Manganese in Health and Disease

Issues in Toxicology

Series Editors:
Professor Diana Anderson, *University of Bradford, UK*
Dr Michael Waters, *Integrated Laboratory Systems Inc, NC, USA*
Dr Timothy C Marrs, *Edentox Associates, Kent, UK*

Advisor to the Board:
Alok Dhawan, *Ahmedabad University, India*

Titles in the Series:

How to obtain future titles on publication:
A standing order plan is available for this series. A standing order will bring delivery of each new volume immediately on publication.

For further information please contact:
Book Sales Department, Royal Society of Chemistry, Thomas Graham House, Science Park, Milton Road, Cambridge, CB4 0WF, UK
Telephone: +44 (0)1223 420066, Fax: +44 (0)1223 420247
Email: booksales@rsc.org
Visit our website at www.rsc.org/books

Manganese in Health and Disease

Edited by

Lucio G. Costa
University of Washington, Seattle, WA, USA
Email: lgcosta@u.washington.edu

Michael Aschner
Albert Einstein College of Medicine, Bronx, NY, USA
Email: michael.aschner@einstein.yu.edu

THE QUEEN'S AWARDS
FOR ENTERPRISE:
INTERNATIONAL TRADE
2013

Issues in Toxicology No. 22

Print ISBN: 978-1-84973-943-6
PDF eISBN: 978-1-78262-238-3
ISSN: 1757-7179

A catalogue record for this book is available from the British Library

Published by The Royal Society of Chemistry,
Thomas Graham House, Science Park, Milton Road,
Cambridge CB4 0WF, UK

Registered Charity Number 207890

For further information see our web site at www.rsc.org

Printed and bound by CPI Group (UK) Ltd, Croydon, CR0 4YY

Preface

Manganese (Mn) is the twelfth most abundant element in the Earth's crust and present in drinking water and in food. As an essential trace element, Mn is required for multiple biochemical and cellular reactions, and is a necessary component for numerous metallo-enzymes, such as Mn superoxide dismutase, arginase, phosphoenol-pyruvate decarboxylase, and glutamine synthase, to name a few.

Despite its essentiality, exposures to high levels of Mn from occupational, iatrogenic, medical, and environmental exposures may contribute to human morbidity. Excessive Mn accumulation in the brain, primarily in basal ganglia, may cause clinical signs and morphological lesions analogous to those seen in Parkinson's disease (PD). Other tissues may be affected as well.

Mn intoxication cases were originally described over two centuries ago. Manganism, resulting from exposure to exceedingly high levels of this metal, was originally described by James Couper (1837), providing insight into the adverse neurological effects in five Scottish men employed in grinding Mn dioxide ore. As Mn began to be used more widely in the steel alloy industry, more cases were recognized, with stronger epidemiological evidence implicating Mn in a number of neurological diseases. Contemporary exposures to Mn at levels described by Couper are rare, yet concerns about the health effects of Mn remain, given its abundant occurrence and the potential exposures throughout various life-stages.

This book, to our knowledge, is the first multidisciplinary scientific endeavor to address the health effects of Mn. It aims to provide state-of-the-art information and deepen the understanding of Mn's adverse health effects. It commences with a description on various pathways for Mn absorption (lung, gastrointestinal tract, olfactory pathway), followed by its nutritional needs, toxicokinetics and toxicodynamics. A large section of the book is devoted to its adverse effects, emphasizing cellular and molecular mechanisms of

Issues in Toxicology No. 22
Manganese in Health and Disease
Edited by Lucio G. Costa and Michael Aschner
© The Royal Society of Chemistry 2015
Published by the Royal Society of Chemistry, www.rsc.org

toxicity in a host of tissues and organs, particularly the nervous system, with emphasis on sensitivity to Mn at various life-stages. We conclude with a list of research needs that will further improve our understanding of the role of Mn both in health and disease.

We called upon internationally recognized experts on Mn to address and facilitate the understanding of its role in health and disease, making a valiant attempt to provide as broad and multidisciplinary approach as possible. Our goal was to assemble a series of chapters that advance the latest developments and scientific breakthroughs in this fast-paced research area, and to provide information that should be of interest to risk assessors, neurobiologists, and neurotoxicologists, as well as metal and trace element biologists. We are hopeful that the book offers the reader appreciation and renewed sense on contemporary issues in Mn research. We are indebted to the authors for their contributions and hope that, as a reader, whether you are a novice or a seasoned Mn researcher, the knowledge amassed herein will stimulate and transform your novel ideas into better understanding on the role of this unique metal in health and disease.

Michael Aschner
Lucio G. Costa

Contents

Issues in Toxicology No. 22
Manganese in Health and Disease
Edited by Lucio G. Costa and Michael Aschner
© The Royal Society of Chemistry 2015
Published by the Royal Society of Chemistry, www.rsc.org

Chapter 8 Manganese and Oxidative Stress **199**

Daiana Silva Ávila, Marcelo Farina,
João Batista Teixeira da Rocha and Michael Aschner

Chapter 9 Mutual Neurotoxic Mechanisms Controlling Manganism and Parkisonism **221**

Jerome A. Roth

Chapter 10 Mechanism of Manganese-Induced Impairment of Astrocytic Glutamate Transporters **258**

Pratap Karki, Keisha Smith, Michael Aschner and Eunsook Lee

Chapter 16   Developmental Effects of Manganese   **426**
Scott M. Langevin and Erin N. Haynes

**Chapter 17   The Effects of Manganese on Female Pubertal
Development**   **437**
William L. Dees, Jill K. Hiney and Vinod K. Srivastava

Huajun Jin, Dilshan S. Harischandra, Christopher Choi,
Dustin Martin, Vellareddy Anantharam, Arthi Kanthasamy
and Anumantha G. Kanthasamy

CHAPTER 1

Manganese Transport, Trafficking and Function in Invertebrates

AMORNRAT NARANUNTARAT JENSEN[a] AND
LARAN T. JENSEN*[b]

[a] Department of Pathobiology, Faculty of Science, Mahidol University,
Bangkok, Thailand; [b] Department of Biochemistry, Faculty of Science,
Mahidol University, Bangkok, Thailand
*Email: laran.jen@mahidol.ac.th

1.1 Introduction

Manganese is a biologically important trace metal and is required for the growth and survival of most, if not all, living organisms. It is perhaps best known for its prominent role as a redox-active cofactor in free radical de-toxifying enzymes.[1-8] However, the utilization of manganese in biological systems is substantially more diverse. The uptake and distribution of manganese is critical for proper function of manganese-requiring enzymes; however, this same metal can have deleterious effects in biological systems if homeostasis is disrupted.[9-12] In order to prevent toxicity, cells maintain manganese under tight homeostatic control. Adding complexity to the cellular control of manganese homeostasis is the presence of multiple types of manganese transporter that participate in the specific transport of manganese or in general divalent metal ion transport.

Issues in Toxicology No. 22
Manganese in Health and Disease
Edited by Lucio G. Costa and Michael Aschner
© The Royal Society of Chemistry 2015
Published by the Royal Society of Chemistry, www.rsc.org

Cells appear to transport manganese solely as the divalent cation and several classes of manganese transporters have been characterized. These include Nramp H^+-manganese transporters,[13–16] ATP-binding cassette (ABC) manganese permeases,[17–21] manganese transporting P-type ATPases,[22,23] cation diffusion facilitators (CDFs),[24–26] and inorganic phosphate transporters with high affinity for $Mn–HPO_4$ complexes.[27–29] Bacteria typically contain one or more of these types of transporter, and these classes of transporter are also present in eukaryotic cells.[30,31] These transporters comprise both high and low affinity manganese uptake systems and the transporter utilized depends on the concentration of manganese in the environment. The homeostatic range for manganese is quite wide, with cellular levels of manganese between 0.04 and 2.0 mM under optimal growth conditions.[10,16,29,32,33] Cells rarely experience optimal environmental levels of manganese and often face extreme conditions of either manganese deficiency or excess.[34] Cells activate stress response mechanisms in an attempt to return manganese levels to the homeostatic range. The response typically results in the upregulation or downregulation of cell surface and intracellular transport systems. The regulation of manganese uptake, distribution, and efflux can occur at both the transcriptional and post-translational levels, although the specific route of regulation varies in different organisms.

1.2 Function of Manganese in Biological Systems

1.2.1 Manganese Metalloenzymes

Manganese metalloenzymes are involved in a wide range of cellular functions, including detoxification of reactive oxygen species, protein glycosylation, polyamine biosynthesis, DNA biosynthesis, nucleic acid degradation, phospholipid biosynthesis and processing, polysaccharide biosynthesis, protein catabolism, the urea cycle, photosynthesis, and sugar catabolism.[2,35–45] Manganese-dependent enzymes that participate in these processes typically utilize manganese in Lewis acid–base reactions or as a reduction/oxidation center to facilitate catalysis. These types of reaction are exemplified by arginase (Lewis acid) and Mn superoxide dismutase (reduction/oxidation),[2,3,46,47] and the role of manganese in these reactions is shown in Figure 1.1.

1.2.2 Non-Protein Manganese Antioxidants

The importance of manganese in biological systems is not limited to enzyme-mediated catalysis. Non-enzymatic manganese is involved in the formation of bacterial products, including secreted antibiotics,[48] and contributes to the stabilization of bacterial cell walls.[49] In addition, the accumulation of non-protein complexes of manganese can function in the removal of reactive oxygen species (ROS), especially superoxide.[50–53] These

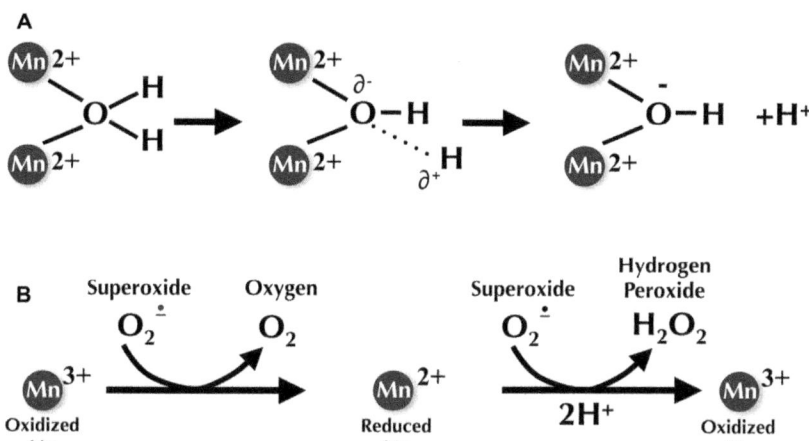

Figure 1.1 Typical chemistry performed by manganese in enzymes. (A) Di-nuclear manganese center of arginase. The manganese cofactor of arginase does not participate in redox reactions but instead functions as a Lewis acid to accept a pair of electrons from the bound water molecule, allowing deprotonation and increasing its reactivity. (B) Catalytic detoxification of superoxide anions by manganese superoxide dismutase (Mn-Sod) enzymes. The catalytic cycle for Mn-Sod has been called a "ping–pong" reaction in which the manganese cofactor alternates between the oxidized and reduced forms.

Mn-antioxidants are divalent manganese complexes of small metabolites, and while the nature of the intracellular Mn-complexes has not been clearly defined, phosphate and lactate Mn-complexes have been shown to display the capacity to react efficiently with superoxide *in vitro*.[52,54,55] Complexes of both iron and copper exhibit superoxide scavenging activity, however these metal ions also exhibit pro-oxidant activity.[56–59] In contrast, manganese ions react poorly with hydrogen peroxide and do not generate the highly toxic hydroxyl radical, providing a beneficial antioxidant activity without the pro-oxidant side effects of other redox active metals.[50,60]

It appears that Mn-antioxidants can serve to enhance oxidative stress protection when enzymatic antioxidants are insufficient in various organisms.[53,61–63] A critical role for Mn-antioxidants has been demonstrated in *Deinococcus radiodurans*, a bacterium that is extremely resistant to radiation and desiccation. In this organism, survival under extreme exposure to radiation and other oxidative stress conditions is not dependent on antioxidant enzymes but instead relies on the accumulation of millimolar concentrations of manganese and the subsequent formation of Mn-antioxidants.[50,51,60,63] Interestingly, *Lactobacillus plantarum*, while resistant to oxidative stress, does not express the antioxidant enzyme superoxide dismutase.[64,65] Indeed, *L. plantarum* appears to rely exclusively on Mn-antioxidants for protection against oxidative stress,[54,66,67] highlighting the power of this alternative ROS detoxification pathway.

The majority of the information on manganese antioxidants has come from investigation of bacterial and yeast systems; however, it is also likely that these complexes are present in multicellular organisms. Elevated manganese accumulation in the nematode *Caenorhabditis elegans* enhances thermotolerance and oxidative stress resistance, and extends life span.[68,69] The mechanism of the enhanced stress resistance due to manganese supplementation in *C. elegans* has not been fully elucidated but is suspected to involve elevated antioxidant activity.

1.2.3 Manganese and Bacterial Virulence

Manganese is either known or proposed to be important for virulence in bacterial species such as *Salmonella enterica*, *Mycobacterium tuberculosis*, *Staphylococcus aureus*, *Yersinia pestis*, and *Streptococcus pneumoniae*.[19,30,31] Invasion and initial survival within host cells is not dependent on manganese; however, extended survival appears to require the element.[70,71] The expression of manganese transporters is required to enhance bacterial survival when challenged by host defenses.[15,18,70,72–74] Whether different classes of manganese transporter are redundant or involved at different stages of infection is not known. Models have been proposed in which manganese transporters, as well as iron transporters, are essential for virulence because of competition between the infecting bacterium and host cells for metal ions.[31,71] The need for manganese in bacterial virulence appears to go beyond its role as a cofactor in ROS detoxifying enzymes such as Mn-superoxide dismutase and catalase. Enterobacteria are capable of rapidly increasing uptake of manganese in response to stress, and can accumulate millimolar levels of manganese.[17] This concentration of manganese far exceeds the level needed to supply Mn-superoxide dismutase with its cofactor. It appears that the formation of non-protein Mn-antioxidant complexes may also be an important virulence factor in some bacterial species. The additional protection against reactive oxygen species generated by the host cells may allow invading bacteria to survive the initial stages of infection, and thus promote colonization.

1.3 Manganese Transport in Bacteria

1.3.1 Bacterial Manganese Uptake Systems

In prokaryotic cells, which lack internal compartmentalization, metal ion homeostasis is maintained primarily by tight regulation of metal cation flux across the cytoplasmic membrane. Manganese uptake in bacteria predominantly involves members of two transporter families, Nramp (MntH) and cation-transporting ABC permeases (MntABCD and related), with many species containing both types of transport system.[17,20,70,73,75,76] In addition, utilization of other transport systems for manganese, such as a P-type adenosine triphosphatase (ATPase) by *Lactobacillus* species (MntP) and

Mycobacterium tuberculosis (CtpC), has also been observed.[23,77] Exposure to excess manganese leads to repression of these dedicated manganese transport systems.[20,74] However, the tight control of manganese influx can be bypassed *via* other transporters that are capable of facilitating the uptake of manganese but escape regulation by this metal. An example of this is PitA, an inorganic phosphate transporter with high affinity for Mn–HPO$_4$ complexes that appears to be a major source of manganese uptake during conditions of excess.[28]

1.3.1.1 Bacterial Nramp Manganese Transporter, MntH

Members of the Nramp (natural resistance-associated macrophage protein) transporter family were first identified in yeast and mammalian cells and subsequently found to play a major role in metal ion homeostasis.[78–80] Nramp proteins function in general metal ion transport, and members of this transporter family have been shown to facilitate the movement of divalent metal ions including manganese, zinc, copper, iron, cadmium, nickel, cobalt, and lead.[14,81–85] Transport of metal ions through Nramp is energized by the symport of protons (Figure 1.2).[15,86]

The majority of bacterial Nramp1 homologues, typically designated as MntH,[15,20,23,80,87] appear to function in manganese homeostasis.[15,73,74] The MntH transporters are commonly found in bacterial species, although examples of bacteria lacking Nramp transporters have been described.[30] Metal accumulation studies revealed that overexpression of *Staphylococcus*

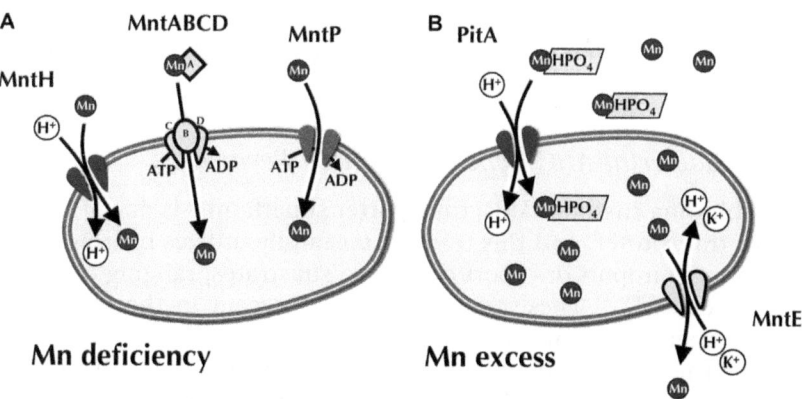

Figure 1.2 Typical manganese transporters in bacterial cells. (A) During conditions of manganese deficiency the high affinity transporters MntH, MntABCD, and MntP facilitate manganese uptake. However, these manganese transporters are not present in all bacterial species. (B) Manganese excess inhibits expression of the high affinity transporters and induces the manganese efflux protein MntE. Uptake of manganese–phosphate complexes may be a source of manganese when cells are exposed to toxic concentrations of this metal.

aureus MntH resulted in increased cell-associated manganese but not calcium, copper, iron, magnesium or zinc, indicating that this Nramp1 transporter was selective for the uptake of manganese.[20] Consistent with these observations, mutants of *mntH* in *Bacillus subtilis* exhibited impaired growth in metal-depleted media that could be rescued by the addition of manganese.[74] Direct transport assays also indicated a preference for manganese in MntH from *Salmonella enterica* serovar Typhimurium and *Escherichia coli*. The affinity for manganese far exceeds that for iron in these MntH proteins, demonstrating the role of Nramp transporters in bacterial manganese uptake.[30]

Species differences in MntH metal ion specificity have been observed, with some MntH homologues appearing to function in the transport of other metals in addition to manganese. While *S. enterica*, *E. coli*, and *B. subtilis* MntH exhibit a strong preference for manganese,[30,74] the *M. tuberculosis* MntH homologue, Mramp, appears to transport not only manganese but also significant amounts of iron and zinc.[88] Roles for Nramp transporters in the uptake of other metal ions, especially iron, have been documented in both prokaryotic and eukaryotic organisms.[16,73,80,89–91] Multiple Nramp isoforms can be present in a single species, and these Nramp transporters, although highly similar, may have divergent metal ion preferences. *Pseudomonas aeruginosa* expresses two distinct Nramp transporters capable of transporting manganese, and multiple Nramp isoforms are present in *Burkholderia* species, although the metal ion preferences of these transporters have not been determined.[30] While the most physiologically relevant substrate for the majority of bacterial MntH transporters appear to be manganese, it is clear that these transporters have the capacity to facilitate the uptake of other metals when they are present in excess. This broad metal ion selectivity in Nramp transporters also appears to enhance the uptake of toxic metal ions, such as cadmium and lead.[74,85,92]

1.3.1.2 Bacterial ABC-Type Manganese Permeases

The ATP-binding cassette (ABC) transporter superfamily is one of the largest classes of transporter, and this transporter family utilizes hydrolysis of ATP to facilitate the import or export of diverse substrates, ranging from ions to macromolecules.[93–95] These transporters are present in the plasma membrane or inner membrane of Gram-negative bacteria,[93,95,96] and are well known for their involvement in multi-drug resistance in both prokaryotic and eukaryotic cells by enhancing the export of toxins and drugs.[97,98] However, ABC transporters functioning as importers have only been described in prokaryotic systems.[93,95,96] Metal ion transporting ABC permeases have been identified with important roles in manganese acquisition.[19,70,76,99,100] The cation selectivity of manganese ABC-type permeases extends to other divalent metal ions including iron, zinc, cobalt, nickel, molybdenum, and cadmium; however, the typical affinities for these metal ions are 10- to 100-fold lower than for manganese.[100–103]

Examples of bacterial ABC transporters involved in manganese import include, but are not limited to, MntABCD (*Bacillus subtilis, Staphylococcus aureus*), SitABCD (*Shigella flexneri*), PsaABCD (*Streptococcus pneumoniae*), and YfeABCD (*Yersinia pestis*),[17,18,21,31,70,74,104,105] and these transporters exhibit similar subunit organization and function. The manganese transporter complex MntABCD (see Figure 1.2) consists of three subunits: MntC and MntD are integral membrane proteins that form the permease subunit and mediate cation import; MntB is the ATPase subunit; and MntA functions as a cation binding protein that delivers manganese to the permease complex.[17,19,93,95,106] MntA is present as a soluble periplasmic protein in Gram-negative bacteria.[94] In Gram-positive bacteria MntA is a lipoprotein anchored to the extracellular side of the plasma membrane,[31,74,107] because these bacteria do not possess an outer membrane. Similar organization is also present in the operons of other manganese ABC transporters such as *sitABCD*, *yfeABCD*, and *psaABCD*.[17,19,21,31,70]

1.3.1.3 Bacterial P-Type Manganese Transporting ATPases

P-type ATPases form a large superfamily of cation and lipid pumps and are distinct from the ABC class of ATPases in that ATP hydrolysis is coupled to transport within a single protein chain.[108] A manganese/cadmium transporting P-type ATPase, MntP (also known as MntA, although distinct from MntABCD) from *Lactobacillus plantarum*, was identified and proposed to be the major source of manganese for this organism.[23] Subsequent analysis of the *L. plantarum* genome revealed the presence of three Nramp transporters as well as a manganese ABC transporter.[109] Mutations of *L. plantarum mntP* or the Nramp and ABC transporters did not alter intracellular manganese concentrations under either manganese deficiency or excess.[109] A primary role for MntP in manganese acquisition in *L. plantarum* is not certain; however, Nramp and manganese ABC transporters were also not essential for manganese uptake. It appears that *L. plantarum* is highly adaptive in maintaining manganese uptake even in the absence of known transporters and additional, yet uncharacterized, transporters may participate in manganese accumulation. Three additional putative P-type calcium/manganese ATPases are present in *L. plantarum* and have been proposed as possible sources of manganese uptake in this bacterium.[109]

1.3.1.4 Bacterial Transport of Manganese–Phosphate Complexes

In *Salmonella* lacking both the Nramp and manganese ABC transporters, manganese uptake activity has been observed, although at low levels.[30] The proposed source of this residual manganese uptake is PitA, a low affinity phosphate transporter.[27,110] The substrate for PitA is a neutral metal phosphate (metal–HPO_4) complex, and this transporter has a preference for phosphate complexes of magnesium, calcium, cobalt, and manganese.[27] In environments rich in metals and phosphate, PitA and related transporters have been proposed to be major suppliers of divalent metal cations[111] and may contribute to metal ion toxicity. Experimental evidence for manganese uptake in intact cells through PitA or other phosphate transporters is limited. However, stimulation of manganese uptake was produced in *L. plantarum* and *B. subtilis* by the addition of phosphate.[112,113]

1.3.2   Bacterial Manganese Efflux

A manganese efflux transporter, MntE, showing homology with the cation diffusion facilitator family (CDF) has been identified in several bacterial species.[26,114] Members of the CDF family are found in most prokaryotic and eukaryotic cells and typically function in metal tolerance by exporting cations from the cytoplasm to the cell exterior.[25,114,115] The most likely transport mechanism for MntE is an antiport cycle consisting of the efflux of manganese with the uptake of hydrogen and potassium ions.[25]

Cells lacking functional MntE exhibit sensitivity to manganese but not other metal ions (cadmium, cobalt, copper, iron, nickel, and zinc) and accumulate three times the intracellular manganese seen in the wild-type strain.[26] The high levels of intracellular manganese in *mntE* mutants increased resistance to oxidative stress but did not lead to enhanced virulence. Bacteria lacking MntE were actually less pathogenic than wild-type cells,[26] indicating that control of manganese homeostasis is critical for both survival and virulence.

In addition to MntE, the P-type ATPase CtpC, from *M. tuberculosis* and *M. smegmatis*, also appears to facilitate manganese efflux. Deletion of *ctpC* leads to sensitivity to oxidative stress and elevated accumulation of cytosolic manganese.[77] Mutations that increase cytosolic manganese commonly enhance resistance to oxidative stress; however, this is not the case for cells lacking CtpC. This discrepancy appears to be explained by the function of CtpC in providing manganese for incorporation into secreted proteins, including Mn-superoxide dismutase enzymes. Thus the primary function of CtpC may not be simply to remove excess manganese from cells, but also to provide the manganese cofactor for secreted enzymes.[77]

The only other example of a bacterial manganese efflux protein is YebN from *Xanthomonas oryzae pv. oryzae*. YebN does not belong to any known transporter family, although cells lacking YebN are manganese sensitive and accumulate high concentrations of intracellular manganese.[116,117] YebN may represent an uncharacterized class of manganese transporter, or alternatively the effect of this protein on manganese efflux may be indirect.

1.3.3   Regulation of Bacterial Manganese Transport

The uptake of essential transition metals, such as manganese, must be regulated in order to respond to changes in environmental conditions. Cells facilitate uptake when faced with deficiency and prevent import under conditions of metal excess. In bacteria this regulation is mediated primarily at the level of transcription.[30,31] The principal bacterial transcription factor involved in manganese-dependent gene expression is MntR, a manganese-specific member of the diphtheria toxin repressor family (DtxR).[74,106,118] MntR largely controls intracellular manganese levels in bacteria through regulated expression of *mntH* and *mntABCD*. Regulation of *mntH* and *mntABCD* is different under conditions of manganese deficiency and manganese excess, and variation among bacterial species has also been observed. In *B. subtilis* cells experiencing manganese deficiency, MntR activates expression of *mntABCD* although *mntH* is expressed independently of MntR.[74] However, under conditions of manganese sufficiency or excess, *B. subtilis* MntR functions as a typical repressor and inhibits the expression of both *mntH* and *mntABCD*,[74] thereby limiting manganese uptake. MntR dependent inhibition of both *mntH* and *mntABCD* expression has also been observed in *E. coli* and *S. enterica*.[119–121] In *S. aureus*, MntR acts as a negative regulator of *mntABCD* in the presence of excess manganese, similar to what is observed in *B. subtilis*. However, *mntH* levels in *S. aureus* are not decreased by excess manganese and positive expression of *mntH* appears to require MntR.[20] The manganese efflux systems *mntE*, *ctpC*, and *yebN* all exhibit positive regulation by MntR,[26,77,117] as would be expected to facilitate removal during condition of manganese excess. MntR homologues from different bacterial species often exhibit a low level of sequence identity,[31] which may account for the contrasting modes of regulation observed.

In addition to MntR, the transcription factors Fur, PerR, and OxyR have been found to have roles in regulating the expression of *mntH* and *mntABCD*. Fur is well characterized for its role in the regulation of iron uptake genes in response to cellular iron status.[122] Expression of *mntH* and *mntABCD* is also regulated by Fur in some bacterial species including *E. coli*, *S. meliloti*, and *S. enterica*, but not in *B. subtilis*.[107,120,121] The ability of manganese transporters to facilitate the uptake of iron when present at high concentrations[15,31,101] may explain the dual regulation of *mntH* and *mntABCD* by manganese and iron in some bacterial species, although the physiological relevance of this regulation has not been established.

The principle roles of PerR and OxyR are in the regulation of genes in response to oxidative stress.[122,123] However, these factors are also involved in manganese homeostasis in several bacterial species including *B. subtilis*, *S. aureus*, *S. pneumoniae*, and *S. enterica*.[31,120,124] The regulation of *mntH* and *mntABCD* by PerR and OxyR is likely to be related to the involvement of manganese in defense against oxidative stress. Enhanced manganese uptake facilitates incorporation of manganese into antioxidant enzymes, such as Mn-superoxide dismutase, as well as production of non-protein antioxidant complexes.[33,61–63,125,126] However, when sufficient cellular antioxidant capacity is obtained, PerR and OxyR may sense the reduction in oxidative stress and limit manganese uptake to prevent toxicity from excessive manganese accumulation.

Limited information is available regarding the regulation of P-type manganese transporting ATPases (MntP) and cation-phosphate transporters

(PitA). Expression of *mntP* from *L. plantarum* is increased under manganese deficiency,[23] consistent with regulation by MntR; however, regulators of *L. plantarum mntP* in response to manganese limitation or excess have not been determined. In the case of *pitA*, there is no report of manganese-mediated transcriptional control but regulation by zinc ions has been observed. The expression of *pitA* was increased approximately two-fold in response to elevated zinc concentrations and is also increased by phosphate limitation,[127] although the transcription factors involved in *pitA* regulation have not been identified. It has been proposed that the addition of excess zinc may compete with magnesium for binding to phosphate, reducing phosphate influx, because $ZnHPO_4$ complexes are less favored for transport by PitA than $MgHPO_4$.[127] This model suggests that *pitA* is regulated by alterations in intracellular phosphate levels and not directly by zinc ions. Excess manganese is not expected to limit phosphate uptake because $MnHPO_4$ complexes are preferred substrates for PitA,[28] thus this type of *pitA* regulation appears unlikely under conditions of manganese excess.

1.4   Manganese Transport in Yeast

1.4.1   Yeast High Affinity Manganese Uptake, Smf1p and Smf2p

The yeast NRAMP transporters Smf1p and Smf2p form the primary manganese uptake system utilized by yeast experiencing manganese deficiency.[16,128–130] The Smf1p and Smf2p transporters have specific roles in the acquisition of manganese and are localized either at the cell surface or in intracellular vesicles (Figure 1.3). Regardless of their localiztion, the Smf1p and Smf2p transporters move manganese across membranes toward the cytosol.[13,34,78,131,132] A third NRAMP homologue, Smf3p, is also present in yeast but this transporter appears to function in the movement of iron from vacuolar stores.[133,134] Similar to other members of the Nramp family, yeast SMFs are capable of facilitating the translocation of divalent metals in addition to manganese, including iron, cobalt, copper, zinc, and cadmium, and may contribute to the toxicity of these metals when present in excess.[13,34,78,79,81–83,89,132,133]

Saccharomyces cerevisiae Smf1p and Smf2p were originally identified on the basis of their ability when overexpressed to bypass a mitochondrial protein-processing defect (*S*uppressor of *M*itochondria import *F*unction).[135] Subsequently Smf1p was characterized as a high affinity manganese transporter,[81] revealing that the role of Smf1p in mitochondrial import was due to the requirement for manganese in mitochondrial protein processing.[135] Despite the high degree of similarity between Smf1p and Smf2p (50% identity and 67% similarity), these transporters are not functionally redundant and exhibit distinct roles in manganese uptake and trafficking.[16,33,34,125,128,134,136] Smf1p is localized at the cell surface under conditions of manganese deficiency; however, the majority of Smf1p is present in intracellular

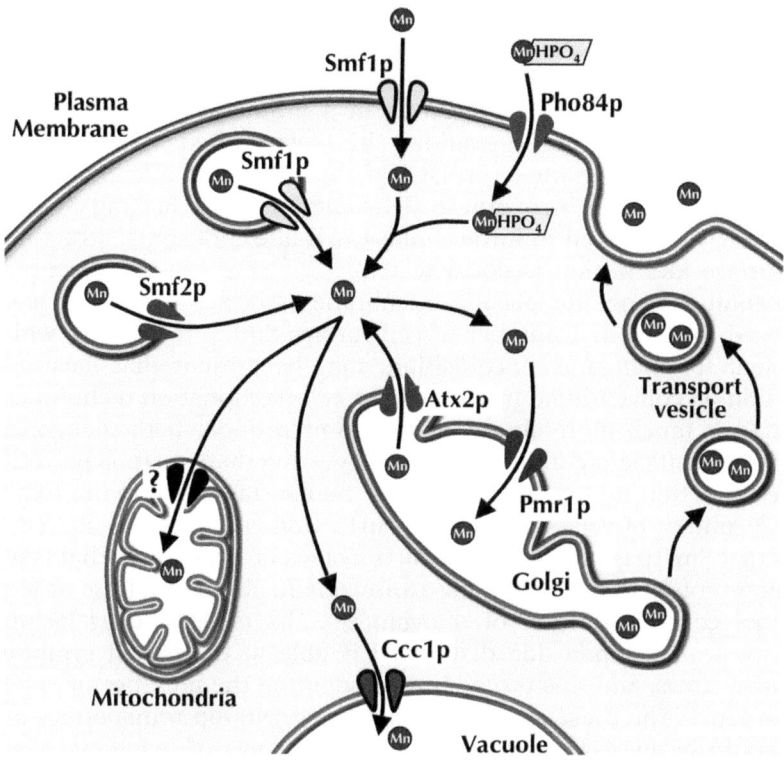

Figure 1.3 Manganese trafficking in the yeast *S. cerevisiae*. The eukaryotic yeast cell requires manganese uptake systems as well as transporters to move manganese in and out of intracellular compartments. The high affinity manganese uptake system is comprised of Smf1p and Smf2p, with Pho84p only functioning as a manganese importer during conditions of manganese excess. Intracellular transporters have been identified for the Golgi apparatus and vacuole. Pmr1p imports manganese into the Golgi as well as facilitating exocytic efflux of excess manganese. Atx2p appears to mobilize manganese from the Gogli back to the cytosol. Ccc1p is a vacuolar manganese importer that functions to limit cytosolic manganese concentrations during conditions of excess. Vacuolar manganese exporters and mitochondrial manganese importers have not been identified.

vesicles, thought to be endosomes, when manganese levels are sufficient.[34,81,128–130,137] Based on changes in Smf1p localization in response to manganese status it appears that Smf1p is functioning in manganese transport only under deficiency conditions and has no role in manganese uptake and distribution during manganese sufficiency. Consistent with this proposal, cellular manganese levels in yeast lacking *SMF1* grown under manganese-sufficient conditions were unchanged when compared with the wild-type strain.[81] In addition, no change in the activity of manganese-dependent enzymes, such as mannosyl-transferases and mitochondrial superoxide dismutase, was observed in cells lacking Smf1p.[125]

In contrast to Smf1p, the second yeast Nramp transporter, Smf2p, exhibits many characteristics of a cell-surface manganese transporter and has a significant role in manganese homeostasis during manganese sufficient conditions. Deletion of *SMF2* results in a dramatic reduction in cellular manganese accumulation, resulting in lowered activity of manganese-dependent enzymes in the cytosol, Golgi apparatus, and mitochondria.[125,138] Despite the cell-wide manganese deficiency in yeast lacking Smf2p, this transporter is observed in intracellular Golgi-like endosomal structures and cell surface localization is not detected.[16,34,125] Even in the absence of experimental evidence for plasma membrane associated Smf2p, it has been proposed that a small number of cell-surface Smf2p molecules, which are sufficient for manganese acquisition, may be present that have escaped detection by conventional microscopy or cell-fractionation techniques.[130]

Smf1p is much more abundant than Smf2p under both manganese deficient and sufficient conditions,[82,128] suggesting that Smf1p is performing a needed function in cellular manganese homeostasis. While the total manganese content of cells lacking the Smf1p transporter is not altered, it appears that Smf1p is important for the transport of manganese that is utilized as a non-proteinaceous manganese antioxidant, similar to those observed in bacterial cells.[33] Addition of manganese salts to yeast cells lacking the cytosolic Cu/Zn superoxide dismutase is able to reverse all symptoms of oxidative stress, and this rescue is dependent on the presence of Smf1p but not Smf2p.[33] The presence of two manganese Nramp transporters in yeast appears to be necessary to provide separate pools of manganese, at least under manganese-sufficient conditions. Smf2p provides manganese for manganese-requiring enzymes while Smf1p plays a critical role in oxidative stress resistance by supplying manganese for non-proteinaceous manganese antioxidant molecules.

1.4.2   Manganese and Phosphate Coupled Uptake in Yeast, Pho84p

The Smf1p and Smf2p transporters are involved in the uptake of manganese that is needed for the activity of manganese enzymes and oxidative stress protection. However, other transporters participate in the uptake of manganese under conditions of manganese surplus. The primary low affinity manganese transporter in yeast cells is Pho84p, a member of the major facilitator superfamily (MFS). Pho84p is a high affinity phosphate/proton symporter[29] and appears to be analogous to the bacterial PitA.[27] Phosphate transport in yeast is mediated by at least six transporters; however, only Pho84p has a major role in manganese uptake.[29] The substrate for Pho84p is a divalent metal phosphate complex and *in vitro* reconstitution studies have shown a high preference for the transport of manganese and cobalt.[28] Manganese transport by Pho84p accounts for the majority of excess manganese accumulated when cells are exposed to manganese surplus, and yeast lacking Pho84p are resistant to manganese toxicity through reduced manganese accumulation.[29,33] In the absence of an excess of manganese, Pho84p appears to transport a magnesium phosphate complex. Magnesium can effectively inhibit manganese uptake through Pho84p,[29] and the higher concentrations of magnesium in yeast growth media compared to manganese probably facilitate the uptake of magnesium phosphate complexes. However, when yeast experience conditions in which a manganese surplus is present, Pho84p-mediated transport of $MnHPO_4$ complexes is favored.[28,29] The uptake of manganese through Pho84p appears to be unintended, and is a consequence of the presence, under manganese surplus conditions, of $MnHPO_4$ complexes that mimic the natural $MgHPO_4$ substrate. Manganese transported by Pho84p is capable of causing toxicity but does not appear to have a role as an essential nutrient. Cells lacking Pho84p do not exhibit defects in manganese-requiring enzymes[29] or in the formation of non-protein manganese antioxidants.[33]

1.4.3 Intracellular Manganese Distribution in Yeast

While much is known about the cellular uptake of manganese in both prokaryotic and eukaryotic systems, it is not entirely clear how cells target manganese to the correct recipient proteins.[34,131,139] The presence of multiple organelles in eukaryotic cells that contain manganese-requiring proteins increases the complexity of manganese trafficking. Proteins involved in the vectoral delivery of manganese inside cells have not been identified; however, manganese is correctly targeted under manganese-sufficient conditions, suggesting that a concerted process for manganese delivery exists. Movement of manganese to specific organelles within yeast cells is mediated by the action of intracellular transporters. Currently, manganese transporters for the Golgi complex[22,140,141] and vacuole[142,143] have been characterized. However, the identity of the manganese transporter for mitochondria is not known, although it is possible that manganese is taken up through the action of several transporters or transporters with dual specificity for other divalent metal ions.

1.4.3.1 *Golgi Manganese Transport, Pmr1p and Atx2p*

One of the major functions of the Golgi apparatus is protein glycosylation, and manganese serves as an essential cofactor for several mannosyl-transferase enzymes present in this organelle.[22,144] Manganese transport from the cytosolic compartment into the Golgi is mediated by Pmr1p, a P-type calcium and manganese-transporting ATPase.[22,34,141,144,145] In the absence of Pmr1p, the sorting of proteins and glycosylation in the Golgi apparatus is impaired.[125,140,144] In addition, loss of function mutations in *PMR1* lead to hyperaccumulation of manganese in the cytosol and severe manganese sensitivity.[145,146] The excess accumulation of manganese in cells lacking Pmr1p reveals that Pmr1p is not only important for supplying the manganese needed for proper activity of mannosyl-transferase enzymes, but it also plays an important role in manganese detoxification. Surplus manganese that has the potential to be toxic is transported into the secretory pathway by Pmr1p and is ultimately released from the cell through exocytic transport (see Figure 1.3).[34,141,145–147] Pmr1p-dependent manganese efflux appears to be the primary mechanism for removal of unwanted manganese from yeast cells, although a dedicated manganese efflux transporter has not been identified. The only example of a manganese efflux transporter in yeast cells is Hip1p, a high affinity histidine permease that is also involved in the transport of manganese;[148] nevertheless, the transport cycle that facilitates manganese efflux is not known. However, yeast lacking functional Hip1p do not display the severe manganese sensitivity seen in cells with mutations in *PMR1*,[148] indicating that the primary manganese efflux pathway is Pmr1p-dependent exocytic transport from the Golgi.

Manganese transported into the Golgi by Pmrp1 can also be redirected back to the cytosol through the action of Atx2p,[32] a putative transporter with similarity to the ZIP (Zrt1, Irt1-like Protein) family. The precise function of Atx2p is not clear but this protein has a role in maintaining the cytosolic concentration of manganese. Overexpression of Atx2p results in increased cellular manganese content, whereas cells lacking Atx2p exhibit decreased cytosolic manganese levels.[32] It appears that Atx2p and Pmr1p work in opposite directions to control manganese levels in both the cytosol and Golgi compartments.

1.4.3.2   Vacuolar Manganese Transport, Ccc1p

In addition to the Pmr1p-mediated cellular export of manganese through the secretory pathway, vacuolar sequestration is another mechanism used to limit the toxicity of manganese. The yeast vacuole, a lysosome-like compartment, is a major site for metal ion storage and detoxification, and yeast strains defective in vacuolar function exhibit sensitivity to several transition metals including manganese.[149,150] Manganese is sequestered in the vacuole through the action of Ccc1p,[34,142,143] a manganese and iron transporter with similarity to the vacuolar iron transporter (VIT) family (see Figure 1.3). While the transport mechanism for the VIT class of transporters is not known, they are predicted to function by an H^+ antiport carrier-type mechanism.[151] Consistent with a role in manganese sequestration, cells lacking Ccc1p display enhanced sensitivity to manganese, and overexpression of Ccc1p reduces manganese toxicity in cells lacking Pmr1p.[142] A transporter that facilitates vacuolar efflux of manganese has not been identified and there is currently no evidence that manganese is released from the vacuole. It is possible that Ccc1p functions to sequester manganese in the vacuole to limit its toxicity, but not for storage and later use by the cell. Alternatively, vacuolar transporters typically associated with other metals, such as iron or calcium, may facilitate vacuolar manganese efflux under specific conditions.

1.4.3.3 Competition between Manganese and Iron in Yeast Mitochondria

The mitochondrial manganese transporter has yet to be identified; however, the search for this transporter has revealed other pathways important for incorporation of manganese into the mitochondrial superoxide dismutase, Sod2p. Movement of manganese to its proper destination does not always occur unhindered, and alterations in both manganese and iron metabolism can affect the delivery of manganese to Sod2p. The mitochondrial carrier protein Mtm1p was identified as an important molecule for manganese insertion into Sod2p.[138] Characterization of Mtm1p revealed that it was not the mitochondrial manganese transporter; instead, Mtm1p participates in mitochondrial iron metabolism, and dysregulation of mitochondrial iron alters manganese binding to Sod2p.[125,152,153] Iron atoms outnumber manganese by nearly two orders of magnitude in mitochondria; however, iron does not bind to Sod2p under these conditions.[152,153] The misincorporation of iron into Sod2p in yeast mutants with defects in mitochondrial iron metabolism is not caused by a global change in the chemical environment of mitochondrial iron. Instead, in the case of mutations that result in mitochondrial iron overload, failure to deliver iron to the proper recipient proteins results in accumulation of iron sulfur precursors on assembly proteins. Iron derived from these stalled iron cluster assembly scaffolds is highly reactive with Sod2p and prevents the proper binding of manganese.[153] A small pool of this Sod2p reactive iron appears always to be present, but under normal conditions the abundance of manganese is sufficient to promote proper insertion of manganese into Sod2p. However, when manganese levels are reduced, this reactive iron can effectively compete with manganese for binding to Sod2p.[153] These observations suggest that, similar to bacterial MnSOD, mitochondria Sod2p captures manganese on the basis of the differential availability of manganese and iron, and they do not support a requirement for a specific carrier protein to deliver manganese to Sod2p.[147]

1.4.4 Regulation of Yeast Manganese Transporters

1.4.4.1 Regulation of Smf1p and Smf2p by Manganese

The regulation of many transition metal transporters, including those for copper, iron, and zinc, occurs primarily at the transcriptional level.[154] However, regulation of the Nramp transporters, Smf1p and Smf2p, occurs principally post-translationally in response to changes in manganese concentrations (Figure 1.4).[16,128,155] Smf1p and Smf2p exhibit multiple levels of post-translational regulation in response to changes in manganese from deficient, through sufficient, to toxic concentrations. When cells are deficient in manganese, the amount of Smf1p and Smf2p increases, allowing enhanced manganese accumulation.[16,128–130] Conditions of sufficient manganese redirect the majority of Smf1p and Smf2p to the vacuole for

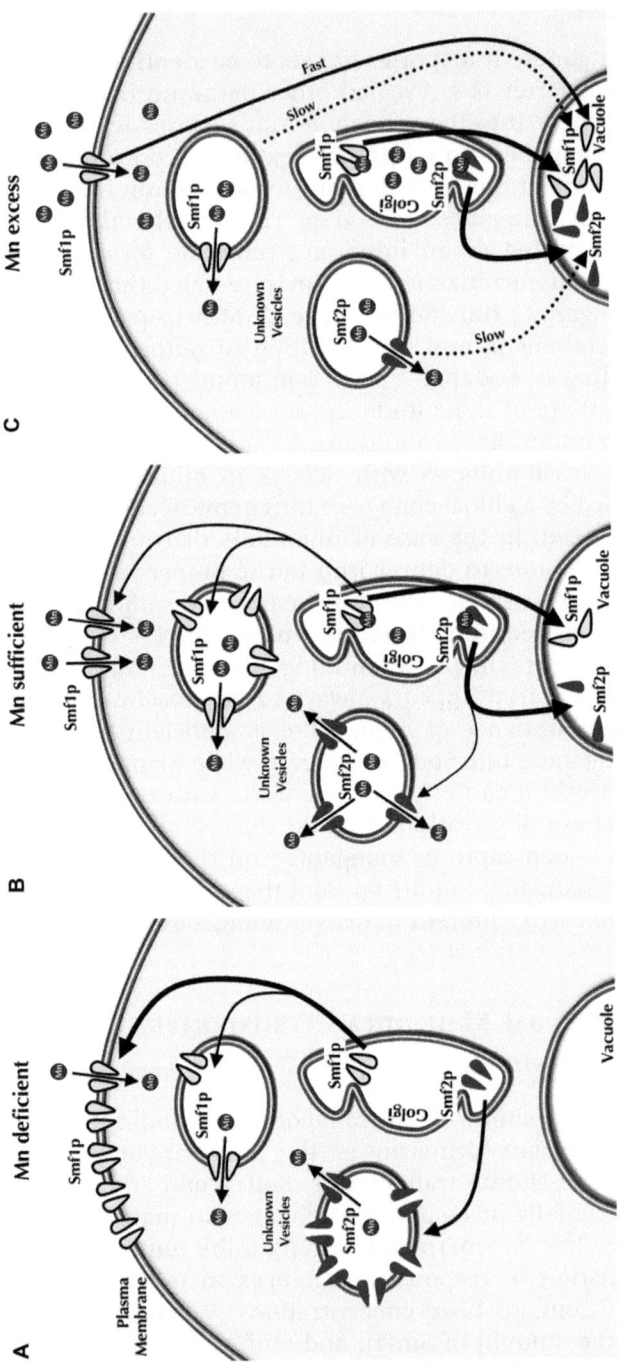

Figure 1.4   Post-translational regulation of *S. cerevisiae* Smf1p and Smf2p. Curved arrows indicate the direction of Smf1p and Smf2p trafficking and straight arrows show Mn transport. Thickness of the arrows shows the relative proportion of proteins that are targeted to the indicated intracellular location. (A) Smf1p and Smf2p are highly abundant when manganese is deficient for cellular needs, and they facilitate manganese uptake. (B) When environmental manganese concentrations are sufficient to provide adequate metal for the cell the majority of Smf1p and Smf2p is degraded in the vacuole. Newly synthesized Smf1p and Smf2p are directed from the secretory pathway (Golgi) toward the vacuole for degradation. (C) Excessive or toxic concentrations of manganese result in the removal of the residual Smf1p and Smf2p proteins. Smf1p that is present at the plasma membrane is rapidly internalized and degraded. However, Smf1p and Smf2p in intracellular vesicles are slowly directed to the vacuole for degradation.

degradation, limiting the uptake of manganese, although a substantial fraction of these transporters is not degraded.[16,128–130] Exposure to toxic concentrations of manganese leads to increased vacuolar degradation of Smf1p and Smf2p, virtually eliminating these transporters from cells.[16,128–130]

1.4.4.1.1 Smf1p and Smf2p Regulation in Manganese Sufficient Conditions.

When manganese levels are within a range that is sufficient for cellular needs but below the toxic threshold, the majority of newly synthesized Smf1p and Smf2p is directly targeted to the vacuole for degradation.[82,128] The constitutive degradation maintains Smf1p and Smf2p at levels that are sufficient to supply the cell with manganese but are not in an excess that would promote manganese overload. Vacuolar targeting of Smf1p and Smf2p during constitutive degradation occurs in the exocytic pathway, in contrast to most cell surface transporters that are targeted to the vacuole through endocytosis. As Smf1p and Smf2p transit the Golgi, the bulk of these proteins are conjugated with ubiquitin, marking them for vacuolar targeting through the multivesicular body (MVB) pathway.[130,137,156] The ubiquitin ligase Rsp5p is responsible for ubiquitination of Smf1p and Smf2p but requires adaptor proteins in order to interact with these transporters. One of these adaptor proteins, Bsd2p, interacts with both Rsp5p and the Smf1p and Smf2p transporters and thereby links Smf1p and Smf2p to Rsp5p, facilitating ubiquitin conjugation.[128,130,157] In addition to Bsd2p, two other adaptor proteins, Tre1p and Tre2p (transferin receptor-like), are also required for Rsp5p-mediated ubiquitination of Smf1p and Smf2p.[137,158] Cells lacking the adaptor protein Bsd2p, or both Tre1p and Tre2p, or containing mutations in *RSP5* accumulate upwards of 10 times more Smf1p and Smf2p, and these transporters display localizations similar to those observed in manganese deficient conditions.[128,130,137] Bsd2p and Rsp5p are thought to form a complex first that is then bound by the Tre1p and Tre2p proteins. The Tre1/2p–Bsd2p–Rsp5 complex then interacts with and mediates the ubiquitination of Smf1p or Smf2p.[130] Subsequently, ubiquitinated Smf1p and Smf2p are directed through the MVB pathway for their eventual degradation in the vacuole.[129,130,137,157]

1.4.4.1.2 Stabilization of Smf1p and Smf2p during Manganese Deficiency.

Manganese deficiency enhances the stability of Smf1p and Smf2p, resulting in increased accumulation of these transporters.[128,134,155] Under manganese deficient conditions, the Smf1p and Smf2p proteins do not interact with the Tre1/2p–Bsd2p–Rsp5 complex and vacuolar degradation is blocked.[128,137,158] Instead, these transporters are delivered either to the cell surface in the case of Smf1p or to intracellular vesicles for Smf2p.[16,128–130] The interaction of the Tre1/2p–Bsd2p–Rsp5 complex with Smf1p and Smf2p appears to be linked to the concentration of manganese in the Golgi.[155] Limiting manganese transport into the Golgi by deletion of *PMR1* promotes the stabilization of Smf1p,[155] even though these conditions result in elevated cytosolic manganese.[141,145] How Smf1p and Smf2p detect manganese deficiency is not known but it has been proposed that these transporters are directly sensing the manganese status of the Golgi lumen.[155]

1.4.4.1.3 Enhanced Degradation of Smf1p in Response to Manganese Toxicity.

In contrast to manganese sufficient conditions, the vacuolar degradation of Smf1p is enhanced to eliminate the residual levels of this transporter when intracellular manganese concentrations exceed the toxic threshold.[155] Degradation of Smf1p in response to toxic concentrations of manganese utilizes trafficking through the MVB pathway, similar to the case of sufficient manganese. However Smf1p vacuolar targeting in response to toxic manganese is independent of Bsd2p, Tre1p, and Tre2p.[155] Smf1p proteins localized to both the plasma membrane and intracellular vesicles are targeted for degradation following exposure to toxic concentrations of manganese. In contrast to conditions of manganese sufficiency, Smf1p appears to sense cytosolic manganese during chronic manganese toxicity. Plasma membrane associated Smf1p is rapidly internalized and delivered to the vacuole for degradation upon exposure to toxic concentrations of manganese. The endocytosis of Smf1p, while independent of Bsd2p, Tre1p, and Tre2p, does require the Rsp5p ubiquitin ligase.[155] However, Rsp5p adaptors that participate in the endocytosis of Smf1p in response to toxic manganese have not been defined. Smf1p localized to intracellular vesicles is slowly trafficked toward the vacuole with prolonged exposure to toxic concentrations of manganese. The slow degradation of Smf1p located in the intracellular vesicles during chronic exposure to toxic manganese does not appear to require Rsp5p or other known ubiquitin ligases. Smf2p is similarly slowly trafficked to the vacuole for degradation in response to chronic manganese toxicity. The slow degradation of the Smf1p and Smf2p proteins present in intracellular vesicles in response to manganese toxicity is unusual. It appears that cells do not immediately degrade these transporters during manganese exposure and Smf1p and Smf2p may be retained in an intracellular pool that can be rapidly recycled in the event of changes in the manganese status of the environment.[155]

1.4.4.2 Regulation of Low Affinity and Intracellular Manganese Transporters

In contrast to the robust regulation of the Smf1p and Smf2p transporters by manganese status, other manganese transporters show limited or no manganese-dependent regulation. The phosphate content of the cell tightly regulates Pho84p, the major low affinity manganese transporter; however, transcription, localization, and stability of Pho84p do not appear to be regulated by manganese.[29] There is also no indication that expression or stability of the intracellular manganese transporters Pmr1p, Atx2p, and Ccc1p are modulated by manganese concentrations. Ccc1p is regulated by the iron status of the cell,[159] and it is possible that Pmr1p and Atx2p are similarly regulated by other environmental cues instead of the level of manganese in the environment.

1.5 Manganese Transport in the nematode *Caenorhabditis elegans*

1.5.1 Nramp Manganese Transporters in *C. elegans*

In eukaryotic organisms, the uptake and trafficking of manganese appear to be mediated largely through the action of Nramp transporters. Among multicellular microorganisms, manganese homeostasis has been best characterized in the nematode *Caenorhabditis elegans*, which expresses three distinct Nramp proteins that function as cation transporters: SMF-1, SMF-2, and SMF-3.[160–162] Each of the *C. elegans* SMF proteins can function in the uptake of manganese, although it appears that they have overlapping but not identical substrate specificity and also participate in iron transport.[161] In addition, each of the SMF proteins has specific functions and localization in intact worms.[160] Intestinal epithelial cells are the major site of manganese accumulation and contain the highest concentrations of manganese in the worm,[163] suggesting that ingestion is a likely route of manganese absorption. SMF-1 and SMF-3 are expressed in partially overlapping regions of intestinal epithelium and primarily localize to the apical plasma membrane; however, localization to intracellular vesicles has also been observed.[160,162,164] The overlapping yet distinct localization pattern for SMF-1 and SMF-3 suggests that these transporters have non-redundant roles in the uptake of intestinal manganese.[160] SMF-3 appears to be the primary manganese uptake transporter, with a minor role in iron uptake. Conversely, SMF-1 appears to have a limited role in manganese uptake and plays a major role in the import of iron.[160]

In contrast to SMF-1 and SMF-3, SMF-2 is restricted to specialized pharyngeal cells and it exhibits a cytoplasmic localization.[160] In addition to functioning as a metal ion transporter, SMF-2 has been proposed to be involved in sensing environmental levels of metal ions.[160] This role for SMF-2 was inferred in part from the phenotypes of *smf* mutants and the localization of the SMF-2 protein. Worms with mutations in *smf-1* or *smf-3* display increased tolerance of manganese and decreased cellular manganese content following exposure to excess manganese, consistent with the role of the gene products in manganese uptake. Surprisingly, mutation of *smf-2* resulted in enhanced sensitivity to manganese and increased manganese accumulation, indicating that SMF-2 may be involved in protecting against toxic manganese exposure.[160] SMF-2 may protect against excess manganese by enhancing excretion or sequestration of this metal; alternatively, SMF-2 may modulate manganese uptake by SMF-1 and SMF-2.[160] Expression of SMF-2 in the pharynx places this transporter at the entry point of manganese into the intestine, a reasonable site for sensing manganese concentrations in food. In this model, SMF-2 facilitates the uptake of manganese into pharyngeal cells, resulting in attenuation of a downstream signaling pathway that inhibits manganese uptake from SMF-1 and SMF-3. Possible mechanisms for SMF-2 mediated changes in manganese uptake by SMF-1 and SMF-2 include altering SMF-1/SMF-3 transporter activity, enhancing excretion through other transporters, or reducing pharyngeal pumping, resulting in reduced nutrient and manganese intake.[160]

1.5.2 Regulation of *C. elegans* Nramp Transporters by Manganese

Transcriptional and post-translational regulation of Nramp transporters occurs in various organisms, and both modes of regulation appear to be present in *C. elegans*. In wild-type worms, the transcriptional response to manganese appears to be a minor component of SMF regulation. Exposure to non-toxic, micromolar levels of manganese results in a modest increase in transcription of the *smf-1* and *smf-2* genes. However, exposure to concentrations of manganese that result in toxicity produces a small decrease in *smf-1* and *smf-3* transcription. Expression of *smf-2* appears to be insensitive to manganese at both low and high concentrations.[160]

Among the *C. elegans* SMF proteins, only SMF-3 exhibits post-translational regulation in response to manganese. Both protein levels and intracellular localization of SMF-1 and SMF-2 are unchanged following exposure of worms to excess manganese. In contrast, manganese exposure results in the rapid translocation of SMF-3 to apical vesicular compartments and the eventual degradation of this protein.[160] SMF-3 appears to be internalized to endosomal compartments in response to manganese excess,[160] through a process that may be mechanistically similar to the vacuolar targeting of Smf1p and Smf2p in yeast under conditions of sufficient or excess manganese.[16,128,155] Consistent with its role as the primary manganese transporter, the post-translational regulation of SMF-3 in response to excess manganese appears to be the principle means utilized by *C. elegans* to limit the unwanted accumulation of this metal.

1.5.3 Intracellular Manganese Transporters in *C. elegans*

Currently, information regarding intracellular manganese transport in *C. elegans* is limited. A functional homologue of the yeast Pmr1p has been identified in *C. elegans* and was found localized to the Golgi. The *C. elegans* PMR-1 appears to perform a similar role to that observed for yeast Pmr1p in facilitating transport of manganese from the cytosol to the Golgi, as well as in protecting against manganese toxicity by enhancing exocytosis of manganese.[68,140,145,165] Intracellular manganese transporters identified in

S. cerevisiae such as the Golgi efflux pump Atx2p[32] and vacuolar importer Ccc1p[142,143] have not been described in *C. elegans*. It is noteworthy that analysis of the *C. elegans* genome does not indicate the presence of an orthologue for yeast Ccc1p; however, a potential orthologue for yeast Atx2p is present.

1.6 Conclusions

Manganese is utilized as a cofactor for metalloenzymes and as part of non-protein antioxidant complexes in both prokaryotic and eukaryotic cells.[33,69,126] The uptake and efflux of manganese in prokaryotes and eukaryotes exhibit some overlap but are substantially distinct. Two major classes of manganese uptake transporter, the Nramp MntH and ABC-type manganese permeases MntABCD/SitABCD, have been characterized in prokaryotes.[15,17,73] Manganese efflux is mediated in most bacterial species by the cation diffusion facilitator (CDF) MntE.[26] Eukaryotes do not contain the ABC-type manganese permeases and rely on Nramp transporters (SMFs) for manganese acquisition under physiological conditions.[81,134,160,162] Intracellular distribution of manganese in eukaryotes utilizes several classes of transporter some of which, such as the Golgi transporter Pmr1p, are conserved across species.[68,140,165]

In prokaryotic cells, manganese transporters are regulated primarily at the level of transcription through the MntR transcription factor.[74,106,119] In contrast, the regulation of proteins involved in manganese uptake in eukaryotes appears to be principally post-translational, through targeted degradation in response to sufficient or excess manganese.[128,155,160] It is likely that the processes of manganese uptake, distribution, and elimination identified in bacteria, yeasts, and nematodes are conserved, at least in part, in more complicated invertebrate organisms.

References

1. H. M. Hassan, Biosynthesis and regulation of superoxide dismutases, *Free Radical Biol. Med.*, 1988, **5**, 377–385.
2. I. Fridovich, Superoxide dismutases, *Annu. Rev. Biochem.*, 1975, **44**, 147–159.
3. J. M. McCord, Iron- and manganese-containing superoxide dismutases, *Adv. Exp. Med. Biol.*, 1976, **74**, 540–550.
4. H. M. Hassan, Microbial superoxide dismutases, *Adv. Genet.*, 1989, **26**, 65–97.
5. B. H. Robinson, The role of manganese superoxide dismutase in health and disease, *J. Inherited Metab. Dis.*, 1998, **21**, 598–603.
6. V. C. Culotta, M. Yang and T. V. O'Halloran, Activation of superoxide dismutases, *Biochim. Biophys. Acta*, 2006, **1763**, 747–758.
7. J. W. Whittaker, Metal uptake by manganese superoxide dismutase, *Biochim. Biophys. Acta*, 2010, **1804**, 298–307.

8. M. J. Horsburgh, S. J. Wharton, M. Karavolos and S. J. Foster, Manganese: elemental defence for a life with oxygen, *Trends Microbiol.*, 2002, **10**, 496–501.

9. J. Crossgrove and W. Zheng, Manganese toxicity upon overexposure, *NMR Biomed.*, 2004, **17**, 544–553.

10. M. Yang, L. T. Jensen, A. J. Gardner and V. C. Culotta, Manganese toxicity and *Saccharomyces cerevisiae* Mam3p, *Biochem. J.*, 2005, **386**, 479–487.

11. E. Delhaize, T. Kataoka, D. M. Hebb, R. G. White and P. R. Ryan, Genes encoding proteins of the cation diffusion facilitator family that confer manganese tolerance, *Plant Cell*, 2003, **15**, 1131–1142.

12. A. B. Santamaria, Manganese exposure, essentiality & toxicity, *Indian J. Med. Res.*, 2008, **128**, 484–500.

13. M. Cellier, G. Prive, A. Belouchi, T. Kwan, V. Rodrigues, W. Chia and P. Gros, Nramp defines a family of membrane proteins, *Proc. Natl. Acad. Sci. U. S. A.*, 1995, **92**, 10089–10093.

14. S. M. Vidal, E. Pinner, P. Lepage, S. Gauthier and P. Gros, Natural resistance to intracellular infections, *J. Immunol.*, 1996, **157**, 3559–3568.

15. D. G. Kehres, M. L. Zaharik, B. B. Finlay and M. E. Maguire, The NRAMP proteins of *Salmonella typhimurium* and *Escherichia coli* are selective manganese transporters involved in the response to reactive oxygen, *Mol. Microbiol.*, 2000, **36**, 1085–1100.

16. M. E. Portnoy, X. F. Liu and V. C. Culotta, *Saccharomyces cerevisiae* expresses three functionally distinct homologues of the nramp family of metal transporters, *Mol. Cell. Biol.*, 2000, **20**, 7893–7902.

17. D. G. Kehres, A. Janakiraman, J. M. Slauch and M. E. Maguire, SitABCD is the alkaline $Mn(2+)$ transporter of *Salmonella enterica* serovar Typhimurium, *J. Bacteriol.*, 2002, **184**, 3159–3166.

18. M. L. Zaharik, V. L. Cullen, A. M. Fung, S. J. Libby, S. L. Kujat Choy, B. Coburn, D. G. Kehres, M. E. Maguire, F. C. Fang and B. B. Finlay, The *Salmonella enterica* serovar typhimurium divalent cation transport systems MntH and SitABCD are essential for virulence in an Nramp1G169 murine typhoid model, *Infect. Immun.*, 2004, **72**, 5522–5525.

19. S. W. Bearden and R. D. Perry, The Yfe system of *Yersinia pestis* transports iron and manganese and is required for full virulence of plague, *Mol. Microbiol.*, 1999, **32**, 403–414.

20. M. J. Horsburgh, S. J. Wharton, A. G. Cox, E. Ingham, S. Peacock and S. J. Foster, MntR modulates expression of the PerR regulon and superoxide resistance in *Staphylococcus aureus* through control of manganese uptake, *Mol. Microbiol.*, 2002, **44**, 1269–1286.

21. L. J. McAllister, H. J. Tseng, A. D. Ogunniyi, M. P. Jennings, A. G. McEwan and J. C. Paton, Molecular analysis of the psa permease complex of *Streptococcus pneumoniae*, *Mol. Microbiol.*, 2004, **53**, 889–901.

22. A. Antebi and G. R. Fink, The yeast $Ca(2+)$-ATPase homologue, *PMR1*, is required for normal Golgi function and localizes in a novel Golgi-like distribution, *Mol. Biol. Cell*, 1992, **3**, 633–654.

23. Z. Hao, S. Chen and D. B. Wilson, Cloning, expression, and characterization of cadmium and manganese uptake genes from *Lactobacillus plantarum*, *Appl. Environ. Microbiol.*, 1999, **65**, 4746–4752.

24. B. Montanini, D. Blaudez, S. Jeandroz, D. Sanders and M. Chalot, Phylogenetic and functional analysis of the Cation Diffusion Facilitator (CDF) family, *BMC Genomics*, 2007, **8**, 107.

25. A. A. Guffanti, Y. Wei, S. V. Rood and T. A. Krulwich, An antiport mechanism for a member of the cation diffusion facilitator family: divalent cations efflux in exchange for K^+ and H^+, *Mol. Microbiol.*, 2002, **45**, 145–153.

26. J. W. Rosch, G. Gao, G. Ridout, Y. D. Wang and E. I. Tuomanen, Role of the manganese efflux system mntE for signalling and pathogenesis in *Streptococcus pneumoniae*, *Mol. Microbiol.*, 2009, **72**, 12–25.

27. H. W. van Veen, T. Abee, G. J. Kortstee, W. N. Konings and A. J. Zehnder, Translocation of metal phosphate via the phosphate inorganic transport system of *Escherichia coli*, *Biochemistry*, 1994, **33**, 1766–1770.

28. U. Fristedt, R. Weinander, H. S. Martinsson and B. L. Persson, Characterization of purified and unidirectionally reconstituted Pho84 phosphate permease of *Saccharomyces cerevisiae*, *FEBS Lett.*, 1999, **458**, 1–5.

29. L. T. Jensen, M. Ajua-Alemanji and V. C. Culotta, The *Saccharomyces cerevisiae* high affinity phosphate transporter encoded by *PHO84* also functions in manganese homeostasis, *J. Biol. Chem.*, 2003, **278**, 42036–42040.

30. D. G. Kehres and M. E. Maguire, Emerging themes in manganese transport, biochemistry and pathogenesis in bacteria, *FEMS Microbiol. Rev.*, 2003, **27**, 263–290.

31. K. M. Papp-Wallace and M. E. Maguire, Manganese transport and the role of manganese in virulence, *Annu. Rev. Microbiol.*, 2006, **60**, 187–209.

32. S. J. Lin and V. C. Culotta, Suppression of oxidative damage by *Saccharomyces cerevisiae ATX2*, which encodes a manganese-trafficking protein that localizes to Golgi-like vesicles, *Mol. Cell. Biol.*, 1996, **16**, 6303–6312.

33. A. R. Reddi, L. T. Jensen, A. Naranuntarat, L. Rosenfeld, E. Leung, R. Shah and V. C. Culotta, The overlapping roles of manganese and Cu/Zn SOD in oxidative stress protection, *Free Radical Biol. Med.*, 2009, **46**, 154–162.

34. V. C. Culotta, M. Yang and M. D. Hall, Manganese transport and trafficking: lessons learned from *Saccharomyces cerevisiae*, *Eukaryotic Cell*, 2005, **4**, 1159–1165.

35. D. W. Christianson, Structural chemistry and biology of manganese metalloenzymes, *Prog. Biophys. Mol. Biol.*, 1997, **67**, 217–252.

36. J. W. Whittaker, Non-heme manganese catalase–the 'other' catalase, *Arch. Biochem. Biophys.*, 2012, **525**, 111–120.

37. T. R. Morgan, J. A. Shand, S. M. Clarke and J. J. Eaton-Rye, Specific requirements for cytochrome c-550 and the manganese-stabilizing protein in photoautotrophic strains of *Synechocystis* sp. PCC 6803, *Biochemistry*, 1998, **37**, 14437–14449.

38. Y. P. Chao, R. Patnaik, W. D. Roof, R. F. Young and J. C. Liao, Control of gluconeogenic growth by pps and pck in *Escherichia coli*, *J. Bacteriol.*, 1993, **175**, 6939–6944.

39. B. Mukhopadhyay, S. F. Stoddard and R. S. Wolfe, Purification, regulation, and molecular and biochemical characterization of pyruvate carboxylase from *Methanobacterium thermoautotrophicum* strain ΔH, *J. Biol. Chem.*, 1998, **273**, 5155–5166.

40. M. Chander, B. Setlow and P. Setlow, The enzymatic activity of phosphoglycerate mutase from gram-positive endospore-forming bacteria requires Mn^{2+} and is pH sensitive, *Can. J. Microbiol.*, 1998, **44**, 759–767.

41. J. E. Seemann and G. E. Schulz, Structure and mechanism of L-fucose isomerase from *Escherichia coli*, *J. Mol. Biol.*, 1997, **273**, 256–268.

42. C. F. Yocum and V. L. Pecoraro, Recent advances in the understanding of the biological chemistry of manganese, *Curr. Opin. Chem. Biol.*, 1999, **3**, 182–187.

43. N. J. Kuhn, S. Ward and W. S. Leong, Submicromolar manganese dependence of Golgi vesicular galactosyltransferase, *Eur. J. Biochem.*, 1991, **195**, 243–250.

44. T. Igarashi, Y. Kono and K. Tanaka, Molecular cloning of manganese catalase from *Lactobacillus plantarum*, *J. Biol. Chem.*, 1996, **271**, 29521–29524.

45. G. S. Johnson, C. R. Adler, J. J. Collins and D. Court, Role of the *spoT* gene product and manganese ion in the metabolism of guanosine 5′-diphosphate 3′-diphosphate in *Escherichia coli*, *J. Biol. Chem.*, 1979, **254**, 5483–5487.

46. C. Reyero and F. Dorner, Purification of arginases from human-leukemic lymphocytes and granulocytes: study of their physicochemical and kinetic properties, *Eur. J. Biochem.*, 1975, **56**, 137–147.

47. M. Leopoldini, N. Russo and M. Toscano, Determination of the catalytic pathway of a manganese arginase enzyme through density functional investigation, *Chemistry*, 2009, **15**, 8026–8036.

48. F. Archibald, Manganese: its acquisition by and function in the lactic acid bacteria, *Crit. Rev. Microbiol.*, 1986, **13**, 63–109.

49. R. J. Doyle, How cell walls of gram-positive bacteria interact with metal ions, in *Metal Ions and Bacteria*, ed. T. J. Beveridge and R. J. Doyle, Wiley, New York, 1989, pp. 275-293.

50. M. J. Daly, A new perspective on radiation resistance based on *Deinococcus radiodurans*, *Nat. Rev. Microbiol.*, 2009, **7**, 237–245.

51. M. J. Daly, E. K. Gaidamakova, V. Y. Matrosova, J. G. Kiang, R. Fukumoto, D. Y. Lee, N. B. Wehr, G. A. Viteri, B. S. Berlett and R. L. Levine, Small-molecule antioxidant proteome-shields in *Deinococcus radiodurans*, *PLoS One*, 2010, **5**, e12570.

52. K. Barnese, E. B. Gralla, D. E. Cabelli and J. S. Valentine, Manganous phosphate acts as a superoxide dismutase, *J. Am. Chem. Soc.*, 2008, **130**, 4604–4606.

53. E. C. Chang and D. J. Kosman, Intracellular Mn (II)-associated superoxide scavenging activity protects Cu, Zn superoxide dismutase-deficient *Saccharomyces cerevisiae* against dioxygen stress, *J. Biol. Chem.*, 1989, **264**, 12172–12178.

54. F. S. Archibald and I. Fridovich, Investigations of the state of the manganese in *Lactobacillus plantarum, Arch. Biochem. Biophys.*, 1982, **215**, 589–596.

55. K. Barnese, E. B. Gralla, J. S. Valentine and D. E. Cabelli, Biologically relevant mechanism for catalytic superoxide removal by simple manganese compounds, *Proc. Natl. Acad. Sci. U. S. A.*, 2012, **109**, 6892–6897.

56. T. Offer, A. Russo and A. Samuni, The pro-oxidative activity of SOD and nitroxide SOD mimics, *FASEB J.*, 2000, **14**, 1215–1223.

57. G. Lupidi, F. Marchetti, N. Masciocchi, D. L. Reger, S. Tabassum, P. Astolfi, E. Damiani and C. Pettinari, Synthesis, structural and spectroscopic characterization and biomimetic properties of new copper, manganese, zinc complexes, *J. Inorg. Biochem.*, 2010, **104**, 820–830.

58. A. Horn, Jr., G. L. Parrilha, K. V. Melo, C. Fernandes, M. Horner, C. Visentin Ldo, J. A. Santos, M. S. Santos, E. C. Eleutherio and M. D. Pereira, An iron-based cytosolic catalase and superoxide dismutase mimic complex, *Inorg. Chem.*, 2010, **49**, 1274–1276.

59. J. D. Aguirre and V. C. Culotta, Battles with iron: manganese in oxidative stress protection, *J. Biol. Chem.*, 2012, **287**, 13541–13548.

60. M. J. Daly, Death by protein damage in irradiated cells, *DNA Repair*, 2012, **11**, 12–21.

61. M. Al-Maghrebi, I. Fridovich and L. Benov, Manganese supplementation relieves the phenotypic deficits seen in superoxide dismutase null *Escherichia coli, Arch. Biochem. Biophys.*, 2002, **402**, 104–109.

62. A. Anjem, S. Varghese and J. A. Imlay, Manganese import is a key element of the OxyR response to hydrogen peroxide in *Escherichia coli, Mol. Microbiol.*, 2009, **72**, 844–858.

63. M. J. Daly, E. K. Gaidamakova, V. Y. Matrosova, A. Vasilenko, M. Zhai, A. Venkateswaran, M. Hess, M. V. Omelchenko, H. M. Kostandarithes, K. S. Makarova, L. P. Wackett, J. K. Fredrickson and D. Ghosal, Accumulation of Mn(II) in *Deinococcus radiodurans* facilitates gamma-radiation resistance, *Science*, 2004, **306**, 1025–1028.

64. E. M. Gregory and I. Fridovich, Oxygen metabolism in *Lactobacillus plantarum, J. Bacteriol.*, 1974, **117**, 166–169.

65. F. Gotz, E. F. Elstner, B. Sedewitz and E. Lengfelder, Oxygen utilization by *Lactobacillus plantarum, Arch. Microbiol.*, 1980, **125**, 215–220.

66. F. S. Archibald and I. Fridovich, Manganese and defenses against oxygen toxicity in *Lactobacillus plantarum, J. Bacteriol.*, 1981, **145**, 442–451.

67. M. Watanabe, S. van der Veen, H. Nakajima and T. Abee, Effect of respiration and manganese on oxidative stress resistance of *Lactobacillus plantarum* WCFS1, *Microbiology*, 2012, **158**, 293–300.

68. J. H. Cho, K. M. Ko, G. Singaravelu and J. Ahnn, *Caenorhabditis elegans* PMR1, a P-type calcium ATPase, is important for calcium/manganese homeostasis and oxidative stress response, *FEBS Lett.*, 2005, **579**, 778–782.

69. Y. T. Lin, H. Hoang, S. I. Hsieh, N. Rangel, A. L. Foster, J. N. Sampayo, G. J. Lithgow and C. Srinivasan, Manganous ion supplementation accelerates wild type development, enhances stress resistance, and rescues the life span of a short-lived *Caenorhabditis elegans* mutant, *Free Radical Biol. Med.*, 2006, **40**, 1185–1193.

70. E. Boyer, I. Bergevin, D. Malo, P. Gros and M. F. Cellier, Acquisition of Mn(II) in addition to Fe(II) is required for full virulence of *Salmonella enterica* serovar Typhimurium, *Infect. Immun.*, 2002, **70**, 6032–6042.

71. A. Janakiraman and J. M. Slauch, The putative iron transport system SitABCD encoded on SPI1 is required for full virulence of *Salmonella typhimurium*, *Mol. Microbiol.*, 2000, **35**, 1146–1155.

72. N. S. Jakubovics and H. F. Jenkinson, Out of the iron age: new insights into the critical role of manganese homeostasis in bacteria, *Microbiology*, 2001, **147**, 1709–1718.

73. H. Makui, E. Roig, S. T. Cole, J. D. Helmann, P. Gros and M. F. Cellier, Identification of the *Escherichia coli* K-12 Nramp orthologue (MntH) as a selective divalent metal ion transporter, *Mol. Microbiol.*, 2000, **35**, 1065–1078.

74. Q. Que and J. D. Helmann, Manganese homeostasis in *Bacillus subtilis* is regulated by MntR, a bifunctional regulator related to the diphtheria toxin repressor family of proteins, *Mol. Microbiol.*, 2000, **35**, 1454–1468.

75. M. H. Saier, Jr., A functional-phylogenetic classification system for transmembrane solute transporters, *Microbiol. Mol. Biol. Rev.*, 2000, **64**, 354–411.

76. L. J. Runyen-Janecky, S. A. Reeves, E. G. Gonzales and S. M. Payne, Contribution of the *Shigella flexneri* Sit, Iuc, and Feo iron acquisition systems to iron acquisition *in vitro* and in cultured cells, *Infect. Immun.*, 2003, **71**, 1919–1928.

77. T. Padilla-Benavides, J. E. Long, D. Raimunda, C. M. Sassetti and J. M. Arguello, A novel P(1B)-type Mn2 + -transporting ATPase is required for secreted protein metallation in mycobacteria, *J. Biol. Chem.*, 2013, **288**, 11334–11347.

78. N. Nelson, Metal ion transporters and homeostasis, *EMBO J.*, 1999, **18**, 4361–4371.

79. J. R. Forbes and P. Gros, Divalent-metal transport by NRAMP proteins at the interface of host-pathogen interactions, *Trends Microbiol.*, 2001, **9**, 397–403.

80. D. Agranoff, L. Collins, D. Kehres, T. Harrison, M. Maguire and S. Krishna, The Nramp orthologue of *Cryptococcus neoformans* is a

pH-dependent transporter of manganese, iron, cobalt and nickel, *Biochem. J.*, 2005, **385**, 225–232.

81. F. Supek, L. Supekova, H. Nelson and N. Nelson, A yeast manganese transporter related to the macrophage protein involved in conferring resistance to mycobacteria, *Proc. Natl. Acad. Sci. U. S. A.*, 1996, **93**, 5105–5110.

82. X. F. Liu, F. Supek, N. Nelson and V. C. Culotta, Negative control of heavy metal uptake by the *Saccharomyces cerevisiae BSD2* gene, *J. Biol. Chem.*, 1997, **272**, 11763–11769.

83. X. Z. Chen, J. B. Peng, A. Cohen, H. Nelson, N. Nelson and M. A. Hediger, Yeast SMF1 mediates H(+)-coupled iron uptake with concomitant uncoupled cation currents, *J. Biol. Chem.*, 1999, **274**, 35089–35094.

84. Y. Nevo and N. Nelson, The mutation F227I increases the coupling of metal ion transport in DCT1, *J. Biol. Chem.*, 2004, **279**, 53056–53061.

85. D. I. Bannon, R. Abounader, P. S. Lees and J. P. Bressler, Effect of DMT1 knockdown on iron, cadmium, and lead uptake in Caco-2 cells, *Am. J. Physiol.: Cell Physiol.*, 2003, **284**, C44–50.

86. H. Gunshin, B. Mackenzie, U. V. Berger, Y. Gunshin, M. F. Romero, W. F. Boron, S. Nussberger, J. L. Gollan and M. A. Hediger, Cloning and characterization of a mammalian proton-coupled metal-ion transporter, *Nature*, 1997, **388**, 482–488.

87. I. Reeve, D. Hummel, N. Nelson and J. Voss, Overexpression, purification, and site-directed spin labeling of the Nramp metal transporter from *Mycobacterium leprae*, *Proc. Natl. Acad. Sci. U. S. A.*, 2002, **99**, 8608–8613.

88. D. Agranoff, I. M. Monahan, J. A. Mangan, P. D. Butcher and S. Krishna, *Mycobacterium tuberculosis* expresses a novel pH-dependent divalent cation transporter belonging to the Nramp family, *J. Exp. Med.*, 1999, **190**, 717–724.

89. A. Sacher, A. Cohen and N. Nelson, Properties of the mammalian and yeast metal-ion transporters DCT1 and Smf1p expressed in *Xenopus laevis* oocytes, *J. Exp. Biol.*, 2001, **204**, 1053–1061.

90. N. Jabado, A. Jankowski, S. Dougaparsad, V. Picard, S. Grinstein and P. Gros, Natural resistance to intracellular infections: natural resistance-associated macrophage protein 1 (Nramp1) functions as a pH-dependent manganese transporter at the phagosomal membrane, *J. Exp. Med.*, 2000, **192**, 1237–1248.

91. V. Picard, G. Govoni, N. Jabado and P. Gros, Nramp 2 (DCT1/DMT1) expressed at the plasma membrane transports iron and other divalent cations into a calcein-accessible cytoplasmic pool, *J. Biol. Chem.*, 2000, **275**, 35738–35745.

92. S. Thomine, R. Wang, J. M. Ward, N. M. Crawford and J. I. Schroeder, Cadmium and iron transport by members of a plant metal transporter family in *Arabidopsis* with homology to Nramp genes, *Proc. Natl. Acad. Sci. U. S. A.*, 2000, **97**, 4991–4996.

93. C. F. Higgins, ABC transporters: from microorganisms to man, *Annu. Rev. Cell Biol.*, 1992, **8**, 67–113.
94. G. F. Ames, Bacterial periplasmic transport systems: structure, mechanism, and evolution, *Annu. Rev. Biochem.*, 1986, **55**, 397–425.
95. D. C. Rees, E. Johnson and O. Lewinson, ABC transporters: the power to change, *Nat. Rev. Mol. Cell Biol.*, 2009, **10**, 218–227.
96. J. S. Klein and O. Lewinson, Bacterial ATP-driven transporters of transition metals, *Metallomics*, 2011, **3**, 1098–1108.
97. M. M. Gottesman and I. Pastan, Biochemistry of multidrug resistance mediated by the multidrug transporter, *Annu. Rev. Biochem.*, 1993, **62**, 385–427.
98. M. N. Alekshun and S. B. Levy, Molecular mechanisms of antibacterial multidrug resistance, *Cell*, 2007, **128**, 1037–1050.
99. T. Kitten, C. L. Munro, S. M. Michalek and F. L. Macrina, Genetic characterization of a *Streptococcus mutans* LraI family operon and role in virulence, *Infect. Immun.*, 2000, **68**, 4441–4451.
100. P. E. Kolenbrander, R. N. Andersen, R. A. Baker and H. F. Jenkinson, The adhesion-associated sca operon in *Streptococcus gordonii* encodes an inducible high-affinity ABC transporter for Mn^{2+} uptake, *J. Bacteriol.*, 1998, **180**, 290–295.
101. M. Sabri, S. Leveille and C. M. Dozois, A SitABCD homologue from an avian pathogenic *Escherichia coli* strain mediates transport of iron and manganese and resistance to hydrogen peroxide, *Microbiology*, 2006, **152**, 745–758.
102. V. V. Bartsevich and H. B. Pakrasi, Manganese transport in the cyanobacterium *Synechocystis* sp. PCC 6803, *J. Biol. Chem.*, 1996, **271**, 26057–26061.
103. G. Kuan, E. Dassa, W. Saurin, M. Hofnung and M. H. Saier, Jr., Phylogenetic analyses of the ATP-binding constituents of bacterial extracytoplasmic receptor-dependent ABC-type nutrient uptake permeases, *Res. Microbiol.*, 1995, **146**, 271–278.
104. H. F. Jenkinson, Cell surface protein receptors in oral *streptococci*, *FEMS Microbiol. Lett.*, 1994, **121**, 133–140.
105. H. J. Tseng, A. G. McEwan, J. C. Paton and M. P. Jennings, Virulence of *Streptococcus pneumoniae*: PsaA mutants are hypersensitive to oxidative stress, *Infect. Immun.*, 2002, **70**, 1635–1639.
106. M. P. Schmitt, Analysis of a DtxR-like metalloregulatory protein, MntR, from *Corynebacterium diphtheriae* that controls expression of an ABC metal transporter by a Mn(2 +)-dependent mechanism, *J. Bacteriol.*, 2002, **184**, 6882–6892.
107. R. Platero, L. Peixoto, M. R. O'Brian and E. Fabiano, Fur is involved in manganese-dependent regulation of *mntA* (*sitA*) expression in *Sinorhizobium meliloti*, *Appl. Environ. Microbiol.*, 2004, **70**, 4349–4355.
108. M. G. Palmgren and P. Nissen, P-type ATPases, *Annu. Rev. Biophys.*, 2011, **40**, 243–266.

109. M. N. Groot, E. Klaassens, W. M. de Vos, J. Delcour, P. Hols and M. Kleerebezem, Genome-based *in silico* detection of putative manganese transport systems in *Lactobacillus plantarum* and their genetic analysis, *Microbiology*, 2005, **151**, 1229–1238.
110. S. J. Beard, R. Hashim, G. Wu, M. R. Binet, M. N. Hughes and R. K. Poole, Evidence for the transport of zinc(II) ions via the pit inorganic phosphate transport system in *Escherichia coli*, *FEMS Microbiol. Lett.*, 2000, **184**, 231–235.
111. A. Kirsten, M. Herzberg, A. Voigt, J. Seravalli, G. Grass, J. Scherer and D. H. Nies, Contributions of five secondary metal uptake systems to metal homeostasis of *Cupriavidus metallidurans* CH34, *J. Bacteriol.*, 2011, **193**, 4652–4663.
112. F. S. Archibald and M. N. Duong, Manganese acquisition by *Lactobacillus plantarum*, *J. Bacteriol.*, 1984, **158**, 1–8.
113. W. W. Kay and O. K. Ghei, Inorganic cation transport and the effects on C4 dicarboxylate transport in *Bacillus subtilis*, *Can. J. Microbiol.*, 1981, **27**, 1194–1201.
114. D. H. Nies, Efflux-mediated heavy metal resistance in prokaryotes, *FEMS Microbiol. Rev.*, 2003, **27**, 313–339.
115. G. Grass, M. Otto, B. Fricke, C. J. Haney, C. Rensing, D. H. Nies and D. Munkelt, FieF (YiiP) from *Escherichia coli* mediates decreased cellular accumulation of iron and relieves iron stress, *Arch. Microbiol.*, 2005, **183**, 9–18.
116. C. Li, J. Tao, D. Mao and C. He, A novel manganese efflux system, YebN, is required for virulence by *Xanthomonas oryzae pv. oryzae*, *PLoS One*, 2011, **6**, e21983.
117. L. S. Waters, M. Sandoval and G. Storz, The *Escherichia coli* MntR miniregulon includes genes encoding a small protein and an efflux pump required for manganese homeostasis, *J. Bacteriol.*, 2011, **193**, 5887–5897.
118. S. A. Lieser, T. C. Davis, J. D. Helmann and S. M. Cohen, DNA-binding and oligomerization studies of the manganese(II) metalloregulatory protein MntR from *Bacillus subtilis*, *Biochemistry*, 2003, **42**, 12634–12642.
119. J. S. Ikeda, A. Janakiraman, D. G. Kehres, M. E. Maguire and J. M. Slauch, Transcriptional regulation of *sitABCD* of *Salmonella enterica* serovar Typhimurium by MntR and Fur, *J. Bacteriol.*, 2005, **187**, 912–922.
120. D. G. Kehres, A. Janakiraman, J. M. Slauch and M. E. Maguire, Regulation of *Salmonella enterica* serovar Typhimurium *mntH* transcription by $H(2)O(2)$, Fe(2 +), and Mn(2 +), *J. Bacteriol.*, 2002, **184**, 3151–3158.
121. S. I. Patzer and K. Hantke, Dual repression by Fe(2 +)-Fur and Mn(2 +)-MntR of the *mntH* gene, encoding an NRAMP-like Mn(2 +) transporter in *Escherichia coli*, *J. Bacteriol.*, 2001, **183**, 4806–4813.
122. K. Hantke, Iron and metal regulation in bacteria, *Curr. Opin. Microbiol.*, 2001, **4**, 172–177.

123. N. Bsat, A. Herbig, L. Casillas-Martinez, P. Setlow and J. D. Helmann, *Bacillus subtilis* contains multiple Fur homologues: identification of the iron uptake (Fur) and peroxide regulon (PerR) repressors, *Mol. Microbiol.*, 1998, **29**, 189–198.

124. M. J. Horsburgh, M. O. Clements, H. Crossley, E. Ingham and S. J. Foster, PerR controls oxidative stress resistance and iron storage proteins and is required for virulence in *Staphylococcus aureus*, *Infect. Immun.*, 2001, **69**, 3744–3754.

125. E. E. Luk and V. C. Culotta, Manganese superoxide dismutase in *Saccharomyces cerevisiae* acquires its metal co-factor through a pathway involving the Nramp metal transporter, Smf2p, *J. Biol. Chem.*, 2001, **276**, 47556–47562.

126. V. C. Culotta and M. J. Daly, Manganese complexes: diverse metabolic routes to oxidative stress resistance in prokaryotes and yeast, *Antioxid. Redox Signaling*, 2013, **19**, 933–944.

127. R. J. Jackson, M. R. Binet, L. J. Lee, R. Ma, A. I. Graham, C. W. McLeod and R. K. Poole, Expression of the PitA phosphate/metal transporter of *Escherichia coli* is responsive to zinc and inorganic phosphate levels, *FEMS Microbiol. Lett.*, 2008, **289**, 219–224.

128. X. F. Liu and V. C. Culotta, Post-translation control of Nramp metal transport in yeast. Role of metal ions and the *BSD2* gene, *J. Biol. Chem.*, 1999, **274**, 4863–4868.

129. X. F. Liu and V. C. Culotta, Mutational analysis of *Saccharomyces cerevisiae* Smf1p, a member of the Nramp family of metal transporters, *J. Mol. Biol.*, 1999, **289**, 885–891.

130. J. A. Sullivan, M. J. Lewis, E. Nikko and H. R. Pelham, Multiple interactions drive adaptor-mediated recruitment of the ubiquitin ligase rsp5 to membrane proteins *in vivo* and *in vitro*, *Mol. Biol. Cell*, 2007, **18**, 2429–2440.

131. V. C. Culotta, Manganese transport in microorganisms, in *Metal ion in biological systems*, ed. A. Sigel and H. Sigel, Marcel Dekker, New York, 2000, vol. 37, pp. 35–53.

132. N. C. Andrews, Iron homeostasis: insights from genetics and animal models, *Nat. Rev. Genet.*, 2000, **1**, 208–217.

133. A. Cohen, H. Nelson and N. Nelson, The family of SMF metal ion transporters in yeast cells, *J. Biol. Chem.*, 2000, **275**, 33388–33394.

134. M. E. Portnoy, L. T. Jensen and V. C. Culotta, The distinct methods by which manganese and iron regulate the Nramp transporters in yeast, *Biochem. J.*, 2002, **362**, 119–124.

135. A. H. West, D. J. Clark, J. Martin, W. Neupert, F. U. Hartl and A. L. Horwich, Two related genes encoding extremely hydrophobic proteins suppress a lethal mutation in the yeast mitochondrial processing enhancing protein, *J. Biol. Chem.*, 1992, **267**, 24625–24633.

136. E. Pinner, S. Gruenheid, M. Raymond and P. Gros, Functional complementation of the yeast divalent cation transporter family

SMF by NRAMP2, a member of the mammalian natural resistance-associated macrophage protein family, *J. Biol. Chem.*, 1997, **272**, 28933–28938.

137. H. E. Stimpson, M. J. Lewis and H. R. Pelham, Transferrin receptor-like proteins control the degradation of a yeast metal transporter, *EMBO J.*, 2006, **25**, 662–672.

138. E. Luk, M. Yang, L. T. Jensen, Y. Bourbonnais and V. C. Culotta, *J. Biol. Chem.*, 2005, **280**, 22715–22720.

139. E. Luk, L. T. Jensen and V. C. Culotta, The many highways for intracellular trafficking of metals, *J. Biol. Inorg. Chem.*, 2003, **8**, 803–809.

140. G. Durr, J. Strayle, R. Plemper, S. Elbs, S. K. Klee, P. Catty, D. H. Wolf and H. K. Rudolph, The medial-Golgi ion pump Pmr1 supplies the yeast secretory pathway with Ca2+ and Mn2+ required for glycosylation, sorting, and endoplasmic reticulum-associated protein degradation, *Mol. Biol. Cell*, 1998, **9**, 1149–1162.

141. D. Mandal, T. B. Woolf and R. Rao, Manganese selectivity of Pmr1, the yeast secretory pathway ion pump, is defined by residue gln783 in transmembrane segment 6. Residue Asp778 is essential for cation transport, *J. Biol. Chem.*, 2000, **275**, 23933–23938.

142. P. J. Lapinskas, S. J. Lin and V. C. Culotta, The role of the *Saccharomyces cerevisiae CCC1* gene in the homeostasis of manganese ions, *Mol. Microbiol.*, 1996, **21**, 519–528.

143. L. Li, O. S. Chen, D. McVey Ward and J. Kaplan, CCC1 is a transporter that mediates vacuolar iron storage in yeast, *J. Biol. Chem.*, 2001, **276**, 29515–29519.

144. H. K. Rudolph, A. Antebi, G. R. Fink, C. M. Buckley, T. E. Dorman, J. LeVitre, L. S. Davidow, J. I. Mao and D. T. Moir, The yeast secretory pathway is perturbed by mutations in PMR1, a member of a Ca2+ ATPase family, *Cell*, 1989, **58**, 133–145.

145. P. J. Lapinskas, K. W. Cunningham, X. F. Liu, G. R. Fink and V. C. Culotta, Mutations in *PMR1* suppress oxidative damage in yeast cells lacking superoxide dismutase, *Mol. Cell. Biol.*, 1995, **15**, 1382–1388.

146. V. K. Ton, D. Mandal, C. Vahadji and R. Rao, Functional expression in yeast of the human secretory pathway Ca(2+), Mn(2+)-ATPase defective in Hailey-Hailey disease, *J. Biol. Chem.*, 2002, **277**, 6422–6427.

147. A. R. Reddi, L. T. Jensen and V. C. Culotta, Manganese homeostasis in *Saccharomyces cerevisiae*, *Chem. Rev.*, 2009, **109**, 4722–4732.

148. I. C. Farcasanu, M. Mizunuma, D. Hirata and T. Miyakawa, Involvement of histidine permease (Hip1p) in manganese transport in *Saccharomyces cerevisiae*, *Mol. Gen. Genet.*, 1998, **259**, 541–548.

149. D. J. Eide, J. T. Bridgham, Z. Zhao and J. R. Mattoon, The vacuolar H(+)-ATPase of *Saccharomyces cerevisiae* is required for efficient

copper detoxification, mitochondrial function, and iron metabolism, *Mol. Gen. Genet.*, 1993, **241**, 447–456.

150. L. M. Ramsay and G. M. Gadd, Mutants of *Saccharomyces cerevisiae* defective in vacuolar function confirm a role for the vacuole in toxic metal ion detoxification, *FEMS Microbiol. Lett.*, 1997, **152**, 293–298.

151. S. A. Kim, T. Punshon, A. Lanzirotti, L. Li, J. M. Alonso, J. R. Ecker, J. Kaplan and M. L. Guerinot, Localization of iron in *Arabidopsis* seed requires the vacuolar membrane transporter VIT1, *Science*, 2006, **314**, 1295–1298.

152. M. Yang, P. A. Cobine, S. Molik, A. Naranuntarat, R. Lill, D. R. Winge and V. C. Culotta, The effects of mitochondrial iron homeostasis on cofactor specificity of superoxide dismutase 2, *EMBO J.*, 2006, **25**, 1775–1783.

153. A. Naranuntarat, L. T. Jensen, S. Pazicni, J. E. Penner-Hahn and V. C. Culotta, The interaction of mitochondrial iron with manganese superoxide dismutase, *J. Biol. Chem.*, 2009, **284**, 22633–22640.

154. J. C. Rutherford and A. J. Bird, Metal-responsive transcription factors that regulate iron, zinc, and copper homeostasis in eukaryotic cells, *Eukaryotic Cell*, 2004, **3**, 1–13.

155. L. T. Jensen, M. C. Carroll, M. D. Hall, C. J. Harvey, S. E. Beese and V. C. Culotta, Down-regulation of a manganese transporter in the face of metal toxicity, *Mol. Biol. Cell*, 2009, **20**, 2810–2819.

156. L. Eguez, Y. S. Chung, A. Kuchibhatla, M. Paidhungat and S. Garrett, Yeast Mn2+ transporter, Smf1p, is regulated by ubiquitin-dependent vacuolar protein sorting, *Genetics*, 2004, **167**, 107–117.

157. E. H. Hettema, J. Valdez-Taubas and H. R. Pelham, Bsd2 binds the ubiquitin ligase Rsp5 and mediates the ubiquitination of transmembrane proteins, *EMBO J.*, 2004, **23**, 1279–1288.

158. E. Nikko, J. A. Sullivan and H. R. Pelham, Arrestin-like proteins mediate ubiquitination and endocytosis of the yeast metal transporter Smf1, *EMBO Rep.*, 2008, **9**, 1216–1221.

159. L. Li, D. Bagley, D. M. Ward and J. Kaplan, Yap5 is an iron-responsive transcriptional activator that regulates vacuolar iron storage in yeast, *Mol. Cell. Biol.*, 2008, **28**, 1326–1337.

160. C. Au, A. Benedetto, J. Anderson, A. Labrousse, K. Erikson, J. J. Ewbank and M. Aschner, SMF-1, SMF-2 and SMF-3 DMT1 orthologues regulate and are regulated differentially by manganese levels in *C. elegans*, *PLoS One*, 2009, **4**, e7792.

161. C. Au, A. Benedetto and M. Aschner, Manganese transport in eukaryotes: the role of DMT1, *Neurotoxicology*, 2008, **29**, 569–576.

162. J. Bandyopadhyay, H. O. Song, B. J. Park, G. Singaravelu, J. L. Sun, J. Ahnn and J. H. Cho, Functional assessment of Nramp-like metal transporters and manganese in *Caenorhabditis elegans*, *Biochem. Biophys. Res. Commun.*, 2009, **390**, 136–141.

163. G. McColl, S. A. James, S. Mayo, D. L. Howard, C. G. Ryan, R. Kirkham, G. F. Moorhead, D. Paterson, M. D. de Jonge and A. I. Bush, *Caenorhabditis elegans* maintains highly compartmentalized cellular distribution of metals and steep concentration gradients of manganese, *PLoS One*, 2012, **7**, e32685.

164. R. Hunt-Newbury, R. Viveiros, R. Johnsen, A. Mah, D. Anastas, L. Fang, E. Halfnight, D. Lee, J. Lin, A. Lorch, S. McKay, H. M. Okada, J. Pan, A. K. Schulz, D. Tu, K. Wong, Z. Zhao, A. Alexeyenko, T. Burglin, E. Sonnhammer, R. Schnabel, S. J. Jones, M. A. Marra, D. L. Baillie and D. G. Moerman, High-throughput in vivo analysis of gene expression in *Caenorhabditis elegans*, *PLoS Biol.*, 2007, **5**, e237.

165. K. Van Baelen, J. Vanoevelen, L. Missiaen, L. Raeymaekers and F. Wuytack, The Golgi PMR1 P-type ATPase of *Caenorhabditis elegans*, *J. Biol. Chem.*, 2001, **276**, 10683–10691.

CHAPTER 2

Nutritional Requirements for Manganese

JEANNE H. FREELAND-GRAVES,* TAMARA Y. MOUSA AND
NAMRATA SANJEEVI

Department of Nutritional Sciences, University of Texas, Austin, Texas,
USA
*Email: jfg@mail.utexas.edu

2.1   Introduction

Manganese (Mn) is a ubiquitous trace mineral that is essential for all living
organisms. Mn has the atomic number 25 and the most biologically
important oxidation state[1] is Mn^{2+}. This trace mineral has a significant
role as a cofactor for several critical enzymes. For example, Mn superoxide
dismutase (MnSOD) is an antioxidant enzyme that is critical for the prevention and/or reduction of oxidative stress, facilitating the control of
chronic diseases such as diabetes mellitus. Arginase is a Mn-dependent
enzyme that regulates urea production in the liver, and nitric oxide synthase
in smooth muscle cells.[2] Other enzymes that utilize Mn as a cofactor or
activator are involved in the metabolism of carbohydrates, amino acids and
cholesterol, the formation of bone and cartilage, and wound healing.[3] They
include pyruvate carboxylase, transferases, hydrolases, kinases,[3] lyases,
oxido-reductases, isomerases, ligases, glutamine synthetase, and phosphoenolpyruvate decarboxylase.[4,5] Another Mn-dependent enzyme is the
reverse transcriptase that some retroviruses use to multiply by transcription
of the genetic material.[6] Finally, Mn assumes a significant role in digestion,
development, reproduction, immune function, and regulation of cellular

Issues in Toxicology No. 22
Manganese in Health and Disease
Edited by Lucio G. Costa and Michael Aschner
© The Royal Society of Chemistry 2015
Published by the Royal Society of Chemistry, www.rsc.org

energy and blood glucose. Thus, an understanding of the nutrient requirement for Mn is important for achievement of optimal health in humans.

2.2   Food Sources

The primary sources of Mn in the diet are whole grains and cereals. Other good sources are brown rice, pineapples, green leafy vegetables, dried legumes, nuts and tea. For example, the Mn content in one cup of tea ranges from 0.4 to 1.3 mg, and one-half cup of pineapple contains 0.77 mg. In contrast, meats, fat, sugar and white flour are poor sources of this mineral, as illustrated in Figure 2.1.

In the past, diets were based on abundant amounts and varieties of whole grains and cereals, foods that are rich in Mn. But the nutrition transition that occurred in the United States (US) and other developed countries has shifted diets more towards those high in processed meats and energy-dense, high-fat foods, all of which are low in this element. Thus, the quantity of Mn in the diet has been reduced substantially in recent years. Manganese also may be derived from drinking water, with concentrations varying between 1 and 100 $\mu g\,l^{-1}$, depending on its source.[7] Finally, other significant contributors of Mn may be supplements and infant formulas.

The Mn content in human breast milk is 3–8 $\mu g\,l^{-1}$, but this amount is much less than that present in infant formulas containing either cow's milk (50–100 $\mu g\,l^{-1}$) or soy (200–300 $\mu g\,l^{-1}$).[8] Cockell and Belonjea (2004)[9] also reported Mn levels in milk protein formulas (90 ± 50 $\mu g\,l^{-1}$) to be less than in those which are soy based (310 ± 80 $\mu g\,l^{-1}$). It has been postulated that the

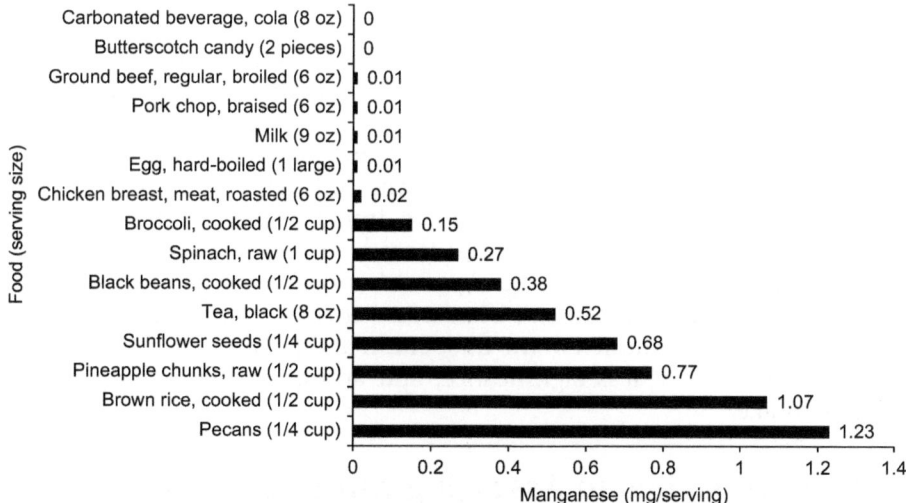

Figure 2.1   Manganese content of common food items.
(USDA food composition database 2014.)

elevated levels of Mn found in soy formula may increase the susceptibility of an infant to Mn toxicity.

2.3   Absorption, Transport and Excretion

The proportion of Mn absorbed from food sources in humans is believed to approximate 5% or less. In meals labeled with ^{54}Mn, mean absorption was 1.71%, 2.16%, 3.81% and 5.20% from sunflower seeds, wheat, spinach and lettuce, respectively.[10] It is believed that homeostatic mechanisms may be employed to increase fecal excretion of this mineral in response to higher dietary levels. Thus, the uptake of Mn may be a highly regulated process that is dependent on the amount of Mn, other dietary components and physiological variants.

The amount of Mn stored in the body of a 70 kg adult man[11] is approximately 12–20 mg. Manganese is absorbed into the intestinal cell by the binding of Mn^{3+} to the transferrin–transferrin receptor (Tfr) complex. Endocytosis brings about internalization of the complex, Mn^{3+}–transferrin–Tfr,[12] with the subsequent dissociation of Mn^{3+} into Mn^{2+}. The resultant Mn^{2+} and proton are then transported into the cytosol by the divalent metal transporter-1 (DMT-1).[13] The protein, ferroportin, which is responsible for export of Fe from the cell, has been implicated in cytoplasmic Mn efflux as well.[14] Subsequently, Mn is released into the circulation, where the proteins involved in the transport of circulating Mn to peripheral tissues include albumin and β-globulin.[15] Recently, Herrera and Bartnikas explored the role of transferrin in Mn distribution in mice. This protein was not observed to be involved in Mn transport to body cells, yet it had an indirect effect on Mn deficiency. Low levels of transferrin were documented to affect hepcidin negatively and increase Fe metabolism, thereby contributing to the low Mn concentration. The inverse association observed between Fe and Mn absorption was due apparently to sharing of the same transporter.[16]

The distribution of Mn in the body is dependent on the mitochondrial content of tissues, with the greatest deposition in mitochondrial-rich tissues such as bone, liver, kidneys, pituitary gland, and pancreas.[3] The liver plays a major role in the excretion of surplus Mn, and helps in maintenance of Mn homeostasis. Excess Mn is secreted into the bile by the liver, and subsequently excreted through the feces.[17] Thus, hepatic dysfunction may result in Mn imbalance, thereby causing deficiency or toxicity of this mineral.[18] Alternatively, Mn can be excreted through pancreatic juices and be reabsorbed in the lumen of the duodenum and jejunum. Excretion of Mn from the urine is of less significance, and appears to be independent of dietary intake.[19]

2.4   Approaches to Assessing Mn Requirements

Approaches to assessing Mn requirements include metabolic balance, blood levels of Mn, and other biomarkers. Biomarkers that have been used to reflect Mn status other than blood are feces, urine, hair,[17] and activities of Mn-dependent enzymes such as MnSOD and arginase. When sufficient data are lacking, a crude method to estimate status and/or requirements in a healthy population is extrapolation to usual dietary intakes.[20]

2.4.1 Metabolic Balance

Metabolic balance has been used in past investigations to set preliminary dietary recommendations for numerous minerals. This method has been replaced largely with newer isotope distribution studies utilizing stable isotopes. Ideally, Mn could be assessed *via* isotopes to measure requirements, yet this approach is precluded because of the existence of only one stable isotope, ^{55}Mn, that does not create any decay products. Thus, below is a brief review of a number of metabolic balance studies that have been useful in the determination of Mn requirements in humans.

2.4.1.1 Adult Diets

An early study by North *et al.* (1960) found that college women fed conventional diets, with a mean Mn intake of 3.7 mg per day, exhibited Mn balance of 1.54 mg per day and retention of 41% (Table 2.1).[21] In college men who were vegetarians and consumed their self-selected diets, Lang *et al.* (1965) found that a higher Mn intake of 7.07 mg per day resulted in a positive Mn balance of 3.34 mg per day and retention of 47%.[22] In 1984, Patterson *et al.* examined Mn dietary levels over the course of one year in 28 young adult men and women. A diet was collected for two days during each of the four seasons of a year. The dietary intake of Mn averaged 3.0 mg per day, with a mean negative metabolic balance of −0.16, −0.16, −0.12 and −0.21 mg per day for Spring, Summer, Fall and Winter, respectively. These negative balances suggested that a dietary level of 3 mg was inadequate to permit positive balance.[23]

The author of this chapter documented Mn metabolic balance in young men fed diets of conventional foods for 105 days. The diets varied in mean Mn concentrations, at 2.89, 2.06, 1.21, 3.79 and 2.65 mg per day; these intakes corresponded to Mn balances of −0.083, −0.018, −0.088, 0.657 and 0.136 mg per day, respectively. Mn retentions of −2.90 to 17.34% were calculated,[19] and this indicated that the levels of dietary Mn influenced both the percentage absorbed and that retained. The authors concluded that the mean sum of endogenous and exogenous losses would be 392 µg per day when dietary levels were theoretically zero. The incorporation of the positive retention would mean a dietary requirement of 3–5 mg per day or 50 µg kg^{-1}. Figure 2.2 illustrates this human requirement based on consumption of mixed diets, in comparison with values recommended for other mammals.

When Finley *et al.* fed a mixed Western diet to 20 men (5.43 mg Mn per day) and women (4.01 mg Mn per day) for 70 days, Mn balances averaged 0.27 and −0.12 mg per day, respectively (Table 2.1).[24] The men showed a

Table 2.1 Characteristics of metabolic balance studies that investigated manganese (Mn) requirements in adults.

Reference	Design	Age (years)	Gender[a]	Sample (n)	Country of origin	Type of diet	Mean Mn intake (mg per day)	Mean Mn balance (mg per day)	Retention (%)
Finley et al., 1994[24]	Controlled trial	18–40	M	20	United States	Conventional foods	5.43	0.27	4.97
		18–40	F	20			4.01	−0.12	−2.99
Freeland-Graves et al., 1988[19]	Controlled trial	19–20	M	5	United States	Conventional foods	2.89	−0.08	−2.90
							2.06	−0.02	−0.88
							1.21	−0.09	−7.40
							3.79	0.66	17.33
							2.60	0.14	5.12
Freidman et al., 1987[25]	Metabolic balance study	19–22	M	7	United States	Minimal Mn	0.11	−0.02	−19.9
							1.53	+0.84	+54.7
							2.55	+1.02	+39.8
Hunt et al., 1998[111]	Crossover	20–42	F	21	United States	Lactoovo-vegetarian	5.90	0.60	10.16
						Non-vegetarian	2.50	0.10	4.00
Hunt et al., 1995[133]	Crossover	51–70	F	14	United States	High meat	3.66	0.11	3.01
						Low meat	3.48	0.04	1.14
						Low meat + K, P, Fe, Mg, Zn	3.63	0.17	4.68
Johnson et al., 1991[10]	Randomized, factorially arranged	27	F	14	United States	Varied Ca	5.66	0.10	1.77
						1.32	5.52	0.30	5.43
						0.59	0.95	−0.01	−1.05
						1.35	0.94	0.06	6.38
						0.57			

Reference	Study design	Age	Sex	N	Country	Diet			
Kelsay et al., 1988[110]	Observational	20–53	F	82	United States	Vegetarian, Asian Indian	3.57	−0.49	−13.72
		24–58	M			Asian Indian	−4.87	−0.37	−7.60
		22–48	F			American	4.06	−1.11	−27.30
		21–49	M			American	5.06	−1.09	−21.50
		22–48	F			Conventional foods, American	2.58	−0.23	−8.91
Lang et al., 1965[22]	Extra period Latin square	21–49	M	8	United States	American	3.27	0	0
		20–29	M			Vegetarian	7.07	3.34	47
North et al., 1960[21]	Randomized crossover trial	18–21	F	9	United States	Conventional foods	3.70	1.54	41
Nishimuta et al., 2012[156]	Metabolic balance studies (13, 1986–2007)	18–26	M	131	Japan	Japanese diet	2.63–3.73[b,c]	−0.02 to −0.03[b]	
Patterson et al., 1984[23]	1 year study; 4 metabolic balance periods	20–35	M	12	United States	Spring	2.8	−0.16	
						Summer	3.0	−0.16	
			F	16		Fall	2.9	−0.21	
						Winter	3.2	−0.12	
Sandberg et al., 1994[114]	Case-control study; ileostomic patients	43.5 ± 18.5	M	2	Sweden	Low-fiber	2.56	0.34	
		47.5 ± 19.7	F	4		Low-fiber + alginate	2.55	0.18	

[a]M, male; F, female.
[b]Milligrams per day based on a mean body weight of 51.9 kg.
[c]Median intake.

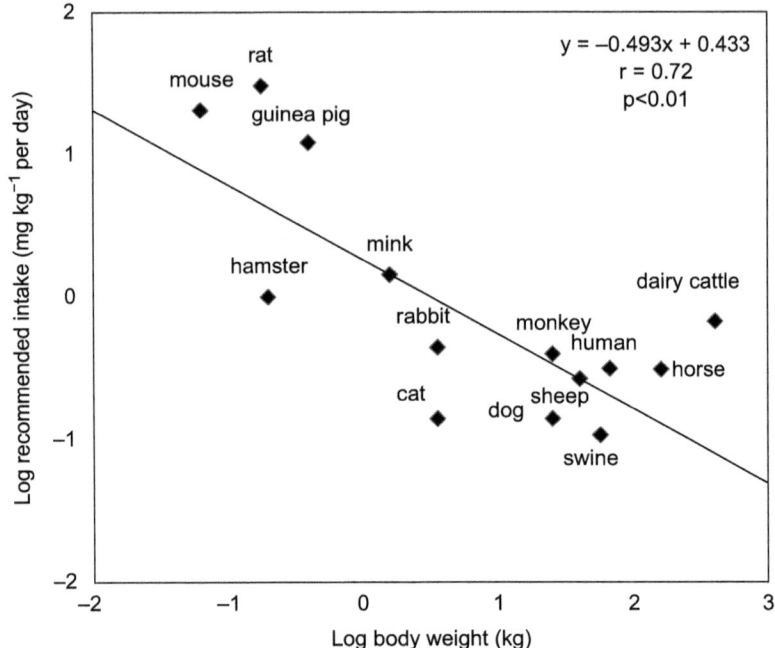

Figure 2.2 Full logarithmic plot of published values of recommended manganese intakes for mammals compared with value for humans.
(Freeland-Graves J. H., Behmardi F., Bales C. W., Dougherty V., Lin P. H., Crosby J. B. and Trickett P. C., Metabolic balance of manganese in young men consuming diets containing five levels of dietary manganese. *J. Nutr.*, 1988, **118**, 764–773. With permission).

positive retention of 4.97% Mn, but the women had a mean negative value of −2.99%. Men were thought to absorb less Mn than the women owing to their higher Fe status, as indicated by elevated serum ferritin concentrations. Another explanation for the gender differences could have been the negative relationships between body weight and percentage absorption of Mn.[24] The authors proposed that women would be expected to retain less Mn than Fe, because the biological half-life of Mn in women is shorter than that in men (34 *vs.* 48 days).

In 1991 Johnson and colleagues fed mixed diets of whole foods to young women in a randomized controlled trial for four 39 day periods. Dietary levels of Mn and Ca were 5.66, 5.52, 0.95, and 0.94 mg per day and 1.32, 0.59, 1.35 and 0.57 mg per day, respectively. The resultant balances were primarily positive, and averaged 0.1, 0.3, −0.01 and 0.06 mg per day.[10] Thus, the influence of dietary Ca on Mn balance appeared to be negligible in this report.

The only study that utilized semi-purified diets to induce a Mn deficiency with the factorial approach was conducted by the author of this chapter.[25] In young men aged 19–22 years, a Mn adequate diet (2.59 mg per day) was fed for three weeks to establish baseline data. This baseline period was followed by 39 days of a deficient diet, 0.11 mg Mn per day, and two five-day periods of repletion (1.53 and 2.55 mg Mn per day). When the intake of Mn was minimal (0.11 mg per day), and coupled with endogenous losses, the Mn requirement was estimated to be 0.74 mg per day. It should be noted that this experiment was based on semi-purified diets free of inhibitory components such as phytates. Thus, the authors suggested that the Mn intake would be increased to 2.11 mg per day if one consumed a typical diet of whole foods.[25] But this value may be artificially low, because copious amounts of dietary components that affect bioavailability are often present. To date, the Mn requirements of humans are unclear and subsequent research has not been found.

2.4.1.2   Infant Diets

Manganese concentrations in breast milk are low, measured in $\mu g\,l^{-1}$, rather than the mg quantities present in foods. Intakes in infants who were solely breast-fed were reported to range from ~ 2.8 to 6.4 μg Mn per day (0.4 to 1.6 $\mu g\,kg^{-1}$ per day).[26] Manganese absorption from cow's milk and infant formulas is lower than that from human milk, regardless of the retention rate.[27] For instance, the average percentage of Mn absorbed from soy formula, cow's milk and human milk averaged 0.7%, 2.4% and 8.2%, respectively.[28] Accordingly, Dörner *et al.* (1989)[27] showed that mean Mn retention levels in infants who were fed breast milk or a cow-based milk formula were 0.43 ± 0.65 $\mu g\,kg^{-1}$ and 2.8 ± 4.8 $\mu g\,kg^{-1}$, respectively. This discrepancy indicates a much greater Mn retention from the formula, as compared to that fed from the breast. In preterm infants, retention rates were even higher in those who were formula-fed, averaging 10.06 ± 5.87 μg Mn kg^{-1}. These elevated retention rates, coupled with higher intakes, from soy formula presumably predispose prenatals to Mn toxicity.[27] The usually high retention among preterm neonates has been attributed to a diminished secretion of Mn into the bile, resulting from the immaturity of the liver and kidneys.[29]

The bioavailability of Mn from breast milk and infant formulas is influenced by several factors including vitamin C, citric acids,[30] casein, and iron (Fe),[31] as well as parenteral nutrition.[32] Both ascorbate and citrate have a positive influence on Mn absorption by facilitating its uptake from human milk and milk formulas.[30] In contrast, casein from cow's milk and cow's milk formulas diminishes Mn absorption as a result of its binding to the protein.[31] This is particularly true for Fe, as it may use the same intestinal transporters, divalent metal transporter-1 (DMT-1) and transferrin receptor (TfR).[33]

In preterm infants parenteral nutrition may adversely affect Mn status. For instance, Mn balance was investigated in 13 very low birth weight preterm infants (mean gestational age 27 weeks) receiving total parenteral nutrition containing 43 ± 4 μg Mn kg^{-1} per day. The retention was extraordinarily high, averaging 88% (*i.e.*, 38 ± 6 μg per day) of the Mn ingested.[32] Again, the lack of regulation for this trace mineral was attributed to the immaturity of body organs.

2.4.2   Blood Levels of Mn

Blood levels of Mn have been utilized as simple biomarkers for Mn status. The mean concentration of Mn in whole blood, where it is usually bound to β1-globulin,[11] was $8.44 \pm 2.73\ \mu g\ l^{-1}$. In a study conducted by the author,[25] 10 young men were fed a baseline diet of conventional foods containing 2.59 mg Mn per day for three weeks. The mean Mn levels for whole blood, serum and plasma averaged 9.57, 1.22 and 0.73 $\mu g\ l^{-1}$ Mn per day, respectively. After the diet was switched to one that was semi-purified with a minimal level of Mn (0.11 mg per day), whole blood, serum and plasma Mn concentrations were 6.01, 1.12 and 1.10 $\mu g\ l^{-1}$ Mn per day. Upon repletion of Mn (2.55 mg per day) serum and plasma Mn concentrations decreased insignificantly to 1.04 and 0.83 $\mu g\ l^{-1}$, respectively, as compared to baseline. Whole blood Mn concentration increased to 6.99 $\mu g\ l^{-1}$, but the change was not significant. Thus, these substantial variations observed in the response of whole blood Mn eliminates its use as a reliable indicator of status. The variation in values of plasma Mn, when measurements were repeated on the same and different subjects several times, is illustrated in Figure 2.3. As expected, the variance was greater between subjects, and levels did not vary significantly within subjects.

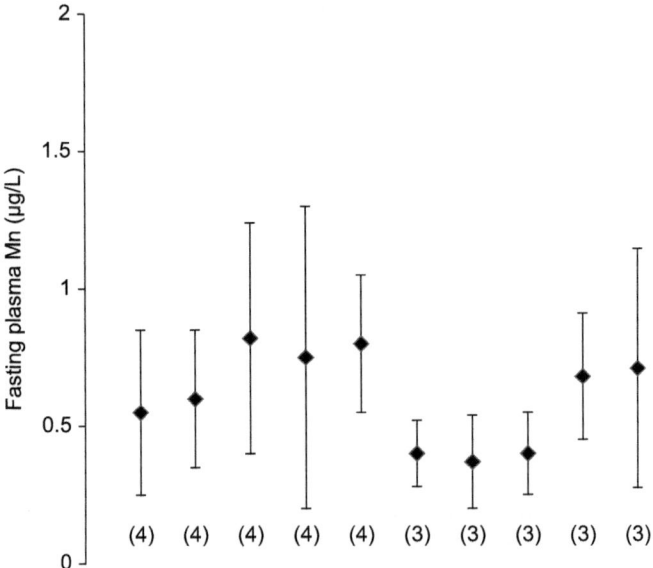

Figure 2.3   Variations in fasting concentrations of plasma manganese within and among normal subjects.
(Bales, Connie, Jeanne Freeland-Graves, Pao-Hwa Lin, Jan Stone and Virginia Dougherty. Plasma uptake of manganese: response to dose and dietary factors. In *Nutritional Bioavailability of Manganese*, C. Kies, ed. Washington, D.C.: American Chemical Society. pp. 112–122, 1987. With permission).

In a study of 47 young women supplemented with 15 mg Mn per day for 125 days, serum Mn concentrations increased to 0.3 µg l^{-1}, and were significantly higher than in those who did not receive Mn.[34] Yet Greger *et al.* reported that serum Mn in young men was not associated with dietary levels either at baseline (0.2 µg l^{-1}) or after 7 days of supplementation with 15 mg Mn (0.1 µg l^{-1}).[35] Thus, blood levels of Mn appear to be too variable or unresponsive to changes in Mn status to be utilized as definitive markers.

2.4.3 Other Biomarkers

Fecal losses of Mn were initially considered as a biomarker, as these are reflective of quantities in the diet. When young men were fed a semi-purified Mn depleted diet for 39 days, mean Mn losses in feces and urine were 86% and 0.32% of the total measured losses.[25] These values increased to 98% and 0.59% when men were supplemented with 1.39 or 2.39 mg Mn, respectively, for five days each.[25] Greger *et al.* also reported that fecal Mn varied with diet, but no effect was observed on urinary excretion of the mineral.[35]

Hair Mn has been utilized as a parameter of Mn by a number of studies. When postmenopausal vegetarian and non-vegetarian women were consuming diets containing 4.4 mg and 2.6 mg Mn per day, respectively, median hair Mn concentrations were significantly higher in vegetarians than in non-vegetarians (0.40 and 0.26 µg g^{-1}, respectively).[36] In younger men who consumed minimal Mn intakes for 39 days,[25] hair Mn concentration declined from 10.4 µg g^{-1} to 9.6 µg g^{-1}. However, hair Mn content may vary according to the rate of hair growth and possible contamination from shampoos and chemical treatments such as permanent colorings. Nonetheless, hair Mn may be useful for screening populations for quick assessment of Mn status under certain circumstances such as contamination in drinking water.

Manganese-dependent enzymes would appear to be ideal candidates as biomarkers of Mn status; these enzymes include MnSOD, arginase, glutamine synthetase, and phosphophenyl pyruvate decarboxylase. Others such as the glycosyl transferases and xylosyltransferases have been observed to respond to Mn, with effects on bone.[37] Finally, pyruvate carboxylase, an enzyme critical in digestion, also is activated by Mn, but the Mn can be substituted by magnesium.[40] Of the above, only the first three have been investigated as parameters of Mn status. In young women supplemented with 15 mg Mn per day, MnSOD activity significantly increased from 0.00174 to 0.70174 units mg^{-1} protein after 124 days.[34] Thus, MnSOD activity may be utilized as a measure of Mn exposure in humans, but measurements have not been made for deficiency states. Moreover, a host of other factors such as mitochondrial dysfunction and disease states (*i.e.*, cancer) may affect the activity of Mn-dependent enzymes.[38]

Arginase activity in the liver[5] and endothelial cells of the aorta[39] has been reported to be reflective of Mn deficiency in animal models. In rats fed for 21 days with a Mn deficient diet, liver arginase activity was lower than that of controls (1.12 *vs.* 1.55 mmol ornithine g^{-1} hepatic protein min^{-1}).[5] In 2004 Ensunsa *et al.* fed two groups of rats low (0.5 µg g^{-1} diet) and high (45 µg g^{-1} diet) Mn diets. Manganese-deficient rodents showed inhibited arginase activity in the endothelial cells of the aorta, producing vasorelaxation.[39] In 2006 Häberle and colleagues reported that Mn had a non-significant effect on glutamine synthetase activity when they detected malformations in the brains of two infants, as a result of inborn errors of amino acid metabolism.[40]

A sensitive, but expensive, method to estimate Mn status is magnetic resonance imaging (MRI). Reynolds *et al.* (1998)[41] showed that three of nine adult patients had Mn deposition in the brain upon MRI scanning. These brain lesions resulted from the high and presumably toxic Mn levels obtained *via* parenteral nutrition formula (27 nmol l^{-1}). Upon removal of Mn from a parenteral nutrition solution, the whole blood Mn level declined from 51 µg l^{-1} to 34 µg l^{-1}. This reduction was accompanied by disappearance of the high intensity lesions in the globus pallidus, as recorded by MRI.[42] The Mn deposition was attributed to the bypassing of homeostatic mechanisms that regulate absorption during oral intake. Thus, MRI may be a good indicator of changes in Mn status, particularly with regards to toxicity. Manganese accumulation in the brain is believed to cause neuronal damage.[43] In a meta-analysis of eight studies with 281 workers occupationally exposed to Mn, Mn accumulation in the brain of the exposed group as measured by the pallidal index was significantly greater than in the non-exposed group (weighted mean difference of 7.76, confidence interval of 4.86–10.65).[44] A positive correlation between the pallidal index and Mn deposition existed in the brain ($r = 0.42$, $p < 0.05$). Thus, MRI may be a successful approach to documenting changes in Mn concentrations in the brain.

2.4.4 Extrapolation to Usual Diet Intake

A lack of sufficient data has precluded the establishment of a Recommended Dietary Allowance for Mn. To date the usual dietary intake that is found in healthy populations has been set on the basis of extrapolation of an Adequate Intake to give the Reference Dietary Intake. The current Adequate Intake (AI) of Mn is based on values derived from the 1966 Total Diet Study by Pennington *et al.* A total of 260 typical food components in the US diet were analyzed for Mn content, and median intakes were calculated based on usual diets.[45] Consequently, the values of 1.8 mg Mn per day and 2.3 mg Mn per day were established as the usual intakes of Mn for adult women and men, respectively.[46]

Requirements for other age groups were then created by extrapolation of the quantities obtained from adult diets in the above study.[46] Whether the values were based on diets that incorporated copious quantities of high fat, high sugar, or refined foods low in Mn owing to the nutrition transition remains to be elucidated.

2.5 Deficiencies

The first experimental Mn deficiency[1] was created by Doisy in 1974, who had prepared a vitamin K deficient diet to be fed to two men (Table 2.2). One of the men developed a scaly, fleeting dermatitis, hypycholesterolemia, a minimal reddening of the hair, and low vitamin K-dependent clotting factors. Supplemental vitamin K was then provided, but the symptoms were not mitigated until a normal diet was reinstated. At that point Doisey discovered that he had inadvertently forgotten to add Mn to the investigational diet and had induced a deficiency.

The second report of an induced Mn deficiency in humans was from the authors' laboratory, in young men undergoing a metabolic balance study.[25] A nutritionally adequate diet of conventional foods was fed for 21 days to establish baseline values of physiological parameters. Then a semi-purified diet that contained only 0.11 mg Mn per day was provided. After 35 days, five of the seven men developed a dermatitis, characterized as *miliaria crystallina* by a consultant dermatologist Stephen D. Houston. Small blisters filled with clear fluid appeared, due to obstruction of sweat glands near the surface of the skin. Physical exercise exacerbated the condition, with breakage of blisters producing a dry, flaky dermis. The finely scaling dermatitis was minimally erythematous, located primarily on the upper body but also affecting areas of the groin and peripheral extremities. The study was stopped at 39 days, following by 10 days of repletion. The dermatitis quickly cleared soon after supplementation with Mn.

The mechanism by which Mn deficiency led to the dermatitis observed in both of the studies above was believed to be related to Mn-dependent enzymes. Gulberti and others (2003)[47] showed that Mn^{2+} is a critical trace element that functions as a substrate for mammalian glycosyltransferase. The binding of Mn^{2+} to this enzyme activates glycosylation and formation of glycoproteins, which maintains cell–cell adhesion in tissues. In particular, this Mn-dependent enzyme induces the production of glycosaminoglycans. These glycans are major constituents of collagen, a wound-healing protein, in the dermis. Consequently, collagen dissociates by prolidase, another Mn-dependent enzyme. Thus, Mn deficiency has adverse effects on skin health and wound healing.[3]

The Mn deficiency experiment conducted by the author[25] also resulted in increased levels of serum calcium, phosphorus and alkaline phosphatase (an enzyme involved with active biosynthesis of bone). Elevated serum calcium is of significance as this mineral rarely fluctuates to any great extent owing to homeostatic controls. In agreement, Strause *et al.* (1986) also documented increased serum calcium and phosphorus in rodents fed long term a diet deficient in Mn. In addition, bones that had diminished Mn concentrations were perceived to exhibit an osteoporotic structure.[48]

Subsequently, the author measured bone density and plasma Mn in 23 osteoporotic and 17 healthy postmenopausal women.[49] Respective values for bone density by dual photon absorptiometry and number of fractures

Table 2.2 Human studies on manganese (Mn) deficiency.

Reference	Design	Age (years)	Gender[a]	Sample (n)	Type of diet	Disorder associated with reduced Mn status
Doisy, 1972[1]	Experimental trial	—	M	2	Vitamin K deficient	Dermatitis, hypocholesterolemia, reddening of the hair
Friedman et al., 1987[25]	Metabolic balance	19–22	M	7	Semi-purified	Dermatitis, hypocholesterolemia, increased serum Ca, P, and alkaline phosphatase
Wu et al., 2006[58]	Case-control	—	F	68 breast cancer (25 malignant, 43 benign), 26 healthy controls	—	Benign and malignant breast cancer
Afridi et al., 2009[54]	Case-control	Mothers aged 30–40 and neonates	F	76 diabetic mothers and infants, 68 controls	—	Type 2 diabetes
Forte et al., 2013[66]	Case-control	>18	M, F	192 type 1 diabetes, 68 type 2 diabetes, 59 controls	—	Type 1 and type 2 diabetes
Li et al., 2013[55]	Cross-sectional	18–65	M, F	221 metabolic syndrome, 329 controls	Conventional	Elevated number of metabolic syndrome components
Patel et al., 2006[59]	Nested case-control	45–75	M, F	515 asthma, 515 controls	Conventional	Symptomatic asthma
Bhang et al., 2013[63]	Cross-sectional	8–11	M, F	1089	Conventional	Lower cognitive processing

[a]M, male; F, female.

(independent of trauma) for these subjects were 0.88 and 1.29 g cm^{-2} and 15 and 0, respectively ($p < 0.0001$). In these same women, plasma Mn was significantly lower in those exhibiting osteoporosis *vs.* those in good health (0.71 $\pm$ 0.04 *vs.* 1.00 $\pm$ 0.06 mg l^{-1}, $p > 0.001$). It has been hypothesized that Mn is critical for the incorporation of Ca into bones.[50] Thus, the elevated serum Ca found in Mn depletion may have been the consequence of the mobilization of Mn and concomitant release of Ca that was sequestered in the bone. Skeletal abnormalities also have been found in animal studies relating to Mn status. Chickens, puppies, pigs, and ducks with low blood levels of Mn suffered from poor formation of extremities, resulting in shortened bones and enlargement of joints.[3] Consequently, these malformations often impair mobility.

A deficiency of Mn was reported as a consequence of total parenteral nutrition (TPN) before Mn was added as a nutrient to solutions. Patients suffered from gastrointestinal problems, dermatitis and weight loss. The low level provided coupled with the high body demands due to the physical condition of the patient were presumably the etiologic factors.[11]

Low dietary intakes of Mn have been associated with a number of health conditions such as skin photo-aging,[51] dyslipidaemia,[52,53] diabetes,[54] metabolic syndrome,[55] hypertension,[56] epilepsy and seizures,[57] cancer,[58] asthma,[59] and inborn metabolic errors of metabolism such as phenylketonuria.[60] Other problems associated with a possible Mn deficiency include inflammatory diseases,[61] altered exocrine and endocrine activities,[3] and problems with fetal growth[62] and cognitive function.[63]

A common basis may be related to diminished activity of MnSOD2, an essential antioxidant enzyme. MnSOD2 has a critical role in reducing oxidative stress by combating the formation of reactive oxygen species (ROS). Scheurmann *et al.*[51] postulated that this Mn-dependent enzyme is a key element in skin anti-aging, through its role in reducing the inflammatory pathways associated with ROS. The mechanism of aging was examined in a study conducted by Jouihan *et al.*,[64] who demonstrated a decrease in Mn uptake by mitochondria in Hfe (haemochromatosis) mutated mice. Subsequently, the decline in Mn-dependent superoxide dismutase activity was associated with lipid peroxidation and oxidative damage of mitochondria.

Deficiency was first associated with dyslipidemia in the initial experiments of Doisy *et al.*[1] and the author described above. Both reported the development of hypercholesterolemia in response to a Mn-depleted diet. Manganese acts as an activator for mevalonate kinase and farnesyl pyrophosphate synthase; the latter initiates the formation of a cholesterol precursor, squalene, in the biosynthesis of cholesterol.[49] In 2011, Bae and colleagues explored the influence of Mn on cholesterol synthesis in Ca-deficient ovariectomized rats. In animals fed a low-Ca diet for 12 weeks, Mn counteracted the effects of decreased cholesterol and low-density lipoprotein levels in blood.[53]

Low blood levels of Mn are associated with dyslipidemia in diabetics. Burtlet *et al.* (2013) found that Mn^{2+} diminished monocyte binding to endothelial cells of veins by inactivating the production of intercellular adhesive molecules and ROS, inhibiting clot formation.[65] Provision of Mn^{2+} to diabetic rats significantly diminished the concentrations of intercellular adhesion molecules and cholesterol in the blood by 17–28% and 25%, respectively.[65] Therefore, Mn appears to be a protective element against endothelial dysfunction in diabetic patients.

Other investigations have shown an inverse relationship between type 2 diabetes and Mn status. In 2009, Afridi *et al.* examined Mn concentrations in the blood of 76 diabetic mothers and their neonates. Mean Mn levels in whole blood of the mothers (40.2 µg l^{-1}) and their infants (41.0 µg l^{-1}) were significantly lower than those of the controls (46.6 µg l^{-1} mothers; 47.5 µg l^{-1} infants, respectively). Additionally, Mn urinary loss was greater in the diabetic mothers (1.9 µg l^{-1}) and infants (1.6 µg l^{-1}), as compared to controls (1.2 µg l^{-1}, 1.3 µg l^{-1}), respectively.[54] Similarly, low Mn was reported in the blood of 260 adult patients with type 1 (9.3 µg l^{-1} in females, 8.6 µg l^{-1} in males) or type 2 diabetes (12.9 µg l^{-1} in women and 9.9 µg l^{-1} in men) *vs.* 59 healthy adult subjects (15.4 µg l^{-1} in women and 12.5 µg l^{-1} in men).[66] It is plausible that low blood Mn concentrations may be linked to diabetes through its unusually high urinary losses.

The relationship of Mn status to metabolic syndrome in China was assessed in 221 patients with symptoms and in 329 adult controls.[55] Manganese intake was negatively associated with the number of metabolic syndrome components. Those who exhibited hypertriglyceridemia, hypercholesterolemia, and hyperlipidemia linked with metabolic syndrome had significantly lower dietary Mn (2.8 mg) than did controls (4.1 mg). The conclusion was that adequate dietary levels of Mn could facilitate reduction of the occurrence of metabolic syndrome.

Hypertension has been reported to be related to altered MnSOD status as well. In 2013 Jin and Vaziri[56] observed that high dietary salt intake induced inflammation through ROS formation, hypertension and urinary albumin loss in mice deficient in MnSOD. However, inflammatory reactions associated with salt intake were insignificant in mice with normal levels of MnSOD. Thus, maintenance of adequate intakes of Mn and its dependent antioxidant enzyme may be important to protect against oxidative stress, subsequent to the development of high blood pressure.

Manganese deficiency is associated with neurologic problems such as epilepsy *via* its association with oxidative stress and MnSOD. Children diagnosed with neurologic problems including epilepsy were observed to have lower blood Mn levels than healthy children (0.85 *vs.* 1.45 µg l^{-1}).[57] Furthermore, whole-blood Mn measurements in 44 epileptic patients were lower than those observed in normal subjects (0.84 µg l^{-1} *vs.* 1.19 µg l^{-1}). Patients with trauma had significantly greater whole blood Mn values than did those whose epilepsy was caused by unknown factors (0.94 µg l^{-1} *vs.* 0.66 µg l^{-1}).[67] The authors concluded that Mn deficiency is linked to epilepsy, but to a lower extent than that observed in patients with a history of trauma.

The influence of Mn-associated enzymes in the liver (arginase) and brain (glutamine synthetase) of mice was explored. Episodes of seizure were seen with low Mn in whole blood, but not in the brain. Compared with controls, these episodes augmented the activity of arginase in the liver but not glutamine synthetase in the brains of epileptic patients. It is conceivable that Mn is related to epileptic seizures by affecting arginase activity in the liver.[68] In contrast, Critchfield and others[69] investigated the effect of Mn supplementation on seizure development in the offspring of two groups of pregnant mice fed 45 (control) and 1000 µg Mn g^{-1} diet, respectively. The high intakes of Mn did not affect glutamine synthetase activity in the brain or seizure occurrence. Nonetheless, Liang *et al.* (2012)[61] supported the findings of Carl *et al.*[68] when they examined the role of Mn deficiency in inducing epilepsy in mice. Animals with low MnSOD activity exhibited increased risk for epileptic seizures that was attributed to oxidative stress, with subsequent mitochondrial dysfunction. Thus, maintaining optimal levels of this antioxidant element may reduce the risk of developing diseases associated with oxidative stress.

Thyroid gland secretions, particularly thyroid (T4) and thyroxine (T3), and thyroid stimulating hormone (TSH), have been reported to be affected by Mn status. For instance, Badiei *et al.* (2008) showed, after 8 weeks of supplementing sheep with 5 mg Mn kg^{-1} per day, that serum levels of T4, T3 and TSH were significantly reduced (67.22 nmol l^{-1}, 0.76 nmol l^{-1}, and 0.33 mIU l^{-1} *vs.* 47.26 nmol l^{-1}, 0.67 nmol l^{-1} and 0.31 mIU l^{-1}, respectively).[70]

Manganese SOD has been linked further to the development of cancer.[71] In an investigation by Wu *et al.* (2006), patients with either benign or malignant cancer had statistically lower serum Mn levels, as compared with healthy subjects (7.47 µg l^{-1} and 5.50 µg l^{-1} *vs.* 9.09 µg l^{-1}, respectively).[58] Thus, Mn may have potential as a protective antioxidant against cancer.

Asthma is another health condition associated with low blood Mn. A nested case-control study measured the relationship between dietary antioxidants and asthma in 515 adults with asthma and 515 matched controls (45–75 years old). From food diaries collected for one week, Mn was inversely and independently associated with symptomatic [odds ratio (OR) 0.85; 95% confidence interval (CI) 0.74–0.98] and diagnosed asthma (OR 0.86; 95% CI 0.77–0.95). The symptomatic asthma was prevalent in those with low dietary intake of fruits and Mn.[59] These findings suggest that diet may be a potentially modifiable risk factor for the development of asthma, and increasing Mn intake could be advantageous.

Low blood Mn levels are observed to be associated with impaired cognitive function in children. A cross-sectional study found that Korean children (8–11 years old) with low blood Mn values (<8.5 µg l^{-1}) had reduced color scores on the Stroop test (≤52.0) (stating the color of a presented word).[63] Thus, maintenance of optimal Mn status in schoolchildren is critical to cognitive health and, presumably, school performance.

Finally, Mn deficiency may be associated with the risk of poor birth outcomes. Hansen and others[62] investigated the effect of Mn dietary intake on fetal growth in pregnant heifers that were followed for 276 days. At birth, whole-blood Mn levels (35.02 mg l^{-1}) and weight of calves (39 kg) born to heifers supplemented with 50 mg k^{-1}g Mn were greater than those born to controls (24.04 mg l^{-1} and 32 kg, respectively). Calves of controls showed signs of Mn deficiency such as dwarfism and swollen joints, while these symptoms were not seen in the supplemented group.

In humans, Eum and colleagues (2014)[72] examined the role of maternal blood Mn levels on birth outcome in 331 mother–infant pairs. Low birth weight (<3 kg) was observed in infants of mothers who exhibited reduced Mn levels in their blood (16.9 µg l^{-1}). Thus, maternal Mn deficiency may be a potential risk factor for reduced natal size. Similarly, Mn concentration in whole (54.9 µg l^{-1}) and umbilical cord (78.8 µg l^{-1}) blood of 125 pregnant mothers had a positive association with birth outcome;[73] the mean weight of an infant at birth was <3.25 kg. Therefore, maintaining optimal Mn blood levels during pregnancy may be critical to positive birth outcomes.

The above suggests that low levels of Mn are associated with numerous health conditions such as dermatitis, osteoporosis, dyslipidemia, diabetes, metabolic syndrome, hypertension, epilepsy, cancer, asthma, cognitive function, and poor birth outcomes.

2.6   Nutritional Recommendations for Mn

The AIs for Mn based on age and gender in the United States (US)[46] are shown in Figure 2.4. These median Mn intakes were derived from the Total Diet Study,[74] as previously discussed. The highest intake values were used to designate the AI for Mn on the basis that dietary assessment methodologies (records, recalls and food frequencies) may underestimate actual daily intake.[46]

Factors that influence Mn requirements include life stage and gender (infants, children and adolescents, pregnancy, lactation, infant nutrition), bioavailability (fiber, phytates, mineral interactions, polyphenolic compounds) and international considerations.

2.6.1   Life Stage and Gender

The estimated safe and adequate daily dietary intake of Mn for those older than 10 years of age was set originally (in 1989) as 2–5 mg per day.[75] In 2001 the Institute of Medicine (IOM), Washington DC, USA, recommended specific intakes of Mn for adults that are still in place.[46] These optimal intakes of Mn were established based on prevention of clinical symptoms of deficiency and maintenance of optimal stores in order to avoid depletion during periods of inadequate consumption.

Between 1982 and 1991, the average intakes of Mn in the US for 25–30 year-old adult females and males (1.11 and 1.43 mg per day), and for 60–65 year-old elderly females and males (1.15 and 1.33 mg per day), were lower

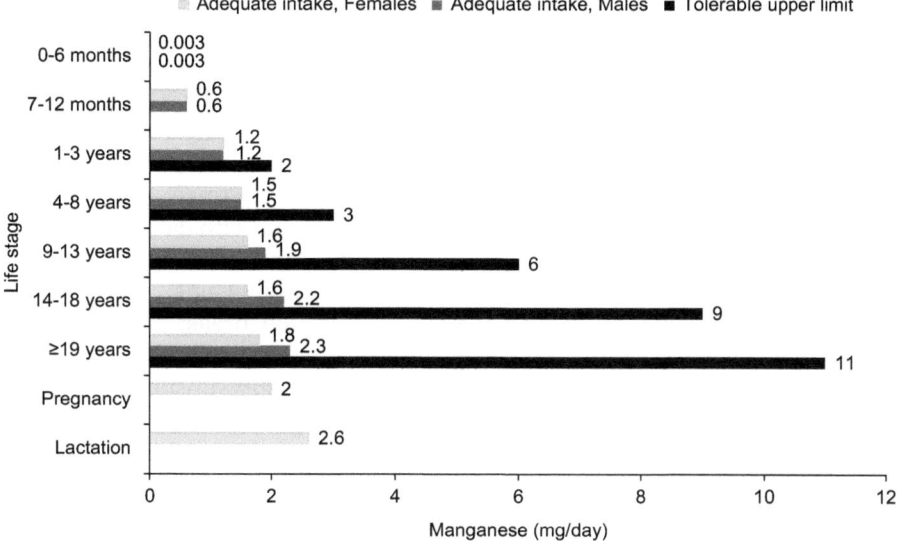

Figure 2.4 shows the legend: Adequate intake, Females ■ Adequate intake, Males ■ Tolerable upper limit

* Tolerable upper limit has not been established for infants aged 0–12 months

Figure 2.4   Adequate intakes (AI) and tolerable upper limit of manganese based on life stage and gender.
(Institute of Medicine (IOM) of the National Academies, 2001.)

than the recommended intakes.[74] A decade later, the Institute of Medicine recommended that adult men and women ≥19 years old should consume at least 2.3 and 1.8 mg per day, respectively, of Mn.[46] In Canadian men and women aged ≤40 years, the Mn median intake (2.6 mg and 2.7 mg, respectively) was lower than that in adults >40 years old (3.4 mg and 4.1 mg, respectively). Thus, Mn intake may not decline with age,[76] but the US has not established any differential values for AIs for the elderly.

2.6.2   Infants

In the US, dietary intake assessment in 1974–1982 and 1982–1991 showed that Mn intakes by infants (6–11 months old) were 1.57 and 1.78 mg per day, respectively. Note that these dietary levels were higher than the recommendations.[74] In 2001 the dietary standards for the first six months of age were changed to 0.3–0.6 mg per day.[46] These values are higher than the amount that would result from ingestion of human milk.[77] Nonetheless, the Mn content in other milk formula sources is higher than that reported in human milk: 50–100 µg l^{-1} in cow formula, and 200–300 µg l^{-1} in soy-based formula.[78]

The effect of dietary Fe, copper (Cu), zinc (Zn) and Mn on the trace element content of human milk was examined in 15 breast-feeding mothers. Seven-day food records and 24 hour milk samples were collected at 6–8 weeks and 17–22 weeks postpartum.[79] The estimated average daily intake of Mn during lactation in the mothers was 5 mg. The mean Mn concentration in milk samples was 4.5 ± 1.8 and 4.0 ± 1.5 µg l^{-1} for the early and late postpartum periods, respectively.[79] A significant positive correlation was observed between maternal Mn intake and its content in human milk in the latter period ($p < 0.01$).

Dörner *et al.* (1989)[27] examined Mn intake in 26 male infants, aged 6–12 weeks, and in preterm infants with a gestational age of 34–36 weeks. Infants were fed breast milk, cow's milk formula without trace element supplementation, or cow's milk formula supplemented with Fe, Cu, Zn, and iodine (I). The Mn contents in breast milk, unsupplemented formula, and supplemented formula were 6.2 µg l^{-1}, 77 µg l^{-1}, and 99 µg l^{-1}, respectively. The mean daily Mn intakes of full-term infants fed breast milk or infant formula, as well as formula-fed preterm infants, were 1.06 ± 0.43, 14.2 ± 3.1, and 15.0 ± 2.2 µg kg^{-1} per day, respectively. Notice the substantially higher levels of Mn in infants fed formula, which could possibly lead to Mn toxicity. A recent report suggested that preterm and very low birth weight infants should be supplemented with 1–15 µg kg^{-1} Mn per day to prevent adverse health effects associated with the deficiency of this trace element.[80]

Infant consumption of Mn was estimated by analyzing 16 milk protein and 14 soy-based formulas. Daily mean Mn intake from milk protein formulas (70.3 ± 36.0 µg) was lower than from those that were soy-based (242 ± 65.0 µg). Cockell *et al.* (2004)[81] calculated an upper limit (UL) for infants' intake of Mn, ranging from 50 to 90 µg 100 kcal^{-1} or 0.35 to 0.6 µg ml^{-1}. In Japan, Tsutie *et al.* (2010)[82] reported inadequate Mn intakes in 9–11 month-old weaned infants, as compared to those who were 7–8 months old. The Mn content of the daily weaning formulas did not satisfy the Japanese dietary reference intakes or energy needs for either age group. Manganese intake from hospital weaning formulas was lower than the recommended intake (*i.e.*, 1.2 mg per day) by 24.2–72.5% for the younger infants, and by 61.7–97.5% for the older group.

The California State Assembly Committee on Public Safety proposed that Mn neurotoxicity from soy formula could be linked to attention deficit hyperactivity disorder.[83] Despite the suggested upper limit[46] for Mn of 0.6 mg l^{-1}, infant formulas were observed to contain about 3.5 mg l^{-1} Mn.[75] This amount could result in ingestion of 0.5 mg Mn per 100 kcal, which is 1000 times higher than the amount consumed by a breast-fed infant (Casey, 1985).[84] The safe amount of Mn to be consumed by infants[75] is 10–20 µg per 100 kcal. Therefore, there appears to be a need to balance Mn inadequacy and potentially toxic intakes from infant formulas.

2.6.3   Children and Adolescents

The reported average Mn dietary intake in 2 year-old children was 1.42 mg per day and 1.47 mg per day in 1974–1982 and 1982–1991, respectively.[74] These quantities were higher than the recommended intake of 1.2 mg per day. In contrast, daily Mn consumption in teenagers (14–16 years old) was 1.44 mg in 1974–1982, and 0.90 and 1.38 mg in females and males aged 15–20 years, respectively, in 1982–1991. These levels in adolescents were lower than the recommendation of 1.6 mg per day.[74] The current AIs for adolescents, 9–13 and 14–18 years old, are 1.9 and 2.2 mg per day for males, and 1.6 mg per day for both age groups of females.

The influence of Mn levels on childhood behavior was investigated by a longitudinal study that measured Mn intake in 27 children at the 20th and 62–64th weeks of gestation, and then followed until the third grade.[85] Overabsorption of Mn, secondary to gestational Fe deficiency anemia, resulted in Mn deposition in tooth enamel. This deposition was correlated with externalizing behavior in first ($r = 0.42$) and third graders ($r = 0.58$), as well as with disruptive disorder symptoms in the third grade ($r = 0.42$) ($p < 0.05$).[85] An interaction between Fe and Mn is believed to induce manganism in children, causing behavioral disorders. However, no study has reported on Mn *deficiency* and behaviour.[85]

Dietary intake of Mn was evaluated in 257 Korean boys and girls aged 11 to 12 years. The Mn intake in boys (4.6 mg per day) was slightly higher than in girls (4.03 mg per day) ($p < 0.001$), and both of these exceeded the recommended intakes by about 17%. The high Mn consumption of this population group could be attributed to the characteristics of the Korean diet. Half of the diet consists of cereals and seeds, which are good sources of Mn.[53] Thus, it appears that more investigation is needed to assess Mn intakes of children and its relationships to health and cognitive function and other consequences.

Although an adequate intake level for adolescent girls is 1.6 mg per day, a negative balance of 0.40 mg per day was observed in adolescent females fed 3 mg per day of Mn and 13.4 mg per day of Zn.[86] An additional study, conducted by Greger *et al.* (1978), reported Mn retention to be 0.44 and 0.17 mg per day in adolescents with a mean Mn intake of 3.2 mg per day. However, the retention rates did not differ significantly.[87] Thus, future clinical trials are warranted for this age group.

2.6.4 Pregnancy

Manganese requirements in pregnancy are estimated to be 2 mg per day.[46] These are affected by maternal dietary intake, degree of Mn absorption, utero-placental blood flow, placental transfer and fetal uptake. Physiological changes that accompany pregnancy include greater dietary absorption of minerals, and alterations in plasma volume, urinary losses, and maternal tissue protein, as well as the number of erythrocytes and leukocytes.[88] Manganese crosses the placenta, as detected by its presence in the body of the fetus where it is required for normal growth and development.[89]

The Mn status of pregnant women, particularly Mn deficiency, has been reported to be linked to a reduction in both intrauterine growth and birth weight of the neonate.[90] Vigeh and colleagues[91] examined the effect of Mn in umbilical cord blood and maternal whole blood on intrauterine growth retardation in 410 postpartum mothers and their infants. Growth retardation in the newborn was associated with low Mn levels in the blood of the mother, but negatively associated with that in the umbilical cord blood ($p = 0.001$).

Rahman and others[92] studied the effect of Mn intake from drinking water during pregnancy on birth outcome in Bangladesh. Concentrations of Mn in water that were obtained from groundwater were between 3 and 6550 µg l^{-1}. Remarkably, the high Mn levels (median Mn $= 1292$ µg l^{-1}) were reported to be protective against abortion and death by 65% and 69%, respectively. In contrast, Zota *et al.*[93] reported a non-linear U-shaped relation between Mn concentrations in the maternal and cord blood. The birth weight of the infant increased when the maternal blood Mn estimates were ≤ 3.1 µg l^{-1}. However it started to decline with increasing values of maternal Mn content. This U-shaped association is consistent with two studies that explored the influence of maternal Mn on birth weight, in Korea[72] and China.[94] The birth weight of 331 infants had a curvilinear association with the blood Mn levels of the mother. Low natal weight (<3 kg) was significantly correlated[72] with maternal blood Mn concentrations of ≤ 17 µg l^{-1} and ≥ 27 µg l^{-1}, and peaked when Mn blood estimates reached 30 and 35 µg l^{-1}.

Takser *et al.*[95] examined the influence of lifestyle on Mn status during pregnancy and at birth in 149 pregnant women (15–39 years old) from Southwest Quebec.[95] Manganese blood levels in the mother increased during the three trimesters of pregnancy (9.0, 9.9, and 16.3 µg l^{-1}, respectively) and were unaffected by age. Manganese estimates in cord blood and placenta were 34.3 and 0.06 µg l^{-1}, respectively. At delivery, mothers who smoked and resided in underdeveloped villages had significantly reduced Mn concentrations in blood (<15 µg l^{-1}). Mothers who were exposed to pesticides had significantly higher Mn levels at delivery than those who did not (18 *vs.* 15 µg l^{-1}, $p < 0.05$).

In conclusion, few human studies have assessed the effect of maternal Mn on that of the infant and additional research is desirable.

2.6.5 Lactation

Manganese requirements for lactation have been set as 2.6 mg per day.[46] The Mn content in breast milk is dependent on maternal intake and the subsequent deposition of this mineral in mother's milk.[79] In premature infants, the time period between 3 and 6 months represents a critical window for the development of Mn deficiency.[89] This is attributed to a decline in the Mn concentration of breast milk during lactation,[79] malabsorption due to immature organs,[29] and the lack of introduction of mixed diets in 3–6 month-old infants.[89]

Ljung and others reported that Mn levels in the drinking water of 408 Bangladeshi women were not associated with those in breast milk. In less-fewer than half of the water samples tested, Mn concentrations were greater than World Health Organization (WHO) recommendations of 400 µg l^{-1}. Values of blood Mn of the mother at the 14th week of gestation, the infant at

6 months of age, and in breast milk averaged 0.022, 0.024, and 0.01 mg kg^{-1}, respectively.[96] Despite these findings, the researchers found that the Mn intakes of mothers from water had no significant influence on of the level in breast milk or infant blood. This lack of association was attributed to a genetic predisposition of Bangladeshi women that might have affected Mn bioavailability, in particular their ability to absorb and retain Mn.

2.6.6 International Variability of Requirements and Dietary Levels for Mn

The recommendations for Mn across different countries vary considerably, as shown in Table 2.3. In Europe, the lowest Mn intakes for men and women were observed in the United Kingdom (3.32 mg per day and 2.69 mg per day),[97] presumably due to less abundant fruits and vegetables in the diet. It is noteworthy to mention that beverages are the most important contributors of Mn in the British diet, accounting for about 41% of Mn consumption.[98] Mean Mn intakes in Austria for men and women were 4.30 mg per day and 4.40 mg per day,[99] and in Germany 6.60 mg per day and 5.50 mg per day,[100] respectively. Thus, the Mn intakes obtained from the National Dietary Survey exceed the AI levels in the United Kingdom, Germany and Austria. In Spain, the daily amount of dietary minerals consumed was examined by analyzing Mn content in 420 food and drink samples found in local markets. The daily dietary values of Mn averaged 2.37 mg, with cereals serving as significant contributors of Mn (48%) in these diets.[101]

Table 2.3 International variability in recommendations for manganese intake.

Country/Union	Year	Adults Men (mg per day)	Women (mg per day)	Pregnancy (mg per day)	Lactation (mg per day)	Reference
Australia/New Zealand	2005	5.5	5.0	5.0	5.0	149
Bangladesh	2011	2.3	1.8	—	—	46
Brazil	2012	1.8	1.8	—	—	105
China	2001	3.5	3.5	3.5	3.5	150
European Union	2013	3.0	3.0	3.0	3.0	99
Germany/Austria/ Switzerland	2003	2.0–5.0	2.0–5.0	2.0–5.0	2.0–5.0	100
India	2012	2.0–5.0	2.0–5.0	2.0–5.0	2.0–5.0	151
Japan	2009	4.0	3.5	3.5	3.5	152
Korea	2012	4.0	3.5	3.5	3.5	153
Lebanon	2010	2.3	1.8	—	—	104
Philippines	2002	2.3	1.8	2.0	2.6	154
Spain	2009	2.0–5.0	2.0–5.0	2.0–5.0	2.0–5.0	101
United Kingdom	2000	1.4	1.4	1.4	1.4	155
United States of America	2010	2.3	1.8	2.0	2.6	46

The 23rd Australian Total Diet Study observed that the mean Mn dietary level was 5.0 and 4.5 mg per day in adult men and women, respectively.[102] Cereal and grain based foods were the highest contributors of Mn in this population. In New Zealand, the National Nutrition Survey documented similar high median Mn dietary estimates of 5.09–5.18 mg per day and 3.91–4.39 mg per day in adult males and females, respectively.[103]

Other international reports of usual dietary intakes exhibit a wide range of findings. For example, in Lebanon, the dietary intake of micronutrients was calculated from 1215 food records of adults.[104] The maximum amounts of Mn were obtained from seeds and pulses (4.4 mg kg^{-1}), as well as cereals and bread (3 mg kg^{-1}); together these formed less than half of the dietary Mn. A recommendation for Mn was set as 2.04 mg per day, which was higher than the values in the US suggested by the IOM for females (1.8 mg per day), but not for males (2.3 mg per day). In Brazil, micronutrient intake was explored in 1663 participants (mean age 40.5 years) using a 24 hour dietary recall.[105] Approximately, 3.0 mg per day of Mn was ingested, which was greater than the recommended value of 1.8 mg. The Brazilian diet is comprised primarily of beans, rice, beef, juice and soft drinks; both beans and rice are significant sources of Mn.

In Asia a study conducted by Kim *et al.* demonstrated that Mn intakes in Korean adults were adequate.[106] Total Mn intake was reported as 5.2 mg per day and 4.1 mg per day in adult males and females, which exceeded recommendations by 1.2 and 0.6 mg per day, respectively. The high intakes are due presumably to a diet rich in rice, cereals and pulses.[53] In Japan, cereals, pulses, vegetables and seaweed provide about 70% of the dietary Mn, but the daily amount of Mn for adults ranges from 2.7 to 2.9 mg per day.[107] A recent study reported a mean value of 2.72 mg per day for Mn intake of adults in Tokyo.[108] These levels are lower than the recommended values in Japan, which are 4.0 and 3.5 mg per day for men and women, respectively.

The highest Mn intake in the world reported to date is found in Bangladesh, with a total amount of 20.3 mg per day.[109] This value exceeds the Provisional Maximum Tolerable Daily Intake of 9.52 mg per day,[109] as well as the upper tolerable level (UL) of Mn for adults (11 mg per day) set by the US.[46] This exceptional amount is due to the consumption of tea infusions and steamed rice, as well as Mn exposure from drinking water that comprises about 90% of the diet.[109] To a lesser extent, the chewing of betel quids (a combination of betel leaf, arcea nut, slaked lime and tobacco)[109] which is a common practice among Bangladeshis, provides about 8%.

2.7   Influence of Bioavailability

Numerous factors affect the bioavailability of Mn, including dietary components such as fiber, phytate, Fe, calcium, cadmium, fats, proteins, and polyphenols. The amount that is absorbed varies from 1 to 5%, according to the variety of inhibition or accelerator components in the diet.

2.7.1 Fiber and Phytate

Both soluble and insoluble fiber, as well as phytate, negatively influence the bioavailability of Mn. An example of the diminished absorption or enhanced excretion from the inclusion of these components is seen with consumption of vegetarian diets. The total Mn content of these diets is high owing to the abundance of plants, nuts and seeds that are naturally rich in this trace element. Despite high intakes of 2.9 to 4.6 mg per day, negative Mn balances (−0.4 to −1.1 mg per day) have been documented in vegetarians (see Table 2.1). Presumably, the negative amounts are the result of a reduced bioavailability of the mineral.[110] However, greater quantities of dietary Mn (7 mg per day) in vegetarian diets increased mean Mn balance to a positive 3.34 mg per day.[22] In 1998 Hunt *et al.* systematically investigated the effects of lacto-ovovegetarian or omnivorous diets in young women for 8 weeks in a cross-over design. Although the lacto-ovovegetarian diet had double the amount of Mn when compared with the meat-based one (5.9 *vs.* 2.5 mg per day), the respective Mn balances (0.6 *vs.* 0.1 mg per day) did not differ significantly.[111]

A negative Mn balance was observed in men who consumed diets containing an insoluble type of fiber, carboxymethylcellulose.[112] However, some studies have not demonstrated significant changes in Mn balance in humans consuming high-fiber diets.[110,113] In contrast, the effect of soluble fiber on excretion of sterols and nutrients in six ileostomy patients was tested by the consumption of a constant low-fiber diet, with and without 7.5 g sodium alginate per day (see Table 2.1). After two weeks, the alginate increased excretion of fat by 140%, but reduced excretion of bile acids by 12%. A non-significant decline of apparent absorption of Mn in five of the six subjects was observed (from 6.1 ± 0.5 to 3.3 ± 1.7 µmol per day). Since almost no digestion of sodium alginate occurred in the stomach and small intestine, it was suggested that alginate may have diminished the binding of endogenous minerals.[114] Thus, soluble fibers may have quite different effects on Mn; these fibers include pectins, gums, resistant starches, lactulose, oligofructose and inulin.

In rats, the addition of soluble fiber to the diet improved the viscosity of gut content, which enhanced fermentation and the formation of volatile fatty acids in the cecum.[115] In turn, these effects increased serum enteroglucagon concentrations, gastric emptying time and absorption of numerous minerals including Mn. Thus, the amount of Mn absorbed and retained may depend on whether the fiber is soluble or insoluble, naturally present in the diet, or supplemented.

The author of this chapter found similar results when measuring the Mn plasma intake in response to an isolated insoluble fiber, alpha cellulose. In a Mn-tolerance test, a significant reduction in the plasma levels of the mineral was observed one-hour post dose (Figure 2.5). It is possible that isolated fibers exhibit effects on minerals that differ from those present in foods naturally abundant in fiber, such as wheat, bran, sunflower and vegetables.[116]

Phytate (the salt of insoluble hexa bisphosphate) is the storage form of phosphorus in plant tissues such as bran and seeds. This compound binds a

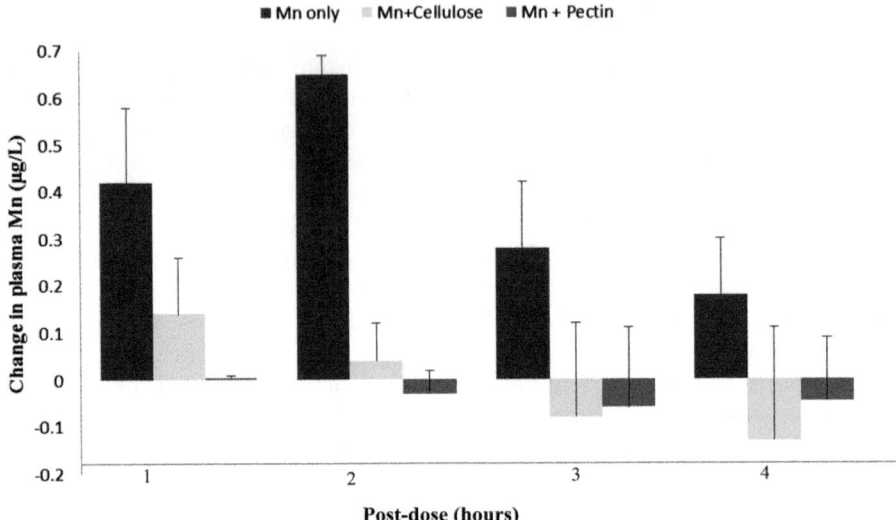

Figure 2.5 Effect of 15 g cellulose or high-methoxyl pectin on plasma uptake of 40 mg manganese.
(Bales, Connie, Jeanne Freeland-Graves, Pao-Hwa Lin, Jan Stone and Virginia Dougherty. Plasma uptake of manganese: response to dose and dietary factors. In *Nutritional Bioavailability of Manganese*, C. Kies, ed. Washington, DC: American Chemical Society. pp. 112–122, 1987. With permission.)

variety of minerals *via* chelation to the oxygen in the phosphate moiety.[117] Davidsson *et al.* (1995) documented the influence of phytate and ascorbic acid on Mn absorption using labeled ^{54}Mn in 16 young adults fed a dephytinized diet. The removal of phytate significantly increased Mn absorption from 0.7% to 1.6% ($p < 0.05$), but this percentage absorption was still quite low. The doubling of the amount of ascorbic acid in the phytinized diet from 625 μmol l^{-1} to 1250 μmol l^{-1} had no influence on Mn absorption.[118] Therefore, despite the fact that whole grains, nuts and seeds are excellent sources of dietary Mn, the presence of fiber and phytates may limit bioavailability to a substantial degree. Accordingly, the phytate in soy formula is presumably a major factor contributing to its lower absorption of Mn, as compared to that of other infant formulas and breast milk.

2.7.2 Mineral Interactions

A variety of mineral interactions significantly affect both manganese requirements and metabolism. Minerals reported to exert adverse effects when ingested concomitantly with Mn include Fe, Ca and cadmium (Cd).

2.7.2.1 Iron

It is well established that Mn status is influenced strongly by other transition minerals, such as inorganic Fe. Manganese negatively affects Fe absorption, presumably because Fe and Mn share a common intestinal transport system[33] including DMT-1 and TfR.[119] Lack of regulation of one mineral may result in imbalance of the other. Consequently, excessive Fe supplementation has the potential to lead to a Mn deficiency.[120] Park and coauthors[121] investigated Mn status in 31 Fe-deficient infants (6–24 months of age). As predicted, mean Mn blood levels of infants deficient in Fe were greater than those of the healthy controls (25.5 *vs.*14.9 µg l^{-1}). However, these estimates were reduced to 20.45 µg l^{-1} upon Fe supplementation to the deficient group (6 mg kg^{-1} Fe^{3+}), who were followed for 1–6 months. These findings were attributed to the incremental increase in Mn absorption that resulted from Fe deficiency.

Finley (1999) assessed Mn absorption and retention in 26 healthy adult women who were fed diets supplemented with 0.7 or 9 mg per day of Mn for 60 days. Absorption was greatest in subjects who consumed a low Mn diet that was coupled with a low ferritin concentration (4.86%), as compared to those with a high ferritin concentration (0.97%). The subjects with low serum ferritin fed a high Mn diet had a higher metabolic balance than those with a high ferritin, (1.53 *vs.* 0.59 mg per day, respectively).[122] Moreover, the half-life of ^{54}Mn was longer in those who had high ferritin concentrations and consumed a low Mn diet (36.6 days).[122] In the investigation conducted in six ileostomy patients who consumed Indian vegetarian diets, Mn intakes ranged between 1 and 3.6 mg per 1000 kcal. The level of Mn in the diet was positively associated with plasma zinc ($r = 0.25$), copper ($r = 0.57$), and Mn absorption ($r = 0.21$) ($p < 0.05$), and negatively correlated with Fe absorption ($r = -0.89$, $p < 0.01$) post surgery.[123] Finally, blood levels of Mn in Fe-deficient anemic patients were higher than in those who were non-anemic, averaging 2.05 *vs.* 1.28 µg dl^{-1}, respectively.[124] This result is similar to that of a study by Smith *et al.* which reported median blood Mn concentrations to be higher in Fe deficient anemic children when compared to controls (1.64 *vs.* 1.1 µg dl^{-1}).[125]

A mouse model of hemochromatosis, a condition whereby loss of the HFE gene leads to unregulated intestinal Fe absorption, was used to demonstrate decreased Mn absorption due to Fe overload. Mice with HFE deficiency displayed significantly higher whole blood Fe and lower whole blood Mn, as compared to normal mice.[126] In women with the unaffected HFE alleles, mean whole blood Mn concentrations were 17.1, compared with 13.5 µg l^{-1} for those with a HFE variant allele.[119] These data suggest that women with hemochromatosis may have increased Mn requirements.

2.7.2.2 Calcium

In a series of plasma tolerance tests in young adults, the addition of Ca, as either calcium carbonate (CaCO$_3$) or in milk, diminished Mn absorption in plasma.[116] Likewise the administration of 40 mg Mn as manganese chloride essentially blocked the uptake of 800 mg Ca, as shown in Figure 2.6. The concomitant administration of 2 mg Cu depressed Mn uptake a great deal,

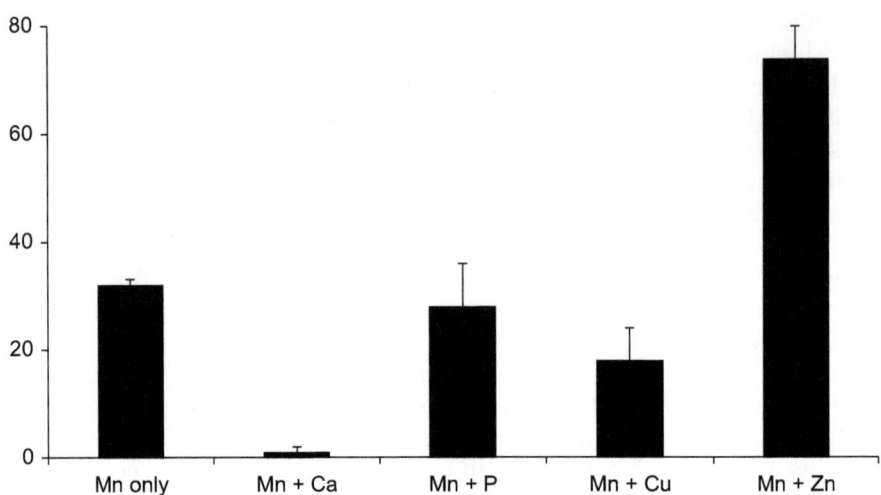

Figure 2.6 Mean areas under the curve for plasma manganese response in humans when the following oral loads were administered: 40 mg manganese, 40 mg manganese + 800 mg calcium, 40 mg manganese + 800 mg phosphorus, 40 mg manganese + 2 mg copper, and 40 mg manganese + 50 mg zinc. (Freeland-Graves JH, Lin PH: Plasma uptake of manganese as affected by oral loads of manganese, calcium, milk, phosphorus, copper, and zinc. *J. Am. Coll. Nutr.* 1991, **10**:38–43. With permission.)

but the change was not significant. In contrast the provision of 50 mg Zn with the 40 mg Mn produced a rapid rise in the area under the curve (124%, $p < 0.05$) for plasma Mn.

In subjects with heartburn, Ca supplements elevated esophageal pH, producing an alkaline environment in the gastrointestinal tract which may have subsequently diminished Mn absorption.[127] However, Johnson *et al.* reported no differences in Mn absorption in young women who consumed diets consisting of 587 or 1336 mg Ca per day.[128] Thus, the impact of Ca on Mn uptake may be dependent on dose and pharmacological administration.

2.7.2.3 Cadmium

Cadmium (Cd) is a toxic trace element that exists in a +2 oxidative state and utilizes a transport system similar to Mn.[129] Ingested Cd accumulates in the kidneys, eyes, and other tissues. It is believed that this accumulation results from the efficient absorption and systemic transport of Cd through multiple carriers used to transport Mn, as well as Ca, Fe, and Zn.[130] An investigation by Kippler *et al.*[130] observed that the concentration of Cd in the breast milk of pregnant Bangladeshi women was 0.14 μg kg^{-1}, and positively associated with levels in breast milk and erythrocytes. Furthermore, the ratio of Cd in breast milk to that in plasma was approximately 3 : 4, indicating the lack of a barrier preventing Cd transport from plasma to breast milk. Cadmium in breast milk also was positively associated with Mn in breast milk $(r = 0.56; p < 0.01)$, suggesting that Cd shares common transporters with Mn when transported to mammary glands and could induce a deficiency.[130] Levels of Cd as low as <1 μg l^{-1} in breast milk have been reported to cause neurobehavioral and endocrine problems that reduce infant suckling, and decrease the Mn intake from breast milk.

2.7.3 Fat and Protein

The effect of dietary fat on mineral utilization was investigated by Kies *et al.* (1988), who provided a diet with two levels of total fat (30% and 40% of total calories), and cholesterol (300 and 600 mg per day) to 30 subjects. The Mn absorption was less negative when participants received the moderately high-fat diet (-28.63% to -30.60%), as compared to the low-fat (-7.11% to -7.80%) group. Thus, fat was believed to enhance Mn absorption,[131] perhaps due to the increased viscosity which created a slower transport through the intestine.[115]

Greger and Snedeker examined the influence of protein and phosphorus (P) on Mn retention. Eight adult men were fed diets that varied in levels either of P or of protein. The variety of treatments did not affect the apparent absorption and retention of Mn.[132] Subsequently, the influence of the quantity of meat intake on minerals such as Mn was examined in 14 postmenopausal women, aged 51–70 years (see Table 2.1).[133] Participants consumed three different diets on a two-day rotation for seven weeks. Metabolic balances for the high- and low-meat and low-meat mineral supplemented diets, on Mn intakes of 3.66, 3.48 and 3.63 mg per day, were 0.11, 0.04 and 0.17 mg per day, respectively. However, these differences were not significant.[133] Thus, it appears that dietary protein is not a significant factor influencing Mn absorption.

2.7.4 Polyphenolic Compounds

Tea is an excellent source of Mn; however, its bioavailability is diminished with the great abundance of polyphenolic compounds naturally present. An investigation of the Mn intake of tea drinkers reported intakes of 5.5–10 mg per day, as compared to 3.2 mg per day in non-tea drinkers.[134] However, indices of Mn status such as whole blood and plasma Mn, and superoxide dismutase expression, did not vary between the two groups. Thus, the abundant quantities of Mn in tea did not appear to translate to increased absorption and/or retention by tea drinkers, presumably owing to the presence of polyphenolic compounds.

2.8 Toxicity

Ingestion of quantities of Mn up to 10 mg per day from foods has not been reported to have any adverse health effects.[20] Therefore, the US Environmental Protection Agency (EPA)[75] set 10 mg per day (0.14 mg kg^{-1} per day) as the Reference Dose for Mn in 2007. The upper tolerable level of Mn for adults has been established as 11 mg per day, according to the IOM.[46] However, no undesirable effect of excessive Mn intake from food or supplements has been reported.[135]

However, the amount of Mn supplied by drinking water may have significant deleterious effects. In 362 Canadian schoolchildren, Bouchard and others[136] examined the consequences of drinking tap water on intellectual functioning. Hair content of Mn was positively associated with Mn content in water ($p < 0.001$), but not with the diet. The content of Mn in the drinking water ranged from 1.0 to 2700 µg l^{-1}, with a median of 34 µg l^{-1}. Of great concern was that the concentrations of Mn in both hair and water were negatively associated with intelligent quotient (IQ) scores. An increase in Mn water concentration by 10-fold was associated with a 2.4-fold decrease in IQ scores ($p < 0.01$). It was concluded that the high Mn content in the groundwater was associated with intellectual impairment in children.

Manganese toxicity is known to result from occupational exposure such as working in mines,[137] air pollution,[138] disease conditions associated with organ dysfunction, such as liver diseases,[139] mineral interactions,[140] and parenteral nutrition.[141] The result is a neurodegenrative effect that contributes to manganism and parkinsonism at levels >1 mg m^{-3} and 100 ng m^{-3}, respectively.[139] Deposition of Mn has been observed in the central nervous system, resulting in adverse outcomes on motor and cognitive functions.[142] These effects include manganism, idiopathic Parkinson's disease,[42] hepatic dysfunction, and cholestasis,[139] attention-deficit hyperactivity disorder and, possibly, autism.[138]

In 2013 Bhang and colleagues[143] explored the role of Mn obtained from the environment on cognitive function in 1089 Korean school children aged 8–11 years. The results showed that the median Mn level in blood was 14.14 µg l^{-1}, ranging between 8.15 and 21.45 µg l^{-1}. After controlling for confounders, elevated Mn levels were significantly associated with attention deficit syndrome. This syndrome is expressed as low cognitive functioning and academic performance, including thinking, reading and computational processes. Thus, monitoring the Mn status of young children may be an important component of effective public health screening.

2.8.1 Parenteral Nutrition

The current recommended Mn level for adults on parenteral nutrition formulas is 0.06–0.1 mg per day. Higher and lower levels are associated with toxicity and deficiency, respectively.[144] In the past, TPN created nutritional imbalances, including Mn accumulation that led to parkinsonism. In a patient receiving a TPN solution containing 10 µmol of Mn per day,[145] whole blood Mn was elevated to 135 µg l^{-1}. Upon removal of Mn from the solution, the whole blood Mn level then declined to 20 µg l^{-1}. This reduction was accompanied by disappearance of high intensity lesions in the globus pallidus, as determined by MRI.[42]

Manganese levels were examined in 21 patients receiving TPN who were provided 21.9 mg per 100 ml per day of Mn gluconate, as compared to 10 healthy controls (26–75 years old) for 3 to 132 months.[146] Serum levels of Mn in TPN patients were more than doubled when compared to controls (1.96 *vs.* 0.81 mg l^{-1}; $p = 0.001$) and were positively related to the amount of Mn in the TPN supplied (21.9 mg per 100 ml).

Patients suffering from inflammation also showed marginal signs of neurologic problems that influence motor activities. It was suggested that sustained inflammation might have induced hypermanganesemia through manifestation of: (1) cholestatic liver disease that reduced Mn biliary excretion, (2) increased nutritional requirements and demand for Mn, and/ or (3) modification of the metabolism or distribution of Mn in the body.[146]

In 2008 Klein *et al.*[147] assessed trace element loss through urine in 12 TPN patients (18–62 years old) exhibiting traumatic renal dysfunction. The TPN formula provided 0.3 mg Mn per day. Since only a minimal amount of Mn was lost in the urine (4.2–12.0 μg per day), it was believed that potential toxicity could have been induced by the high Mn retention. Recently, Abdalian *et al.*[147] found the mean amount of Mn in the TPN formula given to 16 subjects to average 400 μg per day, and Mn values were higher than the upper normal blood levels by about 1.4 times. Deposits of Mn in the brain, a reflection of Mn neurotoxicity, were detected in 81% of the subjects *via* MRI, and 15% suffered from symptoms of parkinsonism such as depression and memory loss.

Manganism also has developed in infants given TPN. Amounts present in the solution ranged from 5.6 to 8.9 μg l^{-1}, levels that are 100 times higher than those found in breast milk.[148] The safe TPN dose of Mn given to preterm infants should be 1 μg kg^{-1} per day.[144] It is recommended that preterm infants weighing less than 3 kg, and full-term infants weighing 3–10 kg, should receive 1 μg kg^{-1} per day.[144]

Today, the trace element composition of commercial TPN formulas available in the US is far from ideal. Current recommendations[144] for daily Mn dose of TPN formulas to provide adequate intake are 60 to 100 μg.

2.9 Conclusions

Manganese is an essential trace element that is found primarily in vegetable products such as whole grains, cereals and nuts. Dietary requirements of Mn differ by age and gender, and it is essential to maintain balanced intake of Mn. Negative balances have been reported in numerous investigations where mean Mn intake was greater than the recommended AI. The adequacy of current dietary recommendations for Mn is unclear.

Deficiency of Mn has been documented in several studies to be associated with adverse health effects including dermatitis, osteoporosis, dyslipidemia, type II diabetes, metabolic syndrome, hypertension, neurological disorders such as epilepsy, thyroid gland secretions, cancer, decreased cognitive function, and asthma. Many of these complications were related to the dysfunction of MnSOD, a Mn-dependent antioxidant enzyme that combats ROS and attenuates oxidative stress.

The bioavailability of Mn is relatively low ($\leq$5%), as it is affected by dietary fiber, phytates, mineral interactions, polyphenolic compounds, type of diet, such as vegetarianism or infant formulas, and route of feeding (*e.g.*, TPN). Toxic levels of Mn may be a potential problem with contaminated drinking water, in individuals with hepatic dysfunction, and in infants drinking soy-based formula. Nonetheless, clear evidence linking Mn toxicity to these conditions is lacking, with the exception of TPN. In sum, understanding the nutrient requirements of Mn is critical for optimal health.

References

1. E. Doisy, Effects of deficiency of manganese upon plasma levels of clotting proteins and cholesterol in man, in *Trace Elements Metabolism in Animals*, ed. W. Hoekstra, J. S and M. Gantner, University Park Press, Baltimore, MD, 1974, pp. 668–70.
2. S. Sarban, U. Isikan, Y. Kocabey and A. Kocyigit, Relationship between synovial fluid and plasma manganese, arginase, and nitric oxide in patients with rheumatoid arthritis, *Biol. Trace Elem. Res.*, 2007, **115**, 97–106.
3. J. Freeland-Graves, T. Bose and A. Karbassian, Manganese Metal-lotherapeutics, in *Metallotherapeutic Drugs and Metal-Based Diagnostic Agents: The Use of Metals in Medicine*, ed. M. Gielen, and E. Tiekink, West Sussex, United Kingdom, 2005, pp. 159–178.
4. J. L. Aschner and M. Aschner, Nutritional aspects of manganese homeostasis, *Mol. Aspects Med.*, 2005, **26**, 353–362.
5. A. A. Brock, S. A. Chapman, E. A. Ulman and G. Wu, Dietary manganese deficiency decreases rat hepatic arginase activity, *J. Nutr.*, 1994, **124**(Mar), 340–4.
6. M. Wilhelm, J. Fishman, R. Pontikis, A. Aubertin and F. Wilhelm, Susceptibility of recombinant porcine endogenous retrovirus reverse transcriptase to nucleoside and non-nucleoside inhibitors, *Cell. Mol. Life Sci.*, 2002, **59**, 2184–2190.
7. M. Johnson and J. Edmonds, Copper, iron, zinc, and manganese in dietary supplements, infant formulas, and ready-to-eat breakfast cereals, *Am. J. Clin. Nutr.*, 1998, **67**, 1035S–40S.
8. B. Lönnerdal, Nutritional aspects of soy formula, *Acta Paediatr. Suppl.*, 1994, **402**, 105–8.
9. K. Cockell, G. Bonacci and B. Belonjea, Manganese content of soy or rice beverages is high in comparison to infant formulas, *J. Am. Coll. Nutr.*, 2004, **23**, 124–130.
10. P. E. Johnson, G. I. Lykken and E. D. Korynta, Absorption and biological half-life in humans of intrinsic and extrinsic 54Mn tracers from foods of plant origin, *J. Nutr.*, 1991, **121**(May), 711–7.

11. R. Allinson, Plasma trace elements during total parenteral nutrition, *JPEN, J. Parenter. Enteral Nutr.*, 1978, **2**, 35–40.
12. T. Gunter, B. Gerstner, K. Gunter, J. Malecki, R. Gelein, W. Valentine, M. Aschner and D. Yule, Manganese transport *via* the transferring mechanism, *Neurotoxicology*, 2013, **34**, 118–127.
13. S. Gruenheid, F. Canonne-Hergaux, S. Gauthier, D. J. Hackam, S. Grinstein and P. Gros, The iron transport protein NRAMP2 is an integral membrane glycoprotein that colocalizes with transferrin in recycling endosomes, *J. Exp. Med.*, 1999, **189**(Mar), 831–41.
14. Z. Yin, H. Jiang, E. S. Lee, M. Ni, K. M. Erikson, D. Milatovic, A. B. Bowman and M. Aschner, Ferroportin is a manganese-responsive protein that decreases manganese cytotoxicity and accumulation, *J. Neurochem.*, 2010, **112**(Mar), 1190–8.
15. C. D. Davis, L. Zech and J. L. Greger, Manganese metabolism in rats: an improved methodology for assessing gut endogenous losses, *Proc. Soc. Exp. Biol. Med.*, 1993, **202**, 103–8.
16. C. Herrera, M. Pettiglio and T. Bartnikas, Investigating the role of transferrin in the distribution of iron, manganese, copper, and zinc, *J. Biol. Inorg. Chem.*, 2014, **19**, 869–877.
17. J. Bresson, A. Flynn, M. Heinonen, K. Hulshof, H. Korhonen, P. Lagiou, M. Løvik, R. Marchelli, A. Martin, B. Moseley, H. Przyrembel, S. Salminen, S. Strain, S. Strobel, I. Tetens, H. Berg, H. Loveren and H. Verhagen, EFSA Panel on Dietetic Products NaAN, Scientific Opinion on the substantiation of health claims related to manganese and protection of DNA, proteins and lipids from oxidative damage (ID 309), maintenance of bone (ID 310), energy-yielding metabolism (ID 311), and cognitive function (ID 340) pursuant to Article 13(1) of Regulation (EC) No 1924/20061, *EFSA J.*, 2009, 7, 1217–1233.
18. J. Crossgrove and W. Zheng, Manganese toxicity upon overexposure, *NMR Biomed.*, 2004, **17**, 544–553.
19. J. H. Freeland-Graves, F. Behmardi, C. W. Bales, V. Dougherty, P. H. Lin, J. B. Crosby and P. C. Trickett, Metabolic balance of manganese in young men consuming diets containing five levels of dietary manganese, *J. Nutr.*, 1988, **118**, 764–73.
20. J. H. Freeland-Graves and J. R. Turnlund, Deliberations and evaluations of the approaches, endpoints and paradigms for manganese and molybdenum dietary recommendations, *J. Nutr.*, 1996, **126**, 2435S–2440S.
21. B. B. North, J. M. Leichsenring and L. M. Norris, Manganese metabolism in college women, *J. Nutr.*, 1960, 72, 217–23.
22. V. M. Lang, B. B. North and L. M. Morse, Manganese metabolism in college men consuming vegetarian diets, *J. Nutr.*, 1965, **85**, 132–8.
23. K. Y. Patterson, J. T. H, J. E. Bodner, J. L. Kelsay, J. C. Smith Jr and C. Veillon, Zinc, copper, and manganese intake and balance for adults consuming self-selected diets, *Am. J. Clin. Nutr.*, 1984, **40**, S1397–S403.

24. J. W. Finley, P. E. Johnson and L. K. Johnson, Sex affects manganese absorption and retention by humans from a diet adequate in manganese, *Am. J. Clin. Nutr.*, 1994, **60**, 949–55.

25. B. J. Friedman, J. H. Freeland-Graves, C. W. Bales, F. Behmardi, R. L. Shorey-Kutschke, R. A. Willis, J. B. Crosby, P. C. Trickett and S. D. Houston, Manganese balance and clinical observations in young men fed a manganese-deficient diet, *J. Nutr.*, 1987, **117**, 133–43.

26. B. Lönnerdal, Effects of milk and milk components on calcium, magnesium, and trace element absorption during infancy, *Physiol. Rev.*, 1997, **77**, 643–69.

27. K. Dörner, S. Dziadzka, A. Höhn, E. Sievers, H. D. Oldigs, G. Schulz-Lell and J. Schaub, Longitudinal manganese and copper balances in young infants and preterm infants fed on breast-milk and adapted cow's milk formulas, *Br. J. Nutr.*, 1989, **61**, 559–572.

28. L. Davidsson, A. Cederblad, B. Lonnerdal and B. Sandstrom, Manganese absorption from humanmilk, cows milk, and infant formulas in humans, *Am. J. Dis. Child.*, 1989, **143**, 823–827.

29. B. Koletzko, S. Baker, G. Cleghorn, F. Neto, S. Gopalan, O. Hernell, Q. Hock, P. Jirapinyo, B. Lonnerdal, P. Pencharz, H. Pzyrembel, J. Ramirez-Mayans, R. Shamir, D. Turck, Y. Yamashiro and D. Zong-Yi, Global standard for the composition of infant formula: recommendations of an ESPGHAN coordinated international expert group, *J. Pediatr. Gastroenterol. Nutr.*, 2005, **41**, 584–599.

30. M. Nicar and C. Pack, Calcium bioavailability from calcium carbonate and calcium citrate, *J. Clin. Endocrinol. Metab.*, 1985, **61**, 391–393.

31. B. Lönnerdal, Trace element absorption in infants as a foundation to setting upper limits for trace elements in infant formulas, *J. Nutr.*, 1989, **119**, 1839S–45S.

32. J. K. Friel, S. Penney, D. W. Reid and W. L. Andrews, Zinc, copper, manganese, and iron balance of parenterally fed very low birth weight preterm infants receiving a trace element supplement, *JPEN, J. Parenter. Enteral Nutr.*, 1988, **12**, 382–386.

33. S. J. Garcia, K. Gellein, T. Syversen and M. Aschner, A manganese-enhanced diet alters brain metals and transporters in the developing rat, *Toxicol. Sci.*, 2006, **92**, 516–525.

34. C. D. Davis and J. L. Greger, Longitudinal changes of manganese-dependent superoxide dismutase and other indexes of manganese and iron status in women, *Am. J. Clin. Nutr.*, 1992, **55**, 747–52.

35. J. L. Greger, C. D. Davis, J. W. Suttie and B. J. Lyle, Intake, serum concentrations, and urinary excretion of manganese by adult males, *Am. J. Clin. Nutr.*, 1990, **51**, 457–61.

36. R. S. Gibson, B. M. Anderson and J. H. Sabry, The trace metal status of a group of post-menopausal vegetarians, *J. Am. Diet. Assoc.*, 1983, **82**, 246–50.

37. J. Bresson, A. Flynn, M. Heinonen, K. Hulshof, H. Korhonen, P. Lagiou, M. Løvik, R. Marchelli, A. Martin, B. Moseley, H. Przyrembel, S. Salminen, S. Strain, S. Strobel, I. Tetens, H. Berg, H. Loveren and

H. Verhagen, EFSA Panel on Dietetic Products NaAN, Scientific Opinion on the substantiation of health claims related to manganese and protection of DNA, proteins and lipids from oxidative damage (ID 309), maintenance of bone (ID 310), energy-yielding metabolism (ID 311), and cognitive function (ID 340) pursuant to Article 13(1) of Regulation (EC) No 1924/20061, *EFSA J.*, 2009, 7, 1212–1233.

38. M. Razandi, A. Pedram, V. C. Jordan, S. Fuqua and E. R. Levin, Tamoxifen regulates cell fate through mitochondrial estrogen receptor beta in breast cancer, *Oncogene*, 2013, **32**, 3274–85.

39. J. L. Ensunsa, J. D. Symons, L. Lanoue, H. R. Schrader and C. L. Keen, Reducing arginase activity *via* dietary manganese deficiency enhances endothelium-dependent vasorelaxation of rat aorta, *Exp. Biol. Med.*, 2004, **229**, 1143–1153.

40. J. Häberle, B. Görg, A. Toutain, F. Rutsch, J. Benoist, A. Gelot, A. Suc, H. Koch, F. Schliess and D. Häussinger, Inborn error of amino acid synthesis: Human glutaminesynthetase deficiency, *J. Inherited Metab. Dis.*, 2006, **29**, 352–358.

41. N. Reynolds, A. Blumsohn, J. P. Baxter, G. Houston and C. R. Pennington, Manganese requirement and toxicity in patients on home parenteral nutrition, *Clin. Nutr.*, 1998, **17**, 227–30.

42. S. Nagatomo, F. Umehara, K. Hanada, Y. Nobuhara, S. Takenaga, K. Arimura and M. Osame, Manganese intoxication during total parenteral nutrition: report of two cases and review of the literature, *J. Neurol. Sci.*, 1999, **162**, 102–5.

43. E. Harris, *Managanese*, Minerals in Foods Nutrition, Metabolism, Bioactivity, DEStech Publications, Lancaster, PA, 2014, pp. 227–37.

44. S. J. Li, L. Jiang, X. Fu, S. Huang, Y. N. Huang, X. R. Li, J. W. Chen, Y. Li, H. L. Luo, W. Fang, O. Shi-Yan and J. Yue-Ming, Pallidal index as biomarker of manganese brain accumulation and associated with manganese levels in blood: a meta-analysis, *PLoS One*, 2014, **9**, 1–7.

45. J. A. Pennington, Total diet studies: the identification of core foods in the United States food supply, *Food Addit. Contam.*, 1992, **9**, 253–64.

46. Food and Nutrition Board, Institute of Medicine, *Manganese*. Dietary reference intakes for vitamin A, vitamin K, boron, chromium, copper, iodine, iron, manganese, molybdenum, nickel, silicon, vanadium, and zinc, National Academy Press, Washington, D.C., 2001, pp. 394–419.

47. S. Gulberti, S. Fournel-Gigleux, G. Mulliert, A. Aubry, P. Netter, J. Magdalou and M. Ouzzine, The functional glycosyltransferase signature sequence of the human beta 1,3-glucuronosyltransferase is a XDD motif, *J. Biol. Chem.*, 2003, **278**, 32219–32226.

48. L. Strause, J. Hegenaur, P. Saltman, R. Cone and D. Resnick, Effects of long term dietary manganese and copper deficiency on rat skeleton, *J. Nutr.*, 1986, **116**, 135–41.

49. J. Freeland-Graves and C. Llanes, Models to study manganese deficiency, in *Manganese in Health and Disease*, ed. D. Klimis-Tavantzis and B. Racon, CRC Press, FL, 1993, pp. 59–86.

50. T. Landete-Castillejos, I. Molina-Quilez, J. Estevez, F. Ceacero, A. Garcia and L. Gallego, Alternative hypothesis for the origin of osteoporosis: the role of Mn, *Front. Biosci., Elite Ed.*, 2012, **4**, 1385–1390.

51. J. Scheurmann, N. Treiber, C. Weber, A. C. Renkl, D. Frenzel, F. Trenz-Buback, A. Ruess, G. Schulz, K. Scharffetter-Kochanek and J. M. Weiss, Mice with heterozygous deficiency of manganese superoxide dismutase (SOD2) have a skin immune system with features of "inflamm-aging", *Arch. Dermatol. Res.*, 2014, **306**, 143–55.

52. B. J. Friedman, J. H. Freeland-Graves, C. W. Bales, F. Behmardi, R. L. Shorey-Kutschke, R. A. Willis, J. B. Crosby, P. C. Trickett and D. S. Houston, Manganese balance and clinical observations in young men fed a manganese-deficient diet, *J. Nutr.*, 1987, **117**, 133–143.

53. Y. Bae, M. Choi and M. Kim, Manganese supplementation reduces the blood cholesterol levels in Ca-deficient ovariectomized rats, *Biol. Trace Elem. Res.*, 2011, **141**, 224–231.

54. H. Afridi, T. Kazi, N. Kazi, J. Baig, M. Jamali, M. Arain, R. Sarfraz, H. Sheikh, G. Kandhro and A. Shah, Status of essential trace metals in biological samples of diabetic mother and their neonates, *Gynecol. Obstet.*, 2009, **280**, 415–23.

55. Y. Li, G. Hongwei, M. Wu and M. Liu, Serum and dietary antioxidant status is associated with lower prevalence of the metabolic syndrome in a study in Shanghai, China, *Asia Pac. J. Clin. Nutr.*, 2013, **22**, 60–68.

56. K. Jin and N. D. Vaziri, Salt-sensitive hypertension in mitochondrial superoxide dismutase deficiency is associated with intra-renal oxidative stress and inflammation, *Clin. Exp. Nephrol.*, 2013, **18**, 445–452.

57. T. Eid, M. Thomas, D. Spencer, E. Rundén-Pran, J. Lai, G. Malthankar, J. Kim, N. Danbolt, O. Ottersen and N. de Lanerolle, Loss of glutamine synthetase in the human epileptogenic hippocampus: possible mechanism for raised extracellular glutamate in mesial temporal lobe epilepsy, *Lancet*, 2004, **363**, 28–37.

58. H. Wu, S. Chou, D. Chen and H. Kuo, Differentiation of serum levels of trace elements in normal and malignant breast patients, *Biol. Trace Elem. Res.*, 2006, **113**, 9–18.

59. B. D. Patel, A. A. Welsh, S. A. Bingham, R. N. Luben, N. E. Day, K. T. Khaw, D. A. Lomas and N. J. Wareham, Dietary antioxidants and asthma in adults, *Thorax*, 2006, **61**, 388–393.

60. A. MacDonald, P. Lee, P. Davies, A. Daly, M. Lilburn, H. Gokmen Ozel, M. A. Preece, C. Hendriksz and A. Chakrapani, Long-term compliance with a novel vitamin and mineral supplement in older people with PKU, *J. Inherited Metab. Dis.*, 2008, **31**, 718–723.

61. L. Liang, S. Waldbaum, S. Rowley, T. Huang, B. Day and M. Patel, Mitochondrial oxidative stress and epilepsy in SOD2 deficient mice: attenuation by a lipophilic metalloporphyrin, *Neurobiol. Dis.*, 2012, **45**, 1068–1076.

62. S. Hansen, J. Spears, K. Lloyd and C. Whisnant, Feeding a low manganese diet to heifers during gestation impairs fetal growth and development, *J. Dairy Sci.*, 2006, **89**, 4305–4311.

63. S. Y. Bhang, S. C. Cho, J. W. Kim, Y. C. Hong, M. S. Shin, H. J. Yoo, I. H. Cho, Y. Kim and B. N. Kim, Relationship between blood manganese levels and children's attention, cognition, behavior, and academic performance–a nationwide cross-sectional study, *Environ. Res.*, 2013, **126**, 9–16.

64. H. A. Jouihan, P. A. Cobine, R. C. Cooksey, E. A. Hoagland, S. Boudina, E. D. Abel, D. R. Winge and D. A. McClain, Iron-mediated inhibition of mitochondrial manganese uptake mediates mitochondrial dysfunction in a mouse model of hemochromatosis, *Mol. Med.*, 2008, **14**, 98–108.

65. E. Burtlet, Manganese supplementation reduces high glucose-induced monocyte adhesion to endothelial cells and endothelial dysfunction in Zucker diabetic fatty rats, *J. Biol. Chem.*, 2013, **288**, 6409–16.

66. G. Forte, B. Bocca, A. Peruzzu, F. Tolu, Y. Asara, C. Farace, R. Oggiano and R. Madeddu, Blood metals concentration in type 1 and type 2 diabetics, *Biol. Trace Elem. Res.*, 2013, **156**, 79–90.

67. G. Carl, C. Keen, B. Gallagher, M. Clegg, W. Littleton, D. Flannery and L. Hurley, Association of low blood manganese concentrations with epilepsy, *Neurology*, 1986, **36**, 1584–1587.

68. G. Carl, L. Blackwell, F. Barnett, L. Thompson, C. Rissinger, K. Olin, J. Critchfield, C. Keen and B. Gallagher, Manganese and epilepsy: brain glutamine synthetase and liver arginase activities in genetically epilepsy prone and chronically seizured rats, *Epilepsia*, 1993, **34**, 441–446.

69. J. Critchfield, G. Carl and C. Keen, The influence of manganese supplementation on seizure onset and severity, and brain monoamines in the genetically epilepsy prone rat, *Epilepsy Res.*, 1993, **14**, 3–10.

70. K. Badiei, K. Mostaghni and E. Gorjizadeh, Effect of manganese on thyroid function in sheep, *Comp. Clin. Pathol.*, 2008, **17**, 259–262.

71. K. Zabłocka-Słowińska and H. Grajeta, The role of manganese in etiopathogenesis and prevention of selected diseases, *Postepy Hig. Med. Dosw.*, 2012, **66**, 549–53.

72. J. Eum, H. Cheong, E. Ha, M. Ha, Y. Kim, Y. Hong, H. Park and N. Chang, Maternal blood manganese level and birth weight: a MOCEH birth cohort study, *Environ. Health*, 2014, **13**, 31–37.

73. H. Guan, M. Wang, X. Li, F. Piao, Q. Li, L. Xu, F. Kitamura and K. Yokoyama, Manganese concentrations in maternal and umbilical cord blood: related to birth size and environmental factors, *Eur. J. Public Health*, 2014, **24**, 150–157.

74. J. A. Pennington, Intakes of minerals from diets and foods: is there a need for concern?, *J. Nutr.*, 1996, **126**, S2304–S8.

75. Toxicity and Exposure Assessment for Children's Health (TEACH). In: (EPA). USEPA, editor. http://www.epa.gov/teach/chem_summ/manganese_summary.pdf. Accessed Dec. 5, 2013.; October 29, 2007.

76. H. Kuhnlein, O. Receveur, R. Soueida and P. Berti, Unique patterns of dietary adequacy in three cultures of Canadian Arctic indigenous peoples, *Public Health Nutr.*, 2008, **11**, 349–360.

77. P. A. Pleban, B. S. Numerof and F. H. Wirth, Trace element metabolism in the fetus and neonate, *Clin. Endocrinol. Metab.*, 1985, **14**, 545–566.

78. B. Lönnerdal, C. L. Keen, M. Ohtake and T. Tamura, Iron, zinc, copper, and manganese in infant formulas, *Am. J. Dis. Child.*, 1983, **137**(May), 433–7.

79. E. Vuori, S. M. Makinen, R. Kara and P. Kuitunen, The effects of the dietary intakes of copper, iron, manganese, and zinc on the trace element content of human milk, *Am. J. Clin. Nutr.*, 1980, **33**, 227–231.

80. M. Domellöf, Nutritional care of premature infants: microminerals, *World Rev. Nutr. Diet.*, 2014, **110**, 121–39.

81. K. Cockell, G. Bonacci and B. Belonjea, Manganese content of soy or rice beverages is high in comparison to infant formulas, *J. Am. Coll. Nutr.*, 2004, **23**, 124–130.

82. S. Tsutie, N. Kurihara, A. Sasaki, A. Takagi, H. Seguti and T. Inatome, Formulas providing adequate pantothenic acid, vitamin D, manganese, iron and vitamin A for infants fed with mother's milk (aged 6–11 months) according to the Japanese Dietary Reference Intakes prepared by the Ministry of Health, Labour and Welfare (2005 edition), *Matern. Child Nutr.*, 2010, **6**, 147–14l.

83. M. Leno, in *Is there a relationship between elevated manganese levels and violent behavior?* ed. C. Sacramento, California Assembly Committee on Public Safety, USA, 2004.

84. C. Casey, K. Hambidge and M. Mebille, Studies in human lactation: Zinc, copper, manganese and chromium in human milk in the first month of lactation, *Am. J. Clin. Nutr.*, 1985, **41**, 1193–1200.

85. J. E. Ericson, F. M. Crinella, K. A. Clarke-Stewart, V. D. Allhusen, T. Chan and R. T. Robertson, Prenatal manganese levels linked to childhood behavioral disinhibition, *Neurotoxicol. Teratol.*, 2007, **29**, 181–187.

86. J. L. Greger, S. C. Zaikis, R. P. Abernathy, O. A. Bennett and J. Huffman, Zinc, nitrogen, copper, iron, and manganese balance in adolescent females fed two levels of zinc, *J. Nutr.*, 1978, **108**, 1449–56.

87. J. L. Greger, P. Baligar, R. P. Abernathy, O. A. Bennett and T. Peterson, Calcium, magnesium, phosphorus, copper, and manganese balance in adolescent females, *Am. J. Clin. Nutr.*, 1978, **31**, 117–21.

88. D. M. Campbell, Trace element needs in human pregnancy, *Proc. Nutr. Soc.*, 1988, **47**, 45–53.

89. J. C. Shaw, Trace metal requirements of preterm infants, *Acta Paediatr. Scand., Suppl.*, 1982, **296**, 93–100.

90. R. Wood, Manganese and birth outcome, *Nutr. Rev.*, 2009, **67**, 416–20.

91. M. Vigeh, K. Yokoyama, F. Ramezanzadeh, M. Dahaghin, E. Fakhriazad, Z. Seyedaghamiri and S. Araki, Blood manganese concentrations and intrauterine growth restriction, *Reprod. Toxicol.*, 2008, **25**, 219–223.

92. S. Rahman, A. Akesson, M. Kippler, M. Grandér, J. Hamadani, P. Streatfield, L.-A. Persson, S. El Arifeen and M. Vahter, Elevated Manganese Concentrations in Drinking Water May Be Beneficial for Fetal Survival, *Plos One*, 2013, **8**, e74119–e74126.

93. A. Zota, A. Ettinger, M. Bouchard, C. Amarasiriwardena, J. Schwartz, H. Hu and R. Wright, Maternal Blood Manganese Levels and Infant Birth Weight, *Epidemiology*, 2009, **20**, 367–373.

94. L. Chen, G. Ding, Y. Gao, P. Wang, R. Shi, H. Huang and Y. Tian, Manganese concentrations in maternal–infant blood and birth weight, *Environ. Sci. Pollut. Res.*, 2014, **21**, 6170–6175.

95. L. Takser, J. Lafond, M. Bouchard, G. St-Amour and D. Mergler, Manganese levels during pregnancy and at birth: relation to environmental factors and smoking in a Southwest Quebec population, *Environ. Res.*, 2004, **95**, 119–125.

96. K. Ljung, M. Kippler, W. Goessler, G. Grandér, B. Nermell and M. Maternal, Early Life Exposure to Manganese in Rural Bangladesh, *Environ. Sci. Technol.*, 2009, **43**, 2595–2601.

97. L. Henderson, K. Irving, J. Gregory, The National Diet & Nutrition Survey: adults aged 19 to 64 years, 2003, pp. 3–10.

98. M. Rose, M. Baxter, N. Brereton and C. Baskaran, Dietary exposure to metals and other elements in the 2006 UK Total Diet Study and some trends over the last 30 years, *Food Addit. Contam., Part A*, 2010, **27**(Oct), 1380–404.

99. C. Agostoni, R. Canani, S. Fairweather-Tait, M. Heinonen, H. Korhonen, S. La Vieille, R. Marchelli, A. Martin, A. Naska, M. Neuhäuser-Berthold, G. Nowicka, Y. Sanz, A. Siani, A. Sjödin, M. Stern, S. Strain, I. Tetens, D. Tomé, D. Turck and H. Verhagen, Scientific Opinion on Dietary Reference Values for manganese, *EFSA J.*, 2013, **11**, 3419–3430.

100. G. B. Mensink and R. Beitz, Food and nutrient intake in East and West Germany, 8 years after the reunification–The German Nutrition Survey 1998, *Eur. J. Clin. Nutr.*, 2004, **58**, 1000–10.

101. C. Rubio, A. J. Guttierrez, C. Revert, J. I. Reguera, A. Burgos and A. Hardisson, Daily dietary intake of iron, copper, zinc and manganese in a Spanish population, *Int. J. Food Sci. Nutr.*, 2009, **60**, 590–600.

102. The 23rd Australian Total Diet Study, 2011.

103. Key results of the 1997 National Nutrition Survey. Ministry of Health, 1999.

104. L. Nasreddine, O. Nashalian, F. Naja, L. Itani, D. Parent-Massin, M. Nabhani-Zeidan and N. Hwalla, Dietary exposure to essential and toxic trace elements from a Total diet study in an adult Lebanese urban population, *Food Chem. Toxicol.*, 2010, **84**, 1262–1269.

105. J. M. Morimoto, D. M. Marchioni, C. L. Cesar and R. M. Fisberg, Statistical innovations improve prevalence estimates of nutrient risk populations: applications in São Paulo, Brazil, *J. Acad. Nutr. Diet.*, 2012, **112**, 1614–1618.

106. E. Y. Kim, Y. J. Bae, S. J. Kim and M. K. Choi, Estimation of manganese daily intake among adults in Korea, *Nutr. Res. Pract*, 2008, **2**, 22–5.

107. Y. Murakami, Y. Suzuki, T. Yamagata and N. Yamagata, Chromium and manganese in Japanese diet, *J. Radiat. Res.*, 1965, **6**, 105–110.
108. N. N. Aung, J. Yoshinaga and J. I. Takahashi, Dietary intake of toxic and essential trace elements by the children and parents living in Tokyo Metropolitan Area, Japan, *Food Addit. Contam.*, 2006, **23**, 883–94.
109. S. W. Al-Rmalli, R. O. Jenkins and P. I. Haris, Betel quid chewing elevates human exposure to arsenic, cadmium and lead, *J. Hazard. Mater.*, 2011, **190**, 69–74.
110. J. Kelsay, C. Frazier, E. Prather, J. Canary, W. Clark and A. Powell, Impact of variation in carbohydrate intake on mineral utilization by vegetarians, *Am. J. Clin. Nutr.*, 1988, **48**, 875–879.
111. J. Hunt, L. Matthys and L. Johnson, Zinc absorption, mineral balance, and blood lipids in women consuming controlled lactoovovegetarian and omnivorous diets for 8 wk, *Am. J. Clin. Nutr.*, 1998, **67**, 421–430.
112. K. M. Behall, D. J. Scholfield, K. Lee, A. S. Powell and P. B. Moser, Mineral balance in adult men: effect of four refined fibers, *Am. J. Clin. Nutr.*, 1987, **46**, 307–14.
113. J. Hallfrisch, A. Powell, C. Carafelli, S. Reiser and E. S. Prather, Mineral balances of men and women consuming high fiber diets with complex or simple carbohydrate, *J. Nutr.*, 1987, **117**, 48–55.
114. A. S. Sandberg, H. Andersson, I. Bosaeus, N. G. Carlsson, K. Hasselblad and M. Härröd, Alginate, small bowel sterol excretion, and absorption of nutrients in ileostomy subjects, *Am. J. Clin. Nutr.*, 1994, **60**, 751–756.
115. N. Delzenne, J. Aertssens, H. Verplaetse, M. Roccaro and M. Roberfroid, Effect of fermentable fructo-oligosaccharides on mineral, nitrogen and energy digestive balance in the rat, *Life Sci.*, 1995, **57**, 1579–1587.
116. J. H. Freeland-Graves and P. H. Lin, Plasma uptake of manganese as affected by oral loads of manganese, calcium, milk, phosphorus, copper, and zinc, *J. Am. Coll. Nutr.*, 1991, **10**, 38–43.
117. J. R. Zhou and J. W. Erdman, Phytic acid in health and disease, *Crit. Rev. Food Sci. Nutr.*, 1995, **35**, 495–508.
118. L. Davidsson, A. Almgren, M. A. Juillerat and R. F. Hurrell, Manganese absorption in humans: the effect of phytic acid and ascorbic acid in soy formula, *Am. J. Clin. Nutr.*, 1995, **62**, 984–987.
119. B. Claus Henn, J. Kim, M. Wessling-Resnick, M. M. Téllez-Rojo, I. Jayawardene, A. S. Ettinger, V. A. Fitsanakis, N. Zhang, S. Garcia and M. Aschner, Manganese (Mn) and iron (Fe): interdependency of transport and regulation, *Neurotoxic. Res.*, 2010, **18**, 124–31.
120. K. Thompson, R. M. Molina, T. Donaghey, J. E. Schwob, J. D. Brain and M. Wessling-Resnick, Olfactory uptake of manganese requires DMT1 and is enhanced by anemia, *FASEB J.*, 2007, **21**, 223–30.
121. S. Park, C. S. Sim, H. Lee and Y. Kim, Blood manganese concentration is elevated in infants with iron deficiency, *Biol. Trace Elem. Res.*, 2013, **155**, 184–9.

122. J. W. Finley, Manganese absorption and retention by young women is associated with serum ferritin concentration, *Am. J. Clin. Nutr.*, 1990, 70, 37–43.

123. V. Agte, M. Jahagirdar and S. Chiplonkar, Apparent absorption of eight micronutrients and phytic acid from vegetarian meals in ileostomized human volunteers, *Nutrition*, 2005, 21, 678–685.

124. Y. Kim, J. K. Park, Y. Choi, C. I. Yoo, C. R. Lee, H. Lee, J. H. Lee, S. R. Kim, T. H. Jeong, C. S. Yoon and J. H. Park, Blood manganese concentration is elevated in iron deficiency anemia patients, whereas globus pallidus signal intensity is minimally affected, *Neurotoxicology*, 2005, 26, 107–111.

125. E. A. Smith, P. Newland, K. G. Bestwick and N. Ahmed, Increased whole blood manganese concentrations observed in children with iron deficiency anaemia, *J. Trace Elem. Med. Biol.*, 2013, 27, 65–9.

126. J. E. Levy, L. K. Montross and N. C. Andrews, Genes that modify the hemochromatosis phenotype in mice, *J. Clin. Invest.*, 2000, 105, 1209–16.

127. D. L. Decktor, M. Robinson, P. N. Maton, F. L. Lanza and S. Gottlieb, Effects of Aluminum/Magnesium Hydroxide and Calcium Carbonate on Esophageal and Gastric pH in Subjects with Heartburn, *Am. J. Ther.*, 1995, 2, 546–52.

128. P. E. Johnson and G. I. Lykken, Manganese and calcium absorption and balance in young women fed diets with varying amounts of manganese and calcium, *J. Trace Elem. Exp. Med.*, 1991, 4, 19–35.

129. S. Satarug, S. H. Garrett, M. A. Sens and D. A. Sens, Cadmium, environmental exposure, and health outcomes, *Environ. Health Perspect.*, 2010, 118, 182–190.

130. M. Kippler, B. Lonnerdal, W. Goessler, E. C. Ekström, S. E. Arifeen and M. Vahter, Cadmium interacts with the transport of essential micronutrients in the mammary gland – a study in rural Bangladeshi women, *Toxicology*, 2008, 257, 64–69.

131. C. V. Kies, Mineral utilization of vegetarians: impact of variation in fat intake, *Am. J. Clin. Nutr.*, 1988, 48, 884S–7S.

132. J. L. Greger and S. M. Snedeker, Effect of dietary protein and phosphorus levels on the utilization of zinc, copper and manganese by adult males, *J. Nutr.*, 1980, 111, 2243–2253.

133. J. Hunt, S. Gallagher, L. Johnson and G. Lykken, High- versus low-meat diets: effects on zinc absorption, iron status, and calcium, copper, iron, magnesium, manganese, nitrogen, phosphorus, and zinc balance in postmenopausal women, *Am. J. Clin. Nutr.*, 1995, 62, 621–632.

134. S. Hope, K. Daniel, K. L. Gleason, S. Comber, M. Nelson and J. J. Powell, Influence of tea drinking on manganese intake, manganese status and leucocyte expression of MnSOD and cytosolic aminopeptidase P, *Eur. J. Clin. Nutr.*, 2006, 60, 1–8.

135. C. Mullholand and D. Benford, What is known about the safety of multivitamin-multimineral supplements for the generally healthy

population? Theoretical basis for harm, *Am. J. Clin. Nutr.*, 2007, **85**, 318S–322S.

136. M. F. Bouchard, S. Sauvé, B. Barbeau, M. Legrand, M. È. Brodeur, T. Bouffard, E. Limoges, D. C. Bellinger and D. Mergler, Intellectual impairment in school-age children exposed to manganese from drinking water, *Environ. Health Perspect.*, 2011, **119**, 138–148.

137. D. Wang, W. Zhou, S. Wang and W. Zheng, Occupational exposure to manganese in welders and associated neurodegenerative diseases in China, *Toxicologist*, 1998, **48**(Suppl. 1), 24.

138. P. Grandjean, Neurobehavioural effects of developmental toxicity, *Lancet*, 2014, **13**, 330–8.

139. R. G. Lucchini, From manganism to manganese-induced parkinsonism: a conceptual model based on the evolution of exposure, *NeuroMol. Med.*, 2009, **11**, 311–21.

140. J. W. Finley and C. D. Davis, Manganese deficiency and toxicity: Are high or low dietary amounts of manganese cause for concern?, *BioFactors*, 1999, **10**, 15–24.

141. R. N. Dickerson, Manganese intoxication and parenteral nutrition, *Nutrition*, 2001, **17**, 689–693.

142. X. Liu, K. Sullivan, J. Madl, M. Legare and R. Tjalkens, Manganese-induced neurotoxicity: the role of astroglial-derived nitric oxide in striatal interneuron degeneration, *Toxicol. Sci.*, 2006, **91**, 521–531.

143. S. Bhang, S. Cho, J. Kim, Y. Hong, M. Shin, H. Yoo, I. Cho, Y. Kim and B. Kim, Relationship between blood manganese levels and children's attention, cognition, behavior, and academic performance–a nationwide cross-sectional study, *Environ. Res.*, 2013, **126**, 9–16.

144. V. Vanek, P. Borum, A. Buchman, T. Fessler, L. Howard, K. Jeejeebhoy, M. Kochevar, A. Shenkin and C. Valentine, Novel Nutrient Task Force and American Society for Parenteral and Enteral Nutrition (A.S.P.E.N.) Board of Directors. A.S.P.E.N. Position paper: recommendations for changes in commercially available parenteral multivitamin and multi-trace element products, *Nutr. Clin. Pract.*, 2012, **27**, 440–491.

145. J. Ono, K. Harada, R. Kodaka, K. Sakurai, H. Tajiri, Y. Takagi, T. Nagai, T. Harada, A. Nihei and A. Okada, Manganese deposition in the brain during long-term total parenteral nutrition, *JPEN, J. Parenter. Enteral Nutr.*, 1995, **19**, 310–2.

146. J. M. Reimund, J. L. Dietemann, J. M. Warter, R. Baumann and B. Duclos, Factors associated to hypermanganesemia in patients receiving home parenteral nutrition, *Clin. Nutr.*, 2000, **19**, 343–348.

147. R. Abdalian, O. Saqui, G. Fernandes and J. Allard, Effects of manganese from a commercial multi-trace element supplement in a population sample of Canadian patients on long-term parenteral nutrition, *JPEN, J. Parenter. Enteral Nutr.*, 2013, **37**, 538–543.

148. D. C. Wilson, T. R. Tubman, H. L. Halliday and D. McMaster, Plasma manganese levels in the very low birth weight infant are high in early life, *Biol. Neonate*, 1992, **61**, 42–6.

149. *Nutrient Reference Values for Australia and New Zealand. Including recommended dietary intakes*, Australian Government, National Health and Medical Research Council, Canberra, 2005.
150. Chinese Nutrition Association, *Chinese dietary reference intakes*, Chinese Light Industry Press, Beijing, 2001.
151. *Dietary Guidelines for Indians*, National Institute of Nutrition, Hyderabad, 2nd edn, 2012.
152. Ministry of Health. Labour and Welfare of Japan, *2010 dietary reference intakes for Japanese*, Daiichi Shuppan Publishing Co., Ltd., Tokyo, 2009.
153. *Dietary Reference Intakes for Koreans 2010*, Korea Institute of Nutrition, Seoul, 2010.
154. *Recommended Energy and Nutrient Intakes (RENI) Philippines, 2002*, RENI Philippines, Manilla, 2002.
155. J. Buttriss, *Nutrient requirements and optimisation of intakes*, British Nutrition Foundation, 2000.
156. M. Nishimuta, N. Kodama, M. Shimada, Y. Yoshitake, N. Matsuzaki and E. Morikuni, Estimated equilibrated dietary intakes for nine minerals (Na, K, Ca, Mg, P, Fe, Zn, Cu, and Mn) adjusted by mineral balance medians in young Japanese females, *J. Nutr. Sci. Vitaminol. (Tokyo)*, 2012, **58**, 118–28.

Manganese: Toxicokinetics and Toxicodynamics

Manganese Superoxide Dismutase

KINSLEY K. KININGHAM

Belmont University College of Pharmacy, Department of Pharmaceutical, Social and Administrative Sciences, Nashville, TN, USA
Email: kelley.kiningham@belmont.edu

3.1 Introduction

Since the initial discovery of the superoxide dismutase (SOD) enzymes by McCord and Fridovich over 45 years ago, numerous investigators have studied their role in various disease states with the ultimate goal of modifying expression to improve prognosis.[1] The primary function of this class of enzymes is to catalyze the dismutation (disproportionation) of superoxide anion $(O_2^{\bullet-})$ into hydrogen peroxide (H_2O_2) and molecular oxygen (O_2) according to the following scheme:

$$O_2^{\bullet-} + 2H^+ \rightarrow H_2O_2 + O_2 \tag{3.1}$$

Low levels of reactive oxygen species (ROS) such as $O_2^{\bullet-}$ and H_2O_2 contribute to normal physiological mechanisms; however, excessive production can lead to the accumulation of alkoxyl radicals $(RO^{\bullet})$, lipid peroxyl $(ROO^{\bullet})$ radicals, or reactive nitrogen species (RNS) such as peroxynitrite $(ONOO^-)$, all of which can lead to cellular damage. Cytotoxicity mediated through $O_2^{\bullet-}$ generation involves disruption of iron–sulfur (Fe–S) clusters in metal-containing proteins such as aconitase as well as inducing genetic mutations. As seen in Table 3.1, numerous sources of $O_2^{\bullet-}$ exist endogenously.

Issues in Toxicology No. 22
Manganese in Health and Disease
Edited by Lucio G. Costa and Michael Aschner
© The Royal Society of Chemistry 2015
Published by the Royal Society of Chemistry, www.rsc.org

Table 3.1 Sources of superoxide anion.

Source of $O_2^{\bullet-}$	Cellular location
Complex I (NADH: ubiquinone oxidoreductase)	Inner mitochondrial membrane
Complex II (succinate-ubiquinone oxidoreductase)	Inner mitochondrial membrane
Complex III (ubiquinol:cytochrome *c* oxidoreductase	Inner mitochondrial membrane
Cyclooxygenase-2	Smooth endoplasmic reticulum, nuclear membrane
Cytochrome P450 mediated reactions	Smooth endoplasmic reticulum
Dihydroorotate dehydrogenase	Inner mitochondrial membrane
Lipoxygenase	Nuclear membrane
Monoamine oxidase	Outer mitochondrial membrane
NADPH oxidase	Plasma membrane, phagosome membrane
β-Oxidation of fatty acids	Peroxisomes
Tricarboxylic acid cycle	Mitochondria matrix
Xanthine/xanthine oxidase	Cytosol

The individual SOD isoforms are localized where they can immediately serve a defensive role as the radical is generated.

Transcribed from three separate gene products, the eukaryotic SOD enzymes have been identified in different cellular compartments, all with catalytic centers dependent on either copper (Cu) or manganese (Mn) for activity (Table 3.2). These redox active transition metals are required in the active site of the SOD enzymes to dismutate $O_2^{\bullet-}$ according to the following scheme:

$$(\text{Cu or Mn})^{ox}\text{SOD} + O_2^{\bullet-} \rightarrow (\text{Cu or Mn})^{red}\text{SOD} + O_2 \qquad (3.2)$$

$$(\text{Cu or Mn})^{red}\text{SOD} + O_2^{\bullet-} \rightarrow (\text{Cu or Mn})^{ox}\text{SOD} + H_2O_2 \qquad (3.3)$$

$$2O_2^{\bullet-} + 2H^+ \rightarrow H_2O_2 + O_2 \qquad (3.4)$$

The Cu or Mn cation catalyzes both a one-electron oxidation (eqn 3.2) and a one-electron reduction (eqn 3.3) of two individual superoxide anions in the dismutation reaction.[2] The catalytic cycle consists of an oscillating ping–pong mechanism between the redox states of the respective transition metal (*i.e.* Mn^{3+}/Mn^{2+}).

Copper, zinc superoxide dismutase (Cu, ZnSOD; SOD1) is primarily localized in the cytosol, although expression has been reported in the mitochondrial intermembrane space as well as other cellular locations (Table 3.2). Manganese-containing superoxide dismutase (MnSOD; SOD2) is located in the mitochondrial matrix closely associated with the inner mitochondrial membrane. Extracellular superoxide dismutase (ECSOD; SOD3) is secreted and then tethered to the plasma membrane and/or extracellular matrix through interactions with heparin sulfate proteoglycan, fibulin-5

Table 3.2 Characteristics of eukaryotic SOD enzymes.

Chromosomal location in humans	Isoform	Cellular location	Molecular weight (kDa)	Subunit conformation	Metal ion(s)
21q22	SOD1	Intermembrane space of mitochondria, cytoplasm, nucleus, lysosomes, peroxisomes	32	Homodimer	Cu^{2+} (catalytic) Zn^{2+} (stability)
6q25.3	SOD2	Mitochondrial matrix	88	Homotetramer	Mn^{3+} (catalytic)
4p-q21	SOD3	Extracellular matrix, plasma membrane, plasma, lymph, ascites, cerebrospinal fluid	135	Homotetramer	Cu^{2+} (catalytic) Zn^{2+} (stability)

Table 3.3 Summarized genomic structure and organization of *SOD* genes.

Gene	Chromosome	Exons	Introns	TATA box in basal promoter	CAAT box in basal promoter	GC rich region in basal promoter
SOD1	21q22	5	4	Yes	1	Yes
SOD2	6q25.3	5	4	No	0	Yes
SOD3	4p-q21	3	2	No	2	Yes

and/or collagen[3–5] (Table 3.2). Based on cellular location, SOD1 is primarily responsible for cytosolic $O_2^{\bullet-}$ removal whereas SOD2 detoxifies mitochondrial levels of $O_2^{\bullet-}$, estimated at 10^{-11}–10^{-12} M, which are predominantly generated as a result of leakage of electrons during oxidative phosphorylation. SOD3 works extracellularly to dismutate $O_2^{\bullet-}$.

Interestingly, these three enzymes show little similarity in primary sequence or tertiary structure. A 2009 review by Miao and St Clair describes the genomic sequences of the three genes and the specific consensus elements of each involved with either constitutive and/or inducible expression.[6] Table 3.3 highlights the genomic structure and organization of the *SOD* genes.

In addition to eukaryotes, the Mn-containing SOD, which is the focus of this review, is found in a variety of organisms including *Escherichia coli* B,[7] *Saccharomyces cerevisiae*,[8] *Porphyridium cruentum*,[9] *Caenorhabditis elegans*,[10] *Callinectes sapidus*[11] and *Candida albicans*.[12] Structure and location of the prokaryotic enzyme may vary in comparison to the eukaryotic form, which is only found in the mitochondrial matrix. Several of the prokaryotic models have helped further our understanding of the enzyme's role within cells as well as elucidate its interaction with Mn.

3.2   Manganese Incorporation into SOD2

Metal ion binding *in vivo* can be incorporated into proteins during synthesis, folding, or following apoprotein subunit assembly. Currently, the exact mechanism by which apo-MnSOD acquires a Mn ion in the active site is un-known; however, key experiments have identified cellular components that contribute to the sequence of events leading to activation of the SOD2 enzyme.

Studies in prokaryotes suggest that various forms of SOD are present with either Mn or Fe in the active site; however, only the Mn-containing form is active.[13] While eukaryotes contain only MnSOD as a tetramer, prokaryotes, such as bacteria, express SOD dimers with tetrameric Mn- or Fe-containing SOD present in hyperthermophiles.[2]

Although MnSOD resides in the mitochondrial matrix, it is actually synthesized on ribosomes in the cytosol and then transported under the direction of an N-terminal signal peptide into the mitochondria. Studies in *S. cerevisiae* suggest a close association between the location of ribosomes involved in MnSOD translation and the outer mitochondrial membrane.[14] Transport of the apo-enzyme into the matrix involves sequential transloca-tion across both the outer and inner mitochondrial membranes followed by cleavage of the signal sequence. A recent study by Magnoni *et al.* suggests an association between heat shock protein 60 (Hsp60) and MnSOD, which is key for proper folding of the apo-enzyme once inside mitochondria.[15] Following protein entry into the mitochondrial matrix, Mn binds the enzyme, leading to its activation.

Deletion analysis by Luk *et al.* suggests that Mtm1p may be essential for proper insertion of Mn into the apo-enzyme within the mitochondrial matrix of yeast.[16] Studies have shown that Mtm1p, a member of the mitochondrial carrier family of transporters, preserves SOD2 activity by preventing its as-sociation with iron.[2] In addition, the Mn transporter Smf2p is also important in activation of yeast SOD2.[2] One of two Nramp metal transporters, Smf2p is crucial in Mn uptake at the cell surface as well as trafficking in *S. cerevisiae*.[2]

According to Whittaker and Whittaker each subunit of the tetrameric structure of eukaryotic MnSOD adopts a two-domain arrangement with an N-terminal α-hairpin domain and a C-terminal mixed α/β-domain with the metal binding site buried at the interface of these two domains.[17] The Mn ion is stabilized in a distorted trigonal-bipyramidal geometry and coordinated by one aspartate, three histidine residues and one –OH or water molecule.

3.3   Manganese Superoxide Dismutase is Essential for Life

Within the SOD family of isoenzymes, MnSOD has been shown to be es-sential for normal cellular development. In a CD1 murine model, in which the *SOD2* gene was knocked out, neonatal lethality occurred as a result of cardiovascular complications. The animals had an average life span of 5.4 days and exhibited dilated cardiomyopathy, resulting in part from reduced ventricular wall thickness and myocardial hypertrophy. Other reported findings from this model included accumulation of lipid in liver and skeletal muscle, metabolic acidosis and ketosis as well as a reduction of heart succinate dehydrogenase (complex II) and aconitase enzyme activities.[18]

In a subsequent study, which utilized a mixed genetic background of C57BL/6 and 129/Sv, Lebovitz *et al.* reported a neurodegenerative phenotype in mice lacking MnSOD.[19] In this model, where both exons 1 and 2 were deleted from the *SOD2* gene, mice lived approximately 21 days. Neurons in the basal ganglia and brainstem were affected with progressive motor abnormalities including weakness, fatigue and circling behavior. Reduced hematopoiesis contributed to the development of anemia in the model and cardiovascular abnormalities including cardiac dilation and thinning of the ventricular wall were also reported.

Utilizing a B6D2F1 background, *SOD2*(−/−) animals developed ataxia within 11 days of birth with progressive neurological abnormalities including frequent seizures, degeneration of the motor cortex as well as thalamic and hippocampal damage. According to Lynn *et al.* differences in genetic backgrounds suggest the presence of modifiers that may contribute to the differences in tissue sensitivity in SOD2 mutant mice.[20] Collectively these studies distinguished MnSOD from both the Cu, ZnSOD and ECSOD isoforms. Knock-out of either the *SOD1*[21] or *SOD3*[22] gene did not result in neonatal lethality in murine models, but led to cellular deficits when challenged under altered redox conditions.

Copin *et al.* overexpressed *SOD1* in a *SOD2* deficient murine model to determine if the neonatal lethality described by Li and others could be ameliorated; however, the lethality could not be prevented by expression of cytosolic Cu, ZnSOD.[23] This suggested that mitochondrial localization of the SOD enzyme was essential to aerobic life.

Studies by Zhang and Van Remmen utilized fibroblasts isolated from the murine *SOD2*(−/−) model to identify potential mechanisms of the neonatal lethality.[24] Studies showed an alteration in proliferation of the cells compared to wildtype (WT). Key signaling molecules that were altered in the *SOD2*(−/−) model included decreased growth stimulatory function of mTOR signaling as well as increased GSK-3β signaling. Also suppressed in the *SOD2*(−/−) murine fibroblasts were the G-protein-coupled receptor-mediated intracellular calcium signal transduction pathways. Impaired oxygen consumption with a reduction in cellular adenosine triphosphate (ATP) levels, enhanced $O_2^{\bullet-}$, and reduced expression of peroxiredoxin 3 (Prdx3) were also noted in the isolated cells.

Heterozygote murine models (*SOD2* (+/−)) with a 50% reduction in MnSOD activity showed increased sensitivity to apoptosis, accumulating oxidative damage in both nuclear and mitochondrial DNA, and impaired mitochondrial respiration when compared to WT counterparts.[25–29] The noted difference in

DNA was the accumulation of 8-oxo-2-deoxyguanosine (8oxodG) in both the nucleus and mitochondria of all tissues as age increased. The increased 8oxodG was correlated with an increase in the number of mice with tumors. Interestingly the reduction in SOD2 expression did not shorten the life span (mean and maximum survival) of the heterozygotes. Markers of aging including cataract formation, immune changes and formation of glycoxidation products were consistent between both *SOD2* (+/−) and WT animals.

To better ascertain the effect of MnSOD deficiency in specific tissues, various groups have utilized the Cre-loxP system to reduce SOD2 expression in liver,[30] brain,[31] heart,[32] skeletal muscle[33–35] or kidney.[36,37] Several labs have reported decreased complex II and aconitase enzyme activities, enhanced lipid peroxidation, reduced ATP production and morphological changes with the Cre-loxP system approach to reducing SOD2 expression (Table 3.4). These changes were most evident in tissues known to be rich in mitochondria such as neurons, cardiac tissue and muscle fibers. In a recent study using the Cre-loxP system, Parajuli *et al.* described the development of a renal-specific MnSOD deficient murine model with reduced enzyme expression in the distal tubules, collecting ducts and Loops of Henle.[36] Initial

Table 3.4 Conditional knock-out models of SOD2 induce tissue specific injury with altered mitochondrial function.

Tissues affected	Significant findings	Reference(s)
Liver	No change in liver morphology or biomarkers of oxidative stress	30
Postnatal neurons	Injury-induced disorganization of distal nerve axons	31
Heart/muscle	Progressive congestive heart failure ↑ $O_2^{\bullet-}$ and lipid peroxidation ↓ Complex II activity, ↓ ATP	32
Skeletal muscle fibers	Centralized nuclei in muscle fibers No atrophic changes in skeletal muscle ↑ $O_2^{\bullet-}$, ↑ F2 isoprostanes ↓ Aconitase, ↓ Complex II activity, ↓ respiration, ↓ ATP ↑ Glycolytic muscle damage ↓ Contractile force of muscles ↓ Aerobic exercise capacity	33–35
Kidney	No change in renal function or life span ↓ Body weight ↑ Tyrosine nitration Epithelial cellular swelling, cast formation, dilation of the tubular lumen Induction of cell survival signals: ↑ Autophagosome formation ↑ Mitochondrial biogenesis ↑ DNA replication/repair	36, 37

findings described a reduction in body weight with normal organ size; however, there was no change in either renal function or lifespan when compared to WT animals, consistent with a study by Van Remmen *et al.*[27] However, the knock-out (KO) mice showed a significant increase in oxidative stress as evidenced by tyrosine nitration in a gene-dose dependent manner. Other noted pathological changes included epithelial cellular swelling, tubular dilation and cast formation within the tubular lumen. A subsequent paper in 2013 described increased autophagosome formation, mitochondrial biogenesis and DNA replication/repair in the KO model that proved protective against subsequent ischemia/reperfusion (I/R) injury.[37] Interestingly, renal function and tubular cell death were found to be similar between WT and KO animals following I/R, suggesting a compensatory mechanism in the animals with reduced SOD2.

These studies have significantly advanced the understanding of MnSOD expression in specific tissues and provide a clear role of the enzyme in preventing mitochondrial dysfunction in numerous disease states. The expanded use of this technology in additional tissues will provide even more insight into how modification of the enzyme's activity can impact disease prognosis.

3.4 Post-Translational Modification of MnSOD

Manganese superoxide dismutase can be regulated post-translationally through nitration, phosphorylation and acetylation reactions resulting in altered expression, activity, localization and/or stability of the enzyme. Nitration of MnSOD has been extensively studied in various disease states and following a variety of pharmacological treatments. These studies have advanced our understanding of the enzyme's role in disease and how it may be pharmacologically targeted to improve prognosis.

3.4.1 Nitration of MnSOD Compromises Mitochondrial Function in Various Disease States

Manganese superoxide dismutase can be inactivated post-translationally through a peroxynitrite-mediated nitration reaction on tyrosine 34 (Tyr34). According to Yamakura and Kawasaki, the modified residue is located within 6 Å from the active site.[38] Radi suggests that the nitration specifically occurs on Tyr34 *via* a Mn-catalyzed process involving three potential redox steps:[39]

$$Mn^{3+}SOD + ONOO^- \rightarrow O{=}Mn^{4+}SOD + {}^{\bullet}NO_2 \tag{3.5}$$

$$Tyr34 + O{=}Mn^{4+} \rightarrow {}^{\bullet}Tyr34 + Mn^{3+} + OH^- \tag{3.6}$$

$${}^{\bullet}Tyr34 + {}^{\bullet}NO_2 \rightarrow NO_2Tyr34 \tag{3.7}$$

Upon nitration of MnSOD, mitochondrial redox status is disrupted, furthering oxidative damage, and causing greater inactivation of the enzyme with subsequent cellular damage and/or death. Nitration of MnSOD is thought to reduce enzymatic function by restricting both access and binding of $O_2^{\bullet-}$ to the active site.[40] In addition, the nitrated tyrosine residue is thought to weaken hydrogen bonding, thereby affecting proton transfer during $O_2^{\bullet-}$ dismutation.[41]

MacMillan-Crow *et al.* first described the nitration and oxidation of MnSOD in chronic rejection of human renal allografts.[42] In the transplanted organs, MnSOD protein was increased; however, activity was significantly decreased. Subsequent reports in different disease state models have noted nitration of MnSOD, thereby contributing to enhanced $O_2^{\bullet-}$ formation and mitochondrial dysfunction.

Anantharaman *et al.* identified MnSOD as a target of nitration in the $APP^{NLh/NLh} \times PS\text{-}1^{P264L/P264L}$ double knock-in murine model of Alzheimer's disease (AD).[43] There was an age-dependent accumulation of amyloid-β (Aβ) peptide deposits in the brain that was associated with an accelerated decline in mitochondrial function. This finding confirmed a report by Sompol *et al.* that had described decreased MnSOD expression and an increase in the colocalization of 3-nitrotyrosine (3-NT) with the enzyme in the $APP^{NLh/NLh} \times PS\text{-}1^{P264L/P264L}$ model.[44] These findings are consistent with reports of protein oxidation[45] and elevated 3-NT levels in brains of AD patients.[46,47]

Prenatal ductal ligation-induced persistent pulmonary hypertension (PPHN) in a fetal lamb model is associated with increased oxidative stress, which contributes to impaired pulmonary vasodilation. A recent study by Afolayan *et al.* showed that adenoviral expression of MnSOD was protective in the PPHN model by effectively scavenging $O_2^{\bullet-}$.[48] The expressed MnSOD improved the relaxation response of the pulmonary arteries. Additional insight into the mechanism of pulmonary hypertension identified nitration of MnSOD as a key contributor to the increased oxidative stress associated with the hypertensive state. PPHN is known to be associated with enhanced $O_2^{\bullet-}$ production in pulmonary arteries leading to depletion of the vasodilator, nitric oxide ($^{\bullet}NO$), with subsequent production of $ONOO^-$. In a separate, but related report, rat kidneys infused with angiotensin II demonstrated hypertension secondary to nitration of the MnSOD enzyme.[49]

Additional drug-induced toxicity models have identified $ONOO^-$ formation with subsequent nitration/inactivation of MnSOD to be critical in the mechanism of cellular damage (Table 3.5). Cytotoxicity secondary to the use of the immunosuppressant cyclosporine A (CsA) has been reported to involve generation of ROS and endothelial derived RNS, specifically $ONOO^-$. Cardiovascular complications as a result of CsA administration, such as hypertension and vascular injury, involve the post-translational nitration of key proteins including MnSOD.[50]

Agarwal *et al.* identified MnSOD as a target of nitration in a murine model of acetaminophen (APAP)-induced hepatotoxicity.[51] Glutathione (GSH), which detoxifies $ONOO^-$, is severely depleted in APAP overdose *via* the production of the metabolite, *N*-acetyl-*p*-benzoquinone imine (NAPQI). Loss

Table 3.5 Nitration of MnSOD contributes to pathophysiology of drug-induced disease.

Drug	Significant findings	Reference(s)
Angiotensin-II	Hypertension	49
Cyclosporin A (CsA)	Hypertension and vascular injury	50
Acetaminophen	Necrosis of hepatocytes	51, 52
Adriamycin	Memory loss and inability to perform complex tasks	53, 54
Opioids	Hyperalgesia and antinociceptive tolerance	55, 56, 57
Streptozotocin	Impaired would healing in a model of diabetes mellitus	59

of GSH results in NAPQI-mediated covalent modification of proteins within the liver. MnSOD was shown to be nitrated and inactivated within 1 h of APAP administration, suggesting an early compromise to mitochondrial function, followed by GSH depletion in both the cytosol and mitochondrial fractions within 2 h. Serum alanine aminotransferase (ALT) and serum aspartate aminotransferase (AST) release increased within 6 h. Subsequent studies in neuronal NOS (nNOS)-KO animals showed protection against APAP-induced hepatotoxicity with no early alteration in MnSOD activity.[52] These studies suggest production of both ROS/RNS upon APAP administration. Loss of MnSOD activity coupled with GSH depletion results in a shift of antioxidant to oxidant imbalance within hepatocytes, resulting in mitochondrial damage and subsequent necrosis.

Tangpong *et al.* identified brain mitochondria as a key target of adriamycin (ADR)-induced $^{\bullet}$NO-mediated injury.[53] Adriamycin, an anthracycline used in the treatment of multiple tumor types can lead to transient memory loss and inability to handle complex tasks in patients. Collectively these symptoms are often referred to as chemobrain. Inducible nitric oxide synthase (iNOS)-KO animals were less susceptible to alterations in brain mitochondrial function with ADR treatment compared to WT animals. Circulating tumor necrosis factor-alpha (TNF-α) produced from brain microglia was shown to mediate iNOS production of $^{\bullet}$NO in WT animals, resulting in MnSOD nitration in brain with a subsequent compromise in mitochondrial respiration.[54]

Long-term use of opioid therapies can lead to the development of hyperalgesia and antinociceptive tolerance.[55,56] Recent findings identify a role for NADPH oxidase (NOX) enzyme activity in spinal cord.[57] More specifically, a murine model in which the NOX2 complex was deleted ($Nox2(-/-)$), identified its contribution to proinflammatory cytokine production, $ONOO^-$ formation and MnSOD nitration.[58] Knock out of the NOX2 complex attenuated morphine-induced antinociceptive tolerance. These findings support a critical role of both ROS/RNS in the development of pain and identify MnSOD as a potential pharmacological target to protect opioid analgesia.

Analysis of kidney tissue from a streptozotocin-induced diabetic murine model indicated nitrated MnSOD. Tie *et al.* have reported that *Ganoderma lucidum* polysaccharide can accelerate healing in the streptozotocin model by suppression of nitration of cutaneous MnSOD.[59]

Both *in vitro* experiments and *in vivo* disease modeling have furthered our understanding of the post-translational inactivation of MnSOD through nitration. These studies lay the foundation for intervention strategies in diseases such as amyotrophic lateral sclerosis (ALS), AD, Parkinson's disease (PD), acquired immune deficiency syndrome (AIDS), diabetes and esophagitis where tissue samples have shown the presence of nitrated MnSOD.[60-63]

From a pathophysiological perspective, excess $^\bullet$NO, generated as a result of enhanced NOS activity, can easily diffuse across cellular membranes. Upon entrance into mitochondria, components of the electron transport chain such as cytochrome oxidase become susceptible to inactivation, resulting in enhanced $O_2^{\bullet-}$ production. Nitric oxide effectively out competes MnSOD for $O_2^{\bullet-}$, thereby generating $ONOO^-$ which then nitrates and inactivates numerous proteins including SOD2. Ultimately this cycle of events in the organelle causes cumulative oxidative and nitrative damage, resulting in mitochondrial dysfunction and either apoptosis or necrosis of cells.

3.4.2   Phosphorylation of MnSOD can Enhance Activity and Stability

Early studies regarding phosphorylation events of MnSOD were reported in plants[64] and swine heart;[65] however, no activity data was reported in either study. More recently, a phosphorylation site on serine 82 (Ser82) was identified in rodent mitochondrial MnSOD; however, effects on stability and/or activity were not described.[66] Candas *et al.* have identified a cell cycle-dependent kinase (Cdk1) phosphorylation site in human MnSOD and have shown Ser106 to be modified upon radiation-induced cyclin B1/Cdk1 activation.[67] Protein modification was shown to be reversible in both *in vivo* and *in vitro* studies. Interestingly, upon phosphorylation MnSOD activity and stability increased, leading to improved mitochondrial function and resistance to radiation-induced apoptosis. This finding is key in introducing a novel regulatory effect on MnSOD activity and a mechanism by which cells enhance their own survival under stressful conditions.

3.4.3   Acetylation of MnSOD Reduces Enzymatic Activity

Recent studies by Tao *et al.* suggest that post-translational acetylation of MnSOD regulates enzymatic activity in response to changes in cellular nutrient status or ionizing radiation.[68] In an *in vitro* deacetylation assay, recombinant sirtuin-3, a mitochondrial deacetylase, was able to directly deacetylate purified MnSOD, restoring enzymatic activity. These studies complemented work by Qiu *et al.* who also identified sirtuin-3 (Sirt3) as a regulator of MnSOD activity.[69] Between the two studies several lysine residues were identified as targets of acetylation in the SOD protein. Tao reported the deacetylation of lysine 122 by Sirt3 and Chen *et al.* reported lysines 59 and 89 as potential targets of acetylation in the antioxidant enzyme. Utilizing Sirt3(−/−) mice, Tao reported increases in mitochondrial $O_2^{\bullet-}$ levels and altered mitochondrial metabolism.

3.5 MnSOD and Redox Signaling

Changes in expression of MnSOD within tissues can alter levels of both $O_2^{\bullet-}$ and H_2O_2, thereby altering various signaling cascades known to be influenced by their generation in cells. Numerous studies have identified a role for ROS, including $O_2^{\bullet-}$ and H_2O_2, in diverse physiological processes such as proliferation, angiogenesis, autophagy, migration, cell survival, invasion/metastasis and differentiation, often through changes in activity levels of redox sensitive kinases, phosphatases and/or transcription factors.[70–75]

An increase in H_2O_2 with enhanced activity of SOD2 has been reported in various eukaryotic cell lines.[76–83] These findings suggest an expanded role for the primary antioxidant in modulating overall cellular redox status. According to Buettner, both $O_2^{\bullet-}$ and H_2O_2 affect cellular signaling processes through different types of chemical reactions. Superoxide anion can be either a one-electron oxidizing or reducing agent, whereas H_2O_2 is primarily a two-electron oxidizing agent.[84] MnSOD is a key cellular switch in that it dismutates $O_2^{\bullet-}$ to form H_2O_2 as shown in Figure 3.1. Therefore, its expression is critical in both $O_2^{\bullet-}$ and H_2O_2-mediated signaling cascades.

Utilizing an inducible retroviral human *SOD2* expression system in mouse NIH/3T3 cells, Kim *et al.* reported alterations in the mitochondrial redox state resulting in elevations of both murine SOD2 and thioredoxin 2 mRNA and protein levels.[85] The authors hypothesized that changes in the antioxidant proteins were adaptive responses to alterations in mitochondrial redox homeostasis elicited by human SOD2 overexpression. They tested this by treating the cells with the mitochondrial specific antioxidant, MitoQ, following induction of the exogenous MnSOD. MitoQ was able to block the increase in murine SOD2 and thioredoxin 2 mRNA levels. Extending these studies to an *in vivo* model, Kim *et al.* showed that liver-specific expression of SOD2 led to an increase in the mitochondrial thioredoxin 2 protein as well as reduced proliferative cell nuclear antigen (PCNA) expression following a partial hepatectomy.[85] The *in vitro* study by Kim *et al.* was consistent with previous reports, suggesting H_2O_2 can induce SOD2 expression.[86–87] Collectively, the reports by Kim *et al.* suggest both induction and inhibition of various genes with SOD2 generated H_2O_2.[85]

MnSOD has been shown to influence cellular processes such as angiogenesis through its dismutation of $O_2^{\bullet-}$. In a study by Connor *et al.,* H_2O_2 was shown to oxidize and then inactivate the tumor suppressor phosphatase and tensin homolog (PTEN), leading to an increase in the second messenger phosphatidylinositol 3,4,5-triphosphate as well as activation of Akt and modulation of its downstream targets.[88] SOD2 overexpression increased blood vessel formation through increased H_2O_2 generation in a chicken chorioallantoic membrane assay. Mitochondrial production of H_2O_2, secondary to MnSOD overexpression, promoted endothelial cell sprouting in an *in vitro* angiogenesis assay. This was a catalase reversible process, which revealed the relationship between SOD2 and H_2O_2-mediated activation of angiogenic signaling pathways.

Differentiation agents such as nerve growth factor (NGF),[89] phorbol esters[90,91] and all-*trans* retinoic acid (ATRA)[92] have been shown to induce expression of MnSOD in various cellular models (Figure 3.1). Upregulation of the enzyme may promote a differentiated phenotype by preventing oxidative damage and ensuring energy production. The contribution of MnSOD to differentiation also involves modulation of intracellular redox status, leading to changes in gene expression important in generation and sustainability of the differentiated phenotype. We have previously shown that overexpression of MnSOD in a murine fibrosarcoma cell line results in a decrease in the activity of Jun associated transcription factors, resulting in down-regulation of target genes.[93] Additionally, numerous studies have shown that MnSOD transgenic cell lines have significantly reduced proliferative rates, decreased colony formation and exhibit changes consistent with a differentiated phenotype compared to WT.[94–97]

Cassano showed that overexpression of SOD2 in PC12 cells mediated an increase in H_2O_2 upon treatment with NGF, leading to sustained mitogen activated protein kinase (MAPK) signaling that was critical for neuronal

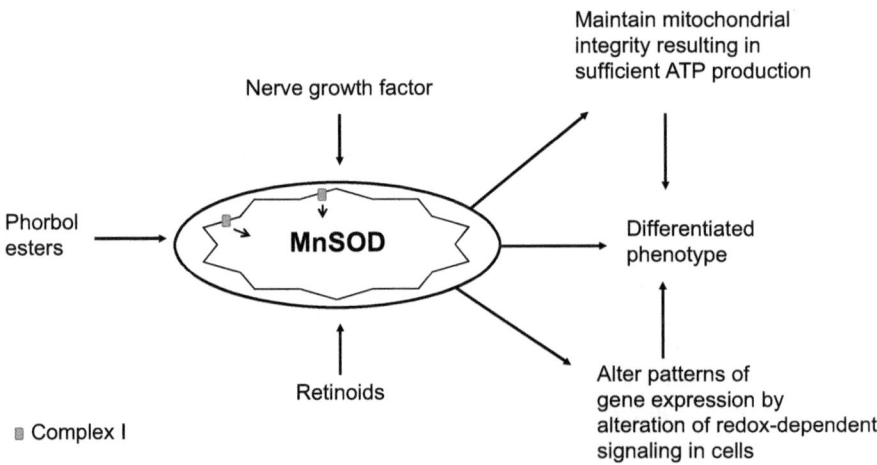

Figure 3.1 MnSOD enhances cellular differentiation. Phorbol ester, nerve growth factor and all-*trans* retinoic acid induce expression of MnSOD, leading to a differentiated phenotype. MnSOD protects mitochondria, allowing for ATP production and through dismutation of $O_2^{\bullet-}$ and generation of H_2O_2 can alter redox dependent pathways, resulting in expression of genes important in differentiation.

differentiation.[98] In another study, complex I derived $O_2^{\bullet-}$ was subsequently dismutated by SOD, leading to a differentiated muscle phenotype through generation of the signaling messenger H_2O_2. MnSOD was shown to be upregulated *via* the redox sensitive transcription factor, nuclear factor kappa binding protein (NFκB) upon complex I generation of $O_2^{\bullet-}$. Mitochondrial targeted catalase and MitoQ attenuated ROS production and suppressed muscle differentiation.[99]

SOD2 has also been shown to contribute to the pathogenesis of pulmonary arterial hypertension (PAH) through H_2O_2-mediated activation of the hypoxia inducible factor-1 alpha (H1F1α)-O_2-sensitive Kv1.5 channel pathway. Studies by Archer *et al.* have shown that reduced activity of MnSOD limits mitochondrial H_2O_2 production, leading to the activation of the hypoxia responsive pathway, which is followed by proliferative changes as well as a loss of apoptotic ability.[100] These events culminate in the lethal syndrome of PAH, which is defined by vascular obstruction and right ventricular failure. Recent studies in models of PAH using dichloroacetate, a mitochondrial pyruvate dehydrogenase kinase inhibitor, revealed that the disease is reversible and that H_2O_2 levels can be restored, thereby limiting H1Fα signaling.[100]

Early studies indicated reduced expression of SOD2 in numerous tumor types compared to normal cellular counterparts, leading to the speculation that MnSOD might be serving as a tumor suppressor gene.[101,102] Recent studies have identified some tumors that actually have increased activity of SOD2. Mechanistic studies suggest that in these tumors MnSOD expression correlates with invasiveness and an increased metastatic phenotype.[103–105] Connor *et al.* utilized transgenic MnSOD HT1080 fibrosarcoma and 253J bladder cells and showed a promigratory and invasive phenotype that was H_2O_2 dependent.[82] Targeted expression of the catalase (CAT) enzyme, which converts H_2O_2 to H_2O, reversed the phenotype. Furthermore, overexpression of MnSOD correlated with upregulation of matrix metalloproteinase-1 (MMP-1), which is known to alter the tumor/stromal microenvironment and enhance metastatic disease. This study complemented earlier work by Zhang *et al.*, which showed MMP-2 expression in MCF-7 breast cancer cells to be dependent on H_2O_2 produced in SOD2 overexpressing cells.[79] Adenoviral-mediated expression of either CAT or glutathione peroxidase (GPx) prevented the upregulation of MMP-2. A report by Liu *et al.* showed that SOD2-mediated production of H_2O_2 contributed to migration and invasion of tongue squamous cell carcinoma (TSCC) *via* the Snail signaling pathway. Increased levels of Snail, MMP-1 and pERK 1/2 protein levels correlated with metastatic potential and higher SOD2 activity in the study.[106] Also reported was an inverse relationship between Snail, MnSOD and H_2O_2 levels with expression of E-cadherin, an adhesion molecule associated with invasion and metastasis. Reduced expression of E-cadherin promotes cell migration and invasion.[107] As in previously mentioned studies, co-expression of a peroxide removal system, such as CAT, decreased the migration and invasion potential of the TSCC cell lines. Chen *et al.* reported a positive correlation between MnSOD expression, forkhead box M1 (FoxM1) and MMP2 expression with tumor aggressiveness in lung cancer.[108] Chromatin immunoprecipitation assays indicated that lung cancer cells, which over-expressed MnSOD, promoted binding of the E2F1 and Sp-1 transcription factors to their promoter-binding sites. Utilizing a luciferase reporter system, MnSOD was shown to activate FoxM1 reporter activity. Collectively, these studies suggest both induction and inhibition of transcriptional events with alterations in cellular redox status mediated by SOD2 activity, leading to alterations in both signaling mediators $O_2^{\bullet-}$ and H_2O_2.

3.6 Transcriptional Regulation of MnSOD Expression

MnSOD expression has been shown to be regulated transcriptionally through activation of redox sensitive transcription factors. Elucidating the signaling pathways upstream of the transcription factors involved in the regulation of the enzyme is important to identifying potential pharmaco-logical targets for the treatment of redox-related diseases such as cancer, hypertension, atherosclerosis, inflammation, diabetes and neurodegenera-tive disorders. Alterations in the cellular redox status by agents such as ROS,[86,87] retinoic acid,[92] manganese,[109–112] ionizing radiation,[113,114] TNF-α,[115,116] interleukins 1, 4 and 6,[117–119] lipopolysaccharide,[118,120] interferon-γ,[121] dinitrophenol,[122] and paraquat[123] have been shown to upregulate *SOD2* gene transcription.

The entire genomic structure for human *SOD2* has been published[124,125] and shows conservation across species when compared to the rat, mouse and bovine genes.[126] The human MnSOD gene consists of five exons and four introns with a typical splice junction, resulting in two mRNA transcripts of 1 and 4 kb. Regulatory elements of the *SOD*2 gene have been identified in both the noncoding and coding regions (Figure 3.2). Although TATA and CAAT boxes are absent in the basal *SOD2* promoter, several SP-1 and AP-2 consensus sequences are localized among GC-rich islands upstream of the initiation site.[125] Additional redox sensitive transcription factors, AP-1, NFκB and FOXO3a, bind consensus sequences further upstream of the SP-1 and AP-2 binding sites.[6]

SP-1, which contains three zinc finger motifs in the DNA binding domain, is important in both constitutive and inducible expression of the *SOD2* gene through binding to GC-rich consensus sequences. AP-2 plays a repressive role by suppressing SP-1-dependent constitutive expression in the basal promoter.[127–129] Zhu *et al.* reported that methylation of an AP-1 consensus sequence within the basal promoter limits AP-1 binding, thereby reversing the repression on SP-1.[130] Redox sensitive p53 competes with SP-1 for binding to the promoter and therefore negatively regulates gene transcription.[131,132]

Although an NF-κB consensus sequence has been identified in the pro-moter of the human *SOD2* gene, the one localized in the second intron is more responsive to stimuli such as cytokines, retinoic acid and phorbol

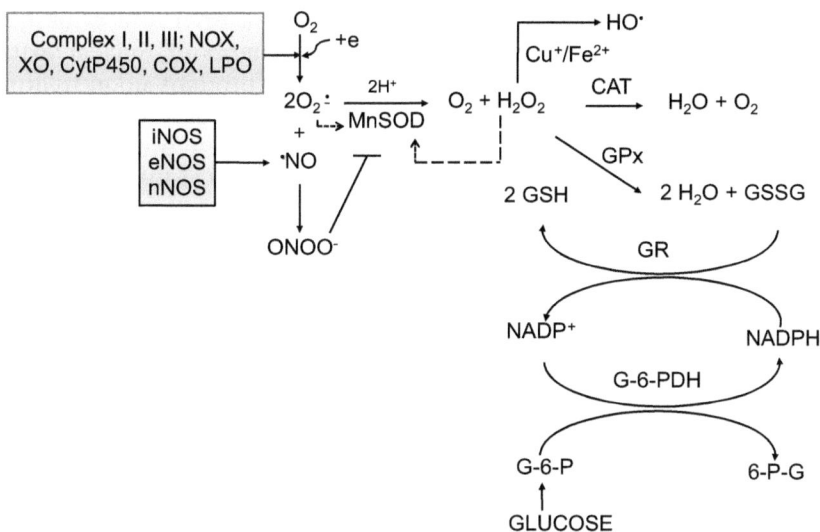

Figure 3.2 Generation and metabolism of reactive oxygen species. Superoxide anion is generated through complex enzyme leakage of electrons (e′), xanthine oxidase (XO), NADPH oxidase (NOX), cytochrome P450 mediated reactions (Cyp450), cyclooxygenase (COX) and lipooxygenase (LPO). Superoxide is converted by MnSOD to H_2O_2, which is reduced to H_2O by glutathione peroxidase (GPx) and catalase (CAT). GPx requires secondary enzymes glutathione reductase (GR) and glucose-6-phosphate dehydrogenase (G-6-PDH) and cofactors glutathione (GSH), NADPH and glucose-6-phosphate (G-6-P). In the presence of transition metal (Cu^+/Fe^{2+}), H_2O_2 can be converted to hydroxyl radical $^\bullet OH$. Nitric oxide ($^\bullet NO$), produced by endothelial, inducible or neuronal nitric oxide synthase (e-, i-, nNOS) can be converted to $ONOO^-$ through reaction with $O_2^{\bullet-}$. Dashed line indicates inducer of MnSOD protein expression and activity.

esters.[90,92,133–135] Proteins p65 and RelB among the NF-κB family members positively regulate *SOD2* gene transcription, whereas p50 represses expression of the gene.[136,137]

Studies by Dhar *et al.* identified a unique loop structure within the *SOD2* gene that promotes the interaction of SP-1 proteins in the basal promoter and NFκB in the second intron through interactions with the ribonuclear protein nucleophosmin.[132,138,139] Disruption of the loop attenuated both constitutive and inducible expression of the *SOD2* gene.[140] Dhar and St Clair suggested that this loop structure is a common site for interactions among transcription factors and other regulatory proteins, which either collectively induce or repress transcription.[141]

Treatment of SV40 transformed fibroblast (VA-13) cells with phorbol ester enhanced NFκB DNA binding activity. This activation was attenuated with either pretreatment of the PKC inhibitor GF109203X or coexpression of dominant negative constructs for either PKCα or PKCβI.[142] In addition, a CCAAT-enhancer-binding protein (C/EBP) consensus sequence within the second intron was shown to contribute to phorbol ester and cytokine induction of *SOD2*.[90] Immunoprecipitation experiments of nuclear extracts revealed physical interactions of C/EBP and NFκB proteins upon treatment with a combination of cytokines and phorbol ester.[90] Studies by Sebastian *et al.* confirmed a co-regulatory role for C/EBP when overexpression of C/EBPβ was unable to induce MnSOD transcription in C/EBPβ(−/−) cells.[143]

Kim *et al.* identified a cAMP-responsive element (CRE)-like sequence in the 5′-flanking region of the *SOD2* gene, located between −1292 and −1202, and demonstrated that CREB/ATF-1 bound to the consensus sequence after phorbol ester PKCα-mediated phosphorylation in human lung adenocarcinoma A549 cells.[144] Small interfering RNA experiments indicated that knockdown of NADPH oxidase components, Rac1, p22(phox), p67(phox), and NOXO1 attenuated phorbol ester-induced MnSOD expression in the A549 cells, indicating a potential role for $O_2^{\bullet-}$. In a subsequent study, Chung *et al.* reported a PKC-dependent decrease in Akt phosphorylation upon phorbol ester treatment of A549 cells with downstream dephosphorylation of FOXO3a, leading to transcriptional upregulation of the *SOD2* gene.[145] This study is consistent with that of Kops *et al.*, who showed that FOXO3a was able to bind to the promoter of the *SOD2* gene at position −1249 (GTAACAA) and upregulate expression during oxidative stress.[146]

In addition to phorbol esters, anticancer agents paclitaxel, vincristine and vinblastine activate the protein kinase C pathway leading to induction of MnSOD in A549 cells in a time- and dose-dependent manner.[147] Phorbol ester effectively caused the movement of PKC isoforms α, β, δ and μ to the A549 membrane; however, the anticancer drugs specifically stimulated the translocation of PKCδ, suggesting a role for this isoform in MnSOD induction. Collectively these studies suggest that different PKC isoforms contribute to induction of the antioxidant and that this may be stimulus and/or cell dependent.

Maehara described early growth-responsive-1 (Egr-1) dependent transcription of *SOD2* upon treatment of NIH3T3 cells with platelet-derived growth factor (PDGF).[148] Both the redox-sensitive mitogen-activated protein kinase kinase-1 (MEK1) and extracellular signal-regulated kinases 1 and 2 (ERK1/2) pathways were identified to be involved in Egr-1 activation by PDGF.

Negoro *et al.* reported signal transducer and activator of transcription 3 (STAT3) protected cardiomyocytes from oxidative injury induced by hypoxia/reoxygenation (H/R) through increased expression and enzymatic activity of MnSOD.[149] Mice that overexpressed STAT3 showed protection against ADR-induced damage. This study complemented the work of Yen *et al.* who utilized MnSOD transgenic animals and showed protection against acute ADR cardiac toxicity, in part, through maintenance of complex 1 enzyme activity.[150,151] Negoro identified three potential STAT3 binding sites in the *SOD2* promoter spanning from −2505 to −1104. Both interleukins and interferon-γ are known to signal through the Janus family tyrosine kinases (JAK)/STAT pathway, are associated with H/R injury and have been reported to induce transcription of MnSOD in some cell types.[152] Additional studies will be required to determine which of the three consensus sequences are crucial for induction of the *SOD2* gene.

Progestin stimulated MnSOD in T47D human breast cancer cells at the mRNA, protein and activity level.[153] Use of RU486, an antiprogestin, and U0126, a MAP kinase kinase inhibitor, attenuated the increase in MnSOD and inhibited migration/invasion in the T47D cell line. Use of U0126 inhibited progestin-mediated phosphorylation of the MAPK target proteins, ERK 1/2, suggesting involvement in the regulation of MnSOD and tumor invasiveness.

While some heavy metals have a negative effect on transcriptional regulation of MnSOD, manganese has been shown to induce MnSOD in *Escherichia coli* B,[109] lactobacilli,[110] human breast cancer Hs578T cells[111] and HeLa cells.[112] In *E. coli* B, Mn^{2+} was shown to induce MnSOD in an O_2-dependent fashion. Increased expression of MnSOD in Hs578T cells was inhibited at the mRNA level by the metal chelator, tiron. Pyruvate, a H_2O_2 scavenger, also blocked Mn-dependent induction at both the mRNA and protein levels in breast cancer. Expression of MnSOD in HeLa cells was correlated with an increase in cellular ROS levels and more specifically peroxides upon Mn^{2+} treatment. Although the mechanism by which Mn^{2+} induces MnSOD expression is unknown, the metal has been shown to activate nuclear factor-E2-related transcription factor (Nrf2) in PC12 cells leading to an adaptive response *via* the antioxidant response element (ARE).[154] Nrf2 has been reported to increase MnSOD expression in many studies *via* the ARE.[155–157]

3.7   MnSOD and Disease

3.7.1   Cancer

Initial focus on MnSOD's role in disease was aimed at understanding its potential as a tumor suppressor gene based on the report of reduced expression and activity in a variety of tumors by Oberley and Buettner.[101] Early studies where MnSOD was overexpressed led to restoration of normal growth rates and suppressed tumorgenicity in melanoma, breast and other cell lines. In addition, numerous studies explored the contribution of the enzyme to inducing differentiation, thereby slowing proliferation and decreasing the risk of a more aggressive phenotype. More recently, elevated MnSOD expression and activity has been reported to be associated with different tumor types (pancreatic, gastric, *etc.*).[158] It is now accepted that low MnSOD expression is an early event in tumor progression and that without sufficient cellular antioxidant capacity, ROS levels can significantly increase, induce mitochondrial and cellular dysfunction and propagate metastatic disease. MnSOD is one of several primary antioxidant enzymes whose activity is critical in maintaining nanomolar levels of cellular ROS (Figure 3.3). Any imbalance between antioxidant levels and oxidant generation stresses the cellular system, resulting in alterations of signaling cascades with

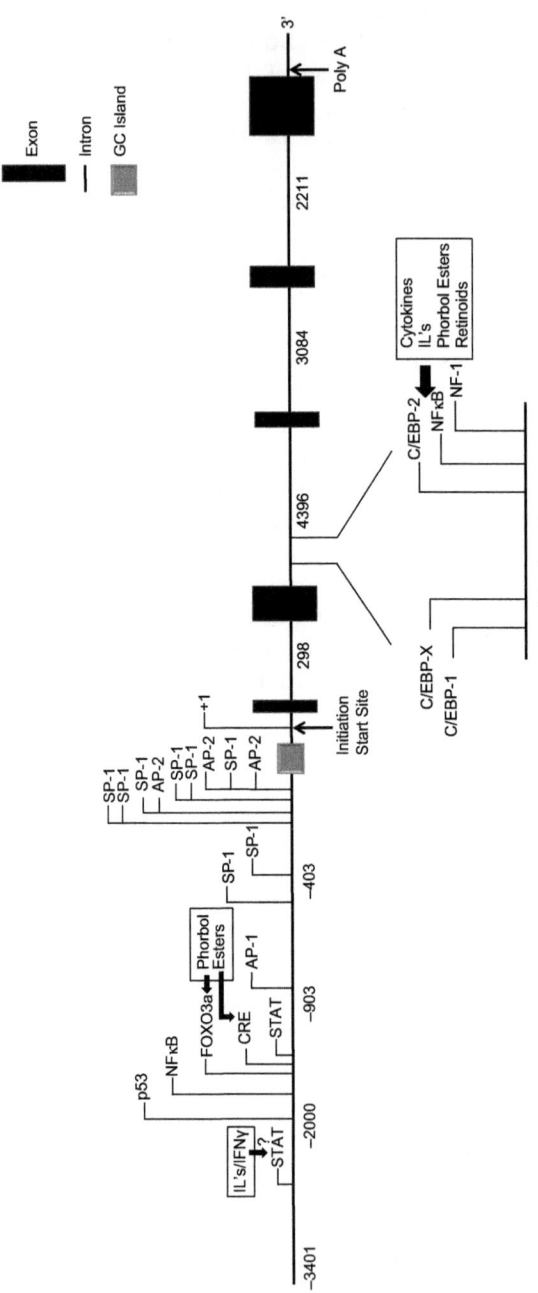

Figure 3.3 Organization of the human *SOD2* gene. The human *SOD2* gene consists of 5 exons and 4 introns. The regulatory elements in the 5′ flanking region with expansion of those in the second intron are shown. Corresponding negative numbers indicate their location relative to the transcription start site, which is designated as +1. The size of each intron is shown. Induction agents of MnSOD expression are shown with the respective consensus sequences for transcription factors they have been reported to activate. Abbreviations: interleukins (ILs) and interferon-γ (IFN-γ).

subsequent modulation of gene expression, leading to a more metastatic phenotype. In a recent review by Miriyala *et al.,* the authors suggest that the oncogenic potential of MnSOD is complex with contributing factors such as enhanced activity of ROS generators, such as NADPH oxidase, as well as deficiencies in their removal systems such as CAT or GPx.[159] Studies by St Clair's group suggests that these changes in redox balance ultimately contribute to the regulation of MnSOD expression.[159] In an *in vivo* model of melanoma, transcription of the *SOD2* gene was suppressed early by a decrease in SP-1 DNA binding. With low MnSOD expression, oxidative stress ensued, leading to a loss of p53 DNA binding activity and subsequent restoration of MnSOD during later stages of tumorigenesis.[160] Prior studies by Fojta *et al.* have described the susceptibility of p53 to oxidative modification with subsequent alterations in the transcription of target genes.[161,162] Dhar *et al.* suggested that the increase in MnSOD activity protected against oxidative injury to mitochondria, thereby promoting a growth advantage to the cancer cells.[163] Miriyala suggested that even though the SOD2 protein is upregulated with loss of p53 DNA binding ability, additional oxidative modifications may attenuate enzymatic activity and suppress the antioxidant's metastatic potential in some tumors. The authors stated that the paradoxical views of MnSOD as a tumor suppressor or oncogene may be explained based on the differences in cellular redox status between normal, transformed and metastatic phenotypes.

3.7.2 Cardiovascular Disease

Studies utilizing knock-out technology highlighted the essential role of MnSOD for aerobic life and focused attention on the importance of the enzyme in overall mitochondrial homeostasis, particularly in tissues with high energy demands, such as heart and brain. Excellent reviews by Fukai and Holley describe the role of MnSOD in pathophysiologies such as atherosclerosis, hypertension, angiogenesis and drug-induced cardiomyopathy.[164,165]

Studies show the protection of mitochondrial DNA, complex enzyme activities and ATP production with overexpression of the SOD2 enzyme in cardiac tissue.[141,142] In apolipoprotein deficient mice (ApoE(−/−)) mitochondrial dysfunction secondary to a *SOD2* deficiency accelerated atherosclerosis, which correlated with mitochondrial DNA damage.[166] Numerous studies suggest an association between increased ROS, SOD expression in various tissues and the development of hypertension. The use of a SOD mimetic has been shown to improve hypertension with an associated reduction in oxidation markers. Mechanistically it has been reported that the improvement includes modulation of various physiological changes: vasodilation, vasoconstriction, vascular remodeling and cardiac hypertrophy among others. One key component in the development of vascular dysfunction in multiple models of hypertension involves the removal of $^{\bullet}$NO by $O_2^{\bullet-}$ to generate $ONOO^-$, thereby decreasing vasorelaxation. The rate of this reaction, $k > 6.7 \times 10^9$ M^{-1} s^{-1}, is greater than that of the dismutation reaction catalyzed by MnSOD, therefore favoring the formation of ONOO⁻ when both species are present. Peroxynitrite can then decay at physiological pH to generate more potent and oxidizing radicals such as $^\bullet$OH resulting in cytotoxicity.

Cole *et al.* investigated the role of decreased $^\bullet$NO production in exacerbation of ADR-induced cardiotoxicity *in vivo* using iNOS(−/−) mice.[167] Cardiac troponin, creatine phosphokinase and lactate dehydrogenase levels were significantly increased after ADR treatment in iNOS(−/−) mice when compared to WT animals. When the iNOS(−/−) mice were crossed with MnSOD overexpressing animals, biomarkers of cardiac injury as well as mitochondrial damage were reduced to that of WT animals.

Changes in mRNA and/or protein expression of SOD2 have been shown to accompany CHF as well as myocardial infarction; however, neither condition results in enhanced SOD2 activity, suggesting post-translational modification of the enzyme.[165] Studies by Van Remmen[28] highlighted the importance of cardioprotection by MnSOD showing that cardiomyocytes are more likely to undergo apoptosis when SOD2 activity is decreased, and cells are exposed to either calcium or *t*-butylhydroperoxide.

3.7.3 Neurodegenerative Disorders

In addition to cardiovascular damage in SOD2 KO animals, neurodegeneration was noted by both Lebovitz[19] and Lynn[20] in different genetic backgrounds. Damage to regions such as the basal ganglia, brain stem, thalamus and motor cortex were reported to contribute to ataxia and a lowered seizure threshold. *In vitro* studies have identified MnSOD to be cytoprotective when exposing neuronal cultures to mutant SOD1,[168] iron,[169] NMDA,[170] β-amyloid,[171] $^\bullet$NO,[170] cystine deprivation[172] or hypoxia.[173] MnSOD transgenic animals have showed protection when exposed to methamphetamine,[174] 1-methyl-4-phenyl-1,2,3,6-tetrahydropyridine[175] or following traumatic brain injury[176] when compared to WT animals.

Immunocytochemical studies performed by Marcus *et al.* in AD patients indicated an elevation of MnSOD immunoreactivity in the CA2/3 and CA4 hippocampal regions when compared to non-AD patients,[177] suggesting an adaptive response against oxidative stress. Li *et al.* crossed the AD murine model, APP(−/−), with a heterozygous KO MnSOD model (SOD2(+/−)) and showed an increase in Aβ levels as well as plaque formation in the mice.[178] A subsequent study by Dumont *et al.* showed that overexpression of MnSOD in the APP −/− model reduced plaque burden by 33%, decreased oxidized proteins by 50% with restoration of memory to that of WT littermates.[179] Collectively, these studies suggest that mitochondrial antioxidant levels improve resistance to Aβ toxicity and attenuate the phenotype of a transgenic AD mouse model.

Amyotrophic lateral sclerosis (ALS) is an age-associated, terminal neurodegenerative disease. Mutations in the *SOD1* gene contribute to 25% of familial cases and between 1–4% of sporadic cases. Transgenic animal models have been constructed to represent some of the *SOD1* mutations identified in the patient population. In the mSOD1^{G93A} murine model, human SOD1 is expressed with the glycine residue at amino acid 93 mutated to an alanine. When this murine model was crossed with the SOD2(+/−) model, disease progression was accelerated, indicating a critical role for mitochondrial redox status in the progression of disease in patients with this specific genetic mutation.[180] Additional information on the role of MnSOD in other neurodegenerative diseases such as Parkinson's disease, stroke, aging and normal age-related cognitive decline can be found in a review by Flynn and Melov.[181]

According to Miriyala[159] antioxidant mimetics show promise in the treatment of various disease states including brain related disorders such as tumors,[182] epilepsy[183] and stroke.[184] Currently both synthetic and natural compounds able to mimic MnSOD's antioxidant capacity are being tested in disease models to determine efficacy. Studies utilizing lipophilic Mn porphyrins, which mimic the localization and activity of MnSOD, have shown promise in treating disorders of the central nervous system where mitochondria are known to play a role. These studies, along with more insight into identifying key molecules in signaling pathways associated with the cytoprotective mechanism of SOD2, may prove key in improving the prognosis of cancer, brain and cardiovascular associated disorders.

3.7.4 MnSOD Polymorphisms and Disease

Various disease states have been studied to determine the presence of *SOD2* polymorphisms. Several known single-nucleotide polymorphisms (SNPs) have been reported in the noncoding and coding regions of the *SOD2* gene.[185] In the noncoding region three mutations have been described that alter expression and lead to decreased enzymatic activity: a C-to-T transition at −102, an A insertion at −93 and a C-to-G transversion at −38. In addition, a C-to-T SNP in the mitochondrial targeting sequence is reported to affect localization and efficiency of transport into the mitochondria.[186] The SNP results in conversion of valine (Val) to alanine (Ala) at amino acid 16. Sutton *et al.* reported than the Ala variant leads to greater enzymatic activity than the Val containing protein by as much as 40%.[187] Structural differences between the two variants lead to the Val16 variant predominantly embedded in the inner mitochondrial membrane. In addition, the Val variant has been investigated in various tumor types and reportedly associated with a greater risk of pancreatic, breast and non-small-cell-lung carcinoma; however, other studies suggest the Ala16Val polymorphism to be associated with an increased risk for cancer.[188–191] Woodson *et al.* suggested an association of the AA genotype with increased risk for prostate cancer[192] and Olson *et al.* reported a similar finding in breast carcinoma.[193] The Ala16Val polymorphism has also been studied in cardiovascular disease and the frequency of the Val/Val allele was found to be higher in patients with either coronary artery disease[194] or vasospastic angina.[195]

Reduced enzyme activity was reported by Hernandez-Saavedra and McCord to result from a Leu60Phe SNP in a Jurkat human T cell leukemia cell line when compared to normal human peripheral blood lymphocytes.[196] In addition, a G1677T polymorphism within the second intron of the *SOD2* gene was reported to generate a consensus sequence for the glucocorticoid receptor. Hernandez-Saavedra and McCord reported that the T/T genotype was associated with a reduced risk of lung cancer.[197]

An additional polymorphism resulting in isoleucine (Ile) instead of threonine (Thr) at amino acid 58 in the MnSOD protein has been reported to result in greater enzymatic activity and tumor suppressor capability in MCF-7 cells.[198] Amino acid 58 is located at the interface of the four monomers that comprise the functional enzyme.[199] The Ile58Thr polymorphism destabilizes the interface of individual monomeric subunits resulting in the formation of dimeric SOD2. A recent analysis of this polymorphism in prostate cancer suggested no association.[200] This finding was consistent with other studies that showed lack of a connection between specific disease states and MnSOD mutations.[201–203]

3.8   Future Directions

Over the past 30 years tremendous progress has been made in understanding the significance of the expression of the mitochondrial MnSOD isoform. Unlike *SOD1* and *SOD3*, *SOD2* is essential to aerobic life. Studies have identified regulatory control at both the transcriptional and post-translational level in addition to epigenetic control of expression. Some investigators have hypothesized that post-translational modifications of the enzyme may be reversible; however, additional studies will be needed to confirm this hypothesis.

Future studies will be needed to further our understanding of additional regulatory mechanisms at the genomic level. Also important will be studies identifying how MnSOD influences signaling cascades by altering cellular redox status, both inside and outside the mitochondria. These studies, in addition to those at the genomic level, may provide pharmacological targets specific to disease states that will ultimately improve prognosis in specific patient populations.

Continued exploration of the role of mitochondria in carcinogenesis is needed. MnSOD is critical in maintaining mitochondrial homeostasis to allow for energy needs during cell growth.[158] Dhar has speculated that lowered MnSOD expression could lead to compromised mitochondria and a shift from oxidative phosphorylation to glycolysis to support energy needs during early stages of tumorigenesis. As cellular conditions change, and MnSOD expression increases, this may result in a shift back to oxidative phosphorylation for energy needs of metastatic cells. Systematic studies of these possibilities, as well as MnSOD regulation during the cancer development process, are essential to developing additional treatment options for patients.

Abbreviation List

Å	Angstrom
Aβ	Amyloid β peptide
AD	Alzheimer's disease
ADR	Adriamycin
AIDS	Acquired immune deficiency syndrome
Ala	Alanine
ALS	Amyotrophic lateral sclerosis
ALT	Serum alanine aminotransferase
ARE	Antioxidant response element
AST	Serum aspartate aminotransferase
ATP	Adenosine triphosphate
Cdk1	Cell cycle dependent kinase
CHF	Congestive heart failure
CsA	Cyclosporine
Cu	Copper
CuZnSOD	Copper, zinc superoxide dismutase
ECSOD	Extracellular superoxide dismutase
Fe	Iron
FoxM1	Forkhead box M1
GSH	Glutathione
Hsp60	Heat shock protein 60
H_2O_2	Hydrogen peroxide
HIFα	Hypoxia inducible factor-alpha
H/R	Hypoxia reoxygenation
iNOS	Inducible nitric oxide synthase
I/R	Ischemia reperfusion
Ile	Isoleucine
JAK	Janus family tyrosine kinases
KO	Knock-out
MAPK	Mitogen activated protein kinase
MMP	Matrix metalloproteinase
Mn	Manganese
MnSOD	Manganese superoxide dismutase
NAPQI	*N*-acetyl-*p*-benzoquinone imine
Nrf2	Nuclear factor-E2-related transcription factor
NGF	Nerve growth factor
nNOS	Neuronal nitric oxide synthase
•NO	Nitric oxide
3-NT	3-Nitrotyrosine
PAH	Pulmonary arterial hypertension
PCNA	Proliferative cell nuclear antigen
PD	Parkinson's disease
PDGF	Platelet derived growth factor
PPHN	Persistent pulmonary hypertension

PTEN	Phosphatase and tensin homolog
RNS	Reactive nitrogen species
$RO^\bullet$	Alkoxyl radical
$ROO^\bullet$	Peroxyl radical
ROS	Reactive oxygen species
$ONOO^-$	Peroxynitrite
$O_2^{\bullet-}$	Superoxide anion
Ser	Serine
Sirt3	Sirtuin-3
SNPs	Single nucleotide polymorphisms
SOD	Superoxide dismutase
STAT3	Signal transducer and activator of transcription 3
Thr	Threonine
TNF-α	Tumor necrosis factor-α
Tyr	Tyrosine
Val	Valine
WT	Wildtype

References

1. J. M. McCord and I. Fridovich, Superoxide dismutase. An enzymic function for erythrocuprein (hemocuprein), *J. Biol. Chem.*, 1969, **244**, 6049.
2. V. C. Culotta, M. Yang and T. V. O'Halloran, Activation of superoxide dismutases: putting the metal to the pedal, *Biochim. Biophys. Acta*, 2006, **1763**, 747.
3. T. Fukai, R. J. Folz, U. Landmesser and D. G. Harrison, Extracellular superoxide dismutase and cardiovascular disease, *Cardiovasc. Res.*, 2002, **55**, 239.
4. A. D. Nguyen, S. Itoh, V. Jeney, H. Yanagisawa, M. Fujimoto, M. Ushio-Fukai and T. Fukai, Fibulin-5 is a novel binding protein for extracellular superoxide dismutase, *Circ. Res.*, 2004, **95**, 1067.
5. S. V. Petersen, T. D. Oury, L. Ostergaard, Z. Valnickova, J. Wegrzyn, I. B. Thøgersen, C. Jacobsen, R. P. Bowler, C. L. Fattman, J. D. Crapo and J. J. Enghild, Extracellular superoxide dismutase (EC-SOD) binds to type i collagen and protects against oxidative fragmentation, *J. Biol. Chem.*, 2004, **279**, 13705.
6. L. Miao and D. K. St Clair, Regulation of superoxide dismutase genes: implications in disease, *Free Radicals Biol. Med.*, 2009, **47**, 344.
7. B. B. Keele Jr, J. M. McCord and I. Fridovich, Superoxide dismutase from escherichia coli B. A new manganese-containing enzyme, *J. Biol. Chem.*, 1970, **245**, 6176.
8. S. D. Ravindranath and I. Fridovich, Isolation and characterization of a manganese-containing superoxide dismutase from yeast, *J. Biol. Chem.*, 1975, **250**, 6107.

9. H. P. Misra and I. Fridovich, Purification and properties of superoxide dismutase from a red alga, Porphyridium cruentum, *J. Biol. Chem.*, 1977, **252**, 6421.

10. T. Hunter, W. H. Bannister and G. J. Hunter, Cloning, expression, and characterization of two manganese superoxide dismutases from Caenorhabditis elegans, *J. Biol. Chem.*, 1997, **272**, 28652.

11. M. Brouwer, T. Hoexum Brouwer, W. Grater and N. Brown-Peterson, Replacement of a cytosolic copper/zinc superoxide dismutase by a novel cytosolic manganese superoxide dismutase in crustaceans that use copper (haemocyanin) for oxygen transport, *Biochem. J.*, 2003, **374**, 219.

12. C. Lamarre, J. D. LeMay, N. Deslauriers and Y. Bourbonnais, Candida albicans expresses an unusual cytoplasmic manganese-containing superoxide dismutase (SOD3 gene product) upon the entry and during the stationary phase, *J. Biol. Chem.*, 2001, **276**, 43784.

13. W. F. Beyer Jr and I. Fridovich, In vivo competition between iron and manganese for occupancy of the active site region of the manganese-superoxide dismutase of Escherichia coli, *J. Biol. Chem.*, 1991, **266**, 303.

14. P. Marc, A. Margeot, F. Devaux, C. Blugeon, M. Corral-Debrinski and C. Jacq, Genome-wide analysis of mRNAs targeted to yeast mito-chondria, *EMBO Rep.*, 2002, **3**, 159.

15. R. Magnoni, J. Palmfeldt, J. Hansen, J. H. Christensen, T. J. Corydon and P. Bross, The Hsp60 folding machinery is crucial for manganese superoxide dismutase folding and function, *Free Radical Res.*, 2014, **48**, 168.

16. E. Luk, M. Carroll, M. Baker and V. C. Culotta, Manganese activation of superoxide dismutase 2 in Saccharomyces cerevisiae requires MTM1, a member of the mitochondrial carrier family, *Proc. Natl. Acad. Sci. U. S. A*, 2003, **100**, 10353.

17. M. M. Whittaker and J. W. Whittaker, In vitro metal uptake by re-combinant human manganese superoxide dismutase, *Arch. Biochem. Biophys.*, 2009, **491**, 69.

18. Y. Li, T. T. Huang, E. J. Carlson, S. Melov, P. C. Ursell, J. L. Olson, L. J. Noble, M. P. Yoshimura, C. Berger, P. H. Chan, D. C. Wallace and C. J. Epstein, Dilated cardiomyopathy and neonatal lethality in mutant mice lacking manganese superoxide dismutase, *Nat. Genet.*, 1995, **11**, 376.

19. R. M. Lebovitz, H. Zhang, H. Vogel, J. Cartwright Jr, L. Dionne, N. Lu, S. Huang and M. M. Matzuk, Neurodegeneration, myocardial injury, and perinatal death in mitochondrial superoxide dismutase-deficient mice, *Proc. Natl. Acad. Sci. U. S. A*, 1996, **93**, 9782.

20. S. Lynn, E. J. Huang, S. Elchuri, M. Naeemuddin, Y. Nishinaka, J. Yodoi, D. M. Ferriero, C. J. Epstein and T. T. Huang, Selective neuronal vul-nerability and inadequate stress response in superoxide dismutase mutant mice, *Free Radicals Biol. Med.*, 2005, **38**, 817.

21. T. Kondo, A. G. Reaume, T. T. Huang, E. Carlson, K. Murakami, S. F. Chen, E. K. Hoffman, R. W. Scott, C. J. Epstein and P. H. Chan,

Reduction of CuZn-superoxide dismutase activity exacerbates neuronal cell injury and edema formation after transient focal cerebral ischemia, *J. Neurosci.*, 1997, **17**, 4180.

22. L. M. Carlsson, J. Jonsson, E. Edlund and S. L. Marklund, Mice lacking extracellular superoxide dismutase are more sensitive to hyperoxia, *Proc. Natl. Acad. Sci. U. S. A*, 1995, **92**, 6264.

23. J. C. Copin, Y. Gasche and P. H. Chan, Overexpression of copper/zinc superoxide dismutase does not prevent neonatal lethality in mutant mice that lack manganese superoxide dismutase, *Free Radicals Biol. Med.*, 2000, **28**, 1571.

24. Y. Zhang, H. M. Zhang, Y. Shi, M. Lustgarten, Y. Li, W. Qi, B. X. Zhang and H. Van Remmen, Loss of manganese superoxide dismutase leads to abnormal growth and signal transduction in mouse embryonic fibroblasts, *Free Radicals Biol. Med.*, 2010, **49**, 1255.

25. M. D. Williams, H. Van Remmen, C. C. Conrad, T. T. Huang, C. J. Epstein and A. Richardson, Increased oxidative damage is correlated to altered mitochondrial function in heterozygous manganese superoxide dismutase knockout mice, *J. Biol. Chem.*, 1998, **273**, 28510.

26. H. Van Remmen, C. Salvador, H. Yang, T. T. Huang, C. J. Epstein and A. Richardson, Characterization of the antioxidant status of the heterozygous manganese superoxide dismutase knockout mouse, *Arch. Biochem. Biophys.*, 1999, **363**, 91.

27. H. Van Remmen, Y. Ikeno, M. Hamilton, M. Pahlavani, N. Wolf, S. R. Thorpe, N. L. Alderson, J. W. Baynes, C. J. Epstein, T. T. Huang, J. Nelson, R. Strong and A. Richardson, Life-long reduction in MnSOD activity results in increased DNA damage and higher incidence of cancer but does not accelerate aging, *Physiol. Genomics*, 2003, **16**, 29.

28. H. Van Remmen, M. D. Williams, Z. Guo, L. Estlack, H. Yang, E. J. Carlson, C. J. Epstein, T. T. Huang and A. Richardson, Knockout mice heterozygous for Sod2 show alterations in cardiac mitochondrial function and apoptosis, *Am. J. Physiol. Heart Circ. Physiol.*, 2001, **281**, H1422.

29. M. Strassburger, W. Bloch, S. Sulyok, J. Schüller, A. F. Keist, A. Schmidt, J. Wenk, T. Peters, M. Wlaschek, J. Lenart, T. Krieg, M. Hafner, A. Kümin, S. Werner, W. Müller and K. Scharffetter-Kochanek, Heterozygous deficiency of manganese superoxide dismutase results in severe lipid peroxidation and spontaneous apoptosis in murine myocardium in vivo, *Free Radicals Biol. Med.*, 2005, **38**, 1458.

30. T. Ikegami, Y. Suzuki, T. Shimizu, K. Isono, H. Koseki and T. Shirasawa, Model mice for tissue-specific deletion of the manganese superoxide dismutase (MnSOD) gene, *Biochem. Biophys. Res. Commun.*, 2002, **296**, 729.

31. H. Misawa, K. Nakata, J. Matsuura, Y. Moriwaki, K. Kawashima, T. Shimizu, T. Shirasawa and R. Takahashi, Conditional knockout of Mn superoxide dismutase in postnatal motor neurons reveals resistance to mitochondrial generated superoxide radicals, *Neurobiol. Dis.*, 2006, **23**, 169.

32. H. Nojiri, T. Shimizu, M. Funakoshi, O. Yamaguchi, H. Zhou, S. Kawakami, Y. Ohta, M. Sami, T. Tachibana, H. Ishikawa, H. Kurosawa, R. C. Kahn, K. Otsu and T. Shirasawa, Oxidative stress causes heart failure with impaired mitochondrial respiration, *J. Biol. Chem.*, 2006, **281**, 33789.
33. M. S. Lustgarten, Y. C. Jang, Y. Liu, F. L. Muller, W. Qi, M. Steinhelper, S. V. Brooks, L. Larkin, T. Shimizu, T. Shirasawa, L. M. McManus, A. Bhattacharya, A. Richardson and H. Van Remmen, Conditional knockout of Mn-SOD targeted to type IIB skeletal muscle fibers increases oxidative stress and is sufficient to alter aerobic exercise capacity, *Am. J. Physiol.: Cell Physiol.*, 2009, **297**, C1520.
34. M. S. Lustgarten, Y. C. Jang, Y. Liu, W. Qi, Y. Qin, P. L. Dahia, Y. Shi, A. Bhattacharya, F. L. Muller, T. Shimizu, T. Shirasawa, A. Richardson and H. Van Remmen, MnSOD deficiency results in elevated oxidative stress and decreased mitochondrial function but does not lead to muscle atrophy during aging, *Aging Cell*, 2011, **10**, 493.
35. H. Kuwahara, T. Horie, S. Ishikawa, C. Tsuda, S. Kawakami, Y. Noda, T. Kaneko, S. Tahara, T. Tachibana, M. Okabe, J. Melki, R. Takano, T. Toda, D. Morikawa, H. Nojiri, H. Kurosawa, T. Shirasawa and T. Shimizu, Oxidative stress in skeletal muscle causes severe disturbance of exercise activity without muscle atrophy, *Free Radicals Biol. Med.*, 2010, **48**, 1252.
36. N. Parajuli, A. Marine, S. Simmons, H. Saba, T. Mitchell, T. Shimizu, T. Shirasawa and L. A. Macmillan-Crow, Generation and characterization of a novel kidney-specific manganese superoxide dismutase knockout mouse, *Free Radicals Biol. Med.*, 2011, **51**, 406.
37. N. Parajuli and L. A. MacMillan-Crow, Role of reduced manganese superoxide dismutase in ischemia-reperfusion injury: a possible trigger for autophagy and mitochondrial biogenesis?, *Am. J. Physiol. Renal Physiol.*, 2013, **304**, F257.
38. F. Yamakura and H. Kawasaki, Post-translational modifications of superoxide dismutase, *Biochim. Biophys. Acta*, 2010, **1804**, 318.
39. R. Radi, Protein tyrosine nitration: biochemical mechanisms and structural basis of functional effects, *Acc. Chem. Res.*, 2013, **46**, 550.
40. D. M. Morena, M. A. Martí, P. M. De Biase, D. A. Estrin, V. Demicheli, R. Radi and L. Boechi, Exploring the molecular basis of human manganese superoxide dismutase inactivation mediated by tyrosine 34 nitration, *Arch. Biochem. Biophys.*, 2011, **507**, 304.
41. P. Quint, R. Reutzel, R. Mikulski, R. McKenna and D. N. Silverman, Crystal structure of nitrated human manganese superoxide dismutase: mechanism of inactivation, *Free Radicals Biol. Med.*, 2006, **40**, 453.
42. L. A. MacMillan-Crow, J. P. Crow, J. D. Derby, J. S. Beckman and J. A. Thompson, Nitration and inactivation of manganese superoxide dismutase in chronic rejection of human renal allografts, *Proc. Natl. Acad. Sci. U. S. A*, 1996, **93**, 11853.

43. M. Anantharaman, J. Tangpong, J. N. Keller, M. P. Murphy, W. R. Markesbery, K. K. Kiningham and D. K. St Clair, Beta-amyloid mediated nitration of manganese superoxide dismutase: implication for oxidative stress in a APPNLH/NLH X PS-1P264L/P264L double knock-in mouse model of Alzheimer's disease, *Am. J. Pathol.*, 2006, **168**, 1608.

44. P. Sompol, W. Ittarat, J. Tangpong, Y. Chen, I. Batinic-Haberle, H. M. Abdul, D. A. Butterfield and D. K. St. Clair, A neuronal model of Alzheimer's disease: an insight into the mechanisms of oxidative stress-mediated mitochondrial injury, *Neuroscience.*, 2008, **153**, 120.

45. D. A. Butterfield and C. M. Lauderback, Lipid peroxidation and protein oxidation in Alzheimer's disease brain: potential causes and consequences involving amyloid beta-peptide-associated free radical oxidative stress, *Free Radicals Biol. Med.*, 2002, **32**, 1050.

46. M. A. Smith, P. L. Richey Harris, L. M. Sayre, J. S. Beckman and G. Perry, Widespread peroxynitrite-mediated damage in Alzheimer's disease, *J. Neurosci.*, 1997, **17**, 2653.

47. A. Castegna, V. Thongboonkerd, J. B. Klein, B. Lynn, W. R. Markesbery and D. A. Butterfield, Proteomic identification of nitrated proteins in Alzheimer's disease brain, *J. Neurochem.*, 2003, **85**, 1394.

48. A. J. Afolayan, A. Eis, R. J. Teng, I. Bakhutashvili, S. Kaul, J. M. Davis and G. G. Konduri, Decreases in manganese superoxide dismutase expression and activity contribute to oxidative stress in persistent pulmonary hypertension of the newborn, *Am. J. Physiol.: Lung Cell. Mol. Physiol.*, 2012, **303**, L870.

49. W. Guo, T. Adachi, R. Matsui, S. Xu, B. Jiang, M. H. Zou, M. Kirber, W. Lieberthal and R. A. Cohen, Quantitative assessment of tyrosine nitration of manganese superoxide dismutase in angiotensin II-infused rat kidney, *Am. J. Physiol. Heart Circ. Physiol.*, 2003, **285**, H1396.

50. M. Redondo-Horcajo, N. Romero, P. Martínez-Acedo, A. Martínez-Ruiz, C. Quijano, C. F. Lourenco, N. Movilla, J. A. Enríquez, F. Rodríguez-Pascual, E. Rial, R. Radi, J. Vázquez and S. Lamas, Cyclosporine A-induced nitration of tyrosine 34 MnSOD in endothelial cells: role of mitochondrial superoxide, *Cardiovasc. Res.*, 2010, **87**, 356.

51. R. Agarwal, L. A. MacMillan-Crow, T. M. Rafferty, H. Saba, D. W. Roberts, E. K. Fifer, L. P. James and J. A. Hinson, Acetaminophen-induced hepatotoxicity in mice occurs with inhibition of activity and nitration of mitochondrial manganese superoxide dismutase, *J. Pharmacol. Exp. Ther.*, 2011, **337**, 110.

52. R. Agarwal, L. Hennings, T. M. Rafferty, L. G. Letzig, S. McCullough, L. P. Janes, L. A. MacMillan-Crow and J. A. Hinson, Acetaminophen-induced hepatotoxicity and protein nitration in neuronal nitric-oxide synthase knockout mice, *J. Pharmacol. Exp. Ther.*, 2012, **340**, 134.

53. J. Tangpong, M. P. Cole, R. Sultana, S. Estus, M. Vore, W. St Clair, S. Ratanachaiyavong, D. K. St Clair and D. A. Butterfield, Adriamycin-mediated nitration of manganese superoxide dismutase in the central

nervous system: insight into the mechanism of chemobrain, *J. Neurochem.*, 2007, **100**, 191.

54. J. Tangpong, P. Sompol, M. Vore, W. St Clair, D. A. Butterfield and D. K. St Clair, Tumor necrosis factor alpha-mediated nitric oxide production enhances manganese superoxide dismutase nitration and mitochondrial dysfunction in primary neurons: an insight into the role of glial cells, *Neuroscience*, 2008, **151**, 622.

55. T. W. Vanderah, M. H. Ossipov, J. Lai, T. P. Malan Jr and F. Porreca, Mechanisms of opioid-induced pain and antinociceptive tolerance: descending facilitation and spinal dynorphin, *Pain*, 2001, **92**, 5.

56. S. Mitra, Opioid-induced hyperalgesia: pathophysiology and clinical implications, *J. Opioid Manag.*, 2008, **4**, 123.

57. T. Doyle, L. Bryant, C. Muscoli, S. Cuzzocrea, E. Esposito, Z. Chen and D. Salvemini, Spinal NADPH oxidase is a source of superoxide in the development of morphine-induced hyperalgesia and antinociceptive tolerance, *Neurosci. Lett.*, 2010, **483**, 85.

58. T. Doyle, E. Esposito, L. Bryant, S. Cuzzocrea and D. Salvemini, NADPH-oxidase 2 activation promotes opioid-induced antinociceptive tolerance in mice, *Neuroscience*, 2013, **241**, 1.

59. L. Tie, H. Q. Yang, Y. An, S. Q. Liu, J. Han, Y. Xu, M. Hu, W. D. Li, A. F. Chen, Z. B. Lin and X. J. Li, Ganoderma lucidum polysaccharide accelerates refractory wound healing by inhibition of mitochondrial oxidative stress in type 1 diabetes, *Cell. Physiol. Biochem.*, 2012, **29**, 583.

60. K. Aoyama, K. Matsubara, Y. Fujikawa, Y. Nagahiro, K. Shimizu, N. Umegae, N. Hayase, H. Shiono and S. Kobayashi, Nitration of manganese superoxide dismutase in cerebrospinal fluids is a marker for peroxynitrite-mediated oxidative stress in neurodegenerative diseases, *Ann. Neurol.*, 2000, **47**, 524.

61. S. R. Mallery, P. Pei, D. J. Landwehr, C. M. Clark, J. E. Bradburn, G. M. Ness and F. M. Robertson, Implications for oxidative and nitrative stress in the pathogenesis of AIDS-related Kaposi's sarcoma, *Carcinogenesis*, 2004, **25**, 597.

62. P. Jiménez, E. Piazuelo, M. T. Sánchez, J. Ortego, F. Soteras and A. Lanas, Free radicals and antioxidant systems in reflux esophagitis and Barrett's esophagus, *World J. Gastroenterol.*, 2005, **11**, 2697.

63. S. Xu, J. Ying, B. Jiang, W. Guo, T. Adachi, V. Sharov, H. Lazar, J. Menzoian, T. V. Knyushko, D. Bigelow, C. Schöneich and R. A. Cohen, Detection of sequence-specific tyrosine nitration of manganese SOD and SERCA in cardiovascular disease and aging, *Am. J. Physiol. Heart Circ. Physiol.*, 2006, **290**, H2220.

64. N. V. Bykova, H. Egsgaard and I. M. Møller, Identification of 14 new phosphoproteins involved in important plant mitochondrial processes, *FEBS Lett.*, 2003, **540**, 141.

65. R. K. Hopper, S. Carroll, A. M. Aponte, D. T. Johnson, S. French, R. F. Shen, F. A. Witzmann, R. A. Harris and R. S. Balaban,

Mitochondrial matrix phosphoproteome: effect of extra mitochondrial calcium, *Biochemistry*, 2006, **45**, 2524.

66. I. Castellano, F. Cecere, A. De Vendittis, R. Cotugno, A. Chambery, A. Di Maro, A. Michniewicz, G. Parlato, M. Masullo, E. V. Avvedimento, E. De Vendittis and M. R. Ruocco, Rat mitochondrial manganese superoxide dismutase: amino acid positions involved in covalent modifications, activity, and heat stability, *Biopolymers*, 2009, **91**, 1215.

67. D. Candas, M. Fan, D. Nantajit, A. T. Vaughan, J. S. Murley, G. E. Woloschak, D. J. Grdina and J. J. Li, CyclinB1/Cdk1 phosphorylates mitochondrial antioxidant MnSOD in cell adaptive response to radiation stress, *J. Mol. Cell Biol.*, 2013, **5**, 166.

68. R. Tao, M. C. Coleman, J. D. Pennington, O. Ozden, S. H. Park, H. Jiang, H. S. Kim, C. R. Flynn, S. Hill, W. Hayes McDonald, A. K. Olivier, D. R. Spitz and D. Gius, Sirt3-mediated deacetylation of evolutionarily conserved lysine 122 regulates MnSOD activity in response to stress, *Mol. Cell*, 2010, **40**, 893.

69. X. Qiu, K. Brown, M. D. Hirschey, E. Verdin and D. Chen, Calorie restriction reduces oxidative stress by SIRT3-mediated SOD2 activation, *Cell Metab.*, 2010, **12**, 662.

70. G. Groeger, C. Quiney and T. G. Cotter, Hydrogen peroxide as a cell-survival signaling molecule, *Antioxid. Redox Signaling*, 2009, **11**, 2655.

71. A. Bindoli and M. P. Rigobello, Principles in redox signaling: from chemistry to functional significance, *Antioxid. Redox Signaling*, 2013, **18**, 1557.

72. R. Bretón-Romero and S. Lamas, Hydrogen peroxide signaling mediator in the activation of p38 MAPK in vascular endothelial cells, *Methods Enzymol.*, 2013, **528**, 49.

73. Y. M. Janssen-Heininger, M. E. Poynter and P. A. Baeuerle, Recent advances towards understanding redox mechanisms in the activation of nuclear factor kappaB, *Free Radicals Biol. Med.*, 2000, **28**, 1327.

74. J. J. Haddad, Science review: redox and oxygen-sensitive transcription factors in the regulation of oxidant-mediated lung injury: role for hypoxia-inducible factor-1alpha, *Crit. Care*, 2003, 7, 47.

75. Y. Sun and L. W. Oberley, Redox regulation of transcriptional activators, *Free Radicals Biol. Med.*, 1996, **21**, 335.

76. S. Li, T. Yan, J. Q. Yang, T. D. Oberley and L. W. Oberley, The role of cellular glutathione peroxidase redox regulation in the suppression of tumor cell growth by manganese superoxide dismutase, *Cancer Res.*, 2000, **60**, 3927.

77. A. M. Rodríquez, P. M. Carrico, J. E. Mazurkiewicz and J. A. Meléndez, Mitochondrial or cytosolic catalase reverses the MnSOD-dependent inhibition of proliferation by enhancing respiratory chain activity, net ATP production, and decreasing the steady state levels of H(2)O(2), *Free Radicals Biol. Med.*, 2000, **29**, 801.

78. K. H. Kim, A. M. Rodríguez, P. M. Carrico and J. A. Meléndez, Potential mechanisms for the inhibition of tumor cell growth by manganese superoxide dismutase, *Antioxid. Redox Signaling*, 2001, **3**, 361.

79. H. J. Zhang, W. Zhao, S. Venkataraman, M. E. Robbins, G. R. Buettner, K. C. Kregel and L. W. Oberley, Activation of matrix metalloproteinase-2 by overexpression of manganese superoxide dismutase in human breast cancer MCF-7 cells involves reactive oxygen species, *J. Biol. Chem.*, 2002, **277**, 20919.

80. S. Venkataraman, X. Jiang, C. Weydert, Y. Zhang, H. J. Zhang, P. C. Goswami, J. M. Ritchie, L. W. Oberley and G. R. Buettner, Manganese superoxide dismutase overexpression inhibits the growth of androgen-independent prostate cancer cells, *Oncogene*, 2005, **24**, 77.

81. G. R. Buettner, C. F. Ng, M. Wang, V. G. Rodgers and F. Q. Schafer, A new paradigm: manganese superoxide dismutase influences the production of H2O2 in cells and thereby their biological state, *Free Radicals Biol. Med.*, 2006, **41**, 1338.

82. K. M. Connor, N. Hempel, K. K. Nelson, G. Dabiri, A. Gamarra, J. Belarmino, L. VanDe Water, B. M. Mian and J. A. Meléndez, Manganese superoxide dismutase enhances the invasive and migratory activity of tumor cells, *Cancer Res.*, 2007, **67**, 10260.

83. S. Venkataraman, B. A. Wagner, X. Jiang, H. P. Wang, F. Q. Schafer, J. M. Ritchie, B. C. Patrick, L. W. Oberley and G. R. Buettner, Overexpression of manganese superoxide dismutase promotes the survival of prostate cancer cells exposed to hyperthermia, *Free Radical Res.*, 2004, **38**, 1119.

84. G. R. Buettner, Superoxide dismutase in redox biology: the roles of superoxide and hydrogen peroxide, *Anti-Cancer Agents Med. Chem.*, 2011, **11**, 341.

85. A. Kim, W. Zhong and T. D. Oberley, Reversible modulation of cell cycle kinetics in NIH/3T3 mouse fibroblasts by inducible overexpression of mitochondrial manganese superoxide dismutase, *Antioxid. Redox Signaling*, 2004, **6**, 489.

86. B. B. Warner, L. Stuart, S. Gebb and J. R. Wispé, Redox regulation of manganese superoxide dismutase, *Am. J. Physiol.*, 1996, **271**, L150.

87. E. Röhrdanz and R. Kahl, Alterations of antioxidant enzyme expression in response to hydrogen peroxide, *Free Radicals Biol. Med.*, 1998, **24**, 27.

88. K. M. Connor, S. Subbaram, K. J. Regan, K. K. Nelson, J. E. Mazurkiewicz, P. J. Bartholomew, A. E. Aplin, Y. T. Tai, J. Aguirre-Ghiso, S. C. Flores and J. A. Melendez, Mitochondrial H2O2 regulates the angiogenic phenotype via PTEN oxidation, *J. Biol. Chem.*, 2005, **280**, 16916.

89. X. M. Li, A. V. Juorio, J. Qi and A. A. Boulton, L-deprenyl potentiates NGF-induced changes in superoxide dismutase mRNA in PC12 cells, *J. Neurosci. Res.*, 1998, **53**, 235.

90. K. K. Kiningham, Y. Xu, C. Daosukho, B. Popova and D. K. St Clair, Nuclear factor kappaB-dependent mechanisms coordinate the synergistic effect of PMA and cytokines on the induction of superoxide dismutase 2, *Biochem. J.*, 2001, **353**, 147.

91. S. Porntadavity, Y. Xu, K. Kiningham, V. M. Rangnekar, V. Prachayasittikul and D. K. St Clair, TPA-activated transcription of the

human MnSOD gene: role of transcription factors Sp-1 and Egr-1, *DNA Cell Biol.*, 2001, **20**, 473.

92. K. K. Kiningham, Z. A. Cardozo, C. Cook, M. P. Cole, J. C. Stewart, M. Tassone, M. C. Coleman and D. R. Spitz, All-trans-retinoic acid induces manganese superoxide dismutase in human neuroblastoma through NF-kappaB, *Free Radicals Biol. Med.*, 2008, **44**, 1610.

93. K. K. Kiningham and D. K. St Clair, Overexpression of manganese superoxide dismutase selectively modulates the activity of Jun-associated transcription factors in fibrosarcoma cells, *Cancer Res.*, 1997, **57**, 5265.

94. J. J. Li, L. W. Oberley, D. K. St Clair, L. A. Ridnour and T. D. Oberley, Phenotypic changes induced in human breast cancer cells by overexpression of manganese-containing superoxide dismutase, *Oncogene*, 1995, **10**, 1989.

95. S. L. Church, J. W. Grant, L. A. Ridnour, L. W. Oberley, P. E. Swanson, P. S. Meltzer and J. M. Trent, Increased manganese superoxide dismutase expression suppresses the malignant phenotype of human melanoma cells, *Proc. Natl. Acad. Sci. U. S. A*, 1993, **90**, 3113.

96. T. Yan, L. W. Oberley, W. Zhong and D. K. St Clair, Manganese-containing superoxide dismutase overexpression causes phenotypic reversion in SV40-transformed human lung fibroblasts, *Cancer Res.*, 1996, **56**, 2864.

97. P. A. Amstad, H. Liu, M. Ichimiya, I. K. Berezesky and B. F. Trump, Manganese superoxide dismutase expression inhibits soft agar growth in JB6 clone41 mouse epidermal cells, *Carcinogenesis*, 1997, **18**, 479.

98. S. Cassano, S. Agnese, V. D'Amato, M. Papale, C. Garbi, P. Castagnola, M. R. Ruocco, I. Castellano, E. De Vendittis, M. Santillo, S. Amente, A. Porcellini and E. V. Avvedimento, Reactive oxygen species, Ki-Ras, and mitochondrial superoxide dismutase cooperate in nerve growth factor-induced differentiation of PC12 cells, *J. Biol. Chem.*, 2010, **285**, 24141.

99. S. Lee, E. Tak, J. Lee, M. A. Rashid, M. P. Murphy, J. Ha and S. S. Kim, Mitochondrial H2O2 generated from electron transport chain complex I stimulates muscle differentiation, *Cell Res.*, 2011, **21**, 817.

100. S. L. Archer, M. Gomberg-Maitland, M. L. Maitland, S. Rich, J. G. Garcia and E. K. Weir, Mitochondrial metabolism, redox signaling, and fusion: a mitochondria-ROS-HIF-1alpha-Kv1.5 O2-sensing pathway at the intersection of pulmonary hypertension and cancer, *Am. J. Physiol. Heart Circ. Physiol.*, 2008, **294**, H570.

101. L. W. Oberley and G. R. Buettner, Role of superoxide dismutase in cancer: a review, *Cancer Res.*, 1979, **39**, 1141.

102. Y. Sun, Free radicals, antioxidant enzymes, and carcinogenesis, *Free Radicals Biol. Med.*, 1990, **8**, 583.

103. E. W. Lam, R. Zwacka, E. A. Seftor, D. R. Nieva, B. L. Davidson, J. F. Engelhardt, M. J. Hendrix and L. W. Oberley, Effects of antioxidant enzyme overexpression on the invasive phenotype of hamster cheek pouch carcinoma cells, *Free Radicals Biol. Med.*, 1999, **27**, 572.

104. M. Malafa, J. Margenthaler, B. Webb, L. Neitzel and M. Christophersen, MnSOD expression is increased in metastatic gastric cancer, *J. Surg. Res.*, 2000, **88**, 130.

105. F. Li, H. Want, C. Huang, J. Lin, G. Zhu, R. Hu and H. Feng, Hydrogen peroxide contributes to the manganese superoxide dismutase promotion of migration and invasion in glioma cells, *Free Radical Res.*, 2011, **45**, 1154.

106. Z. Liu, S. Li, Y. Cai, A. Wang, Q. He, C. Zheng, T. Zhao, X. Ding and X. Zhou, Manganese superoxide dismutase induces migration and invasion of tongue squamous cell carcinoma via H2O2-dependent Snail signaling, *Free Radicals Biol. Med.*, 2012, **53**, 44.

107. R. G. Hardy, C. Vicente-Duenas, I. González-Herrero, C. Anderson, T. Flores, S. Hughes, C. Tselepis, J. A. Ross and I. Sánchez-Gaciá, Snail family transcription factors are implicated in thyroid carcinogenesis, *Am. J. Pathol.*, 2007, **171**, 1037.

108. P. M. Chen, T. C. Wu, S. H. Shieh, Y. H. Wu, M. C. Li, G. T. Sheu, Y. W. Cheng, C. Y. Chen and H. Lee, MnSOD promotes tumor invasion via upregulation of FoxM1-MMP2 axis and related with poor survival and relapse in lung adenocarcinomas, *Mol. Cancer Res.*, 2013, **11**, 261.

109. S. Y. Pugh, J. L. DiGuiseppi and I. Fridovich, Induction of superoxide dismutases in Escherichia coli by manganese and iron, *J. Bacteriol.*, 1984, **160**, 137.

110. S. N. González, M. C. Apella, N. Romero, A. A. Pesce de Ruiz Holgado and G. Oliver, Superoxide dismutase activity in some strains of lactobacilli: induction by manganese, *Chem. Pharm. Bull.*, 1989, **37**, 3026.

111. J. Thongphasuk, L. W. Oberley and T. D. Oberley, Induction of superoxide dismutase and cytotoxicity by manganese in human breast cancer cells, *Arch. Biochem. Biophys.*, 1999, **365**, 317.

112. H. Oubrahim, E. R. Stadtman and P. B. Chock, Mitochondria play no roles in Mn(II)-induced apoptosis in HeLa cells, *Proc. Natl. Acad. Sci. U. S. A*, 2001, **98**, 9505.

113. M. Akashi, M. Hachiya, R. L. Paquette, Y. Osawa, S. Shimizu and G. Suzuki, Irradiation increases manganese superoxide dismutase mRNA levels in human fibroblasts. Possible mechanisms for its accumulation, *J. Biol. Chem.*, 1995, **270**, 15864.

114. J. Eastgate, J. Moreb, H. S. Nick, K. Suzuki, N. Taniguchi and J. R. Zucali, A role for manganese superoxide dismutase in radioprotection of hematopoietic stem cells by interleukin-1, *Blood*, 1993, **81**, 639.

115. G. H. Wong and D. V. Goeddel, Induction of manganous superoxide dismutase by tumor necrosis factor: possible protective mechanism, *Science*, 1988, **242**, 941.

116. G. H. Wong, J. H. Elwell, L. W. Oberley and D. V. Goeddel, Manganous superoxide dismutase is essential for cellular resistance to cytotoxicity of tumor necrosis factor, *Cell*, 1989, **58**, 923.

117. A. Masuda, D. L. Longo, Y. Kobayashi, E. Appella, J. J. Oppenheim and K. Matsushima, Induction of mitochondrial manganese superoxide dismutase by interleukin 1, *FASEB J.*, 1988, **15**, 3087.

118. G. A. Visner, W. C. Dougall, J. M. Wilson, I. A. Burr and H. S. Nick, Regulation of manganese superoxide dismutase by lipopolysaccharide, interleukin-1, and tumor necrosis factor. Role in the acute inflammatory response, *J. Biol. Chem.*, 1990, **265**, 2856.

119. W. C. Dougall and H. S. Nick, Manganese superoxide dismutase: a hepatic acute phase protein regulated by interleukin-6 and glucocorticoids, *Endocrinology*, 1991, **129**, 2376.

120. K. Asayama, R. L. Janco and I. M. Burr, Selective induction of manganous superoxide dismutase in human monocytes, *Am. J. Physiol.*, 1985, **249**, C393.

121. C. A. Harris, K. S. Derbin, B. Hunte-McDonough, M. R. Krauss, K. T. Chen, D. M. Smith and L. B. Epstein, Manganese superoxide dismutase is induced by IFN-gamma in multiple cell types. Synergistic induction by IFN-gamma and tumor necrosis factor or IL-1, *J. Immunol.*, 1991, **147**, 149.

122. S. E. Dryer, R. L. Dryer and A. P. Autor, Enhancement of mitochondrial, cyanide-resistant superoxide dismutase in the livers of rats treated with 2,4-dinitrophenol, *J. Biol. Chem.*, 1980, **255**, 1054.

123. J. Krall, A. C. Bagley, G. T. Mullenbach, R. A. Hallewell and R. E. Lynch, Superoxide mediates the toxicity of paraquat for cultured mammalian cells, *J. Biol. Chem.*, 1988, **263**, 1910.

124. N. Zhang, Characterization of the 5′ flanking region of the human MnSOD gene, *Biochem. Biophys. Res. Commun.*, 1996, **220**, 171.

125. X. S. Wan, M. N. Devalaraja and D. K. St Clair, Molecular structure and organization of the human manganese superoxide dismutase gene, *DNA Cell Biol.*, 1994, **13**, 1127.

126. I. N. Zelko, T. J. Mariani and R. J. Folz, Superoxide dismutase multigene family: a comparison of the CuZn-SOD (SOD1), Mn-SOD (SOD2), and EC-SOD (SOD3) gene structures, evolution, and expression, *Free Radicals Biol. Med.*, 2002, **3**, 337.

127. Y. Xu, S. Porntadavity and D. K. St Clair, Transcriptional regulation of the human manganese superoxide dismutase gene: the role of specificity protein 1 (Sp1) and activating protein-2 (AP-2), *Biochem. J.*, 2002, **362**, 401.

128. C. Zhu, Y. Huang, C. J. Weydert, L. W. Oberley and F. E. Domann, Constitutive activation of transcription factor AP-2 is associated with decreased MnSOD expression in transformed human lung fibroblasts, *Antioxid. Redox Signaling*, 2001, **3**, 387.

129. C. H. Zhu, Y. Huang, L. W. Oberley and F. E. Domann, A family of AP-2 proteins down-regulate manganese superoxide dismutase expression, *J. Biol. Chem.*, 2001, **276**, 14407.

130. C. H. Zhu, Y. Huang, M. T. Broman and F. E. Domann, Expression of AP-2 alpha in SV40 immortalized human lung fibroblasts is associated with a distinct pattern of cytosine methylation in the AP-2 alpha promoter, *Biochim. Biophys. Acta*, 2001, **1519**, 85.

131. S. K. Dhar, Y. Xu and D. K. St Clair, Nuclear factor kappaB- and specificity protein 1-dependent p53-mediated bi-directional regulation of the human manganese superoxide dismutase gene, *J. Biol. Chem.*, 2010, **285**, 9835.

132. S. K. Dhar, Y. Xu, Y. Chen and D. K. St Clair, Specificity protein 1-dependent p53-mediated suppression of human manganese superoxide dismutase gene expression, *J. Biol. Chem.*, 2006, **281**, 21698.

133. Y. Xu, K. K. Kiningham, M. N. Devalaraja, C. C. Yeh, H. Majima, E. J. Kasarskis and D. K. St Clair, An intronic NF-kappaB element is essential for induction of the human manganese superoxide dismutase gene by tumor necrosis factor-alpha and interleukin-1beta, *DNA Cell Biol.*, 1999, **18**, 709.

134. Z. Guo, G. H. Boekhoudt and J. M. Boss, Role of the intronic enhancer in tumor necrosis factor-mediated induction of manganous superoxide dismutase, *J. Biol. Chem.*, 2003, **278**, 23570.

135. R. J. Rogers, S. E. Chesrown, S. Kuo, J. M. Monnier and H. S. Nick, Cytokine-inducible enhancer with promoter activity in both the rat and human manganese-superoxide dismutase genes, *Biochem. J.*, 2000, **347**, 233.

136. S. Josson, Y. Xu, F. Fang, S. K. Dhar, D. K. St Clair and W. H. St Clair, RelB regulates manganese superoxide dismutase gene and resistance to ionizing radiation of prostate cancer cells, *Oncogene*, 2006, **25**, 1554.

137. S. K. Dhar, Y. Xu, T. Noel and D. K. St Clair, Chronic exposure to 12-O-tetradecanoylphorbol-13-acetate represses sod2 induction *in vivo*: the negative role of p50, *Carcinogenesis*, 2007, **28**, 2605.

138. S. K. Dhar, B. C. Lynn, C. Daosukho and D. K. St Clair, Identification of nucleophosmin as an NF-kappaB co-activator for the induction of the human SOD2 gene, *J. Biol. Chem.*, 2004, **279**, 28209.

139. Y. Xu, F. Fang, S. K. Dhar, W. H. St Clair, E. J. Kasarskis and D. K. St Clair, The role of a single-stranded nucleotide loop in transcriptional regulation of the human sod2 gene, *J. Biol. Chem.*, 2007, **282**, 15981.

140. Y. Xu, A. Krishna, X. S. Wan, H. Majima, C. C. Yeh, G. Ludewig, E. J. Kasarskis and D. K. St Clair, Mutations in the promoter reveal a cause for the reduced expression of the human manganese superoxide dismutase gene in cancer cells, *Oncogene*, 1999, **18**, 93.

141. S. K. Dhar and D. K. St Clair, Nucleophosmin blocks mitochondrial localization of p53 and apoptosis, *J. Biol. Chem.*, 2009, **284**, 16409.

142. K. K. Kiningham, C. Daosukho and D. K. St Clair, IkappaBalpha (inhibitory kappaBalpha) identified as labile repressor of MnSOD (manganese superoxide dismutase) expression, *Biochem. J.*, 2004, **384**, 543.

143. T. Sebastian and P. F. Johnson, RasV12-mediated down-regulation of CCAAT/enhancer binding protein beta in immortalized fibroblasts requires loss of p19Arf and facilitates bypass of oncogene-induced senescence, *Cancer Res.*, 2009, **69**, 2588.

144. H. P. Kim, J. H. Roe, P. B. Chock and M. B. Yim, Transcriptional activation of the human manganese superoxide dismutase gene mediated by tetradecanoylphorbol acetate, *J. Biol. Chem.*, 1999, **274**, 37455.
145. Y. W. Chung, H. K. Kim, I. Y. Kim, M. B. Yim and P. B. Chock, Dual function of protein kinase C (PKC) in 12-O-tetradecanoylphorbol-13-acetate (TPA)-induced manganese superoxide dismutase (MnSOD) expression: activation of CREB and FOXO3a by PKC-alpha phosphorylation and by PKC-mediated inactivation of Akt, respectively, *J. Biol. Chem.*, 2011, **286**, 29681.
146. G. J. Kops, T. B. Dansen, P. E. Polderman, I. Saarloos, K. W. Wirtz, P. J. Coffer, T. T. Huang, J. L. Bos, R. H. Medema and B. M. Burgering, Forkhead transcription factor FOXO3a protects quiescent cells from oxidative stress, *Nature*, 2002, **419**, 316.
147. K. C. Das, X. L. Guo and C. W. White, Protein kinase Cdelta-dependent induction of manganese superoxide dismutase gene expression by microtubule-active anticancer drugs, *J. Biol. Chem.*, 1998, **273**, 34639.
148. K. Maehara, K. Oh-Hashi and K. I. Isobe, Early growth-responsive-1-dependent manganese superoxide dismutase gene transcription mediated by platelet-derived growth factor, *FASEB J.*, 2001, **15**, 2025.
149. S. Negoro, K. Kunisada, Y. Fujio, M. Funamoto, M. I. Darville, D. L. Eizirik, T. Osugi, M. Izumi, Y. Oshima, Y. Nakaoka, H. Hirota, T. Kishimoto and K. Yamauchi-Takihara, Activation of signal transducer and activator of transcription 3 protects cardiomyocytes from hypoxia/reoxygenation-induced oxidative stress through the upregulation of manganese superoxide dismutase, *Circulation*, 2001, **104**, 979.
150. H. C. Yen, T. D. Oberley, S. Vichitbandha, Y. S. Ho and D. K. St Clair, The protective role of manganese superoxide dismutase against adriamycin-induced acute cardiac toxicity in transgenic mice, *J. Clin. Invest.*, 1996, **98**, 1253.
151. H. C. Yen, T. D. Oberley, C. G. Gairola, L. I. Szweda and D. K. St Clair, Manganese superoxide dismutase protects mitochondrial complex I against adriamycin-induced cardiomyopathy in transgenic mice, *Arch. Biochem. Biophys.*, 1999, **362**, 59.
152. K. Imada and W. J. Leonard, The Jak-STAT pathway, *Mol. Immunol.*, 2000, **37**, 1.
153. A. K. Holley, K. K. Kiningham, D. R. Spitz, D. P. Edwards, J. T. Jenkins and M. R. Moore, Progestin stimulation of manganese superoxide dismutase and invasive properties in T47D human breast cancer cells, *J. Steroid Biochem. Mol. Biol.*, 2009, **117**, 23.
154. H. Lin, S. Wu, N. Shi, S. Lian and W. Lin, Nrf2/HO-1 pathway activation by manganese is associated with reactive oxygen species and ubiquitin-proteasome pathway, not MAPKs signaling, *J. Appl. Toxicol.*, 2011, **31**, 690.
155. M. K. Kwak, K. Itoh, M. Yamamoto, T. R. Sutter and T. W. Kensler, Role of transcription factor Nrf2 in the induction of hepatic phase 2 and

antioxidative enzymes in vivo by the cancer chemoprotective agent, 3H-1, 2-dimethiole-3-thione, *Mol. Med.*, 2001, 7, 135.

156. P. Gong, C. S. Li, R. Hua, H. Zhao, Z. R. Tang, X. Mei, M. Y. Zhang and J. Cui, Mild hypothermia attenuates mitochondrial oxidative stress by protecting respiratory enzymes and upregulating MnSOD in a pig model of cardiac arrest, *PLoS One*, 2012, 7, e35313.

157. X. S. Huang, H. P. Chen, H. H. Yu, Y. F. Yan, Z. P. Liao and Q. R. Huang, Nrf2-dependent upregulation of antioxidative enzymes: a novel pathway for hypoxic preconditioning-mediated delayed cardioprotection, *Mol. Cell. Biochem.*, 2014, **385**, 33.

158. N. Hempel, P. M. Carrico and J. A. Melendez, Manganese superoxide dismutase (Sod2) and redox-control of signaling events that drive metastasis, *Anti-Cancer Agents Med. Chem.*, 2011, **11**, 191.

159. S. Miriyala, I. Spasojevic, A. Tovmasyan, D. Salvemini, Z. Vujaskovic, D. K. St Clair and I. Batinic-Haberle, Manganese superoxide dismutase, MnSOD and its mimics, *Biochim. Biophys. Acta*, 2012, **1822**, 794.

160. S. K. Dhar, J. Tangpong, L. Chaiswing, T. D. Oberley and D. K. St Clair, Manganese superoxide dismutase is a p53-regulated gene that switches cancers between early and advanced stages, *Cancer Res.*, 2011, **71**, 6684.

161. M. Fojta, T. Kubicárová, B. Vojtěsek and E. Palecek, Effect of p53 protein redox states on binding to supercoiled and linear DNA, *J. Biol. Chem.*, 1999, **274**, 25749.

162. M. Fojta, H. Pivonkova, M. Brazdova, K. Nemcova, J. Palecek and B. Vojtesek, Investigations of the supercoil-selective DNA binding of wild type p53 suggest a novel mechanism for controlling p53 function, *Eur. J. Biochem.*, 2004, **271**, 3865.

163. S. K. Dhar and D. K. St Clair, Manganese superoxide dismutase regulation and cancer, *Free Radicals Biol. Med.*, 2012, **52**, 2209.

164. T. Fukai and M. Ushio-Fukai, Superoxide dismutases: role in redox signaling, vascular function, and diseases, *Antioxid. Redox Signaling*, 2011, **15**, 1583.

165. A. K. Holley, V. Bakthavatchalu, J. M. Velez-Roman and D. K. St Clair, Manganese superoxide dismutase: guardian of the powerhouse, *Int. J. Mol. Sci.*, 2011, **12**, 7114.

166. S. W. Ballinger, C. Patterson, C. A. Knight-Lozano, D. L. Burow, C. A. Conklin, Z. Hu, J. Reuf, C. Horaist, R. Lebovitz, G. C. Hunter, K. McIntyre and M. S. Runge, Mitochondrial integrity and function in atherogenesis, *Circulation*, 2002, **106**, 544.

167. M. P. Cole, L. Chaiswing, T. D. Oberley, S. E. Edelmann, M. T. Piascik, S. M. Lin, K. K. Kiningham and D. K. St Clair, The protective roles of nitric oxide and superoxide dismutase in adriamycin-induced cardiotoxicity, *Cardiovasc. Res.*, 2006, **69**, 186.

168. S. W. Flanagan, R. D. Anderson, M. A. Ross and L. W. Oberley, Overexpression of manganese superoxide dismutase attenuates neuronal death in human cells expressing mutant (G37R) Cu/Zn-superoxide dismutase, *J. Neurochem.*, 2002, **81**, 170.

169. J. N. Keller, M. S. Kindy, F. W. Holtsberg, D. K. St Clair, H. C. Yen, A. Germeyer, S. M. Steiner, A. J. Bruce-Keller, J. B. Hutchins and M. P. Mattson, Mitochondrial manganese superoxide dismutase prevents neural apoptosis and reduces ischemic brain injury: suppression of peroxynitrite production, lipid peroxidation, and mitochondrial dysfunction, *J. Neurosci.*, 1998, **18**, 687.

170. M. Gonzalez-Zulueta, L. M. Ensz, G. Mukhina, R. M. Lebovitz, R. M. Zwacka, J. F. Engelhardt, L. W. Oberley, V. L. Dawson and T. M. Dawson, Manganese superoxide dismutase protects nNOS neurons from NMDA and nitric oxide-mediated neurotoxicity, *J. Neurosci.*, 1998, **18**, 2040.

171. P. Sompol, Y. Xu, W. Ittarat, C. Daosukho and D. K. St Clair, NF-κB-associated MnSOD induction protects against beta-amyloid-induced neuronal apoptosis, *J. Mol. Neurosci.*, 2006, **29**, 279.

172. O. Baud, R. F. Haynes, H. Wang, R. D. Folkerth, J. Li, J. J. Volpe and P. A. Rosenberg, Developmental up-regulation of MnSOD in rat oligodendrocytes confers protection against oxidative injury, *Eur. J. Neurosci.*, 2004, **20**, 29.

173. X. Shan, L. Chi, Y. Ke, C. Luo, S. Qian, D. Gozal and R. Liu, Manganese superoxide dismutase protects mouse cortical neurons from chronic intermittent hypoxia-mediated oxidative damage, *Neurobiol. Dis.*, 2007, **28**, 205.

174. W. F. Maragos, R. Jakel, D. Chesnut, C. B. Pocernich, D. A. Butterfield, D. K. St Clair and W. A. Cass, Methamphetamine toxicity is attenuated in mice that overexpress human manganese superoxide dismutase, *Brain Res.*, 2000, **878**, 218.

175. P. Klivenyi, D. K. St Clair, M. Wermer, H. C. Yen, T. Oberley, L. Yang and M. Flint Beal, Manganese superoxide dismutase overexpression attenuates MPTP toxicity, *Neurobiol. Dis.*, 1998, **5**, 253.

176. Y. Xiong, F. S. Shie, J. Zhang, C. P. Lee and Y. S. Ho, Prevention of mitochondrial dysfunction in post-traumatic mouse brain by superoxide dismutase, *J. Neurochem.*, 2005, **95**, 732.

177. D. L. Marcus, J. A. Strafaci and M. L. Freedman, Differential neuronal expression of manganese superoxide dismutase in Alzheimer's disease, *Med. Sci. Monit.*, 2006, **12**, BR8.

178. F. Li, N. Y. Calingasan, F. Yu, W. M. Mauck, M. Toidze, C. G. Almeida, R. H. Takahashi, G. A. Carlson, M. Flint Beal, M. T. Lin and G. K. Gouras, Increased plaque burden in brains of APP mutant MnSOD heterozygous knockout mice, *J. Neurochem.*, 2004, **89**, 1308.

179. M. Dumont, E. Wille, C. Stack, N. Y. Calingasan, M. F. Beal and M. T. Lin, Reduction of oxidative stress, amyloid deposition, and memory deficit by manganese superoxide dismutase overexpression in a transgenic mouse model of Alzheimer's disease, *FASEB J.*, 2009, **23**, 2459.

180. F. L. Muller, Y. Liu, A. Jernigan, D. Borchelt, A. Richardson and H. Van Remmen, MnSOD deficiency has a differential effect on disease

progression in two different ALS mutant mouse models, *Muscle Nerve*, 2008, **38**, 1173.

181. J. M. Flynn and S. Melov, SOD2 in mitochondrial dysfunction and neurodegeneration, *Free Radicals Biol. Med.*, 2013, **62**, 4.

182. S. T. Keir, M. W. Dewhirst, J. P. Kirkpatrick, D. D. Bigner and I. Batinic-Haberle, Cellular redox modulator, ortho Mn(III) meso-tetrakis(N-n-hexylpyridinium-2-yl)porphyrin, MnTnHex-2-PyP(5+) in the treatment of brain tumors, *Anti-Cancer Agents Med. Chem.*, 2011, **11**, 202.

183. L. P. Liang, S. Waldbaum, S. Rowley, T. T. Huang, B. J. Day and M. Patel, Mitochondrial oxidative stress and epilepsy in SOD2 deficient mice: attenuation by a lipophilic metalloporphyrin, *Neurobiol. Dis.*, 2012, **45**, 1068.

184. H. F. Huang, F. Guo, Y. Z. Cao, W. Shi and Q. Xia, Neuroprotection by manganese superoxide dismutase (MnSOD) mimics: antioxidant effect and oxidative stress regulation in acute experimental stroke, *CNS Neurosci. Ther.*, 2012, **18**, 811.

185. D. St Clair and E. Kasarskis, Genetic polymorphism of the human manganese superoxide dismutase: what difference does it make?, *Pharmacogenetics*, 2003, **13**, 129.

186. J. S. Rosenblum, N. B. Gilula and R. A. Lerner, On signal sequence polymorphisms and diseases of distribution, *Proc. Natl. Acad. Sci. U. S. A*, 1996, **93**, 4471.

187. A. Sutton, H. Khoury, C. Prip-Buus, C. Cepanec, D. Pessayre and F. Degoul, The Ala16Val genetic dimorphism modulates the import of human manganese superoxide dismutase into rat liver mitochondria, *Pharmacogenetics*, 2003, **13**, 145.

188. G. Liu, W. Zhou, S. Park, L. I. Wang, D. P. Miller, J. C. Wain, T. J. Lynch, L. Su and D. C. Christiani, The SOD2 Val/Val genotype enhances the risk of nonsmall cell lung carcinoma by p53 and XRCC1 polymorphisms, *Cancer*, 2004, **101**, 2802.

189. D. G. Cox, R. M. Tamimi and D. J. Hunter, Gene x Gene interaction between MnSOD and GPX-1 and breast cancer risk: a nested case-control study, *BMC Cancer*, 2006, **31**, 217.

190. P. Wheatley-Price, K. Asomaning, A. Reid, R. Zhai, L. Su, W. Zhou, A. Zhu, D. P. Ryan, D. C. Christiani and G. Liu, Myeloperoxidase and superoxide dismutase polymorphisms are associated with an increased risk of developing pancreatic adenocarcinoma, *Cancer*, 2008, **112**, 1037.

191. C. G. Bica, L. L. de Moura da Silva, N. V. Toscani, I. B. da Cruz, G. Sá, M. S. Graudenz and C. G. Zettler, MnSOD gene polymorphism association with steroid-dependent cancer, *Pathol. Oncol. Res.*, 2009, **15**, 19.

192. K. Woodson, J. A. Tangrea, T. A. Lehman, R. Modali, K. M. Taylor, K. Snyder, P. R. Taylor, J. Virtamo and D. Albanes, Manganese superoxide dismutase (MnSOD) polymorphism, alpha-tocopherol supplementation and prostate cancer risk in the alpha-tocopherol, beta-carotene cancer prevention study (Finland), *Cancer Causes Control*, 2003, **14**, 513.

193. S. H. Olson, M. D. Carlson, H. Ostrer, S. Harlap, A. Stone, M. Winters and C. B. Ambrosone, Genetic variants in SOD2, MPO, and NQO1, and risk of ovarian cancer, *Gynecol. Oncol.*, 2004, **93**, 615.
194. H. Fujimoto, J. Taguchi, Y. Imai, S. Ayabe, H. Hashimoto, H. Kobayashi, K. Ogasawara, T. Aizawa, M. Yamakado, R. Nagai and M. Ohno, Manganese superoxide dismutase polymorphism affects the oxidized low-density lipoprotein-induced apoptosis of macrophages and coronary artery disease, *Eur. Heart J.*, 2008, **29**, 1267.
195. H. Fujimoto, H. Kobayashi, K. Ogasawara, M. Yamakado and M. Ohno, Association of the manganese superoxide dismutase polymorphism with vasospastic angina pectoris, *J. Cardiol.*, 2010, **55**, 205.
196. D. Hernandez-Saavedra and J. M. McCord, Paradoxical effects of thiol reagents on Jurkat cells and a new thiol-sensitive mutant form of human mitochondrial superoxide dismutase, *Cancer Res.*, 2003, **63**, 159.
197. D. Hernandez-Saavedra and J. M. McCord, Association of a new intronic polymorphism of the SOD2 gene (G1677T) with cancer, *Cell Biochem. Funct.*, 2009, **27**, 223.
198. H. J. Zhang, T. Yan, T. D. Oberley and L. W. Oberley, Comparison of effects of two polymorphic variants of manganese superoxide dismutase on human breast MCF-7 cancer cell phenotype, *Cancer Res.*, 1999, **59**, 6276.
199. G. E. Borgstahl, H. E. Parge, M. J. Hickey, W. F. Beyer Jr, R. A. Hallewell and J. A. Tainer, The structure of human mitochondrial manganese superoxide dismutase reveals a novel tetrameric interface of two 4-helix bundles, *Cell*, 1992, **71**, 107.
200. T. Tefik, C. Kucukgergin, O. Sanli, T. Oktar, S. Seckin and C. Ozsoy, Manganese superoxide dismutase Ile58Thr, catalase C-262T and myeloperoxidase G-463A gene polymorphisms in patients with prostate cancer: relation to advanced and metastatic disease, *BJU Int.*, 2013, **112**, E406.
201. L. I. Holla, K. Kankova and A. Vasku, Functional polymorphism in the manganese superoxide dismutase (MnSOD) gene in patients with asthma, *Clin. Biochem.*, 2006, **39**, 299.
202. J. C. Ho, J. C. Mak, S. P. Ho, M. S. Ip, K. W. Tsang, W. K. Lam and M. Chan-Yeung, Manganese superoxide dismutase and catalase genetic polymorphisms, activity levels, and lung cancer risk in Chinese in Hong Kong, *J. Thorac. Oncol.*, 2006, **1**, 648.
203. M. Singh, A. J. Khan, P. P. Shah, R. Shukla, V. K. Khanna and D. Parmar, Polymorphism in environment responsive genes and association with Parkinson disease, *Mol. Cell. Biochem.*, 2008, **312**, 131.

CHAPTER 4

Olfactory Transport of Manganese: Implications for Neurotoxicity

DAVID C. DORMAN* AND MELANIE L. FOSTER

College of Veterinary Medicine, North Carolina State University, Raleigh, NC, USA
*Email: david_dorman@ncsu.edu

4.1 Introduction

An association between manganese inhalation and neurotoxicity was first recognized in the mid-19th century in workers at an ore-grinding plant where "black oxide of manganese" was processed.[1] Most epidemiologic research on manganese conducted during the late-20th century focused on occupational exposures associated with inhalation. The results of these studies clearly established a causal association between chronic manganese inhalation and a clinical syndrome known as manganism that mimics Parkinson's disease. Hallmarks of manganism include behavioral changes, extrapyramidal motor dysfunction, and neurochemical and neuropathological changes in the basal ganglia and globus pallidus.[2,3] Manganese-induced cognitive deficitsx have also been reported in workers with exposure to manganese-based welding fumes.[4,5]

Manganese neurotoxicity following inhalation requires prolonged exposure to high airborne concentrations of manganese that result in excessive accumulation of this metal in the brain.[2] There are three important routes by which inhaled manganese can gain access to the brain: (a) direct anterograde

Issues in Toxicology No. 22
Manganese in Health and Disease
Edited by Lucio G. Costa and Michael Aschner
© The Royal Society of Chemistry 2015
Published by the Royal Society of Chemistry, www.rsc.org

axonal delivery *via* olfactory or trigeminal nerve endings located in the nasal cavity; (b) transport across the pulmonary epithelial lining and its subsequent systemic distribution from blood; and/or (c) mucociliary elevator clearance from the lung and subsequent delivery and absorption from the gastro-intestinal tract. This chapter will provide an update of our current under-standing of the first of these routes, namely direct "nose-to-brain" transport (also referred to as olfactory transport) of manganese. No attempt has been made to provide an exhaustive review of each of the issues addressed, but wherever possible references to review articles are provided. Topics addressed in this chapter include a brief description of the olfactory system with a special emphasis on the anatomical basis for olfactory transport; scientific evidence in support of olfactory transport; the use of manganese as a contrast agent for magnetic resonance imaging of the olfactory system; and the toxicological significance of this route of delivery.

4.2 Anatomical Features of the Olfactory System

The past few decades of research have yielded an important discovery that certain chemicals can undergo direct "nose-to-brain" or olfactory transport. This route of direct delivery bypasses the blood–brain-barrier and takes advantage of the unique relationship between the sensory neurons found in the nasal cavity and their projections to the brain.[6,7] The olfactory system begins with a collection of specialized neurons involved in the sense of smell (Figure 4.1).

Olfactory neurons are located in the olfactory epithelium, a patch of pseudostratified cells that line the dorsal or dorso-posterior aspect of the nasal cavity. The olfactory epithelium is composed of olfactory receptor neurons, glia-like supporting sustentacular cells, and basal cells that function as stem cells.[8] Unlike most post-embryonic neurons, olfactory receptor neurons have a lifetime of two to four weeks and are continually replaced by replication from the basal cells.[9,10] The olfactory receptor neuron stem cell (most likely a subpopulation of globose basal cells) may also give rise to the glia-like sustentacular cells and the olfactory ensheathing cells that envelope the olfactory nerve.[11,12]

Olfactory sensory cells are bipolar neurons, with a short apical dendrite and a relatively long, thin (~ 0.2 µm diameter) unmyelinated axon that passes into the underlying lamina propria to form prominent olfactory nerve bundles. The dendritic portions of these neurons extend above the epithelial surface and terminate into a bulbous olfactory knob from which protrude on average 10–15 immotile cilia.[13] These cilia provide an extensive surface area for reception of odorants and are a possible target for other xenobiotics. Bundles of sensory axons, supported by olfactory ensheathing cells, pass through the lamina propria and the cribriform plate and terminate in the olfactory bulb. In the olfactory bulb, the receptor cell axons synapse with mitral cells, the most prominent cell type of the olfactory bulb. The olfactory bulb is the terminal nucleus of the olfactory nerve. Receptor cell axons,

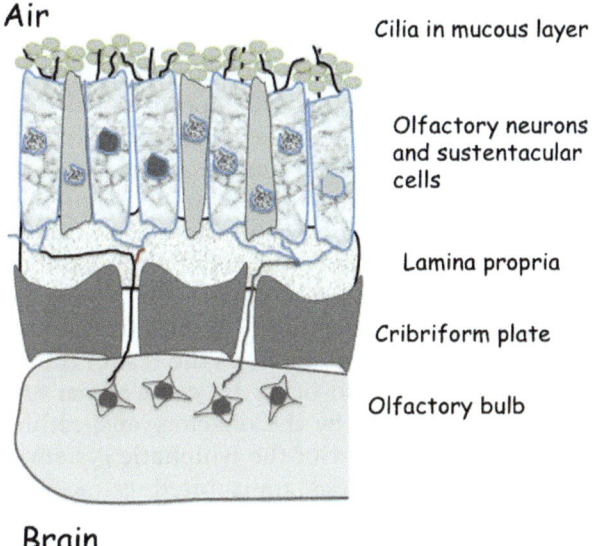

Figure 4.1 Simplified schematic diagram of the olfactory system showing neuronal projections between sensory (olfactory) neurons and the brain's olfactory bulb. Manganese and certain other xenobiotics can travel along these neuronal projections from the olfactory neuron to the olfactory bulb.

mitral cell dendrites, and smaller, tufted cells form brush-like terminals known as olfactory glomeruli. About 25 000 axons enter each glomerulus and synapse with approximately 25 mitral cells that send signals further to the brain.[14] The axons of the tufted and mitral cells largely make up the olfactory tract to the cerebrum. Unlike the other sensory systems, olfactory impulses reach the cerebral cortex without relay through the thalamus. Similar to other sensory systems, the olfactory system has an area of associated neocortex (olfactory cortex).

As one would expect, inhaled manganese is transported in the air from the external environment to the nasal cavity where a portion of the inspired air reaches the olfactory epithelial surface. The proportion of air that reaches the olfactory epithelium is primarily determined by nasal anatomy and the inspiratory airflow rate.[15,16] Prominent species differences in nasal anatomy, airflow, and even the size of the olfactory bulb exist among mammals, for example rodents are obligate nose breathers and have large patches of olfactory epithelium and robust olfactory bulb structures.

4.3 Scientific Evidence in Support of Olfactory Transport of Manganese

Regional delivery of manganese from the olfactory mucosa to the brain could occur *via* blood-borne delivery or direct olfactory transport. Absorption of certain xenobiotics (*e.g.* diazepam, progesterone) from the nasal epithelium with localized delivery to the nose blood vasculature is a well-known mechanism of reaching the central nervous system. Mucosal absorption with blood delivery takes advantage of the rich arterial and venous plexus found in the nasal cavity. For rats and mice, approximately 1% of the cardiac output is delivered to the nasal passages[17] with some currently unknown fraction thereof reaching the olfactory mucosa. Substances that are transported *via* the vasculature directly to the brain demonstrate an initially higher ratio of carotid arterial blood concentration to brain concentration through a process of counter-current transfer.[18] Although the concentration in the olfactory bulb may initially be higher than the rest of the brain following uptake by the nasal vasculature, it rapidly declines within less than an hour and becomes equivalent to levels in the rest of the brain. Olfactory transport of manganese may also occur by extracellular transport in the perineurial spaces encompassed by the olfactory ensheathing cells or the perivascular spaces which are part of the lymphatic system and important for drainage of CSF.[18] Xenobiotics transported extracellularly following intranasal administration (*e.g.* insulin, cocaine) are present in the olfactory bulb within less than 30 minutes.[18] A working theory is that manganese is transported intracellularly within the olfactory sensory neurons. Neuronal uptake may occur by passive diffusion, receptor-mediated uptake or adsorptive endocytosis, followed by axonal transport, which takes several hours to days for manganese to reach the olfactory bulb and other regions of the CNS.[19]

Initial studies examining direct nose-to-brain transport of manganese relied on intranasal instillation of an aqueous solution of radiolabeled manganese chloride ($^{54}MnCl_2$) into one side of the nasal cavity of animals whereas the opposite portion of the nasal cavity would receive the aqueous vehicle control. One of the first studies was performed by Tjälve *et al.*[20] and used freshwater pike as the animal model. In these studies, gamma spectrometry and autoradiography were used to determine the fate of the intranasally instilled manganese. These experiments demonstrated the presence of ^{54}Mn in the olfactory bulb on the same (*i.e.* ipsilateral) side as the instilled nostril.[21] Qualitatively similar intranasal instillation studies performed in the same laboratory that used rats were able to replicate these initial findings.[22] Moreover, rats given an intraperitoneal injection of $^{54}MnCl_2$ failed to accumulate elevated olfactory bulb manganese concentrations when compared with levels seen in the rest of the brain.[22] Brenneman and coworkers[23] extended this general approach to an inhalation route of exposure. Brenneman used a rat animal model in which one nostril was physically occluded using a plastic plug and tissue glue. Pilot studies with inhaled hydrogen sulfide confirmed that the nasal plugging procedure restricted airflow to the occluded nostril. Inhalation exposures to $^{54}MnCl_2$ resulted in olfactory transport of manganese to the side of the rat brain ipsilateral to the patent nostril. In these studies, direct delivery along the olfactory route accounted for nearly all of the ^{54}Mn found in the olfactory bulb and tract of the rat brain following acute manganese inhalation.

Elder and coworkers[24] used this nasal occlusion model to evaluate the distribution of inhaled manganese oxide nanoparticle aerosols. Elder used short-term exposures to demonstrate that manganese accumulated only in the olfactory bulb on the side of the head with a patent nostril.

Transport of manganese is not restricted to the olfactory system, for example translocation of manganese along the optic nerve has also been observed following intravitreal manganese injection.[25,26] Likewise, Lewis and coworkers[27] reported that the rat trigeminal nerve may also absorb and deliver manganese from the nasal cavity to the brain. The available data suggests that manganese ions that enter nerve cells are transported along the axons in an anterograde direction and also trans-synaptically, which allows for investigation of neural pathways and interneuronal connections in the olfactory system and elsewhere.[28]

Since manganese is paramagnetic, it increases the signal intensity seen with T1-weighted magnetic resonance imaging (MRI). Changes in the T1-weighted image correlate with manganese tissue concentration[29] and allow for visual inspection of the body for manganese accumulation. Cross *et al.* used unilateral intranasal $MnCl_2$ instillation in rats and brain MRI to show significant transport of manganese in the olfactory bulb, lateral olfactory tract, and olfactory tubercle ipsilateral to site of manganese administration.[30] Dorman *et al.*[29] also used brain MRI and measurement of tissue manganese concentrations to demonstrate presumed olfactory transport of inhaled manganese sulfate in rhesus monkeys. Visual evaluation of the manganese-enhanced neuronal tracts illustrated changes in the olfactory bulb but failed to demonstrate evidence for direct translocation of manganese from the olfactory bulb to the globus pallidus.[29] Sen and colleagues[31] used brain MRI to evaluate region-specific manganese accumulation in seven welders without obvious neurological deficits. When compared with age- and gender-matched controls, the manganese-exposed welders had significantly higher T1 relaxation rates in the olfactory bulb as well as evidence of manganese accumulation in the frontal white matter, globus pallidus, and putamen. Although the results of the Dorman and Sen studies are consistent with animal studies evaluating olfactory transport of manganese, these studies do not provide direct evidence that olfactory transport of manganese occurs in humans and other primates. Despite this limitation, evidence in favor of olfactory transport of metals in people is available for thallium, a metal that, like manganese, undergoes olfactory transport in rodents.[32] Shiga *et al.*[33] assessed the transport of nasally-administered thallium (as ^{201}Tl) to the human brain in healthy volunteers using a combination of single photon emission computed tomography (SPECT), X-ray computed tomography (CT), and MRI. These studies demonstrated appreciable movement of thallium to the human olfactory bulb with transport kinetic properties that were most consistent with delivery *via* the olfactory nerve. These observations suggest that the mechanisms involved in olfactory transport of metals are conserved across different species.

4.3.1   Manganese Transport Kinetics

Olfactory transport of manganese demonstrates dose-dependent kinetics.[19,34] Tjälve *et al.*[20] reported that manganese was transported along the primary olfactory neurons into the fish (freshwater pike) brain at a maximal velocity of approximately 3 mm h^{-1} at an ambient water temperature of 10 °C. Other investigators have reported that manganese is transported at a rate of 1 to 6 mm h^{-1} in axons so that typically 24 to 72 h is usually needed for manganese to be transported in sufficient quantities to target regions within the central nervous system.[26,28,35] Smaller ($\leq$1.5 μm) manganese particles more efficiently undergo olfactory transport when compared with larger ($\sim$10 μm) aerosol particles.[24,36] Olfactory transport is also favored when soluble manganese particles are used.[19,37] Leavens and coworkers[19] developed a pharmacokinetic model to describe the olfactory transport and blood delivery of manganese in rats following acute manganese inhalation. Tissue compartments in the model included the olfactory mucosa, olfactory bulb, olfactory tract and tubercle, and striatum. Model simulations demonstrated that direct olfactory transport provided the majority of manganese delivered to the olfactory system whereas only a small fraction ($<$3%) is predicted to be delivered to the rat striatum.

Several recent studies have investigated the cellular mechanisms involved with the olfactory transport of manganese. Thompson *et al.*[38] evaluated the role of the divalent metal transporter-1 (DMT1) in the olfactory transport of manganese. Thompson studied manganese pharmacokinetics in Belgrade rats, an animal model with significant defects in both iron and manganese metabolism due to a glycine-to-arginine substitution in their DMT1 gene product. Thompson observed decreased absorption of intranasally instilled ^{54}Mn in Belgrade rats when compared with control animals. Several investigators have also evaluated the olfactory transport of manganese in iron-sufficient and iron-deficient rodents with mixed results.[39–41] Iron deficiency did not affect the olfactory transport of manganese after the inhalation of manganese-containing welding fumes.[41] Other studies have shown an association between olfactory transport of manganese and iron deficiency[39] or hyperferremia (HFE) status.[40]

Thompson *et al.*[38] also used immunohistochemical approaches to show that DMT1 was localized to both the lumen microvilli and end feet of the sustentacular cells of the olfactory epithelium. Follow-up experiments performed in the same laboratory further evaluated whether sustentacular cells play a role in the olfactory transport of manganese. Thompson and co-workers[42] pre-exposed rats to methyl bromide, a known olfactory epithelial toxicant. They then intranasally instilled ^{54}MnCl$_2$ at different times after methyl bromide exposure and assessed whether olfactory transport of manganese was affected. Their study also showed that early olfactory epithelial injury (*e.g.* 2 to 4 days after methyl bromide exposure) was associated with decreased blood ^{54}Mn concentration, suggesting that impaired systemic (blood) absorption of ^{54}Mn into the blood occurred. Systemic absorption returned to normal by day 7, a time that coincided with re-establishment of sustentacular cell foot processes with the epithelial basal lamina. They also found that olfactory transport of manganese was initially impaired when the olfactory epithelium was injured, but olfactory transport became re-established once neuronal regeneration and olfactory bulbar reinnervation occurred, demonstrating the need for intact axonal projections from the nasal cavity.

4.4 Manganese-Enhanced Magnetic Resonance Imaging (MEMRI) of the Olfactory System

As mentioned earlier, accumulation of the paramagnetic manganese ion in tissues can improve MRI contrast between anatomical structures.[43] The past 10 years or so have seen scientists apply manganese-enhanced magnetic resonance imaging (MEMRI) to noninvasively identify functional neural connections in animals.[44] Micro-MRI methods have increasingly become an alternative to more traditional histological and microscopic neuroanatomical studies. Pautler and colleagues[28] were the first to detect enhancement in the mammalian olfactory tract after intranasal manganese administration. Gutman and coworkers[45] used three-dimensional MEMRI before and after bilateral intranasal administration of manganese in mice. The resulting imaging studies clearly show the existence of neuroanatomical pathways between the olfactory bulb, amygdala, piriform cortex, caudate putamen, and olfactory cortex. Other investigators have reported similar MEMRI findings.[30,46] These studies have also identified methodological issues associated with intranasal administration of manganese in rodents. Following intranasal manganese administration, MRI evidence of suspected manganese accumulation has been observed in the pituitary gland and other brain regions outside the expected neuronal pathway associated with the olfactory system.[28,30] These results raise the question of whether following intranasal injection, manganese can also gain access to the systemic circulation *via* the nasal blood vasculature.

MEMRI has also been used to image brain structures that are activated in animals following odorant stimulation. For example, Lehallier *et al.*[47] showed that rats presented with either an aversive odor (male fox feces) or an appetent one (chocolate flavored cereals) developed different patterns of manganese enhancement in the rat olfactory cortex. Kivity and colleagues[48] also used MEMRI to characterize dysosmia associated with experimental neuropsychiatric lupus in mice.

4.5 Toxicological Significance of Olfactory Transport of Manganese

The toxicological significance of olfactory transport of manganese remains incompletely understood. Although olfactory transport rapidly delivers manganese to brain structures in the olfactory pathway, it appears to be relatively slow (and perhaps inefficient) in delivering inhaled manganese to the rat striatum or primate globus pallidus and other more distant brain structures that are typically associated with manganism.[19,29,45]

4.5.1 Olfactory System Pathology

A number of chemicals are known to induce olfactory epithelial damage in laboratory animals. Dorman *et al.*[49] characterized the subchronic nasal toxicity of inhaled manganese sulfate and manganese phosphate in rats. They assessed nasal pathology, brain glial fibrillary acidic protein (GFAP) levels, and brain manganese concentrations. They found elevated olfactory bulb, striatum, and cerebellum manganese concentrations in manganese-exposed rats. Manganese exposure, however, did not affect olfactory bulb, cerebellar, or striatal GFAP concentrations. Exposure to manganese sulfate was also associated with reversible inflammation within the nasal respiratory epithelium, while the olfactory epithelium was unaffected by manganese inhalation. This laboratory also showed that subchronic manganese sulfate inhalation was not associated with olfactory epithelial injury in rhesus monkeys.[50]

Villalobos *et al.*[51] showed that the mouse olfactory bulb develops neuron degeneration and myelin sheath disorganization following high dose intraperitoneal manganese injection. Manganese-induced ultrastructural changes have been reported in rodents following manganese inhalation. Colin-Barenque and colleagues[52] reported that manganese inhalation was associated with ultrastructural changes including disruption of organelle membranes, cytoplasm vacuolation, increased lipofuscin deposits and cell death in the olfactory bulb granule cell.

4.5.2 Biochemical Effects

Henriksson and Tjälve[53] found that intranasal instillation of manganese results in alterations in olfactory bulb expression of GFAP and S-100b in rats. These proteins are known markers of damage to astrocytes, an important support cell found within the central nervous system. As mentioned earlier, Dorman *et al.*[49] however did not observe similar changes in rat olfactory bulb GFAP concentrations following subchronic inhalation to manganese, even though olfactory bulb manganese concentrations were increased approximately 3.5-fold *versus* air-exposed controls. Moberly and coworkers[54] showed that intranasal instillation of manganese caused a dose-dependent reduction in odorant-evoked glutamatergic neurotransmitter release in mice as a result of centrally, not peripherally mediated effects. Monkeys exposed subchronically to inhaled manganese had decreased olfactory cortical glutamate transporter-1 (GLT-1) and glutamate/aspartate transporter (GLAST) protein levels.[55]

4.5.3 Olfactory Function

Disorders of olfactory function in people are associated with diminished quality of life, depression, and other consequences.[56] Epidemiological studies show that loss in odor sensitivity is common in both general and clinical populations and is a common finding in people with Alzheimer's disease, Parkinson's disease, and certain other neurodegenerative diseases.[57] The known association between manganese exposure, olfactory transport, and development of a Parkinsonian syndrome has sparked an interest in whether manganese-induced changes in the sense of smell occur in people.[58] Several studies have shown an association between manganese exposure and olfactory deficits (increased olfactory threshold, reduced olfactory discrimination and scent identification) in adults[59] and children.[60] Studies performed in rodents provide experimental evidence in support of an association between manganese exposure and olfactory deficits. Lehallier and coworkers[61] observed reduced olfactory perception in rats following intranasal administration of manganese. These functional changes occurred in the presence of increased MRI contrast enhancement in the olfactory cortex.

4.5.4 Species Differences

The relevance of nasal uptake studies conducted in rodents to human manganese inhalation exposure is incompletely understood. As mentioned earlier, significant interspecies differences in nasal and brain anatomy and physiology exist.[62]

In the rat, the olfactory bulb accounts for a relatively large portion of the CNS, and the nasal olfactory mucosa covers approximately 50% of the total nasal epithelium. These structures are proportionately smaller in primates, for example the olfactory mucosa covers approximately 5% of the total nasal epithelium in humans. In addition, total airflow to the olfactory mucosa is slightly lower in humans than in rats. These anatomical and physiological differences argue that olfactory transport may be less important in humans when compared to the rat.

4.6 Conclusions

A variety of experimental animal studies have confirmed the importance of the "nose–brain" transport pathway as a delivery route for manganese to the central nervous system. This pathway appears to be highly conserved and has been documented in fish and mammalian animal models. Pharmacokinetic studies have shown this pathway is highly efficient; indeed, in rodents the majority of inhaled manganese that reaches the olfactory bulb results from this transport mechanism. The past decade has seen an explosion of our understanding of this pathway and the use of MEMRI and other imaging modalities has elucidated the anatomical substrates involved in olfactory transport. MEMRI has also been used by basic neuroscientists to interrogate olfactory function in animals following scent delivery. Additional applications of MEMRI to neurosciences are also emerging. The toxicological role of olfactory transport and manganese-induced neurotoxicity remains incompletely understood. Further research is needed to explain the cellular and molecular basis for manganese transport, mechanisms of toxicity, and the prevalence of manganese-induced effects on olfaction especially in children, the elderly, and other potentially susceptible subpopulations.

References

1. J. Couper, On the effects of black oxide of manganese when inhaled into the lungs, *Br. Ann. Med. Pharm. Vital. Stat. Gen. Sci.*, 1837, **1**, 41.
2. H. A. Roels, R. M. Bowler, Y. Kim, B. Claus Henn, D. Mergler, P. Hoet, V. V. Gocheva, D. C. Bellinger, R. O. Wright, M. G. Harris, Y. Chang, M. F. Bouchard, H. Riojas-Rodriguez, J. A. Menezes-Filho and M. M. Téllez-Rojo, Manganese exposure and cognitive deficits: a growing concern for manganese neurotoxicity, *Neurotoxicology*, 2012, **33**, 872.
3. J. A. Roth, Homeostatic and toxic mechanisms regulating manganese uptake, retention, and elimination, *Biol. Res.*, 2006, **39**, 45.
4. R. M. Bowler, S. Gysens, E. Diamond, S. Nakagawa, M. Drezgic and H. A. Roels, Manganese exposure: neuropsychological and neurological symptoms and effects in welders, *Neurotoxicology*, 2006, **27**, 315.
5. D. G. Ellingsen, R. Konstantinov, R. Bast-Pettersen, L. Merkurjeva, M. Chashchin, Y. Thomassen and V. Chashchin, A neurobehavioral study of current and former welders exposed to manganese, *Neurotoxicology*, 2008, **29**, 48.
6. F. W. Merkus and M. P. van den Berg, Can nasal drug delivery bypass the blood-brain barrier?: questioning the direct transport theory, *Drugs RD*, 2007, **8**, 133.
7. H. Wu, K. Hu and X. Jiang, From nose to brain: understanding transport capacity and transport rate of drugs, *Expert Opin. Drug Delivery*, 2008, **5**, 1159.
8. J. R. Harkema, S. A. Carey and J. G. Wagner, The nose revisited: a brief review of the comparative structure, function, and toxicologic pathology of the nasal epithelium, *Toxicol. Pathol.*, 2006, **34**, 252.
9. C. L. Beites, S. Kawauchi, C. E. Crocker and A. L. Calof, Identification and molecular regulation of neural stem cells in the olfactory epithelium, *Exp. Cell Res.*, 2005, **306**, 309.
10. A. Mackay-Sim, Stem cells and their niche in the adult olfactory mucosa, *Arch. Ital. Biol.*, 2010, **148**, 47.
11. X. Chen, H. Fang and J. E. Schwob, Multipotency of purified, transplanted globose basal cells in olfactory epithelium, *J. Comp. Neurol.*, 2004, **469**, 457.

12. Z. Su and C. He, Olfactory ensheathing cells: biology in neural development and regeneration, *Prog. Neurobiol.*, 2010, **92**, 517.
13. R. Elsaesser and J. Paysan, The sense of smell, its signalling pathways, and the dichotomy of cilia and microvilli in olfactory sensory cells, *BMC Neurosci.*, 2007, **8**(Suppl 3), S1.
14. C. A. Greer, Structural organization of the olfactory system, in *Smell and Taste in Health and Disease*, ed. T. V. Getchell, R. L. Doty, L. M. Bartoshuk and J. B. J. Snow, Raven Press, New York, 1991, pp. 65–81.
15. D. J. Doorly, D. J. Taylor, A. M. Gambaruto, R. C. Schroter and N. Tolley, Nasal architecture: form and flow, *Philos. Trans. R. Soc., A*, 2008, **366**, 3225.
16. J. S. Kimbell, Nasal dosimetry of inhaled gases and particles: where do inhaled agents go in the nose?, *Toxicol. Pathol.*, 2006, **34**, 270.
17. W. T. Stott, M. D. Dryzga and J. C. Ramsey, Blood-flow distribution in the mouse, *J. Appl. Toxicol.*, 1983, **3**, 310.
18. S. V. Dhuria, L. R. Hanson and W. H. Frey II, Intranasal delivery to the central nervous system: mechanisms and experimental considerations, *J. Pharm. Sci.*, 2010, **99**, 1654.
19. T. L. Leavens, D. Rao, M. E. Andersen and D. C. Dorman, Dorman DC, Evaluating transport of manganese from olfactory mucosa to striatum by pharmacokinetic modeling, *Toxicol. Sci.*, 2007, **97**, 265.
20. H. Tjälve, C. Mejàre and K. Borg-Neczak, Uptake and transport of manganese in primary and secondary olfactory neurones in pike, *Pharmacol. Toxicol.*, 1995, **77**, 23.
21. H. Tjälve and J. Henriksson, Uptake of metals in the brain via olfactory pathways, *Neurotoxicology*, 1999, **20**, 181.
22. H. Tjälve, J. Henriksson, J. Tallkvist, B. S. Larsson and N. G. Lindquist, Uptake of manganese and cadmium from the nasal mucosa into the central nervous system via olfactory pathways in rats, *Pharmacol. Toxicol.*, 1996, **79**, 347.
23. K. A. Brenneman, B. A. Wong, M. A. Buccellato, E. R. Costa, E. A. Gross and D. C. Dorman, Direct olfactory transport of inhaled manganese (^{54}MnCl$_2$) to the rat brain: toxicokinetic investigations in a unilateral nasal occlusion model, *Toxicol. Appl. Pharmacol.*, 2000, **169**, 238.
24. A. Elder, R. Gelein, V. Silva, T. Feikert, L. Opanashuk, J. Carter, R. Potter, A. Maynard, Y. Ito, J. Finkelstein and G. Oberdörster, Translocation of inhaled ultrafine manganese oxide particles to the central nervous system, *Environ. Health Perspect.*, 2006, **114**, 1172.
25. Ø. Olsen, A. Kristoffersen, M. Thuen, A. Sandvig, C. Brekken, O. Haraldseth and P. E. Goa, Manganese transport in the rat optic nerve evaluated with spatial- and time-resolved magnetic resonance imaging, *J. Magn. Reson. Imaging*, 2010, **32**, 551.
26. T. Watanabe, T. Michaelis and J. Frahm, Mapping of retinal projections in the living rat using high-resolution 3D gradient-echo MRI with Mn^{2+}-induced contrast, *Magn. Reson. Med.*, 2001, **46**, 424.

27. J. Lewis, G. Bench, O. Myers, B. Tinner, W. Staines, E. Barr, K. K. Divine, W. Barrington and J. Karlsson, Trigeminal uptake and clearance of inhaled manganese chloride in rats and mice, *Neurotoxicology*, 2005, **26**, 113.

28. R. G. Pautler, A. C. Silva and A. P. Koretsky, *In vivo* neuronal tract tracing using manganese-enhanced magnetic resonance imaging, *Magn. Reson. Med.*, 1998, **40**, 740.

29. D. C. Dorman, M. F. Struve, B. A. Wong, J. A. Dye and I. D. Robertson, Correlation of brain magnetic resonance imaging changes with pallidal manganese concentrations in rhesus monkeys following subchronic manganese inhalation, *Toxicol. Sci.*, 2006, **92**, 219.

30. D. J. Cross, S. Minoshima, Y. Anzai, J. A. Flexman, B. P. Keogh, Y. Kim and K. R. Maravilla, Statistical mapping of functional olfactory connections of the rat brain *in vivo*, *NeuroImage*, 2004, **23**, 1326.

31. S. Sen, M. R. Flynn, G. Du, A. I. Tröster, H. An and X. Huang, Manganese accumulation in the olfactory bulbs and other brain regions of "asymptomatic" welders, *Toxicol. Sci.*, 2011, **121**, 160.

32. Y. Kanayama, S. Enomoto, T. Irie and R. Amano, Axonal transport of rubidium and thallium in the olfactory nerve of mice, *Nucl. Med. Biol.*, 2005, **32**, 505.

33. H. Shiga, J. Taki, M. Yamada, K. Washiyama, R. Amano, Y. Matsuura, O. Matsui, S. Tatsutomi, S. Yagi, A. Tsuchida, T. Yoshizaki, M. Furukawa, S. Kinuya and T. Miwa, Evaluation of the olfactory nerve transport function by SPECT-MRI fusion image with nasal thallium-201 administration, *Mol. Imaging Biol.*, 2011, **13**, 1262.

34. J. Henriksson, J. Tallkvist and H. Tjälve, Transport of manganese via the olfactory pathway in rats: dosage dependency of the uptake and subcellular distribution of the metal in the olfactory epithelium and the brain, *Toxicol. Appl. Pharmacol.*, 1999, **156**, 119.

35. K. S. Saleem, J. M. Pauls, M. Augath, T. Trinath, B. A. Prause, T. Hashikawa and N. K. Logothetis, Magnetic resonance imaging of neuronal connections in the macaque monkey, *Neuron*, 2002, **34**, 685.

36. L. D. Fechter, D. L. Johnson and R. A. Lynch, The relationship of particle size to olfactory nerve uptake of a non-soluble form of manganese into brain, *Neurotoxicology*, 2002, **23**, 177.

37. D. C. Dorman, K. A. Brenneman, A. M. McElveen, S. E. Lynch, K. C. Roberts and B. Wong, Olfactory transport: a direct route of delivery of inhaled manganese phosphate to the rat brain, *J. Toxicol. Environ. Health, Part A*, 2002, **65**, 1493.

38. K. Thompson, R. M. Molina, T. Donaghey, J. E. Schwob, J. D. Brain and M. Wessling-Resnick, Wessling-Resnick M, Manganese uptake and distribution in the brain after methyl bromide-induced lesions in the olfactory epithelia, *FASEB J.*, 2007, **21**, 223.

39. J. Kim, Y. Li, P. D. Buckett, M. Böhlke, K. J. Thompson, M. Takahashi, T. J. Maher and M. Wessling-Resnick, Iron-responsive olfactory uptake

of manganese improves motor function deficits associated with iron deficiency, *PLoS One*, 2012, 7, e33533.

40. J. Kim, P. D. Buckett and M. Wessling-Resnick, Absorption of manganese and iron in a mouse model of hemochromatosis, *PLoS One*, 2013, 8, e64944.
41. J. D. Park, K. Y. Kim, D. W. Kim, S. J. Choi, B. S. Choi, Y. H. Chung, J. H. Han, J. H. Sung, I. H. Kwon, J. H. Mun and I. J. Yu, Tissue distribution of manganese in iron-sufficient or iron-deficient rats after stainless steel welding-fume exposure, *Inhalation Toxicol.*, 2007, 19, 563.
42. K. J. Thompson, R. M. Molina, T. Donaghey, S. Savaliya, J. E. Schwob and J. D. Brain, Manganese uptake and distribution in the brain after methyl bromide-induced lesions in the olfactory epithelia, *Toxicol. Sci.*, 2011, 120, 163.
43. A. C. Silva, J. H. Lee, I. Aoki and A. P. Koretsky, Manganese-enhanced magnetic resonance imaging (MEMRI): methodological and practical considerations, *NMR Biomed.*, 2004, 17, 532.
44. J. Herberholz, S. H. Mishra, D. Uma, M. W. Germann, D. H. Edwards and K. Potter, Non-invasive imaging of neuroanatomical structures and neural activation with high-resolution MRI, *Front. Behav. Neurosci.*, 2011, 5, 16.
45. D. A. Gutman, M. Magnuson, W. Majeed, O. P. Keifer Jr, M. Davis, K. J. Ressler and S. Keilholz, Mapping of the mouse olfactory system with manganese-enhanced magnetic resonance imaging and diffusion tensor imaging, *Brain Struct. Funct.*, 2013, 218, 527.
46. K. H. Chuang and A. P. Koretsky, Accounting for nonspecific enhancement in neuronal tract tracing using manganese enhanced magnetic resonance imaging, *Magn. Reson. Imaging*, 2009, 27, 594.
47. B. Lehallier, O. Rampin, A. Saint-Albin, N. Jérôme, C. Ouali, Y. Maurin and J. M. Bonny, Brain processing of biologically relevant odors in the awake rat, as revealed by manganese-enhanced MRI, *PLoS One*, 2012, 7, e48491.
48. S. Kivity, G. Tsarfaty, N. Agmon-Levin, M. Blank, D. Manor, E. Konen, J. Chapman, M. Reichlin, C. Wasson, Y. Shoenfeld and T. Kushnir, Abnormal olfactory function demonstrated by manganese-enhanced MRI in mice with experimental neuropsychiatric lupus, *Ann. N. Y. Acad. Sci.*, 2010, 1193, 70.
49. D. C. Dorman, B. E. McManus, C. U. Parkinson, C. A. Manuel, A. M. McElveen and J. I. Everitt, Nasal toxicity of manganese sulfate and manganese phosphate in young male rats following subchronic (13-week) inhalation exposure, *Inhalation Toxicol.*, 2004, 16, 481.
50. D. C. Dorman, M. F. Struve, E. A. Gross, B. A. Wong and P. C. Howroyd, Sub-chronic inhalation of high concentrations of manganese sulfate induces lower airway pathology in rhesus monkeys, *Respir. Res.*, 2005, 6, 121.
51. V. Villalobos, E. Bonilla, A. Castellano, E. Novo, R. Caspersen, D. Giraldoth and S. Medina-Leenderz, Ultrastructural changes of the olfactory bulb in manganese-treated mice, *Biocell.*, 2009, 33, 187.

52. L. Colin-Barenque, L. M. Souza-Gallardo and T. I. Fortoul, Toxic effects of inhaled manganese on the olfactory bulb: an ultrastructural approach in mice, *J. Electron Microsc.*, 2011, **60**, 73.
53. J. Henriksson and H. Tjälve, Manganese taken up into the CNS via the olfactory pathway in rats affects astrocytes, *Toxicol. Sci.*, 2000, **55**, 392.
54. A. H. Moberly, L. A. Czarnecki, J. Pottackal, T. Rubinstein, D. J. Turkel, M. D. Kass and J. P. McGann, Intranasal exposure to manganese disrupts neurotransmitter release from glutamatergic synapses in the central nervous system *in vivo*, *Neurotoxicology*, 2012, **33**, 996.
55. K. M. Erikson, D. C. Dorman, L. H. Lash and M. Aschner, Manganese inhalation by rhesus monkeys is associated with brain regional changes in biomarkers of neurotoxicity, *Toxicol. Sci.*, 2007, **97**, 459.
56. S. Nordin and A. Brämerson, Complaints of olfactory disorders: epidemiology, assessment and clinical implications, *Curr. Opin. Allergy Clin. Immunol.*, 2008, **8**, 10.
57. R. M. Patel and J. M. Pinto, Olfaction: anatomy, physiology, and disease, *Clin. Anat.*, 2014, **27**, 54.
58. S. Zoni, G. Bonetti and R. Lucchini, Olfactory functions at the intersection between environmental exposure to manganese and Parkinsonism, *J. Trace Elem. Med. Biol.*, 2012, **26**, 179.
59. M. Guarneros, N. Ortiz-Romo, M. Alcaraz-Zubeldia, R. Drucker-Colín and R. Hudson, Nonoccupational environmental exposure to manganese is linked to deficits in peripheral and central olfactory function, *Chem. Senses*, 2013, **38**, 783.
60. R. G. Lucchini, S. Guazzetti, S. Zoni, F. Donna, S. Peter, A. Zacco, M. Salmistraro, E. Bontempi, N. J. Zimmerman and D. R. Smith, Tremor, olfactory and motor changes in Italian adolescents exposed to historical ferro-manganese emission, *Neurotoxicology*, 2012, **33**, 687.
61. B. Lehallier, G. Coureaud, Y. Maurin and J. M. Bonny, Effects of manganese injected into rat nostrils: implications for in vivo functional study of olfaction using MEMRI, *Magn. Reson. Imaging*, 2012, **30**, 62.
62. D. C. Dorman, Olfactory system in *Nervous System and Behavioral Toxicology*, ed. C. A. McQueen, Comprehensive Toxicology, Elsevier, Kidlington, 2nd edn, 2010, Nervous System and Behavioral Toxicology, ch. 13.15, vol. 13 , pp. 263–276.

CHAPTER 5

Manganese Transport Across the Pulmonary Epithelium

KHRISTY J. THOMPSON,[a] JONGHAN KIM[b] AND
MARIANNE WESSLING-RESNICK*[a]

[a] Harvard School of Public Health, Boston, MA, USA; [b] Northeastern
University, Boston, MA, USA
*Email: wessling@hsph.harvard.edu

5.1 The Air–Blood Barrier

Lung epithelium provides a protective barrier against foreign substances
while allowing efficient transport of oxygen and other nutrients. In addition,
the lungs play an important role in triggering inflammatory and immune
response to external stress and pathogen infections. Along with other environmental exposures, metals in inhaled particles exacerbate conditions
like asthma and chronic obstructive pulmonary disease (COPD).[1,2] When
exposed, the lungs play a critical role in detoxification from metal-induced
oxidative stress.[3] Thus, our pulmonary epithelium is at the interface between
environmental exposures to protect the body and to initiate its response in a
manner that significantly affects both the pharmacokinetics and pharmacodynamics of uptake of airborne metals.

5.1.1 Microanatomy of the Lungs

Air entering the trachea follows into the right or left bronchus. Throughout
this region goblet cells and ciliated epithelium are present to provide mu-
cous to capture particulates, and to sweep particles and cellular debris to the

Issues in Toxicology No. 22
Manganese in Health and Disease
Edited by Lucio G. Costa and Michael Aschner
© The Royal Society of Chemistry 2015
Published by the Royal Society of Chemistry, www.rsc.org

pharynx. Gas exchange does not occur until reaching the respiratory bronchioles that contain the alveoli infused with capillaries. Over 300 million alveoli exist in the human lung, all covered by an extensive network of capillaries from the pulmonary arteries. The respiratory zone constitutes most of the lung. The bronchioles divide into alveolar ducts leading into alveolar sacs. The walls of the alveoli are composed primarily of alveolar epithelial type I cells. Less abundant are alveolar epithelial type II cells. Type II cells secrete surfactant that keeps the alveoli open by reducing the surface tension of the interface between alveolar surfaces during respiration. The combined alveolar wall, basement membrane, and capillary wall form the respiratory membrane. Alveolar pores connect adjacent alveoli allowing for equalized pressure. Alveolar macrophages are present to clear infectious microorganisms as well as particulate matter. Debris and dead cells are also removed from the alveolar region by the bronchial ciliary current. An important component of the lung's immune system is bronchus-associated lymphoid tissue (BALT), which initiates a local immune response to antigens entering the airways.[4] Inflammatory response factors, such as neutrophils and cytokines, can be measured in the lung fluid collected by bronchoalveolar lavage (BAL).

5.1.2   Lung Manganese Exposures

Metal-containing particles originate from natural sources like soil as well as man-made sources like automobile exhaust. For example, recent concerns about manganese exposures were generated from its use in the fuel additive methylcyclopentadienyl manganese tricarbonyl (MMT).[5] Occupational exposures to manganese can be chronic; in particular, welders and miners often develop manganism, a Parkinson's-like neurological disorder due to excessive manganese accumulation in the brain.[6] In urban settings, multiple transition metals, including manganese, are found in particulate matter.[7] Although insoluble larger particles can be cleared from the airways by mucociliary transport, those penetrating deeper into the lungs and beyond ciliated bronchial cells must either be dissolved or taken up by lung macrophages.[8] Such inhaled particles are known to induce inflammatory responses.[9] Metals like manganese that are solubilized from particles can be subsequently transported by the pulmonary epithelium across the "air–blood barrier".

Manganese from a variety of airborne sources enters the bloodstream from the lungs and is pumped to the heart and the entire body before traveling to the liver for detoxification. Manganese homeostasis appears to be tightly maintained by the liver, which is responsible for excreting excess metal into bile.[10,11] The circulatory pathway of manganese across the lung air–blood barrier is quite different than that taken by ingested manganese, which passes by hepatobiliary circulation to the liver so that any excess can be directly cleared from the body. Airborne manganese that penetrates the air–blood barrier escapes first-pass clearance, and therefore can travel directly to the brain where it exerts toxic effects. Thus, the brain is particularly vulnerable to inhalation exposures to manganese and its associated neurotoxicity. For these reasons, it is important to understand the mechanisms underlying pulmonary absorption of manganese. While the general form of airborne manganese from our environment is particulate manganese dioxide or tetraoxide rather than manganese solutions or vapors, most *in vivo* and *in vitro* studies have been conducted with metal salts such as $MnCl_2$. Evidence characterizing these processes using such approaches is discussed next.

5.2 Overview of Manganese Transport

Current models of manganese transport across alveolar epithelium are guided by information gained from studies of dietary absorption across the duodenal epithelium. In the gut, iron and manganese are taken up across the apical surface by a common transporter called Divalent Metal Transporter-1 (DMT1). Transfer of both metals across the basolateral surface is most likely mediated by an export protein called ferroportin (Fpn). It is important to note that both transporters, referred to as SLC11A2 and SLC40A1, respectively, are regulated by iron status. Therefore, absorption from the diet is increased by iron deficiency and decreased in iron overload states. The interplay between manganese and iron is complex, but the fact that manganese levels in the body correlate with iron status is well established. Iron deficiency results in increased dietary manganese absorption[12-16] and increased brain manganese levels.[13,17,18] This relationship underscores the use of a common set of transporters and their regulation by iron. Thus, the iron-responsive nature of lung manganese uptake has been tested. The influence of iron status on the expression of DMT1 and Fpn in the lungs has also been studied. Once transferred to circulation, some 80% of manganese is bound to β-globulin and albumin as Mn^{+2}; some portion may be oxidized by ceruloplasmin and bound to transferrin (Tf) as trivalent Mn^{+3}.[19] Tf and the Tf receptor are also iron-regulated species postulated to play a role in manganese metabolism.

5.2.1 Divalent Metal Transporter-1 (Slc11a2)

DMT1 (DCT1, Slc11a2, Nramp2) has been shown to transport multiple metals.[20] Physiologically, this transporter plays a significant role in uptake of both iron and manganese across the intestinal mucosa as essential minerals. The overlap in transport activity explains why increased manganese uptake is associated with iron deficiency. DMT1 is regulated by iron status at both transcriptional and post-transcriptional levels. At least two, and in some tissues up to four splice variants are transcribed from the DMT1 gene.[21,22] A fifth isoform is transcribed from an alternate promoter.[23] The tissue-specific nature of DMT1 expression and mRNA splicing is not fully understood but can be defined by the presence or absence of an iron-responsive element (IRE) in the $3'$ untranslated region. This difference not only leads to different 18–21 terminal amino acids, but also confers regulation by iron-responsive proteins (IRPs) that stabilize the transcript to permit increased translation under low iron conditions (see later). The Belgrade (*b*) rat and the microcytic anemia (*mk*) mouse carry the same G185R point mutation in DMT1, which limits the transporter's activity.[24] The Belgrade rat displays decreased iron and manganese uptake by kidney, brain, bone marrow, and erythroblasts,[25] and this animal model has been used to probe the possible role of DMT1 in pulmonary manganese absorption, as discussed later.

5.2.2   Ferroportin (Slc40a1)

The pathway for iron assimilation across the intestinal mucosa also includes export of the metal across the basolateral surface of the intestinal epithelium into circulation. This step is carried out by Fpn (IREG1, Slc40a1, MTP1).[26–28] Fpn expression is also regulated by iron and subject to both translational and post-translational control.[29] Iron depletion diminishes Fpn expression and iron loading increases expression. Increased expression results from translational control because Fpn transcripts harbor an IRE in the $5'$-untranslated region.[30] Under conditions of high iron, IRPs are released from the IRE and Fpn expression is increased.[31]

It has been determined that systemic iron efflux *via* Fpn is also regulated post-translationally through its interaction with hepcidin.[32] Hepcidin is considered to be the principal hormone involved in iron regulation.[32,33] Hepcidin synthesis is upregulated in response to iron loading and inflammation and is suppressed by anemia and hypoxia.[32,34,35] Circulating hepcidin binds to Fpn and leads to its internalization and degradation through endocytic trafficking to the lysosome;[32] therefore, conditions leading to high levels of hepcidin result in the retention of intracellular iron. Conversely, reduction of hepcidin results in a diminution of intracellular iron through export into the plasma.[36]

Accumulating evidence from *in vitro* studies supports the idea that Fpn exports manganese in addition to iron. For example, *Xenopus* oocytes expressing this transporter export more ^{54}Mn than controls.[37] Studies of manganese transfer across Caco-2 cell monolayers, an *in vitro* model for intestinal absorption, are consistent with DMT1 and Fpn mediating apical uptake and basolateral export of manganese.[38] In other studies, inducible expression of Fpn in HEK293T cells reduced manganese accumulation and toxicity.[39] Treatment with manganese[40,41] and other metals[42] has been shown to regulate Fpn expression. In macrophages, Fpn expression is altered by exposure to manganese,[43] Hepcidin is also present in alveolar macrophages,[38,44] but the possible role for Fpn in manganese transport, its speculative function in the respiratory uptake of the metal and/or its detoxification, and the potential role for hepcidin in the regulation of these processes remain to be fully explored.

5.2.3 Transferrin/Transferrin Receptor (CD71)

Transferrin (Tf) is synthesized in the liver as the major plasma iron-binding protein for delivery to peripheral tissues and is upregulated by iron deficiency. Although this carrier has high affinity for Fe^{+3} and will bind Mn^{+3}, its affinity for manganese is significantly less than iron.[45] The half-life of injected Mn–Tf is greater than free Mn^{+2}/Mn^{+3} because unbound manganese is rapidly cleared by the liver.[46] Thus, the physiological relevance of Tf interactions with manganese is not clear. In biological systems, Mn^{+2} appears to be more stable as a divalent cation, requiring oxidation to Mn^{+3} before Tf binding. Early studies involved prolonged incubation of sera with aeration, conditions that could allow for manganese oxidation, thereby promoting Tf interactions.[46–48] While one study identified Mn–Tf in plasma after oral or intravenous administration,[49] other work showed *in vitro* formation of Mn–Tf when conditions supported oxidation.[50] In addition to these uncertainties, whether manganese binds to Tf in pulmonary fluid following exposure to the metal is uncertain.[51]

Transferrin receptor (CD71) is also expressed in an iron-responsive manner. Similar to regulation of DMT1 and Fpn, this effect is exerted through IREs under IRP control to stabilize the receptor's mRNA under low iron conditions.[23] Tf receptors would interact with Fe–Tf or Mn–Tf, and thus could be involved in manganese uptake from the lung. In addition to its role in metal uptake across the plasma membrane, DMT1 appears to be essential in the Tf/TfR pathway as the transporter necessary for exit of iron from the endosome.[24] DMT1 would have an analogous role in Mn uptake by the Tf/TfR pathway or after dissolution of internalized particles within acidic phagocytic organelles.

5.2.4 Other Manganese Transporters: Non-Selective Ion Channels

Amongst other transporters, voltage-gated calcium channels (VGCs) and store-operated calcium channels (SOCs) are known to transport iron and manganese. In fact, many studies use Mn^{+2} as a proxy for Ca^{+2}.[52,53] Iron uptake by the heart and Fe^{+2} and Mn^{+2} uptake by cardiomyocytes are sensitive to L-type VGC inhibitors.[54,55] Similar analysis of transport across the blood–brain barrier demonstrated Mn^{+2} uptake is sensitive to SOC inhibitors and activators.[56] Transient receptor potential (TRP) channels, especially the TRPC subfamily, have broad substrate specificity that includes Mn^{+2}. Notably, TRPC1, TRPC3, TRPC4, TRPC5, TRPC6, and TRPC7 are permeable to Mn^{+2}.[57–60] Pulmonary epithelial cells express many different iron channels including VGCs, SOCs and TRP channels.[61–63] These channels, and others, are thought to maintain fluid balance in lungs, but potential roles in the transport of manganese and other metals are possible as well.[64]

5.2.5 Hypothetical Model for Pulmonary Manganese Transport

Pulmonary epithelial cells express transferrin receptors, both IRE- and nonIRE-isoforms of DMT1, Fpn and a variety of channels. Tf is an abundant protein in bronchoalveolar lavage fluid, consisting $\sim 5\%$ of the total protein[65] and is synthesized by the lungs.[66] In cultured bronchial epithelia cells, Tf levels decreased in response to residual oil fly ash (ROFA), an effect reversible upon iron chelation suggesting regulation.[67] Ceruloplasmin is also present in pulmonary fluid and therefore could potentially oxidize Mn^{+2} for loading onto Tf.[68] Tf receptor expression in the lungs appears to respond to inflammation,[69,70] and exposure of lungs to ferric ammonium citrate upregulates DMT1 in an isoform specific manner.[71] As described earlier, all of these factors could play a role in clearance of inhaled manganese from the lungs. Figure 5.1 presents a theoretical model for uptake of manganese across the pulmonary epithelium. After phagocytosis by alveolar macrophages, Mn^{+2} released from solubilized particles is mobilized by DMT1. Mn^{+2} could cross the apical surface of the pulmonary epithelium *via* DMT1, manganese-permeable ion channels, or other pathways. Alternatively, manganese oxidized to Mn^{+3} might interact with Tf in pulmonary fluid, and then associate with Tf receptors in a complex to be taken up by endocytosis. DMT1 would presumably also play a role in this pathway, transporting manganese to the cytoplasm. In all of these scenarios, export and/or detoxification pathways mediated by Fpn might be envisioned. Some of these transport elements have been tested while others await further study. It should be emphasized that Figure 5.1 is a "working model" illustrating the multiple mechanisms that are implicated in pulmonary manganese absorption and providing a theoretical framework to evaluate the evidence.

5.3 Interplay Between Manganese and Iron Status

Understanding the interplay between manganese and iron is important when considering exposures to mixtures of metals and exposures to metals under conditions of altered iron homeostasis, for example iron deficiency and loading. As a transition metal, manganese has characteristics similar to iron and can therefore directly interact with proteins that bind iron.[72] These include several of the transport factors for iron discussed earlier. A common theme amongst this set of transporters is their regulation by an iron-sensing regulatory mechanism. An mRNA element called the iron responsive element (IRE) confers cell-based control over iron status where low iron conditions engage this system to increase proteins of iron import (DMT1 and Tf receptor) and decrease iron export mechanisms (Fpn). This mechanism allows for the adequate uptake and distribution of iron as well as the reduction in export when sufficiency is achieved. This is achieved by the iron regulatory proteins IRP1 and IRP2, which act to prevent translation of proteins of iron export

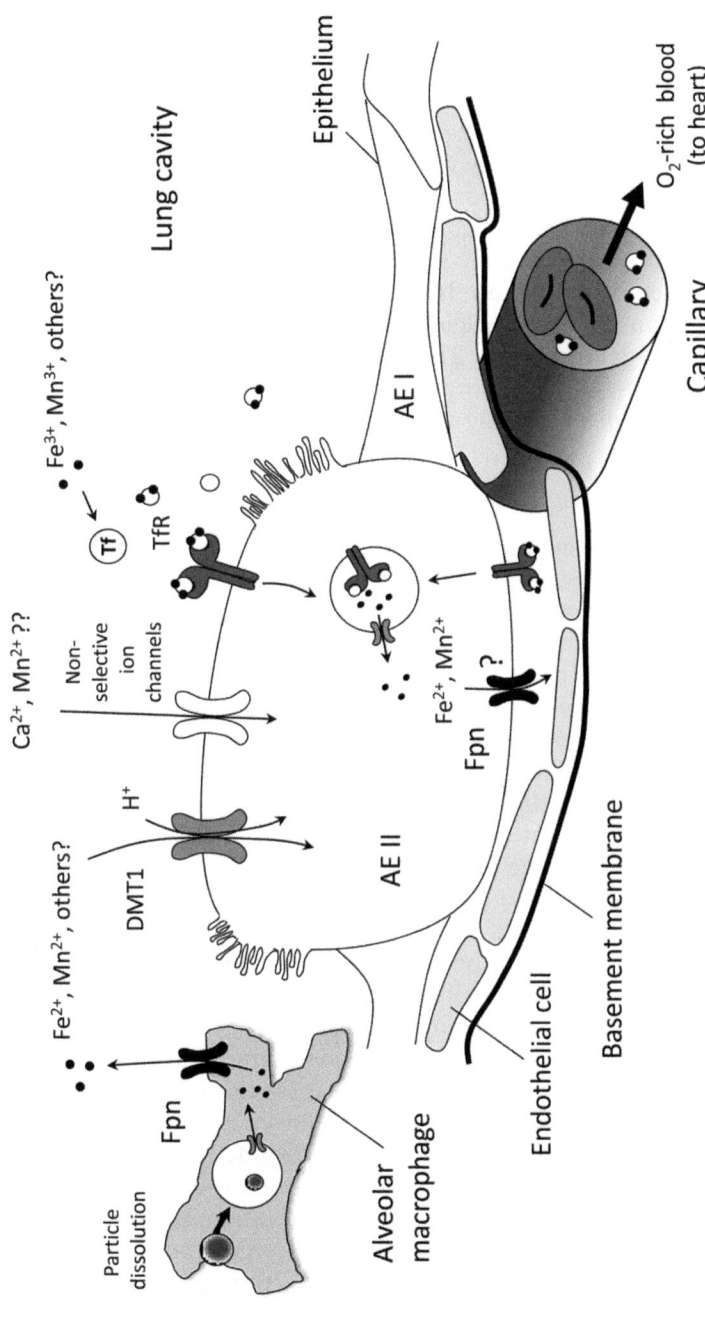

Figure 5.1 A working model of pulmonary manganese transport. Manganese present in the lung cavity (alveolar space) may be transported by alveolar epithelial type II cells (AE II) *via* several mechanisms including Divalent Metal Transporter-1 (DMT1), ferroportin (Fpn), non-selective ion channels, and the transferrin/transferrin receptor (Tf/TfR) endosomal pathway. Alveolar macrophages contribute to the metal content in the alveolar space by engulfing and breaking down particulates and releasing metals through DMT1 and Fpn. Alveolar type I cells (AE I) are very abundant but nevertheless their role in metal transport is understudied. Adapted from Heilig.[124]

(Fpn) and stabilize messages of iron import (DMT1) during periods of low iron. Manganese can influence this iron regulatory mechanism. Mn^{+2} (like Fe^{+2}) has been shown to interact with the IRE structure and alter the interaction of IRP1 with this RNA element.[73] The physiological implications of this interaction have yet to be tested, but the result would mimic high iron conditions. Table 5.1 summarizes the iron regulatory control of transport factors in various cell types of the lungs and their control by IRE/IRP binding.

During periods of iron deficiency when iron import is increased, manganese utilization of the iron transport mechanisms can potentially increase manganese toxicity. Several investigators have noted increased brain manganese levels associated with iron deficiency.[13,17,18] The iron-responsive nature of manganese uptake can have profound impact on vulnerable populations. For example, iron deficient children[74,75] and women[76,77] have increased blood manganese levels. The established interplay between iron metabolism and manganese absorption has been used to explore the potential pathways involved in lung metal absorption.

5.3.1   Pulmonary Manganese Uptake and Iron Deficiency

Pharmacokinetic studies using tracer levels of radioactive ^{59}Fe and ^{54}Mn have been used to compare aspects of pulmonary transport of these two metals in the search for common elements. In studies from our lab, rats were placed on iron deficient diet or repetitively phlebotomized to promote the development of anemia.[51,64,78] Radioactive ^{59}Fe or ^{54}Mn was then instilled intratracheally so that transport could be examined by measuring blood isotope levels over time. Following intratracheal instillation, ^{59}Fe was cleared more rapidly from the lungs to the blood in rats made iron-deficient, either by diet or phlebotomy, compared to controls.[64,78] ^{54}Mn levels increased in the blood over time, but followed a completely different pharmacokinetic time course. Overall, ^{54}Mn and ^{59}Fe tissue levels in the lung were not altered. Since lung non-heme iron was not altered significantly, the lack of iron-responsive lung tissue uptake of the metals most likely reflects local iron status, which did not change. Consistent with this observation, both IRE and nonIRE DMT1 expression, which was localized to airway epithelium (typed II cells) and alveolar macrophages, was unaltered in the lung by iron deficiency.

DMT1 is known to transport other metals, including manganese, and to investigate its specific role in lung manganese transport our group also studied the Belgrade rat. This animal model carries a point mutation in DMT1 (G185R) leading to protein misfolding and degradation; Belgrade rats therefore are iron deficient.[25] Previous studies showed the Belgrade rat has reduced clearance of iron.[71] In intratracheal instillation experiments, the transport of ^{54}Mn to the blood was unaltered.[51] However, intranasal instillation of ^{54}Mn into Belgrade rats did reveal that olfactory uptake of ^{54}Mn was DMT1-dependent.[79] Surprisingly, ^{54}Mn also failed to associate with Tf present in BAL from control rats.[64] This combined evidence completely rules out a role for DMT1, suggesting the existence of additional transport pathways for manganese in the lungs.

Table 5.1 Iron-responsive metal transport factors in lung.

Name	Cell type	Normal	Iron deficient	Iron loading	Metal specificity	IRE/IRP Regulation
Transferrin	Bronchial epithelium	+	+	++ (mRNA)	Fe^{+3}, Mn^{+3}	3′ IRE; message stabilized by IRPs
	BALT (focal)	+	+	+ (protein)		
	Alveolar macrophages	+	+			
Transferrin receptor	Bronchial epithelium	+	+	+++ (mRNA but not protein)	Determined by Tf	3′ IRE; message stabilized by IRPs
	BALT	+	+			
	Alveolar epithelium	−	++			
	Alveolar macrophages	−	−			
DMT1	Bronchial epithelium	+	+	+	Fe^{+2}, Mn^{+2}, and others	3′ IRE; message stabilized by IRPs
	BALT	+	+	+		
	Alveolar epithelium	+/−	+/−	+		
	Alveolar macrophages	+	+	+++		
Fpn	Alveolar macrophages	+	?	+	Fe^{+2}, Mn^{+2}	5′ IRE; translation suppressed *via* IRPs
	Airway epithelium (apical surface)	+		+		

5.3.2   Iron Overload and Lung Manganese Absorption

Manganese absorption by the lungs under iron-overload conditions has also been studied using dietary manipulations and iron oxide exposures. In rats fed a 1% carbonyl iron diet,[80] both liver and lung non-heme iron levels increased. Although Tf mRNA levels increased in bronchial and alveolar epithelial cells and BALT, changes in protein levels were not observed. DMT1 was also found in airway epithelium and BALT, and iron-loading resulted in decreased expression in BALT. Intratracheal instillation and intravenous injection studies using ^{54}Mn showed lower absorption from the lungs to the blood, but that once in the blood, ^{54}Mn was cleared more rapidly in iron-loaded rats. Combined, these data demonstrate that manganese absorption from the lung to the blood is affected by iron loading through diet.[80]

A similar study was conducted using intratracheal instillation of iron oxide particles to load iron directly into the lung. DMT1 levels were increased in iron oxide-exposed rats at focal regions where particles were observed, along with a recruitment of macrophages near the particles.[78] Upregulation of DMT1 in BALT was also observed. Intratracheal instillation of iron oxide was associated with reduced transport of ^{54}Mn from the lung to the blood. ^{59}Fe transport also was reduced, as would be expected based on competition with the instilled metal.[78] Importantly, local upregulation of DMT1 mRNA was observed around the iron oxide particles in the lung. Wang *et al.*[71] have shown that non-IRE DMT1 protein is upregulated in rat lung epithelium after instillation of ferric ammonium citrate. Such evidence supports a role for DMT1, however the observations that neither ^{54}Mn nor ^{59}Fe uptake across the pulmonary epithelium was enhanced in iron oxide-exposed animals argues that this transporter would not be rate-limiting and/or that solubilized iron from the oxide particles blocked uptake of the radioisotopes by this pathway. It is possible that other lung functions are modified, for example, solubilization of metal in the phagolysosome of alveolar macrophages. DMT1 could also promote iron sequestration and storage in pulmonary epithelium because increased ferritin and hemosiderin are observed in epithelia following iron oxide inhalation.[81]

5.3.3   Roles for Tf and the Tf Receptor in the Lungs

Tf and TfR could also affect the pulmonary clearance of inhaled manganese and iron. Both are upregulated by iron deficiency. Tf receptor expression is post-transcriptionally regulated by IRP activation (Table 5.1). Both Tf and Tf receptor mRNA and protein levels have been studied in iron-deficient rats and rats exposed to iron oxide particles. As reported by Yang *et al.*,[66] Tf message was found in bronchial epithelium, type II cells, alveolar macrophages, and BALT, but the expression of mRNA and protein did not change upon iron deficiency.[51] These results demonstrate that Tf levels in BAL fluid must be locally controlled by lungs because plasma Tf is markedly increased in the iron-deficient rats. Tf receptor was found in BALT and bronchial epithelium, and Tf receptor message levels were increased in Type II cells by iron deficiency. Protein levels were increased overall in the lung in iron-deficient rats. Despite this increase, ^{54}Mn was not found to be associated with Tf in lung fluid. Interestingly, iron-deficient rats had increased ^{54}Mn in the brain suggesting that iron status can modify the potential toxicity of metal taken up by pulmonary absorption, potentially through plasma transport pathways.[64]

5.4 Non-Selective Ion Channels

Studies on the Belgrade rat, which is deficient in DMT1 activity yet displays no defects in pulmonary ^{54}Mn absorption, the failure of ^{54}Mn to show significant binding by Tf present in lung fluid, and the demonstration that ^{54}Mn is more rapidly taken up by the lung than ^{59}Fe represent independent lines of evidence suggesting that factors other than or in addition to DMT1, Tf, and the Tf receptor must be involved in pulmonary manganese absorption. These observations prompted *in vitro* studies using the Type II alveolar A549 cell line. Characterization of A549 cell manganese transport showed time- and temperature-dependency of a saturable process.[64] Although iron uptake was enhanced by low pH, manganese transport was unaffected. This observation indicates proton-coupled DMT1 functions to take up iron but that other transporters must mediate uptake of manganese by A549 cells. Among known non-selective ion channels, L-type voltage-gated channels, TRPC1, TRPC3, TRPC6, TRPC7 and TRPM7 were identified to be expressed in this Type II alveolar cell line. A series of pharmacological experiments identified possible roles for L-type voltage-gated channels, TRPM7, and TRPC6 in manganese uptake by A549 cells.[64] These and other newly identified transporters must be explored as potential gatekeepers for pulmonary manganese absorption. Since iron loading by diet influences lung manganese transport[80] and iron deficiency enhances brain iron uptake after pulmonary absorption,[78] such transport pathways may also process pulmonary manganese absorption in parallel to known iron transport factors (see Table 5.2, for example).

Table 5.2 Iron-responsive manganese transport.

	Iron response	Manganese transport
DMT1	↑ Fe depletion	↑ import into cell
	↓ Fe overload	↓ import into cell
TfR	↑ Fe depletion	↑ import into cell
	↓ Fe overload	↓ import into cell
Fpn	↓ Fe depletion	↓ export out of cell
	↑ Fe overload	↑ export out of cell
Hepcidin	↓ Fe depletion	↑ export out of cell
	↑ Fe overload	↓ export out of cell

5.5 Toxic Effects of Manganese on Lung Epithelial Cells

Manganese exposure is associated with lung toxicity and increased susceptibility to infections and inflammatory response of both humans and animals, resulting in impaired lung function.[82] Studies of the Belgrade rat suggest that DMT1 is involved in pulmonary inflammation[83,84] as well as metal-induced injury,[85,86] emphasizing an important detoxification pathway that may transport and sequester iron and manganese after inhalation exposures.

Lung impairments due to manganese exposure include cough, bronchitis, pneumonitis, and manganic pneumonia,[82,87–89] however, these effects may result from acute inhalation toxicity because other metal particulates induce similar manifestations.[90] Nonetheless, several lines of evidence have revealed that manganese inhalation is associated with pulmonary inflammation[91] and its inflammatory potential is comparable to or greater than other transition metals.[92] Recent investigations have demonstrated that manganese promotes pulmonary toxicity by several mechanisms, including upregulation of hypoxia inducible factor (HIF) and vascular endothelial growth factor (VEGF),[93,94] inflammatory response in the small airway,[95] and induction of apoptosis.[96]

Cyclooxygenase 2 (Cox-2), an inducible inflammatory enzyme, is involved in pathogenesis of lung inflammation mediated by manganese. A549 human epithelial cells exposed to $MnCl_2$ increased Cox-2 expression in a dose-dependent manner *via* transcriptional activation of signaling molecules, including ERKs, p38 MAPK, JNKs, ATF-2, and PKB, but not NF-κB.[97] Other human airway cells like H292 bronchial and Hep2 laryngeal cells also responded to manganese-mediated Cox-2 induction. Another group showed similar results using SAEC normal human small airway epithelial cells, but without a change in p38 MAPK phosphorylation.[98] Following phosphorylation of signaling molecules, inflammatory cytokines IL-6 and IL-8, but not TNF-α were released.[99] Manganese also depletes intracellular glutathione levels.[97] Moreover, these changes were abrogated by *N*-acetylcysteine and by signaling inhibitors.[97] Such results indicate a critical role of inhaled manganese in the progression of pulmonary inflammation and altered redox potential in the lung.

Manganese also transactivates HIF-1α,[94] which senses cellular oxygen availability and promotes erythropoiesis and angiogenesis.[100] While HIF-1α is induced by many stimuli, such as IL-1β, TGF-1β and IGF-1, Shin *et al.*[101] demonstrated that its protein level was post-transcriptionally elevated upon $MnCl_2$ exposure in Hep2 laryngeal cells, but not in H292 bronchial or A549 lung cells. Li *et al.*[102] showed that $MnCl_2$ increased A549 cell IRP-1 activity and stabilized HIF-1α protein, both of which were associated with depletion of intracellular iron as evidenced by reduced ferritin levels. They further demonstrated that $MnCl_2$ treatment activated hypoxia-response element (HRE) reporter gene expression and increased the protein levels of Cap43, an HIF-1α responsive protein. Along the same lines, Han *et al.*[94] showed that manganese inhibited HIF prolyl hydroxylase, an enzyme that leads to HIF inactivation, resulting in elevated expression of VEGF by an HIF-dependent manner. Manganese-mediated HIF induction was attenuated by iron supplementation, indicating a deleterious effect of manganese on iron homeostasis during pulmonary inflammation.[94] In addition, manganese can induce VEGF expression independent of HIF activation based on the finding that deletion of the HRE from the VEGF promoter did not diminish manganese-induced promoter activity.[93] The concentration of manganese exposure may determine the degree of HIF dependency on VEGF induction. It is also notable that manganese promotes caspase-9-mediated apoptosis in lung epithelium,[96] which could also contribute to pathophysiology of pulmonary inflammation.

Several *in vivo* studies have characterized the inflammatory potential and toxicity of manganese and other metals in the lung. Mice exposed to 2 mg Mn m^{-3}, 6 h per day for 5 days displayed a 2-fold increase in pulmonary VEGF mRNA levels with no significant pulmonary inflammation.[93] These authors speculated that manganese-associated pathological changes in the lung gradually become detectable over time, similar to manganese-induced neurotoxic effects.[93] Rice *et al.*[92] exposed rats to six transition metals, including vanadium, nickel, iron, copper, manganese, and zinc, by a single intratracheal instillation with two different doses (0.1 or 1 μmol kg^{-1}) and determined biochemical and inflammatory markers in BAL fluid at different time points (4, 16, or 48 h) post-instillation. There were dose- and time-dependent changes in lung injury; manganese elevated BAL lactate dehydrogenase levels at 1 μmol kg^{-1} but not 0.1 μmol kg^{-1} as early as 4 hours post-instillation, whereas the numbers of both neutrophils and eosinophils significantly increased 16 h after instillation only at higher dose.[92] Moreover, manganese exposure induced mRNA levels of macrophage inflammatory proteins (MIP1 and 2) at the high dose,[92] indicating a pivotal role of inhaled manganese in lung inflammation and injury. In rabbit lungs, inhalation of aerosolized $MnCl_2$ (median aerodynamic diameter of 1 μm) did not cause abnormalities.[103] In another study, monkeys were subjected to subchronic manganese exposure *via* inhalation at 0.06, 0.3, or 1.5 mg Mn m^{-3} as $MnSO_4$, 6 h per day, 5 days per week.[95] Although administration of the higher doses for ≥15 days induced mild subacute bronchiolitis, alveolar duct inflammation and proliferation of bronchus-associated lymphoid tissue, these lesions disappeared 45 days after cessation of exposure, suggesting manganese-induced lung toxicity is reversible at this dose range.[95] These authors also found that a lower dose (≤0.3 mg Mn m^{-3}), a dose similar to 8 h occupational threshold limit value, did not elicit respiratory tract pathology.

5.6 Infection and Manganese in the Lungs

Although the importance of iron withdrawal as a host response to infection has been well studied, there is increasing awareness that manganese also plays a key role in host-pathogen interactions.[104,105] Amongst its functions in microorganisms, manganese-dependent defenses against oxidative stress, the role of manganese as a cofactor in signaling and metabolism, and the metal's regulatory role in virulence gene expression are thought to influence infection.[106] Thus, strategies for the host to limit the access of lung pathogens to manganese are as critical as the need for iron limitation to combat infection. Host-pathogen interactions are particularly relevant to the cellular physiology of pneumococcus because manganese appears to be critical to the survival of *Streptococcus pneumoniae*. Adkins *et al.*[107] showed increased susceptibility in mice to *Streptococcal pyogenes* after manganese inhalation. Manganese dust plus pneumocci types I and II cause bronchopneumonia in rabbits.[107] *S. pneumoniae* is the cause of infections such as meningitis, bacteremia, pneumonia, upper respiratory infections, and otitis media.[108–111] The PSA genes of *S. pneumoniae* encode a manganese uptake system for this organism,[112] and it is clear that manganese is essential for the virulence and survival. Interestingly, manganese also protected *S. pneumonia* by competing with iron for transport into the bacterium in a model of mouse infection.[113] The SloABC family also transports both Mn^{2+} and Fe^{3+}, suggesting interactions between the metals can cause competitive effects.[113,114]

Lipopolysaccharide (LPS) has been shown to induce hepcidin expression in mouse alveolar macrophages, resulting in reduced Fpn expression and altered iron release.[44] DMT1 expression was induced, suggesting that under infection and inflammation, these transporters could act to take up and sequester manganese as a protective host response. The net effect of such alterations would be to reduce levels of manganese available in the lung to limit survival of pathogens like *S. pneumonia*. Recently, the function of the S100 family of calcium binding proteins has also been implicated in host defense against pathogens at epithelial surfaces. Calprotectin, a heterodimer of S100A8 and S100A9, is an antimicrobial proposed to act by binding manganese to limit bacterial and fungal growth.[115,116] It represents up to 50% of the protein content of neutrophils, and it most likely sequesters manganese in the lung during the inflammatory response.

5.7   Future Directions

Pulmonary pharmacokinetics and pharmacodynamics of manganese are controlled by many physicochemical and biological/physiological factors, including solubility, charge, size (of particle), dose, exposure period, route of exposure, animal species, nutritional status, disease state, and genetic polymorphism/mutation. More systematic and controlled studies are warranted to better understand the underlying mechanism of manganese-associated pulmonary epithelial injury. Likewise, the potential role of metal-binding agents, such as calprotectin, must be explored in order to better understand the pathologies associated with lung infection and possibly develop therapies to mitigate infection or injury of the lung.

Among yet-to-be explored manganese transporters are members of the zinc transport family of ZIP proteins, which have been recently characterized to take up manganese.[117] Both ZIP8 (SLC39A8) and ZIP14 (SLC39A14) are expressed in the lungs and, unlike DMT1, display a pH optimum of 7.5 and therefore are attractive candidates for transport of manganese across the apical surface of pulmonary epithelia cells. DMT1, which acts as a H + symporter and functions at lower pH, might have a more defined role in uptake metals after particle dissolution in acidic endocytic compartments. Zip8 and Zip14 expression can be upregulated under inflammatory conditions, further suggesting a role in lung pathology.[117] The potential functions of these and other members of the ZIP family in pulmonary manganese uptake, and their possible role in metal detoxification, need to be more fully examined.

Lung manganese export pathways also must be better defined. Manganese absorbed by the intestine enters hepatobiliary circulation and is transported to the liver where excess metal is removed from the blood and excreted into bile.[10,11] Recently, mutations in the transporter SLC30A10 have been associated with hypermanganesemia with accumulation of manganese in the liver.[118,119] This suggests the possible function of SLC30A10 in pulmonary manganese uptake should be tested. Whether human mutations in this gene affect transfer of manganese across the air–blood barrier could also be investigated in clinical studies. A second transporter, the SPCA Ca^{+2}/Mn^{+2} pump, has been shown to play a role in uptake of manganese by the Golgi apparatus, serving to limit intracellular damage due to toxic levels of intracellular manganese.[120] Overexpression of SPCA *in vitro* protects against manganese toxicity,[121] thus this factor should also be examined as a potential metal handling player by the pulmonary epithelium. Finally, emerging evidence that the ATP12A2 (PARK9) plays a role in manganese toxicity[122] further suggests this P-type ATPase possibly interacts in lung metal homeostasis.

Perhaps one of the more intriguing questions that remains unanswered is what role does hepcidin play in the regulation of manganese uptake by the lungs? Nguyen *et al.*[44] have shown that alveolar macrophages express this regulatory protein and the levels of Fpn decrease in response to its induction by LPS. This same group has identified expression of Fpn in airway epithelia but its putative localization on the apical surface is difficult to reconcile with transfer of metals across this barrier and better supports a role in detoxification.[123] Because emerging evidence indicates that Fpn also acts as a manganese exporter, its role in pulmonary metal absorption should be tested along with the other manganese exporters described earlier. Moreover, Fpn regulation by the local synthesis of hepcidin in response to inflammation merits further study due to the high therapeutic value of potential interventions that might help to limit manganese exposure and/or assist in metal detoxification.

These newly discovered manganese-binding proteins, transport pathways, and their regulation exemplify just a few recent developments that could help to improve our understanding of manganese trafficking across the air–blood barrier. There is much yet to learn about the mechanisms and molecules involved. Importantly, this research can potentially yield new targets and therapeutic approaches to ameliorate or prevent manganese neurotoxicity through respiratory pathways.

Acknowledgements

The authors wish to thank Dr Elizabeth Heilig for her contributions to the figure and tables in this chapter. This work is supported in part by NIH R00 ES017781 and P30 ES000002 (JK) and R01 ES014638 (MWR).

References

1. D. L. Costa and K. L. Dreher, Bioavailable transition metals in particulate matter mediate cardiopulmonary injury in healthy and compromised animal models, *Environ. Health Perspect.*, 1997, **105**(Suppl 5), 1053–1060.
2. S. H. Gavett and H. S. Koren, The role of particulate matter in exacerbation of atopic asthma, *Int. Arch. Allergy Immunol.*, 2001, **124**, 109–112.
3. J. L. Turi, F. Yang, M. D. Garrick, C. A. Piantadosi and A. J. Ghio, The iron cycle and oxidative stress in the lung, *Free Radical Biol. Med.*, 2004, **36**, 850–857.
4. T. D. Randall, Bronchus-associated lymphoid tissue (BALT) structure and function, *Adv. Immunol.*, 2010, **107**, 187–241.
5. K. Blumberg and M. P. Walsh, *Status Report Concerning the Use of MMT in Gasoline*, International Council on Clean Transportation, Washington, DC, 2004.
6. G. C. Cotzias, K. Horiuchi, S. Fuenzalida and I. Mena, Chronic manganese poisoning. Clearance of tissue manganese concentrations with persistance of the neurological picture, *Neurology*, 1968, **18**, 376–382.
7. A. K. Prahalad, J. M. Soukup, J. Inmon, R. Willis, A. J. Ghio, S. Becker and J. E. Gallagher, Ambient air particles: effects on cellular oxidant radical generation in relation to particulate elemental chemistry, *Toxicol. Appl. Pharmacol.*, 1999, **158**, 81–91.
8. J. B. West, *Respiratory Physiology: The Essentials*, Lippincott Williams & Wilkins, Baltimore, MD, 1995.
9. Z. T. Handzel, Effects of environmental pollutants on airways, allergic inflammation, and the immune response, *Rev. Environ. Health*, 2000, **15**, 325–336.
10. A. J. Bertinchamps, S. T. Miller and G. C. Cotzias, Interdependence of routes excreting manganese, *Am. J. Physiol.*, 1966, **211**, 217–224.
11. P. S. Papavasiliou, S. T. Miller and G. C. Cotzias, Role of liver in regulating distribution and excretion of manganese, *Am. J. Physiol.*, 1966, **211**, 211–216.

12. S. V. Chandra and G. S. Shukla, Role of iron deficiency in inducing susceptibility to manganese toxicity, *Arch. Toxicol.*, 1976, **35**, 319–323.
13. A. C. Chua and E. H. Morgan, Effects of iron deficiency and iron overload on manganese uptake and deposition in the brain and other organs of the rat, *Biol. Trace Elem. Res.*, 1996, **55**, 39–54.
14. J. W. Finley, Manganese absorption and retention by young women is associated with serum ferritin concentration, *Am. J. Clin. Nutr.*, 1999, **70**, 37–43.
15. I. Mena, K. Horiuchi, K. Burke and G. C. Cotzias, Chronic manganese poisoning. Individual susceptibility and absorption of iron, *Neurology*, 1969, **19**, 1000–1006.
16. K. Yokoi, M. Kimura and Y. Itokawa, Effect of dietary iron deficiency on mineral levels in tissues of rats, *Biol. Trace Elem. Res.*, 1991, **29**, 257–265.
17. K. M. Erikson, Z. K. Shihabi, J. L. Aschner and M. Aschner, Manganese accumulates in iron-deficient rat brain regions in a heterogeneous fashion and is associated with neurochemical alterations, *Biol. Trace Elem. Res.*, 2002, **87**, 143–156.
18. K. M. Erikson, T. Syversen, E. Steinnes and M. Aschner, Globus pallidus: a target brain region for divalent metal accumulation associated with dietary iron deficiency, *J. Nutr. Biochem.*, 2004, **15**, 335–341.
19. M. Aschner, T. R. Guilarte, J. S. Schneider and W. Zheng, Manganese: recent advances in understanding its transport and neurotoxicity, *Toxicol. Appl. Pharmacol.*, 2007, **221**, 131–147.
20. A. C. Illing, A. Shawki, C. L. Cunningham and B. Mackenzie, Substrate profile and metal-ion selectivity of human divalent metal-ion transporter-1, *J. Biol. Chem.*, 2012, **287**, 30485–30496.
21. N. Hubert and M. W. Hentze, Previously uncharacterized isoforms of divalent metal transporter (DMT)-1: implications for regulation and cellular function, *Proc. Natl. Acad. Sci. U. S. A.*, 2002, **99**, 12345–12350.
22. A. Shawki, P. B. Knight, B. D. Maliken, E. J. Niespodzany and B. Mackenzie, H(+)-coupled divalent metal-ion transporter-1: functional properties, physiological roles and therapeutics, *Curr. Top. Membr.*, 2012, **70**, 169–214.
23. M. W. Hentze, M. U. Muckenthaler, B. Galy and C. Camaschella, Two to tango: regulation of Mammalian iron metabolism, *Cell*, 2010, **142**, 24–38.
24. M. D. Fleming, M. A. Romano, M. A. Su, L. M. Garrick, M. D. Garrick and N. C. Andrews, Nramp2 is mutated in the anemic Belgrade (b) rat: evidence of a role for Nramp2 in endosomal iron transport, *Proc. Natl. Acad. Sci. U. S. A.*, 1998, **95**, 1148–1153.
25. A. C. Chua and E. H. Morgan, Manganese metabolism is impaired in the Belgrade laboratory rat, *J. Comp. Physiol., B*, 1997, **167**, 361–369.
26. A. Donovan, A. Brownlie, Y. Zhou, J. Shepard, S. J. Pratt, J. Moynihan, B. H. Paw, A. Drejer, B. Barut, A. Zapata, T. C. Law, C. Brugnara, S. E. Lux, G. S. Pinkus, J. L. Pinkus, P. D. Kingsley, J. Palis,

M. D. Fleming, N. C. Andrews and L. I. Zon, Positional cloning of zebrafish ferroportin1 identifies a conserved vertebrate iron exporter, *Nature*, 2000, **403**, 776–781.

27. A. T. McKie, P. Marciani, A. Rolfs, K. Brennan, K. Wehr, D. Barrow, S. Miret, A. Bomford, T. J. Peters, F. Farzaneh, M. A. Hediger, M. W. Hentze and R. J. Simpson, A novel duodenal iron-regulated transporter, IREG1, implicated in the basolateral transfer of iron to the circulation, *Mol. Cell*, 2000, **5**, 299–309.

28. S. Abboud and D. J. Haile, A novel mammalian iron-regulated protein involved in intracellular iron metabolism, *J. Biol. Chem.*, 2000, **275**, 19906–19912.

29. A. Pietrangelo, The ferroportin disease, *Blood Cells, Mol., Dis.*, 2004, **32**, 131–138.

30. H. Mok, J. Jelinek, S. Pai, B. M. Cattanach, J. T. Prchal, H. Youssoufian and A. Schumacher, Disruption of ferroportin 1 regulation causes dynamic alterations in iron homeostasis and erythropoiesis in polycythaemia mice, *Development*, 2004, **131**, 1859–1868.

31. A. Lymboussaki, E. Pignatti, G. Montosi, C. Garuti, D. J. Haile and A. Pietrangelo, The role of the iron responsive element in the control of ferroportin1/IREG1/MTP1 gene expression, *J. Hepatol.*, 2003, **39**, 710–715.

32. E. Nemeth, M. S. Tuttle, J. Powelson, M. B. Vaughn, A. Donovan, D. M. Ward, T. Ganz and J. Kaplan, Hepcidin regulates cellular iron efflux by binding to ferroportin and inducing its internalization, *Science*, 2004, **306**, 2090–2093.

33. T. Ganz, Hepcidin, a key regulator of iron metabolism and mediator of anemia of inflammation, *Blood*, 2003, **102**, 783–788.

34. C. Pigeon, G. Ilyin, B. Courselaud, P. Leroyer, B. Turlin, P. Brissot and O. Loreal, A new mouse liver-specific gene, encoding a protein homologous to human antimicrobial peptide hepcidin, is overexpressed during iron overload, *J. Biol. Chem.*, 2001, **276**, 7811–7819.

35. G. Nicolas, C. Chauvet, L. Viatte, J. L. Danan, X. Bigard, I. Devaux, C. Beaumont, A. Kahn and S. Vaulont, The gene encoding the iron regulatory peptide hepcidin is regulated by anemia, hypoxia, and inflammation, *J. Clin. Invest.*, 2002, **110**, 1037–1044.

36. I. De Domenico, D. M. Ward, G. Musci and J. Kaplan, Iron overload due to mutations in ferroportin, *Haematologica*, 2006, **91**, 92–95.

37. M. S. Madejczyk and N. Ballatori, The iron transporter ferroportin can also function as a manganese exporter, *Biochim. Biophys. Acta*, 2012, **1818**, 651–657.

38. X. Li, J. Xie, L. Lu, L. Zhang, Y. Zou, Q. Wang, X. Luo and S. Li, Kinetics of manganese transport and gene expressions of manganese transport carriers in Caco-2 cell monolayers, *Biometals*, 2013, **26**, 941–953.

39. Z. Yin, H. Jiang, E. S. Lee, M. Ni, K. M. Erikson, D. Milatovic, A. B. Bowman and M. Aschner, Ferroportin is a manganese-responsive protein that decreases manganese cytotoxicity and accumulation, *J. Neurochem.*, 2010, **112**, 1190–1198.

40. F. Aydemir, S. Jenkitkasemwong, S. Gulec and M. D. Knutson, Iron loading increases ferroportin heterogeneous nuclear RNA and mRNA levels in murine J774 macrophages, *J. Nutr.*, 2009, **139**, 434–438.

41. M. B. Troadec, D. M. Ward, E. Lo, J. Kaplan and I. De Domenico, Induction of FPN1 transcription by MTF-1 reveals a role for ferroportin in transition metal efflux, *Blood*, 2010, **116**, 4657–4664.

42. J. Chung, D. J. Haile and M. Wessling-Resnick, Copper-induced ferroportin-1 expression in J774 macrophages is associated with increased iron efflux, *Proc. Natl. Acad. Sci. U. S. A.*, 2004, **101**, 2700–2705.

43. B. Y. Park and J. Chung, Effects of various metal ions on the gene expression of iron exporter ferroportin-1 in J774 macrophages, *Nutr. Res. Pract.*, 2008, **2**, 317–321.

44. N. B. Nguyen, K. D. Callaghan, A. J. Ghio, D. J. Haile and F. Yang, Hepcidin expression and iron transport in alveolar macrophages, *Am. J. Physiol.: Lung Cell. Mol. Physiol.*, 2006, **291**, L417–425.

45. P. Aisen, R. Aasa and A. G. Redfield, The chromium, manganese, and cobalt complexes of transferrin, *J. Biol. Chem.*, 1969, **244**, 4628–4633.

46. R. A. Gibbons, S. N. Dixon, K. Hallis, A. M. Russell, B. F. Sansom and H. W. Symonds, Manganese metabolism in cows and goats, *Biochim. Biophys. Acta*, 1976, **444**, 1–10.

47. R. C. Keefer, A. J. Barak and J. D. Boyett, Binding of manganese and transferrin in rat serum, *Biochim. Biophys. Acta*, 1970, **221**, 390–393.

48. A. M. Scheuhammer and M. G. Cherian, Binding of manganese in human and rat plasma, *Biochim. Biophys. Acta*, 1985, **840**, 163–169.

49. L. Davidsson, B. Lonnerdal, B. Sandstrom, C. Kunz and C. L. Keen, Identification of transferrin as the major plasma carrier protein for manganese introduced orally or intravenously or after in vitro addition in the rat, *J. Nutr.*, 1989, **119**, 1461–1464.

50. A. A. Moshtaghie, A. Badii and T. Hsanazadeh, Role of cerulplasmin and ethanolamine in manganese binding to human serum apo-transferrin, *Iran. J. Sci. Technol.*, 1997, **21**, 157–167.

51. E. A. Heilig, K. J. Thompson, R. M. Molina, A. R. Ivanov, J. D. Brain and M. Wessling-Resnick, Manganese and iron transport across pulmonary epithelium, *Am. J. Physiol.: Lung Cell. Mol. Physiol.*, 2006, **290**, L1247–1259.

52. C. Fasolato, M. Hoth, G. Matthews and R. Penner, Ca2+ and Mn2+ influx through receptor-mediated activation of nonspecific cation channels in mast cells, *Proc. Natl. Acad. Sci. U. S. A.*, 1993, **90**, 3068–3072.

53. J. E. Merritt, R. Jacob and T. J. Hallam, Use of manganese to discriminate between calcium influx and mobilization from internal stores in stimulated human neutrophils, *J. Biol. Chem.*, 1989, **264**, 1522–1527.

54. H. Masumiya, H. Tsujikawa, N. Hino and R. Ochi, Modulation of manganese currents by 1, 4-dihydropyridines, isoproterenol and forskolin in rabbit ventricular cells, *Pflugers Archiv.*, 2003, **446**, 695–701.

55. R. G. Tsushima, A. D. Wickenden, R. A. Bouchard, G. Y. Oudit, P. P. Liu and P. H. Backx, Modulatj.ion of iron uptake in heart by L-type Ca2 + channel modifiers: possible implications in iron overload, *Circ. Res.*, 1999, **84**, 1302–1309.

56. J. S. Crossgrove and R. A. Yokel, Manganese distribution across the blood-brain barrier. IV. Evidence for brain influx through store-operated calcium channels, *Neurotoxicology*, 2005, **26**, 297–307.

57. J. Chen and G. J. Barritt, Evidence that TRPC1 (transient receptor potential canonical 1) forms a Ca(2+)-permeable channel linked to the regulation of cell volume in liver cells obtained using small interfering RNA targeted against TRPC1, *Biochem. J.*, 2003, **373**, 327–336.

58. T. Hofmann, A. G. Obukhov, M. Schaefer, C. Harteneck, T. Gudermann and G. Schultz, Direct activation of human TRPC6 and TRPC3 channels by diacylglycerol, *Nature*, 1999, **397**, 259–263.

59. M. K. Monteilh-Zoller, M. C. Hermosura, M. J. Nadler, A. M. Scharenberg, R. Penner and A. Fleig, TRPM7 provides an ion channel mechanism for cellular entry of trace metal ions, *J. Gen. Physiol.*, 2003, **121**, 49–60.

60. J. Mwanjewe and A. K. Grover, Role of transient receptor potential canonical 6 (TRPC6) in non-transferrin-bound iron uptake in neuronal phenotype PC12 cells, *Biochem. J.*, 2004, **378**, 975–982.

61. R. L. Corteling, S. Li, J. Giddings, J. Westwick, C. Poll and I. P. Hall, Expression of transient receptor potential C6 and related transient receptor potential family members in human airway smooth muscle and lung tissue, *Am. J. Respir. Cell Mol. Biol.*, 2004, **30**, 145–154.

62. A. Riccio, A. D. Medhurst, C. Mattei, R. E. Kelsell, A. R. Calver, A. D. Randall, C. D. Benham and M. N. Pangalos, mRNA distribution analysis of human TRPC family in CNS and peripheral tissues, *Mol. Brain Res.*, 2002, **109**, 95–104.

63. H. H. Xue, D. M. Zhao, T. Suda, C. Uchida, T. Oda, K. Chida, A. Ichiyama and H. Nakamura, Store depletion by caffeine/ryanodine activates capacitative Ca(2+) entry in nonexcitable A549 cells, *J. Biochem.*, 2000, **128**, 329–336.

64. E. Heilig, R. Molina, T. Donaghey, J. D. Brain and M. Wessling-Resnick, Pharmacokinetics of pulmonary manganese absorption: evidence for increased susceptibility to manganese loading in iron-deficient rats, *Am. J. Physiol.: Lung Cell. Mol. Physiol.*, 2005, **288**, L887–893.

65. D. Y. Bell, J. A. Haseman, A. Spock, G. McLennan and G. E. Hook, Plasma proteins of the bronchoalveolar surface of the lungs of smokers and nonsmokers, *Am. Rev. Respir. Dis.*, 1981, **124**, 72–79.

66. F. Yang, W. E. Friedrichs and J. J. Coalson, Regulation of transferrin gene expression during lung development and injury, *Am. J. Physiol.*, 1997, **273**, L417–426.

67. A. J. Ghio, J. D. Carter, J. M. Samet, W. Reed, J. Quay, L. A. Dailey, J. H. Richards and R. B. Devlin, Metal-dependent expression of ferritin and lactoferrin by respiratory epithelial cells, *Am. J. Physiol.*, 1998, **274**, L728–736.

68. E. R. Pacht and W. B. Davis, Role of transferrin and ceruloplasmin in antioxidant activity of lung epithelial lining fluid, *J. Appl. Physiol.*, 1988, **64**, 2092–2099.

69. A. J. Ghio, J. D. Carter, J. H. Richards, L. D. Richer, C. K. Grissom and M. R. Elstad, Iron and iron-related proteins in the lower respiratory tract of patients with acute respiratory distress syndrome, *Crit. Care Med.*, 2003, **31**, 395–400.

70. R. L. Upton, Y. Chen, S. Mumby, J. M. Gutteridge, P. B. Anning, A. G. Nicholson, T. W. Evans and G. J. Quinlan, Variable tissue expression of transferrin receptors: relevance to acute respiratory distress syndrome, *Eur. Respir. J.*, 2003, **22**, 335–341.

71. X. Wang, A. J. Ghio, F. Yang, K. G. Dolan, M. D. Garrick and C. A. Piantadosi, Iron uptake and Nramp2/DMT1/DCT1 in human bronchial epithelial cells, *Am. J. Physiol.: Lung Cell. Mol. Physiol.*, 2002, **282**, L987–995.

72. V. A. Fitsanakis, N. Zhang, S. Garcia and M. Aschner, Manganese (Mn) and iron (Fe): interdependency of transport and regulation, *Neurotoxic. Res.*, 2010, **18**, 124–131.

73. J. Ma, S. Haldar, M. A. Khan, S. D. Sharma, W. C. Merrick, E. C. Theil and D. J. Goss, Fe2+ binds iron responsive element-RNA, selectively changing protein-binding affinities and regulating mRNA repression and activation, *Proc. Natl. Acad. Sci. U. S. A.*, 2012, **109**, 8417–8422.

74. M. A. Rahman, B. Rahman and N. Ahmed, High blood manganese in iron-deficient children in Karachi, *Public Health Nutr.*, 2013, **16**, 1677–1683.

75. E. A. Smith, P. Newland, K. G. Bestwick and N. Ahmed, Increased whole blood manganese concentrations observed in children with iron deficiency anaemia, *J. Trace Elem. Med. Biol.*, 2013, **27**, 65–69.

76. H. M. Meltzer, A. L. Brantsaeter, B. Borch-Iohnsen, D. G. Ellingsen, J. Alexander, Y. Thomassen, H. Stigum and T. A. Ydersbond, Low iron stores are related to higher blood concentrations of manganese, cobalt and cadmium in non-smoking, Norwegian women in the HUNT 2 study, *Environ. Res.*, 2010, **110**, 497–504.

77. Y. Kim and B. K. Lee, Iron deficiency increases blood manganese level in the Korean general population according to KNHANES 2008, *Neurotoxicology*, 2011, **32**, 247–254.

78. J. D. Brain, E. Heilig, T. C. Donaghey, M. D. Knutson, M. Wessling-Resnick and R. M. Molina, Effects of iron status on transpulmonary transport and tissue distribution of Mn and Fe, *Am. J. Respir. Cell Mol. Biol.*, 2006, **34**, 330–337.

79. K. Thompson, R. M. Molina, T. Donaghey, J. E. Schwob, J. D. Brain and M. Wessling-Resnick, Olfactory uptake of manganese requires DMT1 and is enhanced by anemia, *FASEB J.*, 2007, **21**, 223–230.

80. K. Thompson, R. Molina, T. Donaghey, J. D. Brain and M. Wessling-Resnick, The influence of high iron diet on rat lung manganese absorption, *Toxicol. Appl. Pharmacol.*, 2006, **210**, 17–23.

81. A. Y. Watson and J. D. Brain, Uptake of iron aerosols by mouse airway epithelium, *Lab. Invest.*, 1979, **40**, 450–459.

82. H. Roels, R. Lauwerys, J. P. Buchet, P. Genet, M. J. Sarhan, I. Hanotiau, M. de Fays, A. Bernard and D. Stanescu, Epidemiological survey among workers exposed to manganese: effects on lung, central nervous system, and some biological indices, *Am. J. Ind. Med.*, 1987, **11**, 307–327.

83. J. Kim, R. M. Molina, T. C. Donaghey, P. D. Buckett, J. D. Brain and M. Wessling-Resnick, Influence of DMT1 and iron status on inflammatory responses in the lung, *Am. J. Physiol.: Lung Cell. Mol. Physiol.*, 2011, **300**, L659–665.

84. X. Wang, M. D. Garrick, F. Yang, L. A. Dailey, C. A. Piantadosi and A. J. Ghio, IFN-gamma, and endotoxin increase expression of DMT1 in bronchial epithelial cells, *Am. J. Physiol.: Lung Cell. Mol. Physiol.*, 2005, **289**, L24–33.

85. A. J. Ghio, C. A. Piantadosi, X. Wang, L. A. Dailey, J. D. Stonehuerner, M. C. Madden, F. Yang, K. G. Dolan, M. D. Garrick and L. M. Garrick, Divalent metal transporter-1 decreases metal-related injury in the lung, *Am. J. Physiol.: Lung Cell. Mol. Physiol.*, 2005, **289**, L460–467.

86. A. J. Ghio, J. L. Turi, M. C. Madden, L. A. Dailey, J. D. Richards, J. G. Stonehuerner, D. L. Morgan, S. Singleton, L. M. Garrick and M. D. Garrick, Lung injury after ozone exposure is iron dependent, *Am. J. Physiol.: Lung Cell. Mol. Physiol.*, 2007, **292**, L134–143.

87. F. Akbar-Khanzadeh, Short-term respiratory function changes in relation to workshift welding fume exposures, *Int. Arch. Occup. Environ. Health*, 1993, **64**, 393–397.

88. M. M. Boojar and F. Goodarzi, A longitudinal follow-up of pulmonary function and respiratory symptoms in workers exposed to manganese, *J. Occup. Environ. Med.*, 2002, **44**, 282–290.

89. T. A. Davies, Manganese pneumonitis, *Br. J. Ind. Med.*, 1946, **3**, 111–135.

90. ATSDR, *Toxicological Profile for Manganese*. U.S. Department of Health and Human Services Public Health Service, http://www.atsdr.cdc.gov/toxprofiles. Assessed October 2013.

91. J. M. Antonini, M. D. Taylor, A. T. Zimmer and J. R. Roberts, Pulmonary responses to welding fumes: role of metal constituents, *J. Toxicol. Environ. Health, Part A*, 2004, **67**, 233–249.

92. T. M. Rice, R. W. Clarke, J. J. Godleski, E. Al-Mutairi, N. F. Jiang, R. Hauser and J. D. Paulauskis, Differential ability of transition metals to induce pulmonary inflammation, *Toxicol. Appl. Pharmacol.*, 2001, **177**, 46–53.

93. S. Bredow, M. M. Falgout, T. H. March, C. M. Yingling, S. P. Malkoski, J. Aden, E. J. Bedrick, J. L. Lewis and K. K. Divine, Subchronic inhalation of soluble manganese induces expression of hypoxia-associated angiogenic genes in adult mouse lungs, *Toxicol. Appl. Pharmacol.*, 2007, **221**, 148–157.

94. J. Han, J. S. Lee, D. Choi, Y. Lee, S. Hong, J. Choi, S. Han, Y. Ko, J. A. Kim, Y. M. Kim and Y. Jung, Manganese (II) induces chemical

hypoxia by inhibiting HIF-prolyl hydroxylase: implication in manganese-induced pulmonary inflammation, *Toxicol. Appl. Pharmacol.*, 2009, **235**, 261–267.

95. D. C. Dorman, M. F. Struve, E. A. Gross, B. A. Wong and P. C. Howroyd, Sub-chronic inhalation of high concentrations of manganese sulfate induces lower airway pathology in rhesus monkeys, *Respir. Res.*, 2005, **6**, 121.

96. L. Zhang, H. Sang, Y. Liu and J. Li, Manganese activates caspase-9-dependent apoptosis in human bronchial epithelial cells, *Hum. Exp. Toxicol.*, 2013, **32**, 1155–1163.

97. B. C. Jang, Induction of COX-2 in human airway cells by manganese: role of PI3K/PKB, p38 MAPK, PKCs, Src, and glutathione depletion, *Toxicol. In Vitro*, 2009, **23**, 120–126.

98. D. M. Tessier and L. E. Pascal, Pascal, Activation of MAP kinases by hexavalent chromium, manganese and nickel in human lung epithelial cells, *Toxicol. Lett.*, 2006, **167**, 114–121.

99. L. E. Pascal and D. M. Tessier, Cytotoxicity of chromium and manganese to lung epithelial cells in vitro, *Toxicol. Lett.*, 2004, **147**, 143–151.

100. M. Ivan, K. Kondo, H. Yang, W. Kim, J. Valiando, M. Ohh, A. Salic, J. M. Asara, W. S. Lane and W. G. Kaelin, Jr., HIFalpha targeted for VHL-mediated destruction by proline hydroxylation: implications for O2 sensing, *Science*, 2001, **292**, 464–468.

101. H. J. Shin, M. S. Choi, N. H. Ryoo, K. Y. Nam, G. Y. Park, J. H. Bae, S. I. Suh, W. K. Baek, J. W. Park and B. C. Jang, Manganese-mediated up-regulation of HIF-1alpha protein in Hep2 human laryngeal epithelial cells via activation of the family of MAPKs, *Toxicol. In Vitro*, 2010, **24**, 1208–1214.

102. Q. Li, H. Chen, X. Huang and M. Costa, Effects of 12 metal ions on iron regulatory protein 1 (IRP-1) and hypoxia-inducible factor-1 alpha (HIF-1alpha) and HIF-regulated genes, *Toxicol. Appl. Pharmacol.*, 2006, **213**, 245–255.

103. P. Camner, T. Curstedt, C. Jarstrand, A. Johannsson, B. Robertson and A. Wiernik, Rabbit lung after inhalation of manganese chloride: a comparison with the effects of chlorides of nickel, cadmium, cobalt, and copper, *Environ. Res.*, 1985, **38**, 301–309.

104. T. E. Kehl-Fie and E. P. Skaar, Nutritional immunity beyond iron: a role for manganese and zinc, *Curr. Opin. Chem. Biol.*, 2010, **14**, 218–224.

105. M. I. Hood and E. P. Skaar, Nutritional immunity: transition metals at the pathogen-host interface, *Nat. Rev. Microbiol.*, 2012, **10**, 525–537.

106. M. L. Zaharik and B. B. Finlay, Mn2+ and bacterial pathogenesis, *Front. Biosci.*, 2004, **9**, 1035–1042.

107. B. Adkins, Jr., G. H. Luginbuhl, F. J. Miller and D. E. Gardner, Increased pulmonary susceptibility to streptococcal infection following inhalation of manganese oxide, *Environ. Res.*, 1980, **23**, 110–120.

108. M. Kalin, Pneumococcal serotypes and their clinical relevance, *Thorax*, 1998, **53**, 159–162.

109. A. Marra, S. Lawson, J. S. Asundi, D. Brigham and A. E. Hromockyj, In vivo characterization of the psa genes from Streptococcus pneumoniae in multiple models of infection, *Microbiology*, 2002, **148**, 1483–1491.

110. H. Y. Reynolds, Defense mechanisms against infections, *Curr. Opin. Pulm. Med.*, 1999, **5**, 136–142.

111. D. A. Watson, D. M. Musher and J. Verhoef, Pneumococcal virulence factors and host immune responses to them, *Eur. J. Clin. Microbiol. Infect. Dis.*, 1995, **14**, 479–490.

112. A. Dintilhac, G. Alloing, C. Granadel and J. P. Claverys, Competence and virulence of Streptococcus pneumoniae: Adc and PsaA mutants exhibit a requirement for Zn and Mn resulting from inactivation of putative ABC metal permeases, *Mol. Microbiol.*, 1997, **25**, 727–739.

113. C. L. Ong, A. J. Potter, C. Trappetti, M. J. Walker, M. P. Jennings, J. C. Paton and A. G. McEwan, Interplay between manganese and iron in pneumococcal pathogenesis: role of the orphan response regulator RitR, *Infect. Immun.*, 2013, **81**, 421–429.

114. S. Paik, A. Brown, C. L. Munro, C. N. Cornelissen and T. Kitten, The sloABCR operon of Streptococcus mutans encodes an Mn and Fe transport system required for endocarditis virulence and its Mn-dependent repressor, *J. Bacteriol.*, 2003, **185**, 5967–5975.

115. B. D. Corbin, E. H. Seeley, A. Raab, J. Feldmann, M. R. Miller, V. J. Torres, K. L. Anderson, B. M. Dattilo, P. M. Dunman, R. Gerads, R. M. Caprioli, W. Nacken, W. J. Chazin and E. P. Skaar, Metal chelation and inhibition of bacterial growth in tissue abscesses, *Science*, 2008, **319**, 962–965.

116. S. M. Damo, T. E. Kehl-Fie, N. Sugitani, M. E. Holt, S. Rathi, W. J. Murphy, Y. Zhang, C. Betz, L. Hench, G. Fritz, E. P. Skaar and W. J. Chazin, Molecular basis for manganese sequestration by calprotectin and roles in the innate immune response to invading bacterial pathogens, *Proc. Natl. Acad. Sci. U. S. A.*, 2013, **110**, 3841–3846.

117. S. Jenkitkasemwong, C. Y. Wang, B. Mackenzie and M. D. Knutson, Physiologic implications of metal-ion transport by ZIP14 and ZIP8, *Biometals*, 2012, **25**, 643–655.

118. M. Quadri, A. Federico, T. Zhao, G. J. Breedveld, C. Battisti, C. Delnooz, L. A. Severijnen, L. Di Toro Mammarella, A. Mignarri, L. Monti, A. Sanna, P. Lu, F. Punzo, G. Cossu, R. Willemsen, F. Rasi, B. A. Oostra, B. P. van de Warrenburg and V. Bonifati, Mutations in SLC30A10 cause parkinsonism and dystonia with hypermanganesemia, polycythemia, and chronic liver disease, *Am. J. Hum. Genet.*, 2012, **90**, 467–477.

119. K. Tuschl, P. T. Clayton, S. M. Gospe, Jr., S. Gulab, S. Ibrahim, P. Singhi, R. Aulakh, R. T. Ribeiro, O. G. Barsottini, M. S. Zaki, M. L. Del Rosario, S. Dyack, V. Price, A. Rideout, K. Gordon, R. A. Wevers, W. K. Chong and P. B. Mills, Syndrome of hepatic cirrhosis, dystonia, polycythemia, and hypermanganesemia caused by mutations in SLC30A10, a manganese transporter in man, *Am. J. Hum. Genet.*, 2012, **90**, 457–466.

120. W. He and Z. Hu, The role of the Golgi-resident SPCA Ca(2)(+)/ Mn(2)(+) pump in ionic homeostasis and neural function, *Neurochem. Res.*, 2012, **37**, 455–468.

121. S. Leitch, M. Feng, S. Muend, L. T. Braiterman, A. L. Hubbard and R. Rao, Vesicular distribution of Secretory Pathway Ca(2)+-ATPase isoform 1 and a role in manganese detoxification in liver-derived po-larized cells, *Biometals*, 2011, **24**, 159–170.

122. A. D. Gitler, A. Chesi, M. L. Geddie, K. E. Strathearn, S. Hamamichi, K. J. Hill, K. A. Caldwell, G. A. Caldwell, A. A. Cooper, J. C. Rochet and S. Lindquist, Alpha-synuclein is part of a diverse and highly conserved interaction network that includes PARK9 and manganese toxicity, *Nat. Genet.*, 2009, **41**, 308–315.

123. F. Yang, D. J. Haile, X. Wang, L. A. Dailey, J. G. Stonehuerner and A. J. Ghio, Apical location of ferroportin 1 in airway epithelia and its role in iron detoxification in the lung, *Am. J. Physiol.: Lung Cell. Mol. Physiol.*, 2005, **289**, L14–23.

124. E. A. Heilig, Ph.D. Dissertation, Harvard University, 2005.

CHAPTER 6

Are There Distinguishable Roles for the Different Oxidation States of Manganese in Manganese Toxicity?

THOMAS E. GUNTER

Department of Biochemistry and Biophysics, University of Rochester
Medical Center, Rochester, NY, USA
Email: thomas_gunter@urmc.rochester.edu

6.1 Introduction

Manganese (Mn) is an essential biological element and a cofactor in a number of important biochemical reactions. Nevertheless, excessive accumulation of Mn in certain regions of the brain can lead to neurodegeneration. This neurodegenerative syndrome, known as "manganism," shows characteristics similar to those of Parkinson's disease (PD).

Over many years of observation, exposure to excessive Mn has been associated clinically with psychiatric and neurological symptoms.[1-3] These include emotional instability, compulsive sometimes violent behavior, hallucinations, dystonia, bradykinesia, slurring or stuttering speech at a lower volume, and a gait disturbance sometimes known as "cock walk."[1-3] These symptoms correlate with increased Mn concentration in the striatum, globus pallidus, caudate-pitamen and subthalamic nuclei.[3-5] There is a significant resemblance between the symptoms of Mn toxicity and idiopathic PD.

Issues in Toxicology No. 22
Manganese in Health and Disease
Edited by Lucio G. Costa and Michael Aschner
© The Royal Society of Chemistry 2015
Published by the Royal Society of Chemistry, www.rsc.org

There is also evidence that increased Mn in these brain areas may be a risk factor for PD.[6] At the organ and cellular level, accumulation of Mn in these brain areas correlate with cell death in the internal segment of the globus pallidus, decrease of dopamine in the striatum, and gliosis and swelling in the posterior limb of the internal capsule.[3] The striatum and globus pallidus seem to be areas of the brain most sensitive to damage by Mn.[7–10]

Most of the earlier work focused on understanding the biological transport of Mn, the inhibitions and effects of excessive Mn, and the possible mechanisms through which the initial biochemical effects of Mn are translated into the observed signs and symptoms of Mn toxicity used the more stable Mn^{2+} ion.[10] Far too many inhibitory effects of Mn, primarily Mn^{2+}, are reported to detail here. To summarize, the most frequently proposed mechanisms leading to Mn toxicity can be categorized into two general classes: (1) mitochondrial effects, including impaired energy metabolism and ATP production, production of reactive oxygen species, and apoptotic effects including activation of proteases;[11–21] and (2) pathological changes in the levels of neurotransmitters such as dopamine, γ-aminobutyric acid (GABA), and glutamate.[3,6,10,22–25] Some would prefer to split the mitochondrial effects from the apoptotic effects;[26] however, because of the central role played by mitochondria in apoptosis,[27–29] these effects could prove difficult to separate in some cases. There are now hundreds to thousands of papers in the scientific literature relevant to Mn toxicity. For a broad background in the general characteristics of Mn toxicity, the reader is referred to the many excellent reviews available today,[6,10,26,30–32] while we restrict our focus here to those issues important to distinguishing effects of the different ionic states of Mn or Mn speciation in Mn toxicity.

6.2 A Brief Review of the Inorganic Chemistry of Mn^{2+} and Mn^{3+}

Mn is the twelfth most abundant element in the crust of the earth and shows seven different oxidation states $(0, 2+, 3+, 4+, 5+, 6+, \text{and } 7+)$.[33] Mn^{2+} is by far the most stable state.[33–35] Mn^{2+} and Mn^{3+} have been found in animal cells,[36] small amounts of Mn^{4+} in the form of MnO_2 have been reported in plant root tissue,[37,38] and both Mn^{3+} and Mn^{4+} have been found to be associated with the process of oxygen generation in photosynthesis (see Armstrong[33] and references therein). Mn^{2+} behaves quite a bit like Mg^{2+} and Ca^{2+} (Group 2 elements) in that it is soluble in aqueous medium, complexes weakly like Group 2 elements, and is relatively redox inactive in aqueous media.[33] Mn^{2+} has a half-filled 3d shell giving it spherical symmetry like that of Mg^{2+} and Ca^{2+}, and its ionic radius is intermediate between those of Mg^{2+} and Ca^{2+} (Mg^{2+} – 86 pm, Mn^{2+} – 97 pm, and Ca^{2+} – 114 pm).[33] This is undoubtedly why Mn^{2+} behaves biologically so much like Mg^{2+} or Ca^{2+}. Mg^{2+} is a common cofactor for numerous biochemical reactions, and Mn^{2+} will often substitute for Mg^{2+} in this role. Ca^{2+} is perhaps the most common second messenger – Mn^{2+} often binds like Ca^{2+} and is often transported over Ca^{2+} transport mechanisms.[26,39,40]

Mn^{3+}, on the other hand is relatively insoluble. Mn^{3+} compounds prepared from water solutions are only slightly soluble or slightly dissociated.[33] Linus Pauling in his book General Chemistry says that "manganic ion, Mn^{3+}, is a strong oxidizing agent and its salts are unimportant,"[35]; however, there can be no doubt that Mn^{3+} has the potential to induce damage to biological tissue though oxidation upon exposure. Therefore, the issue of how much biological damage is caused by Mn^{3+} and how much by Mn^{2+} becomes, in part, an issue of where and to what extent cellular components are exposed to Mn^{2+} or Mn^{3+}. This suggests that both cellular and intracellular transport of both of these ionic states is important to the causes of the biological damage induced by Mn. It also suggests that we explore the question of whether or not oxidation of the more stable Mn^{2+} to the more reactive Mn^{3+} occurs within living cells or their organelles.

There are relatively stable Mn^{3+} complexes, such as Mn^{3+} transferrin (Tf) and Mn^{3+} pyrophosphate,[9,41–43] which can be put into aqueous solution as soluble complexes; however, the concentration of free Mn^{3+} in aqueous solution is always very small.[33–35] In the presence of oxygen, Mn^{3+} readily disproportionates into Mn^{2+} and MnO_2 (a very dark precipitate).[33–35]

The transition metal, Mn, is adjacent to iron (Fe) in the periodic table and shares properties and binding sites with Fe. Like Fe, it can transition between the $2+$ and $3+$ oxidation states as part of its biological function.[33]

6.3   Current Physical Techniques Useful in Mn Speciation

6.3.1   UV-Visible Spectroscopy

It has long been known that solutions of transition metals show color changes upon changes of oxidation state. The exact color shown generally depends on the compound. Mn^{2+} solutions are very light pink, while Mn^{3+} solutions take on a much deeper hue (often a dark purplish-red, indicating a greater absorbance). Mn^{4+} is a very dark gray/black. Fe^{2+} is often a light blue/green color, while Fe^{3+} is a darker yellowish-red to dark red. Cu^{2+} is blue, but Cu^+ compounds are generally colorless because Cu^+ has a completed 3d shell. UV–visible spectroscopy is not only useful in identifying the species of transition metal ions present in the solution, but it is also available in almost every laboratory. Therefore, UV–visible spectroscopy is a convenient source of information on the species of Mn ions present.

In any spectroscopy, if the components of a solution are independent, that is, if the presence of one component does not interact with and change the spectrum of another, one should be able to add the contributions of each component to get the absorption spectrum of the final solution. This might be called "the addition rule." In practice, this is difficult to do with the UV–visible spectroscopy of Mn, particularly with biological samples. First, because many different chromaphores absorb light in the same UV–visible range as Mn ions, making the final spectrum a very complicated combination of known and unknown components in which the Mn-related part may be very small. Although background spectra can be subtracted out (*i.e.* the biological sample without the added Mn ions) interactions between components often invalidate the addition rule. Finally, the size of many biological organelles and debris usually add intense Rayleigh scattering to the final UV–visible spectrum, which further complicates the use of the addition rule and the interpretation of the Mn components in the spectrum. This means that UV–visible spectroscopy is most often useful in demonstrating the presence of Mn^{3+} within a simple chemical component before it is introduced into a biological system.

Information on preparation of purified samples of some of these soluble Mn^{3+} complexes, their UV–visible spectra, and extinction coefficients that can be used to measure the concentrations of purified samples are available in the chemical literature. For example, the absorption maximum of Mn^{3+} pyrophosphate is $\lambda = 258$ nm $(\varepsilon = 6.2$ mM^{-1} cm^{-1}),[42,44,45] the absorption maximum of Mn^{3+} citrate is $\lambda = 340$ nm $(\varepsilon = 310$ mM^{-1} cm^{-1}),[42,45] and the absorption maximum of Mn^{3+}Tf is $\lambda = 420$ nM.[41,46]

6.3.2 XANES Spectroscopy

XANES stands for X-ray absorption near edge structure. Energetic X-rays can promote electrons from the lower energy orbitals of an atom either to unoccupied higher orbitals or to the continuum, which is the band of energies above the highest bound state orbital of the atom. Absorption of X-rays in promotion of electrons to unoccupied higher orbitals gives absorption lines in the X-ray spectrum while their promotion to the continuum gives an absorption edge. This spectrum looks like an edge or a very wide band of energies rather than a narrow band, as with an absorption line, because all the energies of the continuum are available. Each element has absorption edges that relate to promotion of s electrons, p electrons, *etc.* to the continuum. Usually the edge at the highest energy representing a transition from the 1s shell to the continuum (the K edge) is used because that edge generally gives the highest sensitivity. Several characteristics make XANES spectroscopy very useful for speciation of Mn or any other element which has a high enough atomic number to give sufficient sensitivity (usually this means heavier than calcium (Ca)):

(1) The absorption edges of the elements are tabulated and the corresponding K edges of successively heavier elements are separated by wide differences in energy. Therefore, the edge for each element is in its own energy region usually free from interference from absorption due to other elements.

(2) At the K edge for Mn near 6.550 keV, the background, from anything other than Mn absorption, is almost completely Compton scattering

from scattering of the incident X-rays by the electrons in the sample. This can be measured accurately using a similar sample that contains no Mn and subtracted from the total absorption of the sample containing Mn, leaving the edge absorption.

(3) All of the Mn complexes present in the sample show an absorbance proportional to their concentration in the sample.

(4) The absorption edges of each oxidation state present fall at a slightly different energy, with the edges for the higher oxidation states at a higher energy. For example, the difference in the position of the K alpha absorption edges of Mn^{2+} and Mn^{3+} is about 3 eV, which can be easily measured.

(5) There are spectroscopic differences in the XANES absorption of each Mn complex that can be determined using model compounds and used to identify the complexes present in the sample. This can be done either with the absorption spectra or with the first derivatives of the absorption spectra, which often show more structure.[36,47–49]

Let's go over the physical reasons for each of these useful characteristics. In points (1) and (2), the energy separation between the 1s level and the continuum is due to the Coulomb attraction between these innermost electrons (*i.e.* the 1s electrons) and the nucleus. As one goes from one element to the next highest in atomic number, the charge on the nucleus goes up by one proton, the attraction becomes progressively stronger, and the position of the corresponding edge moves to higher energy. These Coulomb energies are large with respect to the range of energy that should be scanned in XANES spectroscopy. For example, the separation between the K edges of chromium (6.005 keV) and manganese (6.550 keV), and of manganese (6.550 keV) and iron (7.125 keV), are hundreds of eV, and so there should be no overlap between the XANES spectra of these elements. The same argument can be made for the energy separation between each progressively higher electron shell and the continuum (*i.e.* for other sets of absorption edges). Although it is possible for the K edge of a lighter element to lie near another edge (say the L edge or the M edge of a heavier element), this is not often a problem because the width of the X-ray spectrum is huge when compared with that of the UV–visible spectrum and the concentration of these heavier elements, which could lead to such problems, is usually completely negligible.

In point (3), XANES is not troubled by the problem seen in some other spectroscopies such as magnetic resonance, in which there can be relaxation times for states that are so rapid that the absorption lines are broadened to invisibility. If a complex is present in the sample, it will contribute its share toward the total absorbance.

Point (4) illustrates a very useful characteristic of XANES spectra in that the absorption edge for successively higher oxidation states of an element falls at successively higher energies. The cause of this effect is again related to charge–charge or Coulomb interaction, but the energy shift is not nearly so great as that seen when the positive charge of the nucleus is increased. The issue is the effect on the energy of a 1s electron, for example, when an outer electron is removed from its orbital (increase in oxidation state). The effect is that the nucleus will attract the remaining electrons a little tighter, slightly lowering the energy of the 1s state and moving the XANES K edge to a slightly higher energy.

Point (5) shows that XANES spectra have structure. That's what the "near edge structure" in the term XANES refers to. This structure is primarily determined by the symmetries (*i.e.* geometries) and intensities of scattering from the ligands surrounding the ion of interest in its complex. While there is theory that can help us interpret this structural information, it's much easier to simply set up the complexes of interest as model compounds and experimentally determine their spectra. Then, using the addition rule, obtain information on the complexes present in a biological sample by trying to represent the experimental spectrum of the sample as a sum of those of the model compounds. This can be done using either the experimental absorption spectra or their first derivatives. The first derivatives often provide a wider range of structural differences.

The only practical difficulty in using XANES spectroscopy is that in order to get the broad X-ray spectrum and high X-ray intensity necessary to obtain good spectra, an electron synchrotron storage ring is required. Facilities with good detection sensitivity are available in the United States, Europe, and Japan. Almost all of these facilities have programs through which scientists from outside institutions can get beam time for independent experiments. A more complete description of how to carry out XANES experiments and reduce XANES data is given in earlier publications.[36,47–49]

6.3.3 Electron Paramagnetic Resonance Spectroscopy (EPR)

EPR is a field of spectroscopy open for study of paramagnetic ions and molecules such as transition metal ions and free radicals. Mn^{2+}, Mn^{3+}, and Mn^{4+}, and many other species are paramagnetic and under the right conditions can be studied using EPR. Several of the references discussed next use EPR to obtain information in studies comparing the observed effects in biological systems exposed to free Mn^{2+} or Mn^{3+} complexes. Unfortunately, EPR is a very complex spectroscopic field that can't be adequately covered in a few paragraphs. Consequently, this section will be restricted to pointing out a few caveats for use of EPR, which could impact speciation studies similar to those discussed.

Mn^{2+} has five 3d electrons in a half-filled 3d shell and a nuclear spin of 5/2. This gives the free ion spherical symmetry like that of Ca^{2+} and Mg^{2+} and is one of the reasons it binds so readily to their binding sites. In high symmetry complexes, as in the octahedral symmetry found in free Mn^{2+} (the inner hydration shell has six waters located at equidistance along the plus and minus *x*, *y*, and *z* axes), the relaxation times (T1 and T2) are long, and there is no strong relaxation broadening of the six line hyperfine spectrum, even in liquid water at 25 °C.[39,50–55] However, a major problem in using EPR to quantify the amount of Mn^{2+} present is that the spectra of most Mn^{2+} complexes are more complex than the simple six line spectrum of free Mn^{2+} and have broader lines, and the accuracy of quantifying the amount of Mn^{2+} responsible for the spectrum can be very poor. The physical reasons that quantization using EPR is generally less accurate than quantization using other fields of spectroscopy lie in the properties of EPR detection. To obtain good sensitivity, EPR uses phase sensitive detection. This type of detection gives a first derivative absorption output that requires integrating the signal twice to obtain the area under the absorption curve, which is related to the amount of material generating a specific signal. Phase sensitive detection is also sensitive to several types of phase-coupled artifacts that distort the background and make acquiring accurate absorption information in the tails of EPR spectra very difficult. At the same time, the necessity of integrating the signal twice makes an accurate result very dependent on accurate information of absorption in the tails of the EPR signals. When care is taken with this quantization, EPR can be used as described in the following references; however, even small changes in line with high symmetry spectra, like that of free Mn^{2+}, due to spin–spin interactions or to changes of viscosity in liquid samples can lead to significant quantization error.[55]

Usually, EPR spectra of Mn^{3+} are not seen except perhaps in some cases at very low temperatures (such as 4.2 °K). There are two primary reasons for this. There may be no low-lying energy level split by the magnetic field, and large zero field splittings (due to crystal fields, *etc.*) may move spectra out of the detectible field range. Usually, relaxation times are so fast that lines that might otherwise be seen are relaxation broadened into invisibility, unless the work is done at very low temperature.[55]

In the hands of spectroscopists who understand EPR, the technique can be useful in Mn speciation, but the work should be overseen by experts.

6.4 Studies Most Relevant to Mn Speciation

There are many issues and questions involved with determining whether an event that leads to pathology is caused by Mn^{2+} or Mn^{3+}. We already know that Mn^{2+} is by far the more stable of these two ionic forms of Mn as free ions, especially in the pH range where biological processes occur. Correspondingly, Mn^{3+} is a strong oxidizing agent and is much more reactive than Mn^{2+}. It would be reasonable to expect that exposure to free Mn^{3+} (*i.e.* Mn^{3+} surrounded by only its hydration shells and not stabilized by complexation to Tf, citrate, pyrophosphate, *etc.*) could be much more damaging to a cell or to intracellular or intramitochondrial components than exposure to Mn^{2+}. Because of the instability of Mn^{3+}, it is not possible to expose cells to high free Mn^{3+} concentrations; however, they can be exposed to high concentrations of stable Mn^{3+} complexes. In such exposures, it might be useful to keep in mind that the induced damage could be related to the instability of the complex used. It is not hard to imagine the strong oxidizing agent, Mn^{3+}, oxidizing some cell component and being converted to Mn^{2+} inside a cell or mitochondrion, but whether conversion of Mn^{2+} to Mn^{3+} occurs in one of these locations is another very important question. Since the concentration of free Mn^{3+} is very small, if significant transport of Mn^{3+} occurs, it must be as a complex in which the Mn^{3+} is stabilized, such as Mn^{3+} citrate or Mn^{3+}Tf, and not as the free Mn^{3+} ion. The only transport of a Mn^{3+} complex that is well documented in the literature is transport of Mn^{3+}Tf.[41,56,57]

Although Mn can be important biologically because it functions as a cofactor in a number of biochemical reactions, it is also toxicologically important because it can bind to almost all Ca^{2+} and Mg^{2+} binding sites and to many iron (Fe) binding sites. Ca^{2+} is perhaps the most common second messenger, Mg^{2+} is an important cofactor in many biochemical reactions, and Fe is central to the function of hemes. Because of the important biological roles of Ca^{2+}, Mg^{2+}, and Fe, this translates into many possible sites at which Mn could inhibit or otherwise interfere with the processes in which Ca^{2+}, Mg^{2+} or Fe participate. Mn^{2+}, Mn^{3+} or Mn^{4+} should therefore be expected to interfere with or inhibit many biological processes in the striatum or globus pallidus. When Mn interferes by binding to Ca^{2+} or Mg^{2+} binding sites, it must do so in the Mn^{2+} form because only in this form does it have the ionic radius and preferences in binding symmetry to substitute for Ca^{2+} or Mg^{2+}. The relevance of this to Mn speciation is that it may allow us to identify a process in which the Mn-induced injury may occur strictly by a process involving only the Mn^{2+} form. A likely example of this type of process is the observation, mentioned earlier, that Mn^{2+} inhibits ATP production within the coupled reactions of oxidative phosphorylation in mitochondria. A complicating factor is that not all Mn-induced interferences or inhibitions are toxicologically relevant. Suppose, for example, that Mn inhibits two steps in the metabolic processes leading to ATP production, one at very low Mn concentrations and the other at much higher concentrations. We'll probably never see biological effects from the inhibition that requires the higher Mn concentration. Another complicating factor with mitochondrial processes is that mitochondria from different tissues contain different relative amounts of the enzymes catalyzing the different steps of oxidative phosphorylation, so that a step that may be rate-limited by Mn^{2+} in one type of mitochondria may not be rate-limited in another type of mitochondria. It is very important to understanding Mn toxicity to be able to identify the toxicologically relevant initial effects of Mn on the brain, particularly on the striatum and globus pallidus. For the complex case of oxidative phosphorylation in mitochondria, an experimental process has been identified that does allow us to identify the Mn^{2+}-coupled, rate-limiting steps of ATP production and therefore identify the important steps of a Mn^{2+}-linked inhibition.

Finally, Mn toxicity in humans often occurs following occupational exposure to small Mn-containing particles that may contain a range of Mn oxidation states. The effects of these Mn nanoparticles on cells should be of great importance to toxicology.

Unfortunately, the ionic state in which Mn exerts its most important effects on toxicity generally has received relatively little study, probably because assessing its ionic state within biological systems requires equipment, such as that for XANES spectroscopy, rarely available in academic labs.

6.5   Studies of Biological Effects of Exposure to Mn^{2+} or Mn^{3+} Complexes

An important issue that should be addressed is whether at the isolated cell, tissue, or intact animal level, the treatment of a biological sample with Mn^{3+} complexes shows more or less toxicity than treatment with similar amounts of Mn^{2+} and, if so, whether the effects observed are caused by higher toxicity of Mn^{3+} complexes or higher uptake of Mn^{3+} complexes. We might expect Mn^{3+} complexes to be more dangerous because Mn^{3+} is a strong oxidizing agent and is so much more reactive than Mn^{2+}. As discussed earlier, although Mn^{2+} is relatively stable as a free ion in aqueous solution at neutral pH, Mn^{3+} is not and can only be stabilized in such solutions in specific complexes. Stated simply, what this means is that while the concentration of free Mn^{3+} is not zero in these solutions, it is very small. The Mn^{3+} needs the stabilizing effect of the ligands and geometries of specific complexes around it even in solution to be stable. We cannot easily calculate exactly how low the concentration of this free Mn^{3+} is in these solutions, particularly in biological solutions, because of incomplete information on the necessary solubility products, dissociation constants, and other thermodynamic parameters for free Mn^{3+} in the literature. To give a rough estimate for the upper limits of free or "aqueous" Mn^{3+} concentration in aqueous solution, Klewicki and Morgan[45] estimated it to be about 8 nM, where free Mn^{2+} concentration was about 1 mM, total Mn^{3+} concentration in a soluble complex, such as Mn^{3+} pyrophosphate, was about 0.1 mM at pH 2, where free Mn^{3+} is significantly more stable than at pH 7. Since Mn^{3+} can only be held in any significant concentration in these complexes, we might also expect the mechanisms of transport of Mn^{2+} and Mn^{3+} complexes to be very different.

A few studies have focused on comparing the effects of treatment of cells or animals with solutions containing Mn^{2+} with those containing Mn^{3+} complexes. Chen et al.[58] compared several characteristics following uptake from $Mn^{2+}Cl_2$ or Mn^{3+} acetate into PC12 cells. They dissolved $Mn^{2+}Cl_2$ in H_2O and Mn^{3+} acetate in a 1 : 1 mixture of ethanol and DMSO, which they found gave no precipitate when added to culture medium. After treatment of PC12 cells with one of these Mn solutions, they monitored aconitase activity, mitochondrial complex I activity, amount of transferrin receptor (TfR), cell growth, and amount of mitochondrial DNA (mtDNA) in samples.[58]

The results showed that the Mn^{3+} complex inactivated aconitase almost 10 times more strongly than Mn^{2+} (at a similar dose), that reactivation of aconitase activity with Fe^{2+} was more complete following Mn^{3+} treatment, that both Mn species inhibited complex I activity roughly equally well, that Mn^{3+} caused a greater stimulation of TfR, that Mn^{3+} was more effective in decreasing cell growth, and that Mn^{2+} treatment caused a slightly greater decrease in mtDNA than Mn^{3+}.[58] They concluded that Mn^{3+} was more toxic to neuron-like cells than Mn^{2+}. Furthermore, they speculated that the reason was that Mn^{3+} was better able than Mn^{2+} to cause loss of an Fe ion from iron–sulfur clusters with the substitution of Mn for the Fe and a decrease in complex activity.[58]

Smith, Reaney, and coworkers have published a series of papers comparing uptake and toxicity of Mn^{2+} and Mn^{3+} compounds from the cellular to the whole animal levels.[9,42,43] Mn^{3+} acetate, Mn^{3+} citrate, and Mn^{3+} pyrophosphate were studied for suitability for uptake and toxicity studies in PC12 cells using light scattering to show precipitation and atomic absorption measurements after centrifugation to indicate a decrease in the Mn^{3+} complex with time in solution. Of these three complexes only Mn^{3+} pyrophosphate was considered stable enough to use in experiments.[42] Mn uptake into PC12 cells following treatment with either $Mn^{2+}Cl_2$ or Mn^{3+} pyrophosphate was found to be significantly greater for cells treated with the Mn^{3+} pyrophosphate containing medium than with an equimolar amount of the $Mn^{2+}Cl_2$ containing medium.[43] Assays were also carried out measuring EPR measurable Mn, cellular ATP content, lactate dehydrogenase (LDH) activity (a cytotoxicity assay), dopamine (DA) content, 5-hydroxytryptophan (5-HT) content (a measure of seratonin), H-ferritin content, TfR, and MnSOD content as a function of Mn concentration during treatment.[43] When compared with controls, ATP content fell slightly less than 20% (from zero to 150 to 200 µM Mn concentration with both Mn^{2+} and Mn^{3+} treatment), while LDH activity increased a bit more than 20%. DA content fell as much as 40% within the same dose range and, interestingly, fell more for Mn^{2+} treatment than for Mn^{3+} treatment. Seratonin content fell precipitously for Mn^{3+} treatment but not nearly as much for Mn^{2+} treatment. H-ferritin content fell precipitously, particularly for Mn^{2+} treatment, while TfR increased 35–40% in the same dose range. MnSOD fell nearly 40% with Mn^{2+} treatment but increased 20% with Mn^{3+} treatment.

In a later paper,[9] this same laboratory studied uptake from $Mn^{2+}Cl_2$ and Mn^{3+} pyrophosphate following intraperitoneal injection of rats in the dose range 0, 30, and 90 mg kg^{-1} body weight (total injected Mn) over a 5 week period. Slides were taken of the striatum, globus pallidus (GP), thalamus, and cortex and [Mn] measured using proton-induced X-ray emission (PIXE) to measure Mn, Fe, Zn, and Cu in these brain locations.[9] Tissue was taken from the other hemisphere of the brain to measure DA (and its oxidation products), GABA, TfR, H-ferritin, and tubulin.[9]

Results showed that blood uptake of Mn was almost twice as high from the Mn^{3+} complex than from Mn^{2+}; however, liver uptake didn't change much except at the highest level of Mn^{3+} where it was up 25%.[9] The Mn uptake in all four brain regions was about the same with a smaller (15–20%) uptake from the Mn^{3+} complex than from the Mn^{2+}-containing solution. Uptake into the brain regions went up sharply with blood concentrations to somewhere around 100 ppb (blood) and 4 ppb (tissue), where it roughly saturated.[9] In the GP, GABA increased a little at the highest Mn doses while DA decreased for Mn^{2+} between 30 and 90 mg kg^{-1}, but increased for Mn^{3+} at 90 mg kg^{-1}.[9] TfR increased a little with dose both in the GP and the average of the tissues, and increased a little more rapidly for Mn^{3+} than Mn^{2+}. H-ferritin increased slightly more with dose in the GP. Mean plasma prolactin decreased significantly with dose at about the same rate for Mn^{2+} and Mn^{3+}.[9]

Most of the increased effect of Mn^{3+} over Mn^{2+} could be accounted for by the higher blood levels.[9] The data suggest that the apparent susceptibility of the GP and striatum to toxicity by Mn must be due to their inherent sensitivity and not greater transport because uptake in all the brain tissues studied was similar.

More recently, others have used different cells or tissues and focused on the results of different types of measurement in an approach somewhat like that of Chen *et al.*[58] and Reaney *et al.*[9,42,43] in which they investigated a range of effects of simultaneous measurements after treatment with both Mn^{2+} and Mn^{3+} to draw their conclusions. Hernandez *et al.*[59] studied the effects of $Mn^{2+}Cl_2$, Mn^{2+} citrate, Mn^{3+} citrate, and Mn^{3+} pyrophosphate on two primary neuronal cultures of both mature and immature mouse cerebellar granular neurons (CGC) and neocortical neurons (CTX). Effects on cell viability, mitochondrial function, and glutathione levels were monitored following treatment in which the cells were attached to the bottoms of multiwell plates and exposed to a range of Mn concentration for 120 h. The effects of the neuroprotective agents ascorbate and lactate were also studied. Both UV–visible spectroscopy and EPR measurable Mn were used to follow Mn^{2+} and Mn^{3+} content of the initial solutions. The MTT assay and the propidium iodide assay were used to determine mitochondrial metabolic impairment and cell viability and rhodamine 123 fluorescence was used to monitor mitochondrial membrane potential. Mn accumulation was followed using inductively coupled plasma mass spectroscopy (ICP-MS), and glutathione, a protectant against oxidative damage, was estimated using the 5,5′-dithio-bis-2-nitrobenzoic acid (DTNB) method.[59]

LC_{50} (the doses at which viability was 50% of control) were lowest for the $Mn^{2+}Cl_2$ and Mn^{3+}citrate-treated samples, suggesting that these were the more toxic of the four different types of Mn treatment used. Uptake into CGC neurons was perhaps 15% higher following Mn^{3+} citrate treatment than Mn^{2+} Cl_2 treatment, which was itself higher than following Mn^{2+} citrate treatment. Based on the portion of cells showing mitochondrial disfunction (percent of control) with the MTT assay, CGC neurons were almost an order of magnitude more susceptible to Mn toxicity than CTX neurons. Glutathione levels increased perhaps as much as 25–30% with dosage up to the LC_{50}, and then decreased for both cell types. A significant fall in mitochondrial membrane potential correlated with the loss of cell viability. The MTT assay suggested that mitochondrial disfunction also correlated with decreasing cell viability. Both ascorbic acid and lactate were effective in decreasing the toxic effects of all Mn treatments, and a combination of ascorbic acid and lactate were even more effective. There was actually not much difference between the toxicity of Mn^{2+} or Mn^{3+} treatments. The authors concluded that cellular Mn toxicity was caused by mitochondrial energy metabolism impairment and oxidative stress.[59]

Huang *et al.*[60] studied the effects of $Mn^{2+}Cl_2$ and Mn^{3+} acetate treatments on rat liver tissue. Rats were injected intraperitoneally daily with 2 mg Mn per kg body weight or with vehicle (saline) for 90 days. Following the injection period, liver nuclei and cell membranes were isolated and studied and the activities of SOD, glutathione peroxidase (GPx), and the Na^+/K^+ ATPase were followed. The levels of glutathione (GSH) and malondialdehyde (MDA) (to follow lipid peroxidation/oxidative stress) were measured. The samples were also studied using both light microscopy and electron microscopy.[60]

Results showed a decrease in SOD and GPx activity, a decrease in GSH and an increase in MDA following injection of either form of Mn with all of these effects being greater following Mn^{2+} injection than Mn^{3+} injection. There was also an increase in MDA and a decrease in Na^+/K^+ ATPase activity. Liver sections from Mn^{2+} injected rats showed more hydropic degeneration around the hepatic lobule and infiltration of inflammatory cells than in either the control or Mn^{3+} injected liver sections. Using electron microscopy, mitochondria in Mn^{2+} injected livers showed more extensive swelling and hyperplasia than mitochondria in either control or Mn^{3+} injected livers. The authors concluded that more toxicity is shown following Mn^{2+} injection than following Mn^{3+} injection.[60]

Ducic *et al.*[61] have recently used the elegant new microscopic X-ray fluorescence and micro XANES techniques to show the distribution of Fe and Mn in midbrain dopaminergic (MDN) and nondopaminergic neurons. and to attempt some XANES analysis of the results. Midbrain neurons from transgenic mice, which expressed green fluorescent protein (GFP) in 90% of the dopaminergic cells, were cultured. These cultured neurons were treated for 3 h with either vehicle or with 50 µM (final concentration) of either $FeCl_2$, $FeCl_3$, Mn^{3+} pyrophosphate, or with 500 µM $MnCl_2$. After being placed in the sample chamber of the X-ray fluorescence/microXANES equipment, visible light fluorescence was used to determine which neurons were dopaminergic and which were nondopaminergic.[61] Both types of neurons were then studied using the X-ray fluorescence technique to determine the distribution within the neurons of Fe and Mn and a series of other elements.[61]

Results showed that a low basal amount of Fe and Mn was distributed homogeneously within the cell soma. Following treatment with Fe^{2+}, the Fe signal was seen to localize with the signal for phosphorus, indicating colocalization primarily within the cell soma.[61] Uptake following Fe^{3+} treatment was smaller than with Fe^{2+} and no clear cut intracellular distribution of Fe was observed. XANES analysis following the Fe^{2+} treatment suggested that most of the Fe was in the Fe^{3+} state or a ferritin-bound state.[61] Following treatment with Mn^{2+}, most of the Mn was located in a perinuclear location,[61] which is similar to the distribution of the mitochondrial network.[62] Following treatment with Mn^{3+}, the signal was weak and the distribution hard to determine definitively. Because of low signal strength, definitive XANES analysis was not possible from the intracellular Mn.[61]

6.6   Is Mn^{2+} Oxidized to Mn^{3+} within Cells or Mitochondria?

As discussed earlier, Mn^{3+} is a strong oxidizing agent while Mn^{2+} is not. Archibald and Tyree[63] make a strong case for superoxide radical being able to oxidize Mn^{2+} to Mn^{3+} *in vitro* using chemical systems, and propose that the toxicity of Mn is due to oxidation of dopamine by Mn^{3+}. It is a bit surprising that superoxide, which despite its name is not a particularly strong oxidizing agent,[64] can oxidize Mn^{2+} to Mn^{3+}; however, Archibald and Tyree[63] suggest that this reaction can proceed *via* Fenton chemistry. A strong caveat to concluding that this is the mechanism of Mn toxicity, however, is that the same authors also show that superoxide dismutase, present in both the cytosol of cells and mitochondria, strongly inhibits this conversion of Mn^{2+} to Mn^{3+}.[63] The mitochondrial electron transport system produces well over 90% of the superoxide produced in the typical cell, most of it within the mitochondrial matrix.[65,66] Furthermore, Mn^{2+} can diffuse across the outer mitochondrial membrane and is rapidly transported across the inner membrane *via* the mitochondrial Ca^{2+} uniporter.[62,67] If Mn^{2+} can be oxidized to Mn^{3+} *via* the superoxide-requiring mechanism proposed by Archibald and Tyree anywhere in the cell, then the mitochondrial matrix would be the first place to seek to confirm this at the cellular or mitochondrial levels.

This issue has been addressed using XANES spectroscopy at both the mitochondrial and cellular levels.[36,47–49] In these experiments, isolated liver, heart, and brain mitochondria, rat brain astrocytes, and neuron-like cells, such as PC12 cells, nerve growth factor-induced PC12 cells, and NC2 cells were exposed to a range of Mn^{2+} concentrations in medium. The concentrations with which the samples were treated varied from no added Mn^{2+} to concentrations calculated to be in the pathological range. Although the concentration of free Mn^{3+} produced by oxidation of Mn^{2+} in biological tissue might not be expected to reach levels where detection by XANES might be possible, there is plenty of citrate, pyrophosphate, and acetate present, particularly in the mitochondria, to form stable complexes with any Mn^{3+} formed. The resultant XANES spectra were fit to a set of XANES spectra of model compounds chosen to be Mn compounds likely to be found inside the mitochondria or cells or to Mn compounds believed to have similar local binding sites to compounds likely to be present because similar local binding sites yield similar XANES structure. Model compounds of Mn^{2+}, Mn^{3+}, and Mn^{4+} ionic state were used in every analysis. Because of the shift in absorption edge with ionic state with XANES spectroscopy, it was easy to separate Mn^{2+}, Mn^{3+}, and Mn^{4+} spectra present in the biological samples.

None of the XANES measurements showed any sign of the presence of Mn^{4+} compounds within any of the mitochondrial or cell samples.[36,47–49] One of the model compounds used was Mn superoxide dismutase (MnSOD), which showed a XANES spectrum that could be interpreted as a sum of Mn^{2+} and Mn^{3+} compounds. This was expected because the function of MnSOD induces changes between the Mn^{2+} and Mn^{3+} ionic states in the compound. The spectrum of endogenous Mn in both the mitochondrial and the cell samples could easily be seen using XANES spectroscopy. The spectra, particularly in the mitochondrial samples, where the Mn concentration was higher, showed a very strong resemblance to the spectrum of MnSOD.[36,47–49] No other spectra showing Mn^{3+} content were seen.[36,47–49] As the Mn content of the mitochondria or cells increased, XANES spectra similar to those of other Mn^{2+} compounds increased. No spectra showing any Mn^{3+} character were seen other than the MnSOD-like spectra, which did not increase as the total Mn concentration increased. The only things that increased with total Mn concentration were the spectra of Mn^{2+} compounds.[36,47–49] This was true even in those mitochondrial samples in which additions, such as antimycin A, were made to increase superoxide production.[36,47–49] These results suggest that the amount of any Mn^{3+} compound produced by the Fenton-like mechanism suggested by Archibald and Tyree *in vivo* is very small.[36,47–49,63]

6.7 Transport of Mn^{3+} *via* the Transferrin Mechanism

Fe uptake *via* the Tf mechanism has been thoroughly studied and is currently fairly well understood. First, the more stable ionic state of Fe, Fe^{3+}, binds to Tf in the blood, and the Fe^{3+}Tf complex then binds to the transferrin receptor (TfR) on the outside of the cell membrane.[68] Then the membrane invaginates forming an endosome with the TfR and its associated Fe^{3+}Tf complex still bound on the inside of the vesicle.[68] The endosome moves away from the membrane toward the perinuclear area where most of the mitochondrial network is found.[68,69] The inside of the endosome is acidified, which aids in releasing the Fe^{3+} and Tf from the TfR, and then the Fe^{3+} is reduced to Fe^{2+} by Steap proteins present within the endosomes on the inside of the endosome.[69–71] Finally, the Fe^{2+} is transported out of the endosome by DMT1.[72] Most of the Fe^{2+} is taken up by mitochondria for incorporation in hemes.[73]

The corresponding Mn^{3+} uptake experiments used confocal microscopy and fluorescently-labeled (*i.e.* Alexa green labeled) Tf to follow the Mn^{3+}Tf uptake in both cultured striatal and hippocampal neurons. The fluorescently labeled $Mn^{3+}Tf$ molecules were seen to first form a bright outline of the cell as if they were binding to the membrane. Then very small, bright green sources of fluorescence, similar to the size range of endosomes, were seen moving into the cell cytosol across the cell membrane and later moving toward the mitochondrial network, which was labeled by Mitotracker red.[41] What was seen was exactly what was expected if transport of Mn^{3+} by the Tf mechanism is a direct analog to the transport of Fe^{3+} by the same mechanism.[41] This strongly suggests that the Mn reaching the cytosol and mitochondria *via* the Tf uptake mechanism is in the Mn^{3+} oxidation state within the endosome but is converted to the Mn^{2+} oxidation state before reaching the cytosol or the mitochondria.

The very effective Vivaspin 15 purification procedure used to purify the fluorescent $Mn^{3+}Tf$ probe could also be used to purify nonfluorescently labeled $Mn^{3+}Tf$. The purified $Mn^{3+}Tf$ was then used in quantitative uptake experiments. The $Mn^{3+}Tf$ was purified to 99.8% by this procedure and then used to compare the uptake rates of $Mn^{3+}Tf$ with those of Mn^{2+} in solution using atomic absorption.[41] These experiments showed that there was clearly significant Mn uptake *via* the Tf mechanism; however, uptake of Mn^{2+} was about 8.7 times faster than that of $Mn^{3+}Tf$ in both types of neuronal cells.[41] Further uptake experiments also showed that uptake of Mn^{3+} *via* the Tf mechanism was strongly inhibited by chlorpromazine or dynasore, known inhibitors of Fe^{3+} transport *via* the Tf mechanism.[41] It appears that uptake of Mn^{3+} *via* the Tf mechanism is not as rapid as transport of the same amount of Mn^{2+} *via* other mechanisms, suggesting that perhaps alternative mechanisms for transporting Mn^{3+} complexes may exist as suggested by the data of Reaney *et al.*[9,42,43]

6.8 The Toxicologically Important Steps of a Mn^{2+}-Inhibited Process

Although it is certainly possible that several separate initial effects of Mn on the important target areas of the brain contribute to the observed effects of Mn toxicity, in sequential processes, such as energy metabolism and ATP production, steps at which Mn inhibition becomes rate limiting for the overall process can be identified. These steps would therefore be the most important steps for any toxicology based on disruption of energy metabolism. Mitochondrial oxidation rate has been shown to be a good measure of the rate of ATP production.[74-76] Therefore, Gunter *et al.*[77] used the oxidation rates of isolated brain, liver, and heart mitochondria under state 3 (rapid phosphorylation conditions) using different sets of substrates over a range of Mn^{2+} concentrations to investigate where Mn^{2+} was inhibiting the overall process of oxidative phosphorylation. The different sets of mitochondrial substrates, succinate, glutamate plus malate, and pyruvate plus malate energize different portions of the Krebs cycle and, to a lesser extent, different portions of the electron transport chain. This advantage of being able to use different portions of the coupled reactions of oxidative phosphorylation through use of different sets of substrates allowed the identification of the loci within the processes which were rate limited by Mn^{2+}. These authors first showed that Mn^{2+} clearly inhibits α-ketoglutarate dehydrogenase (α KGDH), one of the sites at which Ca^{2+} activates oxidative phosphorylation. The oxidation rate results, however, showed that, consistent with the earlier discussion, Mn inhibition of α KGDH does not play a role in Mn^{2+} inhibition of the overall rate of oxidative phosphorylation. Results also showed that in liver and heart mitochondria, Mn^{2+} inhibition of the F_1F_0 ATP synthase did inhibit the overall rate of ATP production, whereas in brain mitochondria, this inhibition did not affect the overall rate. In brain mitochondria, Mn^{2+} inhibited the overall rate of oxidative phosphorylation about three times more strongly than it did in liver mitochondria, but the two primary sites of inhibition were first either complex II or fumarase and second either the glutamate aspartate exchanger or amino transferase.[77] This is consistent with the earlier work of Zwingmann *et al.*[18] whose work using nuclear magnetic resonance spectroscopy found complex II to be a rate-limiting step.

The Mn-induced toxicological effect studied here and inhibition of ATP production by oxidative phosphorylation appears to be induced strictly by Mn^{2+}-linked processes because (1) the inhibitory effects can be induced by addition of Mn^{2+} to mitochondria,[11] (2) XANES spectroscopy carried out under relatively similar conditions shows no evidence for the presence of any Mn^{3+} complexes other than MnSOD,[47] and (3) some of the relevant toxicological steps (*e.g.* inhibition of the F_1F_0ATP synthase) are known to be Ca^{2+}-activated.[78] Only the 2 + oxidation state of Mn has the necessary ionic radius and habits of binding symmetry to bind to Ca^{2+} and Mg^{2+} sites.

6.9 Effects of Exposure to Nanoparticles Containing a Range of Mn Oxidation States

Mn toxicity is often caused by exposure to small Mn-containing particles that interact with the nasal passages and the lungs in work environments. The sources of these small particles include welding fumes, dust from metal working, mining dust, agricultural toxins, and other work-related processes. Uptake of Mn-containing nanoparticles (NP) (40 nm) by type I alveolar epithelial cells (R3-1 cells) was compared with uptake of titanium dioxide (TiO_2) (25 nm), gold (Au) (20 nm), silver (Ag) (39 nm), and copper (Cu) (40 and 60 nm) nanoparticles in the same cell line as a model of the interactions of these NP with the initial cells they might encounter in either the lungs or nasal passages. The sizes given by each type of NP represent averages. The Mn NP used were generated by electric arc discharge in an argon-filled chamber between two opposing Mn metal rods in a Palas generator (Palas GmbH, Karlsruhe, Germany), while the other NP used were obtained commercially. These Palas-generated NP contained Mn metal, Mn^{2+}, and Mn^{3+}.

Uptake times and location of each type of NP and the per cent of condensed nuclei within the R3-1 cells were determined using transmission electron microscopy (TEM) and elemental analysis (EDS) using an electron dispersive X-ray spectrometer. The generation of H_2O_2 in the cells and medium by these NP was measured using amplex red fluorescence. Dosing of the cells with each type of NP was based on the mass of particles per available surface area in growth dishes (1.2 μg cm^{-2}, which is approximately 3–6 μg ml^{-1} in cell medium) for the TEM studies and a smaller amount (0.4 μg cm^{-2} or approximately 1 μg ml^{-1} in cell medium) for the amplex red assay.[79]

Results showed that the Mn, TiO_2, and Au NP were found in relatively large numbers (up to hundreds per cell) in the cytosol and particularly within the lysosomes within minutes to hours. Ag NP were widely distributed throughout the cells within minutes of exposure, while the Cu NP were not found within the cells. There was no H_2O_2 production above background when the TiO_2 or Au NP were used, even after 48 h of exposure. The Mn NP caused a significant increase in H_2O_2 production and hyper-activation of the lysosomes (termed heterolysosomes) at longer exposures. Condensation of the cell nuclei also occurred in parallel with the hyper-activation of the lysosomes. Exposure of the cells to Cu NP caused a large burst of H_2O_2 formation simultaneous with extensive damage to the cell membranes and cell death, indicated by the finding of condensed nuclei within a short time as the Cu NP dissolved. At longer times after the disappearance of the Cu NP, the surviving cells recovered. It should be noted that the free energy for the reaction:

$$Cu(s) + O_2 + 2H^+ = Cu^{2+} + H_2O_2 \qquad (6.1)$$

is -66.57 kjoules mole^{-1} and for the similar Mn reaction:

$$Mn + O_2 + 2H^+ = Mn^{2+} + H_2O_2 \qquad (6.2)$$

is -330.15 kjoules mole^{-1}, suggesting that the H_2O_2 is being produced by these reactions. In both cases, cell death and the appearance of condensed nuclei suggested apoptotic death occurred in parallel with H_2O_2 production. The similarity of the uptake times and distribution of Mn, TiO_2, and Au NP within the cells suggested a common process of entry into the cells that probably involved binding to the cell membrane and endocytosis. H_2O_2 production, particularly within the lysosomes, seemed to relate to the mechanism of cell death.[79]

6.10 Conclusions

Clearly, different responses and effects can be evoked by treatment of animals or cells with Mn^{2+} or with Mn^{3+} complexes. Many of the results suggest that there is not too much difference in toxicity between treatment with Mn^{2+} or treatment with Mn^{3+} complexes; however, many other results do suggest that one treatment or the other is more harmful. A major reason for the diversity of results lies in the large differences in types of cells used or cells *versus* live animals to obtain different results. Another lies in the many different assays used to follow the results. It is useful to keep in mind that in this study of the available literature comparing uptake and toxicity of free Mn^{2+} with those of Mn^{3+} complexes, we are not really comparing the effects of free Mn^{2+} with those of free Mn^{3+}. We don't know the concentration of free Mn^{3+} only that it's very small. We can't calculate it because we also don't know the concentrations or even the thermodynamic parameters of the many ligands present with which it can associate. The results suggest that progress in two areas, transport, and speciation will contribute the most toward our understanding of the effects described earlier. A more direct knowledge of speciation and of how the timing of changes in speciation correlate with toxic effects could be very important.

First, there seems to be a considerable difference between the transport properties of Mn^{2+} and the Mn^{3+} complexes. There are also transport differences between cell types, even between different types of neuron-like cells. A good example of how a small difference in cell type can make a large difference in transport is when Gunter and coworkers[48] found that non-induced PC12 cells sequestered Mn^{2+} roughly 6.6 times faster than the more neuron-like nerve growth factor-induced PC12 cells. The fact that we know far too little about how the Mn^{3+} complexes are transported and how that transport varies between different cell types has been highlighted by both Reaney *et al.*[9,43] and by Quintanar.[26] Although we know something about the biological mechanisms that transport Mn^{2+} and Mn^{3+}, there has still been only very limited success in distinguishing between the effects of one mechanism and others while they're functioning within the biological system.[41] A better understanding of the inhibitors of the different mechanisms of transport could be useful in this area of the work.

Second, it would greatly aid efforts to understand what is happening in experiments involving Mn^{3+} complexes if we knew when and into what these Mn^{3+} complexes were transformed during experiments. Carrying out XANES experiments in conjunction with experiments similar to those described earlier could significantly improve our understanding.

Finally, there has been considerable work associating Mn toxicity with inhalation of small particles produced in conjunction with welding, metal working, Mn mining, and agricultural toxins. It is clear that NP, somewhat similar to these small Mn-containing particles, can get into cells like those found in the lining of the lung and nasal passages by a process probably involving binding to the cell membrane and entering the cell inside endosomes.[79] This can kill these cells through a process involving hyper-activation of the lysosomes. Since the Mn NP dissolve slowly inside the cells, it can also provide Mn^{2+} or Mn^{3+} complexes that can later enter astrocytes or neurons in the more sensitive areas of the brain.

The study of Mn toxicity is important both to improve safety in the workplace and because of its relevance to Parkinsonism. Currently, there is strong evidence that Mn^{2+} can inhibit oxidative phosphorylation and there are reasons to suspect that uncharged Mn and Mn^{3+} might cause additional damage. However, additional information, particularly in the areas of transport, speciation, and timing of speciation changes, is necessary to determine which of the differences observed in damage caused by Mn^{2+} and Mn^{3+} complexes are due to differences in transport, which are due to oxidation by Mn^{3+}, and which are due to other effects. All of these things should be studied further.

Acknowledgements

The author wishes to thank Dr Karlene Gunter and Dr Richard Watson for reading and criticizing the manuscript and for discussions on its contents. He wishes to thank Kerstin Navik for copy editing the manuscript and for help in improving its wording. He also wishes to thank Dr Karlene Gunter for help with the references and for help in putting the manuscript into the form necessary for submission.

References

1. A. Barbeau, Manganese and extrapyramidal disorders, *Neurotoxicology*, 1984, **5**, 13–35.
2. D. B. Calne, N.-S. Chu, C.-C. Huang, C.-S. Lu and W. Olanow, Manganism and idiopathic parkinsonism: similarities and differences, *Neurology*, 1994, **44**, 1583–1586.
3. M. Aschner, K. E. Vrana and W. Zheng, Manganese uptake and distribution in the central nervous system (CNS), *Neurotoxicology*, 1999, **20**, 173–180.
4. M. C. Newland, T. L. Ceckler, J. H. Kordower and B. Weiss, Visualizing manganese in the primate basal ganglia with magnetic resonance imaging, *Exp. Neurol.*, 1989, **106**, 251–258.
5. H. Eriksson, J. Tedroff, K. A. Thuomas, S. M. Aquilonius, P. Hartvig, K. J. Fasth, P. Bjurling, B. Langstrom, K. G. Hedstrom and E. Heilbronn, Manganese induced brain lesions in Macaca fascicularis as revealed by positron emission tomography and magnetic resonance imaging, *Arch. Toxicol.*, 1992, **66**, 403–407.
6. A. Benedetto, C. Au and M. Aschner, Mangnese-induced dopamineergic neurodegeneration: Insight into mechanisms and genetics shared with Parkinson's disease, *Chem. Rev.*, 2009, **109**, 4862–4884.
7. Y. Suzuki, T. Mouri, Y. Suzuki, K. Nishiyama, N. Fujii and H. Yano, Study of subacute toxicity of manganese dioxide in monkeys, *Tokushima J. Exp. Med.*, 1975, **22**, 5–10.
8. C. E. Gavin, *A Role for the Mitochondrion in Manganese Toxicity*, University of Rochester, Rochester, NY, 1991, p. 99.
9. S. H. Reaney, G. Bench and D. R. Smith, Brain accumulation and toxicity of Mn(II) and Mn(III) exposures, *Toxicol. Sci.*, 2006, **93**, 114–124.

10. M. Aschner, T. R. Guilarte, J. S. Schneider and W. Zheng, Manganese: recent advances in understanding its transport and neurotoxicity, *Toxicol. Appl. Pharmacol.*, 2007, **221**, 131–147.
11. C. E. Gavin, K. K. Gunter and T. E. Gunter, Mn^{2+} sequestration by mitochondria and inhibition of oxidative phosphorylation, *Toxicol. Appl. Pharmacol.*, 1992, **115**, 1–5.
12. E. P. Brouillet, L. Shinobu, U. McGarvey, F. Hochberg and M. F. Beal, Manganese injection into the rat striatum produces excitotoxic lesions by impairing energy metabolism, *Exp. Neurol.*, 1993, **120**, 89–94.
13. P. Galvani, P. Fumagalli and A. Santagostino, Vulnerability of mitochondrial complex I in PC12 cells exposed to manganese, *Eur. J. Pharmacol.*, 1995, **293**, 377–383.
14. P. Du, C. R. Buerstatte, A. W. K. Chan, M. J. Minski, L. Bennett and J. C. K. Lai, Accumulation of manganese in liver can result in decreases in energy metabolism in mitochondria, in *Proceedings of 1997 Conference of Hazardous Wastes and Materials*, 1997, pp. 1–14.
15. J. A. Roth, L. Feng, J. Walowitz and R. W. Browne, Manganese-induced rat pheochromocytoma (PC12) cell death is independent of caspase activation, *J. Neurosci. Res.*, 2000, **61**, 162–171.
16. J. A. Roth, C. Horbinski, D. Higgins, P. Lein and M. D. Garrick, Mechanisms of manganese-induced rat pheochromocytoma (PC12) cell death and cell differentiation, *Neurotoxicology*, 2002, **23**, 147–157.
17. E. A. Malecki, Manganese toxicity is associated with mitochondrial dysfunction and DNA fragmentation in rat primary striatal neurons, *Brain Res. Bull.*, 2001, **55**, 225–228.
18. C. Zwingmann, D. Leibfritz and A. S. Hazell, Energy metabolism in astrocytes and neurons treated with manganese: Relation among cell-specific energy failure, glucose metabolism, and intercellular trafficking using multinuclear NMR-spectroscopic analysis, *J. Cereb. Blood Flow Metab.*, 2003, **23**, 756–771.
19. G. V. Malthankar, B. K. White, A. Bhushan, C. K. Daniels, K. J. Rodnick and J. C. Lai, *Neurochem. Res.*, 2004, **29**, 709–717.
20. C. Zwingmann, D. Leibfritz and A. S. Hazell, Brain Energy Metabolism in a sub-acute model of manganese neurotoxicity: an *ex vivo* nuclear magnetic resonance study using (1–13C)glucose, *Neurotoxicology*, 2004, **25**, 573–587.
21. C. Zwingmann, D. Leibfritz and A. S. Hazell, NMR spectroscopic analysis of regional brain energy metabolism in manganese neurotoxicity, *GLIA*, 2007, **55**, 1610–1617.
22. E. D. Bird, A. H. Anton and B. Bullock, The effect of manganese inhalation on basal ganglia dopamine concentrations in rhesus monkey, *Neurotoxicology*, 1984, **5**, 59–66.
23. D. G. Graham, Catecholamine toxicity: a proposal for the molecular pathogenesis of manganese neurotoxicity and Parkinson's disease, *Neurotoxicology*, 1984, **5**, 83–96.

24. J. Donaldson, The physiopathologic significance of manganese in brain: its relation to schizophrenia and neurodegenerative disorders, *Neurotoxicology*, 1987, **8**, 451–462.

25. E. J. Martinez-Finley, C. E. Gavin, M. Aschner and T. E. Gunter, Manganese neurotoxicology and the role of reactive oxygen species, *Free Radical Biol. Med.*, 2013, **62**, 65.

26. L. Quintanar, Manganese toxicity: A bioinorganic chemist's perspective, *Inorg. Chim. Acta*, 2008, **361**, 875–884.

27. G. Kroemer, Mitochondrial control of apoptosis: An overview, *Biochem. Soc. Symp.*, 1999, **66**, 1–15.

28. G. Kroemer, N. Zamzami and S. A. Susin, Mitochondrial control of apoptosis, *Immunol. Today*, 1997, **18**, 44–51.

29. N. Zamzami, S. A. Susin, P. Marchetti, T. Hirsch, I. Gomez-Monterrey, M. Castedo and G. Kroemer, Mitochondrial control of nuclear apoptosis, *J. Exp. Med.*, 1996, **183**, 1533–1544.

30. J. L. Aschner and M. Aschner, Nutritional aspects of manganese homeostasis, *Mol. Aspects Med.*, 2005, **26**, 353–362.

31. M. Aschner and D. C. Dorman, Manganese: pharmacokinetics and molecular mechanisms of brain uptake, *Toxicol. Rev.*, 2006, **25**, 147–154.

32. M. Aschner, K. M. Erikson, E. Herrero Hernandez and R. Tjalkens, Manganese and its role in Parkinson's disease: from transport to neuropathology, *NeuroMol. Med.*, 2009, **11**, 252–266.

33. F. A. Armstrong, Why did nature choose manganese to make oxygen?, *Philos. Trans. R. Soc., B*, 2008, **363**, 1263–1270.

34. W. M. Latimer and J. H. Hildebrand, *Reference Book of Inorganic Chemistry*, The McMillan Company, New York, NY, 1956, p. 625.

35. L. Pauling, in *General Chemistry*, ed. L. Pauling, W. H. Freeman and Co., San Francisco, CA, 1956, pp. 315–328.

36. T. E. Gunter, C. E. Gavin, M. Aschner and K. K. Gunter, Speciation of manganese in cells and mitochondria: A search for the proximal cause of manganese neurotoxicity, *Neurotoxicology*, 2006, **27**, 765–776.

37. D. G. Schulze and P. M. Bertsch, Synchrotron x-ray techniques in soil, plant, and environmental research, *Adv. Agron.*, 1995, **55**, 1–66.

38. D. G. Schulze, T. McCay-Buis, S. R. Sutton and D. M. Huber, Manganese oxidation states in Gaeumannomyces-infested wheat rhizospheres probed by micro-XANES spectroscopy, *Phytopathology*, 1995, **85**, 990–994.

39. T. E. Gunter and J. S. Puskin, Manganous ion as a spin label in studies of mitochondrial uptake of manganese, *Biophys. J.*, 1972, **12**, 625–635.

40. T. E. Gunter and J. S. Puskin, The use of electron paramagnetic resonance in studies of free and bound divalent cation: the measurement of membrane potentials in mitochondria, *Ann. N. Y. Acad. Sci.*, 1975, **264**, 112–123.

41. T. E. Gunter, B. Gerstner, K. K. Gunter, J. Malecki, R. Gelein, W. M. Valentine, A. Aschner and D. I. Yule, Manganese transport via the transferrin mechanism, *Neurotoxicology*, 2013, **34**, 118–127.

42. S. H. Reaney, C. L. Kwik-Uribe and D. R. Smith, Manganese oxidation state and its implications for toxicity, *Chem. Res. Toxicol.*, 2002, **15**, 1119–1126.
43. S. H. Reaney and D. R. Smith, Manganese oxidation state mediates toxicity in PC12 cells, *Toxicol. Appl. Pharmacol.*, 2005, **205**, 271–281.
44. R. H. Kenten and P. J. G. Mann, The oxidation of manganese by illuminated chloroplast preparations, *Biochem. J.*, 1955, **61**, 279–286.
45. J. K. Klewicki and J. J. Morgan, Kinetic behavior of Mn-(III) complexes of pyrophosphate, EDTA, and citrate, *Environ. Sci. Technol.*, 1998, **32**, 2916–2922.
46. P. Aisen, R. Aasa and A. G. Redfield, The chromium, manganese, and cobalt complexes of transferrin, *J. Biol. Chem.*, 1969, **244**, 4628–4633.
47. T. E. Gunter, L. M. Miller, C. E. Gavin, R. Eliseev, J. Salter, L. Buntinas, A. Alexandrov, S. Hammond and K. K. Gunter, Determination of the oxidation states of manganese in brain, liver, and heart mitochondria, *J. Neurochem.*, 2004, **88**, 266–280.
48. K. K. Gunter, M. A. Aschner, L. M. Miller, R. Eliseev, J. Salter, K. Anderson, S. Hammond and T. E. Gunter, Determining the oxidation states of manganese in PC12 and nerve growth factor-induced PC12 cells, *Free Radical Biol. Med.*, 2005, **39**, 164–181.
49. K. K. Gunter, M. Aschner, L. M. Miller, R. Eliseev, J. Salter, K. Anderson and T. E. Gunter, Determining the oxidation states of manganese in NT2 cells and cultured astrocytes, *Neurobiol. Aging*, 2006, **27**, 1816–1826.
50. T. E. Gunter, J. S. Puskin and P. R. Russell, Quantitative magnetic resonance studies of manganese uptake by mitochondria, *Biophys. J.*, 1975, **15**, 319–333.
51. T. E. Gunter, K. K. Gunter, J. S. Puskin and P. R. Russell, Efflux of Ca^{2+} and Mn^{2+} from rat liver mitochondria, *Biochemistry*, 1978, **17**, 339–345.
52. J. S. Puskin and T. E. Gunter, Evidence for the transport of manganous ion against an activity gradient by mitochondria, *Biochim. Biophys. Acta*, 1972, **275**, 302–307.
53. J. S. Puskin and T. E. Gunter, Ion and pH gradients across the transport membrane of mitochondria following Mn^{++} uptake in the presence of acetate, *Biochem. Biophys. Res. Commun.*, 1973, **51**, 797–803.
54. J. S. Puskin, T. E. Gunter, K. K. Gunter and P. R. Russell, Evidence for more than one Ca^{2+} transport mechanism in mitochondria, *Biochemistry*, 1976, **15**, 3834–3842.
55. A. Abragam and B. Bleaney, *Electron Paramagnetic Resonance of Transition Ions*, Clarendon Press, Oxford, UK, 1970, p. 911.
56. M. Aschner and J. L. Aschner, Manganese transport across the blood brain barrier: Relationship to iron homeostasis, *Brain Res. Bull.*, 1990, **24**, 857–860.
57. M. Aschner and M. Gannon, Manganese (Mn) transport across the rat blood-brain barrier: Saturable and transferrin-independent mechanisms, *Brain Res. Bull.*, 1994, **33**, 345–349.

58. J. Y. Chen, G. C. Tsao, Q. Zhao and W. Zheng, Differential cytotoxicity of Mn(II) and Mn(III): special reference to mitochondrial [Fe-S] containing enzymes, *Toxicol. Appl. Pharmacol.*, 2001, **175**, 160–168.

59. R. B. Hernandez, M. Farina, B. P. Esposito, N. C. Souza-Pinto, F. Barbosa, Jr. and C. Sunol, Mechanisms of manganese-induced neurotoxicity in primary neuronal cultures: The role of manganese speciation and cell type, *Toxicol. Sci.*, 2011, **124**, 414–423.

60. P. Huang, G. Li, C. Chen, H. Wang, Y. Han, S. Zhang, Y. Xiao, M. Zhang, N. Liu, J. Chu, L. Zhang and Z. Sun, Differential toxicity of Mn2 + and Mn3 + to rat liver tissues: Oxidative damage, membrane fluidity and histopathological changes, *Exp. Toxicol. Pathol.*, 2012, **64**, 197–203.

61. T. Ducic, E. Barski, M. Salome, J. C. Koch, M. Bahr and P. Lingor, X-ray fluorescence analysis of iron and manganese distribution in primary dopaminergic neurons, *J. Neurochem.*, 2013, **124**, 250–261.

62. T. E. Gunter and S.-S. Sheu, Characteristics and possible functions of mitochondrial Ca^{2+} transport mechanisms, *Biochim. Biophys. Acta*, 2009, **1787**, 1291–1308.

63. F. S. Archibald and C. Tyree, Manganese poisoning and the attack of trivalent manganese upon catecholamines, *Arch. Biochem. Biophys.*, 1987, **256**, 638–650.

64. D. T. Sawyer and J. S. Valentine, How super is superoxide?, *Acc. Chem. Res.*, 1981, **14**, 393–400.

65. M. D. Brand, C. Affourtit, T. C. Esteves, K. Green, A. J. Lambert, S. Miwa, J. L. Pakay and N. Parker, Mitochondrial superoxide: production, biological effects, and activation of uncoupling proteins, *Free Radical Biol. Med.*, 2004, **37**, 755–767.

66. M. D. Brand, The sites and topology of mitochondrial superoxide production, *Exp. Gerontol.*, 2010, **45**, 466–472.

67. T. E. Gunter and D. R. Pfeiffer, Mechanisms by which mitochondria transport calcium, *Am. J. Physiol.*, 1990, **258**, C755–C786.

68. M. W. Hentze, M. U. Muckenthaler and N. C. Andrews, Balancing acts: molecular control of mammalian iron metabolism, *Cell*, 2004, **117**, 285–297.

69. D. S. Richardson and P. Ponka, The molecular mechanisms of the metabolism and transport of iron in normal and neoplastic cells, *Biochim. Biophys. Acta*, 1997, **1331**, 1–40.

70. R. S. Ohgami, D. R. Campagna, E. L. Greer, B. Antiochos, A. McDonald, J. Chen, J. J. Sharp, Y. Fujiwara, J. E. Barker and M. D. Fleming, Identification of a ferrireductase required for efficient transferrin-dependent iron uptake in erythroid cells, *Nat. Genet.*, 2005, **37**, 1264–1269.

71. R. S. Ohgami, D. R. Campagna, A. McDonald and M. D. Fleming, The Steap proteins are metalloreductases, *Blood*, 2006, **108**, 1388–1394.

72. C. Au, A. Benedetto and M. Aschner, Manganese transport in eukaryotes: the role of DMT1, *Neurotoxicology*, 2008, **29**, 569–576.

73. A. D. Sheftel, A.-S. Zhang, C. Brown, O. S. Shirihai and P. Ponka, Direct interorganellar transfer of iron from endosome to mitochondrion, *Blood*, 2007, **110**, 125–132.

74. B. Chance and G. R. Williams, Respiratory enzymes in oxidative phosphorylation. III. The steady state, *J. Biol. Chem.*, 1955, **217**, 409–427.

75. B. Chance, G. R. Williams, W. F. Holmes and J. Higgins, Respiratory enzymes in oxidative phosphorylation. V. A mechanism for oxidative phosphorylation, *J. Biol. Chem.*, 1955, **217**, 439–451.

76. B. Chance and G. R. Williams, The respiratory chain and oxidative phosphorylation, *Adv. Enzymol.*, 1956, **17**, 65–134.

77. T. E. Gunter, B. Gerstner, T. Lester, A. P. Wojtovich, J. Malecki, S. G. Swarts, P. S. Brookes, C. E. Gavin and K. K. Gunter, An analysis of the effects of Mn^{2+} on oxidative phosphorylation in liver, brain, and heart mitochondria using state 3 oxidation rate assays, *Toxicol. Appl. Pharmacol.*, 2010, **249**, 65–75.

78. J. G. McCormack, A. P. Halestrap and R. M. Denton, Role of calcium ions in regulation of mammalian intramitochondrial metabolism, *Physiol. Rev.*, 1990, **70**, 391–425.

79. B. A. Van Winkle, K. L. Bentley, J. Malecki, K. K. Gunter, I. M. Evans, A. Elder, J. N. Finkelstein, G. Oberdorster and T. E. Gunter, Nanoparticle (NP) uptake by type I alveolar epithelial cells and their oxidant stress response, *Nanotoxicology*, 2009, **3**, 307–318.

CHAPTER 7

Effect of Manganese on Signaling Pathways

TANARA V. PERES, FABIANO M. CORDOVA, MARK W. LOPES, ANA PAULA COSTA AND RODRIGO BAINY LEAL*

Universidade Federal de Santa Catarina, Centro de Ciências Biológicas, Departamento de Bioquímica, Florianópolis, Brazil
*Email: rbleal@gmail.com

7.1 Introduction

Cell signaling, or signal transduction, can be defined as the process by which cells communicate with their environment and respond temporally to external cues that are sensed by them. The central importance of this phenomenon has been appreciated for a long time because, as described by Bradshaw and Dennis, the dynamic responses of cells to external stimuli are, in essence, a description of the life process itself.[1] In that sense, it is noteworthy that a significant part of the human genome (25 000 genes) codifies cellular signaling elements including 1543 signaling receptors, 518 protein kinases and $\sim$150 protein phosphatases.[2]

One prominent cell signaling mechanism involves reversible post-translational modification of proteins by phosphorylation, which is a versatile mode to regulate protein activity and all aspects of cell function and development.[3] In the central nervous system (CNS), a rich and complex intercellular and intracellular signaling network is involved in the modulation of brain development, synaptic activity, and cognitive functions. Neurotransmitters or trophic molecules, *via* protein–protein interaction and by modulation of intracellular messengers production, can modulate

Issues in Toxicology No. 22
Manganese in Health and Disease
Edited by Lucio G. Costa and Michael Aschner
Published by the Royal Society of Chemistry, www.rsc.org

protein kinases and protein phosphatases activities, which modify the phosphorylation state of target proteins such as enzymes, receptors, ion channels, and transcription factors.[4] All these events direct a diversity of cell fates such as gene expression, neural cell migration, differentiation or proliferation, cell survival or death, and synaptic plasticity.[3,5-7] Regardless of all these aspects, only recently have studies addressed the modulation of intracellular signaling pathways by toxicants and truly become a significant part of the molecular toxicology research.

Manganese (Mn) neurotoxicity was first discovered in 1837 by John Couper, who observed a parkinsonian illness characterized by slowness of movement (bradykinesia), masked facies, and gait impairment (postural instability) in individuals occupationally exposed to Mn. Therefore, it was noted that despite its essentiality, high levels of Mn could be neurotoxic and affect motor function. The mechanisms involved in Mn neurotoxicity are not completely elucidated; however, alterations in dopamine (DA) metabolism,[8,9] mitochondrial dysfunctions, induction of reactive oxygen species (ROS) formation, and nitrosative stress leading to cell death[10-17] have been considered important aspects of Mn action.

The modulation by Mn of intracellular cell signaling elements, as well as the mechanism and cell fates of these effects, is an issue that requires further attention. In this chapter, we will focus on the main cell signaling pathways that have been reported to be altered by Mn exposure. Since only few studies have addressed these aspects *in vivo*, a series of data obtained from cell culture and brain slices models exposed *in vitro* to Mn will also be presented, aiming to help us identify the possible intracellular signaling targets involved in Mn neurotoxicity *in vivo*.

7.2 Manganese may Alter Cell Signaling in the Striatum

Striatum has been reported as a target of Mn neurotoxicity because it expresses higher levels of divalent metal transporter-1 (DMT-1), responsible in part for the transport of Mn, than the cortex and hippocampus. This may explain the higher accumulation of Mn and consequently the most prominent toxic effects in this region.[18,19] Another factor for increased susceptibility in the striatum is its DA content, which contributes to the generation of oxidative stress in the presence of Mn, possibly due to interaction with DA, causing its oxidation, decreasing the level of DA, and forming unstable quinones that causes oxidative stress.[20,21] The selectivity of Mn towards the basal ganglia structures, especially the nigrostriatal dopaminergic system, is confirmed in rodents and non-human primates models exposed *in vivo* to Mn where dysfunction of dopaminergic neurons and decreased DA release with consequent motor alterations were observed.[19,22,23]

The striatum is the main entry station of the basal ganglia, and striatal medium-sized spiny neurons (MSNs) are the major cells and its sole output neurons. MSNs utilize gamma-aminobutyric acid (GABA) as their neurotransmitter, making the striatum a large inhibitory structure. MSNs receive excitatory glutamatergic inputs from the cortex and the thalamus and are a major target of DA projections from the terminals of midbrain dopaminergic neurons.[24] The actions of DA in the striatum are mediated mainly by the D1 and D2 types of DA receptors (D1R and D2R) that are coupled to different G proteins having opposite effects on the production of cAMP. D1Rs are coupled to the striatal-enriched Gαolf that stimulates adenylyl cyclase activity, thus increasing intracellular cAMP levels. D2Rs exert, *via* protein Gi, a tonic inhibitory effect on membrane potential and adenylyl cyclase, thereby opposing other receptors expressed in the same neurons, such A2A adenosine receptors (see Figure 7.1). Noteworthy, D1Rs and D2Rs are remarkably segregated in striatonigral and striatopallidal neurons, respectively, with little overlap.[24,25]

Dopamine- and cAMP-regulated phosphoprotein of Mr 32 kDa (DARPP-32) is selectively enriched in MSNs in the striatum, and plays an essential role in dopaminergic neurotransmission. DA activates dopamine D1 receptors coupled to Gα/olf, leading to an activation of cAMP/protein kinase A (PKA) signaling and the phosphorylation of DARPP-32 at Thr34, known as the PKA-site (see Figure 7.1). When DARPP-32 is phosphorylated on Thr34, it is converted into a potent inhibitor of protein phosphatase-1 (PP1), thereby controlling the phosphorylation state and activity of many downstream physiological effectors, including various neurotransmitter receptors, voltage-gated ion channels, and transcription factors.[25] DARPP-32 is also phosphorylated on Thr-75 by CDK5 in response to glutamatergic excitatory signals from cerebral cortex, which is also crucial to the striatal function. However, in this case DARPP-32 inhibits PKA activity.[25] This way, DARPP-32 acts as a centralized element for reception, integration, and processing of signals from several neurotransmitters and is also a key element in the cortical–striatal–pallidal–thalamic–cortical circuit. It is well documented that DARPP-32 has important roles in the modulation of behavior and motor activity.[26–29] Due to the fact that the striatum is a major target of Mn neurotoxicity[30–33] and that this metal interferes markedly in dopaminergic and glutamatergic neurotransmission,[20,34–37] it is plausible therefore that Mn could interfere with the phosphorylation state of DARPP-32.

Regarding this possibility, Cordova and colleagues[22] demonstrated alteration in DARPP-32 phosphorylation in the striatum using a rat model with five consecutive daily intraperitoneal injections of $MnCl_2$ during postnatal day (PND) 8 to 12, with neurochemical evaluation on PND14. An increase in the phosphorylation of Thr-34 was noted at 5 and 10 mg kg^{-1} Mn. In this protocol, the phosphorylation of Thr-75 site was not modified by exposure to the metal compared to the control. The main consequence of increasing phosphorylation of DARPP-32 at Thr-34 is the inhibition of PP1, which is a protein phosphatase with many possible targets (see Figure 7.2). Therefore, such change might have important pathophysiological implications because it could lead to excessive phosphorylation of various substrates (receptors, transcription factor, and enzymes). Specifically in this study, these alterations take place during a critical period of postnatal development in which

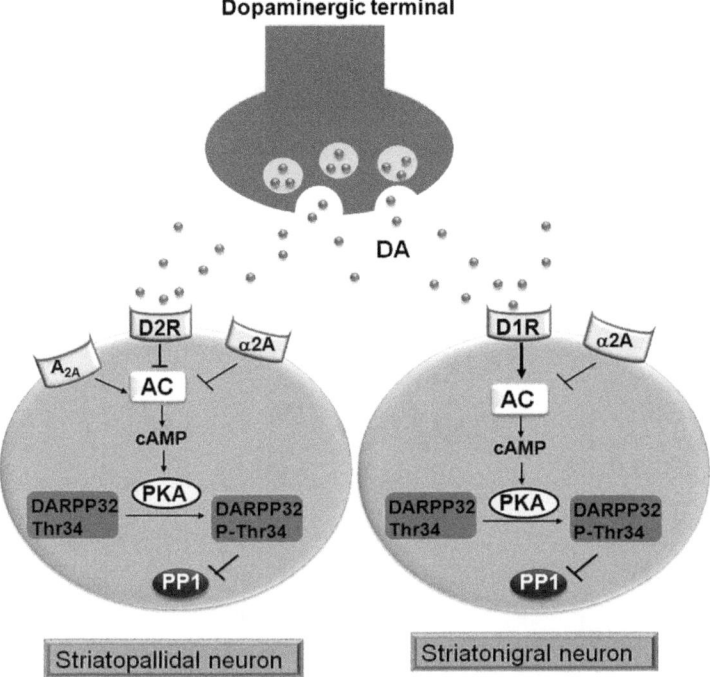

Figure 7.1 The actions of dopamine in the striatum are mediated mainly by the D1 and D2 types of DA receptors (D1R and D2R). D1R and D2R are coupled to different G proteins having opposite effects on the production of cAMP in the medium sized spiny neurons (MSNs). D1Rs are coupled to the striatal-enriched Gαolf that stimulates adenylyl cyclase activity, thus increasing intracellular cAMP levels. The principal target of cAMP is the cAMP-dependent protein kinase (PKA) and the D1R-induced PKA activation modulates multiple voltage- and ligand-gated ion channels, thereby modifying the effectiveness of synaptic inputs. D2Rs exert a tonic inhibitory effect on membrane potential and adenylyl cyclase, thereby opposing other receptors expressed in the same neurons such as A2A adenosine receptors. D1Rs and D2Rs are remarkably segregated in striatonigral and striatopallidal neurons, respectively, with little overlap. DA- and cAMP-regulated phosphoprotein Mr ∼32 000 (DARPP-32) is an endogenous inhibitor of PP1 particularly enriched in striatal MSNs. DARPP-32 activity is modulated by multiple neurotransmitters and is involved in the response to many neurotransmitters and psychoactive drugs. Phosphorylation of specific threonine (Thr) and serine (Ser) residues determines DARPP-32 overall function. For instance, phosphorylation of Thr-34 by PKA or cGMP-dependent protein kinase (PKG) turns DARPP-32 into a very potent inhibitor of PP1c. Inhibition of PP1c by DARPP-32 amplifies PKA signaling by enhancing the phosphorylation of targets that are substrates of PP1, constituting a positive feed-forward mechanism by which the D1R DA signal is amplified (for review, see ref. 24 and 25).

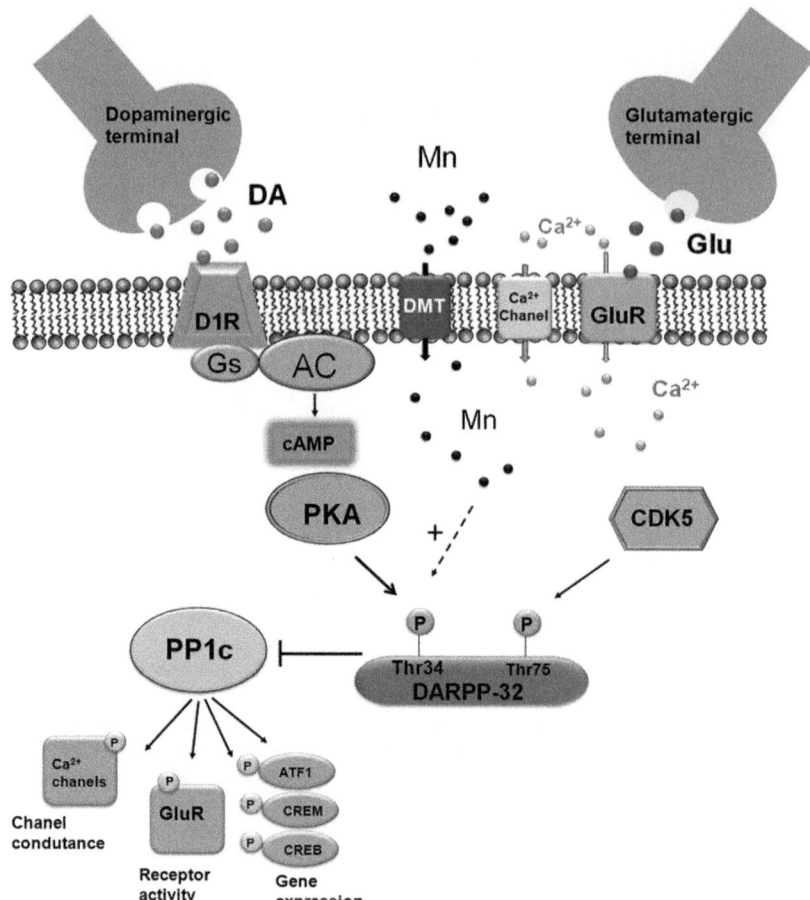

Figure 7.2 DARPP-32 phosphorylation in the striatum may be modulated by Mn exposure. Dopamine acting on D1 receptors stimulates the activity of adenylate cyclase with consequent increase in cAMP production and activation of PKA, which may phosphorylate DARPP-32 at Thr34. Imma-ture rats treated with Mn displayed increased phosphorylation at Thr34 of DARPP-32 when treated with doses 5 and 10 mg per kg Mn. In this protocol the phosphorylation of Thr-75 site was not modified.[22] Cyclin-dependent kinase 5 (CDK5) phosphorylates DARPP-32 at Thr 75 which converts it into a PKA inhibitor. The main consequence of increasing phosphorylation of DARPP-32 at Thr-34 is the inhibition of the catalytic subunit of the phospho-serine/phospho-threonine protein phosphatase 1 (PP1c), which is capable of dephosphorylating a diversity of targets, including receptors, ion channels and transcription factors. Therefore, such Mn effect on DARPP-32 might have important pathophysiological implications, since it could lead to excessive phosphorylation of various substrates.

intense synaptogenesis is occurring. The mechanism involved in the Mn-induced increase in Thr-34 phosphorylation in DARPP-32 is not clear; however, PKA activation by an unknown mechanism in adult rats in response to developmental (PND1-21) Mn exposure has been reported.[38] It is possible, therefore, that a direct and/or indirect action of Mn on striatum reinforces the cAMP/PKA/Thr34-DARPP-32 pathway.

7.3 Manganese Modulation of Tyrosine Hydroxylase Activity

One important aspect that regulates DA metabolism is the rate of synthesis of DA. Tyrosine hydroxylase (TH) is the rate-limiting enzyme in catecholamine synthesis and catalyzes the first step of a biochemical synthetic pathway in which L-tyrosine is converted to L-3,4-dihydroxyphenylalanine (L-DOPA). TH activity can be regulated acutely by protein phosphorylation and chronically by protein synthesis. TH phosphorylation can occur at serine (Ser) residues Ser8, Ser19, Ser31 and Ser40. Phosphorylation represents an important mechanism for maintaining DA levels in tissues immediately after DA secretion.[39] Although other sites of phosphorylation have been recognized in TH, only Ser19, Ser31 and Ser40 are regulated *in vivo*.[39,40] Phosphorylation of Ser40 is considered to be the most important mechanism to activate TH, causing a decrease in the feedback inhibition by DA.[39,41] PKA, protein kinase C (PKC), and protein phosphatase 2A (PP2A) are involved in the phosphorylation and dephosphorylation of Ser40 *in situ*, which has been demonstrated in response to both acute and sustained stimuli (see Figure 7.3).[39,42-44] Protein phosphatase 2A (PP2A) is the major serine/threonine phosphatase that dephosphorylates TH, resulting in reduced TH activity.[45] Moreover, a functional interaction between PKCδ and TH has been documented, in which PKCδ negatively regulates TH activity and DA synthesis by enhancing PP2A activity in DAergic neurons.[46]

Concerning the effect of Mn on TH, Posser and colleagues[47] demonstrated that Mn (100 µM for a period of 1–24 h) stimulated sustained phosphorylation of TH at Ser-40 in rat pheochromocytoma 12 (PC12) DAergic cells without altering protein synthesis of TH or cell viability (see Figure 7.3). The effect of Mn exposure on phosphorylation at Ser40 was not due to oxidative stress, or PKA or PKC activity as demonstrated by the use of an antioxidant compound and kinase inhibitors, respectively.[47] Interestingly, the effect of Mn, which improved TH phosphorylation and activity, was dependent on the exposure period and the concentration of metal applied. In striatal slices exposed to Mn (10–1000 µM) for short periods (3–6 h), this effect was not observed.[48] Zhang and colleagues[49] demonstrated that acute treatment (3 h) of differentiated N27 dopaminergic cells with subtoxic concentrations of $MnCl_2$ (3 µM and 10 µM) induced an increase in TH activity as well as dose-dependent increases in the levels of TH-Ser40 phosphorylation. On the other hand, treatment with low Mn concentrations and for longer period (0.1–1 µM Mn for 24 h), produced a decrease in TH activity without affecting phosphorylation at Ser40. Therefore, considering the *in vitro* data from all these studies, showing that alterations in TH activity may play a role in the action of Mn towards the DAergic system, the effect of Mn on TH activity *in vivo* remains to be determined.

7.4 Alteration in MAPK and AKT Signaling Induced by Manganese

The extracellular regulated protein kinase (ERK1/2), c-jun amino-terminal kinase (JNK1/2/3) and p38[MAPK] are the foremost enzymes studied in the mitogen-activated protein kinase (MAPK) family.[50–53] The ERK1/2 cascade is

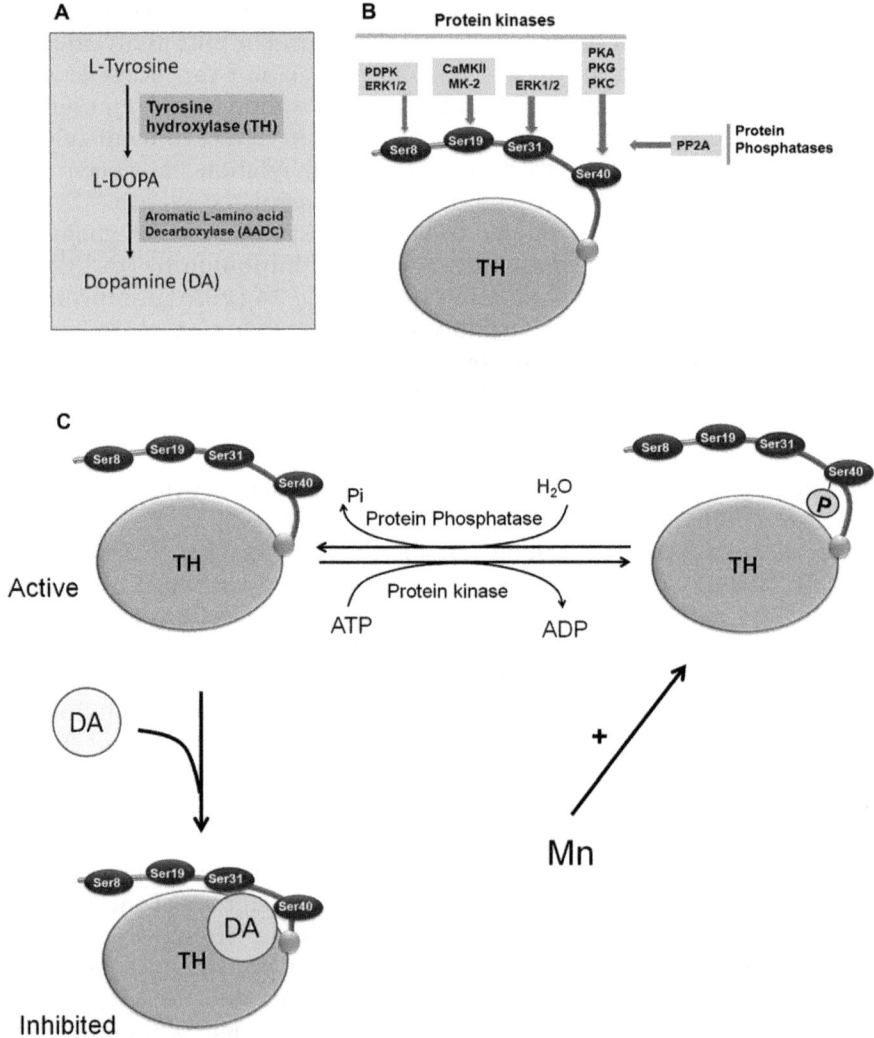

primarily activated by growth factors, regulating gene expression, embryo-genesis, proliferation, cell death/survival, and neuroplasticity.[51,52] The JNK1/2/3 and p38[MAPK] protein kinases, classically recognized as stress-activated protein kinases (SAPKs), are activated by cytokines and cytotoxic insults, and are often related to stress and cell death;[54,55] however, they may also regulate neurodevelopment and neuroplasticity.[53,56] The PI3K/AKT (PKB) pathway can be activated by several growth factors and plays a central role in the regulation of cell growth, proliferation, metabolism, and cell survival, as well as neuroplasticity.[57,58]

Activation of ERK1/2, JNKs, and p38[MAPK] and apoptosis in response to *in vitro* Mn exposure has been reported in astrocytes, microglia, and cell lineages.[20,59–64] Regarding the striatum, it was shown that *in vivo* Mn treatment at PND8-12 or PND8-27 may stimulate AKT phosphorylation at Ser473 by an unknown mechanism, but apparently independent of oxidative stress

---

Figure 7.3   Tyrosine hydroxylase, an enzyme expressed in dopaminergic neurons, regulates dopamine (DA) synthesis and can be modulated by Mn. The activity of TH can be physiologically modulated by medium- to long-term regulation of gene expression (not shown) and short-term regulation of enzyme activity including feedback inhibition, allosteric regulation, and phosphorylation. (A) Tyrosine hydroxylase (TH; tyrosine 3-monooxygenase; E.C. 1.14.16.2) is the first and rate-limiting enzyme in cathecolamine (CA) synthesis catalyzing the hydroxylation of L-tyrosine to L-DOPA. The active form of TH contains Fe^{2+} and uses the cofactor tetrahydrobiopterin (BH_4) and molecular oxygen for catalysis. (B) TH can be phosphorylated at serine residues (Ser) 8, 19, 31 and 40 by a variety of protein kinases. The main protein kinases and the main protein phosphatase (PP) that are able to phosphorylate and dephosphorylate TH at serine residue Ser8, Ser19, Ser31 and Ser40 are shown. These four sites are all phosphorylated *in situ*. TH activation by phosphorylation is the primary mechanism responsible for the maintenance of DA levels in tissue after its secretion. (C) Phosphorylation of Ser40 by PKA, PKC, or PKG, which are the main kinases involved in the phosphorylation of this site, will lead to a small increase in TH activity due to a decrease in the Km. Furthermore, it will maintain TH in the fully active form for a longer time because phosphorylation of TH at Ser40 precludes TH inactivation by catecholamine binding (for review, see ref. 39). Notably, the dopaminergic pheochromocytoma (PC12) cells exposed to Mn (100 µM) for 24 h present an increased level of Ser40 phosphorylation and an increment of TH activity.[47] Conversely, other studies using differentiated N27 dopaminergic cells exposed to low Mn concentration did not demonstrate alteration of Ser40 phosphorylation.[49] However, Mn caused a decrease of TH activity in a manner dependent of PKCδ and PP2A activity.[49] Therefore, it is evident that Mn may affect TH activity; however, the effect of Mn on TH level, activity, and phosphorylation deserves to be studied in dopaminergic neurons from intact tissue exposed *in vivo* in order to clarify the TH modulation *in situ* by Mn. Abbreviations: CaMK, calcium- and calmodulin-stimulated protein kinase; ERK, extracellular signal-regulated protein kinase; MK-2, MAPK-activated protein kinase; PDPK, proline-directed protein kinase; PKA, protein kinase A; PKC, protein kinase C; PKG, protein kinase G; PP2A, protein phosphatase 2A.

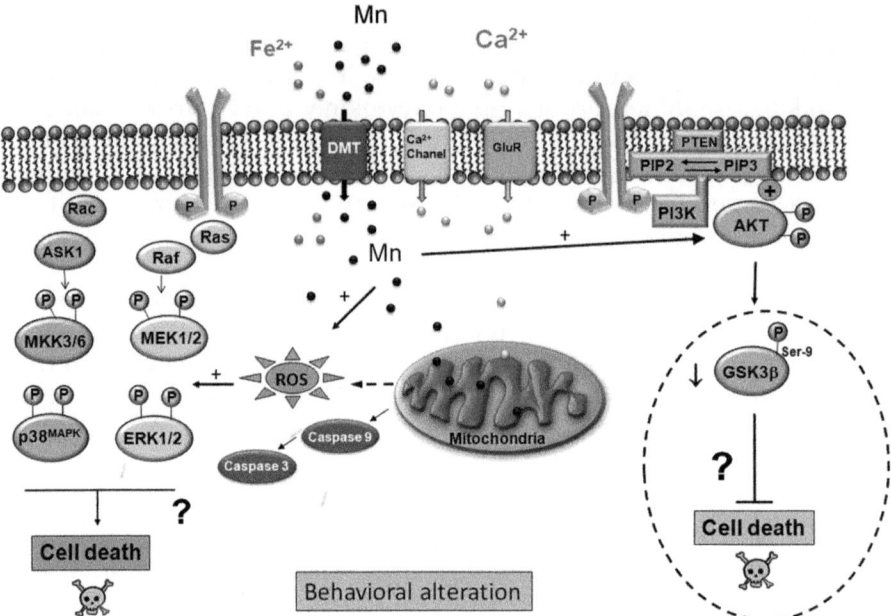

Figure 7.4   Mn can modulate MAPKs and AKT. Immature rats treated at PND8-12 or PND8-27 with 20 mg kg^{-1} Mn displayed an increment of striatal Akt phosphorylation by undefined mechanism but apparently independent of oxidative stress. Moreover, striatal activation of both ERK1/2 in response to Mn exposure at PN8-12 and p38^{MAPK} in response to more prolonged exposure (PND8-27) was dependent on reactive oxygen species (ROS) production, since the antioxidant Trolox abrogated this effect.[22,65] Sustained activation of ERK1/2, p38^{MAPK} and AKT by Mn may trigger long-lasting effects involved in the alteration of cell migration, proliferation, differentiation or survival. Moreover, alterations in behavioral tests were also observed. It is important to state that the striatal cell types involved in the response to Mn intoxication were not determined; however, *in vitro* studies suggest that ERK, p38^{MAPK} and AKT contained into glial cells may be target to Mn modulation and mediate cell death or neuroinflammation.[67,68] The dashed circle shows GSK3β as a potential target for activated AKT. Noteworthy is that phosphorylation at Ser9 site of GSK3β by AKT causes inhibition of the enzyme and could be a neuroprotective event; however, a recent study that analyzed striatum from adult rats exposed *in vivo* and PC12 cells exposed *in vitro* to Mn indicated dephosphorylation and activation of GSK3β probably by the canonical Wnt pathway. The primary consequence of this effect was the phosphorylation of the transcriptional co-factor β-catenin followed by its proteosomal degradation and induction of cell death.[71]

(see Figure 7.4).[22,65] Moreover, striatal activation of both ERK1/2 in response to Mn exposure at PN8-12 and p38^{MAPK} in response to more prolonged exposure (PND8-27), in a manner dependent of oxidative stress production, were reported.[22,65] Consistent with this temporal effect and other *in vitro* and *in vivo* studies,[38,47] it is plausible that *in vivo* Mn exposure may lead to sustained rather than a transient AKT and MAPK activation profile. Given that signaling duration can markedly alter qualitative and quantitative features of downstream elements driving distinct cell fate decisions,[66] the pattern of ERK1/2, p38[MAPK], and AKT phosphorylation may trigger long-lasting effects involved in the alteration of cell migration, proliferation, differentiation, or survival. It is important to state that the striatal cell types involved in these Mn responses were not determined; however, *in vitro* studies in Mn-treated glial cells showed the participation of ERK1/2 and AKT in the expression of iNOS in microglia[67] and COX-2 in astrocytes.[68] In addition, it has been demonstrated that Mn produced apoptotic cell death *via* the ERK1/2 signaling pathway, with caspase-3 activation in PC12 cells.[60] Therefore, Mn-induced ERK1/2, p38[MAPK], and AKT activation may be associated with changes in neuroplasticity and/or cell viability in the immature rat striatum, thus disturbing and impairing neurophysiological functions and neurodevelopment. Moreover, the oxidative stress response may be involved only in the ERK1/2 and p38[MAPK] activation because the effect of Mn on these kinases, but not on AKT, was abrogated by the antioxidant Trolox.[22]

7.5 Manganese Action on GSK-3β and the Canonical Wnt/β-Catenin Pathway

Wnts are a family of glycoproteins that regulate development and synaptic formation. Wnts and their downstream effectors modulate synaptogenesis, synaptic transmission, and plasticity. They act through a diverse number of pathways downstream of the seven-pass transmembrane Frizzled receptors. The canonical Wnt/β-catenin pathway is implicated in various cellular processes, such as cell proliferation, differentiation, survival, and motility. The canonical pathway consists of the stabilization of β-catenin in its dephosphorylated state in the cytoplasm, from where it can translocate to the nucleus and activate the transcription of pro-survival targets, such as the protein survivin.[69,70] Glycogen synthase kinase 3β (GSK-3β) plays a crucial role in the stabilization of β-catenin because it can phosphorylate this protein and mark it for ubiquitin-proteasome degradation. Deregulation of Wnt/β-catenin signaling contributes to the action of several neurotoxins, including β-amyloid, 6-hydroxydopamine and 1-methyl-4phenyl-1,2,3,4-tetrahydropyridine (MPTP). To test whether it may exert a similar effect in Mn neurotoxicity Jiang *et al.*[71] measured the content of β-catenin and its downstream target survivin, a pro-survival protein, in striatum samples of adult rats exposed to Mn. It was found particularly decreased in neurons. GSK-3β activation was increased, suggesting that down regulation of the Wnt/β-catenin pathway and its pro-survival target, survivin, occurs through GSK-3β action and that these events contribute to Mn-induced neuronal apoptosis.[71]

The hyperphosphorylation of tau protein is an important characteristic of Alzheimer's disease, but it has also been implicated as a mechanism of Mn neurotoxicity.[72] GSK-3β has been reported to contribute to tau hyperphosphorylation in PC12 cells exposed to Mn. Such activation was associated with ERK1/2 activation because the inhibition of ERK1/2 alleviated the activation of GSK-3β in response to Mn. Moreover, an inhibitor of GSK-3β, LiCl, caused inhibition of both the hyperphosphorylation of tau and the cytotoxicity induced by Mn. These findings suggest a role of ERK1/2 and GSK-3β in Mn-induced tau hyperphosphorylation.[72]

7.6   Final Considerations

Both *in vitro* and *in vivo* data have demonstrated that one of the mechanisms by which excessive levels of Mn might disrupt cellular homeostasis is altering the activity of cell signaling elements. Mn affects the dopaminergic system, with the striatum being a major target for Mn accumulation. In this structure, DARPP-32 acts as an important integrator of signaling pathways because it responds to phosphorylation at Thr34 or Thr75, acting as an inhibitor of PP1 or PKA, respectively. In addition, the activity of TH, the enzyme that catalyzes the rate-limiting step for DA synthesis in the dopaminergic neurons, may be modulated by phosphorylation at Ser40, and previous works have shown that Mn may affect the phosphorylation of this site as well as the enzyme activity. The mechanism of these actions is not clear, but it did not seem dependent on oxidative stress. Accordingly, disruption of these pathways may have important consequences for basal ganglia function. Different results from *in vitro* and *in vivo* studies point that oxidative stress and mitochondrial damage, generated by excessive exposure to Mn, may alter MAPKs activity. Moreover, it has been observed that Mn, *in vivo* and *in vitro*, may improve AKT activity, apparently by a mechanism independent of oxidative stress. Recent evidence has also demonstrated activation of GSK3β and induction of neuronal apoptosis in the striatum, possibly *via* modulation of the canonical Wnt/β-catenin pathway. Collectively, the studies presented in this chapter have demonstrated that disturbance in the activity of protein kinases and phosphatases might be part of the mechanism of Mn neurotoxicity that impairs neurophysiological functions and neurodevelopment. This aspect is mainly based in the well-documented roles that these signaling pathways play in regulating gene expression, embryogenesis, cell proliferation, cell death/survival, inflammatory responses, and synaptic plasticity. However, it is important to highlight that further studies will be necessary to disclose the mechanisms involved in the modulation of signaling proteins by Mn. Moreover, the potential role of the Mn-dependent disturbance of each cell signaling pathway, mainly observed *in vitro*, for the pathophysiology of the neurological syndrome induced by Mn overexposure deserves to be addressed *in situ*.

References

1. R. A. Bradshaw and E. A. Dennis, Transmembrane receptor receptors and their signaling properties, in *Functioning of Transmembrane*

Receptors in Cell Signaling, ed. R. A. Bradshaw and E. A. Dennis, Elsevier, San Diego, CA, 1st edn, 2011, ch. 1, vol. 1, p. 456.

2. J. A. Papin, T. Hunter, B. O. Palsson and S. Subramaniam, Reconstruction of cellular signalling networks and analysis of their properties, *Nat. Rev. Mol. Cell Biol.*, 2005, **6**, 99.

3. T. Hunter, The Age of Crosstalk: Phosphorylation, Ubiquitination, and Beyond, *Mol. Cell*, 2007, **28**, 730.

4. A. Yoshii and M. Constantine-Paton, Postsynaptic BDNF-TrkB signaling in synapse maturation, plasticity, and disease, *Dev. Neurobiol.*, 2010, **70**, 304.

5. R. Rodnight, C. Perrett, S. Soteriou, Aspects of Protein Phosphorylation in the Nervous System with Particular Reference to Synaptic Transmission, in *Phosphoproteins in Neuronal Function: Progress in Brain Research*, ed. G. Willem Hendrik and R. Aryeh, Elsevier, San Diego, CA, 1st edn, 1986, ch. 28, vol. 69, p. 394.

6. P. Greengard, The Neurobiology of Slow Synaptic Transmission, *Science*, 2001, **294**, 1024.

7. E. R. Kandel, The Molecular Biology of Memory Storage: A Dialogue Between Genes and Synapses, *Science*, 2001, **294**, 1030.

8. T. R. Guilarte, N. C. Burton, J. L. McGlothan, T. Verina, Y. Zhou, M. Alexander, L. Pham, M. Griswold, D. F. Wong, T. Syversen and J. S. Schneider, Impairment of nigrostriatal dopamine neurotransmission by manganese is mediated by pre-synaptic mechanism(s): implications to manganese-induced parkinsonism, *J. Neurochem.*, 2008, **107**, 1236.

9. T. R. Guilarte, Manganese and Parkinson's disease: a critical review and new findings, *Environ. Health Perspect.*, 2010, **118**, 1071.

10. J. A. Roth and M. D. Garrick, Iron interactions and other biological reactions mediating the physiological and toxic actions of manganese, *Biochem. Pharmacol.*, 2003, **66**, 1.

11. S. Zhang, J. Fu and Z. Zhou, In vitro effect of manganese chloride exposure on reactive oxygen species generation and respiratory chain complexes activities of mitochondria isolated from rat brain, *Toxicol. in Vitro*, 2004, **18**, 71.

12. T. E. Gunter, C. E. Gavin, M. Aschner and K. K. Gunter, Speciation of manganese in cells and mitochondria: A search for the proximal cause of manganese neurotoxicity, *NeuroToxicology*, 2006, **27**, 765.

13. C. Tamm, F. Sabri and S. Ceccatelli, Mitochondrial-mediated apoptosis in neural stem cells exposed to manganese, *Toxicol. Sci.*, 2008, **101**, 310.

14. F. Zhang, Z. Xu, J. Gao, B. Xu and Y. Deng, In vitro effect of manganese chloride exposure on energy metabolism and oxidative damage of mitochondria isolated from rat brain, *Environ. Toxicol. Pharmacol.*, 2008, **26**, 232.

15. D. Milatovic, S. Zaja-Milatovic, R. C. Gupta, Y. Yu and M. Aschner, Oxidative damage and neurodegeneration in manganese-induced neurotoxicity, *Toxicol. Appl. Pharmacol.*, 2009, **240**, 219.

16. D. Milatovic, R. C. Gupta, Y. Yu, S. Zaja-Milatovic and M. Aschner, Protective effects of antioxidants and anti-inflammatory agents against manganese-induced oxidative damage and neuronal injury, *Toxicol. Appl. Pharmacol.*, 2011, **256**, 219.

17. J. A. Moreno, K. M. Streifel, K. A. Sullivan, M. E. Legare and R. B. Tjalkens, Developmental exposure to manganese increases adult susceptibility to inflammatory activation of glia and neuronal protein nitration, *Toxicol. Sci.*, 2009, **112**, 405.

18. C. Au, A. Benedetto and M. Aschner, Manganese transport in eukaryotes: The role of DMT1, *NeuroToxicology*, 2008, **29**, 569.

19. D. S. Ávila, D. Colle, P. Gubert, A. S. Palma, G. Puntel, F. Manarin, S. Noremberg, P. C. Nascimento, M. Aschner, J. B. T. Rocha and F. A. A. Soares, A Possible Neuroprotective Action of a Vinylic Telluride against Mn-Induced Neurotoxicity, *Toxicol. Sci.*, 2010, **115**, 194.

20. K. Prabhakaran, D. Ghosh, G. D. Chapman and P. G. Gunasekar, Molecular mechanism of manganese exposure-induced dopaminergic toxicity, *Brain Res. Bull.*, 2008, **76**, 361.

21. S. C. Sistrunk, M. K. Ross and N. M. Filipov, Direct effects of manganese compounds on dopamine and its metabolite Dopac: An in vitro study, *Environ. Toxicol. Pharmacol.*, 2007, **23**, 286.

22. F. M. Cordova, A. S. Aguiar, Jr., T. V. Peres, M. W. Lopes, F. M. Gonçalves, A. P. Remor, S. C. Lopes, C. Pilati, A. S. Latini, R. D. S. Prediger, K. M. Erikson, M. Aschner and R. B. Leal, In Vivo Manganese Exposure Modulates Erk, Akt and Darpp-32 in the Striatum of Developing Rats, and Impairs Their Motor Function, *PLoS One*, 2012, 7, e33057.

23. T. R. Guilarte, M.-K. Chen, J. L. McGlothan, T. Verina, D. F. Wong, Y. Zhou, M. Alexander, C. A. Rohde, T. Syversen, E. Decamp, A. J. Koser, S. Fritz, H. Gonczi, D. W. Anderson and J. S. Schneider, Nigrostriatal dopamine system dysfunction and subtle motor deficits in manganese-exposed non-human primates, *Exp. Neurol.*, 2006, **202**, 381.

24. M. Matamales and J. A. Girault, Signaling from the cytoplasm to the nucleus in striatal medium-sized spiny neurons, *Front. Neuroanat.*, 2011, 5, 37.

25. P. Svenningsson, A. Nishi, G. Fisone, J.-A. Girault, A. C. Nairn and P. Greengard, DARPP-32: An Integrator of Neurotransmission, *Annu. Rev. Pharmacol. Toxicol.*, 2004, **44**, 269.

26. E. Santini, E. Valjent, A. Usiello, M. Carta, A. Borgkvist, J.-A. Girault, D. Hervé, P. Greengard and G. Fisone, Critical Involvement of cAMP/DARPP-32 and Extracellular Signal-Regulated Protein Kinase Signaling in l-DOPA-Induced Dyskinesia, *J. Neurosci.*, 2007, **27**, 6995.

27. A. Polissidis, O. Chouliara, A. Galanopoulos, G. Rentesi, M. Dosi, T. Hyphantis, M. Marselos, Z. Papadopoulou-Daifoti, G. G. Nomikos, C. Spyraki, E. T. Tzavara and K. Antoniou, Individual differences in the effects of cannabinoids on motor activity, dopaminergic activity and DARPP-32 phosphorylation in distinct regions of the brain, *Int. J. Neuropsychopharmacol.*, 2010, **13**, 1175.

28. K. Botsakis, O. Pavlou, P. D. Poulou, N. Matsokis and F. Angelatou, Blockade of adenosine A2A receptors downregulates DARPP-32 but increases ERK1/2 activity in striatum of dopamine deficient "weaver" mouse, *Neurochem. Int.*, 2010, **56**, 245.
29. M. Lebel, L. Chagniel, G. Bureau and M. Cyr, Striatal inhibition of PKA prevents levodopa-induced behavioural and molecular changes in the hemiparkinsonian rat, *Neurobiol. Dis.*, 2010, **38**, 59.
30. M. Yamada, S. Ohno, I. Okayasu, R. Okeda, S. Hatakeyama, H. Watanabe, K. Ushio and H. Tsukagoshi, Chronic manganese poisoning: A neuropathological study with determination of manganese distribution in the brain, *Acta Neuropathol.*, 1986, **70**, 273.
31. K. M. Erikson, K. Thompson, J. Aschner and M. Aschner, Manganese neurotoxicity: A focus on the neonate, *Pharmacol. Ther.*, 2007, **113**, 369.
32. D. S. Ávila, P. Gubert, R. Fachinetto, C. Wagner, M. Aschner, J. B. T. Rocha and F. A. A. Soares, Involvement of striatal lipid peroxidation and inhibition of calcium influx into brain slices in neurobehavioral alterations in a rat model of short-term oral exposure to manganese, *NeuroToxicology*, 2008, **29**, 1062.
33. J. Roth, Are There Common Biochemical and Molecular Mechanisms Controlling Manganism and Parkisonism, *NeuroMol. Med.*, 2009, **11**, 281.
34. K. Erikson and M. Aschner, Manganese Causes Differential Regulation of Glutamate Transporter (GLAST) Taurine Transporter and Metallothionein in Cultured Rat Astrocytes, *NeuroToxicology*, 2002, **23**, 595.
35. K. M. Erikson, R. L. Suber and M. Aschner, Glutamate/Aspartate Transporter (GLAST), Taurine Transporter and Metallothionein mRNA Levels are Differentially Altered in Astrocytes Exposed to Manganese Chloride, Manganese Phosphate or Manganese Sulfate, *NeuroToxicology*, 2002, **23**, 281.
36. A. Takeda, Manganese action in brain function, *Brain Res. Rev.*, 2003, **41**, 79.
37. M. Sidoryk-Węgrzynowicz, E. Lee, J. Albrecht and M. Aschner, Manganese disrupts astrocyte glutamine transporter expression and function, *J. Neurochem.*, 2009, **110**, 822.
38. S. A. McDougall, T. Der-Ghazarian, C. E. Britt, F. A. Varela and C. A. Crawford, Postnatal manganese exposure alters the expression of D2L and D2S receptor isoforms: Relationship to PKA activity and Akt levels, *Synapse*, 2011, **65**, 583.
39. P. R. Dunkley, L. Bobrovskaya, M. E. Graham, E. I. Von Nagy-Felsobuki and P. W. Dickson, Tyrosine hydroxylase phosphorylation: regulation and consequences, *J. Neurochem.*, 2004, **91**, 1025.
40. J. W. Haycock, Phosphorylation of tyrosine hydroxylase in situ at serine 8, 19, 31, and 40, *J. Biol. Chem.*, 1990, **265**, 11682.
41. H. Fujisawa and S. Okuno, Regulatory mechanism of tyrosine hydroxylase activity, *Biochem. Biophys. Res. Commun.*, 2005, **338**, 271.

42. D. P. Gelain, J. C. F. Moreira, L. R. M. Bevilaqua, P. W. Dickson and P. R. Dunkley, Retinol activates tyrosine hydroxylase acutely by increasing the phosphorylation of serine40 and then serine31 in bovine adrenal chromaffin cells, *J. Neurochem.*, 2007, **103**, 2369.

43. L. Bobrovskaya, D. P. Gelain, C. Gilligan, P. W. Dickson and P. R. Dunkley, PACAP stimulates the sustained phosphorylation of tyrosine hydroxylase at serine 40, *Cell. Signalling*, 2007, **19**, 1141.

44. L. Bobrovskaya, C. Gilligan, E. K. Bolster, J. J. Flaherty, P. W. Dickson and P. R. Dunkley, Sustained phosphorylation of tyrosine hydroxylase at serine 40: a novel mechanism for maintenance of catecholamine synthesis, *J. Neurochem.*, 2007, **100**, 479.

45. J. Haavik, D. L. Schelling, D. G. Campbell, K. K. Andersson, T. Flatmark and P. Cohen, Identification of protein phosphatase 2A as the major tyrosine hydroxylase phosphatase in adrenal medulla and corpus striatum: evidence from the effects of okadaic acid, *FEBS Lett.*, 1989, **251**, 36.

46. D. Zhang, A. Kanthasamy, Y. Yang, V. Anantharam and A. Kanthasamy, Protein Kinase C? Negatively Regulates Tyrosine Hydroxylase Activity and Dopamine Synthesis by Enhancing Protein Phosphatase-2A Activity in Dopaminergic Neurons, *J. Neurosci.*, 2007, **27**, 5349.

47. T. Posser, J. L. Franco, L. Bobrovskaya, R. B. Leal, P. W. Dickson and P. R. Dunkley, Manganese induces sustained Ser40 phosphorylation and activation of tyrosine hydroxylase in PC12 cells, *J. Neurochem.*, 2009, **110**, 848.

48. T. V. Peres, D. Z. Pedro, F. M. de Cordova, M. W. Lopes, F. M. Gonçalves, C. B. N. Mendes-de-Aguiar, R. Walz, M. Farina, M. Aschner and R. B. Leal, In Vitro Manganese Exposure Disrupts MAPK Signaling Pathways in Striatal and Hippocampal Slices from Immature Rats, *BioMed Res. Int.*, 2013, **2013**, 12.

49. D. Zhang, A. Kanthasamy, V. Anantharam and A. Kanthasamy, Effects of manganese on tyrosine hydroxylase (TH) activity and TH-phosphorylation in a dopaminergic neural cell line, *Toxicol. Appl. Pharmacol.*, 2011, **254**, 65.

50. L. Chang and M. Karin, MAP kinase signalling cascades, *Nature*, 2001, **410**, 37.

51. Z. Chen, T. B. Gibson, F. Robinson, L. Silvestro, G. Pearson, B.-e. Xu, A. Wright, C. Vanderbilt and M. H. Cobb, MAP Kinases, *Chem. Rev.*, 2001, **101**, 2449.

52. G. M. Thomas and R. L. Huganir, MAPK cascade signalling and synaptic plasticity, *Nat. Rev. Neurosci.*, 2004, **5**, 173.

53. V. Waetzig and T. Herdegen, Neurodegenerative and physiological actions of c-Jun N-terminal kinases in the mammalian brain, *Neurosci. Lett.*, 2004, **361**, 64.

54. K. Mielke and T. Herdegen, JNK and p38 stresskinases – degenerative effectors of signal-transduction-cascades in the nervous system, *Prog. Neurobiol.*, 2000, **61**, 45.

55. K. J. Cowan and K. B. Storey, Mitogen-activated protein kinases: new signaling pathways functioning in cellular responses to environmental stress, *J. Exp. Biol.*, 2003, **206**, 1107.

56. V. Waetzig, Y. Zhao and T. Herdegen, The bright side of JNKs-Multitalented mediators in neuronal sprouting, brain development and nerve fiber regeneration, *Prog. Neurobiol.*, 2006, **80**, 84.

57. D. P. Brazil, Z. Z. Yang and B. A. Hemmings, Advances in protein kinase B signalling: AKTion on multiple fronts, *Trends Biochem. Sci.*, 2004, **29**, 233.

58. L. P. van der Heide, G. M. Ramakers and M. P. Smidt, Insulin signaling in the central nervous system: learning to survive, *Prog. Neurobiol.*, 2006, **79**, 205.

59. Y. Hirata, K. Furuta, S. Miyazaki, M. Suzuki and K. Kiuchi, Anti-apoptotic and pro-apoptotic effect of NEPP11 on manganese-induced apoptosis and JNK pathway activation in PC12 cells, *Brain Res.*, 2004, **1021**, 241.

60. Y. Ito, K. Oh-hashi, K. Kiuchi and Y. Hirata, p44/42 MAP kinase and c-Jun N-terminal kinase contribute to the up-regulation of caspase-3 in manganese-induced apoptosis in PC12 cells, *Brain Res.*, 2006, **1099**, 1.

61. L. E. Gonzalez, A. A. Juknat, A. J. Venosa, N. Verrengia and M. L. Kotler, Manganese activates the mitochondrial apoptotic pathway in rat astrocytes by modulating the expression of proteins of the Bcl-2 family, *Neurochem. Int.*, 2008, **53**, 408.

62. Z. Yin, J. L. Aschner, A. P. dos Santos and M. Aschner, Mitochondrial-dependent manganese neurotoxicity in rat primary astrocyte cultures, *Brain Res.*, 2008, **1203**, 1.

63. Y. Li, L. Sun, T. Cai, Y. Zhang, S. Lv, Y. Wang and L. Ye, α-Synuclein overexpression during manganese-induced apoptosis in SH-SY5Y neuroblastoma cells, *Brain Res. Bull.*, 2010, **81**, 428.

64. E. J. Park and K. Park, Induction of oxidative stress and inflammatory cytokines by manganese chloride in cultured T98G cells, human brain glioblastoma cell line, *Toxicol. In Vitro*, 2010, **24**, 472.

65. F. Cordova, A. Aguiar, Jr., T. Peres, M. Lopes, F. Gonçalves, D. Pedro, S. Lopes, C. Pilati, R. S. Prediger, M. Farina, K. Erikson, M. Aschner and R. Leal, Manganese-exposed developing rats display motor deficits and striatal oxidative stress that are reversed by Trolox, *Arch. Toxicol.*, 2013, 1.

66. L. O. Murphy and J. Blenis, MAPK signal specificity: the right place at the right time, *Trends Biochem. Sci.*, 2006, **31**, 268.

67. J.-H. Bae, B.-C. Jang, S.-I. Suh, E. Ha, H. H. Baik, S.-S. Kim, M.-y. Lee and D.-H. Shin, Manganese induces inducible nitric oxide synthase (iNOS) expression via activation of both MAP kinase and PI3K/Akt pathways in BV2 microglial cells, *Neurosci. Lett.*, 2006, **398**, 151.

68. S. L. Liao, Y. C. Ou, S. Y. Chen, A. N. Chiang and C. J. Chen, Induction of cyclooxygenase-2 expression by manganese in cultured astrocytes, *Neurochem. Int.*, 2007, **50**, 905.

69. V. Y. Poon, S. Choi and M. Park, Growth factors in synaptic function, *Front. Synaptic Neurosci.*, 2013, **5**, 6.
70. E. M. Dickins and P. C. Salinas, Wnts in action: from synapse formation to synaptic maintenance, *Front. Cell. Neurosci.*, 2013, 7, 162.
71. J. Jiang, S. Shi, Q. Zhou, X. Ma, X. Nie, L. Yang, J. Han, G. Xu and C. Wan, Downregulation of the Wnt/beta-catenin signaling pathway is involved in manganese-induced neurotoxicity in rat striatum and PC12 cells, *J. Neurosci. Res.*, 2014, **92**, 783.
72. T. Cai, H. Che, T. Yao, Y. Chen, C. Huang, W. Zhang, K. Du, J. Zhang, Y. Cao, J. Chen and W. Luo, Manganese Induces Tau Hyperphosphorylation through the Activation of ERK MAPK Pathway in PC12 Cells, *Toxicol. Sci.*, 2011, **119**, 169.

CHAPTER 8

Manganese and Oxidative Stress

DAIANA SILVA ÁVILA,*[a] MARCELO FARINA,[b]
JOÃO BATISTA TEIXEIRA DA ROCHA[c] AND
MICHAEL ASCHNER[d]

[a] Universidade Federal do Pampa-UNIPAMPA Uruguaiana, Uruguaiana, RS, Brazil; [b] Departamento de Bioquímica, Centro de Ciências Biológicas, Universidade Federal de Santa Catarina, Florianópolis, SC, Brazil; [c] Departamento de Química, Universidade Federal de Santa Maria, Santa Maria, RS, Brazil; [d] Department of Molecular Pharmacology, Albert Einstein College of Medicine, Bronx, NY, USA
*Email: avilads1@gmail.com

8.1 Introduction

Reactive oxygen species (ROS) are a set of oxygen-containing molecules that include highly reactive oxygen radicals (*i.e.* hydroxyl radical, $HO^{\bullet-}$, and superoxide anion, $O_2^{\bullet-}$) and relatively stable non-radical oxidants (*i.e.* hydrogen peroxide, H_2O_2). These molecules are physiologically generated in the cytosol, peroxisomes, and membranes.[1–4] However, mitochondria represent the major source of intracellular ROS, where the one-electron reduction of O_2 generates significant amounts of $O_2^{\bullet-}$ (for a review, see ref. 5). Aside from ROS, reactive nitrogen species (RNS), such as nitric oxide (NO) and peroxynitrite (ONOO), are also normally generated under physiological conditions.[6,7]

Issues in Toxicology No. 22
Manganese in Health and Disease
Edited by Lucio G. Costa and Michael Aschner
© The Royal Society of Chemistry 2015
Published by the Royal Society of Chemistry, www.rsc.org

At physiological levels, ROS and RNS may function as signaling molecules through oxidation of redox-sensitive cysteine residues, thus modulating signaling pathways.[8,9] For example, H_2O_2 (usually generated in mitochondria *via* dismutation of $O_2^{\bullet-}$) modulates insulin release and signaling, adipocyte differentiation, regulation of cell cycle, and the hypoxic and immune responses.[10–13] Consequently, ROS signaling has been implicated in several events that can be advantageous to the cell.

On the other hand, excessive ROS/RNS levels are deleterious to cells. In this regard, *oxidative stress* has been defined as "a disturbance in the pro-oxidant–antioxidant balance in favor of the former, leading to potential damage" of biological systems.[14] Oxidative stress is a causal, or at least an auxiliary, factor in the pathogenesis of several human disorders, including cancer, cardiovascular disease, and neurodegeneration.[15–17] Events mediating ROS/RNS-induced cell damage comprise the excessive oxidation of nucleic acids, proteins, and lipids, thus impairing their structure and function.[18,19]

Because of the potential detrimental effects of reactive species, aerobic organisms have evolved a complex antioxidant system, including both enzymatic and non-enzymatic molecules able to reduce oxidative stress by either removal of reactive species or repair of particular forms of oxidation within cellular macromolecules. There are numerous enzymes involved in the antioxidant defense system, not all of which will be discussed here. The main antioxidant enzymes in mammals comprise the family of glutathione peroxidases (GPx),[20] superoxide dismutases (Sod),[21] peroxiredoxins (Prdx),[22] and catalase (Cat).[23] Sod reduces $O_2^{\bullet-}$ levels in the cell by catalyzing its conversion to molecular oxygen and H_2O_2. The principal isoforms of Sod found in mammals are the cytosolic CuZn superoxide dismutase (CuZnSod; Sod1), the mitochondrial Mn superoxide dismutase (MnSod; Sod2), and the extracellular superoxide dismutase (ECSod; Sod3).[24–26] This set of enzymes represents the first line of defense against $O_2^{\bullet-}$ produced by both the mitochondria and other cellular sources, such as NADPH oxidases.

H_2O_2 molecules (including those generated by Sod) are converted into water mainly by Cat, GPx, and Prdx. Cat is an enzyme ubiquitously expressed among mammalian tissues and is predominantly located in the peroxisomes. Its principal catalytic function is the decomposition of H_2O_2 to oxygen and water.[23,27] The GPx family, which comprises eight putative isoforms, can reduce peroxides (including H_2O_2 and lipid hydroperoxides) to less toxic forms, including water and alcohols. Prdxs represent a relatively newly discovered class of antioxidant enzymes with peroxidase activity that can reduce H_2O_2, peroxynitrite, and organic hydroperoxides.[28] Prdxs are sensitive to oxidation, and it is hypothesized that they also act as redox sensors, modulating different cellular events.[29]

Of particular importance, manganese (Mn), the fundamental subject of this book, is known to contribute to the generation of oxidative and nitrosative stresses.[16,30] Although this element is an essential metal for humans and animals,[31] excessive Mn exposure may stimulate the generation of ROS and RNS.[32] Accordingly, evidence exists supporting increased pro-oxidative damage and decreased antioxidant capacity as putative mechanisms mediating Mn-induced toxicity.[33–35] The main events related to such toxicity include mitochondrial dysfunction and energetic failure, increased oxidation of dopamine, production of reactive species and toxic metabolites, direct effects on antioxidant enzymes, and formation of protein aggregates, as discussed later.

8.2 Mechanisms Mediating Mn-Induced Oxidative Stress and Toxicity

8.2.1 Mn and Mitochondria

Mitochondria are dynamic organelles that change in number and morphology in healthy cells.[36] Although mitochondria play pivotal roles in buffering cytosolic Ca^{2+},[37] controlling apoptosis,[38] enabling β-oxidation of fatty acids,[39] and participating in the urea cycle,[40] their primary function is related to oxidative phosphorylation, which contributes to the production of over 90% of the cell's ATP.[41] As already mentioned, mitochondria are the main site for ROS/RNS production.[5] ROS/RNS generated within mitochondria can modulate several mitochondrial events (*i.e.* electron transport chain, tricarboxylic acid cycle, Ca^{2+} buffering), some of which regulate events taking place in other cellular compartments, such as apoptosome formation.[42] Accordingly, it is not surprising that ROS/RNS-induced mitochondrial dysfunction can also affect whole cell homeostasis, culminating in damage and cellular demise.

A growing body of experimental studies (presented later) has demonstrated that mitochondria are a main target for Mn accumulation and toxicity. Mn decreases energy metabolism by affecting the activities of mitochondrial enzymes, membrane potential, and ATP production.[43] This phenomenon is likely related to the Mn's electron structure. The Mn^{2+} ion presents electronic features similar to those of Ca^{2+} and Mg^{2+}, and its ionic radius is intermediate between those of Ca^{2+} and Mg^{2+}.[44] Thus, it is likely able to bind to almost every Ca^{2+} or Mg^{2+} binding-site and can frequently substitute for Ca^{2+} or Mg^{2+} in biological processes or act as an inhibitor of these processes.[32] Approximately four decades ago, experimental evidence showed that Mn^{2+} can be taken up by mitochondria *via* the mitochondrial Ca^{2+} uniporter.[44] Later, Gavin *et al.* investigated both Mn^{2+} and Ca^{2+} mitochondrial efflux kinetics in isolated rat brain mitochondria.[45] The authors observed that Mn^{2+} is not exported out of mitochondria *via* the Na^{2+}-dependent efflux mechanism (the more active mechanism in brain mitochondria), but rather by the Na^{2+}-independent efflux mechanism (much less active in brain mitochondria). Consequently, once Mn^{2+} is sequestered within brain mitochondria, it is very difficult for it to be transported out again, accounting for the relatively long half-life of Mn within this organelle and the brain, in general.[32]

In vitro studies using suspensions of rat liver mitochondria[46] showed that Mn^{2+} inhibited oxygen consumption after the addition of ADP (when either succinate or glutamate/malate was used as substrate). The authors proposed a direct inhibitory effect of Mn^{2+} on oxidative phosphorylation, speculating that Mn^{2+} bound to the $F_1ATPase$. Additional experimental studies based on *in vitro* approaches also pointed to mitochondria as a potential Mn target. Using rat primary neuron cultures, Malecki reported a dose-dependent decrease of the mitochondrial membrane potential and complex II activity following exposure to Mn^{2+}.[47] Based on protocols with rat cortical astrocytes, Gonzalez *et al.* demonstrated that Mn^{2+} targeted mitochondria and caused mitochondrial membrane depolarization followed by cytochrome *c* release to the cytoplasm and induction of effector caspases, which culminated in apoptotic death of astrocytes.[48] At that time, such results were particularly important, taking into account the pivotal role of mitochondria in controlling apoptotic events.[42] In agreement, additional *in vitro* studies with cultured cells showed that Mn exposure stimulated apoptotic-related events, such as caspase-3 activation and dissipation of mitochondrial membrane potential $(\Delta\Psi m)$,[49] as well as chromatin condensation and cell shrinkage, mitochondrial cytochrome c release, and caspase-specific cleavage of the endogenous substrate poly (ADP-ribose) polymerase.[50] These authors also showed that Mn treatment increased the formation of ROS in cultured neural stem cells, as well as showing that the pre-treatment with the antioxidant MnTBAP significantly increased cell viability of C17.2 cells exposed to Mn.[50] Using primary cultures of mouse cortical and cerebellar neurons, Hernández *et al.* showed that Mn treatment caused mitochondrial dysfunction and that this phenomenon was mitigated by the antioxidant ascorbate.[51] Together, these *in vitro* studies with cultured cells strongly suggest that Mn-induced cell death is likely mediated by apoptotic events *via* mitochondrial-dependent pathways, and that ROS might represent primary mediators of this process.[48–51]

With respect to potential molecular targets that mediate Mn-induced dysfunction, a recent study with isolated mitochondria[52] showed that Mn^{2+} inhibited ATP production with very different patterns in liver, brain, and heart mitochondria. In fact, the primary Mn^{2+} inhibition site in liver and heart mitochondria, but not in brain mitochondria, was the F1F0-ATP-synthase. In mitochondria fueled by succinate or glutamate + malate, ATP production was much more strongly inhibited in brain than in liver or heart mitochondria. The authors also observed that Mn^{2+} inhibited two independent sites in brain mitochondria: (i) fumarase or complex II (when succinate was substrate) and (ii) glutamate/aspartate exchanger or aspartate aminotransferase (when glutamate plus malate were the substrates). One of the important results from this study relates to the fact that the potential mitochondrial molecular targets of Mn^{2+} are tissue-dependent and also depend on the available substrates fuelling complex I or II. Mitocondrial aconitase (an important enzyme of the tricarboxylic acid (TCA) cycle) was particularly inhibited by Mn in AF5 cells, which resulted in a 90% increase in intracellular citrate concentrations.[53] The authors proposed that mitochondrial aconitase represents an important target of Mn. Based on the crucial role of this enzyme in supporting the TCA cycle, the energetic failure observed after Mn exposure seems to represent an expected phenomenon. Regardless, as shown in these studies, Mn^{2+} interacts with different molecular targets in the mitochondria (aconitase, F1F0-ATP-synthase, fumarase, complex II, glutamate/aspartate exchanger, and aspartate aminotransferase). These versatile interactions, which are likely related to Mn's similarity to Ca^{2+} and Mg^{2+}, seems to be tissue-dependent and change upon specific metabolic cellular conditions (*i.e.* availability of energetic substrates).[52] The interactions of Mn with the aforementioned mitochondrial targets may lead to energetic failure. However, although ROS is intimately linked to mitochondrial activities and energetic failure, these interactions do not directly explain the increased ROS observed after Mn treatment.

In vivo studies have also pointed to mitochondria as an important Mn target. Zhang *et al.* performed an *ex vivo* protocol to evaluate mitochondrial function in liver and brain of Mn^{2+}-exposed rats.[54] After *in vivo* Mn treatments (i.p. injections for 6 weeks) and mitochondria isolation, the authors observed that Mn accumulated in mitochondria and inhibited calcium efflux, as well as decreased the activity of mitochondrial complexes (I, II, III, and IV), and increased ROS formation in both liver and brain mitochondria.[54] A few years later, the same group (Zhang *et al.*) observed that an analogous exposure protocol changed the expression of mitochondrial proteins involved in energetic metabolism, such as ATP Synthase Beta Chain and Succinate dehydrogenase flavoprotein.[55] In an *in vivo* protocol with Mn-exposed chicken, Shao *et al.* found that subchronic Mn exposure induced damage in chicken hearts, which was related to disruption of mitochondrial metabolism and the alteration in iron (Fe) homeostasis.[56] In an *in vivo* protocol of developmental neurotoxicity with suckling rats, Cordova *et al.* showed that early postnatal exposure to Mn increased ROS formation and stimulated caspase mitochondrial complex I activities in the striatum.[57] These events were paralleled by impairments in behavioral tests related to locomotor activity (open field) and motor performance (rotarod task). Notably, the co-treatment with trolox (an antioxidant vitamin E analog) prevented Mn-induced ROS formation, but did not mitigate the Mn-induced motor deficits. The same researchers also showed that Mn exposure during the weaning period increased caspase activity in the striatum (suggesting the occurrence of a potential apoptotic event) and that trolox significantly decreased Mn-induced caspase activity.[33] This result is particularly important because it indicates that Mn-induced ROS formation under *in vivo* conditions is able to stimulate apoptotic events in the striatum, which represents an important target site of Mn-induced neurotoxicity.

The behavioral profile of animals (mainly rodents) exposed to $Mn^{33,35,57}$ is generally characterized by hypokinesia and motor impairment, in agreement with symptoms observed in "manganism," a human condition characterized by hypokinesia and postural instability.[58] Based on these similarities between experimental models and manganism, one might posit the presence of similar toxicity mechanisms in both conditions, and that the mito-chondria might represent a key target of Mn-induced toxicity in humans. However, this hypothesis has yet to be fully established.

8.2.2 Manganese and Dopamine Oxidation

Mn accumulation was found to lead to symptoms analogous to those observed in Parkinson's disease (PD), thus special attention has been focused on the dopaminergic (DAergic) system and its role in Mn neuro-toxicity. Indeed, chronic exposure to Mn leads to the degeneration of nigrostriatal DAergic neurons.[59,60] In addition, it was shown that autophagy plays a pivotal role on Mn-induced DAergic neurodegeneration.[61] However, whether striatal levels of dopamine (DA) are altered in experimental manganism remains controversial.[62] This is likely associated with differ-ential experimental conditions, the form of Mn (state of oxidation), route of administration, and time of exposure.[16,32,107] DA levels were depleted in striatum of animals exposed to elevated Mn concentration.[16] Others found a biphasic response, depending on Mn dose, resulting in increased DA and its metabolites at low doses and decreased levels at high doses.[16] Similarly, it was found that in rats treated chronically with oral $MnCl_2$, the activity of striatal tyrosine hydroxylase had increased at the beginning of the experiment and decreased after 6 months of treatment.[63] Likewise, a biphasic clinical appearance is also characteristic in Mn-exposed patients, with the first symptoms likely caused by the initial exposure to Mn and the late symptoms by long-term incremental cumulative doses.[63,64]

Recently, it was shown that 8 week Mn exposure in drinking water did not alter striatal dopamine, its metabolites, or the expression of key dopa-mine homeostatic proteins; however, Mn significantly increased striatal 5-hydroxyindoleacetic acid (a serotonin metabolite) levels, without affecting the levels of serotonin itself.[65] The altered glutamatergic and GABAergic functioning may contribute to abnormal striatal DA metabolism.[65-67]

Glutathione S-transferases (GSTs) of the class pi (GSTπ) are phase II detoxification enzymes that conjugate both endogenous and exogenous compounds to glutathione (GSH) to reduce cellular oxidative stress, and their decreased expression has recently been implicated in PD progression.[68] For example, it was demonstrated that a *Caenorhabditis elegans* GSTπ homologue, GST-1, inhibits Mn-induced DAergic degeneration *via* up-regulation of GST-1 gene and protein expression, which was reported to be dependent upon the transcription factor Nrf2/SKN-1. Thus, a reduction in SKN-1 gene expression resulted in a decrease in GST-1 protein expression and an increase in DA neuronal death. These findings are further supported by data showing that decreases in gene expression of the SKN-1 inhibitor, WDR-23, or the GSTp-binding cell death activator JNK/JNK-1, result in increased in resistance to this metal. Additionally, the same authors have shown that the Mn-induced DA neuron degeneration is independent of the dopamine transporter DAT, but is largely dependent upon the caspase CED-3 and the novel caspase CSP-1. Thus, this study provided *in vivo* evidence that a phase II detoxification enzyme may modulate DAergic vulnerability in manganism.[69]

Both Mn^{2+} and Mn^{3+} are known to react with DA *via* the Fenton's reaction, catalyzing its auto-oxidation and generating ROS, thereby contributing to oxidative damage.[59,70–76] However, it is noteworthy that neither Mn^{2+} nor Mn^{3+} can directly generate hydroxyl radicals from hydrogen peroxide and/or superoxide *via* Fenton-type or Haber–Weiss-type reactions.[16,70] Hence, Mn-induced DA auto-oxidation is a complex process involving several steps in which semi-quinone and aminochrome intermediates, L-cysteine or copper (Cu) and NAD(P)H facilitation are implicated.[59,71,75,77–79] Thus, the possible mechanisms underlying semi-quinone and aminochrome-induced damage in Mn-induced neurodegenerative process may include NADH or NADPH depletion, inactivation of enzymes by oxidizing thiol groups or essential amino acids, formation of RS, and lipid peroxidation.

8.2.3 Manganese and Antioxidant Homeostasis

Mn toxicity can be exacerbated by disruption of the antioxidant systems, implying that in addition to generating free radicals by reacting with dopamine, Mn may block the antioxidant system that scavenge these damaging molecules. For example, ROS/RNS can act as potent electrophiles and key activators of nuclear factor erythroid 2-related factor 2 (Nrf2), a master regulator of the specific antioxidant phenotype.[80–83] Casalino *et al.* reported that acute Mn exposure in rats caused increases in both the hepatic level of Nrf2 and its transfer from the cytoplasm to the nucleus where it actively regulates the induction of phase II enzymes, such as alpha-class GST subunit genes.[84] In the case of induction of phase II enzymes by heavy metals, it has been postulated that phosphorylation signaling pathways may contribute to this effect. In this context, it may be postulated that Mn affects the mitogen-activated protein kinases (MAPKs) and protein kinases (PKs) that phosphorylate Nfr2/Keap 1 complex in the cytoplasm, resulting in Nrf2 dissociation and nuclear translocation. In fact, Mn activates PI3K and MAPKs (JNK and p38) in several types of cells.[85,86]

By activating MAPK pathway, Mn also induces iNOS, thus increasing the formation of nitric oxide radical (NO), as observed *in vitro* in microglial cells[87] and in astrocytes.[88] These high levels of NO may render the brain more susceptible to oxidative and nitrosative stress, inducing the formation of reactive nitrogen species (RNS). In agreement, depletion of iNOS in mice reduced formation of 3-nitrotyrosine protein adducts within neurons in the basal ganglia and correlated with protection against Mn-induced neurobehavioral defects.[89]

8.2.4 Manganese and Protein Aggregates

Protein oxidation has been associated to promotion of protein aggregates. When the chaperone and the ubiquitin/proteasome systems are impaired by oxidative stress, misfolded proteins accumulate. These aggregates are physiopathological hallmarks of different neurodegenerative diseases that appear with aging in subjects with genetic predisposition. However, environmental/occupational exposures to toxicants that cause oxidative injury and endoplasmic reticulum (ER) stress, as Mn, have been linked to enhanced and earlier formation of these aggregates.

8.2.4.1   α-Synuclein

α-Synuclein is a highly charged 140-amino acid heat-stable protein that is soluble and natively unfolded.[90] It is a protein with a natural tendency to undergo conformational changes and to aggregate into olygomers called Lewy bodies, which are considered to be a key player in pathophysiology of PD. Mn has been previously shown to induce α-synuclein aggregation.[90–97] Cell lines overexpressing α-synuclein are more vulnerable to Mn toxicity, leading to cell death by apoptosis.[93,94] In addition, it has been reported that Mn up-regulates α-synuclein expression in SH-SY5Y neuroblastoma cells, in brain slices, in *C. elegans*,[92] in transgenic mice, and in the cortex of non-human primates. A compelling evidence of the involvement of oxidative stress in Mn-induced α-synuclein aggregation posits that glutathione (GSH) pre-treatment alleviates its oligomerization whereas H_2O_2 aggravated it.[97]

In a different approach, Xu *et al.* silenced α-synuclein in brain slices and observed that this alleviated Mn-induced ER stress, thus reducing apoptosis in neuronal cells.[96] This provides one putative hypothesis for Mn-induced α-synuclein aggregation because Mn induces damage to ER, impairing the proper unfolded protein response, and thus affecting the degradation of the olygomers. An additional hypothesis was provided from studies in PC12 cells: Mn induces α-synuclein overexpression *via* ERK1/2 activation as it was observed that the inhibition of this pathway by an inhibitor (PD98059) attenuated the aggregates formation and the citotoxicity induced by Mn.[91]

On the other hand, *C. elegans* mutated for several Parkinson's disease-related genes (*parkin, pink-1 and dj-1*) showed enhanced Mn accumulation and oxidative stress that was reduced by α-synuclein protein expression. Moreover, DAergic neurodegeneration, while unchanged with Mn exposure, returned to wild-type (WT) levels for pdr1, but not djr1.1 mutants expressing α-synuclein.[98] This study indicates a protective role for α-synuclein at least in its wildtype form. Studies with mutated forms of α-synuclein have yet to be carried out.

8.2.4.2   β-Amyloid

β-Amyloid is a peptide of 36–43 amino acids that is involved in Alzheimer's disease (AD) as the main component of the amyloid plaques inherent to afflicted subjects. Recently, reports describing that children exposed to elevated levels of Mn depicted impaired intellectual and cognitive functioning, which are domains of frontal cortex and subcortical structures that accumulate very little Mn, led researchers to further investigate the ability of this protein to mediate Mn neurotoxicity.

Gene expression analysis revealed that in the frontal cortex of Mn-exposed non-human primates, 61 genes were increased and 4 genes were decreased relative to controls from a total of 6766 genes.[99] The biological functions of the genes altered by Mn-exposure varied from cholesterol metabolism to proteosome/protein folding/protein turnover. The most highly upregulated gene was amyloid beta (Aβ) precursor-like protein 1 (APLP1), a member of the amyloid precursor protein (APP) family. Immunohistochemistry confirmed that APLP1 protein expression was increased and revealed the presence of diffuse amyloid-β plaques in the frontal cortex of Mn-exposed non-human primates.[100] The same Mn-exposed animals also expressed a significant degree of frontal cortex neuronal degeneration and glial cell activation with white matter involvement.[100] Further, they exhibited significant deficits in working memory with effects on motor function primarily affecting fine motor control.[101] The doses used were non-physiological; nevertheless, the findings were unexpected because the degeneration was found in very young animals that should not have β-amyloid plaques.

8.2.4.3 *PrP*

Prion diseases are neurodegenerative diseases that can be transmitted between individuals. The exact cause of these diseases remains unknown; however, one of the key events associated with the disease is the aggregation of a cellular protein, the prion protein (PrP).

Interestingly, increased Mn content has been observed in the blood and brain of humans infected with Creutzfeldt–Jakob disease (CJD), mice infected with scrapie, and cattle infected with bovine spongiform encephalopathy (BSE).[102–104] Johnson *et al.* have shown that Mn is present in aggregates of prion protein in hamsters infected with transmissible spongiform encephalopathies.[105] *In vitro*, it was demonstrated that Mn is incorporated to recombinant mouse PrP, thus causing conformational changes, increasing its protease resistance and then increasing its neurotoxicity. Additionally, Mn-bound PrPSc can be isolated from both humans and animals infected with prion disease. In agreement, a Mn chelator lengthens the incubation period of prion disease in mice, and reduces the amount of PrP present in their brains.[106] Despite these findings, the role of Mn in the pathogenesis of prion disease is currently unknown.

8.3 Antioxidant Approach against Manganism

As detailed earlier, *in vivo* and *in vitro* exposure to Mn is associated with oxidative stress,[107–112] leading to behavioral impairments and neuroinflammation.[35,107,113] Consequently, the use of antioxidants has been investigated as a potential pharmacological approach to treat the toxicity of Mn. Different classes of antioxidant agents have been reported to protect against Mn toxicity and here we will briefly discuss *in vitro* and *in vivo* studies showing that natural antioxidants found in mammals (for instance, vitamin E) and in plants (complex mixtures of plant extracts containing polyphenols and non-characterized components), iron chelating agents, precursors of glutathione (*N*-acetylcysteine), and synthetic antioxidants can afford protection against manganese-induced cytotoxicity.[107]

Plant extracts have been demonstrated to confer protection against the neurotoxicity of Mn after *in vitro*[114] and *in vivo* exposure in mice.[115] Açaí (*Euterpe oleracea* methanolic extract) protected astrocytes from Mn-induced oxidative stress. The protective effects can be associated to the antioxidant and anti-inflammatory effects of its anthocyanin components.[114] Similarly, crude aqueous extracts of *Melissa officinalis* blunted the Mn-induced striatal and hipocampal lipid peroxidation.[115] Purified flavonoids, for instance, silymarin (obtained from *Silybum marianum*, a plant with hepatoprotecive properties) protected neuroblastoma cells[116] and prevented Mn-induced oxidative stress in brain, liver, and kidney of rats.[117–119] Lycopene has also been reported to decrease the neurotoxicity of this metal in rats.[120]

Mn in its different chemical forms is not expected to cause oxidative stress directly, but it has been suggested that the transition of $Mn(II)$ to $Mn(III)$ could be involved in the pro-oxidative effects of this metal.[121] However, the direct oxidation of biomolecules by $Mn(III)$ has not yet been proved either *in vitro* or *in vivo*.[122] Because Mn shares some chemical properties with Fe, it is also possible that the neurotoxic effects of Mn might be associated with competition with Fe for "safe-non-redox" domains in proteins.[16] Thus, Mn could transiently increase the levels of free intra- or extracellular Fe, which, in turn, would cause oxidative stress either *via* stimulation of Fenton's reaction or reaction with oxygen to form reactive perferryl and related species.[16] Importantly, it was demonstrated that Mn exposure increases the ratio of $Fe(II):Fe(III)$, which can facilitate the Fenton's reaction *in vivo*.[123] Consequently, compounds with Fe chelating properties or those interfering with the Fenton's reaction can be of potential pharmacological importance in the treatment of Mn toxicity. In view of the fact that polyphenol compounds can have powerful antioxidant properties, quench hydroxyl radical and chelate Fe, or form less redox active complexes with Fe species,[124–126] the antioxidant efficacy of natural products should be studied within the context of Mn-induced oxidative stress.

Vitamin E and trolox (hydrophilic analog of vitamin E) have been reported to protect the CNS of rodents and cultured cells from the toxic effects of Mn.[33,127,128] As noted earlier, exposure of lactating rats to Mn caused striatal and hipocampal oxidative stress and motor impairments, which were blunted by trolox administration.[33] *N*-Acetylcysteine (NAC), a precursor of GSH, and reduced GSH can also decrease the toxicity of Mn *in vitro*;[129] however, the protective mechanism involved in NAC and GSH has yet to be fully studied. These compounds most likely serve as indirect antioxidants because they are substrates of Gpx enzymes.

Synthetic antioxidants have also been reported to reduce Mn toxicity. For instance, several organochalcogens (*i.e.* organocompounds containing selenium or tellurium atoms bond to carbon) have been reported to possess antioxidant and anti-inflammatory properties.[130] The protective effects of organoselenide and telluride compounds against Mn induced *in vitro* and *in vivo* neurotoxicity, including ebselen, have been reported.[128] The mechanism may be related to a direct antioxidant property of these compounds against inorganic (H_2O_2) or organic peroxides overproduction stimulated by Fe or by thiol depletion caused by Mn poisoning. Indeed, this class of compounds can imitate the native thiol-peroxidase activity catalyzed by glutathione-peroxidase isoforms.[130] Of particular importance to the treatment of Mn intoxication, ebselen has been used in clinical trials and it exhibits anti-inflammatory properties.[131] Consequently, in addition to counteracting free radicals, ebselen and related compounds could decrease Mn toxicity *via* anti-inflammatory properties. Of note, anti-inflammatory agents have been reported to decrease Mn neurotoxicity *in vitro* and after *in vivo* exposure.[128]

The indirect pro-oxidative effects of Mn have been linked to disruption of synaptic glutamate homeostasis by interfering with glutamate uptake in astrocytes.[132] The increase in extracellular glutamate can cause excitotoxicity, which is associated with oxidative stress in neurons.[132] Furthermore, Mn decreases astrocytic glutamate uptake and expression of the astrocytic glutamate transporter, GLAST[132] *via* disruption of intracellular signaling.[133] Of potential clinical significance, estrogen and tamoxifen have been reported to increase the expression of glutamate transporters (GLAST and GLT-1) in astrocytes, decreasing Mn toxicity.[134–137]

8.4 Concluding Remarks

Oxidative stress plays a determinant role in Mn-induced toxicity. High Mn levels disrupt redox homeostasis, especially in those cells where the metal accumulates. The brain is particularly sensitive to Mn, where accumulation and uptake is high. There, it causes mitochondrial dysfunction and dopamine oxidation and it affects antioxidant pathways, which generate the production of massive reactive species, thus unbalancing oxidative/antioxidant homeostasis within the neuronal cells (see Figure 8.1). In addition, the reactive species can interact with biomolecules, thus causing DNA damage, lipid oxidation, and protein misfolding and aggregation. There may be neuronal and astrocytic impairment, or even death. The damage to the neurons, especially the dopaminergic, causes symptoms of manganism (tremors, gait disturbance, body rigidity, and cognitive alterations), and it is therefore plausible that therapy with antioxidants may minimize the oxidative stress induced by Mn on these cells, thus decreasing cell death and attenuating the symptoms.

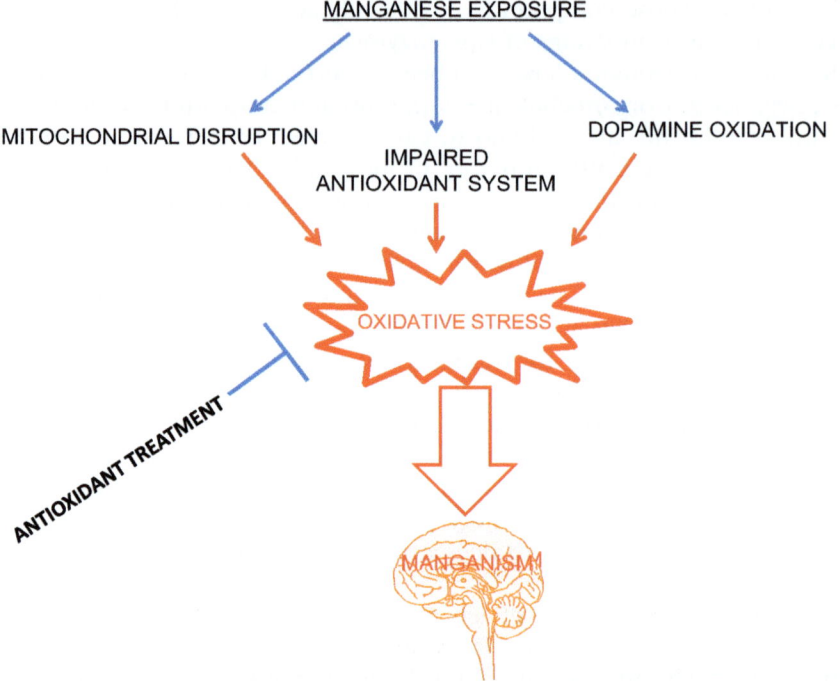

Figure 8.1   Mn exposure disrupts various cell functions, thus triggering oxidative stress. Oxidative stress plays an important role in manganism pathophysiology, and studies provide evidence that antioxidant therapy may attenuate the effects of Mn-induced oxidative stress.

References

1. J. Arnhold and J. Flemmig, Human myeloperoxidase in innate and acquired immunity, *Arch. Biochem. Biophys.*, 2010, **500**, 92–106.
2. S. Dikalov, Cross talk between mitochondria and NADPH oxidases, *Free Radical Biol. Med.*, 2011, **51**, 1289–1301.
3. M. Fransen, M. Nordgren, B. Wang and O. Apanasets, Role of peroxisomes in ROS/RNS-metabolism: implications for human disease, *Biochim. Biophys. Acta*, 2012, **1822**, 1363–1373.
4. T. J. Tavender and N. J. Bulleid, Molecular mechanisms regulating oxidative activity of the Ero1 family in the endoplasmic reticulum, *Antioxid. Redox Signaling*, 2010, **13**, 1177–1187.
5. M. P. Murphy, How mitochondria produce reactive oxygen species, *Biochem. J.*, 2009, **417**, 1–13.
6. J. M. Fukuto, S. J. Carrington, D. J. Tantillo, J. G. Harrison, L. J. Ignarro, B. A. Freeman, A. Chen and D. A. Wink, Small molecule signaling agents: the integrated chemistry and biochemistry of nitrogen oxides, oxides of carbon, dioxygen, hydrogen sulfide, and their derived species, *Chem. Res. Toxicol.*, 2012, **25**, 769–793.

7. J. Lee, S. Giordano and J. Zhang, Autophagy, mitochondria and oxidative stress: cross-talk and redox signalling, *Biochem. J.*, 2012, **441**, 523–540.

8. T. C. Meng, T. Fukada and N. K. Tonks, Reversible oxidation and inactivation of protein tyrosine phosphatases in vivo, *Mol. Cell*, 2002, **9**, 387–399.

9. R. E. Tooker, M. Y. Lipin, V. Leuranguer, E. Rozsa, J. R. Bramley, J. L. Harding, M. M. Reynolds and J. Vigh, Nitric oxide mediates activity-dependent plasticity of retinal bipolar cell output via S-nitrosylation, *J. Neurosci.*, 2013, **33**, 19176–19193.

10. R. B. Hamanaka and N. S. Chandel, Mitochondrial reactive oxygen species regulate cellular signaling and dictate biological outcomes, *Trends Biochem. Sci.*, 2010, **35**, 505–513.

11. K. Loh, H. Deng, A. Fukushima, X. Cai, B. Boivin, S. Galic, C. Bruce, B. J. Shields, B. Skiba, L. M. Ooms, N. Stepto, B. Wu, C. A. Mitchell, N. K. Tonks, M. J. Watt, M. A. Febbraio, P. J. Crack, S. Andrikopoulos and T. Tiganis, Reactive oxygen species enhance insulin sensitivity, *Cell Metab.*, 2009, **10**, 260–272.

12. R. J. Mailloux, A. Fu, C. Robson-Doucette, E. M. Allister, M. B. Wheeler, R. Screaton and M. E. Harper, Glutathionylation state of uncoupling protein-2 and the control of glucose-stimulated insulin secretion, *J. Biol. Chem.*, 2012, **287**, 39673–39685.

13. K. V. Tormos, E. Anso, R. B. Hamanaka, J. Eisenbart, J. Joseph, B. Kalyanaraman and N. S. Chandel, Mitochondrial complex III ROS regulate adipocyte differentiation, *Cell Metab.*, 2011, **14**, 537–544.

14. H. Sies, Oxidative stress: oxidants and antioxidants, *Exp. Physiol.*, 1997, **82**, 291–295.

15. S. K. Choudhari, M. Chaudhary, A. R. Gadbail, A. Sharma and S. Tekade, Oxidative and antioxidative mechanisms in oral cancer and precancer: a review, *Oral Oncol.*, 2014, **50**, 10–18.

16. M. Farina, D. S. Avila, J. B. da Rocha and M. Aschner, Metals, oxidative stress and neurodegeneration: a focus on iron, manganese and mercury, *Neurochem. Int.*, 2013, **62**, 575–594.

17. C. Vetrani, G. Costabile, L. Di Marino and A. A. Rivellese, Nutrition and oxidative stress: a systematic review of human studies, *Int. J. Food Sci. Nutr.*, 2013, **64**, 312–326.

18. E. Cadenas and K. J. Davies, Mitochondrial free radical generation, oxidative stress, and aging, *Free Radical Biol. Med.*, 2000, **29**, 222–230.

19. M. Valko, C. J. Rhodes, J. Moncol, M. Izakovic and M. Mazur, Free radicals, metals and antioxidants in oxidative stress-induced cancer, *Chem.-Biol. Interact.*, 2006, **160**, 1–40.

20. R. Margis, C. Dunand, F. K. Teixeira and M. Margis-Pinheiro, Glutathione peroxidase family – an evolutionary overview, *FEBS J.*, 2008, **275**, 3959–3970.

21. I. N. Zelko, T. J. Mariani and R. J. Folz, Superoxide dismutase multigene family: a comparison of the CuZn-SOD (SOD1), Mn-SOD (SOD2), and

EC-SOD (SOD3) gene structures, evolution, and expression, *Free Radical Biol. Med.*, 2002, **33**, 337–349.

22. J. Fujii and Y. Ikeda, Advances in our understanding of peroxiredoxin, a multifunctional, mammalian redox protein, *Redox Rep.*, 2002, 7, 123–130.

23. A. Deisseroth and A. L. Dounce, Catalase: Physical and chemical properties, mechanism of catalysis, and physiological role, *Physiol. Rev.*, 1970, **50**, 319–375.

24. I. Fridovich, Fundamental aspects of reactive oxygen species, or what's the matter with oxygen?, *Ann. N. Y. Acad. Sci.*, 1999, **893**, 13–18.

25. S. L. Marklund, Extracellular superoxide dismutase in human tissues and human cell lines, *J. Clin. Invest.*, 1984, **74**, 1398–1403.

26. A. Okado-Matsumoto and I. Fridovich, Subcellular distribution of superoxide dismutases (SOD) in rat liver: Cu,Zn-SOD in mitochondria, *J. Biol. Chem.*, 2001, **276**, 38388–38393.

27. B. Halliwell and J. Gutteridge, *Free Radicals in Biology and Medicine*, Oxford University Press, Oxford, 1989, p. 1–20.

28. Z. A. Wood, E. Schroder, J. Robin Harris and L. B. Poole, Structure, mechanism and regulation of peroxiredoxins, *Trends Biochem. Sci.*, 2003, **28**, 32–40.

29. R. A. Poynton and M. B. Hampton, Peroxiredoxins as biomarkers of oxidative stress, *Biochim. Biophys. Acta*, 2014, **1840**, 906–912.

30. J. A. Moreno, K. M. Streifel, K. A. Sullivan, W. H. Hanneman and R. B. Tjalkens, Manganese-induced NF-kappaB activation and nitrosative stress is decreased by estrogen in juvenile mice, *Toxicol. Sci.*, 2011, **122**, 121–133.

31. T. V. Peres, D. Z. Pedro, F. M. de Cordova, M. W. Lopes, F. M. Goncalves, C. B. Mendes-de-Aguiar, R. Walz, M. Farina, M. Aschner and R. B. Leal, In vitro manganese exposure disrupts MAPK signaling pathways in striatal and hippocampal slices from immature rats, *BioMed Res. Int.*, 2013, **2013**, 769295.

32. E. J. Martinez-Finley, C. E. Gavin, M. Aschner and T. E. Gunter, Manganese neurotoxicity and the role of reactive oxygen species, *Free Radical Biol. Med.*, 2013, **62**, 65–75.

33. F. M. Cordova, A. S. Aguiar, Jr., T. V. Peres, M. W. Lopes, F. M. Goncalves, D. Z. Pedro, S. C. Lopes, C. Pilati, R. D. Prediger, M. Farina, K. M. Erikson, M. Aschner and R. B. Leal, Manganese-exposed developing rats display motor deficits and striatal oxidative stress that are reversed by Trolox, *Arch. Toxicol.*, 2013, **87**, 1231–1244.

34. D. Milatovic, S. Zaja-Milatovic, R. C. Gupta, Y. Yu and M. Aschner, Oxidative damage and neurodegeneration in manganese-induced neurotoxicity, *Toxicol. Appl. Pharmacol.*, 2009, **240**, 219–225.

35. D. Santos, D. Milatovic, V. Andrade, M. C. Batoreu, M. Aschner and A. P. Marreilha dos Santos, The inhibitory effect of manganese on acetylcholinesterase activity enhances oxidative stress and neuroinflammation in the rat brain, *Toxicology*, 2012, **292**, 90–98.

36. D. C. Chan, Mitochondrial fusion and fission in mammals, *Annu. Rev. Cell Dev. Biol.*, 2006, **22**, 79–99.

37. T. E. Gunter and D. R. Pfeiffer, Mechanisms by which mitochondria transport calcium, *Am. J. Physiol.*, 1990, **258**, C755–786.

38. G. Kroemer, Mitochondrial control of apoptosis: an overview, *Biochem. Soc. Symp.*, 1999, **66**, 1–15.

39. E. P. Kennedy and A. L. Lehninger, The products of oxidation of fatty acids by isolated rat liver mitochondria, *J. Biol. Chem.*, 1950, **185**, 275–285.

40. H. A. Krebs, R. Hems and P. Lund, Some regulatory mechanisms in the synthesis of urea in the mammalian liver, *Adv. Enzyme Regul.*, 1973, **11**, 361–377.

41. B. Chance and G. R. Williams, The respiratory chain and oxidative phosphorylation, *Adv. Enzymol. Relat. Subj. Biochem.*, 1956, **17**, 65–134.

42. D. R. Green and G. Kroemer, The pathophysiology of mitochondrial cell death, *Science*, 2004, **305**, 626–629.

43. T. E. Gunter, C. E. Gavin and K. K. Gunter, The case for manganese interaction with mitochondria, *Neurotoxicology*, 2009, **30**, 727–729.

44. R. E. Gunter, J. S. Puskin and P. R. Russell, Quantitative magnetic resonance studies of manganese uptake by mitochondria, *Biophys. J.*, 1975, **15**, 319–333.

45. C. E. Gavin, K. K. Gunter and T. E. Gunter, Manganese and calcium efflux kinetics in brain mitochondria. Relevance to manganese toxicity, *Biochem. J.*, 1990, **266**, 329–334.

46. C. E. Gavin, K. K. Gunter and T. E. Gunter, Mn^{2+} sequestration by mitochondria and inhibition of oxidative phosphorylation, *Toxicol. Appl. Pharmacol.*, 1992, **115**, 1–5.

47. E. A. Malecki, Manganese toxicity is associated with mitochondrial dysfunction and DNA fragmentation in rat primary striatal neurons, *Brain Res. Bull.*, 2001, **55**, 225–228.

48. L. E. Gonzalez, A. A. Juknat, A. J. Venosa, N. Verrengia and M. L. Kotler, Manganese activates the mitochondrial apoptotic pathway in rat astrocytes by modulating the expression of proteins of the Bcl-2 family, *Neurochem. Int.*, 2008, **53**, 408–415.

49. Z. Yin, J. L. Aschner, A. P. dos Santos and M. Aschner, Mitochondrial-dependent manganese neurotoxicity in rat primary astrocyte cultures, *Brain Res.*, 2008, **1203**, 1–11.

50. C. Tamm, F. Sabri and S. Ceccatelli, Mitochondrial-mediated apoptosis in neural stem cells exposed to manganese, *Toxicol. Sci.*, 2008, **101**, 310–320.

51. R. B. Hernandez, M. Farina, B. P. Esposito, N. C. Souza-Pinto, F. Barbosa, Jr. and C. Sunol, Mechanisms of manganese-induced neurotoxicity in primary neuronal cultures: the role of manganese speciation and cell type, *Toxicol. Sci.*, 2011, **124**, 414–423.

52. T. E. Gunter, B. Gerstner, T. Lester, A. P. Wojtovich, J. Malecki, S. G. Swarts, P. S. Brookes, C. E. Gavin and K. K. Gunter, An analysis of

the effects of Mn^{2+} on oxidative phosphorylation in liver, brain, and heart mitochondria using state 3 oxidation rate assays, *Toxicol. Appl. Pharmacol.*, 2010, **249**, 65–75.

53. D. R. Crooks, M. C. Ghosh, M. Braun-Sommargren, T. A. Rouault and D. R. Smith, Manganese targets m-aconitase and activates iron regulatory protein 2 in AF5 GABAergic cells, *J. Neurosci. Res.*, 2007, **85**, 1797–1809.

54. S. Zhang, Z. Zhou and J. Fu, Effect of manganese chloride exposure on liver and brain mitochondria function in rats, *Environ. Res.*, 2003, **93**, 149–157.

55. S. Zhang, J. Fu and Z. Zhou, Changes in the brain mitochondrial proteome of male Sprague-Dawley rats treated with manganese chloride, *Toxicol. Appl. Pharmacol.*, 2005, **202**, 13–17.

56. J. J. Shao, H. D. Yao, Z. W. Zhang, S. Li and S. W. Xu, The disruption of mitochondrial metabolism and ion homeostasis in chicken hearts exposed to manganese, *Toxicol. Lett.*, 2012, **214**, 99–108.

57. F. M. Cordova, A. S. Aguiar, Jr., T. V. Peres, M. W. Lopes, F. M. Goncalves, A. P. Remor, S. C. Lopes, C. Pilati, A. S. Latini, R. D. Prediger, K. M. Erikson, M. Aschner and R. B. Leal, In vivo manganese exposure modulates Erk, Akt and Darpp-32 in the striatum of developing rats, and impairs their motor function, *PLoS One*, 2012, 7, e33057.

58. K. Tuschl, P. B. Mills and P. T. Clayton, Manganese and the brain, *Int. Rev. Neurobiol.*, 2013, **110**, 277–312.

59. D. G. Graham, Catecholamine toxicity: a proposal for the molecular pathogenesis of manganese neurotoxicity and Parkinson's disease, *Neurotoxicology*, 1984, 5, 83–95.

60. A. Barbeau, Manganese and extrapyramidal disorders (a critical review and tribute to Dr. George C. Cotzias), *Neurotoxicology*, 1984, 5, 13–35.

61. J. Zhang, R. Cao, T. Cai, M. Aschner, F. Zhao, T. Yao, Y. Chen, Z. Cao, W. Luo and J. Chen, The role of autophagy dysregulation in manganese-induced dopaminergic neurodegeneration, *Neurotoxic. Res.*, 2013, 24, 478–490.

62. S. Rivera-Mancia, C. Rios and S. Montes, Manganese accumulation in the CNS and associated pathologies, *Biometals*, 2011, 24, 811–825.

63. E. Bonilla, L-tyrosine hydroxylase activity in the rat brain after chronic oral administration of manganese chloride, *Neurobehav. Toxicol.*, 1980, 2, 37–41.

64. Y. Finkelstein, N. Zhang, V. A. Fitsanakis, M. J. Avison, J. C. Gore and M. Aschner, Differential deposition of manganese in the rat brain following subchronic exposure to manganese: a T1-weighted magnetic resonance imaging study, *Isr. Med. Assoc. J.*, 2008, **10**, 793–798.

65. S. Krishna, C. A. Dodd, S. K. Hekmatyar and N. M. Filipov, Brain deposition and neurotoxicity of manganese in adult mice exposed via the drinking water, *Arch. Toxicol.*, 2014, **88**, 47–64.

66. M. Carlsson and A. Carlsson, Interactions between glutamatergic and monoaminergic systems within the basal ganglia--implications for schizophrenia and Parkinson's disease, *Trends Neurosci.*, 1990, **13**, 272–276.

67. S. L. Castro and M. J. Zigmond, Stress-induced increase in extracellular dopamine in striatum: role of glutamatergic action via N-methyl-D-aspartate receptors in substantia nigra, *Brain Res.*, 2001, **901**, 47–54.

68. R. Vilar, H. Coelho, E. Rodrigues, M. J. Gama, I. Rivera, E. Taioli and M. C. Lechner, Association of A313 G polymorphism (GSTP1*B) in the glutathione-S-transferase P1 gene with sporadic Parkinson's disease, *Eur. J. Neurol.*, 2007, **14**, 156–161.

69. R. Settivari, N. VanDuyn, J. LeVora and R. Nass, The Nrf2/SKN-1-dependent glutathione S-transferase pi homologue GST-1 inhibits dopamine neuron degeneration in a Caenorhabditis elegans model of manganism, *Neurotoxicology*, 2013, **38**, 51–60.

70. J. Donaldson, The physiopathologic significance of manganese in brain: its relation to schizophrenia and neurodegenerative disorders, *Neurotoxicology*, 1987, **8**, 451–462.

71. R. V. Lloyd, Mechanism of the manganese-catalyzed autoxidation of dopamine, *Chem. Res. Toxicol.*, 1995, **8**, 111–116.

72. M. Parenti, L. Rusconi, V. Cappabianca, E. A. Parati and A. Groppetti, Role of dopamine in manganese neurotoxicity, *Brain Res.*, 1988, **473**, 236–240.

73. S. H. Reaney, C. L. Kwik-Uribe and D. R. Smith, Manganese oxidation state and its implications for toxicity, *Chem. Res. Toxicol.*, 2002, **15**, 1119–1126.

74. S. H. Reaney and D. R. Smith, Manganese oxidation state mediates toxicity in PC12 cells, *Toxicol. Appl. Pharmacol.*, 2005, **205**, 271–281.

75. X. M. Shen and G. Dryhurst, Iron- and manganese-catalyzed autoxidation of dopamine in the presence of L-cysteine: possible insights into iron- and manganese-mediated dopaminergic neurotoxicity, *Chem. Res. Toxicol.*, 1998, **11**, 824–837.

76. A. Y. Sun, W. L. Yang and H. D. Kim, Free radical and lipid peroxidation in manganese-induced neuronal cell injury, *Ann. N. Y. Acad. Sci.*, 1993, **679**, 358–363.

77. T. M. Florence and J. L. Stauber, Stauber, Manganese catalysis of dopamine oxidation, *Sci. Total Environ.*, 1989, **78**, 233–240.

78. R. Graumann, I. Paris, P. Martinez-Alvarado, P. Rumanque, C. Perez-Pastene, S. P. Cardenas, P. Marin, F. Diaz-Grez, R. Caviedes, P. Caviedes and J. Segura-Aguilar, Oxidation of dopamine to aminochrome as a mechanism for neurodegeneration of dopaminergic systems in Parkinson's disease. Possible neuroprotective role of DT-diaphorase, *Pol. J. Pharmacol.*, 2002, **54**, 573–579.

79. O. Terland, T. Flatmark, A. Tangeras and M. Gronberg, Dopamine oxidation generates an oxidative stress mediated by dopamine

semiquinone and unrelated to reactive oxygen species, *J. Mol. Cell. Cardiol.*, 1997, **29**, 1731–1738.

80. B. M. Hybertson, B. Gao, S. K. Bose and J. M. McCord, Oxidative stress in health and disease: the therapeutic potential of Nrf2 activation, *Mol. Aspects Med.*, 2011, **32**, 234–246.

81. M. O. Leonard, N. E. Kieran, K. Howell, M. J. Burne, R. Varadarajan, S. Dhakshinamoorthy, A. G. Porter, C. O'Farrelly, H. Rabb and C. T. Taylor, Reoxygenation-specific activation of the antioxidant transcription factor Nrf2 mediates cytoprotective gene expression in ischemia-reperfusion injury, *FASEB J.*, 2006, **20**, 2624–2626.

82. X. Li, D. Zhang, M. Hannink and L. J. Beamer, Crystal structure of the Kelch domain of human Keap1, *J. Biol. Chem.*, 2004, **279**, 54750–54758.

83. M. McMahon, K. Itoh, M. Yamamoto and J. D. Hayes, Keap1-dependent proteasomal degradation of transcription factor Nrf2 contributes to the negative regulation of antioxidant response element-driven gene expression, *J. Biol. Chem.*, 2003, **278**, 21592–21600.

84. E. Casalino, G. Calzaretti, M. Landriscina, C. Sblano, A. Fabiano and C. Landriscina, The Nrf2 transcription factor contributes to the induction of alpha-class GST isoenzymes in liver of acute cadmium or manganese intoxicated rats: comparison with the toxic effect on NAD(P)H:quinone reductase, *Toxicology*, 2007, **237**, 24–34.

85. O. Dormond, L. Ponsonnet, M. Hasmim, A. Foletti and C. Ruegg, Manganese-induced integrin affinity maturation promotes recruitment of alpha V beta 3 integrin to focal adhesions in endothelial cells: evidence for a role of phosphatidylinositol 3-kinase and Src, *Thromb. Haemostasis*, 2004, **92**, 151–161.

86. Y. Hirata, K. Furuta, S. Miyazaki, M. Suzuki and K. Kiuchi, Anti-apoptotic and pro-apoptotic effect of NEPP11 on manganese-induced apoptosis and JNK pathway activation in PC12 cells, *Brain Res.*, 2004, **1021**, 241–247.

87. J. H. Bae, B. C. Jang, S. I. Suh, E. Ha, H. H. Baik, S. S. Kim, M. Y. Lee and D. H. Shin, Manganese induces inducible nitric oxide synthase (iNOS) expression via activation of both MAP kinase and PI3K/Akt pathways in BV2 microglial cells, *Neurosci. Lett.*, 2006, **398**, 151–154.

88. M. Spranger, S. Schwab, S. Desiderato, E. Bonmann, D. Krieger and J. Fandrey, Manganese augments nitric oxide synthesis in murine astrocytes: a new pathogenetic mechanism in manganism?, *Exp. Neurol.*, 1998, **149**, 277–283.

89. K. M. Streifel, J. A. Moreno, W. H. Hanneman, M. E. Legare and R. B. Tjalkens, Gene deletion of nos2 protects against manganese-induced neurological dysfunction in juvenile mice, *Toxicol. Sci.*, 2012, **126**, 183–192.

90. A. V. Krasnoslobodtsev, J. Peng, J. M. Asiago, J. Hindupur, J. C. Rochet and Y. L. Lyubchenko, Effect of spermidine on misfolding and interactions of alpha-synuclein, *PLoS One*, 2012, **7**, e38099.

91. T. Cai, T. Yao, G. Zheng, Y. Chen, K. Du, Y. Cao, X. Shen, J. Chen and W. Luo, Manganese induces the overexpression of alpha-synuclein in PC12 cells via ERK activation, *Brain Res.*, 2010, **1359**, 201–207.

92. A. D. Gitler, A. Chesi, M. L. Geddie, K. E. Strathearn, S. Hamamichi, K. J. Hill, K. A. Caldwell, G. A. Caldwell, A. A. Cooper, J. C. Rochet and S. Lindquist, Alpha-synuclein is part of a diverse and highly conserved interaction network that includes PARK9 and manganese toxicity, *Nat. Genet.*, 2009, **41**, 308–315.

93. Y. Li, L. Sun, T. Cai, Y. Zhang, S. Lv, Y. Wang and L. Ye, α-Synuclein overexpression during manganese-induced apoptosis in SH-SY5Y neuroblastoma cells, *Brain Res. Bull.*, 2010, **81**, 428–433.

94. C. Pifl, M. Khorchide, A. Kattinger, H. Reither, J. Hardy and O. Hornykiewicz, α-Synuclein selectively increases manganese-induced viability loss in SK-N-MC neuroblastoma cells expressing the human dopamine transporter, *Neurosci. Lett.*, 2004, **354**, 34–37.

95. V. N. Uversky, J. Li and A. L. Fink, Metal-triggered structural transformations, aggregation, and fibrillation of human alpha-synuclein. A possible molecular NK between Parkinson's disease and heavy metal exposure, *J. Biol. Chem.*, 2001, **276**, 44284–44296.

96. B. Xu, F. Wang, S. W. Wu, Y. Deng, W. Liu, S. Feng, T. Y. Yang and Z. F. Xu, Alpha-synuclein is involved in manganese-induced ER stress via PERK signal pathway in organotypic brain slice cultures, *Mol. Neurobiol.*, 2014, **49**, 399–412.

97. B. Xu, S. W. Wu, C. W. Lu, Y. Deng, W. Liu, Y. G. Wei, T. Y. Yang and Z. F. Xu, Oxidative stress involvement in manganese-induced α-synuclein oligomerization in organotypic brain slice cultures, *Toxicology*, 2013, **305**, 71–78.

98. J. Bornhorst, S. Chakraborty, S. Meyer, H. Lohren, S. Grosse Brinkhaus, A. L. Knight, K. A. Caldwell, G. A. Caldwell, U. Karst, T. Schwerdtle, A. Bowman and M. Aschner, The effects of pdr1, djr1.1 and pink1 loss in manganese-induced toxicity and the role of alpha-synuclein in C. elegans, *Metallomics*, 2014, **6**, 476–490.

99. T. R. Guilarte, N. C. Burton, T. Verina, V. V. Prabhu, K. G. Becker, T. Syversen and J. S. Schneider, Increased APLP1 expression and neurodegeneration in the frontal cortex of manganese-exposed non-human primates, *J. Neurochem.*, 2008, **105**, 1948–1959.

100. T. R. Guilarte, APLP1, Alzheimer's-like pathology and neurodegeneration in the frontal cortex of manganese-exposed non-human primates, *Neurotoxicology*, 2010, **31**, 572–574.

101. J. S. Schneider, E. Decamp, K. Clark, C. Bouquio, T. Syversen and T. R. Guilarte, Effects of chronic manganese exposure on working memory in non-human primates, *Brain Res.*, 2009, **1258**, 86–95.

102. S. Hesketh, J. Sassoon, R. Knight and D. R. Brown, Elevated manganese levels in blood and CNS in human prion disease, *Mol. Cell. Neurosci.*, 2008, **37**, 590–598.

103. A. M. Thackray, R. Knight, S. J. Haswell, R. Bujdoso and D. R. Brown, Metal imbalance and compromised antioxidant function are early changes in prion disease, *Biochem. J.*, 2002, **362**, 253–258.

104. B. S. Wong, S. G. Chen, M. Colucci, Z. Xie, T. Pan, T. Liu, R. Li, P. Gambetti, M. S. Sy and D. R. Brown, Aberrant metal binding by prion protein in human prion disease, *J. Neurochem.*, 2001, **78**, 1400–1408.

105. C. J. Johnson, P. U. Gilbert, M. Abrecht, K. L. Baldwin, R. E. Russell, J. A. Pedersen, J. M. Aiken and D. McKenzie, Low copper and high manganese levels in prion protein plaques, *Viruses*, 2013, **5**, 654–662.

106. M. W. Brazier, I. Volitakis, M. Kvasnicka, A. R. White, J. R. Underwood, J. E. Green, S. Han, A. F. Hill, C. L. Masters and S. J. Collins, Manganese chelation therapy extends survival in a mouse model of M1000 prion disease, *J. Neurochem.*, 2010, **114**, 440–451.

107. M. Aschner, K. M. Erikson, E. Herrero Hernandez and R. Tjalkens, Manganese and its role in Parkinson's disease: from transport to neuropathology, *NeuroMol. Med.*, 2009, **11**, 252–266.

108. J. Bornhorst, S. Meyer, T. Weber, C. Boker, T. Marschall, A. Mangerich, S. Beneke, A. Burkle and T. Schwerdtle, Molecular mechanisms of Mn induced neurotoxicity: RONS generation, genotoxicity, and DNA-damage response, *Mol. Nutr. Food Res.*, 2013, **57**, 1255–1269.

109. C. J. Chen and S. L. Liao, Oxidative stress involves in astrocytic alterations induced by manganese, *Exp. Neurol.*, 2002, **175**, 216–225.

110. A. W. Dobson, S. Weber, D. C. Dorman, L. K. Lash, K. M. Erikson and M. Aschner, Oxidative stress is induced in the rat brain following repeated inhalation exposure to manganese sulfate, *Biol. Trace Elem. Res.*, 2003, **93**, 113–126.

111. N. M. Filipov, R. F. Seegal and D. A. Lawrence, Manganese potentiates in vitro production of proinflammatory cytokines and nitric oxide by microglia through a nuclear factor kappa B-dependent mechanism, *Toxicol. Sci.*, 2005, **84**, 139–148.

112. D. Milatovic, Z. Yin, R. C. Gupta, M. Sidoryk, J. Albrecht, J. L. Aschner and M. Aschner, Manganese induces oxidative impairment in cultured rat astrocytes, *Toxicol. Sci.*, 2007, **98**, 198–205.

113. D. Santos, M. C. Batoreu, I. Tavares de Almeida, L. Davis Randall, M. L. Mateus, V. Andrade, R. Ramos, E. Torres, M. Aschner and A. P. Marreilha dos Santos, Evaluation of neurobehavioral and neuroinflammatory end-points in the post-exposure period in rats subacutely exposed to manganese, *Toxicology*, 2013, **314**, 95–99.

114. V. da Silva Santos, E. Bisen-Hersh, Y. Yu, I. S. Cabral, V. Nardini, M. Culbreth, J. B. Teixeira da Rocha, F. Barbosa, Jr. and M. Aschner, Anthocyanin-Rich Acai (Euterpe oleracea Mart.) Extract Attenuates Manganese-Induced Oxidative Stress in Rat Primary Astrocyte Cultures, *J. Toxicol. Environ. Health, Part A*, 2014, **77**, 390–404.

115. E. N. Martins, N. T. Pessano, L. Leal, D. H. Roos, V. Folmer, G. O. Puntel, J. B. Rocha, M. Aschner, D. S. Avila and R. L. Puntel, Protective effect of Melissa officinalis aqueous extract against

Mn-induced oxidative stress in chronically exposed mice, *Brain Res. Bull.*, 2012, **87**, 74–79.

116. Y. Chtourou, K. Trabelsi, H. Fetoui, G. Mkannez, H. Kallel and N. Zeghal, Manganese induces oxidative stress, redox state unbalance and disrupts membrane bound ATPases on murine neuroblastoma cells in vitro: protective role of silymarin, *Neurochem. Res.*, 2011, **36**, 1546–1557.

117. Y. Chtourou, H. Fetoui, M. Garoui el, T. Boudawara and N. Zeghal, Improvement of cerebellum redox states and cholinergic functions contribute to the beneficial effects of silymarin against manganese-induced neurotoxicity, *Neurochem. Res.*, 2012, **37**, 469–479.

118. Y. Chtourou, H. Fetoui, M. Sefi, K. Trabelsi, M. Barkallah, T. Boudawara, H. Kallel and N. Zeghal, Silymarin, a natural antioxidant, protects cerebral cortex against manganese-induced neurotoxicity in adult rats, *Biometals*, 2010, **23**, 985–996.

119. Y. Chtourou, E. Mouldi Garoui, T. Boudawara and N. Zeghal, Protective role of silymarin against manganese-induced nephrotoxicity and oxidative stress in rat, *Environ. Toxicol.*, 2013.

120. M. A. Lebda, M. S. El-Neweshy and Y. S. El-Sayed, Neurohepatic toxicity of subacute manganese chloride exposure and potential chemoprotective effects of lycopene, *Neurotoxicology*, 2012, **33**, 98–104.

121. D. HaMai, A. Campbell and S. C. Bondy, Modulation of oxidative events by multivalent manganese complexes in brain tissue, *Free Radical Biol. Med.*, 2001, **31**, 763–768.

122. K. K. Gunter, M. Aschner, L. M. Miller, R. Eliseev, J. Salter, K. Anderson, S. Hammond and T. E. Gunter, Determining the oxidation states of manganese in PC12 and nerve growth factor-induced PC12 cells, *Free Radical Biol. Med.*, 2005, **39**, 164–181.

123. K. Fernsebner, J. Zorn, B. Kanawati, A. Walker and B. Michalke, Manganese leads to an increase in markers of oxidative stress as well as to a shift in the ratio of Fe(ii)/(iii) in rat brain tissue, *Metallomics*, 2014, **6**, 921–931.

124. Y. Hanasaki, S. Ogawa and S. Fukui, The correlation between active oxygens scavenging and antioxidative effects of flavonoids, *Free Radical Biol. Med.*, 1994, **16**, 845–850.

125. G. Oboh and J. B. Rocha, Hot Pepper (Capsicum spp.) protects brain from sodium nitroprusside- and quinolinic acid-induced oxidative stress in vitro, *J. Med. Food*, 2008, **11**, 349–355.

126. R. P. Pereira, R. Fachinetto, A. de Souza Prestes, R. L. Puntel, G. N. Santos da Silva, B. M. Heinzmann, T. K. Boschetti, M. L. Athayde, M. E. Burger, A. F. Morel, V. M. Morsch and J. B. Rocha, Antioxidant effects of different extracts from Melissa officinalis, Matricaria recutita and Cymbopogon citratus, *Neurochem. Res.*, 2009, **34**, 973–983.

127. A. P. Marreilha dos Santos, D. Santos, C. Au, D. Milatovic, M. Aschner and M. C. Batoreu, Antioxidants prevent the cytotoxicity of manganese in RBE4 cells, *Brain Res.*, 2008, **1236**, 200–205.

128. D. Milatovic, R. C. Gupta, Y. Yu, S. Zaja-Milatovic and M. Aschner, Protective effects of antioxidants and anti-inflammatory agents against manganese-induced oxidative damage and neuronal injury, *Toxicol. Appl. Pharmacol.*, 2011, **256**, 219–226.

129. A. P. Stephenson, J. A. Schneider, B. C. Nelson, D. H. Atha, A. Jain, K. F. Soliman, M. Aschner, E. Mazzio and R. Renee Reams, Manganese-induced oxidative DNA damage in neuronal SH-SY5Y cells: attenuation of thymine base lesions by glutathione and N-acetylcysteine, *Toxicol. Lett.*, 2013, **218**, 299–307.

130. C. W. Nogueira, G. Zeni and J. B. Rocha, Organoselenium and organotellurium compounds: toxicology and pharmacology, *Chem. Rev.*, 2004, **104**, 6255–6285.

131. C. W. Nogueira and J. B. Rocha, Toxicology and pharmacology of selenium: emphasis on synthetic organoselenium compounds, *Arch. Toxicol.*, 2011, **85**, 1313–1359.

132. K. Erikson and M. Aschner, Manganese causes differential regulation of glutamate transporter (GLAST) taurine transporter and metallothionein in cultured rat astrocytes, *Neurotoxicology*, 2002, **23**, 595–602.

133. M. Sidoryk-Wegrzynowicz, E. Lee, N. Mingwei and M. Aschner, Disruption of astrocytic glutamine turnover by manganese is mediated by the protein kinase C pathway, *Glia*, 2011, **59**, 1732–1743.

134. E. Lee, M. Sidoryk-Wegrzynowicz, M. Farina, J. B. Rocha and M. Aschner, Estrogen attenuates manganese-induced glutamate transporter impairment in rat primary astrocytes, *Neurotoxic. Res.*, 2013, **23**, 124–130.

135. E. Lee, M. Sidoryk-Wegrzynowicz, Z. Yin, A. Webb, D. S. Son and M. Aschner, Transforming growth factor-alpha mediates estrogen-induced upregulation of glutamate transporter GLT-1 in rat primary astrocytes, *Glia*, 2012, **60**, 1024–1036.

136. E. S. Lee, M. Sidoryk, H. Jiang, Z. Yin and M. Aschner, Estrogen and tamoxifen reverse manganese-induced glutamate transporter impairment in astrocytes, *J. Neurochem.*, 2009, **110**, 530–544.

137. J. Pawlak, V. Brito, E. Kuppers and C. Beyer, Regulation of glutamate transporter GLAST and GLT-1 expression in astrocytes by estrogen, *Brain Res. Mol. Brain Res.*, 2005, **138**, 1–7.

CHAPTER 9

Mutual Neurotoxic Mechanisms Controlling Manganism and Parkisonism

JEROME A. ROTH

Department of Pharmacology and Toxicology, University at Buffalo, Buffalo, NY, USA
Email: jaroth@buffalo.edu

9.1 Introduction

Although Mn is an essential heavy metal required for normal cellular function, chronic exposure can lead to a disorder known as manganism characterized by severe neurological deficits that often resemble the involuntary extrapyramidal symptoms associated with Parkinson's disease. Numerous review articles over the past several years have attempted to describe the pathophysiological events accountable for the neurotoxicity of Mn as these articles have provided a detailed description of the cellular mechanisms and metabolic pathways responsible for Mn toxicity and have delineated the anatomical sites within the basal ganglia associated with the neurological deficits observed.[1–5] The disorder is most often linked to occupations such as welding, steel and battery manufacturing, and Mn mining in which exposure to abnormally high atmosphere levels is a daily occurrence. In addition, toxicity is also observed in patients with chronic liver disease because hepatic failure precludes the obligatory route required for its elimination.[6,7] Although Mn toxicity was first described in 1837,[8] universal recognition and acceptance of its deleterious actions on human

Issues in Toxicology No. 22
Manganese in Health and Disease
Edited by Lucio G. Costa and Michael Aschner
Published by the Royal Society of Chemistry, www.rsc.org

health has only recently evolved to the forefront with the concern of increased atmospheric levels produced from exhaust emissions with the impending use of methylcyclopentadienyl manganese tricarbonyl (MMT) to boost octane ratings in gasoline.

Although similar to Parkinsonism, the express symptoms produced initially from overexposure to chronic levels of Mn are somewhat unique because manganism is characterized by only mild tremors at rest, a masked-like face, and a more upright standing position. Most notable is significant loss of balance and a distinctive gait often referred to as a cock-like walk. Many of the features of Mn intoxication, however, can overlap or even progress to more Parkinson-like characteristics and, in fact, there is substantial evidence in the literature to suggest that excessive exposure to Mn may predispose a person to develop earlier onset of Parkinson's disease.[9-14] The link between Parkinsonism and manganism is not totally surprising as both alter the signaling output from the basal ganglia. Whereas Mn toxicity is, at least initially, associated with dysfunction of the globus pallidus, Parkinson's lesions mainly occur upstream in the dopaminergic neurons associated with the substantia nigra pars compacta. The subsequent impact from loss of these neurons is reduced release of dopamine in the striatum, which is the primary operative defect resulting in the dystonic movements associated with Parkinsonism. One of the fundamental similarities between manganism and Parkinsonism results from the fact that Mn is capable of suppressing dopamine release in the striatum, which will functionally have the impact to generate symptoms that overlap with those observed in Parkinson's disease.[15-17] Thus, the classical distinguishing features of the two neurological disorders may, in effect, merge based on the relative intensity to which Mn inhibits dopamine release.

Despite the fact that Mn can affect dopamine release in the striatum, it is significant that these dopaminergic neurons, which originate in the substantia nigra pars compacta, remain essentially intact during the initial stages upon chronic exposure to Mn.[18] This is especially remarkable because Mn has also been reported to accumulate both in the substantia nigra[19-22] and the striatum.[23] The fact that these neurons survive and that the dopamine content in the striatum is largely unaffected[16] implies that the action of Mn on dopamine release is likely to be a local response occurring within the presynaptic nerve terminal. The observation that Mn inhibits amphetamine-induced dopamine release implies its actions may be mediated *via* the dopamine transporter (DAT).[16] In support of this are recent studies demonstrating that Mn can promote the internalization of DAT and suppress dopamine release in DAT-transfected HEK cells.[15] In all likelihood, Mn-induced disruption in dopamine transmission has the potential to generate a neurological condition resembling that seen in patients with Parkinson's disease and, therefore, is predicted to contribute to the overall pathology seen in manganism. The extent to which inhibition of Mn-induced DA efflux progresses may also impact on the nature of the symptoms expressed and the rate at which the Parkinson-like syndrome develops.

As noted earlier, excess Mn perturbs the highly regulated and organized components of the basal ganglia, resulting in neurological deficits that can resemble those seen in Parkinson's disease. Similar to other neurological disorders, onset, severity, and individual symptoms expressed in Mn toxicity, as well as the progression of the disorder, often deviates unpredictably, implying underlying genetic variability as a potential origin responsible for the divergence in the pathological lesions and features expressed. This is clearly demonstrated in a study by Sadek *et al.*[24] that describes a patient who worked as a welder for a total of 3 years. Within 1 year after beginning employment, symptoms of Mn toxicity were initially perceived by the individual as a disturbance in his gait that, upon continued progression and subsequent neurological testing, was diagnosed as manganism. What makes the history for the development of the neurological deficits in this individual so remarkable is the precipitous rate of onset as opposed to the more typical case that requires longer exposure times, usually in excess of 10 years. The reason for the rapid onset of the neurological deficits in this short a timeframe is unknown, but most likely reflects a genetic predisposition in this individual. Within several years of the initial diagnosis, he developed a unilateral tremor in his right hand more characteristic of Parkinsonism. The contribution of a genetic component responsible for precipitating the onset of manganism, in some individuals, is further substantiated by the reality that not all welders or Mn miners develop manganism, yet exposures are comparable to their fellow workers who acquire the disorder. Clearly, this underscores the potential impact of genetic makeup in contributing to the vulnerability to develop Mn toxicity as well as the symptoms expressed. It is important to stress that although chronic exposure to Mn is not the initial causative agent inducing dopaminergic cell loss, there is compelling evidence in the literature that exposure to Mn may predispose patients to earlier onset of the disorder or a spectrum of symptoms that resemble Parkinsonism.[9–11,25]

In order to evaluate which genes may likely contribute to development of manganism, several issues need to be clarified that relate to both the overall spectrum of symptoms expressed and known contributing factors associated with onset of Parkinson's disease. As noted earlier, manganism is a disorder anatomically and functionally distinct from that of Parkinson's disease, at least in regard to the presumed primary site of injury, leaving us with the question as to why many of the expressed extrapyramidal dysfunctions observed in manganism overlap with those of Parkinsonism. Related to this are the concerns as to whether chronic exposure to Mn can eventually provoke Parkinsonism or a condition resembling Parkinson's disease. Assuming vulnerability to developing manganism is, in part, genetically dependent, then the most obvious correlation linking the two would be the genes associated with early-onset of Parkinson's disease, which include parkin (PARK2), DJ-1 (PARK7), PINK1 (PTEN-induced putative kinase 1, PARK6) and ATP13A2 (PARK9) as well as LRRK2 (leucine-rich repeat kinase 2, PARK8) and VPS35 (vacuolar protein sorting-associated protein 35, PARK17), which

associate with late onset of the disorder.[26–29] Linking these genes to acquisition of manganism may therefore explain why symptoms and features
between the two neurological disorders appear to be interrelated and why
Mn may accelerate onset of Parkinson-like symptoms.

This following review is intended to describe the function and properties
of the most studied genes that have been linked to either early or late onset
of Parkinsonism and to illustrate how the proposed mechanism of action for
each may possibly relate to onset and severity of Mn toxicity.

9.2 Parkin

Parkin is one of over 600 identified E3 ligases that are each responsible for
ubiquitination of a unique and select set of proteins.[30,31] The parkin gene
consists of 12 exons spanning over 1.5 megabases and encodes a protein of
465 amino acids with a molecular weight of approximately 52 kDa. It has an
ubiquitin-like domain at the N-terminal portion and two RING-finger motifs
at the C-terminal portion.[32] It is present in multiple sites within the cell
having been localized to the cytosol, trans-Golgi complex, mitochondria,
and the cytoplasmic surface of secretory vesicles.[33] Parkin catalyzes the
transfer of ubiquitin, a small 76 amino acid protein, from the E2 ubiquitin-
conjugating enzymes, UbcH6, UbcH7, UbcH8 and UbcH13/Uev1, to lysine
residues of a distinct set of substrate proteins. The specific site or sites as
well as the configuration of ubiquitinated products formed serve to regulate
both the biological activity and fate of these substrates.

Mutations of the parkin gene were initially identified from linkage studies
from a Japanese autosomal recessive-juvenile Parkinson (AR-JP) family in
1998.[32] Numerous missense mutations have been identified and shown to
alter parkin function through a number of different mechanisms, resulting
in aberrant ubiquitination and impairment of proteasomal degradation as
well as destabilization of parkin itself. The collective impact of these
malfunctioning biological processes regardless of the mutation position
within the protein all result in dopaminergic cell death. The mechanism
responsible for dopaminergic cell death is likely to be multifaceted and involve a number of dependent systems, all of which are basically contingent
on changes in the operative function of parkin to selectively ubiquitinate
proteins. This relates to parkin's role in the maintenance of cellular redox
balance, which associates with mitochondrial function and survival. Parkin
is normally recruited to depolarize mitochondria in order to foster their
elimination by autophagy (mitophagy).[34] It has also been suggested that
mitochondrial depolarization is the signaling mechanism responsible for
parkin's translocation to the mitochondria.[35] Phosphorylation plays an
essential role in this process because PINK1 kinase activity is required
for recruitment of parkin to depolarized mitochondria and, with the support
of DJ-1, for activation of its ubiquitin ligase activity.[35–40] After parkin
translocates to damaged mitochondria, it selectively ubiquitinates several
outer mitochondrial membrane proteins, resulting in their rapid

degradation through the proteasomal pathway and thus promoting elimination of mitochondria by autophagy.[37,41] Accordingly, parkin is a pivotal mediator of mitophagy, suggesting that impaired clearance of damaged mitochondria may trigger a variety of signaling processes leading to neurodegeneration observed in the Parkinson's disease brain. Along with its role in mitophagy, striatal mitochondrial respiratory chain function, including complex I activity and complex I linked ATP-production, were also found to be much lower in parkin knockout mice.[42] Parkin is also reported to decrease during the normal aging process and may be partially responsible for the accumulation of misfolded proteins in the elderly brain. Interestingly, ubiquitous or neuron-specific overexpression of parkin in adult Drosophila promotes increases both in mean and maximum lifespan.[43]

The consequence of parkin's role in altering mitochondrial function and survival likely results from stimulation of oxidative stress and the ensuing elevation of apoptotic signaling pathways. Consistent with this is the fact that parkin has been reported to protect against dopamine toxicity by decreasing ROS formation and the subsequent activation of apoptotic events, such as the JNK and caspase activation.[44,45] Overexpression of wild-type parkin reduces ROS production in neuronal cells by enhancing mitochondrial membrane potential, whereas overexpression of mutant parkin stimulated ROS formation.[46] These data suggest that parkin's homeostatic management of mitochondria integrity *via* its proteasomal activity may indirectly limit free radical-mediated injury.

This discussion reveals that there is substantial overlap between parkin-induced changes in biochemical processes responsible for promoting dopaminergic cell death and the mechanisms responsible Mn toxicity. Parkin-induced changes in mitochondrial stability in many respects parallel the loss of mitochondrial activity provoked by Mn. Thus, it is not surprising that overexpression of parkin can prevent dopaminergic cell death induced by a number of stress-promoting agents, including Mn.[47,48] For example, overexpression of parkin in human SHSY5Y and CATH.a cells prevent Mn-induced cell death whereas human lymphocytes containing an inactive mutant form of the ligase display greater loss of mitochondrial activity compared to control cells possessing wild-type parkin as assessed by decreases in ATP production.[49] Mn also promoted greater increases in oxidative stress and apoptotic signals in the parkin mutant cell line as indicated by enhanced ROS formation and caspase 3-activation. In another study,[50] human-induced pluripotent stem cells derived from mutant parkin subjects displayed significantly higher levels of ROS generation upon exposure to Mn than cells obtained from control subjects, suggesting that inheritance of a Parkinson's disease genetic risk factor may increase susceptibility to disease relevant Mn toxicity. Thus, it is reasonable to assume that the degree to which mutant parkin results in mitochondrial dysfunction and the age of onset for the critical loss of dopaminergic activity more than likely can impact on the appearance of Parkinson-like symptoms associated with Mn toxicity. In a reciprocal manner, Mn, by promoting degeneration of mitochondria, can likely facilitate mitophagy generated by mutant parkin.

Although by itself, the mutual actions of parkin and Mn on mitochondrial activity are probably sufficient to link some of the expressed symptoms of the two disorders, there is yet another mechanism by which parkin can directly influence Mn toxicity. This relates to the ability of parkin to influence transport of Mn into cells.[48] Several different transport systems have been identified for cellular uptake of Mn, although the general consensus presumes that the primary process utilizes divalent metal transporter 1 (DMT1, also called NRAMP2 and SLC11A2).[2,3] DMT1 has a broad substrate specificity and is generally considered to be the primary transmembrane protein responsible for the uptake of a variety of divalent cations, including iron, manganese, cadmium, cobalt, and nickel.[51] Four isoforms of DMT1 have been identified in mammalian cells encoded by a single gene. The isomers differ both in their N- and C-terminal residues with two mRNA isoforms possessing an iron response element (IRE) motif downstream from the stop codon, which presumably regulates DMT1 message levels *via* binding of the iron response proteins, IRP1 and IRP2. Transcriptionally regulated splice variants, exon 1A and 1B, have also been identified on the proximal N-terminal end of the message.[52] Selective transcription of the 1B species is regulated by several proinflammatory agents and cytokines including HIFα, TNFα, and IFN-γ as well as Fe, Mn, and LPS, all of which are capable of inciting an inflammatory response[53–58] along with activation of NF-κB. In this case, the 1B promoter contains a NF-κB transcription binding site, which is shown to increase synthesis of DMT1 and transport of Mn.[57,58] Interestingly, the inflammatory response produced in the substantia nigra of Parkinson's disease patients may be responsible for the increased levels of DMT1 observed.[59]

Expression of the 1B isoforms is also impacted by post-translational processing *via* the ubiquitin/proteasomal degradative pathway.[56,57,60,61] Studies have reported that parkin is the E3 ligase responsible for the ubiquitination and proteasomal degradation of the 1B isoforms of DMT1.[48] This is consistent with the findings that several different cell lines overexpressing parkin display decreased levels of the transporter, as well as reduced transport and increased toxicity of Mn, whereas cells expressing a native mutant construct of parkin, as well as brains from parkin-knockout rats, display increased levels of the transporter. Thus, mutations in the parkin are expected to lead to increase cellular transport of Mn and facilitate Mn toxicity. Accordingly, mutations in parkin may not only lead to early onset of Parkinson's disease but may also be responsible for increased uptake of Mn and propensity to develop Mn toxicity.

9.3   DJ-1

DJ-1 functionally plays a critical role in the cellular defensive mechanisms in order to preserve and protect normal biological responses against oxidative stress.[62] Deletions and point mutations in DJ-1 (PARK7), which lead to loss of function have been found worldwide and shown to generate autosomal recessive Parkinson's disease.[63] On the basis of the screening analysis of patients with Parkinsonism, DJ-1 mutations are the third most frequent identifiable genetic cause of early onset of the disorder behind that of parkin and PINK1.[64,65] Onset of DJ-1-induced Parkinsonism in the early 30s with gradual disease progression, and good response to levodopa treatment, is common to the other known autosomal recessive forms of the disorder. The physiological role of the gene, however, has yet to be identified although it represents a highly conserved protein, implying a universal function required for normal homeostatic function. It can function as a positive regulator of androgen receptor-dependent transcription,[66] and accumulating evidence suggests that DJ-1 may play a role in the cellular defensive mechanisms regulating and preserving mitochondria function.[67] Accordingly, either a deficiency or malfunction of DJ-1, may innately stimulate increases in ROS formation and oxidative insults, leading to premature degeneration of dopaminergic neurons.

DJ-1 is ubiquitously expressed in both brain and peripheral tissue. DJ-1 is a 20 kDa protein that shows a compact globular domain with an active site, Cys106, particularly sensitive to oxidation. In solution, DJ-1 normally forms a stable dimer and several Parkinson's disease mutations, which hinder dimerization, abolishes its neuroprotective activity, suggesting that dimerization may be essential to its function. Although DJ-1 protein is primarily located in the cytosol, minor portions are also present in the nucleus and mitochondria where it is selectively partitioned in the matrix and intermembrane space.[68] DJ-1 preferential redistributes to the mitochondria under conditions of oxidative stress, where it is presumed to act as a neuroprotective intracellular redox sensor[69] and thus, is important in maintaining mitochondrial homeostasis.[70,71] Several labs have generated DJ-1 −/− mice that exhibit little of the characteristic changes in the nigrostriatal pathway associated with Parkinsonism, although mouse embryonic fibroblasts from these mice contain fragmented mitochondria and defects in mitochondrial oxygen utilization.[72] ATP production and mitochondrial transmembrane potential are also reduced[73] along with opening of the mitochondrial permeability transition pore in isolated primary mouse embryonic fibroblasts from the DJ-1 −/− mice. Despite the lack of apparent pathology detected in the substantia nigra pars compacta of these mice, loss of DJ-1 sensitizes these animals to oxidative stress, as demonstrated by the observation of increased sensitivity to MPTP toxicity compared to wild-type mice.[74] In addition, cell culture models exhibiting loss of DJ-1 similarly display mitochondrial dysfunction and fragmentation associated with autophagy and mitophagy.[75]

There is increasing evidence that DJ-1 activity is regulated by the MAP kinase pathway based on the observation that its expression is increased through activation of ERK 1/2.[76] Thus, increased ROS produced by oxidative stress initiated by mitochondrial damage or oxidation of dopamine itself has the potential to promote phosphorylation of both ERK 1/2, which subsequently has the indirect effect of protecting dopaminergic cells by suppressing oxidative stress *via* upregulation of DJ-1. This mechanism may partially explain how native DJ-1 may function to act as an antioxidant defending cells from oxidative damage and thus possibly explain how modulation of DJ-1 activity has the potential to alter the fundamental neuroprotective component within dopaminergic cells.

In addition to its direct neuroprotective actions, DJ-1 has also been shown to form a complex with parkin and PINK1 to stimulate ubiquitination and degradation of parkin substrates, including parkin itself.[40,77–79] Whether these proteins actually form a distinct complex, however, has been questioned,[75] but decreased expression of either PINK1 or DJ-1 results in reduced ubiquitination of endogenous parkin, which subsequently causes a reduction in the degradation and increased accumulation of other parkin substrates. Thus, DJ-1 may function *in vivo* not only as an antioxidant but also to enhance ubiquitin E3 ligase activity of parkin and, therefore, facilitate the elimination of a variety of proteins linked to Parkinson's disease.

The actual mechanisms by which mutations in DJ-1 alter cellular defensive mechanisms in dopaminergic cell are likely to be multifaceted mediated by several different cellular processes working in synchrony to ultimately produce death *via* apoptosis or possibly other types of programmed cell death, including autophagy or programmed necrosis–all of which are relevant to Parkinson's disease. Whether any of the responses caused by mutations in DJ-1 can parallel and/or, exacerbate the toxic events provoked by Mn and thus potentiate early onset of manganism is not an unlikely scenario given the suspected common biological actions of the two. Equally important, is whether Mn has the potential to augment the neurotoxic processes initiated by loss of DJ-1 activity and thus promote premature development of Parkinson-like symptoms.

As described earlier, one of the acknowledged mechanisms by which Mn elicits its toxic response is *via* its impairment of mitochondrial activity. The fact that both mutant DJ-1 and Mn cause a decrease in ATP production and a decrease in mitochondrial transmembrane potential, as well as increased opening of the mitochondrial permeability transition pore, minimally implies there is likely to be amplification of these deleterious activities when Mn is present in patients exhibiting homozygous mutations in DJ-1.[2,73,80,81] Whereas Mn-induced degeneration of mitochondria function is presumed to be initiated by an increase in Ca^{+2} levels, the precise mechanism for DJ-1 is not known, but unlikely to be caused by a simple build-up of Ca^{+2}. Since DJ-1 is an autosomal recessive gene, it is difficult to assess the extent to which mitochondrial failure may similarly occur in the heterozygous DJ-1 mutated subject in the presence of Mn. Regardless, it is not unreasonable to assume that, under these conditions, the presence of Mn may be adequate to trigger a heightened response that can potentially augment mitochondrial dysfunction. In addition to inhibiting mitochondrial function, Mn has also been shown to affect the MAP kinase pathway because it can result in activation of ERK 1/2, p38, and JNK and thus potentially impact on DJ-1 activity.[82–84] Also, as noted earlier, the proposed interaction of DJ-1 with parkin is likely to have an influence on Mn toxicity, as prior studies have revealed that parkin is the E3 ligase responsible for ubiquitination of the major Mn transporter, DMT1.[48] Thus, if DJ-1 is mutated, even in the presence of normal parkin, DMT1 levels may increase sufficiently resulting in greater cellular uptake of Mn. Mechanistic differences, however, do exist between the actions of the DJ-1 and Mn, as striatal synaptosome preparations from DJ-1 −/− mice exhibit increases in the dopamine transporter, DAT,[85] whereas Mn results in a decrease of surface levels of DAT as well as decreased DA uptake and amphetamine-induced DA efflux in HEK cells transfected with the transporter.[15–17] The reduction in surface levels of DAT by Mn was shown to be caused by internalization of the transporter in these cells.[15] Despite these differences, the majority of the evidence clearly supports the premise that mutations in DJ-1 are likely to exacerbate the neurotoxic actions of Mn and ultimately lead to premature onset of manganism.

9.4 PINK1

PINK1 (PTEN-induced putative kinase 1) is a 581 amino acid protein containing an N-terminal mitochondrial-targeting sequence, a transmembrane domain and a highly conserved serine/threonine kinase sequence that is structurally similar to the Ca^{2+}/calmodulin family of proteins.[86] Studies[87,88] have shown that PINK1 is an integral mitochondrial membrane protein whose kinase domain faces the outer surface accessible to appropriate substrates within the cytoplasmic milieu. Although the precise function of PINK1 has not been elucidated, a number of mitochondrial proteins, including the mitochondrial chaperone protein, TRAP1 (TNF receptor-associated protein 1), the integral membrane protein, Miro, as well as parkin[89] have been identified as downstream substrates because PINK1 directly phosphorylates these proteins resulting in altered mitochondrial function, oxidative damage, and degradation of parkin-dependent E3 ligase-dependent proteins. Thus, by protecting neurons from oxidative stress, PINK1 is presumed to play a major role in preserving mitochondrial morphology and activity[90–92] because PINK1 deficiency results in an age-related loss of neuronal viability and increased sensitivity to stress-induced mitochondrial apoptosis. Morphological changes of mitochondrial function associated with PINK1 deficiency also include the presence of lowered mitochondrial membrane potential and opening of the permeability transition pore, all of which contribute to dopaminergic cell degeneration.[93] The actual mechanisms, however, responsible for changes in mitochondrial membrane potential, as well as oxidative stress that ultimately leads to cell death, remain unknown.

Like the other early onset genes, the average age of inception of PINK1-linked Parkinson's disease is in the early thirties. As noted earlier, the majority of the evidence points to loss of PINK1 activity as causing dopaminergic neuronal cell death primarily due to functional defects in mitochondria. Cultured cells derived from PINK1-deficient mice display increased intracellular calcium levels, along with excess production of ROS, loss of membrane potential, and defects of complex I causing opening of the mitochondrial permeability transition pore.[94] As part of the parkin E3 ligase complex, PINK1 knockdown also leads to proteasomal dysfunction, accompanied by increased α-synuclein aggregates,[95] a primary constituent of Lewy bodies classically observed in dopaminergic neurons associated with Parkinson's disease. The literature, however, is somewhat conflicting as to whether Parkinson subjects with PINK1 mutations can actually form Lewy bodies[96–98] because PINK1-induced Parkinsonism has been reported to represent mitochondrial cytopathies in the substantia nigra distinct from Lewy body Parkinson's disease. The role of parkin in the functioning of PINK1 is evidenced by the observation that Drosophila with phenotypic defects of both PINK1 and parkin are remarkably similar,[99] although mitochondrial dysfunction of PINK1-deficient Drosophila can be rescued by over-expression of parkin, whereas over-expression of PINK1 fails to overcome the defects of parkin-deficiency.[91,92,100,101] Similarly, mitochondrial abnormalities in HeLa cells produced by PINK1 deficiency were also shown to be rescued by overexpression of wild-type parkin but not parkin mutants possessing functionally impaired mutations.[91] These data are consistent with PINK1 function upstream of parkin by phosphorylating and activating the E3 ligase.

These observations lead to the basic question addressed in this review as to whether mutations in PINK1 can potentiate symptoms and possibly promote premature development of manganism. The fact that mutations in PINK1 promote degeneration of mitochondria and induce oxidative stress, definitely support the conclusion that deficiencies in this kinase have the potential to influence onset and progression of Mn toxicity. In addition, by normally stimulating parkin activity, it is likely that functionally inactive PINK1 mutants will cause a decrease in parkin E3 ligase activity, which will have the deleterious consequence of reducing ubiquitination and degradation of the major Mn transporter, DMT1, and thus increasing uptake of Mn.[48] Accordingly, the combined actions of the two processes on mitochondrial function and oxidative stress strongly reinforce the probability that mutations in PINK1 have the potential to facilitate Mn-induced toxicity. At the same time, it is not unreasonable to hypothesize that exposure to Mn can also exacerbate mitochondrial degeneration caused by mutations in PINK1 and thus provide additional stress to dopaminergic cells leading to premature onset of Parkinson-like symptoms.

9.5 ATP13A2

Mutations in ATP13A2/PARK9 have been reported to cause Kufor–Rakeb syndrome, a juvenile recessive multi-systemic neurodegenerative disorder characterized by slowly progressive levodopa-responsive

Parkinsonism.[102–104] Subjects normally display generalized brain atrophy with evidence of diminished nigrostriatal dopaminergic function consistent with the observation that ATP13A2 is particularly enriched in the substantia nigra.[103,105,106] Mutations of ATP13A2 results in its re-localization from intracellular acidic vesicular compartments to the endoplasmic reticulum in mammalian cells where it is subsequently digested *via* the proteasomal ER-associated degradation pathway.[105,107] Homozygous mutations of ATP13A2 have been reported to cause juvenile-onset Parkinsonism (10–22 years) whereas other heterozygous modifications are associated with early-onset Parkinsonism (<50 years),[108–111] suggesting a gene graded response to these mutations. However, the mechanism by which missense mutations cause Parkinson disease is still unclear.

Human ATP13A2 gene encodes an 1180 amino acid transmembrane-spanning domain protein possessing a P5 subfamily of P-type transport ATPases.[112] As noted earlier, mutant forms involved in disease promote retention of the kinase in the endoplasmic reticulum and the subsequent elimination *via* the proteasomal pathway.[105,107] Mutations of ATP13A2 leads to altered lysosomal function, including impaired lysosomal acidification, decreased proteolytic processing of lysosomal enzymes, reduced degradation of lysosomal substrates and diminished lysosomal-mediated clearance of autophagosomes. Thus, disruption of dopaminergic cells by the mutant species of the kinase has been suggested to be elicited by triggering an array of lysosomal insufficiencies, initiating impaired autophagic activity and the subsequent accumulation of toxic α-synuclein aggregates.[106,113,114] Consistent with this is the observation that orthologs of ATP13A2 can protect against cellular toxicity induced by overexpression of α-synuclein in yeast, nematode worm, and primary neurons from midbrain dopaminergic cells.[106] Thus, similar to the other genes related to early onset of Parkinsonism, mutant ATP13A2 disruption of α-synuclein degradation and the subsequent production of α-synuclein aggregates within the dopaminergic cell may represent a common cytotoxic mechanism linking these genes. These findings, however, further suggest that the neurotoxic actions of mutant ATP13A2, unlike the other genes associated with Parkinsonism, are not simply due to stimulation of apoptotic or typical oxidative stress pathways because these processes are still activated in native ATP13A2-expressing cells. In fact, cells expressing non-functional ATP13A2 activity display a remarkable reduction in the accumulated oxidized and damaged proteins implying that ATP13A2 may be required for normal functioning of lysosomes.[115] It should be noted, however, that a recent study raises the possibility that both mitochondrial degeneration and oxidative stress may also contribute to mutant ATP13A2-induced dopaminergic cell death, although this may likely be caused by an indirect consequence of lysosomal dysfunction.[116]

ATP13A2 has been suggested to play a role in the active transport of cations across lysosomal membranes in mammalian cells in an ATP-dependent fashion,[105,106,117,118] although the cation transporting activity and metal selectivity has not been adequately elucidated. This transport function in yeast is significant because deletion of the ATP13A2 ortholog, ykp9, confers sensitivity to growth in the presence of heavy metals, including Cd, Mn, Ni, and Se.[118] Overexpression of Ypk9 is capable of protecting cells from Mn toxicity whereas Parkinson's disease-linked mutations or ATPase-dead mutants fail to prevent metal-dependent cell death.[106,119] The precise mechanism by which Ypk9 prevents heavy metal toxicity is not known, although it has been reported to interact with essential genes involved in cellular trafficking and the cell cycle. Cumulatively, these data suggest that wild-type Ypk9 preserves cell viability in the presence of Mn and other heavy metals possibly *via* its role in sequestration of the metals within vacuoles within cells, whereas the harmful mutant species lack this ability. This possibly results from reduced transport of Mn or disruption of normal trafficking of these vesicles or by suppressing the digestion of essential proteins within the lysosomes. As anticipated, expression of ATP13A2 is capable of suppressing Mn toxicity along with a reduction in the levels of intracellular Mn.[117] This observation was obtained in several different cell lines, although relatively high toxic concentrations of Mn were used in this study to assess the measured levels of Mn within the cells. As to why Mn concentrations decreased is not known, although the concentrations used, even in the absence of excess expression of ATP13A2, are sufficient to promote apoptotic cell death. Interestingly, ATP13A2 expression was upregulated by Mn in this study, which is consistent with the upregulated protein levels observed in surviving substantia nigra dopaminergic neurons from Parkinson's disease brains,[105] suggesting that ATP13A2 may provide an inducible protective function to cells. Recent studies have suggested that the promoter site of ATP13A2 contains a functional hypoxia response element that may be responsible, at least in part, for the induction of ATP13A2 expression because Mn has previously been reported to increase HIF-1α.[120]

In summary, there is compelling evidence to suggest that mutant forms of ATP13A2 have the ability to potentiate Mn toxicity, even though the mechanism by which these aberrant forms of the kinase provoke either early onset of Parkinson's disease or manganism has not been firmly established. Supporting this is a recent study implicating ATP13A2 variants as potential risk factors for the neurotoxic actions of Mn in humans.[121] Based on the overall findings in the literature, ATP13A2 stimulation of Mn toxicity is likely to involve intracellular vesicular cation transport processes, which have the potential to indirectly provoke oxidative stress signals. Since ATP13A2 is selectively localized to the substantia nigra dopaminergic neurons, the actions of the variants will have little direct impact on Mn activity in the globus pallidus, at least in regard to its neurotic actions. Nevertheless, the consequence of this localization is likely to influence the characteristics of the symptoms expressed and possibly progression of Mn toxicity. The combination of excess Mn suppressing dopamine release in the striatum and its cooperative noxious actions in cells expressing mutated ATP13A2 may potentially exacerbate or promote earlier onset of the Parkinson-like symptoms in individuals exposed to high levels of the metal.

9.6   LRRK2

LRRK2 is a widely expressed 2527 amino acid protein that is present in many organs and tissues including the brain.[122] Full-length LRRK2 is mainly present in the cytoplasm, with partial localization to mitochondria and membranes, such as endoplasmic reticulum and synaptic vesicles.[123] It is a relatively complex gene encoding a multi-domain protein that includes an LRR (leucine-rich repeat) region and two catalytically active domains, a serine/threonine kinase with sequence homology to MAPKKK, and a GTPase domain belonging to the Ras-GTPase superfamily of GTPases, more specifically to the ROC (Ras complex proteins) subfamily.[124–126] LRRK2 functions upstream of canonical MAPKK and therefore, acts to phosphorylate several essential MAPKs, including p38 and JNK, which mediate oxidative cell stress, neurotoxicity, and apoptosis.[125,127] Studies with purified LRRK2 from eukaryotic cells have confirmed that LRRK2 binds guanine nucleotides and is capable of catalyzing the hydrolysis of GTP to GDP.[128,129] There is an interplay between the LRRK2 GTPase function and its kinase function, with the majority of the data suggesting ROC sequence functions as an upstream modulator of the kinase domain. Clearly, the ROC domain is important for the functioning of LRRK2 because several different mutations located in this sequence have been found to be linked to Parkinson's disease.[124,128] Thus, GTPase regulation of the kinase activity plays a pivotal function within the LRRK2-mediated signaling cascade. Recent studies have also revealed the presence of a variety of other multiple genetic mutations associated with decreased kinase activity as being linked to susceptibility to develop both familial and sporadic Parkinson's disease.[130–132] These mutations suppress both auto-phosphorylation as well as phosphorylation of LRRK2 substrates.[133]

Because of the structural and functional complexity of LRRK2, the mechanisms by which mutations elicit Parkinsonism are likely to involve multi-interactive signaling systems because variants have been reported to influence mitochondrial function,[134] protein homeostasis,[135] as well as a number of signaling pathways within neurons.[136] Microarray analysis has indicated that down-regulation of LRRK2 can lead to changes in genes involved in actin cytoskeleton-related processes, which are known to influence neurite outgrowth, synaptic plasticity, and axonal/dendritic transport.[136] Other studies have reported that LRRK2 deficiency may also act *via* its inhibition of mitochondrial function[134] or even possibly play an essential role in regulating protein homeostasis during aging, resulting in impairment of the autophagy–lysosomal pathway and leading to α-synuclein accumulation and aggregation.[135] Consistent with this is the observation that Parkinson's disease-associated defects in LRRK2 can cause endolysosomal and Golgi apparatus sorting defects.[137] Dominantly inherited mutations in both

LRRK2 and α-synuclein cause late onset of Parkinson's disease along with formation of Lewy bodies, implying that the two genes may interact as part of a common pathway. Recent studies have, in fact, indicated that LRRK2 can interact with α-synuclein and possibly be responsible for aggregate formation.[135,138–140] LRRK2 can also cause a decrease in the DA transporter, DAT,[141,142] as well as impair DA-stimulated neurotransmission.[143] These latter observations are particularly significant with respect to Mn toxicity because several reports indicate that Mn also perturbs dopaminergic function[144,145] and inhibits amphetamine-induced DA release,[16,17] as well as promotes mitochondria degeneration.[80,146] Thus, LRRK2 displays a number of critical functions, many of which can affect or be affected by Mn.

Several different mutations in LRRK2 have been identified and, although some mutations impair activity, others, in contrast, promote increased activity. For example, LRRK2 mutations, R1441C and R1441G, have been shown to affect the GTPase domain and decrease GTPase activity,[147] whereas the R1941H and G2385R mutations suppress auto-phosphorylation, as well as phosphorylation of myelin basic protein and moesin.[133] In contrast, the most prevalent mutant species, G2019S, actually increases kinase activity.[148,149] Interestingly, a recent study has suggested that difference in the biochemical properties of aggregated α-synuclein is produced in G2019S-linked Parkinson's disease patients when compared to those with idiopathic Parkinson's disease despite a similar histopathological presentation.[150] The reason for this is not known but implies that this mutation may promote phosphorylation of a unique set of sites within α-synuclein or that other proteins are capable of associating with the formed aggregates. What is most significant is the observation that Mn stimulates phosphorylation activity of the G2019S mutant protein, whereas it inhibits wild-type kinase activity normally activated by Mg.[151,152]

Recent studies have confirmed that when LRRK2 is silenced with an shRNA, Mn can potentiate oxidative stress, leading to cell death as indicated by activation of both ROS and JNK.[153] In contrast, Mn was shown to reduced phosphorylation of p38, consistent with the fact that p38 is a likely a downstream substrate of LRRK2.[154] Unlike the other genes associated with Mn toxicity, LRRK2 is an autosomal dominant gene and therefore potentially differentiates a population of individuals that, upon exposure to excess Mn, can develop early symptoms of manganism even in the absence of overt signs of Parkinson's disease.[125,155] Thus, mutations in LRRK2 not only play an essential role in development of Parkinson's disease but also may result in the development of manganism.

9.7 α-Synuclein

Aggregation of α-synuclein in the brain is a central pathological feature of several neurodegenerative disorders classified as synucleinopathies, which include Parkinson's disease.[156] α-Synuclein is present in neurons as well as glial cells where it preferentially localizes to presynaptic terminals and is most abundant in the neocortex, hippocampus, striatum, thalamus, and cerebellum. It has been associated with a diverse array of activities including functioning of the ubiquitin–proteasome system,[157,158] maintenance of synaptic functionality,[159] dopamine-mediated toxicity,[160–162] vesicle trafficking defects,[163] and regulation of oxidative stress.[164] Selective toxicity for dopamine neurons has been suggested to be caused by the actions of reactive oxidative products of dopamine on α-synuclein function and aggregation.[165] Aggregated α-synuclein is a major component of Lewy bodies associated with Parkinsonism and mutations are linked with familial Parkinson's disease.[166–168]

α-Synuclein is a 140 amino acid protein whose primary sequence is divided into three regions: the amphipathic N-terminal region, the hydrophobic central region, and the largely acidic C-terminal region.[169] The central region is a hydrophobic segment that corresponds to the non-amyloid component that regulates the amyloidogenic properties of the protein allowing α-synuclein to form fibrils. Native α-synuclein is normally a monomeric random coil protein that assumes various conformations that hinder the amyloidogenic sequence from forming aggregates. There are long-range interactions between the N- and C-terminal sequences, which inhibit the interactions of the different monomers that are presumed to be responsible for oligomerization and aggregation of the protein.[170] This provides an intrinsic autoinhibitory mechanism that, when disrupted, favor α-synuclein oligomer and fibril formation.[171] The majority of the diseases-associated missense mutations reside in the N-terminal region, although other critical sites are also present in the C-terminal region, all of which likely result in the breakdown of the monomer interactions facilitating formation of aggregates. The presence of abnormally high levels of α-synuclein protein caused by either increased production or decreased degradation foster formation of aggregates and is thought to also lead to dopamine neuronal death in Parkinson's disease.

As briefly mentioned earlier, oxidative stress has also been predicted to participate in the neurotoxic actions of α-synuclein because a number of studies have demonstrated mitochondrial dysfunction as being linked to neuronal cell degeneration in Parkinson's disease.[172] α-Synuclein protein has a noncanonical mitochondrial targeting sequence in its N-terminal portion, presumably facilitating its translocatation to mitochondria.[173,174] The presence of α-synuclein in mitochondria is enhanced in Parkinson's disease brains, implying a responsibility for its neurotoxic actions.[174] α-Synuclein appears to interact directly with mitochondrial membranes to regulate mitochondrial activity[175,176] because it has been reported to interact with complex I and interfere with its function, ultimately promoting the production of ROS.

Clearly, the consequence that mutations of α-synuclein has on the development of aggregates and the subsequent disruption of essential components of the dopaminergic neurons within the substantia nigra pars compacta establishes its role in the generation of Parkinsonism. The fact that the other early and late onset genes associated with Parkinson's disease have the potential to influence aggregate formation implies α-synuclein is a central element associated with dopaminergic cell death (see earlier discussion).

The question addressed in this review is whether alterations in α-synuclein activity or aggregation can provoke early development of manganism and whether exposure to Mn can further stimulate the neurotoxic events induced by altered α-synuclein function in dopaminergic neurons leading to early onset of Parkinsonism. A number of studies have attempted to address these questions using both *in vivo* and various *in vitro* model systems. Mn has been reported to increase expression of both mRNA and α-synuclein protein as well as promote its oligomerization.[177] In PC12 cells, Mn exposure was similarly found to induce overexpression of α-synuclein while siRNA knockdown of α-synuclein reversed Mn-induced cytotoxicity.[178] Additionally, Mn induced the activation of ERK1/2 in PC12 cells, whereas the MEK1 inhibitor, PD98059, which inhibits the activation of ERK, attenuated both the overexpression of α-synuclein and cytotoxicity. Mn-induced oxidative neuronal damage and α-synuclein oligomerization was also shown to be partially alleviated by pretreatment with reduced glutathione and aggravated by H_2O_2 pretreatment.[177] The mechanism for Mn toxicity in rat mesencephalic cells (MES 23.5) overexpressing α-synuclein was suggested to involve the translocation of NF-κB to the nucleus, which was inhibited by both antioxidants and inhibitors of p38 MAP kinase.[179] Mn was also reported to promote increased toxicity in SKN-MC neuroblastoma cells stably expressing the human dopamine transporter when transfected with human α-synuclein when compared to vehicle transfected controls.[180] In addition, α-synuclein-positive cells were found to increase in the gray matter of Cynomolgus macaques in Mn-exposed animals, and some of these neurons displayed loss of Nissl staining along with α-synuclein-positive aggregates.[181] Another study compared the effect of Mn in male transgenic C57BL/6J mice expressing human α-synuclein or a doubly mutated human α-synuclein to their non-transgenic littermates.[145] Results of this study revealed that Mn decreased dopamine turnover in the striatum of mice transgenic for human wild-type α-synuclein, but not in mice expressing two mutatant species, nor in non-transgenic littermates, although Mn failed to induce signs of neurodegeneration of nigrostriatal dopamine neurons.

In conclusion, these results provide substantial evidence that there may be mutual deleterious interactions between Mn and altered α-synuclein activity that are likely to promote dopaminergic cell death and promote early onset of either Parkinsonism or manganism. The overall importance of this is reinforced by the fact that many of the other known genes linked to Parkinsonism influence α-synuclein function and aggregation within the dopaminergic neurons.

9.8   VPS35

One of the newest genes linked with autosomal recessive late onset Parkinsonism is VPS35. This gene encodes a protein that is a component of a large heteropentameric complex, termed the retromer complex, involved in retrograde transport of proteins from endosomes to the trans-Golgi network.[182] The basic structure includes a sorting nexin dimer, comprised of either sorting nexins 1 or 2 and 5 or 6, as well as a cargo-recognition trimer, composed of VPS26, VPS29, and VPS35.[183] It has been suggested that pathogenic mutation in VPS35 may function by disrupting recognition sites required for the binding to essential cargo proteins.[184] Recently, a number of studies have identified patients with mutations in VPS35 that present with late onset autosomal-dominant Parkinson's disease characterized with tremor-predominant symptoms.[184–186] An association of VPS35 component of the retromer complex with Parkinson's disease-associated defects in LRRK2 was recently reported as causing endolysosomal and Golgi apparatus sorting defects and deficiency.[137] Although, there is currently no study that has directly linked VSP35 to Mn toxicity, it has been reported that VSP35 is associated in the recycling of the major Mn transport protein, DMT1.[187] It appears that VPS35 binds exclusively to the cytoplasmic tail of the -IRE species DMT1 because binding is dependent on a specific hydrophobic motif that is necessary for its endosomal recycling. Both the -IRE isoform of DMT1 and VPS35 were shown to colocalize with transferrin receptor-positive endosomes. siRNA-dependent depletion of VPS35 leads to missorting of the transporter to the lysosome-associated membrane protein, LAMP2-positive vesicles, although this did not result in alteration of the levels of DMT1. Thus, it is difficult to predict whether mutations in the gene can actually alter endosomal or surface levels of DMT1 or whether the missorted vesicles display differences in their capacity to transport Mn. Interestingly, prior studies have reported that the major signal for internalization and recycling of DMT1-IRE species resides in its carboxyl terminus and that removal of this signal leads to a default lysosomal targeting.[188] Since four different isoforms of DMT1 exist, it is also difficult to speculate how changes in only the -IRE species will affect the overall uptake of Mn in any given cell type. In a reciprocal fashion, it is not known whether Mn can disrupt the actions of VPS35 on recycling of DMT1. This is an important concern because Mn has recently been reported to promote internalization of the dopamine transporter, DAT, and thus Mn may influence internalization of other surface proteins that may be regulated by VPS35.

9.9 Others

The previous discussion attempts to delineate genes acknowledged to be risk factors for development of Parkinsonism and their potential for overlap with onset and progression of Mn toxicity. It should be noted that there are a number of large case-control genome-wide and meta-analysis studies that have clearly identified polymorphisms in a number of other genes that also significantly correlate with the development of Parkinson's disease. These include GBA (β-glucocerebrosidase),[189] MAPT (microtubule-associated protein tau),[190] GAK (cyclin-G-associated kinase),[191] EIF4G1 (eukaryotic translation initiation factor 4-gamma),[192] HLA (human leukocyte antigen) and the transport protein,[193] SLC30A10,[194] as well as several others. In the case of SLC30A10, it has been suggested that this membrane-bound protein may be involved in the export of Mn and therefore mutations in the gene may not only be responsible for familial predisposition to Parkinsonism but also for hypermanganesemia.[195] The pathogenicity of mutations for this and several of these genes, however, requires additional basic and clinical studies to corroborate the link with expression of Parkinson-like features and Mn toxicity. In some cases, the functionality of these genes has been identified, although for many the physiological actions have yet to be examined and therefore, the triggering process by which they promote dopaminergic degeneration remains unknown. Patients with mutations in several of these genes also present with additional symptoms other than Parkinsonism and therefore, the Parkinson-like symptoms expressed may be distinct from the more classical pattern. Regardless of the characteristics of the symptoms presented, at this point it is difficult to predict *a priori* whether any of these genes will also segregate with progression of manganism, although based on the other risk factors already linked to Parkinsonism, there is a reasonable probability that some, if not all, will also correlate with provocation of Mn toxicity.

9.10   Conclusion

Although mutations in genes identified with promoting neurological deficits resembling Parkinsonism occur with relatively high frequency in specific populations, worldwide these are relatively rare events. The typical person has approximately a 2.5% chance of developing Parkinson's disease in their lifetime, and the risk for people whose close relatives have the disorder is increased to about 6–10%. Thus, the role of genetic mutations in generating Parkinsonism is comparatively small, reflecting the opinion that environmental factors most likely contribute to onset of the disease or symptoms presenting with Parkinson-like movement impairment. In some cases, the aberrant genes not only affect functionality of the basal ganglia, but also other areas of the CNS and therefore, the observed clinical phenotypes expressed are clearly distinguishable from Parkinson's disease. In the case of Mn intoxication, the symptoms are conspicuously distinct from that of Parkinsonism, at least in the early stages of the disorder. Initial and partially reversible neurological symptoms consist of reduced response speed, irritability, intellectual deficits, mood changes, and compulsive behaviors indicative of impairment independent of the actions within the globus pallidus. Upon protracted exposure, the disorder can progress to more prominent and irreversible extrapyramidal dysfunction, which includes a masklike face, limb rigidity, mild tremors, gait disturbance, cock-like walk, slurred speech, excessive salivation, sweating, and a marked disturbance of balance. Although, the primary extrapyramidal symptoms are presumed to be related to loss of function of neurons within the globus pallidus and not the substantia nigra pars compacta, Mn has been reported to also cause loss of dopamine release in the striatum, which would be expected to lead to symptoms associated with Parkinson's disease.

The studies presented in this review have attempted to characterize the functional properties of genes identified as generating Parkinson's disease or Parkinson-like disorders and how mutation of these genes may correlate from a mechanistic perspective to manganese toxicity. Because both neurological disorders are associated with changes in the function and output of the basal ganglia, it is not surprising that symptoms of Mn toxicity often overlap with that of Parkinson's disease. The expressed behavioral and extrapyramidal symptoms of individuals exhibiting Mn toxicity can be quite broad, which probably mirrors a diverse array of lesions within the CNS. In all likelihood, the extent and specificity of these lesions and the subsequent abnormalities expressed are likely to be genetically regulated. What is not known is the extent to which genes that contribute to Parkinsonism can also influence both the characteristics and the emergence of the symptoms observed in manganism.

As noted in this review, there appears to be several common threads linking the two disorders because mutations in genes associated with early and late onset of Parkinsonism produce similar adverse biological responses acknowledged to provoke neuronal cell death. Examination of the data reveal there are four collective themes connecting mutated Parkinson's disease related genes to Mn toxicity: (1) disruption of mitochondrial function leading to oxidative stress; (2) abnormalities in vesicle processing; (3) altered proteasomal and lysosomal protein degradation; and (4) α-synuclein aggregation. Clearly, these are not mutually exclusive events because each has the potential to influence the others' behavior and performance. The functional consequences of deviations in these four biological systems in regard to potentiating manganism is readily apparent because each has previously been implicated as contributing to Mn toxicity. In a reciprocal fashion, because Mn can independently influence each of these processes, further implicates it in possibly potentiating the degenerative actions of mutant forms of Parkinson-related genes and may be considered a risk factor provoking onset of Parkinsonism. It is important to emphasize that chronic exposure to Mn is definitely not the initial causative agent inciting dopaminergic cell loss associated within Parkinsonism, although, as noted earlier, the preponderance of the evidence suggests that persistent exposure to Mn has the potential to supplement the development of dystonic movements associated with Parkinson's disease. The interdependence of the two therefore implies that genetic variance coupled with additional environmental or nutritional factors most likely act in synchrony to contribute to the severity, characteristics, and onset of both disorders.

References

1. T. R. Guilarte, Manganese and Parkinson's disease: a critical review and new findings, *Cien Saude Colet*, 2011, **16**, 4549–4566.

2. J. A. Roth, Homeostatic and toxic mechanisms regulating manganese uptake, retention, and elimination, *Biol. Res.*, 2006, **39**, 45–57.
3. J. A. Roth, Are there common biochemical and molecular mechanisms controlling manganism and parkisonism, *NeuroMol. Med.*, 2009, **11**, 281–296.
4. M. Aschner, K. M. Erikson, E. Herrero Hernandez and R. Tjalkens, Manganese and its role in Parkinson's disease: from transport to neuropathology, *NeuroMol. Med.*, 2009, **11**, 252–266.
5. B. A. Racette, M. Aschner, T. R. Guilarte, U. Dydak, S. R. Criswell and W. Zheng, Pathophysiology of manganese-associated neurotoxicity, *Neurotoxicology*, 2012, **33**, 881–886.
6. S. R. Criswell, J. S. Perlmutter, J. S. Crippin, T. O. Videen, S. M. Moerlein, H. P. Flores, A. M. Birke and B. A. Racette, Reduced uptake of FDOPA PET in end-stage liver disease with elevated manganese levels, *Arch. Neurol.*, 2012, **69**, 394–397.
7. R. A. Hauser and T. A. Zesiewicz, Manganese and chronic liver disease, *Mov. Disord.*, 1996, **11**, 589.
8. J. Couper, On the effects of black oxide of manganese which inhaled into the lungs, *Br. Ann. Med. Pharm.*, 1837, **1**, 41–42.
9. J. M. Gorell, C. C. Johnson, B. A. Rybicki, E. L. Peterson, G. X. Kortsha, G. G. Brown and R. J. Richardson, Occupational exposure to manganese, copper, lead, iron, mercury and zinc and the risk of Parkinson's disease, *Neurotoxicology*, 1999, **20**, 239–247.
10. H. K. Hudnell, Effects from environmental Mn exposures: a review of the evidence from non-occupational exposure studies, *Neurotoxicology*, 1999, **20**, 379–397.
11. Y. Kim, J. M. Kim, J. W. Kim, C. I. Yoo, C. R. Lee, J. H. Lee, H. K. Kim, S. O. Yang, H. K. Chung, D. S. Lee and B. Jeon, Dopamine transporter density is decreased in parkinsonian patients with a history of manganese exposure: what does it mean?, *Mov. Disord*, 2002, **17**, 568–575.
12. P. K. Pal, A. Samii and D. B. Calne, Manganese neurotoxicity: a review of clinical features, imaging and pathology, *Neurotoxicology*, 1999, **20**, 227–238.
13. B. A. Racette, L. McGee-Minnich, S. M. Moerlein, J. W. Mink, T. O. Videen and J. S. Perlmutter, Welding-related parkinsonism: clinical features, treatment, and pathophysiology, *Neurology*, 2001, **56**, 8–13.
14. B. A. Racette, J. A. Antenor, L. McGee-Minnich, S. M. Moerlein, T. O. Videen, V. Kotagal and J. S. Perlmutter, [18F]FDOPA PET and clinical features in parkinsonism due to manganism, *Mov. Disord*, 2005, **20**, 492–496.
15. J. A. Roth, Z. Li, S. Sridhar and H. Khoshbouei, The effect of manganese on dopamine toxicity and dopamine transporter (DAT) in control and DAT transfected HEK cells, *Neurotoxicology*, 2013, **35**, 121–128.
16. T. R. Guilarte, M. K. Chen, J. L. McGlothan, T. Verina, D. F. Wong, Y. Zhou, M. Alexander, C. A. Rohde, T. Syversen, E. Decamp, A. J. Koser,

S. Fritz, H. Gonczi, D. W. Anderson and J. S. Schneider, Nigrostriatal dopamine system dysfunction and subtle motor deficits in manganese-exposed non-human primates, *Exp. Neurol.*, 2006, **202**, 381–390.

17. T. R. Guilarte, N. C. Burton, J. L. McGlothan, T. Verina, Y. Zhou, M. Alexander, L. Pham, M. Griswold, D. F. Wong, T. Syversen and J. S. Schneider, Impairment of nigrostriatal dopamine neurotransmission by manganese is mediated by pre-synaptic mechanism(s): implications to manganese-induced parkinsonism, *J. Neurochem.*, 2008, **107**, 1236–1247.

18. C. W. Olanow, P. F. Good, H. Shinotoh, K. A. Hewitt, F. Vingerhoets, B. J. Snow, M. F. Beal, D. B. Calne and D. P. Perl, Manganese intoxication in the rhesus monkey: a clinical, imaging, pathologic, and biochemical study, *Neurology*, 1996, **46**, 492–498.

19. N. H. Park, J. K. Park, Y. Choi, C. I. Yoo, C. R. Lee, H. Lee, H. K. Kim, S. R. Kim, T. H. Jeong, J. Park, C. S. Yoon and Y. Kim, Whole blood manganese correlates with high signal intensities on T1-weighted MRI in patients with liver cirrhosis, *Neurotoxicology*, 2003, **24**, 909–915.

20. J. Zhang, R. Cao, T. Cai, M. Aschner, F. Zhao, T. Yao, Y. Chen, Z. Cao, W. Luo and J. Chen, The Role of Autophagy Dysregulation in Manganese-Induced Dopaminergic Neurodegeneration, *Neurotoxic. Res.*, 2013, **24**, 478–490.

21. T. Verina, S. F. Kiihl, J. S. Schneider and T. R. Guilarte, Manganese exposure induces microglia activation and dystrophy in the substantia nigra of non-human primates, *Neurotoxicology*, 2010, **32**, 215–226.

22. G. Robison, T. Zakharova, S. Fu, W. Jiang, R. Fulper, R. Barrea, M. A. Marcus, W. Zheng and Y. Pushkar, X-ray fluorescence imaging: a new tool for studying manganese neurotoxicity, *PLoS One*, 2012, **7**, e48899.

23. K. M. Erikson, C. E. John, S. R. Jones and M. Aschner, Manganese accumulation in striatum of mice exposed to toxic doses is dependent upon a functional dopamine transporter, *Environ. Toxicol. Pharmacol.*, 2005, **20**, 390–394.

24. A. H. Sadek, R. Rauch and P. E. Schulz, Parkinsonism due to manganism in a welder, *Internet J. Toxicol.*, 2003, **22**, 393–401.

25. B. A. Racette, A. Perry, G. D'Avossa and J. S. Perlmutter, Late-onset neurodegeneration with brain iron accumulation type 1: expanding the clinical spectrum, *Mov. Disord*, 2001, **16**, 1148–1152.

26. S. Lesage and A. Brice, Parkinson's disease: from monogenic forms to genetic susceptibility factors, *Hum. Mol. Genet.*, 2009, **18**, R48–59.

27. C. Wider and Z. K. Wszolek, Clinical genetics of Parkinson's disease and related disorders, *Parkinsonism Relat. Disord.*, 2007, **13**(Suppl 3), S229–232.

28. Y. X. Yang, N. W. Wood and D. S. Latchman, Molecular basis of Parkinson's disease, *NeuroReport*, 2009, **20**, 150–156.

29. H. Houlden and A. B. Singleton, The genetics and neuropathology of Parkinson's disease, *Acta Neuropathol.*, 2012, **124**, 325–338.

30. K. K. Dev, H. van der Putten, B. Sommer and G. Rovelli, Part I: parkin-associated proteins and Parkinson's disease, *Neuropharmacology*, 2003, **45**, 1–13.

31. K. L. Lim, K. C. Chew, J. M. Tan, C. Wang, K. K. Chung, Y. Zhang, Y. Tanaka, W. Smith, S. Engelender, C. A. Ross, V. L. Dawson and T. M. Dawson, Parkin mediates nonclassical, proteasomal-independent ubiquitination of synphilin-1: implications for Lewy body formation, *J. Neurosci.*, 2005, **25**, 2002–2009.

32. T. Kitada, S. Asakawa, N. Hattori, H. Matsumine, Y. Yamamura, S. Minoshima, M. Yokochi, Y. Mizuno and N. Shimizu, Mutations in the parkin gene cause autosomal recessive juvenile parkinsonism, *Nature*, 1998, **392**, 605–608.

33. H. Shimura, N. Hattori, S. Kubo, M. Yoshikawa, T. Kitada, H. Matsumine, S. Asakawa, S. Minoshima, Y. Yamamura, N. Shimizu and Y. Mizuno, Immunohistochemical and subcellular localization of Parkin protein: absence of protein in autosomal recessive juvenile parkinsonism patients, *Ann. Neurol.*, 1999, **45**, 668–672.

34. D. Narendra, A. Tanaka, D. F. Suen and R. J. Youle, Parkin is recruited selectively to impaired mitochondria and promotes their autophagy, *J. Cell Biol.*, 2008, **183**, 795–803.

35. N. Matsuda, S. Sato, K. Shiba, K. Okatsu, K. Saisho, C. A. Gautier, Y. S. Sou, S. Saiki, S. Kawajiri, F. Sato, M. Kimura, M. Komatsu, N. Hattori and K. Tanaka, PINK1 stabilized by mitochondrial depolarization recruits Parkin to damaged mitochondria and activates latent Parkin for mitophagy, *J. Cell Biol.*, 2010, **189**, 211–221.

36. S. Geisler, K. M. Holmstrom, D. Skujat, F. C. Fiesel, O. C. Rothfuss, P. J. Kahle and W. Springer, PINK1/Parkin-mediated mitophagy is dependent on VDAC1 and p62/SQSTM1, *Nat. Cell Biol.*, 2010, **12**, 119–131.

37. D. P. Narendra, S. M. Jin, A. Tanaka, D. F. Suen, C. A. Gautier, J. Shen, M. R. Cookson and R. J. Youle, PINK1 is selectively stabilized on impaired mitochondria to activate Parkin, *PLoS Biol.*, 2010, **8**, e1000298.

38. C. Vives-Bauza and S. Przedborski, PINK1 points Parkin to mitochondria, *Autophagy*, 2010, **6**, 674–675.

39. M. Lazarou, D. P. Narendra, S. M. Jin, E. Tekle, S. Banerjee and R. J. Youle, PINK1 drives Parkin self-association and HECT-like E3 activity upstream of mitochondrial binding, *J. Cell Biol.*, 2013, **200**, 163–172.

40. H. Xiong, D. Wang, L. Chen, Y. S. Choo, H. Ma, C. Tang, K. Xia, W. Jiang, Z. Ronai, X. Zhuang and Z. Zhang, Parkin, PINK1, and DJ-1 form a ubiquitin E3 ligase complex promoting unfolded protein degradation, *J. Clin. Invest.*, 2009, **119**, 650–660.

41. N. C. Chan and D. C. Chan, Parkin uses the UPS to ship off dysfunctional mitochondria, *Autophagy*, 2011, **7**, 771–772.

42. M. Muftuoglu, B. Elibol, O. Dalmizrak, A. Ercan, G. Kulaksiz, H. Ogus, T. Dalkara and N. Ozer, Mitochondrial complex I and IV activities in

leukocytes from patients with parkin mutations, *Mov. Disord*, 2004, **19**, 544–548.

43. A. Rana, M. Rera and D. W. Walker, Parkin overexpression during aging reduces proteotoxicity, alters mitochondrial dynamics, and extends lifespan, *Proc. Natl. Acad. Sci. U. S. A.*, 2013, **110**, 8638–8643.

44. Y. Wang, Y. Nartiss, B. Steipe, G. A. McQuibban and P. K. Kim, ROS-induced mitochondrial depolarization initiates PARK2/PARKIN-dependent mitochondrial degradation by autophagy, *Autophagy*, 2012, **8**, 1462–1476.

45. A. P. Joselin, S. J. Hewitt, S. M. Callaghan, R. H. Kim, Y. H. Chung, T. W. Mak, J. Shen, R. S. Slack and D. S. Park, ROS-dependent regulation of Parkin and DJ-1 localization during oxidative stress in neurons, *Hum. Mol. Genet.*, 2012, **21**, 4888–4903.

46. D. H. Hyun, M. Lee, B. Halliwell and P. Jenner, Effect of overexpression of wild-type or mutant parkin on the cellular response induced by toxic insults, *J. Neurosci. Res.*, 2005, **82**, 232–244.

47. Y. Higashi, M. Asanuma, I. Miyazaki, N. Hattori, Y. Mizuno and N. Ogawa, Parkin attenuates manganese-induced dopaminergic cell death, *J. Neurochem.*, 2004, **89**, 1490–1497.

48. J. A. Roth, S. Singleton, J. Feng, M. Garrick and P. N. Paradkar, Parkin regulates metal transport via proteasomal degradation of the 1B isoforms of divalent metal transporter 1, *J. Neurochem.*, 2010, **113**, 454–464.

49. J. A. Roth, B. Ganapathy and A. J. Ghio, Manganese-induced toxicity in normal and human B lymphocyte cell lines containing a homozygous mutation in parkin, *Toxicol. In Vitro*, 2012, **26**, 1143–1149.

50. A. A. Aboud, A. M. Tidball, K. K. Kumar, M. D. Neely, K. C. Ess, K. M. Erikson and A. B. Bowman, Genetic risk for Parkinson's disease correlates with alterations in neuronal manganese sensitivity between two human subjects, *Neurotoxicology*, 2012, **33**, 1443–1449.

51. M. D. Garrick, S. T. Singleton, F. Vargas, H. C. Kuo, L. Zhao, M. Knopfel, T. Davidson, M. Costa, P. Paradkar, J. A. Roth and L. M. Garrick, DMTI: Which metals does it transport?, *Biol. Res.*, 2006, **39**, 79–85.

52. N. Hubert and M. W. Hentze, Previously uncharacterized isoforms of divalent metal transporter (DMT)-1: implications for regulation and cellular function, *Proc. Natl. Acad. Sci. U. S. A.*, 2002, **99**, 12345–12350.

53. A. J. Ghio, J. L. Turi, M. C. Madden, L. A. Dailey, J. D. Richards, J. G. Stonehuerner, D. L. Morgan, S. Singleton, L. M. Garrick and M. D. Garrick, Lung injury after ozone exposure is iron dependent, *Am. J. Physiol.: Lung Cell. Mol. Physiol.*, 2007, **292**, L134–143.

54. X. Wang, M. D. Garrick, F. Yang, L. A. Dailey, C. A. Piantadosi and A. J. Ghio, TNF, IFN-gamma, and endotoxin increase expression of DMT1 in bronchial epithelial cells, *Am. J. Physiol.: Lung Cell. Mol. Physiol.*, 2005, **289**, L24–33.

55. A. Lis, P. N. Paradkar, S. Singleton, H. C. Kuo, M. D. Garrick and J. A. Roth, Hypoxia induces changes in expression of isoforms of the

divalent metal transporter (DMT1) in rat pheochromocytoma (PC12) cells, *Biochem. Pharmacol.*, 2005, **69**, 1647–1655.

56. P. N. Paradkar and J. A. Roth, Expression of the 1B isoforms of divalent metal transporter (DMT1) is regulated by interaction of NF-Y with a CCAAT-box element near the transcription start site, *J. Cell. Physiol.*, 2007, **211**, 183–188.

57. P. N. Paradkar and J. A. Roth, Post-translational and transcriptional regulation of DMT1 during P19 embryonic carcinoma cell differentiation by retinoic acid, *Biochem. J.*, 2006, **394**, 173–183.

58. P. N. Paradkar and J. A. Roth, Nitric oxide transcriptionally down-regulates specific isoforms of divalent metal transporter (DMT1) *via* NF-kappaB, *J. Neurochem.*, 2006, **96**, 1768–1777.

59. J. Salazar, N. Mena, S. Hunot, A. Prigent, D. Alvarez-Fischer, M. Arredondo, C. Duyckaerts, V. Sazdovitch, L. Zhao, L. M. Garrick, M. T. Nunez, M. D. Garrick, R. Raisman-Vozari and E. C. Hirsch, Divalent metal transporter 1 (DMT1) contributes to neurodegeneration in animal models of Parkinson's disease, *Proc. Natl. Acad. Sci. U. S. A.*, 2008, **105**, 18578–18583.

60. C. Brasse-Lagnel, Z. Karim, P. Letteron, S. Bekri, A. Bado and C. Beaumont, Intestinal DMT1 cotransporter is down-regulated by hepcidin via proteasome internalization and degradation, *Gastroenterology*, 2011, **140**, 1261–1271, e1261.

61. M. D. Garrick, L. Zhao, J. A. Roth, H. Jiang, J. Feng, N. J. Foot, H. Dalton, S. Kumar and L. M. Garrick, Isoform specific regulation of divalent metal (ion) transporter (DMT1) by proteasomal degradation, *BioMetals*, 2012, **25**, 787–793.

62. M. M. Wilhelmus, P. G. Nijland, B. Drukarch, H. E. de Vries and J. van Horssen, Involvement and interplay of Parkin, PINK1, and DJ1 in neurodegenerative and neuroinflammatory disorders, *Free Radical Biol. Med.*, 2012, **53**, 983–992.

63. V. Bonifati, P. Rizzu, F. Squitieri, E. Krieger, N. Vanacore, J. C. van Swieten, A. Brice, C. M. van Duijn, B. Oostra, G. Meco and P. Heutink, DJ-1(PARK7), a novel gene for autosomal recessive, early onset parkinsonism, *Neurol. Sci.*, 2003, **24**, 159–160.

64. P. M. Abou-Sleiman, D. G. Healy, N. Quinn, A. J. Lees and N. W. Wood, The role of pathogenic DJ-1 mutations in Parkinson's disease, *Ann. Neurol.*, 2003, **54**, 283–286.

65. K. Hedrich, A. Djarmati, N. Schafer, R. Hering, C. Wellenbrock, P. H. Weiss, R. Hilker, P. Vieregge, L. J. Ozelius, P. Heutink, V. Bonifati, E. Schwinger, A. E. Lang, J. Noth, S. B. Bressman, P. P. Pramstaller, O. Riess and C. Klein, DJ-1 (PARK7) mutations are less frequent than Parkin (PARK2) mutations in early-onset Parkinson disease, *Neurology*, 2004, **62**, 389–394.

66. K. Takahashi, T. Taira, T. Niki, C. Seino, S. M. Iguchi-Ariga and H. Ariga, DJ-1 positively regulates the androgen receptor by impairing

the binding of PIASx alpha to the receptor, *J. Biol. Chem.*, 2001, **276**, 37556–37563.

67. T. Taira, Y. Saito, T. Niki, S. M. Iguchi-Ariga, K. Takahashi and H. Ariga, DJ-1 has a role in antioxidative stress to prevent cell death, *EMBO Rep.*, 2004, **5**, 213–218.

68. L. Zhang, M. Shimoji, B. Thomas, D. J. Moore, S. W. Yu, N. I. Marupudi, R. Torp, I. A. Torgner, O. P. Ottersen, T. M. Dawson and V. L. Dawson, Mitochondrial localization of the Parkinson's disease related protein DJ-1: implications for pathogenesis, *Hum. Mol. Genet.*, 2005, **14**, 2063–2073.

69. R. M. Canet-Aviles, M. A. Wilson, D. W. Miller, R. Ahmad, C. McLendon, S. Bandyopadhyay, M. J. Baptista, D. Ringe, G. A. Petsko and M. R. Cookson, The Parkinson's disease protein DJ-1 is neuroprotective due to cysteine-sulfinic acid-driven mitochondrial localization, *Proc. Natl. Acad. Sci. U. S. A.*, 2004, **101**, 9103–9108.

70. A. K. Ashley, W. H. Hanneman, T. Katoh, J. A. Moreno, A. Pollack, R. B. Tjalkens and M. E. Legare, Analysis of targeted mutation in DJ-1 on cellular function in primary astrocytes, *Toxicol. Lett.*, 2009, **184**, 186–191.

71. E. Junn, W. H. Jang, X. Zhao, B. S. Jeong and M. M. Mouradian, Mitochondrial localization of DJ-1 leads to enhanced neuroprotection, *J. Neurosci. Res.*, 2009, **87**, 123–129.

72. G. Krebiehl, S. Ruckerbauer, L. F. Burbulla, N. Kieper, B. Maurer, J. Waak, H. Wolburg, Z. Gizatullina, F. N. Gellerich, D. Woitalla, O. Riess, P. J. Kahle, T. Proikas-Cezanne and R. Kruger, Reduced basal autophagy and impaired mitochondrial dynamics due to loss of Parkinson's disease-associated protein DJ-1, *PLoS One*, 2010, **5**, e9367.

73. E. Giaime, H. Yamaguchi, C. A. Gautier, T. Kitada and J. Shen, Loss of DJ-1 does not affect mitochondrial respiration but increases ROS production and mitochondrial permeability transition pore opening, *PLoS One*, 2012, **7**, e40501.

74. R. H. Kim, P. D. Smith, H. Aleyasin, S. Hayley, M. P. Mount, S. Pownall, A. Wakeham, A. J. You-Ten, S. K. Kalia, P. Horne, D. Westaway, A. M. Lozano, H. Anisman, D. S. Park and T. W. Mak, Hypersensitivity of DJ-1-deficient mice to 1-methyl-4-phenyl-1,2,3,6-tetrahydropyrindine (MPTP) and oxidative stress, *Proc. Natl. Acad. Sci. U. S. A.*, 2005, **102**, 5215–5220.

75. K. J. Thomas, M. K. McCoy, J. Blackinton, A. Beilina, M. van der Brug, A. Sandebring, D. Miller, D. Maric, A. Cedazo-Minguez and M. R. Cookson, DJ-1 acts in parallel to the PINK1/parkin pathway to control mitochondrial function and autophagy, *Hum. Mol. Genet.*, 2011, **20**, 40–50.

76. N. Lev, D. Ickowicz, Y. Barhum, S. Lev, E. Melamed and D. Offen, DJ-1 protects against dopamine toxicity, *J. Neural Transm.*, 2009, **116**, 151–160.

77. D. Sha, L. S. Chin and L. Li, Phosphorylation of parkin by Parkinson disease-linked kinase PINK1 activates parkin E3 ligase function and NF-kappaB signaling, *Hum. Mol. Genet.*, 2010, **19**, 352–363.

78. C. Vives-Bauza, C. Zhou, Y. Huang, M. Cui, R. L. de Vries, J. Kim, J. May, M. A. Tocilescu, W. Liu, H. S. Ko, J. Magrane, D. J. Moore, V. L. Dawson, R. Grailhe, T. M. Dawson, C. Li, K. Tieu and S. Przedborski, PINK1-dependent recruitment of Parkin to mitochondria in mitophagy, *Proc. Natl. Acad. Sci. U. S. A.*, 2010, **107**, 378–383.

79. D. J. Moore, L. Zhang, J. Troncoso, M. K. Lee, N. Hattori, Y. Mizuno, T. M. Dawson and V. L. Dawson, Association of DJ-1 and parkin mediated by pathogenic DJ-1 mutations and oxidative stress, *Hum. Mol. Genet.*, 2005, **14**, 71–84.

80. C. E. Gavin, K. K. Gunter and T. E. Gunter, Manganese and calcium transport in mitochondria: implications for manganese toxicity, *Neurotoxicology*, 1999, **20**, 445–453.

81. D. Milatovic, Z. Yin, R. C. Gupta, M. Sidoryk, J. Albrecht, J. L. Aschner and M. Aschner, Manganese induces oxidative impairment in cultured rat astrocytes, *Toxicol. Sci.*, 2007, **98**, 198–205.

82. Y. Hirata, K. Adachi and K. Kiuchi, Activation of JNK pathway and induction of apoptosis by manganese in PC12 cells, *J. Neurochem.*, 1998, **71**, 1607–1615.

83. J. A. Roth, L. Feng, J. Walowitz and R. W. Browne, Manganese-induced rat pheochromocytoma (PC12) cell death is independent of caspase activation, *J. Neurosci. Res.*, 2000, **61**, 162–171.

84. F. M. Cordova, A. S. Aguiar, Jr., T. V. Peres, M. W. Lopes, F. M. Goncalves, A. P. Remor, S. C. Lopes, C. Pilati, A. S. Latini, R. D. Prediger, K. M. Erikson, M. Aschner and R. B. Leal, *In vivo* manganese exposure modulates Erk, Akt and Darpp-32 in the striatum of developing rats, and impairs their motor function, *PLoS One*, 2012, **7**, e33057.

85. A. B. Manning-Bog, W. M. Caudle, X. A. Perez, S. H. Reaney, R. Paletzki, M. Z. Isla, V. P. Chou, A. L. McCormack, G. W. Miller, J. W. Langston, C. R. Gerfen and D. A. Dimonte, Increased vulnerability of nigrostriatal terminals in DJ-1-deficient mice is mediated by the dopamine transporter, *Neurobiol. Dis.*, 2007, **27**, 141–150.

86. E. M. Valente, P. M. Abou-Sleiman, V. Caputo, M. M. Muqit, K. Harvey, S. Gispert, Z. Ali, D. Del Turco, A. R. Bentivoglio, D. G. Healy, A. Albanese, R. Nussbaum, R. Gonzalez-Maldonado, T. Deller, S. Salvi, P. Cortelli, W. P. Gilks, D. S. Latchman, R. J. Harvey, B. Dallapiccola, G. Auburger and N. W. Wood, Hereditary early-onset Parkinson's disease caused by mutations in PINK1, *Science*, 2004, **304**, 1158–1160.

87. M. M. Muqit, P. M. Abou-Sleiman, A. T. Saurin, K. Harvey, S. Gandhi, E. Deas, S. Eaton, M. D. Payne Smith, K. Venner, A. Matilla, D. G. Healy, W. P. Gilks, A. J. Lees, J. Holton, T. Revesz, P. J. Parker, R. J. Harvey, N. W. Wood and D. S. Latchman, Altered cleavage and localization of

PINK1 to aggresomes in the presence of proteasomal stress, *J. Neurochem.*, 2006, **98**, 156–169.

88. S. Gandhi, M. M. Muqit, L. Stanyer, D. G. Healy, P. M. Abou-Sleiman, I. Hargreaves, S. Heales, M. Ganguly, L. Parsons, A. J. Lees, D. S. Latchman, J. L. Holton, N. W. Wood and T. Revesz, PINK1 protein in normal human brain and Parkinson's disease, *Brain*, 2006, **129**, 1720–1731.

89. J. W. Pridgeon, J. A. Olzmann, L. S. Chin and L. Li, PINK1 protects against oxidative stress by phosphorylating mitochondrial chaperone TRAP1, *PLoS Biol.*, 2007, **5**, e172.

90. M. E. Haque, M. P. Mount, F. Safarpour, E. Abdel-Messih, S. Callaghan, C. Mazerolle, T. Kitada, R. S. Slack, V. Wallace, J. Shen, H. Anisman and D. S. Park, Inactivation of Pink1 gene in vivo sensitizes dopamine-producing neurons to 1-methyl-4-phenyl-1,2,3,6-tetrahydropyridine (MPTP) and can be rescued by autosomal recessive Parkinson disease genes, Parkin or DJ-1, *J. Biol. Chem.*, 2012, **287**, 23162–23170.

91. N. Exner, B. Treske, D. Paquet, K. Holmstrom, C. Schiesling, S. Gispert, I. Carballo-Carbajal, D. Berg, H. H. Hoepken, T. Gasser, R. Kruger, K. F. Winklhofer, F. Vogel, A. S. Reichert, G. Auburger, P. J. Kahle, B. Schmid and C. Haass, Loss-of-function of human PINK1 results in mitochondrial pathology and can be rescued by parkin, *J. Neurosci.*, 2007, **27**, 12413–12418.

92. Y. Yang, S. Gehrke, Y. Imai, Z. Huang, Y. Ouyang, J. W. Wang, L. Yang, M. F. Beal, H. Vogel and B. Lu, Mitochondrial pathology and muscle and dopaminergic neuron degeneration caused by inactivation of Drosophila Pink1 is rescued by Parkin, *Proc. Natl. Acad. Sci. U. S. A.*, 2006, **103**, 10793–10798.

93. S. Gandhi, A. Vaarmann, Z. Yao, M. R. Duchen, N. W. Wood and A. Y. Abramov, Dopamine induced neurodegeneration in a PINK1 model of Parkinson's disease, *PLoS One*, 2012, 7, e37564.

94. B. Heeman, C. Van den Haute, S. A. Aelvoet, F. Valsecchi, R. J. Rodenburg, V. Reumers, Z. Debyser, G. Callewaert, W. J. Koopman, P. H. Willems and V. Baekelandt, Depletion of PINK1 affects mito-chondrial metabolism, calcium homeostasis and energy maintenance, *J. Cell Sci.*, 2011, **124**, 1115–1125.

95. W. Liu, C. Vives-Bauza, R. Acin-Perez, A. Yamamoto, Y. Tan, Y. Li, J. Magrane, M. A. Stavarache, S. Shaffer, S. Chang, M. G. Kaplitt, X. Y. Huang, M. F. Beal, G. Manfredi and C. Li, PINK1 defect causes mitochondrial dysfunction, proteasomal deficit and alpha-synuclein aggregation in cell culture models of Parkinson's disease, *PLoS One*, 2009, **4**, e4597.

96. L. Samaranch, O. Lorenzo-Betancor, J. M. Arbelo, I. Ferrer, E. Lorenzo, J. Irigoyen, M. A. Pastor, C. Marrero, C. Isla, J. Herrera-Henriquez and P. Pastor, PINK1-linked parkinsonism is associated with Lewy body pathology, *Brain*, 2010, **133**, 1128–1142.

97. J. E. Ahlskog, Parkin and PINK1 parkinsonism may represent nigral mitochondrial cytopathies distinct from Lewy body Parkinson's disease, *Parkinsonism Relat. Disord.*, 2009, **15**, 721–727.

98. J. W. Um, H. J. Park, J. Song, I. Jeon, G. Lee, P. H. Lee and K. C. Chung, Formation of parkin aggregates and enhanced PINK1 accumulation during the pathogenesis of Parkinson's disease, *Biochem. Biophys. Res. Commun.*, 2010, **393**, 824–828.

99. L. Pallanck and J. T. Greenamyre, Neurodegenerative disease: pink, parkin and the brain, *Nature*, 2006, **441**, 1058.

100. I. E. Clark, M. W. Dodson, C. Jiang, J. H. Cao, J. R. Huh, J. H. Seol, S. J. Yoo, B. A. Hay and M. Guo, Drosophila pink1 is required for mitochondrial function and interacts genetically with parkin, *Nature*, 2006, **441**, 1162–1166.

101. J. Park, S. B. Lee, S. Lee, Y. Kim, S. Song, S. Kim, E. Bae, J. Kim, M. Shong, J. M. Kim and J. Chung, Mitochondrial dysfunction in Drosophila PINK1 mutants is complemented by parkin, *Nature*, 2006, **441**, 1157–1161.

102. H. Eiberg, L. Hansen, L. Korbo, I. M. Nielsen, K. Svenstrup, S. Bech, L. H. Pinborg, L. Friberg, L. E. Hjermind, O. R. Olsen and J. E. Nielsen, Novel mutation in ATP13A2 widens the spectrum of Kufor-Rakeb syndrome (PARK9), *Clin. Genet.*, 2012, **82**, 256–263.

103. A. Ramirez, A. Heimbach, J. Grundemann, B. Stiller, D. Hampshire, L. P. Cid, I. Goebel, A. F. Mubaidin, A. L. Wriekat, J. Roeper, A. Al-Din, A. M. Hillmer, M. Karsak, B. Liss, C. G. Woods, M. I. Behrens and C. Kubisch, Hereditary parkinsonism with dementia is caused by mutations in ATP13A2, encoding a lysosomal type 5 P-type ATPase, *Nat. Genet.*, 2006, **38**, 1184–1191.

104. J. S. Park, P. Mehta, A. A. Cooper, D. Veivers, A. Heimbach, B. Stiller, C. Kubisch, V. S. Fung, D. Krainc, A. Mackay-Sim and C. M. Sue, Pathogenic effects of novel mutations in the P-type ATPase ATP13A2 (PARK9) causing Kufor-Rakeb syndrome, a form of early-onset parkinsonism, *Hum. Mutat.*, 2011, **32**, 956–964.

105. D. Ramonet, A. Podhajska, K. Stafa, S. Sonnay, A. Trancikova, E. Tsika, O. Pletnikova, J. C. Troncoso, L. Glauser and D. J. Moore, PARK9-associated ATP13A2 localizes to intracellular acidic vesicles and regulates cation homeostasis and neuronal integrity, *Hum. Mol. Genet.*, 2012, **21**, 1725–1743.

106. A. D. Gitler, A. Chesi, M. L. Geddie, K. E. Strathearn, S. Hamamichi, K. J. Hill, K. A. Caldwell, G. A. Caldwell, A. A. Cooper, J. C. Rochet and S. Lindquist, Alpha-synuclein is part of a diverse and highly conserved interaction network that includes PARK9 and manganese toxicity, *Nat. Genet.*, 2009, **41**, 308–315.

107. J. Ugolino, S. Fang, C. Kubisch and M. J. Monteiro, Mutant Atp13a2 proteins involved in parkinsonism are degraded by ER-associated degradation and sensitize cells to ER-stress induced cell death, *Hum. Mol. Genet.*, 2011, **20**, 3565–3577.

108. A. Podhajska, A. Musso, A. Trancikova, K. Stafa, R. Moser, S. Sonnay, L. Glauser and D. J. Moore, Common pathogenic effects of missense mutations in the P-type ATPase ATP13A2 (PARK9) associated with early-onset parkinsonism, *PLoS One*, 2012, 7, e39942.

109. L. Santoro, G. J. Breedveld, F. Manganelli, R. Iodice, C. Pisciotta, M. Nolano, F. Punzo, M. Quarantelli, S. Pappata, A. Di Fonzo, B. A. Oostra and V. Bonifati, Novel ATP13A2 (PARK9) homozygous mutation in a family with marked phenotype variability, *Neurogenetics*, 2010, **12**, 33–39.

110. H. F. Chien, V. Bonifati and E. R. Barbosa, ATP13A2-related neurodegeneration (PARK9) without evidence of brain iron accumulation, *Mov. Disord.*, 2011, **26**, 1364–1365.

111. Y. P. Ning, K. Kanai, H. Tomiyama, Y. Li, M. Funayama, H. Yoshino, S. Sato, M. Asahina, S. Kuwabara, A. Takeda, T. Hattori, Y. Mizuno and N. Hattori, PARK9-linked parkinsonism in eastern Asia: mutation detection in ATP13A2 and clinical phenotype, *Neurology*, 2008, **70**, 1491–1493.

112. P. J. Schultheis, T. T. Hagen, K. K. O'Toole, A. Tachibana, C. R. Burke, D. L. McGill, G. W. Okunade and G. E. Shull, Characterization of the P5 subfamily of P-type transport ATPases in mice, *Biochem. Biophys. Res. Commun.*, 2004, **323**, 731–738.

113. M. Usenovic and D. Krainc, Lysosomal dysfunction in neurodegeneration: the role of ATP13A2/PARK9, *Autophagy*, 2012, **8**, 987–988.

114. M. Usenovic, A. L. Knight, A. Ray, V. Wong, K. R. Brown, G. A. Caldwell, K. A. Caldwell, I. Stagljar and D. Krainc, Identification of novel ATP13A2 interactors and their role in alpha-synuclein misfolding and toxicity, *Hum. Mol. Genet.*, 2012, **21**, 3785–3794.

115. J. P. Covy, E. A. Waxman and B. I. Giasson, Characterization of cellular protective effects of ATP13A2/PARK9 expression and alterations resulting from pathogenic mutants, *J. Neurosci. Res.*, 2012, **90**, 2306–2316.

116. A. M. Gusdon, J. Zhu, B. Van Houten and C. T. Chu, ATP13A2 regulates mitochondrial bioenergetics through macroautophagy, *Neurobiol. Dis.*, 2012, **45**, 962–972.

117. J. Tan, T. Zhang, L. Jiang, J. Chi, D. Hu, Q. Pan, D. Wang and Z. Zhang, Regulation of intracellular manganese homeostasis by Kufor-Rakeb syndrome-associated ATP13A2 protein, *J. Biol. Chem.*, 2011, **286**, 29654–29662.

118. K. Schmidt, D. M. Wolfe, B. Stiller and D. A. Pearce, Cd2+, Mn2+, Ni2+ and Se2+ toxicity to Saccharomyces cerevisiae lacking YPK9p the orthologue of human ATP13A2, *Biochem. Biophys. Res. Commun.*, 2009, **383**, 198–202.

119. A. Chesi, A. Kilaru, X. Fang, A. A. Cooper and A. D. Gitler, The role of the Parkinson's disease gene PARK9 in essential cellular pathways and the manganese homeostasis network in yeast, *PLoS One*, 2012, 7, e34178.

120. H. J. Shin, M. S. Choi, N. H. Ryoo, K. Y. Nam, G. Y. Park, J. H. Bae, S. I. Suh, W. K. Baek, J. W. Park and B. C. Jang, Manganese-mediated

up-regulation of HIF-1alpha protein in Hep2 human laryngeal epithelial cells via activation of the family of MAPKs, *Toxicol. In Vitro*, 2010, **24**, 1208–1214.

121. G. Rentschler, L. Covolo, A. A. Haddad, R. G. Lucchini, S. Zoni and K. Broberg, ATP13A2 (PARK9) polymorphisms influence the neurotoxic effects of manganese, *Neurotoxicology*, 2012, **33**, 697–702.

122. I. F. Mata, W. J. Wedemeyer, M. J. Farrer, J. P. Taylor and K. A. Gallo, LRRK2 in Parkinson's disease: protein domains and functional insights, *Trends Neurosci.*, 2006, **29**, 286–293.

123. S. Biskup, D. J. Moore, F. Celsi, S. Higashi, A. B. West, S. A. Andrabi, K. Kurkinen, S. W. Yu, J. M. Savitt, H. J. Waldvogel, R. L. Faull, P. C. Emson, R. Torp, O. P. Ottersen, T. M. Dawson and V. L. Dawson, Localization of LRRK2 to membranous and vesicular structures in mammalian brain, *Ann. Neurol.*, 2006, **60**, 557–569.

124. A. Zimprich, S. Biskup, P. Leitner, P. Lichtner, M. Farrer, S. Lincoln, J. Kachergus, M. Hulihan, R. J. Uitti, D. B. Calne, A. J. Stoessl, R. F. Pfeiffer, N. Patenge, I. C. Carbajal, P. Vieregge, F. Asmus, B. Muller-Myhsok, D. W. Dickson, T. Meitinger, T. M. Strom, Z. K. Wszolek and T. Gasser, Mutations in LRRK2 cause autosomal-dominant parkinsonism with pleomorphic pathology, *Neuron*, 2004, **44**, 601–607.

125. C. J. Gloeckner, A. Schumacher, K. Boldt and M. Ueffing, The Parkinson disease-associated protein kinase LRRK2 exhibits MAPKKK activity and phosphorylates MKK3/6 and MKK4/7, *in vitro*, *J. Neurochem.*, 2009, **109**, 959–968.

126. E. Greggio, J. M. Taymans, E. Y. Zhen, J. Ryder, R. Vancraenenbroeck, A. Beilina, P. Sun, J. Deng, H. Jaffe, V. Baekelandt, K. Merchant and M. R. Cookson, The Parkinson's disease kinase LRRK2 autophosphorylates its GTPase domain at multiple sites, *Biochem. Biophys. Res. Commun.*, 2009, **389**, 449–454.

127. C. H. Hsu, D. Chan and B. Wolozin, LRRK2 and the stress response: interaction with MKKs and JNK-interacting proteins, *Neurodegener. Dis.*, 2010, 7, 68–75.

128. J. Deng, P. A. Lewis, E. Greggio, E. Sluch, A. Beilina and M. R. Cookson, Structure of the ROC domain from the Parkinson's disease-associated leucine-rich repeat kinase 2 reveals a dimeric GTPase, *Proc. Natl. Acad. Sci. U. S. A.*, 2008, **105**, 1499–1504.

129. D. Korr, L. Toschi, P. Donner, H. D. Pohlenz, B. Kreft and B. Weiss, LRRK1 protein kinase activity is stimulated upon binding of GTP to its Roc domain, *Cell. Signalling*, 2006, **18**, 910–920.

130. S. Bardien, S. Lesage, A. Brice and J. Carr, Genetic characteristics of leucine-rich repeat kinase 2 (LRRK2) associated Parkinson's disease, *Parkinsonism Relat. Disord.*, 2011, **17**, 501–508.

131. O. A. Ross, A. I. Soto-Ortolaza, M. G. Heckman, J. O. Aasly, N. Abahuni, G. Annesi, J. A. Bacon, S. Bardien, M. Bozi, A. Brice, L. Brighina, C. Van Broeckhoven, J. Carr, M. C. Chartier-Harlin, E. Dardiotis,

D. W. Dickson, N. N. Diehl, A. Elbaz, C. Ferrarese, A. Ferraris, B. Fiske, J. M. Gibson, R. Gibson, G. M. Hadjigeorgiou, N. Hattori, J. P. Ioannidis, B. Jasinska-Myga, B. S. Jeon, Y. J. Kim, C. Klein, R. Kruger, E. Kyratzi, S. Lesage, C. H. Lin, T. Lynch, D. M. Maraganore, G. D. Mellick, E. Mutez, C. Nilsson, G. Opala, S. S. Park, A. Puschmann, A. Quattrone, M. Sharma, P. A. Silburn, Y. H. Sohn, L. Stefanis, V. Tadic, J. Theuns, H. Tomiyama, R. J. Uitti, E. M. Valente, S. van de Loo, D. K. Vassilatis, C. Vilarino-Guell, L. R. White, K. Wirdefeldt, Z. K. Wszolek, R. M. Wu and M. J. Farrer, Genetic Epidemiology Of Parkinson's Disease, Association of LRRK2 exonic variants with susceptibility to Parkinson's disease: a case-control study, *Lancet Neurol.*, 2011, **10**, 898–908.

132. X. Wu, K. F. Tang, Y. Li, Y. Y. Xiong, L. Shen, Z. Y. Wei, K. J. Zhou, J. M. Niu, X. Han, L. Yang, G. Y. Feng, L. He and S. Y. Qin, Quantitative assessment of the effect of LRRK2 exonic variants on the risk of Parkinson's disease: a meta-analysis, *Parkinsonism Relat. Disord.*, 2012, **18**, 722–730.

133. M. Jaleel, R. J. Nichols, M. Deak, D. G. Campbell, F. Gillardon, A. Knebel and D. R. Alessi, LRRK2 phosphorylates moesin at threonine-558: characterization of how Parkinson's disease mutants affect kinase activity, *Biochem. J.*, 2007, **405**, 307–317.

134. S. Saha, M. D. Guillily, A. Ferree, J. Lanceta, D. Chan, J. Ghosh, C. H. Hsu, L. Segal, K. Raghavan, K. Matsumoto, N. Hisamoto, T. Kuwahara, T. Iwatsubo, L. Moore, L. Goldstein, M. Cookson and B. Wolozin, LRRK2 modulates vulnerability to mitochondrial dysfunction in Caenorhabditis elegans, *J. Neurosci.*, 2009, **29**, 9210–9218.

135. Y. Tong, H. Yamaguchi, E. Giaime, S. Boyle, R. Kopan, R. J. Kelleher, 3rd and J. Shen, Loss of leucine-rich repeat kinase 2 causes impairment of protein degradation pathways, accumulation of alpha-synuclein, and apoptotic cell death in aged mice, *Proc. Natl. Acad. Sci. U. S. A.*, 2010, **107**, 9879–9884.

136. K. Habig, M. Walter, S. Poths, O. Riess and M. Bonin, RNA interference of LRRK2-microarray expression analysis of a Parkinson's disease key player, *Neurogenetics*, 2008, **9**, 83–94.

137. D. A. MacLeod, H. Rhinn, T. Kuwahara, A. Zolin, G. Di Paolo, B. D. McCabe, K. S. Marder, L. S. Honig, L. N. Clark, S. A. Small and A. Abeliovich, RAB7L1 interacts with LRRK2 to modify intraneuronal protein sorting and Parkinson's disease risk, *Neuron*, 2013, **77**, 425–439.

138. E. Greggio, M. Bisaglia, L. Civiero and L. Bubacco, Leucine-rich repeat kinase 2 and alpha-synuclein: intersecting pathways in the pathogenesis of Parkinson's disease?, *Mol. Neurodegener.*, 2011, **6**, 6–10.

139. K. Kondo, S. Obitsu and R. Teshima, alpha-Synuclein aggregation and transmission are enhanced by leucine-rich repeat kinase 2 in human neuroblastoma SH-SY5Y cells, *Biol. Pharm. Bull.*, 2011, **34**, 1078–1083.

140. X. Lin, L. Parisiadou, X. L. Gu, L. Wang, H. Shim, L. Sun, C. Xie, C. X. Long, W. J. Yang, J. Ding, Z. Z. Chen, P. E. Gallant,

J. H. Tao-Cheng, G. Rudow, J. C. Troncoso, Z. Liu, Z. Li and H. Cai, Leucine-rich repeat kinase 2 regulates the progression of neuro-pathology induced by Parkinson's-disease-related mutant alpha-synuclein, *Neuron*, 2009, **64**, 807–827.

141. J. R. Adams, H. van Netten, M. Schulzer, E. Mak, J. McKenzie, A. Strongosky, V. Sossi, T. J. Ruth, C. S. Lee, M. Farrer, T. Gasser, R. J. Uitti, D. B. Calne, Z. K. Wszolek and A. J. Stoessl, PET in LRRK2 mutations: comparison to sporadic Parkinson's disease and evidence for presymptomatic compensation, *Brain*, 2005, **128**, 2777–2785.

142. R. Nandhagopal, E. Mak, M. Schulzer, J. McKenzie, S. McCormick, V. Sossi, T. J. Ruth, A. Strongosky, M. J. Farrer, Z. K. Wszolek and A. J. Stoessl, Progression of dopaminergic dysfunction in a LRRK2 kindred: a multitracer PET study, *Neurology*, 2008, **71**, 1790–1795.

143. Y. Tong, A. Pisani, G. Martella, M. Karouani, H. Yamaguchi, E. N. Pothos and J. Shen, R1441C mutation in LRRK2 impairs dopa-minergic neurotransmission in mice, *Proc. Natl. Acad. Sci. U. S. A.*, 2009, **106**, 14622–14627.

144. K. Sriram, G. X. Lin, A. M. Jefferson, J. R. Roberts, O. Wirth, Y. Hayashi, K. M. Krajnak, J. M. Soukup, A. J. Ghio, S. H. Reynolds, V. Castranova, A. E. Munson and J. M. Antonini, Mitochondrial dysfunction and loss of Parkinson's disease-linked proteins contribute to neurotoxicity of manganese-containing welding fumes, *FASEB J.*, 2010, **24**, 4989–5002.

145. T. M. Peneder, P. Scholze, M. L. Berger, H. Reither, G. Heinze, J. Bertl, J. Bauer, E. K. Richfield, O. Hornykiewicz and C. Pifl, Chronic exposure to manganese decreases striatal dopamine turnover in human alpha-synuclein transgenic mice, *Neuroscience*, 2011, **180**, 280–292.

146. T. E. Gunter, C. E. Gavin and K. K. Gunter, The case for manganese interaction with mitochondria, *Neurotoxicology*, 2009, **30**, 727–729.

147. G. Santpere and I. Ferrer, LRRK2 and neurodegeneration, *Acta Neuro-pathol.*, 2009, **117**, 227–246.

148. J. M. Bravo-San Pedro, M. Niso-Santano, R. Gomez-Sanchez, E. Pizarro-Estrella, A. Aiastui-Pujana, A. Gorostidi, V. Climent, R. Lopez de Maturana, R. Sanchez-Pernaute, A. Lopez de Munain, J. M. Fuentes and R. A. Gonzalez-Polo, The LRRK2 G2019S mutant exacerbates basal autophagy through activation of the MEK/ERK pathway, *Cell. Mol. Life Sci.*, 2013, **70**, 121–136.

149. J. P. Covy and B. I. Giasson, Identification of compounds that inhibit the kinase activity of leucine-rich repeat kinase 2, *Biochem. Biophys. Res. Commun.*, 2009, **378**, 473–477.

150. A. Mamais, M. Raja, C. Manzoni, S. Dihanich, A. Lees, D. Moore, P. A. Lewis and R. Bandopadhyay, Divergent alpha-synuclein solubility and aggregation properties in G2019S LRRK2 Parkinson's disease brains with Lewy Body pathology compared to idiopathic cases, *Neu-robiol. Dis.*, 2013, **58**, 183–190.

151. B. Lovitt, E. C. Vanderporten, Z. Sheng, H. Zhu, J. Drummond and Y. Liu, Differential effects of divalent manganese and magnesium on

the kinase activity of leucine-rich repeat kinase 2 (LRRK2), *Biochemistry*, 2010, **49**, 3092–3100.

152. J. P. Covy and B. I. Giasson, The G2019S pathogenic mutation disrupts sensitivity of leucine-rich repeat kinase 2 to manganese kinase inhibition, *J. Neurochem.*, 2010, **115**, 36–46.

153. J. A. Roth and M. Eichhorn, Down-regulation of LRRK2 in control and DAT transfected HEK cells increases manganese-induced oxidative stress and cell toxicity, *Neurotoxicology*, 2013, **37**, 100–107.

154. B. Kim, M. S. Yang, D. Choi, J. H. Kim, H. S. Kim, W. Seol, S. Choi, I. Jou, E. Y. Kim and E. H. Joe, Impaired inflammatory responses in murine Lrrk2-knockdown brain microglia, *PloS ONE*, 2012, **7**, e34693.

155. J. P. Covy and B. I. Giasson, alpha-Synuclein, leucine-rich repeat kinase-2, and manganese in the pathogenesis of Parkinson disease, *Neurotoxicology*, 2011, **32**, 622–629.

156. O. Marques and T. F. Outeiro, Alpha-synuclein: from secretion to dysfunction and death, *Cell Death Dis.*, 2012, **3**, e350.

157. M. C. Bennett, J. F. Bishop, Y. Leng, P. B. Chock, T. N. Chase and M. M. Mouradian, Degradation of alpha-synuclein by proteasome, *J. Biol. Chem.*, 1999, **274**, 33855–33858.

158. Y. Chu, H. Dodiya, P. Aebischer, C. W. Olanow and J. H. Kordower, Alterations in lysosomal and proteasomal markers in Parkinson's disease: relationship to alpha-synuclein inclusions, *Neurobiol. Dis.*, 2009, **35**, 385–398.

159. J. B. Watson, A. Hatami, H. David, E. Masliah, K. Roberts, C. E. Evans and M. S. Levine, Alterations in corticostriatal synaptic plasticity in mice overexpressing human alpha-synuclein, *Neuroscience*, 2009, **159**, 501–513.

160. S. J. Tabrizi, M. Orth, J. M. Wilkinson, J. W. Taanman, T. T. Warner, J. M. Cooper and A. H. Schapira, Expression of mutant alpha-synuclein causes increased susceptibility to dopamine toxicity, *Hum. Mol. Genet.*, 2000, **9**, 2683–2689.

161. M. Periquet, T. Fulga, L. Myllykangas, M. G. Schlossmacher and M. B. Feany, Aggregated alpha-synuclein mediates dopaminergic neurotoxicity *in vivo*, *J. Neurosci.*, 2007, **27**, 3338–3346.

162. A. Abeliovich, Y. Schmitz, I. Farinas, D. Choi-Lundberg, W. H. Ho, P. E. Castillo, N. Shinsky, J. M. Verdugo, M. Armanini, A. Ryan, M. Hynes, H. Phillips, D. Sulzer and A. Rosenthal, Mice lacking alpha-synuclein display functional deficits in the nigrostriatal dopamine system, *Neuron*, 2000, **25**, 239–252.

163. N. M. Bonini and B. I. Giasson, Snaring the function of alpha-synuclein, *Cell*, 2005, **123**, 359–361.

164. E. Junn and M. M. Mouradian, Human alpha-synuclein over-expression increases intracellular reactive oxygen species levels and susceptibility to dopamine, *Neurosci. Lett.*, 2002, **320**, 146–150.

165. C. L. Pham, S. L. Leong, F. E. Ali, V. B. Kenche, A. F. Hill, S. L. Gras, K. J. Barnham and R. Cappai, Dopamine and the dopamine oxidation

product 5,6-dihydroxylindole promote distinct on-pathway and off-pathway aggregation of alpha-synuclein in a pH-dependent manner, *J. Mol. Biol.*, 2009, **387**, 771–785.

166. M. Baba, S. Nakajo, P. H. Tu, T. Tomita, K. Nakaya, V. M. Lee, J. Q. Trojanowski and T. Iwatsubo, Aggregation of alpha-synuclein in Lewy bodies of sporadic Parkinson's disease and dementia with Lewy bodies, *Am. J. Pathol.*, 1998, **152**, 879–884.

167. W. J. Schulz-Schaeffer, The synaptic pathology of alpha-synuclein aggregation in dementia with Lewy bodies, Parkinson's disease and Parkinson's disease dementia, *Acta Neuropathol.*, 2010, **120**, 131–143.

168. M. G. Spillantini, M. L. Schmidt, V. M. Lee, J. Q. Trojanowski, R. Jakes and M. Goedert, Alpha-synuclein in Lewy bodies, *Nature*, 1997, **388**, 839–840.

169. L. Maroteaux, J. T. Campanelli and R. H. Scheller, Synuclein: a neuron-specific protein localized to the nucleus and presynaptic nerve terminal, *J. Neurosci.*, 1988, **8**, 2804–2815.

170. A. Deleersnijder, M. Gerard, Z. Debyser and V. Baekelandt, The remarkable conformational plasticity of alpha-synuclein: blessing or curse?, *Trends Mol. Med.*, 2013, **19**, 368–377.

171. C. W. Bertoncini, Y. S. Jung, C. O. Fernandez, W. Hoyer, C. Griesinger, T. M. Jovin and M. Zweckstetter, Release of long-range tertiary interactions potentiates aggregation of natively unstructured alpha-synuclein, *Proc. Natl. Acad. Sci. U. S. A.*, 2005, **102**, 1430–1435.

172. K. Nakamura, alpha-Synuclein and mitochondria: partners in crime?, *Neurotherapeutics*, 2013, **10**, 391–399.

173. L. Devi and H. K. Anandatheerthavarada, Mitochondrial trafficking of APP and alpha synuclein: Relevance to mitochondrial dysfunction in Alzheimer's and Parkinson's diseases, *Biochim. Biophys. Acta*, 2010, **1802**, 11–19.

174. L. Devi, V. Raghavendran, B. M. Prabhu, N. G. Avadhani and H. K. Anandatheerthavarada, Mitochondrial import and accumulation of alpha-synuclein impair complex I in human dopaminergic neuronal cultures and Parkinson disease brain, *J. Biol. Chem.*, 2008, **283**, 9089–9100.

175. K. Banerjee, M. Sinha, L. Pham Cle, S. Jana, D. Chanda, R. Cappai and S. Chakrabarti, Alpha-synuclein induced membrane depolarization and loss of phosphorylation capacity of isolated rat brain mitochondria: implications in Parkinson's disease, *FEBS Lett.*, 2010, **584**, 1571–1576.

176. K. Nakamura, V. M. Nemani, F. Azarbal, G. Skibinski, J. M. Levy, K. Egami, L. Munishkina, J. Zhang, B. Gardner, J. Wakabayashi, H. Sesaki, Y. Cheng, S. Finkbeiner, R. L. Nussbaum, E. Masliah and R. H. Edwards, Direct membrane association drives mitochondrial fission by the Parkinson disease-associated protein alpha-synuclein, *J. Biol. Chem.*, 2011, **286**, 20710–20726.

177. B. Xu, S. W. Wu, C. W. Lu, Y. Deng, W. Liu, Y. G. Wei, T. Y. Yang and Z. F. Xu, Oxidative stress involvement in manganese-induced

alpha-synuclein oligomerization in organotypic brain slice cultures, *Toxicology*, 2013, **305**, 71–78.

178. T. Cai, T. Yao, G. Zheng, Y. Chen, K. Du, Y. Cao, X. Shen, J. Chen and W. Luo, Manganese induces the overexpression of alpha-synuclein in PC12 cells via ERK activation, *Brain Res.*, 2010, **1359**, 201–207.

179. K. Prabhakaran, G. D. Chapman and P. G. Gunasekar, alpha-Synuclein overexpression enhances manganese-induced neurotoxicity through the NF-kappaB-mediated pathway, *Toxicol. Mech. Methods*, 2011, **21**, 435–443.

180. C. Pifl, M. Khorchide, A. Kattinger, H. Reither, J. Hardy and O. Hornykiewicz, alpha-Synuclein selectively increases manganese-induced viability loss in SK-N-MC neuroblastoma cells expressing the human dopamine transporter, *Neurosci. Lett.*, 2004, **354**, 34–37.

181. T. Verina, J. S. Schneider and T. R. Guilarte, Manganese exposure induces alpha-synuclein aggregation in the frontal cortex of non-human primates, *Toxicol. Lett.*, 2013, **217**, 177–183.

182. J. S. Bonifacino and R. Rojas, Retrograde transport from endosomes to the trans-Golgi network, *Nat. Rev. Mol. Cell Biol.*, 2006, 7, 568–579.

183. J. S. Bonifacino and J. H. Hurley, Retromer, *Curr. Opin. Cell Biol.*, 2008, **20**, 427–436.

184. A. Zimprich, A. Benet-Pages, W. Struhal, E. Graf, S. H. Eck, M. N. Offman, D. Haubenberger, S. Spielberger, E. C. Schulte, P. Lichtner, S. C. Rossle, N. Klopp, E. Wolf, K. Seppi, W. Pirker, S. Presslauer, B. Mollenhauer, R. Katzenschlager, T. Foki, C. Hotzy, E. Reinthaler, A. Harutyunyan, R. Kralovics, A. Peters, F. Zimprich, T. Brucke, W. Poewe, E. Auff, C. Trenkwalder, B. Rost, G. Ransmayr, J. Winkelmann, T. Meitinger and T. M. Strom, A mutation in VPS35, encoding a subunit of the retromer complex, causes late-onset Parkinson disease, *Am. J. Hum. Genet.*, 2011, **89**, 168–175.

185. M. Sharma, J. P. Ioannidis, J. O. Aasly, G. Annesi, A. Brice, L. Bertram, M. Bozi, M. Barcikowska, D. Crosiers, C. E. Clarke, M. F. Facheris, M. Farrer, G. Garraux, S. Gispert, G. Auburger, C. Vilarino-Guell, G. M. Hadjigeorgiou, A. A. Hicks, N. Hattori, B. S. Jeon, Z. Jamrozik, A. Krygowska-Wajs, S. Lesage, C. M. Lill, J. J. Lin, T. Lynch, P. Lichtner, A. E. Lang, C. Libioulle, M. Murata, V. Mok, B. Jasinska-Myga, G. D. Mellick, K. E. Morrison, T. Meitnger, A. Zimprich, G. Opala, P. P. Pramstaller, I. Pichler, S. S. Park, A. Quattrone, E. Rogaeva, O. A. Ross, L. Stefanis, J. D. Stockton, W. Satake, P. A. Silburn, T. M. Strom, J. Theuns, E. K. Tan, T. Toda, H. Tomiyama, R. J. Uitti, C. Van Broeckhoven, K. Wirdefeldt, Z. Wszolek, G. Xiromerisiou, H. S. Yomono, K. C. Yueh, Y. Zhao, T. Gasser, D. Maraganore, R. Kruger and G. consortium, A multi-centre clinico-genetic analysis of the VPS35 gene in Parkinson disease indicates reduced penetrance for disease-associated variants, *J. Med. Genet.*, 2012, **49**, 721–726.

186. C. Vilarino-Guell, C. Wider, O. A. Ross, J. C. Dachsel, J. M. Kachergus, S. J. Lincoln, A. I. Soto-Ortolaza, S. A. Cobb, G. J. Wilhoite, J. A. Bacon,

B. Behrouz, H. L. Melrose, E. Hentati, A. Puschmann, D. M. Evans, E. Conibear, W. W. Wasserman, J. O. Aasly, P. R. Burkhard, R. Djaldetti, J. Ghika, F. Hentati, A. Krygowska-Wajs, T. Lynch, E. Melamed, A. Rajput, A. H. Rajput, A. Solida, R. M. Wu, R. J. Uitti, Z. K. Wszolek, F. Vingerhoets and M. J. Farrer, VPS35 mutations in Parkinson disease, *Am. J. Hum. Genet.*, 2011, **89**, 162–167.

187. M. Tabuchi, T. Yoshimori, K. Yamaguchi, T. Yoshida and F. Kishi, Human NRAMP2/DMT1, which mediates iron transport across endosomal membranes, is localized to late endosomes and lysosomes in HEp-2 cells, *J. Biol. Chem.*, 2000, **275**, 22220–22228.

188. S. Lam-Yuk-Tseung, N. Touret, S. Grinstein and P. Gros, Carboxyl-terminus determinants of the iron transporter DMT1/SLC11A2 isoform II (-IRE/1B) mediate internalization from the plasma membrane into recycling endosomes, *Biochemistry*, 2005, **44**, 12149–12159.

189. M. Swan and R. Saunders-Pullman, The association between ss-glucocerebrosidase mutations and parkinsonism, *Curr. Neurol. Neurosci. Rep.*, 2013, **13**, 368–382.

190. C. M. Lill, J. T. Roehr, M. B. McQueen, F. K. Kavvoura, S. Bagade, B. M. Schjeide, L. M. Schjeide, E. Meissner, U. Zauft, N. C. Allen, T. Liu, M. Schilling, K. J. Anderson, G. Beecham, D. Berg, J. M. Biernacka, A. Brice, A. L. DeStefano, C. B. Do, N. Eriksson, S. A. Factor, M. J. Farrer, T. Foroud, T. Gasser, T. Hamza, J. A. Hardy, P. Heutink, E. M. Hill-Burns, C. Klein, J. C. Latourelle, D. M. Maraganore, E. R. Martin, M. Martinez, R. H. Myers, M. A. Nalls, N. Pankratz, H. Payami, W. Satake, W. K. Scott, M. Sharma, A. B. Singleton, K. Stefansson, T. Toda, J. Y. Tung, J. Vance, N. W. Wood, C. P. Zabetian, 23andMe Genetic Epidemiology of Parkinson's Disease Consortium, International Parkinson's Disease Genomics Consortium, Parkinson's Disease GWAS Consortium, Wellcome Trust Case Control Consortium 2, P. Young, R. E. Tanzi, M. J. Khoury, F. Zipp, H. Lehrach, J. P. Ioannidis and L. Bertram, Comprehensive research synopsis and systematic meta-analyses in Parkinson's disease genetics: The PDGene database, *PLoS Genet.*, 2012, **8**, e1002548.

191. S. L. Rhodes, J. S. Sinsheimer, Y. Bordelon, J. M. Bronstein and B. Ritz, Replication of GWAS associations for GAK and MAPT in Parkinson's disease, *Ann. Hum. Genet.*, 2011, **75**, 195–200.

192. M. C. Chartier-Harlin, J. C. Dachsel, C. Vilarino-Guell, S. J. Lincoln, F. Lepretre, M. M. Hulihan, J. Kachergus, A. J. Milnerwood, L. Tapia, M. S. Song, E. Le Rhun, E. Mutez, L. Larvor, A. Duflot, C. Vanbesien-Mailliot, A. Kreisler, O. A. Ross, K. Nishioka, A. I. Soto-Ortolaza, S. A. Cobb, H. L. Melrose, B. Behrouz, B. H. Keeling, J. A. Bacon, E. Hentati, L. Williams, A. Yanagiya, N. Sonenberg, P. J. Lockhart, A. C. Zubair, R. J. Uitti, J. O. Aasly, A. Krygowska-Wajs, G. Opala, Z. K. Wszolek, R. Frigerio, D. M. Maraganore, D. Gosal, T. Lynch, M. Hutchinson, A. R. Bentivoglio, E. M. Valente, W. C. Nichols, N. Pankratz, T. Foroud, R. A. Gibson, F. Hentati, D. W. Dickson,

A. Destee and M. J. Farrer, Translation initiator EIF4G1 mutations in familial Parkinson disease, *Am. J. Hum. Genet.*, 2011, **89**, 398–406.

193. A. Puschmann, C. Verbeeck, M. G. Heckman, A. I. Soto-Ortolaza, T. Lynch, B. Jasinska-Myga, G. Opala, A. Krygowska-Wajs, M. Barcikowska, R. J. Uitti, Z. K. Wszolek and O. A. Ross, Human leukocyte antigen variation and Parkinson's disease, *Parkinsonism Relat. Disord.*, 2011, **17**, 376–378.

194. M. Quadri, A. Federico, T. Zhao, G. J. Breedveld, C. Battisti, C. Delnooz, L. A. Severijnen, L. Di Toro Mammarella, A. Mignarri, L. Monti, A. Sanna, P. Lu, F. Punzo, G. Cossu, R. Willemsen, F. Rasi, B. A. Oostra, B. P. van de Warrenburg and V. Bonifati, Mutations in SLC30A10 cause parkinsonism and dystonia with hypermanganesemia, polycythemia, and chronic liver disease, *Am. J. Hum. Genet.*, 2012, **90**, 467–477.

195. M. R. DeWitt, P. Chen and M. Aschner, Manganese efflux in Parkinsonism: insights from newly characterized SLC30A10 mutations, *Biochem. Biophys. Res. Commun.*, 2013, **432**, 1–4.

CHAPTER 10

Mechanism of Manganese-Induced Impairment of Astrocytic Glutamate Transporters

PRATAP KARKI,[a] KEISHA SMITH,[a] MICHAEL ASCHNER[b] AND EUNSOOK LEE*[a]

[a] Department of Physiology, Meharry Medical College, Nashville, TN, USA;
[b] Department of Molecular Pharmacology, Albert Einstein College of Medicine, Bronx, NY, USA
*Email: elee@mmc.edu

10.1 Introduction

Manganese (Mn) is an essential element in the human tissues, playing a critical role in blood clotting, blood sugar homeostasis, digestion, reproduction, bone growth, immune responsiveness, and ATP generation.[1] Mn is also an integral component of multiple metalloenzymes, including glutamine synthetase,[2] mitochondrial superoxide dismutase,[3] arginase, and pyruvate carboxylase.[4] However, chronic exposure to excessive levels of Mn leads to neurotoxicity in the central nervous system (CNS), causing psychiatric, cognitive and motor abnormalities, and neuropathological conditions referred to as manganism.[5–7] Recent evidences suggest that Mn neurotoxicity is associated with a plethora of neurodegenerative diseases, including Alzheimer's disease (AD), Parkinson's disease (PD), Huntington disease (HD), and

Issues in Toxicology No. 22
Manganese in Health and Disease
Edited by Lucio G. Costa and Michael Aschner

amyotrophic lateral sclerosis (ALS).[8] Several mechanisms such as oxidative stress and mitochondrial impairment have been proposed to be involved in Mn neurotoxicity, but Mn-induced repression of both expression and function of glutamate aspartate transporter (GLAST) (excitatory amino acid transporter [EAAT]1 in human) and glutamate transporter-1 (GLT-1) (EAAT2 in human) has been suggested to be a critical mechanism for Mn neurotoxicity.[9,10]

The astrocytic glutamate transporters are crucial in maintaining optimal glutamate levels in the brain by taking up the excess glutamate from the synaptic cleft, thus preventing glutamate-induced excitotoxic neuronal death.[11] Given that dysregulation of glutamate homeostasis associated with the impairment of astrocytic glutamate transporters contributes to several neurodegenerative diseases, Mn-induced downregulation of the astrocytic glutamate transporters (GLAST and GLT-1) might be an important mechanism in inducing Mn-induced neuropathology. In this review, in addition to the established mechanisms for Mn-induced neurotoxicity, such as oxidative stress and mitochondrial impairment, we will discuss the potential for Mn-induced repression of GLAST and GLT-1 at the transcriptional level to mediate this metal's neurotoxic effects.

10.2   Mn Neurotoxicity

10.2.1   Sources of Human Exposure to Mn and its Transport to the Central Nervous System

Humans absorb Mn from the diet, which is estimated to contain 0.9–10 mg Mn per day.[1] The richest dietary sources of Mn include rice, grain, nuts, legumes, and blueberries, which have Mn levels in excess of 30 mg kg^{-1}.[12] Elevated Mn concentrations in the brain generally result from two major sources–occupational and environmental. Occupational exposures to Mn are well documented in a variety of industries, including welding,[13,14] mining,[15,16] ferroalloy smelting,[17,18] battery,[19] glass and ceramics.[20] Mn released into the environment as a product of industrial activities represents another major public health concern for Mn toxicity. These include the use of Mn-containing gasoline anti-knock additive methylcyclopentadienyl manganese tricarbonyl (MMT) and fungicidal pesticide maneb.[21] In addition, individuals receiving parenteral nutrition[22] and suffering from liver dysfunction[23] tend to have increased levels of Mn in their brain. Additionally, consumption of water and soy-based formulas that have high Mn content[24,25] represents another source of Mn. Recently, Mn neurotoxicity in humans has been reported in young drug abusers who used potassium permanganate in the formulations of homemade ephedron (Methcathinone), leading to clinical features of parkinsonism.[26,27]

Inhalation and gastrointestinal absorption represent the two major routes for Mn absorption into the human body.[28] The absorption of Mn is tightly regulated with approximately 1–5% of ingested Mn being absorbed by the gastrointestinal tract, and 30–40% of inhaled Mn absorbed by the lungs.[29,30]

After absorption into the body, Mn transport to the CNS is facilitated by a number of transporter proteins, including the divalent metal transporter-1(DMT-1),[31] transferrin system,[32] the divalent metal/bicarbonate ion symporters ZIP8 and ZIP14, the magnesium transporter hip14 and the transient receptor potential melastatin 7.[8] Notably, inhaled Mn can be directly transported into the brain *via* the olfactory tract.[33] Activation of *N*-methyl-D-aspartate (NMDA) receptor channel accelerates Mn transport to the brain.[34] Mn is distributed throughout the brain, with the highest levels found in the globus pallidus, substantia nigra, and striatum.[35] In cellular levels, astrocytes are the main cell type to take up most Mn *via* DMT1,[36] and neurons also uptake Mn *via* the transferrin system.[37]

10.2.2   Cellular Mechanisms of Mn Neurotoxicity

10.2.2.1   Mitochondrial Dysfunction

A large portion (60–70%) of intracellularly accumulated Mn is sequestered in mitochondria.[38] Mn entry into mitochondria is mediated by the calcium uniporter. Intramitochondrial Mn is bound to the inner mitochondrial membrane and interacts with proteins involved in oxidative phosphorylation.[39,40] Mn is also an important cofactor for various mitochondrial enzymes and, thus, increased Mn levels in this organelle can directly interfere with ATP synthesis. Mn has been shown to interfere with oxidative phosphorylation by inhibiting F1ATPase at low levels of Mn,[39] and complex I of the electron transport chain at higher Mn concentrations.[41] The involvement of mitochondrial dysfunction has been observed in the neurotoxicity induced by Mn-containing welding fumes in rats.[42] In cultured astrocytes, Mn induces the mitochondrial permeability transition[43] and activates mitochondrial apoptotic pathway.[44]

10.2.2.2   Oxidative Stress

Numerous studies have reported that Mn-induced oxidative stress plays a critical role in Mn-induced neurotoxicity. Mn-induced generation of reactive oxygen species (ROS) causes dopamine oxidation, which leads to dopaminergic cell death.[45] The Mn-induced dopamine oxidation is supported by the results that animals exposed to Mn show increased production of uric acid, a dopamine oxidation product.[46] Mn-exposed rats[47] and primates[48] show the presence of markers of oxidative stress, including the reduction in glutathione (GSH) and increase in metallothionein. The contribution of ROS to Mn neurotoxicity was further confirmed when co-treatment with an antioxidant *N*-acetylcysteine (NAC) prevented the pathological changes observed in Mn-exposed rats.[49] Since Mn accumulates in dopamine-rich regions of the brain, such as globus pallidus, substantia nigra, and striatum, Mn-enhanced dopamine oxidation and consequent dopamine cell death are highly pertinent for Mn neurotoxicity and manganism.[50–52] Mn induced oxidative stress has been shown both in astrocytes[53,54] and neurons.[55] Once taken up into the cells, Mn may induce oxidative stress favorably in mitochondria as the preferential sequestration of Mn in mitochondria interferes with normal energy production pathways, leading to the production of ROS.[39] At the molecular levels, Mn induces oxidative stress by the activation of oxidative stress-related signaling pathways, such as nuclear factor-κB (NF-κB)[56] and activator protein-1 (AP-1).[57] Importantly, the activity of glutamate transporters is regulated by the redox state of their reactive cysteine residues, with a dramatic decrease in activity once the reduced cysteine is oxidized.[58] It has also been reported that glutamate uptake by the recombinant glutamate transporters EAAT1, EAAT2, and EAAT3 was inhibited by peroxynitrite and H_2O_2 and restored upon treatment with the reducing agent, dithithreotol.[59] This indicates that oxidative stress plays a role in the regulation of glutamate transporter function.

10.2.2.3 Apoptosis and Inflammation

Mn induces apoptosis in both neuronal[60] and glial cells[44,61] *via* both tumor necrosis factor (TNF) receptor-mediated extrinsic apoptotic and mitochondria-derived intrinsic apoptotic pathways,[62] by activation of various signaling pathways including protein kinase C-delta (PKC-δ).[63] Mn activates extrinsic apoptotic pathway by increasing FasL levels, caspase-8 activation, and Bid cleavage in C6 astrocytoma cells.[62] Mn also activates intrinsic mitochondrial apoptotic pathways by increasing expression of pro-apoptotic genes, such as caspase 3/7, caspase 6, and Bax, and a ratio of Bcl-X_S/Bcl-X_L in cultured astrocytes.[44] Apoptosis and inflammation are not independent cellular events because TNF-α induces inflammation as well as apoptosis.[64]

TNF-α plays a role in excitotoxic neuronal injury because it impairs glutamate transporters and, importantly, Mn potentiates and increases lipopolysaccharide (LPS)-induced TNF-α release from astrocytes and microglia.[65] Moreover, Mn potentiates the release of several inflammatory molecules such as prostaglandins, cytokines including TNF-α, Interleukin (IL)-6, and IL-1β, and nitric oxide from the activated glial cells.[66–69] Among these inflammatory cytokines, TNF-α and IL-1β have been shown to decrease GLT-1 mRNA and protein expression levels in astrocytes, acting as negative regulators of GLT-1.[70,71] Mn induces nitric oxide synthase 2 (NOS2, an inflammatory gene) *via* NF-κB activation in astrocytes.[72,73] Mn increased expression of NOS2 in glial cells located in the globus pallidus and substantia nigra where Mn mainly damages the tissue in the brain of juvenile C56Bl/6J mice.[74] Mn also induces cyclooxygenase-2 (COX-2) expression at both protein and mRNA levels in astrocytes *via* MAPK/p38, PKC, and AP1 pathways.[75] These results indicate that Mn induces COX-2 by transcriptional upregulation *via* multiple signaling pathways.

10.2.2.4 Excitotoxicity

Elevated extracellular glutamate levels in the brain have been postulated as a critical event in inducing Mn neurotoxicity.[76] Mn decreases expression of astrocytic glutamate transporters in astrocytes,[9,77] leading to reduced glutamate uptake into astrocytes and excitotoxic neuronal death. These *in vitro* results are corroborated by *in vivo* studies showing that MK 801, an NMDA antagonist, blocked the Mn-induced excitotoxic lesions in rat striatum.[76] This excess glutamate-induced excitotoxicity in Mn-induced neurotoxicity is congruent with the results that hyperactivity of neuronal cells, rather than increased postsynaptic α-amino-3-hydroxy-5-methyl-4-isoxazolepropionic acid (AMPA) and NMDA receptor sensitivity to glutamate accounts for Mn-induced excitotoxic mechanisms in the development of Mn-induced neurodegeneration in the striatum.[78]

10.3 Role of Astrocytes in Mn Neurotoxicity

Astrocytes are considered the site of early damage induced by Mn because the metal preferentially accumulates in these cells, at 50–60-fold higher concentration than in neurons.[79] Astrocytes are the most abundant non-neuronal cells in the CNS and they perform numerous essential functions for normal neuronal activity, such as glutamate uptake, glutamine release, K^+ and H^+ buffering and antioxidant defense mechanism.[66,80] Mn alters multiple cellular and molecular functions in astrocytes. Mn exposure in astrocytes has been shown to induce cell swelling, gliosis, and Alzheimer type II astrocytes.[49,81,82] Mn also disrupts the intracellular calcium homeostasis[83] and increases the expression of NOS2 and subsequent release of nitric oxide.[73] Mn decreases the expression of glutamine transporters as well as glutamate transporters in astrocytes.[77,84,85] Particularly, Mn-induced impairment of astrocytic glutamate transporters leads to dysregulation of glutamate homeostasis, which might contribute to Mn-induced excitotoxic neurodegeneration.

10.3.1 Mn-induced astrocyte swelling

The abnormal accumulation of Mn in the brain is associated with the development of hepatic encephalopathy (HE)[86] that causes astrocytic abnormalities, including Alzheimer type II astrocytosis.[87] Astrocyte swelling has been a common feature of brain edema in the acute HE and there is pathological similarities with Mn neurotoxicity. Thus, Norenberg's group examined whether Mn also induces astrocytic swelling and found that Mn induces time and concentration-dependent astrocyte swelling in cultured astrocytes.[82] Pretreatment of astrocytes with vitamin E (antioxidant), L-NAME (nitric oxide synthase inhibitor) and Cyclosporin A (mitochondrial permeability transition inhibitor) blocked Mn-induced astrocyte swelling, indicating the involvement of oxidative/nitrosative stress and mitochondrial dysfunction in this Mn-induced astrocyte swelling process.[82] Furthermore, the same group reported that Mn increased the membrane expression of water channel protein aquaporin-4 and knockdown of aquaporin-4 with siRNA significantly attenuated Mn-induced astrocyte swelling, suggesting that aquaporin-4 mediates Mn-induced astrocyte swelling.[88]

10.3.2 Glial Cell-derived Neuroinflammation in Mn Neurotoxicity

Microglia are considered the primary source of inflammatory cytokines in response to various neurotoxic stimuli, including Mn. Mn upregulates TNF-α and IL-1β proteins in cultured microglia[68] as well as the rat manganism model,[89] and exerts synergic effect on LPS-induced TNF-α release from microglia.[90] At higher, pathophysiologically relevant concentrations, Mn also increases release of TNF-α from astrocytes.[65] These results demonstrate that both astrocytes and microglia play a critical role in Mn-induced neurodegeneration secondary to inflammation. There is abundance of reactive astrocytes in the basal ganglia of PD patients[91] and Mn exposure increases the number of inflammatory cytokines in rat astrocytes.[49] Neuronal injury coincides with increased number of reactive astrocytes expressing NOS2 in Mn-treated mouse brains,[69] and deletion of NOS2 attenuates Mn-induced neurotoxicity,[92] suggesting the critical role of NOS2-mediated inflammation in Mn-induced neurotoxicity. Mn enhances the production and expression of various inflammatory mediators, including prostaglandin E2 and COX-2 following co-treatment of Mn with LPS/Interferon-γ.[75] Mn significantly potentiates LPS-induced release of TNF-α and IL-1β in microglia, effectively inducing the formation of ROS and nitric oxide (NO).[93] The effect of Mn on inflammatory cytokine production in both microglia and astrocytes is closely related to the impairment of astrocytic glutamate transporters as the released TNF-α and IL-1β from both cell types decrease the expression and function of astrocytic glutamate transporter GLT-1.[65,70,94]

10.3.3 Astrocytic Glutamate Transporters in Neurological Disorders

Glutamate is the principal excitatory amino acid neurotransmitter in the brain, playing a critical role in synaptic plasticity, learning, and development.[11] Given that glutamate levels in the synaptic cleft are tightly regulated by glutamate transporters, any disturbances in this regulation can lead to excess glutamate in the synaptic cleft and excitatory neuronal injury.[95] Astrocytic glutamate transporters are mainly responsible for maintaining optimal synaptic glutamate levels and glutamate-induced excitotoxicity,[96,97] as evidenced in various *in vitro* and *in vivo* studies. Genetic deletion of GLT-1 in mice develops lethal seizures and increases susceptibility to acute cortical injury.[98] GLAST-deficient mice also induce loss of motor coordination and increase susceptibility to cerebellar injury.[99] The knockdown of GLT-1 and GLAST in the brain by antisense oligonucleotides administration induces the selective loss of expression and function of GLT-1 and GLAST, respectively, elevating extracellular glutamate levels, excitotoxicity-related neurodegeneration and progressive paralysis.[100] Clinically, the loss of GLT-1 expression has been detected in ALS patients[101] and cultured astrocytes from the cortex of AD patients in parallel with reduced glutamate uptake.[102] The dysfunction of GLT-1 is associated with various neurological disorders including stroke,[103] brain tumors,[104] epilepsy, PD, HD, and cerebral ischemia, indicating that drugs targeting to enhance the expression and function of glutamate transporters would be a rational therapeutic approach for treating these diseases.[97]

10.3.4 Mn Reduces Expression and Function of Astrocytic Glutamate Transporters

Mn causes a reduction of glutamate uptake into astrocytes, leading to elevated extracellular glutamate levels. Hazell and Norenberg reported more than decade ago that Mn decreased glutamate uptake in cultured astrocytes.[105] Mn decreased GLAST mRNA levels in astrocytes[106] and decreased glutamate uptake in Chinese Hamster Ovary (CHO) cells transfected with GLAST and GLT-1 expression vectors.[107] Mn exposure *via* inhalation has also been reported to reduce GLT-1 and GLAST mRNA and protein levels in the non-human primate brain.[48,108] Studies from our laboratory have established that a pathophysiologically relevant concentration of Mn decreases the expression of GLT-1 and GLAST at both mRNA and protein levels with parallel reduction of glutamate uptake, and further, modulates transcriptional regulations in rat primary astrocytes.[77,84] We have also discovered that estrogen and selective estrogen receptor modulators attenuated Mn toxicity on astrocytic glutamate transporters by restoring the expression of GLT-1 and GLAST mRNA and proteins.[77,84]

10.4 Mechanism of Mn-induced Impairment of Astrocytic Glutamate Transporters

Although the precise mechanisms involved in Mn-induced impairment of glutamate transporters at the cellular levels remain to be elucidated, it appears to be closely associated with oxidative stress and neuroinflammation, which are known mechanisms for Mn neurotoxicity. Mn-induced oxidative stress impairs these astrocytic transporters.[9] Mn-induced release of inflammatory cytokines, such as TNF-α and IL-1β also reduces GLT-1 mRNA and protein levels.[65,70] In addition, Mn modulates signaling pathways and transcriptional regulation of glutamate transporters to decrease expression of these transporters, which will be discussed in the following sections.

10.4.1 Mn-activated Signaling Pathways in Astrocytes

Mn activates protein kinase C (PKC) signaling pathway by modulating the glutamate/glutamine cycle in the brain[80,109] by phosphorylation of PKC-α and -δ isoforms in astrocytes.[110] Recent studies revealed that phorbol ester-induced activation of PKC decreased glutamate uptake, and inhibition of PKC reversed the Mn-induced decrease of glutamate uptake as well as GLT-1 and GLAST protein levels.[109] Furthermore, Mn enhanced the interaction of GLT-1 with PKC-δ, indicating that PKC-δ isoform plays a critical role in glutamate turnover in astrocytes.[109] The mechanisms by which PKC decreases glutamate uptake involve the modulation of transporter activity and expression, as well as plasma membrane trafficking. Inhibition of caspase activity also attenuates Mn-induced reduction in glutamate uptake and protein levels of GLT-1 and GLAST, indicating that apoptotic pathways play a significant role in Mn-induced impairment of glutamate uptake and expression of glutamate transporters. Given that caspase-3 is activated in various Mn neurotoxicity models[63,111,112] and inhibition of caspase-3 or PKC-δ attenuates Mn neurotoxicity, caspase-3 dependent PKC-δ activation might be an important signaling event in Mn-induced disruption of glutamate transporters in astrocytes. Mn activates mitogen-activated protein kinase (MAPK), including extracellular signal-regulated kinase (ERK) and c-Jun-N-terminal kinase (JNK), to activate caspase-3 in astrocytes.[61] Mn exposure activates ERK and JNK in striatal and hippocampal slices as an early event in the immature rats.[113] Mn activates ERK in the striatum of developing rats, which is associated with motor dysfunction.[114] Moreover, Mn-induced cell swelling in astrocytes has been linked to the activation of MAPK/ERK and MAPK/p38 that increases the membrane expression of aquaporin-4.[88] In addition, Mn (at 10 μM, near the physiological concentration) potentiates the effects of inflammatory cytokines, such as TNF-α on NF-κB activation in astrocytes, leading to increased NOS2 production.[72]

10.4.2 Mn-induced Transcriptional Regulation of Glutamate Transporter GLT-1: Role of Yin Yang 1

Given that Mn reduces GLT-1 protein and mRNA levels, elucidating the mechanism of Mn-reduced expression of glutatame transporters at the trancritpional level is important to identify the molecular target of Mn action. Numerous studies have reported that NF-κB plays a critical role in mediating the stimulatory effects of various positive regulators of GLT-1, such as neuronal secreting factors,[115] epidermal growth factor, dibutyryl cyclic AMP,[71] ceftriaxone,[116] estrogen and tamoxifen.[117] However, there is no study reported for the negative regulatory mechanism of GLT-1 expression. Interestingly, Mn also activates NF-κB, but decreases GLT-1 expression, suggesting that other mechanisms must compromise with Mn's NF-κB activation to be able to repress GLT-1 promoter activity and subsequent reduction of GLT-1 mRNA and protein levels. Our latest findings revealed that a transcription factor yin yang 1 (YY1) is directly involved in Mn-activated NF-κB and Mn-induced repressiion on GLT-1.[65]

10.4.2.1   YY1 is a Negative Regulator of the GLT-1 Promoter

YY1 is a multifunctional transcription factor that can initiate, activate, or repress gene transcription, depending upon its interaction with available cofactors.[118] YY1 binds to a specific DNA sequence (CGCCATNTT) located in many different promoters[119] and contributes to GLT-1 regulation as a repressor[65] as GLT-1 (EAAT2) contains YY1 binding motifs in the promoter (Figure 10.1). It has been reported that glutamate increases YY1 binding to the GLAST promoter, leading to the decrease in GLAST-mediated glutamate uptake in Bergman glia cells.[120] YY1 is also recruited to the GLT-1 promoter when astrocyte elevated gene-1(AEG-1) induces repression of GLT-1

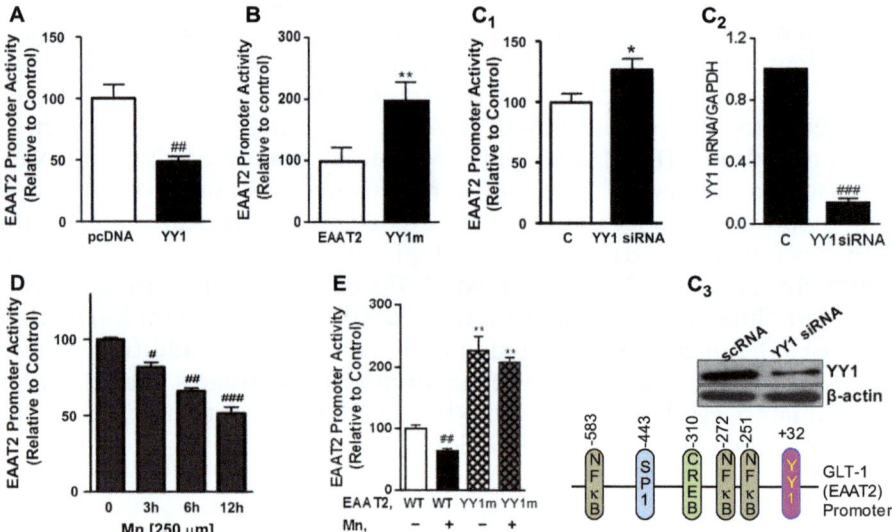

Figure 10.1   YY1 is a negative regulator of EAAT2 (YY1 consensus site is on +32 on the GLT-1 promoter) and it mediates Mn-induced repression of EAAT2. (A) Astrocytes were co-transfected overnight with 0.5 µg of EAAT2 luciferase plasmid and 0.1 µg of either control vector pcDNA or YY1, followed by luciferase assay to determine EAAT2 promoter activity. (B) YY1 consensus site (+34) in the EAAT2 promoter was mutated by site-directed mutagenesis, and the promoter activity of the YY1 mutant of EAAT2 was compared with the wild type EAAT2 by luciferase assay. (C) Astrocytes were transfected with YY1 siRNA or scrambled control siRNA for 48 h, followed by luciferase assay (C-1). The YY1 mRNA levels by qPCR (C-2) and YY1 protein levels by western blot (C-3) were measured to determine the efficiency of YY1 siRNA knockdown. (D) Astrocytes were treated with Mn (250 µM) for indicated time periods, and EAAT2 promoter activity was measured by luciferase assay. (E) After overnight transfection with wild type or YY1 mutant of EAAT2 promoter vectors, astrocytes were treated with Mn (250 µM) for 6 h, followed by luciferase assay. ($^{\#}p < 0.05$, $^{\#\#}p < 0.01$, $^{\#\#\#}p < 0.001$, $^{*}p < 0.05$, $^{**}p < 0.01$; ANOVA followed by Tukey's *post hoc* test; $N = 3$). (Adapted from Karki *et al.*[65])

promoter activity with parallel reduction of glutamate uptake in astrocytes.[121] We have studied the detailed mechanisms by which YY1 regulates GLT-1 promoter activity in relation to the positive regulator NF-κB and an epigenetic modifier histone deacetylases (HDACs). The findings indicate that YY1 is a critical repressor of the GLT-1 promoter as knockdown of YY1 with siRNA and mutation of YY1 binding sites in the GLT-1 promoter increased GLT-1 promoter activity, whereas YY1 overexpression decreased GLT-1 promoter activity.[65]

10.4.2.2 Mn Upregulates YY1 via NF-κB and YY1 Overrides the Positive NF-κB Action on GLT-1

Mn increases YY1 promoter activity as well as its mRNA/protein levels in astrocytes.[65] Intriguingly, Mn-activated NF-κB is a major regulator of YY1 promoter activity within 3 h of Mn exposure, but the positive regulation of GLT-1 *via* NF-κB by the positive regulators such as EGF takes much longer time to increase GLT-1 promoter activity and expression, requiring at least 24 h. These findings suggest that Mn-induced early NF-κB activation is likely attributing to YY1 upregulation. Moreover, under conditions in which both YY1 and NF-κB are activated, YY1 action dominates and overrides NF-κB's positive effect on GLT-1 promoter activity.[65] The proposed mechanism delineating that Mn-activated NF-κB upregulates YY1 expression which, in turn, represses GLT-1 promoter activity is described in Figure 10.2. Several findings indicate that Mn-induced oxidative stress and inflammation impairs glutamate transporters. For example, Mn-induced release of TNF-α from astrocytes likely mediates the Mn effect on YY1, as both Mn and TNF-α increased YY1 promoter activity as well as mRNA and protein levels in a

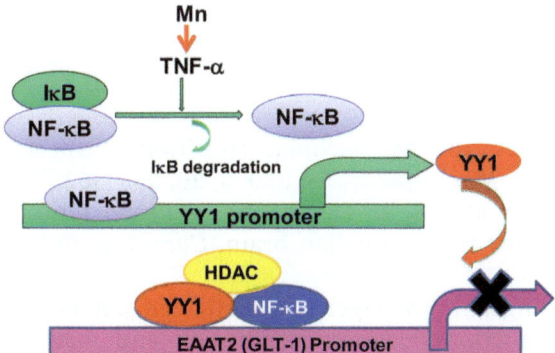

Figure 10.2   Proposed mechanism for Mn-induced repression of EAAT2. TNF-α is released by Mn which activates the NF-κB pathway, followed by YY1 activation. The upregulation of YY1 represses EAAT2 using HDACs as co-repressors. YY1 also physically interacts with NF-κB, inhibiting its positive regulation on EAAT2.
(Adapted from Karki *et al.*[65])

NF-κB dependent manner. Mn-induced oxidative stress may also repress GLT-1 promoter activity because oxidative stress has been reported to be a causal factor for impairment of GLT-1 in experimental and clinical studies.[122–125] YY1 also interacts with an epigenetic modifier HDAC1, which is enhanced by Mn,[65] indicating that HDACs serve as co-repressors of YY1 and inhibition of HDACs with pharmacological inhibitors increases GLT-1 promoter activity[126,127] and attenuates Mn-reduced GLT-1 promoter activity.[65]

10.5 Summary

Since Mn induces neurological disorders similar to PD, understanding the cellular and molecular mechanisms of Mn-induced neurotoxicity is important. Mn induces oxidative stress, mitochondrial impairment, neuroinflammation, apoptosis, and excitotoxicity. In particular, delineating the molecular mechanism of Mn-induced excitotoxic neuronal injury *via* impairment of astrocytic glutamate transporters is crucial in advancing our understanding of the role of impaired glutamate transport in triggering manganism. The findings that YY1 plays a critical role in Mn-induced repression of GLT-1 promoter activity as well as expression, cooperating with epigenetic modifier HDAC at the transcription level is highly valuable for understanding molecualr mechanisms of Mn-induced neurotoxicity. This can also lead to identification of novel molecular targets involved in Mn-induced dysregulation of glutamate transporters leading to neurotoxicity.

Acknowledgements

The present study was supported by NIH grants, NIGMS SC1 089630, NIEHS R01 10563, and NIH UL1 TR000445.

References

1. J. L. Aschner and M. Aschner, Nutritional aspects of manganese homeostasis, *Mol. Aspects Med.*, 2005, **26**, 353–362.
2. F. C. Wedler and R. B. Denman, Glutamine synthetase: the major Mn(II) enzyme in mammalian brain, *Curr. Top. Cell. Regul.*, 1984, **24**, 153–169.
3. W. C. Stallings, A. L. Metzger, K. A. Pattridge, J. A. Fee and M. L. Ludwig, Structure-function relationships in iron and manganese superoxide dismutases, *Free Radical Res. Commun.*, 1991, **12–13**, Pt 1, 259–268.
4. L. A. Bentle and H. A. Lardy, Interaction of anions and divalent metal ions with phosphoenolpyruvate carboxykinase, *J. Biol. Chem.*, 1976, **251**, 2916–2921.
5. J. Couper, On the effects of black oxide of manganese when inhaled into the lungs, *Br. Ann. Med. Pharmacol.*, 1837, **1**, 41–42.

6. C. C. Huang, N. S. Chu, C. S. Lu, J. D. Wang, J. L. Tsai, J. L. Tzeng, E. C. Wolters and D. B. Calne, Chronic manganese intoxication, *Arch. Neurol.*, 1989, **46**, 1104–1106.

7. I. Mena, O. Marin, S. Fuenzalida and G. C. Cotzias, Chronic manganese poisoning. Clinical picture and manganese turnover, *Neurology*, 1967, **17**, 128–136.

8. A. B. Bowman, G. F. Kwakye, E. H. Hernandez and M. Aschner, Role of manganese in neurodegenerative diseases, *J. Trace Elem. Med. Biol.*, 2011, **25**, 191–203.

9. K. Erikson and M. Aschner, Manganese causes differential regulation of glutamate transporter (GLAST) taurine transporter and metallothionein in cultured rat astrocytes, *Neurotoxicology*, 2002, **23**, 595–602.

10. K. M. Erikson and M. Aschner, Manganese neurotoxicity and glutamate-GABA interaction, *Neurochem. Int.*, 2003, **43**, 475–480.

11. N. C. Danbolt, Glutamate uptake, *Prog. Neurobiol.*, 2001, **65**, 1–105.

12. J. W. Finley and C. D. Davis, Manganese deficiency and toxicity: are high or low dietary amounts of manganese cause for concern?, *BioFactors*, 1999, **10**, 15–24.

13. R. M. Bowler, S. Nakagawa, M. Drezgic, H. A. Roels, R. M. Park, E. Diamond, D. Mergler, M. Bouchard, R. P. Bowler and W. Koller, Sequelae of fume exposure in confined space welding: a neurological and neuropsychological case series, *Neurotoxicology*, 2007, **28**, 298–311.

14. M. R. Flynn and P. Susi, Neurological risks associated with manganese exposure from welding operations–a literature review, *Int. J. Hyg. Environ. Health*, 2009, **212**, 459–469.

15. S. Montes, H. Riojas-Rodriguez, E. Sabido-Pedraza and C. Rios, Biomarkers of manganese exposure in a population living close to a mine and mineral processing plant in Mexico, *Environ. Res.*, 2008, **106**, 89–95.

16. Y. Rodriguez-Agudelo, H. Riojas-Rodriguez, C. Rios, I. Rosas, E. Sabido Pedraza, J. Miranda, C. Siebe, J. L. Texcalac and C. Santos-Burgoa, Motor alterations associated with exposure to manganese in the environment in Mexico, *Sci. Total Environ.*, 2006, **368**, 542–556.

17. R. Bast-Pettersen, D. G. Ellingsen, S. M. Hetland and Y. Thomassen, Neuropsychological function in manganese alloy plant workers, *Int. Arch. Occup. Environ. Health*, 2004, 77, 277–287.

18. D. Mergler, G. Huel, R. Bowler, A. Iregren, S. Belanger, M. Baldwin, R. Tardif, A. Smargiassi and L. Martin, Nervous system dysfunction among workers with long-term exposure to manganese, *Environ. Res.*, 1994, **64**, 151–180.

19. M. Bader, M. C. Dietz, A. Ihrig and G. Triebig, Biomonitoring of manganese in blood, urine and axillary hair following low-dose exposure during the manufacture of dry cell batteries, *Int. Arch. Occup. Environ. Health*, 1999, 72, 521–527.

20. A. K. Srivastava, B. N. Gupta, N. Mathur, R. C. Murty, N. Garg and S. V. Chandra, An investigation of metal concentrations in blood of industrial workers, *Vet. Hum. Toxicol.*, 1991, **33**, 280–282.

21. ATSDR, Agency for Toxic Substances and Disease Registry, *Toxicological Profile for Manganese*, Atlanta, GA, 2000, pp. 1–466.
22. G. Alves, J. Thiebot, A. Tracqui, T. Delangre, C. Guedon and E. Lerebours, Neurologic disorders due to brain manganese deposition in a jaundiced patient receiving long-term parenteral nutrition, *JPEN, J. Parenter. Enteral Nutr.*, 1997, **21**, 41–45.
23. R. F. Butterworth, L. Spahr, S. Fontaine and G. P. Layrargues, Manganese toxicity, dopaminergic dysfunction and hepatic encephalopathy, *Metab. Brain Dis.*, 1995, **10**, 259–267.
24. X. G. Kondakis, N. Makris, M. Leotsinidis, M. Prinou and T. Papapetropoulos, Possible health effects of high manganese concentration in drinking water, *Arch. Environ. Health*, 1989, **44**, 175–178.
25. M. Krachler and E. Rossipal, Concentrations of trace elements in extensively hydrolysed infant formulae and their estimated daily intakes, *Ann. Nutr. Metab.*, 2000, **44**, 68–74.
26. K. Sikk, S. Haldre, S. M. Aquilonius and P. Taba, Manganese-Induced Parkinsonism due to Ephedrone Abuse, *Parkinsons Dis.*, 2011, **2011**, 865319.
27. K. Sikk, P. Taba, S. Haldre, J. Bergquist, D. Nyholm, H. Askmark, T. Danfors, J. Sorensen, L. Thurfjell, R. Raininko, R. Eriksson, R. Flink, C. Farnstrand and S. M. Aquilonius, Clinical, neuroimaging and neurophysiological features in addicts with manganese-ephedrone exposure, *Acta Neurol. Scand.*, 2010, **121**, 237–243.
28. N. C. Burton and T. R. Guilarte, Manganese Neurotoxicity: Lessons Learned from Longitudinal Studies in Nonhuman Primates, *Environ. Health Perspect.*, 2009, **117**, 325–332.
29. C. D. Davis, L. Zech and J. L. Greger, Manganese metabolism in rats: an improved methodology for assessing gut endogenous losses, *Proc. Soc. Exp. Biol. Med.*, 1993, **202**, 103–108.
30. I. Mena, The role of manganese in human disease, *Ann. Clin. Lab. Sci.*, 1974, **4**, 487–491.
31. C. Au, A. Benedetto and M. Aschner, Manganese transport in eukaryotes: the role of DMT1, *Neurotoxicology*, 2008, **29**, 569–576.
32. M. Aschner and M. Gannon, Manganese (Mn) transport across the rat blood-brain barrier: saturable and transferrin-dependent transport mechanisms, *Brain Res. Bull.*, 1994, **33**, 345–349.
33. D. C. Dorman, K. A. Brenneman, A. M. McElveen, S. E. Lynch, K. C. Roberts and B. A. Wong, Olfactory transport: a direct route of delivery of inhaled manganese phosphate to the rat brain, *J. Toxicol. Environ. Health, Part A*, 2002, **65**, 1493–1511.
34. K. Itoh, M. Sakata, M. Watanabe, Y. Aikawa and H. Fujii, The entry of manganese ions into the brain is accelerated by the activation of N-methyl-D-aspartate receptors, *Neuroscience*, 2008, **154**, 732–740.
35. D. C. Dorman, M. F. Struve, M. W. Marshall, C. U. Parkinson, R. A. James and B. A. Wong, Tissue manganese concentrations in young

male rhesus monkeys following subchronic manganese sulfate inhalation, *Toxicol. Sci.*, 2006, **92**, 201–210.

36. K. M. Erikson and M. Aschner, Increased manganese uptake by primary astrocyte cultures with altered iron status is mediated primarily by divalent metal transporter, *Neurotoxicology*, 2006, **27**, 125–130.

37. N. Suarez and H. Eriksson, Receptor-mediated endocytosis of a manganese complex of transferrin into neuroblastoma (SHSY5Y) cells in culture, *J. Neurochem.*, 1993, **61**, 127–131.

38. G. Tholey, M. Ledig, P. Mandel, L. Sargentini, A. H. Frivold, M. Leroy, A. A. Grippo and F. C. Wedler, Concentrations of physiologically important metal ions in glial cells cultured from chick cerebral cortex, *Neurochem. Res.*, 1988, **13**, 45–50.

39. C. E. Gavin, K. K. Gunter and T. E. Gunter, Mn2+ sequestration by mitochondria and inhibition of oxidative phosphorylation, *Toxicol. Appl. Pharmacol.*, 1992, **115**, 1–5.

40. C. E. Gavin, K. K. Gunter and T. E. Gunter, Manganese and calcium transport in mitochondria: implications for manganese toxicity, *Neurotoxicology*, 1999, **20**, 445–453.

41. J. Y. Chen, G. C. Tsao, Q. Zhao and W. Zheng, Differential cytotoxicity of Mn(II) and Mn(III): special reference to mitochondrial [Fe-S] containing enzymes, *Toxicol. Appl. Pharmacol.*, 2001, **175**, 160–168.

42. K. Sriram, G. X. Lin, A. M. Jefferson, J. R. Roberts, O. Wirth, Y. Hayashi, K. M. Krajnak, J. M. Soukup, A. J. Ghio, S. H. Reynolds, V. Castranova, A. E. Munson and J. M. Antonini, Mitochondrial dysfunction and loss of Parkinson's disease-linked proteins contribute to neurotoxicity of manganese-containing welding fumes, *FASEB J.*, 2010, **24**, 4989–5002.

43. K. V. Rao and M. D. Norenberg, Manganese induces the mitochondrial permeability transition in cultured astrocytes, *J. Biol. Chem.*, 2004, **279**, 32333–32338.

44. L. E. Gonzalez, A. A. Juknat, A. J. Venosa, N. Verrengia and M. L. Kotler, Manganese activates the mitochondrial apoptotic pathway in rat astrocytes by modulating the expression of proteins of the Bcl-2 family, *Neurochem. Int.*, 2008, **53**, 408–415.

45. A. W. Dobson, K. M. Erikson and M. Aschner, Manganese neurotoxicity, *Ann. N. Y. Acad. Sci.*, 2004, **1012**, 115–128.

46. M. S. Desole, M. Miele, G. Esposito, R. Migheli, L. Fresu, G. De Natale and E. Miele, Dopaminergic system activity and cellular defense mechanisms in the striatum and striatal synaptosomes of the rat subchronically exposed to manganese, *Arch. Toxicol.*, 1994, **68**, 566–570.

47. A. W. Dobson, S. Weber, D. C. Dorman, L. K. Lash, K. M. Erikson and M. Aschner, Oxidative stress is induced in the rat brain following repeated inhalation exposure to manganese sulfate, *Biol. Trace Elem. Res.*, 2003, **93**, 113–126.

48. K. M. Erikson, D. C. Dorman, L. H. Lash and M. Aschner, Manganese inhalation by rhesus monkeys is associated with brain regional changes in biomarkers of neurotoxicity, *Toxicol. Sci.*, 2007, **97**, 459–466.

49. A. S. Hazell, L. Normandin, M. D. Norenberg, G. Kennedy and J. H. Yi, Alzheimer type II astrocytic changes following sub-acute exposure to manganese and its prevention by antioxidant treatment, *Neurosci. Lett.*, 2006, **396**, 167–171.
50. J. Donaldson, F. S. LaBella and D. Gesser, Enhanced autoxidation of dopamine as a possible basis of manganese neurotoxicity, *Neurotoxicology*, 1981, **2**, 53–64.
51. W. N. Sloot, J. Korf, J. F. Koster, L. E. De Wit and J. B. Gramsbergen, Manganese-induced hydroxyl radical formation in rat striatum is not attenuated by dopamine depletion or iron chelation *in vivo*, *Exp. Neurol.*, 1996, **138**, 236–245.
52. A. Benedetto, C. Au and M. Aschner, Manganese-induced dopaminergic neurodegeneration: insights into mechanisms and genetics shared with Parkinson's disease, *Chem. Rev.*, 2009, **109**, 4862–4884.
53. C. J. Chen and S. L. Liao, Oxidative stress involves in astrocytic alterations induced by manganese, *Exp. Neurol.*, 2002, **175**, 216–225.
54. D. Milatovic, Z. Yin, R. C. Gupta, M. Sidoryk, J. Albrecht, J. L. Aschner and M. Aschner, Manganese induces oxidative impairment in cultured rat astrocytes, *Toxicol. Sci.*, 2007, **98**, 198–205.
55. A. P. Stephenson, J. A. Schneider, B. C. Nelson, D. H. Atha, A. Jain, K. F. Soliman, M. Aschner, E. Mazzio and R. Renee Reams, Manganese-induced oxidative DNA damage in neuronal SH-SY5Y cells: attenuation of thymine base lesions by glutathione and N-acetylcysteine, *Toxicol. Lett.*, 2013, **218**, 299–307.
56. G. T. Ramesh, D. Ghosh and P. G. Gunasekar, Activation of early signaling transcription factor, NF-κB following low-level manganese exposure, *Toxicol. Lett.*, 2002, **136**, 151–158.
57. K. Wise, S. Manna, J. Barr, P. Gunasekar and G. Ramesh, Activation of activator protein-1 DNA binding activity due to low level manganese exposure in pheochromocytoma cells, *Toxicol. Lett.*, 2004, **147**, 237–244.
58. D. Trotti, N. C. Danbolt and A. Volterra, Glutamate transporters are oxidant-vulnerable: a molecular link between oxidative and excitotoxic neurodegeneration?, *Trends Pharmacol. Sci.*, 1998, **19**, 328–334.
59. V. J. Miralles, I. Martinez-Lopez, R. Zaragoza, E. Borras, C. Garcia, F. V. Pallardo and J. R. Vina, Na+ dependent glutamate transporters (EAAT1, EAAT2, and EAAT3) in primary astrocyte cultures: effect of oxidative stress, *Brain Res.*, 2001, **922**, 21–29.
60. Y. Hirata, Manganese-induced apoptosis in PC12 cells, *Neurotoxicol. Teratol.*, 2002, **24**, 639–653.
61. Z. Yin, J. L. Aschner, A. P. dos Santos and M. Aschner, Mitochondrial-dependent manganese neurotoxicity in rat primary astrocyte cultures, *Brain Res.*, 2008, **1203**, 1–11.
62. A. Alaimo, R. M. Gorojod and M. L. Kotler, The extrinsic and intrinsic apoptotic pathways are involved in manganese toxicity in rat astrocytoma C6 cells, *Neurochem. Int.*, 2011, **59**, 297–308.

63. C. Latchoumycandane, V. Anantharam, M. Kitazawa, Y. Yang, A. Kanthasamy and A. G. Kanthasamy, Protein kinase Cδ is a key downstream mediator of manganese-induced apoptosis in dopaminergic neuronal cells, *J. Pharmacol. Exp. Ther.*, 2005, **313**, 46–55.

64. M. K. Jang, H. S. Kim and Y. H. Chung, Clinical aspects of Tumor Necrosis Factor-α Signaling in Hepatocellular Carcinoma, *Curr. Pharm. Des.*, 2014, **20**, 2799–2808.

65. P. Karki, A. Webb, K. Smith, J. Johnson, Jr., K. Lee, D. S. Son, M. Aschner and E. Lee, Yin Yang 1 is a Repressor of Glutamate Transporter EAAT2 and it Mediates Manganese-induced Decrease of EAAT2 Expression in Astrocytes, *Mol. Cell. Biol.*, 2014, DOI: 10.1128/mcb.01176-13.

66. N. M. Filipov and C. A. Dodd, Role of glial cells in manganese neurotoxicity, *J. Appl. Toxicol.*, 2012, **32**, 310–317.

67. N. M. Filipov, R. F. Seegal and D. A. Lawrence, Manganese potentiates in vitro production of proinflammatory cytokines and nitric oxide by microglia through a nuclear factor κ B-dependent mechanism, *Toxicol. Sci.*, 2005, **84**, 139–148.

68. M. Liu, T. Cai, F. Zhao, G. Zheng, Q. Wang, Y. Chen, C. Huang, W. Luo and J. Chen, Effect of microglia activation on dopaminergic neuronal injury induced by manganese, and its possible mechanism, *Neurotoxic. Res.*, 2009, **16**, 42–49.

69. X. Liu, K. A. Sullivan, J. E. Madl, M. Legare and R. B. Tjalkens, Manganese-induced neurotoxicity: the role of astroglial-derived nitric oxide in striatal interneuron degeneration, *Toxicol. Sci.*, 2006, **91**, 521–531.

70. R. Sitcheran, P. Gupta, P. B. Fisher and A. S. Baldwin, Positive and negative regulation of EAAT2 by NF-κB: a role for N-myc in TNFα-controlled repression, *EMBO J.*, 2005, **24**, 510–520.

71. Z. Z. Su, M. Leszczyniecka, D. C. Kang, D. Sarkar, W. Chao, D. J. Volsky and P. B. Fisher, Insights into glutamate transport regulation in human astrocytes: cloning of the promoter for excitatory amino acid transporter 2 (EAAT2), *Proc. Natl. Acad. Sci. U. S. A.*, 2003, **100**, 1955–1960.

72. J. A. Moreno, K. A. Sullivan, D. L. Carbone, W. H. Hanneman and R. B. Tjalkens, Manganese potentiates nuclear factor-κB-dependent expression of nitric oxide synthase 2 in astrocytes by activating soluble guanylate cyclase and extracellular responsive kinase signaling pathways, *J. Neurosci. Res.*, 2008, **86**, 2028–2038.

73. M. Spranger, S. Schwab, S. Desiderato, E. Bonmann, D. Krieger and J. Fandrey, Manganese augments nitric oxide synthesis in murine astrocytes: a new pathogenetic mechanism in manganism?, *Exp. Neurol.*, 1998, **149**, 277–283.

74. J. A. Moreno, K. M. Streifel, K. A. Sullivan, M. E. Legare and R. B. Tjalkens, Developmental exposure to manganese increases adult susceptibility to inflammatory activation of glia and neuronal protein nitration, *Toxicol. Sci.*, 2009, **112**, 405–415.

75. S. L. Liao, Y. C. Ou, S. Y. Chen, A. N. Chiang and C. J. Chen, Induction of cyclooxygenase-2 expression by manganese in cultured astrocytes, *Neurochem. Int.*, 2007, **50**, 905–915.
76. E. P. Brouillet, L. Shinobu, U. McGarvey, F. Hochberg and M. F. Beal, Manganese injection into the rat striatum produces excitotoxic lesions by impairing energy metabolism, *Exp. Neurol.*, 1993, **120**, 89–94.
77. E. S. Lee, M. Sidoryk, H. Jiang, Z. Yin and M. Aschner, Estrogen and tamoxifen reverse manganese-induced glutamate transporter impairment in astrocytes, *J. Neurochem.*, 2009, **110**, 530–544.
78. D. Centonze, P. Gubellini, G. Bernardi and P. Calabresi, Impaired excitatory transmission in the striatum of rats chronically intoxicated with manganese, *Exp. Neurol.*, 2001, **172**, 469–476.
79. M. Morello, A. Canini, P. Mattioli, R. P. Sorge, A. Alimonti, B. Bocca, G. Forte, A. Martorana, G. Bernardi and G. Sancesario, Sub-cellular localization of manganese in the basal ganglia of normal and manganese-treated rats An electron spectroscopy imaging and electron energy-loss spectroscopy study, *Neurotoxicology*, 2008, **29**, 60–72.
80. M. Sidoryk-Wegrzynowicz and M. Aschner, Role of astrocytes in manganese mediated neurotoxicity, *BMC Pharmacol. Toxicol.*, 2013, **14**, 23.
81. C. W. Olanow, P. F. Good, H. Shinotoh, K. A. Hewitt, F. Vingerhoets, B. J. Snow, M. F. Beal, D. B. Calne and D. P. Perl, Manganese intoxication in the rhesus monkey: a clinical, imaging, pathologic, and biochemical study, *Neurology*, 1996, **46**, 492–498.
82. K. V. Rama Rao, P. V. Reddy, A. S. Hazell and M. D. Norenberg, Manganese induces cell swelling in cultured astrocytes, *Neurotoxicology*, 2007, **28**, 807–812.
83. R. B. Tjalkens, M. J. Zoran, B. Mohl and R. Barhoumi, Manganese suppresses ATP-dependent intercellular calcium waves in astrocyte networks through alteration of mitochondrial and endoplasmic reticulum calcium dynamics, *Brain Res.*, 2006, **1113**, 210–219.
84. E. Lee, M. Sidoryk-Wegrzynowicz, Z. Yin, A. Webb, D. S. Son and M. Aschner, Transforming growth factor-α mediates estrogen-induced upregulation of glutamate transporter GLT-1 in rat primary astrocytes, *Glia*, 2012, **60**, 1024–1036.
85. M. Sidoryk-Wegrzynowicz, E. Lee, J. Albrecht and M. Aschner, Manganese disrupts astrocyte glutamine transporter expression and function, *J. Neurochem.*, 2009, **110**, 822–830.
86. D. Krieger, S. Krieger, O. Jansen, P. Gass, L. Theilmann and H. Lichtnecker, Manganese and chronic hepatic encephalopathy, *Lancet*, 1995, **346**, 270–274.
87. M. Norenberg, The role of astrocytes in hepatic encephalopathy, *Neurochem. Pathol.*, 1987, **6**, 13–33.
88. K. V. Rao, A. R. Jayakumar, P. V. Reddy, X. Tong, K. M. Curtis and M. D. Norenberg, Aquaporin-4 in manganese-treated cultured astrocytes, *Glia*, 2010, **58**, 1490–1499.

89. F. Zhao, T. Cai, M. Liu, G. Zheng, W. Luo and J. Chen, Manganese induces dopaminergic neurodegeneration via microglial activation in a rat model of manganism, *Toxicol. Sci.*, 2009, **107**, 156–164.

90. P. L. Crittenden and N. M. Filipov, Manganese-induced potentiation of in vitro proinflammatory cytokine production by activated microglial cells is associated with persistent activation of p38 MAPK, *Toxicol. In Vitro*, 2008, **22**, 18–27.

91. P. L. McGeer and E. G. McGeer, Glial reactions in Parkinson's disease, *Mov. Disord.*, 2008, **23**, 474–483.

92. K. M. Streifel, J. A. Moreno, W. H. Hanneman, M. E. Legare and R. B. Tjalkens, Gene deletion of nos2 protects against manganese-induced neurological dysfunction in juvenile mice, *Toxicol. Sci.*, 2012, **126**, 183–192.

93. P. Zhang, K. M. Lokuta, D. E. Turner and B. Liu, Synergistic dopaminergic neurotoxicity of manganese and lipopolysaccharide: differential involvement of microglia and astroglia, *J. Neurochem.*, 2010, **112**, 434–443.

94. H. M. Abdul, M. A. Sama, J. L. Furman, D. M. Mathis, T. L. Beckett, A. M. Weidner, E. S. Patel, I. Baig, M. P. Murphy, H. LeVine, 3rd, S. D. Kraner and C. M. Norris, Cognitive decline in Alzheimer's disease is associated with selective changes in calcineurin/NFAT signaling, *J. Neurosci.*, 2009, **29**, 12957–12969.

95. R. Sattler and M. Tymianski, Molecular mechanisms of glutamate receptor-mediated excitotoxic neuronal cell death, *Mol. Neurobiol.*, 2001, **24**, 107–129.

96. A. Lau and M. Tymianski, Glutamate receptors, neurotoxicity and neurodegeneration, *Pflugers Arch*, 2010, **460**, 525–542.

97. A. L. Sheldon and M. B. Robinson, The role of glutamate transporters in neurodegenerative diseases and potential opportunities for intervention, *Neurochem. Int.*, 2007, **51**, 333–355.

98. K. Tanaka, K. Watase, T. Manabe, K. Yamada, M. Watanabe, K. Takahashi, H. Iwama, T. Nishikawa, N. Ichihara, T. Kikuchi, S. Okuyama, N. Kawashima, S. Hori, M. Takimoto and K. Wada, Epilepsy and exacerbation of brain injury in mice lacking the glutamate transporter GLT-1, *Science*, 1997, **276**, 1699–1702.

99. K. Watase, K. Hashimoto, M. Kano, K. Yamada, M. Watanabe, Y. Inoue, S. Okuyama, T. Sakagawa, S. Ogawa, N. Kawashima, S. Hori, M. Takimoto, K. Wada and K. Tanaka, Motor discoordination and increased susceptibility to cerebellar injury in GLAST mutant mice, *Eur. J. Neurosci.*, 1998, **10**, 976–988.

100. J. D. Rothstein, M. Dykes-Hoberg, C. A. Pardo, L. A. Bristol, L. Jin, R. W. Kuncl, Y. Kanai, M. A. Hediger, Y. Wang, J. P. Schielke and D. F. Welty, Knockout of glutamate transporters reveals a major role for astroglial transport in excitotoxicity and clearance of glutamate, *Neuron*, 1996, **16**, 675–686.

101. J. D. Rothstein, M. Van Kammen, A. I. Levey, L. J. Martin and R. W. Kuncl, Selective loss of glial glutamate transporter GLT-1 in amyotrophic lateral sclerosis, *Ann. Neurol.*, 1995, **38**, 73–84.

102. Z. Liang, J. Valla, S. Sefidvash-Hockley, J. Rogers and R. Li, Effects of estrogen treatment on glutamate uptake in cultured human astrocytes derived from cortex of Alzheimer's disease patients, *J. Neurochem.*, 2002, **80**, 807–814.

103. V. L. Rao, A. Dogan, K. G. Todd, K. K. Bowen, B. T. Kim, J. D. Rothstein and R. J. Dempsey, Antisense knockdown of the glial glutamate transporter GLT-1, but not the neuronal glutamate transporter EAAC1, exacerbates transient focal cerebral ischemia-induced neuronal damage in rat brain, *J. Neurosci.*, 2001, **21**, 1876–1883.

104. Z. C. Ye, J. D. Rothstein and H. Sontheimer, Compromised glutamate transport in human glioma cells: reduction-mislocalization of sodium-dependent glutamate transporters and enhanced activity of cystine-glutamate exchange, *J. Neurosci.*, 1999, **19**, 10767–10777.

105. A. S. Hazell and M. D. Norenberg, Manganese decreases glutamate uptake in cultured astrocytes, *Neurochem. Res.*, 1997, **22**, 1443–1447.

106. K. M. Erikson, R. L. Suber and M. Aschner, Glutamate/aspartate transporter (GLAST), taurine transporter and metallothionein mRNA levels are differentially altered in astrocytes exposed to manganese chloride, manganese phosphate or manganese sulfate, *Neurotoxicology*, 2002, **23**, 281–288.

107. L. Mutkus, J. L. Aschner, V. Fitsanakis and M. Aschner, The in vitro uptake of glutamate in GLAST and GLT-1 transfected mutant CHO-K1 cells is inhibited by manganese, *Biol. Trace Elem. Res.*, 2005, **107**, 221–230.

108. K. M. Erikson, D. C. Dorman, L. H. Lash and M. Aschner, Duration of airborne-manganese exposure in rhesus monkeys is associated with brain regional changes in biomarkers of neurotoxicity, *Neurotoxicology*, 2008, **29**, 377–385.

109. M. Sidoryk-Wegrzynowicz, E. Lee and M. Aschner, Mechanism of Mn(II)-mediated dysregulation of glutamine-glutamate cycle: focus on glutamate turnover, *J. Neurochem.*, 2012, **122**, 856–867.

110. M. Sidoryk-Wegrzynowicz, E. Lee, N. Mingwei and M. Aschner, Disruption of astrocytic glutamine turnover by manganese is mediated by the protein kinase C pathway, *Glia*, 2011, **59**, 1732–1743.

111. H. S. Chun, H. Lee and J. H. Son, Manganese induces endoplasmic reticulum (ER) stress and activates multiple caspases in nigral dopaminergic neuronal cells, SN4741, *Neurosci. Lett.*, 2001, **316**, 5–8.

112. M. Kitazawa, V. Anantharam, Y. Yang, Y. Hirata, A. Kanthasamy and A. G. Kanthasamy, Activation of protein kinase C δ by proteolytic cleavage contributes to manganese-induced apoptosis in dopaminergic cells: protective role of Bcl-2, *Biochem. Pharmacol.*, 2005, **69**, 133–146.

113. T. V. Peres, D. Z. Pedro, F. M. de Cordova, M. W. Lopes, F. M. Goncalves, C. B. Mendes-de-Aguiar, R. Walz, M. Farina, M. Aschner and R. B. Leal, *In vitro* manganese exposure disrupts MAPK signaling pathways in striatal and hippocampal slices from immature rats, *BioMed Res. Int.*, 2013, **2013**, 769295.

114. F. M. Cordova, A. S. Aguiar, Jr., T. V. Peres, M. W. Lopes, F. M. Goncalves, A. P. Remor, S. C. Lopes, C. Pilati, A. S. Latini, R. D. Prediger, K. M. Erikson, M. Aschner and R. B. Leal, *In vivo* manganese exposure modulates Erk, Akt and Darpp-32 in the striatum of developing rats, and impairs their motor function, *PLoS One*, 2012, 7, e33057.

115. M. Ghosh, Y. Yang, J. D. Rothstein and M. B. Robinson, Nuclear factor-κB contributes to neuron-dependent induction of glutamate transporter-1 expression in astrocytes, *J. Neurosci.*, 2011, 31, 9159–9169.

116. J. D. Rothstein, S. Patel, M. R. Regan, C. Haenggeli, Y. H. Huang, D. E. Bergles, L. Jin, M. Dykes Hoberg, S. Vidensky, D. S. Chung, S. V. Toan, L. I. Bruijn, Z. Z. Su, P. Gupta and P. B. Fisher, β-lactam antibiotics offer neuroprotection by increasing glutamate transporter expression, *Nature*, 2005, 433, 73–77.

117. P. Karki, A. Webb, K. Smith, K. Lee, D. S. Son, M. Aschner and E. Lee, cAMP response element-binding protein (CREB) and nuclear factor κB mediate the tamoxifen-induced up-regulation of glutamate transporter 1 (GLT-1) in rat astrocytes, *J. Biol. Chem.*, 2013, 288, 28975–28986.

118. Z. Deng, P. Cao, M. M. Wan and G. Sui, Yin Yang 1: a multifaceted protein beyond a transcription factor, *Transcription*, 2010, 1, 81–84.

119. M. J. Thomas and E. Seto, Unlocking the mechanisms of transcription factor YY1: are chromatin modifying enzymes the key?, *Gene*, 1999, 236, 197–208.

120. S. Rosas, M. A. Vargas, E. Lopez-Bayghen and A. Ortega, Glutamate-dependent transcriptional regulation of GLAST/EAAT1: a role for YY1, *J. Neurochem.*, 2007, 101, 1134–1144.

121. S. G. Lee, K. Kim, T. P. Kegelman, R. Dash, S. K. Das, J. K. Choi, L. Emdad, E. L. Howlett, H. Y. Jeon, Z. Z. Su, B. K. Yoo, D. Sarkar, S. H. Kim, D. C. Kang and P. B. Fisher, Oncogene AEG-1 promotes glioma-induced neurodegeneration by increasing glutamate excitotoxicity, *Cancer Res.*, 2011, 71, 6514–6523.

122. M. Matos, E. Augusto, C. R. Oliveira and P. Agostinho, Amyloid-beta peptide decreases glutamate uptake in cultured astrocytes: involvement of oxidative stress and mitogen-activated protein kinase cascades, *Neuroscience*, 2008, 156, 898–910.

123. M. Hayashi, N. Arai, J. Satoh, H. Suzuki, K. Katayama, K. Tamagawa and Y. Morimatsu, Neurodegenerative mechanisms in subacute sclerosing panencephalitis, *J. Child Neurol.*, 2002, 17, 725–730.

124. M. Hayashi, S. Araki, N. Arai, S. Kumada, M. Itoh, K. Tamagawa, M. Oda and Y. Morimatsu, Oxidative stress and disturbed glutamate transport in spinal muscular atrophy, *Brain Dev.*, 2002, 24, 770–775.

125. M. Hayashi, M. Itoh, S. Araki, S. Kumada, K. Shioda, K. Tamagawa, T. Mizutani, Y. Morimatsu, M. Minagawa and M. Oda, Oxidative stress and disturbed glutamate transport in hereditary nucleotide repair disorders, *J. Neuropathol. Exp. Neurol.*, 2001, 60, 350–356.

126. S. Baltan, S. P. Murphy, C. A. Danilov, A. Bachleda and R. S. Morrison, Histone deacetylase inhibitors preserve white matter structure and function during ischemia by conserving ATP and reducing excitotoxicity, *J. Neurosci.*, 2011, **31**, 3990–3999.
127. M. Itoh, T. Hiroi, N. Nishibori, T. Sagara, S. Her, M. S. Lee and K. Morita, Trichostatin A enhances glutamate transporter GLT-1 mRNA levels in C6 glioma cells via neurosteroid-mediated cell differentiation, *J. Mol. Neurosci.*, 2013, **49**, 21–27.

CHAPTER 11

Impairment of Glutamine/ Glutamate-γ-aminobutyric Acid Cycle in Manganese Toxicity in the Central Nervous System

MARTA SIDORYK-WEGRZYNOWICZ[*a] AND
MICHAEL ASCHNER[b]

[a] Cambridge Centre of Brain Repair, University of Cambridge, Cambridge, UK; [b] Department of Mol Pharmacol, Albert Einstein College of Medicine, Bronx, NY, USA
*Email: martasidoryk@gmail.com

11.1 Glutamine Content and Regional Distribution and its Role in the Central Nervous System

Glutamine (Gln) content in the extracellular fluid (microdialysates) (ECF) or in the cerebrospinal fluid (CSF) ($\sim$0.5–1 mM) exceeds, by at least one order of magnitude, the extracellular contents of other amino acids in these compartments. Gln is the most abundant amino acid in the plasma (at 600–800 µmol) and exhibits extremely rapid cellular turnover rates.[1,2] This important amino acid serves in multiple roles in the mammalian brain – as essential precursor in nucleotide, glucose and amino sugar biosynthesis, glutathione homeostasis, and protein synthesis. Additionally, the growth of proliferating cells, such as fibroblasts, lymphocytes, enterocytes and neoplastic tissue depends on Gln, which provides oxidative energy fuel.[3]

Issues in Toxicology No. 22
Manganese in Health and Disease
Edited by Lucio G. Costa and Michael Aschner
© The Royal Society of Chemistry 2015
Published by the Royal Society of Chemistry, www.rsc.org

In the central nervous system (CNS), Gln supports tissue homeostasis by participation in the intercellular substrate cycles. Gln metabolism generates compounds that serve as direct precursors of the tricarboxylic acid cycle (TCA), thus contributing to the supply of the high-energy demand in the brain. Gln serves as a precursor of the amino acid neuro-transmitters glutamate (Glu), γ-amino butyric acid (GABA).[4] In addition to its importance in neurotransmission, Gln plays a significant role in ammonia assimilation and detoxification within the CNS, where glia-specific enzyme Gln synthetase (GS) catalyzed glutamine from glutamate and ammonia. As GS is an essential part of a complex astrocyte-neuron signaling processes, deregulation of GS dependent metabolites is likely to greatly affect brain function. A number of different pathologies (*e.g.* epilepsy, stroke, Alzheimer's disease (AD), multiple sclerosis, schizophrenia, manganism) are associated with aberrant expression patterns and alternation in activity of GS.[5]

11.2   Glutamine Transporting Systems in General

Glutamine transport across cell membranes has been extensively studied. In the past few years, a number of glutamine carriers have been cloned and their molecular and functional properties have been characterized. Gln transporting systems in mammalian brain are classified into two distinct groups: sodium-dependent and sodium-independent.

In general, four systems have been identified: A (alanine-preferring); L (leucine-preferring); ASC (alanine-, serine-, cysteine-preferring); and N (glutamine-, asparagine-, histidine-preferring). Systems A, ASC and N are sodium-dependent, while system L is sodium-independent. Systems A and N tolerate lithium as a substitute for sodium. All four of the systems have been identified in the brain, indicating that the CNS possesses multiple pathways for Gln transportation across the plasma membrane.[6]

System A transporters SNAT2 and SNAT1 recognize the prototypic substrate methyl-amino-iso-butyric acid (MeAIB) and a broad range of neutral amino acids. Both transporters are pH sensitive. SNAT2 is expressed in neurons and other cells, but has relatively low affinity for Gln compared to other amino acids. SNAT1 expression is restricted to brain and heart, suggesting that its specialized properties are congruent with the function of these organs. Moreover, this transporter has higher affinity for Gln than for other amino acids. The mRNA for SNAT1 and SNAT2 is expressed in cultured astrocytes, but the transporters remain inactive unless stimulated by amino acid depletion.[7] Both SNAT1 and SNAT2 share about 60% homology with SNAT3, a variant of System N.[8]

Among the transporters identified so far, SNAT3 possesses the highest affinity for Gln. Functional analysis demonstrated that SNAT3 transports neutral amino acid and Na^+ in exchange for H^+, and mediates Gln uptake as well as efflux. Gln transport *via* SNAT3 is electroneutral, reversible, and involves the co-transport of one Na^+ and the antiport of one H^+. Efflux of

Gln is particularly favoured at high intracellular Na^+ and low extracellular pH.[9] Structurally and mechanistically, another variant of system N, SNAT5, is closely related to the SNAT3 transporter. The SNAT5 transporter differs in substrate profile; SNAT3 can recognize classic system N substrates such as Gln, asparagine and histidine, while SNAT5 favours serine.[10]

Transporters of ASC system, ASCT1 and ASCT2 variants exhibit Na^+-dependent nature and high affinity for alanine, serine and cysteine with distinct substrate selectivity. In addition to the common substrates of ASC transporters, ASCT2 accepts glutamine and asparagine as high-affinity substrates, while ASCT1 does not. ASCT1 and ASCT2 are obligatory exchangers for amino acids and cannot function in a unidirectional manner. ASCT2 efficiently transports Gln in cell lines and neoplastic tissues. ASCT2 takes up Gln with high affinity and transports a wide panel of other zwitterionic as well as bulky/branch-chain amino acids.[11]

System L, a member of the *SLC7* gene family, is a Na^+-independent transport system for glutamine, although Gln is not recognized as a preferred substrate.[11] Low-affinity, high-capacity uptake of Gln mediated by system L is observed both in astrocytes and neurons. System L functions as an amino acid exchanger and can mediate Gln release in the presence of extracellular amino acid substrates, such as alanine and leucine. Members of this family include LAT1 and LAT2 transporters and both variants require disulfide linkage to 4F2hc, which appears to facilitate their translocation to the plasma membrane.[12] Experiments with rodent and human LAT isoforms revealed that glutamine is recognized with higher efficiency by LAT2 than by LAT1. Both LAT1 and LAT2 mediate Gln transport at the blood–brain barrier (BBB).[13] Other transport systems for glutamine include y + L consisted with y + LAT1 (SLC7A7) and y + LAT2 (SLC7A6) members. Both variants associate with 4F2hc to mediate Na^+-independent transport of cationic amino acids and Na^+-dependent uptake of neutral amino acids. Transport by family members of this system can be inhibited competitively by arginine and lysine.[11]

11.3 Glutamate: Role in Central Nervous System and Transporters

Glutamate is the principal excitatory neurotransmitter in the CNS and mediates up to 80% of synaptic transmission in the brain.[14] Glu is the product of the deamination of glutamine by mitochondrial enzyme phosphate-activated glutaminase. Furthermore, Glu is converted to alpha-ketoglutaric acid (αKG) through deamination by glutamate dehydrogenase or by transamination and metabolized through the TCA to succinate, fumarate, and malate (see Figure 11.1).[15] Glu transporting system is essential for maintaining optimal extracellular Glu concentrations that do not activate Glu transporters and receptors and extracellular accumulation of Glu causes toxicity in the CNS. Impaired uptake of glutamate by astrocytes

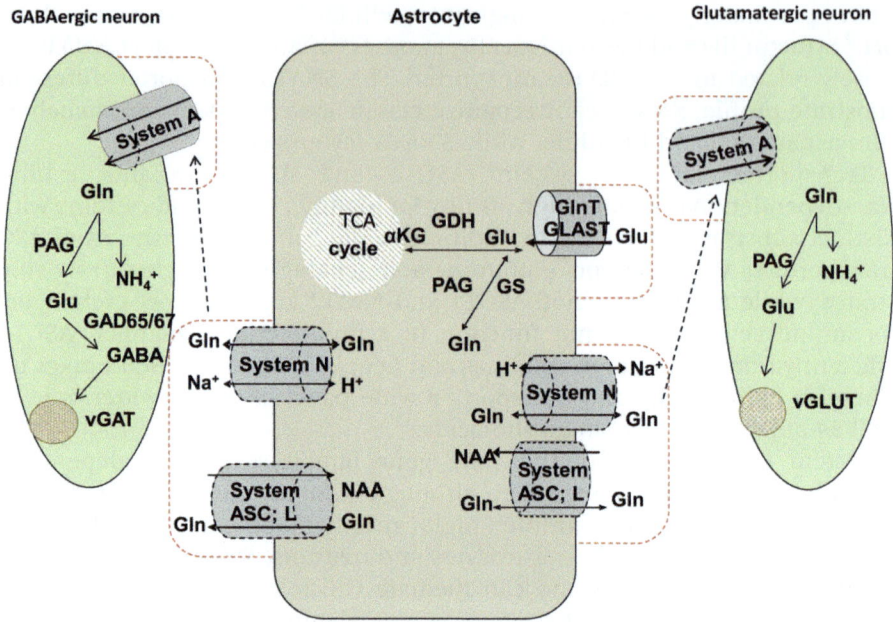

Figure 11.1   Schematic representation showing the amino acid transport systems with suggested Mn involvement in glutamine and glutamate transport related to the glutamine/glutamate-γ-aminobutyric acid cycle. The red dashed frames indicate areas being affected by manganese (discussed in the later sections of the review). Abbreviations: αKG: Alpha-ketoglutaric acid; GABA: γ-amino butyric acid; GAD: glutamic acid decarboxylase; GAD65/67: glutamate decarboxylase isoforms 65/67; GDH: glutamate dehydrogenase; Gln: glutamine; GLAST: glutamate-aspartate transporter: GLT1, glutamate transporter; Glu: glutamate; GS: glutamine synthetase; NAA: neutral amino acids; PAG: phosphate activated glutaminase; VGAT: vesicular GABA transporter; VGLUT: vesicular glutamate transporter.

may slow the synthesis of glutathione and leave the brain vulnerable to oxidative damage.[16] In general, optimal regulation of glutamate transporters is critical for normal CNS function as the dysfunction of these transporters leads to various neurological diseases, such as epilepsy, cerebral ischemia, AD, Parkinson disease (PD), and schizophrenia.[17,18]

Glutamate uptake in mammalian cells is mediated by an amino solute carrier family 1 (SLC1) represented by five high affinity glutamate/aspartate transporters named EAAT (excitatory amino acid transporter) 1–5 (also known as a SLC1A1–A3, SLC1A6, and SLC1A7). The substrate is taken up in co-transport with $3Na^+$ and $1H^+$ and efflux is facilitated by K^+ antiport. Among the five subtypes of human EAATs, two glial subtypes, EAAT1 and EAAT2 (GLAST and GLT-1, respectively, in rodents), are localized primarily in astrocytes and responsible for the clearance of excess glutamate from the synaptic cleft.[19] Notably, astrocytic glutamate transporters are responsible for over 80% of glutamate uptake in the brain, which prevent excitotoxic glutamate elevation and determine neuronal survival.

GLT1 is a high-affinity glutamate transporter expressed in astrocytes and is very abundant in the brain in various brain regions, especially in Cereb Cortex and hippocampus. Studies revealed that glutamate uptake by crude cortical synaptosomes of GLT1-knockout mice are significantly reduced compare to wild-type mice, suggesting that GLT1 is primarily responsible for the removal of glutamate from the synapse. Moreover, histological analysis of GLT1-knockout mice showed selective neuronal degeneration in the hippocampal CA1 region, suggesting a crucial role of GLT1 in neuroprotection.[20]

Another high-affinity glutamate transporter, GLAST, is predominantly concentrated in the cerebellum[21] and is expressed in astrocytes and cerebellar Bergmann glia.[22] The functional relevance of this transporter was confirmed in animal studies where knockout of GLAST induced an increase of cerebellar damage after brain injury and a mild motor discoordination, suggesting an active role of GLAST in preventing excitotoxic damage after acute brain injury.[23]

11.4   The Glutamine/Glutamate-γ-Aminobutyric Acid

Gln functions *via* the glutamine/glutamate-γ-aminobutyric acid cycle (GGC) and the reactions within the cycle are carried out in both astrocytic and neuronal compartments (see Figure 11.1). Glial cell processes are uniquely positioned to receive information from neurons and to signal back to them *via* the release of neurotransmitters. After exocytotic release at synaptic terminals, Glu is taken up by surrounding astrocytes *via* the glia-specific Glu transporter 1 GLT-1 and Glu–aspartate transporter GLAST. In astrocytes, Glu is converted to Gln by the Gln synthetase. In turn, Gln is transferred into the neurons mainly by SNAT1 and SNAT2 transporters belonging to system A. In neurons Gln is catalyzed to Glu *via* phosphate-dependent glutaminase (PAG). Glu is subsequently converted into GABA *via* decarboxylation by glutamic acid decarboxylase (GAD). In both astrocytes and neurons, Glu is used for the synthesis of α-ketoglutaric acid, a substrate for the tricarboxylic acid cycle by oxidative deamination mediated by Glu dehydrogenase (GDH).[4] Additionally, Glu is stored in synaptic vesicles at presynaptic terminals by the vesicular glutamate transporters VGluT1, VGluT2, and VGluT3 (SLC17A6, SLC17A7, and SLC17A8, respectively). Glutamate is released into the synaptic cleft to act on glutamate AMPA and NMDA receptors.

High affinity glutamate transporters play essential roles in removing released glutamate from the synaptic cleft and are major players of the GGC cycle.[24] These transporters are also crucial for maintaining the extracellular glutamate concentration of the cerebrospinal fluid below neurotoxic levels. Disruption of GGC has been reported in numerous pathological conditions, such as epilepsy, cerebral ischemia, AD, PD, and manganism.[4]

11.5 Manganese

11.5.1 Essentiality and Toxicity

Manganese (Mn) is a trace metal commonly found in the environment. It is an essential dietary nutrient and is required for growth, development, and maintenance of normal physiological function. Mn is necessary for a variety of metabolic functions, including lipid, protein, and carbohydrate metabolism. However, exposure to high levels of Mn may lead to a neurological disorder that shares many similarities with PD, and is referred to as manganism.[25] Exposure to Mn in the general population occurs from the automobile combustion of gasoline containing methylcyclopentadienyl manganese tricarbonyl (MMT) or from organic Mn-containing pesticides, such as manganese ethylene-bis-dithiocarbamate.[26] The most frequent cause of Mn neurotoxicity is believed to be chronic occupational exposure to high levels of inhalable manganese (1–5 mg Mn m^3), which is commonly associated with occupations such as mining, smelting, battery manufacturing, and steel production.[27] Further, individuals with some medical conditions like liver dysfunction or total parenteral nutrition exhibit increased Mn blood levels and neurological dysfunctions. The accumulation of Mn in basal ganglia is responsible for the form of parkinsonism with overlapping, but distinct clinical features with those observed in idiopathic Parkinson's disease.[28] Within the central nervous system, exposure to high levels of Mn is accompanied by Mn accumulation in specific brain regions that are highly sensitive to oxidative injury, including the global pallidus (GP), striatum (STR) and substantia nigra (SN).[29,30] Additionally, studies suggest that working memory is affected by chronic Mn exposition, and that this could be associated with the alteration in brain structures such as the frontal and pariental cortex that are responsible for the memory network.[31]

11.5.2 Transporting System

It has been shown that Mn can be transported within the CNS *via* high affinity metal transporters such as calcium (Ca) and iron (Fe) transporters. Some of these transporters include the divalent metal transporter (DMT1), which belongs to the family of natural resistance-associated macrophage protein (NRAMP);[32,33] ZIP-8, a member of the solute carrier-39; voltage regulated and store-operated Ca^{2+} channels; ionotropic glutatmate receptor Ca^{2+} channels or transferrin receptor (TfR).[34] Moreover additional channels/transporters have been identified as potentially Mn-recognizing transporters, including the divalent metal/bicarbonate ion symporters ZIP8 and ZIP14, various calcium channels, the solute carrier-39 (SLC39) family of zinc transporters, park9/ATP13A2, the magnesium transporter hip14, and the transient receptor potential melastatin 7 (TRPM7).[35–38] It is not clear what the relative contribution of each of these transporters in Mn translocation is; however, it is most likely that optimal Mn concentration in normal physiological conditions is maintained by the involvement of all mentioned carriers.

11.5.3 Manganese Effects on Astrocytes Function and Astrocyte–Neuronal Integrity

Astrocytes are essential for normal neuronal function in CNS, such as regulation and maintenance of the blood–brain barrier and neuronal homeostasis. Astrocytes ensure trophic, energy and metabolic support to neurons, and modulate synaptic activity and neuronal plasticity or viability. Glia cells actively participate in neuronal excitability and survival by controlling the extracellular levels of ions, amino acids, and neurotransmitters.[39] In the glutamatergic system, astrocytes secrete glutamate in response to activation, modulate glutamate receptor expression, and remove glutamate from the synaptic cleft by glutamate transporters.[40] It has been demonstrated that an increase in brain ammonia elicits an increase in glutamine synthetase activity and that the glia cells detoxifies ammonia by converting it to glutamine because GS is primarily localized in that cell.[41] As glutamine metabolism is a significant part of complex astrocyte–neuron integration, changes in expression and/or activity of GS may lead to the astrocyte–neuron processes dysfunction. Recently, it has clearly been shown in many neuropathogical conditions (*e.g.* Mn neurotoxicity) that disruption of glutamine metabolism in astrocytes affects neuronal viability and function.

Mn-mediated neuropathological changes are manifested by gliosis and neuronal loss predominately in the globus pallidus and the substantia nigra pars reticulata or the striatum. Astrocytes express high-capacity transporters for Mn, and are therefore considered as an initial target for Mn-induced neurotoxicity.[42] In response to chronic Mn exposure, astrocytes display morphological changes characterized as astrocytosis and adopt a reactive phenotype, which is manifested by enlarged, pale nuclei and margination of chromatin.[43] Furthermore, Mn increases the expression of the glia or astrocytic markers: S100 and glial fibrillary acidic protein (GFAP) in several rat brain structures.[44] The majority of Mn (60–70%) is sequestered in mitochondria, leading to mitochondrial dysfunction and energy depletion.[45] GSH, the main antioxidant constituent, plays an important role in the detoxification of ROS and neutralization of organic hydroperoxides. Notably, GSH levels are lower in neurons than in astrocytes and cysteine derived from astrocytes is essential for the maintenance of stable GSH levels in neurons. In general, neuronal stores of GSH are largely dependent upon astrocytic stores, and neurons are more sensitive than astrocytes to oxidative stress.[46] Several studies demonstrated that Mn promotes failure of astrocytes to maintain antioxidant defence mechanisms *via* disruption of GSH synthesis.[47] *In vivo* interneuron injury in striatal and pallidal regions of Mn-exposured mice is associated with increased number of reactive astrocytes expressing inducible forms of nitric oxide synthase (iNOS) in the same brain areas.[48] Treatment of astrocytes with Mn increases uptake of L-arginine, an iNOS substrate, leading to increased ROS generation as a consequence of nitric oxide (NO) production and deterioration of cellular antioxidant capacity and energy metabolism.

Mammalian brain energy metabolism depends on the oxidation of glucose (Glc). Sodium-dependent translocation of Glu into the astrocytes and activation of Na^+/K^+-ATPase stimulate Glc uptake and glycolytic pathways.[49] In the cerebral, the stoichiometry between oxidative Glc metabolism and Glu cycling is close to $1:1$, which could suggest that the majority of cortical energy production supports functional glutamatergic neuronal activity. Interestingly, both astrocytes and neurons take up and metabolize glucose, but astrocytes are more capable of increasing glycolytic activity and lactate formation than neurons.[50] Glucose metabolism not only produces reduced equivalents *via* activity of the Krebs cycle for subsequent ATP formation, but also leads to the biosynthesis of neurotransmitters, including glutamate, aspartate, and GABA.[51] Furthermore, replenishment of Krebs cycle intermediates *via* anaplerotic processes is necessary for continued ATP synthesis. Astrocytes are uniquely capable of anaplerosis due to their preferential expression of pyruvate carboxylase (PC), which converts pyruvate to oxaloacetate.[52] Because PC is not present in neurons, these cells are incapable of anaplerosis and therefore dependent on astrocytes for glutamate synthesis. Notably, Mn plays an important role in brain energy metabolism and glutamatergic action by affecting PC.[53] Mn mediates changes in major astrocytic metabolic pathways by inhibiting the astrocyte-specific enzyme, GS.[54] Furthermore, studies demonstrated that acute manganese exposure not only disrupts glia cells, but also induces dysfunction of dopaminergic and GABAergic neurons within the basal ganglia.[55] These effects of Mn may cause changes in astrocytic and neuronal integrity and affect astrocyte-mediated neuroprotection. For example, comparative NMR study on co-cultures revealed that Mn disturbs astrocytic function and mediates failure of astrocytes to support neurons with substrates for energy and neurotransmitter metabolism.[53]

11.5.4   Manganese Involvement in PKCδ Signalling

Protein kinase C (PKC) isozymes comprise a family of serine–threonine kinases that regulate diverse cellular function, such as differentiation, proliferation, tumorigenesis, apoptosis, and survival.[56] PKCδ isoform has been shown to be critical regulator of pro-apoptosis. Mn exposure has been demonstrated to mediate the specific phosphorylation of PKCδ isozyme and significantly increase PKC activity. In addition, astrocytes transfected with shRNA against PKCδ are significantly less sensitive to Mn compared to those transfected with control shRNA.[57]

It has been reported that apoptotic signals induce nuclear translocation of PKCδ and activate caspase-3. In the nucleus, caspase-3 mediates proteolytic cleavage of PKCδ, giving rise to the PKCδ catalytic fragment (PKCδ CF). This constitutive PKCδ activation induces chromatin condensation and DNA fragmentation, which supports a role for PKCδ cleavage in the induction of apoptosis.[58] Notably, as observed in several *in vitro* and *in vivo* models, Mn toxicity is associated with caspase-3 activation and PKCδ nuclear accumulation.[59] It has been established that caspase-3-dependent PKCδ activation not only contributes to the neuronal apoptosis, but also has a significant feedback regulatory role in amplification of the apoptotic cascade during neurotoxic stress upon Mn treatment.[60] PKCδ plays a crucial role in the cellular response to ROS and oxidative stress.[61] Notably, as discussed before, oxidative stress, particularly in mitochondria, is a common feature of Mn toxicity.[62] All together, these evidences could represent a link of manganese to PKCδ/caspase3 pathway.

11.6 Manganese and GGC

11.6.1 Manganese and Glutamate Transporting System

Mn has been widely reported to induce the impairment of glutamate transporters, GLAST and GLT1 expression and function (see Table 11.1).[63,64] PKC signalling has been invoked as a potential mechanism leading to transporters inactivation and glial dysfunction in Mn-mediated disruption of Glu transport. Study revealed that PKC stimulation by α-phorbol 12-myristate (PMA) significantly decreases astrocytic Glu uptake, while treatment with the general PKC inhibitor, bisindolylmaleimide II (BIS II) or with rottlerin (ROT) and Gö6976 (specific inhibitors of the PKCδ and PKCα, respectively) protects astrocytes from the Mn-induced downregulation of Glu

Table 11.1 Transporters involved in the glutamine/glutamate-γ-aminobutyric acid cycle and their sensitivity to manganese. Abbreviations: Ala: alanine; Asp: asparagine; Cys: cysteine; Glu: glutamate; Gln: glutamine; His: histidine; Ile: isoleucine; Leu: leucine; Met: methionie; Pro: proline; Trp: tryptophan; Tyr: tyrosine; Val: valine.

SLC	Acronym	System	Substrates	Sensitivity to manganese
SLC1A5	ASCT2	System ASC	Ala, Ser, Cys, Thr, Gln	+
SLC7A5	LAT1/4F2hc	System L	His, Met, Leu, Ile, Val, Phe, Tyr, Trp	+
SLC7A8	LAT2/4F2hc	System L	All neutral amino acids, except Pro	−
SLC38A1	SNAT1	System A	Gly, Ala, Asp, Cys, Glu, His, Met	+
SLC38A2	SNAT2	System A	Gly, Pro, Ala, Ser, Cys, Gln, Asp, His, Met	−
SLC38A3	SNAT3	System N	Gln, Asn, His	+
SLC38A5	SNAT5	System N	Gln, Asn, His, Ala	+
SLC1A2	EAAT2	System X-AG	Asp, Glu	+
SLC1A3	EAAT1	System X-AG	Asp, Glu	+

transport. In addition, inhibition of caspase-3 by Z-VAD-FMK reverses Mn-dependent disruption of Glu transport. Regarding the uptake, Mn decreases GLT1 protein level, an effect that is blocked by Bis II, ROT, Gö6976, Z-VAD-FMK, while Mn-dependent downregulation of GLAST protein level is reversed in the presence of PKCα and casapase-3 inhibitors. Interestingly, co-immunoprecipitation studies demonstrated association of GLT-1, not GLAST, with the PKCδ and PKCα isoforms and Mn-induced specific increasing in PKCδ-GLT-1 interaction.[59]

Several studies revealed contradictory effects of PKCs on GLAST activity,[67] showing that activation of PKC increases, decreases, or has no effect on GLAST expression and/or function.[65] More consistent evidence has been observed in the case of the GLT1 transporter.[6,67] For example, studies on C6 glioma cell line stably transfected with GLT1 or on primary cultures that endogenously express GLT1 showed that activation of PKC rapidly decreases GLT1 cell surface expression. In stably transfected MDCK cells, activation of PKC deregulates GLT1 mediated activity.[66,67] As mentioned earlier, recent study on the role of manganese in glutamate turnover shows that both GLAST and GLT1 are deregulated by PKC in primary culture of astrocytes;[59] however, contribution of the α and δ PKC isoforms for each of these transporters seems to be different.

11.6.2 Manganese and Glutamine Turnover

Research has demonstrated that Mn inhibits the initial net uptake of Gln in a concentration-dependent manner in primary culture of astrocyte and induces deregulation of SNAT3, SNAT2 (Systems N), ASCT2 (System ASC), and LAT2 (Systems L), transporters expression and function (see Table 11.1).[68,69]

The contribution of PKC signalling to Mn-induced dyshomeostasis in Gln transport has been investigated in a primary culture of astrocytes.[70] Research has revealed that PKC inhibition by BIS II blocked the Mn-induced down-regulation in Glu uptake. Treatment of primary astrocyte cultures with a PKC stimulator decreases Gln uptake mediated by Systems ASC and N, and decreases expression of ASCT2 and SNAT3 protein levels in cell lysates and in plasma membranes.[70] It is noteworthy that both Mn-affected transporters contain putative PKC phosphorylation sites, which are conserved in the human, rat, and mouse. For example, few PKC isoforms, including PKCδ and PKCα, phosphorylate SNAT3 at the 52 position. A recent *in situ* study revealed that PKC activation induces phosphorylation of SNAT3 and regulates its membrane trafficking and protein degradation.[71] In addition, internalization of SNAT3 upon PKC-dependent phosphorylation corroborates with the evidence from *Xenopus laevis* oocytes model where PKC activation reduced Vmax of the Glu-mediated Gln uptake activity.[72] Furthermore, increased binding of PKCδ to ASCT2 and SNAT3 upon exposure to Mn has been identified by co-immunoprecipitation research. Taken together, these findings suggest a prominent role for PKCδ in Mn-mediated disruption of Gln turnover,[57] and combined with findings on Mn's influence on Glu transport,[59] are consistent with abnormal GGC cycling function by Mn at two key steps, including Gln and Glu transport, *via* a homologous PKC-dependent pattern.

11.6.3 Manganese Involvement in SNAT3 Expression and Function

Astroglial transporter SNAT3, a major component of GGC cycle, regulates extracellular levels of glutamine and shows dynamic membrane trafficking. Research showed that prolonged PKC activation not only promotes SNAT3 internalization by a caveolin dependent mechanism, but also increases its ubiquitination.[71] Furthermore, both PKC and ubiquitination pathways have been identified as a possible mechanism underlying Mn-mediated SNAT3 degradation.[57,70]

SNAT3 and many members of the SLC6 family are highly regulated, involving multiple protein–protein interactions.[73] These include direct interactions between the transporter and another protein(s) or involve a post-translational modification, such as phosphorylation mediated by PKC(s) or ubiquitination of a transporter, usually at the N- or C-terminus, followed by recognition of the modified sequence by other proteins.[74]

Ubiquitination involves the sequential action of several enzymes, including an ubiquitin-activating enzyme E1, an ubiquitin-conjugating enzyme E2, and an ubiquitin-protein ligase E3. Nedd4-2 (neuronal precursor cell expressed, developmentally downregulated 4-2), a member of the HECT (homology to the E6-associated protein C terminus) family of E3 ubiquitin ligases, is a physiologically important regulator of numerous membrane channels and transporters by promoting their ubiquitination. Poly-ubiquitination triggers transporter sorting and targeting into endosomal compartments, followed by recycling back to the plasma membrane degradation, or activates proteasome-dependent protein degradation machinery. Further, phosphorylation of Nedd4-2 and E3 ligases represents a powerful regulatory mechanism for altering the fate of ion channels and other targets for ubiquitination.[75] It has been demonstrated that SGK1 (serum- and glucocorticoid-regulated kinase 1) decreases the interaction between Nedd4-2 and target proteins by phosphorylation of ubiquitin ligase at three residues.[76] Interestingly, in an *X. laevis* oocyte expression system, SNAT3 is downregulated by Nedd4-2 and this effect is blocked by SGK1.[77] These findings, together with evidence that Mn decreases SGK1 expression and phosphorylation,[70] suggest involvement of Nedd4-2/SGK1 signalling followed by ubiquitination machinery in the Mn-mediated degradation of SNAT3.

PKC isoforms and their substrates regulate a variety of membrane proteins and in particular transporters. Many neurotransmitter transporters are withdrawn from the surface into endosomal compartments after treatment of cells with agents that activate PKC.[74] In most cases, this is accompanied by increases in transporter phosphorylation on the amino-terminus or the carboxyl-terminus. It has been shown that PKC activation regulates dopamine transporter (DAT) function by its internalization and degradation in an ubiquitin-dependent manner. Furthermore, RNA interference (RNAi) analysis shows the essential and specific role of Nedd4-2 in PKC-mediated endocytosis of DAT, where knockdown of Nedd4-2 results in a significant reduction in PKC-dependent ubiquitination of DAT.[78] Research demonstrated that PKC activation induces hyper-ubiquitination and increases the association of SNAT3 with Nedd4-2 in a primary culture of astrocytes. PKC stimulation impairs system N-mediated Gln uptake by astrocytes and significantly decreases SNAT3 protein level in plasma membrane, while, as mentioned before, Mn exposure leads to the PKCs activation.[70] In addition, both prolonged stimulation of PKC and exposition to Mn decreases SNAT3 protein levels.[71] Further, gene expression profiling experiments in non-human primates demonstrated that Mn exposure alternates the expression of genes associated with protein folding and turnover *via* a ubiquitin/proteasome system.[79] Together, these evidences could represent a link of Mn to PKC-regulation in GGC cycle because phosphorylation by PKC is frequently a way to target proteins for ubiquitination and further lysosomal or proteasomal degradation.

11.7 Summary

Glutamine metabolism in the healthy brain initiates a complex chain of metabolic events, including synthesis of the neurotransmitter amino acids glutamate and γ-aminobutyric acid. Accordingly, disrupted glutamine turnover may affect the amino acid neurotransmitter balance, the disturbances of which contribute to the neuropathologic manifestations of manganism. Manganism has been considered as a metabolic syndrome related to impairment of glutamate transport and more recent glutamine/glutamate-γ-aminobutyric acid cycle. *In vivo* and *in vitro* studies demonstrated that Mn evokes mitochondrial abnomalities, oxidative/nitrosative stress, and morphological/functional changes of astrocytes, a major component of the GGC cycle. Mn effectively increases abnormalities in the glutamine metabolism and turnover between glia and neurons. *In vitro* research revealed that Mn significantly decreases the activity of the major carrier of Glu–SNAT3 *via* the ubiquitination-dependent mechanisms. In addition, Mn mediates disruption of glutamate uptake from the synapse increasing the chances of glutamate-mediated excitotoxicity to surrounding neurons. There appear to be common signalling targets of Mn in GGC cycling in glial cells. Namely, the PKC signalling is affected by Mn in glutamine and glutamate transporters expression and function. The evidences described here not only contribute to understanding the mechanism by which Mn disrupts astrocytes function and astrocyte–neurons intercommunication, but may also potentially lead to the development of novel therapeutic interventions in animal models of manganese toxicity.

References

1. M. Erecinska and I. A. Silver, Metabolism and role of glutamate in mammalian brain, *Prog. Neurobiol.*, 1990, **35**, 245–296.
2. D. Darmaun, D. E. Matthews and D. M. Bier, Glutamine and glutamate kinetics in humans, *Am. J. Physiol.*, 1986, **251**, E117–E126.
3. H. Eagle, Nutrition needs of mammalian cells in tissue culture, *Science*, 1955, **122**, 501–514.
4. L. K. Bak, A. Schousboe and H. S. Waagepetersen, The glutamate/GABA-glutamine cycle: aspects of transport, neurotransmitter homeostasis and ammonia transfer, *J. Neurochem.*, 2006, **98**, 641–653.
5. C. F. Rose, A. Verkhratsky and V. Parpura, Astrocyte glutamine synthetase: pivotal in health and disease, *Biochem. Soc. Trans.*, 2013, **41**, 1518–1524.
6. H. Sershen and A. Lajtha, Inhibition pattern by analogs indicates the presence of ten or more transport systems for amino acids in brain cells, *J. Neurochem.*, 1979, **32**, 719–726.
7. M. Melone, F. Quagliano, P. Barbaresi, H. Varoqui, J. D. Erickson and F. Conti, Localization of the glutamine transporter SNAT1 in rat cerebral cortex and neighboring structures, with a note on its localization in human cortex, *Cereb. Cortex*, 2004, **14**, 562–574.
8. R. J. Reimer, F. A. Chaudhry, A. T. Gray and R. H. Edwards, Amino acid transport system A resembles system N in sequence but differs in mechanism, *Proc. Natl. Acad. Sci. U. S. A.*, 2000, **97**, 7715–7720.
9. A. Broer, A. Albers, I. Setiawan, R. H. Edwards, F. A. Chaudhry, F. Lang, C. A. Wagner and S. Broer, Regulation of the glutamine transporter SN1 by extracellular pH and intracellular sodium ions, *J. Physiol.*, 2002, **539**, 3–14.
10. T. Nakanishi, M. Sugawara, W. Huang, R. G. Martindale, F. H. Leibach, M. E. Ganapathy, P. D. Prasad and V. Ganapathy, Structure, function, and tissue expression pattern of human SN2, a subtype of the amino acid transport system N, *Biochem. Biophys. Res. Commun.*, 2001, **281**, 1343–1348.
11. B. P. Bode, Recent molecular advances in mammalian glutamine. transport, *J. Nutr.*, 2001, **131**, 2475S–2485S, discussion 2486S–2477S.
12. M. Pineda, E. Fernandez, D. Torrents, R. Estevez, C. Lopez, M. Camps, J. Lloberas, A. Zorzano and M. Palacin, Identification of a membrane protein, LAT-2, that Co-expresses with 4F2 heavy chain, an L-type amino acid transport activity with broad specificity for small and large zwitterionic amino acids, *J. Biol. Chem.*, 1999, **274**, 19738–19744.
13. J. Xiang, S. R. Ennis, G. E. Abdelkarim, M. Fujisawa, N. Kawai and R. F. Keep, Glutamine transport at the blood-brain and blood-cerebrospinal fluid barriers, *Neurochem. Int.*, 2003, **43**, 279–288.
14. D. A. Coulter and T. Eid, Astrocytic regulation of glutamate homeostasis in epilepsy, *Glia*, 2012, **60**, 1215–1226.
15. A. C. Yu, A. Schousboe and L. Hertz, Metabolic fate of 14C-labeled glutamate in astrocytes in primary cultures, *J. Neurochem.*, 1982, **39**, 954–960.

16. J. Lewerenz, P. Albrecht, M. L. Tien, N. Henke, S. Karumbayaram, H. I. Kornblum, M. Wiedau-Pazos, D. Schubert, P. Maher and A. Methner, Induction of Nrf2 and xCT are involved in the action of the neuroprotective antibiotic ceftriaxone in vitro, *J. Neurochem.*, 2009, **111**, 332–343.

17. A. L. Sheldon and M. B. Robinson, The role of glutamate transporters in neurodegenerative diseases and potential opportunities for intervention, *Neurochem. Int.*, 2007, **51**, 333–355.

18. C. Zoia, T. Cogliati, E. Tagliabue, G. Cavaletti, G. Sala, G. Galimberti, I. Rivolta, V. Rossi, L. Frattola and C. Ferrarese, Glutamate transporters in platelets: EAAT1 decrease in aging and in Alzheimer's disease, *Neurobiol. Aging*, 2004, **25**, 149–157.

19. M. L. Mayer and G. L. Westbrook, The physiology of excitatory amino acids in the vertebrate central nervous system, *Prog. Neurobiol.*, 1987, **28**, 197–276.

20. K. Tanaka, K. Watase, T. Manabe, K. Yamada, M. Watanabe, K. Takahashi, H. Iwama, T. Nishikawa, N. Ichihara, T. Kikuchi, S. Okuyama, N. Kawashima, S. Hori, M. Takimoto and K. Wada, Epilepsy and exacerbation of brain injury in mice lacking the glutamate transporter GLT-1, *Science*, 1997, **276**, 1699–1702.

21. T. Storck, S. Schulte, K. Hofmann and W. Stoffel, Structure, expression, and functional analysis of a Na(+)-dependent glutamate/aspartate transporter from rat brain, *Proc. Natl. Acad. Sci. U. S. A.*, 1992, **89**, 10955–10959.

22. J. D. Rothstein, L. Martin, A. I. Levey, M. Dykes-Hoberg, L. Jin, D. Wu, N. Nash and R. W. Kuncl, Localization of neuronal and glial glutamate transporters, *Neuron*, 1994, **13**, 713–725.

23. K. Watase, K. Hashimoto, M. Kano, K. Yamada, M. Watanabe, Y. Inoue, S. Okuyama, T. Sakagawa, S. Ogawa, N. Kawashima, S. Hori, M. Takimoto, K. Wada and K. Tanaka, Motor discoordination and increased susceptibility to cerebellar injury in GLAST mutant mice, *Eur. J. Neurosci.*, 1998, **10**, 976–988.

24. R. J. Vandenberg and R. M. Ryan, Mechanisms of glutamate transport, *Physiol. Rev.*, 2013, **93**, 1621–1657.

25. M. M. Finkelstein and M. Jerrett, A study of the relationships between Parkinson's disease and markers of traffic-derived and environmental manganese air pollution in two Canadian cities, *Environ. Res.*, 2007, **104**, 420–432.

26. K. S. Crump, Manganese exposures in Toronto during use of the gasoline additive, methylcyclopentadienyl manganese tricarbonyl, *J. Exposure Sci. Environ. Epidemiol.*, 2000, **10**, 227–239.

27. A. B. Santamaria, C. A. Cushing, J. M. Antonini, B. L. Finley and F. S. Mowat, State-of-the-science review: Does manganese exposure during welding pose a neurological risk?, *J. Toxicol. Environ. Health, Part B*, 2007, **10**, 417–465.

28. T. R. Guilarte, Manganese and Parkinson's disease: a critical review and new findings, *Cien. Saude Colet.*, 2011, **16**, 4549–4566.

29. R. G. Lucchini, C. J. Martin and B. C. Doney, From manganism to manganese-induced parkinsonism: a conceptual model based on the evolution of exposure, *NeuroMol. Med.*, 2009, **11**, 311–321.
30. A. M. McKinney, R. W. Filice, M. Teksam, S. Casey, C. Truwit, H. B. Clark, C. Woon and H. Y. Liu, Diffusion abnormalities of the globi pallidi in manganese neurotoxicity, *Neuroradiology*, 2004, **46**, 291–295.
31. J. S. Schneider, C. Williams, M. Ault and T. R. Guilarte, Chronic manganese exposure impairs visuospatial associative learning in non-human primates, *Toxicol. Lett.*, 2013, **221**, 146–151.
32. H. Gunshin, B. Mackenzie, U. V. Berger, Y. Gunshin, M. F. Romero, W. F. Boron, S. Nussberger, J. L. Gollan and M. A. Hediger, Cloning and characterization of a mammalian proton-coupled metal-ion transporter, *Nature*, 1997, **388**, 482–488.
33. M. D. Garrick, K. G. Dolan, C. Horbinski, A. J. Ghio, D. Higgins, M. Porubcin, E. G. Moore, L. N. Hainsworth, J. N. Umbreit, M. E. Conrad, L. Feng, A. Lis, J. A. Roth, S. Singleton and L. M. Garrick, DMT1: a mammalian transporter for multiple metals, *Biometals*, 2003, **16**, 41–54.
34. L. He, K. Girijashanker, T. P. Dalton, J. Reed, H. Li, M. Soleimani and D. W. Nebert, ZIP8, member of the solute-carrier-39 (SLC39) metal-transporter family: characterization of transporter properties, *Mol. Pharmacol.*, 2006, **70**, 171–180.
35. A. Riccio, C. Mattei, R. E. Kelsell, A. D. Medhurst, A. R. Calver, A. D. Randall, J. B. Davis, C. D. Benham and M. N. Pangalos, Cloning and functional expression of human short TRP7, a candidate protein for store-operated Ca^{2+} influx, *J. Biol. Chem.*, 2002, **277**, 12302–12309.
36. S. S. Kannurpatti, P. G. Joshi and N. B. Joshi, Calcium sequestering ability of mitochondria modulates influx of calcium through glutamate receptor channel, *Neurochem. Res.*, 2000, **25**, 1527–1536.
37. L. Davidsson, B. Lonnerdal, B. Sandstrom, C. Kunz and C. L. Keen, Identification of transferrin as the major plasma carrier protein for manganese introduced orally or intravenously or after in vitro addition in the rat, *J. Nutr.*, 1989, **119**, 1461–1464.
38. M. Aschner and M. Gannon, Manganese (Mn) transport across the rat blood-brain barrier: saturable and transferrin-dependent transport mechanisms, *Brain Res. Bull.*, 1994, **33**, 345–349.
39. A. Volterra and J. Meldolesi, Astrocytes, from brain glue to communication elements: the revolution continues, *Nat. Rev. Neurosci.*, 2005, **6**, 626–640.
40. A. Araque, G. Carmignoto and P. G. Haydon, Dynamic signaling between astrocytes and neurons, *Annu. Rev. Physiol.*, 2001, **63**, 795–813.
41. C. Cudalbu, B. Lanz, J. M. Duarte, F. D. Morgenthaler, Y. Pilloud, V. Mlynarik and R. Gruetter, Cerebral glutamine metabolism under hyperammonemia determined in vivo by localized (1)H and (15)N NMR spectroscopy, *J. Cereb. Blood Flow Metab.*, 2012, **32**, 696–708.
42. M. Aschner, M. Gannon and H. K. Kimelberg, Manganese uptake and efflux in cultured rat astrocytes, *J. Neurochem.*, 1992, **58**, 730–735.

43. A. S. Hazell and M. D. Norenberg, Ammonia and manganese increase arginine uptake in cultured astrocytes, *Neurochem. Res.*, 1998, **23**, 869–873.

44. J. Henriksson and H. Tjalve, Manganese taken up into the CNS via the olfactory pathway in rats affects astrocytes, *Toxicol. Sci.*, 2000, **55**, 392–398.

45. C. J. Chen and S. L. Liao, Oxidative stress involves in astrocytic alterations induced by manganese, *Exp. Neurol.*, 2002, **175**, 216–225.

46. A. Y. Shih, D. A. Johnson, G. Wong, A. D. Kraft, L. Jiang, H. Erb, J. A. Johnson and T. H. Murphy, Coordinate regulation of glutathione biosynthesis and release by Nrf2-expressing glia potently protects neurons from oxidative stress, *J. Neurosci.*, 2003, **23**, 3394–3406.

47. S. Weber, D. C. Dorman, L. H. Lash, K. Erikson, K. E. Vrana and M. Aschner, Effects of manganese (Mn) on the developing rat brain: oxidative-stress related endpoints, *Neurotoxicology*, 2002, **23**, 169–175.

48. M. Spranger, S. Schwab, S. Desiderato, E. Bonmann, D. Krieger and J. Fandrey, Manganese augments nitric oxide synthesis in murine astrocytes: a new pathogenetic mechanism in manganism?, *Exp. Neurol.*, 1998, **149**, 277–283.

49. P. J. Magistretti, Regulation of glycogenolysis by neurotransmitters in the central nervous system, *Diabete Metab.*, 1988, **14**, 237–246.

50. H. Marrif and B. H. Juurlink, Astrocytes respond to hypoxia by increasing glycolytic capacity, *J. Neurosci. Res.*, 1999, **57**, 255–260.

51. N. R. Sibson, A. Dhankhar, G. F. Mason, D. L. Rothman, K. L. Behar and R. G. Shulman, Stoichiometric coupling of brain glucose metabolism and glutamatergic neuronal activity, *Proc. Natl. Acad. Sci. U. S. A.*, 1998, **95**, 316–321.

52. R. P. Shank, G. S. Bennett, S. O. Freytag and G. L. Campbell, Pyruvate carboxylase: an astrocyte-specific enzyme implicated in the replenishment of amino acid neurotransmitter pools, *Brain Res.*, 1985, **329**, 364–367.

53. C. Zwingmann, D. Leibfritz and A. S. Hazell, Energy metabolism in astrocytes and neurons treated with manganese: relation among cell-specific energy failure, glucose metabolism, and intercellular trafficking using multinuclear NMR-spectroscopic analysis, *J. Cereb. Blood Flow Metab.*, 2003, **23**, 756–771.

54. K. M. Erikson, D. C. Dorman, V. Fitsanakis, L. H. Lash and M. Aschner, Alterations of oxidative stress biomarkers due to in utero and neonatal exposures of airborne manganese, *Biol. Trace Elem. Res.*, 2006, **111**, 199–215.

55. G. D. Stanwood, D. B. Leitch, V. Savchenko, J. Wu, V. A. Fitsanakis, D. J. Anderson, J. N. Stankowski, M. Aschner and B. McLaughlin, Manganese exposure is cytotoxic and alters dopaminergic and GABAergic neurons within the basal ganglia, *J. Neurochem.*, 2009, **110**, 378–389.

56. Y. Nishizuka, The role of protein kinase C in cell surface signal transduction and tumour promotion, *Nature*, 1984, **308**, 693–698.

57. M. Sidoryk-Wegrzynowicz, E. Lee, N. Mingwei and M. Aschner, Disruption of astrocytic glutamine turnover by manganese is mediated by the protein kinase C pathway, *Glia*, 2011, **59**, 1732–1743.

58. M. F. Denning, Y. Wang, S. Tibudan, S. Alkan, B. J. Nickoloff and J. Z. Qin, Caspase activation and disruption of mitochondrial membrane potential during UV radiation-induced apoptosis of human keratinocytes requires activation of protein kinase C, *Cell Death Differ.*, 2002, **9**, 40–52.

59. M. Sidoryk-Wegrzynowicz, E. Lee and M. Aschner, Mechanism of Mn(II)-mediated dysregulation of glutamine-glutamate cycle: focus on glutamate turnover, *J. Neurochem.*, 2012, **122**, 856–867.

60. C. Latchoumycandane, V. Anantharam, M. Kitazawa, Y. Yang, A. Kanthasamy and A. G. Kanthasamy, Protein kinase Cdelta is a key downstream mediator of manganese-induced apoptosis in dopaminergic neuronal cells, *J. Pharmacol. Exp. Ther.*, 2005, **313**, 46–55.

61. X. Sun, F. Wu, R. Datta, S. Kharbanda and D. Kufe, Interaction between protein kinase C delta and the c-Abl tyrosine kinase in the cellular response to oxidative stress, *J. Biol. Chem.*, 2000, **275**, 7470–7473.

62. S. Rivera-Mancia, C. Rios and S. Montes, Manganese accumulation in the CNS and associated pathologies, *Biometals*, 2011, **24**, 811–825.

63. K. M. Erikson, R. L. Suber and M. Aschner, Glutamate/aspartate transporter (GLAST), taurine transporter and metallothionein mRNA levels are differentially altered in astrocytes exposed to manganese chloride, manganese phosphate or manganese sulfate, *Neurotoxicology*, 2002, **23**, 281–288.

64. L. Mutkus, J. L. Aschner, V. Fitsanakis and M. Aschner, The in vitro uptake of glutamate in GLAST and GLT-1 transfected mutant CHO-K1 cells is inhibited by manganese, *Biol. Trace Elem. Res.*, 2005, **107**, 221–230.

65. M. Conradt and W. Stoffel, Inhibition of the high-affinity brain glutamate transporter GLAST-1 via direct phosphorylation, *J. Neurochem.*, 1997, **68**, 1244–1251.

66. I. M. Gonzalez-Gonzalez, N. Garcia-Tardon, C. Gimenez and F. Zafra, PKC-dependent endocytosis of the GLT1 glutamate transporter depends on ubiquitylation of lysines located in a C-terminal cluster, *Glia*, 2008, **56**, 963–974.

67. M. B. Robinson, Regulated trafficking of neurotransmitter transporters: common notes but different melodies, *J. Neurochem.*, 2002, **80**, 1–11.

68. D. Milatovic, Z. Yin, R. C. Gupta, M. Sidoryk, J. Albrecht, J. L. Aschner and M. Aschner, Manganese induces oxidative impairment in cultured rat astrocytes, *Toxicol. Sci.*, 2007, **98**, 198–205.

69. M. Sidoryk-Wegrzynowicz, E. Lee, J. Albrecht and M. Aschner, Manganese disrupts astrocyte glutamine transporter expression and function, *J. Neurochem.*, 2009, **110**, 822–830.

70. M. Sidoryk-Wegrzynowicz, E. S. Lee, M. Ni and M. Aschner, Manganese-induced downregulation of astroglial glutamine transporter SNAT3 involves ubiquitin-mediated proteolytic system, *Glia*, 2010, **58**, 1905–1912.

71. L. S. Nissen-Meyer, M. C. Popescu, H. Hamdani el and F. A. Chaudhry, Protein kinase C-mediated phosphorylation of a single serine residue on the rat glial glutamine transporter SN1 governs its membrane trafficking, *J. Neurosci.*, 2011, **31**, 6565–6575.

72. S. Balkrishna, A. Broer, A. Kingsland and S. Broer, Rapid downregulation of the rat glutamine transporter SNAT3 by a caveolin-dependent trafficking mechanism in Xenopus laevis oocytes, *Am. J. Physiol.: Cell Physiol.*, 2010, **299**, C1047–C1057.

73. M. Miranda and A. Sorkin, Regulation of receptors and transporters by ubiquitination: new insights into surprisingly similar mechanisms, *Mol. Interventions*, 2007, 7, 157–167.

74. S. Ramamoorthy, T. S. Shippenberg and L. D. Jayanthi, Regulation of monoamine transporters: Role of transporter phosphorylation, *Pharmacol. Ther.*, 2011, **129**, 220–238.

75. C. Debonneville, S. Y. Flores, E. Kamynina, P. J. Plant, C. Tauxe, M. A. Thomas, C. Munster, A. Chraibi, J. H. Pratt, J. D. Horisberger, D. Pearce, J. Loffing and O. Staub, Phosphorylation of Nedd4-2 by Sgk1 regulates epithelial Na(+) channel cell surface expression, *EMBO J.*, 2001, **20**, 7052–7059.

76. D. Rotin and O. Staub, Nedd4-2 and the regulation of epithelial sodium transport, *Front. Physiol.*, 2012, **3**, 212.

77. C. Boehmer, F. Okur, I. Setiawan, S. Broer and F. Lang, Properties and regulation of glutamine transporter SN1 by protein kinases SGK and PKB, *Biochem. Biophys. Res. Commun.*, 2003, **306**, 156–162.

78. T. Sorkina, M. Miranda, K. R. Dionne, B. R. Hoover, N. R. Zahniser and A. Sorkin, RNA interference screen reveals an essential role of Nedd4-2 in dopamine transporter ubiquitination and endocytosis, *J. Neurosci.*, 2006, **26**, 8195–8205.

79. T. R. Guilarte, N. C. Burton, T. Verina, V. V. Prabhu, K. G. Becker, T. Syversen and J. S. Schneider, Increased APLP1 expression and neurodegeneration in the frontal cortex of manganese-exposed non-human primates, *J. Neurochem.*, 2008, **105**, 1948–1959.

CHAPTER 12

Manganese and Neuroinflammation

KELLY A. KIRKLEY[a,b] AND RONALD B. TJALKENS*[a,b]

[a] Center for Environmental Medicine, Colorado State University, Fort Collins, CO, USA; [b] Department of Environmental and Radiological Health Sciences, Colorado State University, Fort Collins, CO, USA
*Email: Ron.Tjalkens@colostate.edu

12.1 Introduction

Neurotoxicity due to excessive exposure to manganese (Mn) has been described as early as 1837.[1] Despite extensive study over the past century, it is only now becoming clear that Mn neurotoxicity involves complex pathophysiological signaling mechanisms between neurons and glial cells. Glial cells are an important target of Mn in the brain, where high levels of the metal accumulate, activating inflammatory signaling pathways that damage neurons through overproduction of numerous reactive oxygen and nitrogen species and inflammatory cytokines. Understanding how these pathways are regulated in glial cells during Mn exposure is critical to determining the mechanisms underlying permanent neurological dysfunction stemming from excess exposure.

Glia represent a diverse class of cells grouped together due to their status as non-excitable neural cells that lack the ability to form an action potential and thus transmit electrical signals.[2] Within the central nervous system (CNS), glia represent 90% of all cells and are classified on the basis of morphology, function, and location consisting of astrocytes, microglia, oligodendrocytes, and ependymal cells.[3] Early descriptions of these cells

Issues in Toxicology No. 22
Manganese in Health and Disease
Edited by Lucio G. Costa and Michael Aschner
© The Royal Society of Chemistry 2015
Published by the Royal Society of Chemistry, www.rsc.org

labeled them as "glue" with a primarily passive structural/supportive role. However, with the advent of patch clamping and fluorescent calcium dye techniques in the late 1980s and early 1990s, research over the past 30 years has found that the role of these cells is much more extensive and complex.[4] Glia are essential for neuronal survival and function, a role that is evolutionary conserved across different phyla, playing prominent roles in both development and pathology of the CNS.[2]

12.2 Astrocytes

12.2.1 Description and Distribution of Astrocytes

Astrocytes accumulate higher levels of Mn than neurons and are therefore considered an important target cell for transport of Mn into the brain, as well as for initiating inflammatory signaling during neuronal stress and injury. The term astrocyte encompasses a heterogeneous population of cells that can have vastly different morphological and physiological character-istics depending on their location with the brain.[3,5] Their morphological forms range from the protoplasmic astrocyte with extensive arborization found in the gray matter to the more rod-like fibrous astrocyte located within the white matter.[6,7] With their extensive processes, they make contacts with neuronal bodies, synapses, axons, blood vessels, and other astrocytes thereby creating a vast network that allows them to serve a multitude of both structural and important physiological roles within the CNS.[2]

Astrocytes are the most prominent type of cell in the CNS, making up 60–70% of all cells in the brain and are also the most prominent glial type with 90% of all glia cells classified as astrocytes.[3] These cells are found throughout the CNS in a contingent, but non-overlapping manner forming distinct microdomains.[7] Astrocytes are morphologically characterized by their classic expression of the intermediate filament proteins glial fibrillary acidic protein (GFAP) and vimentin. Other known markers of astrocytes in the adult brain include glutamine synthetase (GS), S100 calcium binding protein β, and glutamate transporters GLT-1/EAAT2 and GLAST/EAAT1;[8] however, GFAP has been shown to be the most consistent marker in both physiological and pathological states.[9]

12.2.2 Functional Roles of Astrocytes

The first noted function of astrocytes within the adult CNS was purely structural; astrocytes were described as a scaffold to arrange and contain the neuronal circuitry due to their relative abundance and formation of glial scars in disease.[3] Although it is now known that astrocytes have more complex roles, their formation of a continuous syncytium is still important for the structural integrity of the brain. These vast networks help to create specific micro and macro domains and help to create physical barriers between neuronal synapses.[2,7] Furthermore, astrocytic endfeet are an important component of the glial limitans, a barrier that helps to isolate the brain parenchyma from the vasculature and subarachnoid compartments,[3,10] as well as the blood–brain barrier (BBB) through the ensheathing of blood vessels throughout the CNS.[11]

Past their structural roles, astrocytes serve as important facilitators of neuronal homeostasis through nutritive and trophic support. As a primary component of the BBB, astrocytes that surround endothelial cells are enriched in glucose receptors and channels and act as the main vehicle for the movement of glucose and oxygen from the blood to neurons.[2] Astrocytes, not neurons, are capable of storing glucose in the form of glycogen and of *de novo* synthesis of glutamate.[12] Glutamate is the primary excitatory neurotransmitter in the brain and its synaptic concentration is tightly regulated by astrocytes, which rapidly removed glutamate from synapses, where it can be safely transaminated to glutamine for recycling to neurons in the glutamate–glutamine cycle. Eighty percent of glutamate released into the synapse is removed by astrocytes and then converted to glutamine by GS. This glutamine is released and then taken up by neurons that convert glutamine into glutamate and γ-amino butyric acid (GABA). Additionally, production of lactate by astrocytes is used by neurons to produce pyruvate and generate adenosine triphosphate (ATP) *via* the tricarboxcylic acid cycle (TCA).[3] These metabolically coupled support pathways in astrocytes are critical for neuronal survival and are important targets of Mn during neurotoxic exposures.

In addition to being critical for neuronal metabolism, astrocytes are required for normal synaptic transmission through regulation of neurotransmitters, ions, water, and extracellular pH.[3,7] Astrocytes enwrap pre- and post-synaptic membranes forming what is known as the tripartite synapse, which allows astrocytes to not only regulate neurotransmitters, but also actively respond to and modulate synaptic plasticity through the release of gliotransmitters.[4,6,13,14] Astrocytes express a wide assortment of functional neurotransmitters including glutamate, GABA, dopamine, adrenalin/epinephrine, histamine, and glycine, the expression of which varies depending on the local microenvironment to match the physiology of their neuronal neighbors.[3,12] Majority of the neurotransmitter receptors expressed are metabotropic receptors coupled to G-proteins whose activation results in the generation of inositol triphosphate (IP3) and the release of calcium (Ca^{2+}). However, astrocytes express at least three types of ionotrophic receptors: α-amino-3-hydroxy-5-methyl-isoxazole propionate (AMPA), *N*-methyl-D-aspartate (NMDA) types of tetrameric glutamate receptors, and P2X trimeric purinoreceptors.[15] Activation of glia metabotropic and inotropic receptors results in the generation of Ca^{2+} waves within astrocytes that are propagated between astroglial networks through connexin gap junctions and glia release of ATP and glutamate.[4,16] This intercommunication between astrocytes is dynamic and is influenced by the extent of and frequency of neurotransmitter release, which is important in the modulation of synapses in both learning and memory.[14]

Calcium-based communication between astrocytes plays not only a large role in synaptic plasticity, but is vital to the regulation of blood flow in response to neuronal activity known as neurovascular coupling.[7,11] In areas of high neuronal activity, elevations in calcium in astrocytes results in release of vasoactive compounds such as nitric oxide (NO), prostaglandin E2 (PGE$_2$), potassium (K$^+$), and epoxygenase derivatives (EETs) at astrocytic endfeet that results in a dilation or constriction of local vasculature.[10,17] This control of cerebral blood flow is complex and the elucidations of how astrocytes cause specific vasodilation *versus* vasoconstriction in response to neuronal activity is still being fully elucidated. Mn can disrupt ATP-induced Ca^{2+} signaling and intercellular Ca^{2+} waves in astrocytes,[18] which could be detrimental to neuronal trophic support, rendering affected brain regions both focally hypoxic and with insufficient metabolic support.

Astrocytes are thus diverse and important regulators of neuronal metabolism and activity in the developed CNS; likewise, they also play an important role in the developing CNS, through neuronal guidance and synaptogenesis,[3] and in adult neurogenesis.[19] In development, boundaries created by astrocytes help the migration of axons and neuroblasts and release of thrombospondin directs synapse formation. Furthermore, tagging of formed synapses with complement protein, C1q, helps tags synapses for pruning and removal.[20,21] In the adult CNS, neurogenesis within the subventricular zone of the olfactory bulb and the hippocampus is regulated by secretion of astrocytic factors such as Wnt3, interleukin-1β (IL-1β), interleukin-6, and insulin-like growth factor binding protein 6.[12] Additionally, astrocytes themselves are believed to be the source of newly generated neurons determined by labeling based lineage tracking experiments.[19] Thus neuronal generation, function, and continued survival is intimately linked and dependent on the vast and extensive physiology of their astrocytic counterparts.

12.3 Microglia

12.3.1 Description and Distribution of Microglia

Microglia are the resident immunological effector cells of the brain entering the CNS during embryonic development from a monocyte-derived cell type.[2,22] As discussed later, microglia represent an important effector cell type during Mn neurotoxicity that respond rapidly with increased production of neuroinflammatory mediators. In the adult brain microglia have very low rates of division but their numbers can be replenished by perivascular mononuclear phagocytes.[23] They are heterogeneous through the adult brain and constitute 10–15% of all glial cells with greater numbers located within the gray matter.[3] In particular, the highest concentrations of microglia are found within the olfactory bulb, hippocampus, and basal ganglia, with the substantia nigra holding the greatest density of microglia making up 12% of all cells. They exist in three different morphological states: a ramified phenotype found within the neuropil, a rod-like state in fiber tracts, and a macrophage/amoeboid shape in areas with an incomplete BBB.[24] Microglia are never at rest and are constantly migrating; however, these migration patterns are distinct between different cells and do not overlap.[3,23]

12.3.2 Functional Roles of Microglia

Microglial functions within the CNS are still not fully known. As the resident immune cell, the primary role for microglia is immunosurveillance.[22,25,26] Microglia constantly move and sample within their individual domains, clearing up debris *via* their phagocytic function as they migrate.[2,3] They express a variety of both neurotransmitter receptors, pattern recognition receptors (PRRs), and receptors such as P2×7 to sense alterations in brain homeostasis, presence of foreign materials, and neuronal damage.[27] Microglia represent the main class of cell involved in antigen presentation and are important in recruitment of immune cells such at T and B lymphocytes to the sites of injury.[23,25]

More recently, research has determined that microglia may also play integral roles in neuronal development and migration. Amoeboid microglia are implicated in synaptic remodeling and regulation of neuronal apoptosis through the release of soluble factors and phagocytic pruning of synapses in late embryonic development.[2,26] Furthermore, studies have shown microglia to release growth and neurotrophic factors during synaptogenesis.[28] The role of microglia in mediating both trophic and neurotoxic cell–cell interactions during Mn neurotoxicity is still not well known.

12.4 Neuroinflammation

12.4.1 Overview of Neuroinflammation in the CNS

It is now clear that Mn exposure, even early in life, can have lasting effects on the neuroinflammatory status of glial cells.[29] Thus, neuroinflammatory activation of glia may be a fundamental mechanism in determining long-term neurological outcomes from Mn exposure. Astrocytes and microglia serve a multitude of essential functions within the CNS, including integral roles in the innate immune system of the brain.[30] In response to foreign or endogenous signals, both astrocytes and microglia adopt an activated phenotype resulting in the release of proinflammatory mediators.[31] This inflammatory system, known as neuroinflammation, is essential in normal tissue repair and in defense against foreign invasion; however, when sustained, this process can become deleterious through the release of neurotoxic factors that amplify underlying disease.[32–34]

In normal circumstances, the neuroinflammatory reaction has autoregulatory mechanisms in place to limit the extent of activation as the process is neither discriminatory or specific.[30,33] For sustained inflammation to occur there must be failure of self-resolution mechanisms or the presence of endogenous or environmental factors that are perceived as a threat. There are a variety of factors known to elicit activation of both microglia and astrocytes, including products released by injured neurons such as glutamate,[35] ATP,[36] and matrix metalloproteinase-3,[22] cytokines including interferon gamma (IFNγ), interleukin-1β (IL-1β), and interleukin-6 (IL-6); adhesion molecules; growth factors, blood derived factors; ionic imbalances; activation of complement, products from viruses and bacteria; and presence of reactive oxygen species.[7,23,30] Furthermore, new evidence suggests that both microglia and astrocytes express endogenous pattern recognition receptors (PRRs) that respond to a variety of damage-associated molecular patterns (DAMPs) that results in inflammation in a sterile environment and may act as important inducers of inflammation in disease pathology.[33] These PRRs become activated in response to signals released by necrotic neurons or other pathologic products produced during disease, including oxidized proteins and lipids,[37] messenger ribonucleic acid (mRNA), fibronectin, hyaluronic acid, heat shock proteins, amyloid-beta, neuromelanin, and alpha-synuclein.[26,38,39] The production of these products is further increased by the activating glia leading to feed-forward loops and a continued cycle of inflammation, which leads to a release of neurotoxic mediators and tissue injury.

Activated glia release a plethora of factors including cytokines, chemokines, reactive oxygen species (ROS), and nitric oxide (NO) that have shown to be toxic to neurons.[3,22,25] Cytokines such as tumor necrosis factor alpha (TNFα) and interleukin-6 are often upregulated early and have shown to directly lead to neuronal apoptosis and amplify inflammation through recruitment of both innate and adaptive immune cells.[25,38] Released reactive oxygen products are extensible tied to causing lipid peroxidation, mitochondrial dysfunction and subsequent energy failure and apoptosis, protein modifications, and DNA damage in both neurons and glial cells.[32] The formation of peroxynitrite, a by-product of superoxide and NO, is thought to be a major contributor to neuronal-induced cell death through nitration and nitrosylation of tyrosine and serine residues of proteins leading to impairment of normal cellular functions.[40] Mn exposure results in significant increases in protein nitrosylation, indicative of nitrosative stress from NO production by glial cells.[29] Inhibition or deletion of many of these pathways have shown to be neuroprotective, but often the neuroprotection achieved is dependent on the timing of inhibition as often early downregulation of inflammation has actually worsened neuronal injury.[41] However, mice lacking the inducible form of NO synthase (iNOS/NOS2) are protected from Mn neurotoxicity, demonstrating the importance of this glial inflammatory pathway in the mechanism of neuronal injury.[18] Due to the complicated nature of neuroinflammation and the vast majority of implicated factors, systematic and thorough understanding is vital to understanding the implications that may come from targeting this pathway.

12.4.2 Role of Astrocytes in Neuroinflammation

Astrocyte activation is a biological reaction that is documented in most CNS diseases, as measured by upregulation of GFAP expression and alterations in astrocyte morphology, and is an important early indicator of neuropathology.[9,12] Neuroinflammatory activation of astrocytes can be neuroprotective through isolation of damage, glutathione production, BBB repair, and release of neurotrophic factors such as neural growth factor and glial derived growth factor;[7,26,42] however, astrogliosis can also be neurotoxic and promote disease progression. Detrimental consequences of astrogliosis include inhibition of axonal regeneration,[26,43] exacerbation of inflammation *via* cytokine production,[44,45] production of reactive oxygen and nitrogen species,[46–48] and excess glutamate release.[49] Additionally, chronic inflammatory stimulation of astrocytes reduces glial capacity to generate and release neurotrophic mediators and execute normal physiological functions.[12]

The regulation of astrocyte activation is under the control of many factors including cytokines IL6, IFNγ, TNFα, toll-like receptor activators, neurotransmitters, ATP, reactive oxygen species, hypoxia, glucose deprivation, ammonia, and protein aggregates.[7,12] These activators are often by-products of already injured neurons or factors released by activated microglia, which indicate that astrocyte activation is often later in disease progression.[50] However, astrogliosis is often less transitory than microgliosis and is believed to be important in amplifying the inflammatory process thereby inducing greater damage.[51] Moreover, *in vitro* studies have shown that isolated human astrocytes and not microglia are the major source of NO-induced neurotoxicity indicating they may be more significant in neuroinflammatory-induced neuronal death in humans than have been indicated in rodent models.[52]

12.4.3 Role of Microglia in Neuroinflammation

As the resident immune cells, microglia are the effectors of the innate immune response within the CNS with activation occurring early in disease often prior to overt neuropathology.[23,50] Under physiological conditions, microglia exist in a resting, ramified state releasing both anti-inflammatory and neurotrophic factors while surveying their domains.[53] However, in the presence of viral or bacterial products,[33] ATP, changes in ion or neurotransmitter homeostasis,[54] cytokines such as IFNγ and interleukin 4,[23] colony stimulating factors (CSFs),[16] and a list of other pathological products, microglia transform into an activated phenotype, proliferate, and migrate to the site of injury.[3,26] Activation occurs in two stages. The first stage microglia adopt a rod-like shape and increase expression of major histocompatibility complex two (MHCII) and other inflammatory molecules. In the second stage, microglia morph into an amoeboid cell capable of phagocytosis.[16,23]

Once activated, microglia can be both beneficial and deleterious in disease as they release both pro and anti-inflammatory factors.[26] Determining whether microglia neuroinflammatory responses will be helpful or toxic is often predicted by adoption of either the M1 known as "classical activation" or M2, the "alternative activation," phenotype.[55] The M1 phenotype is primarily inflammatory with microglia upregulating MHCII, CD86, CD32, and CD16 with the production of TNFα, IL-1β, and IL-6. The M2 phenotype, on the other hand, is more involved in tissue repair with increased expression of arginase 1 and CD206 and release of brain-derived neurotrophic factor (BDNF), insulin-like growth factor-1 (IGF-1), and interleukin 10 (IL10).[56]

The classical activated or M1 microglia are known to elicit neuronal death and perpetuate inflammation through release of a variety of cytotoxic substances such as NO, superoxide anion, cytokines, glutamate, prostaglandins, and aspartate.[23,28] They appear to be the major initial sensors of foreign or endogenous signals, secreting inflammatory mediators such as TNFα and IL-1β that can act on astrocytes to induce secondary inflammatory responses.[51] Furthermore, prevention of microglia activation *via* pharmacological or genetic approaches often protects against neuroinflammatory pathology thus placing them as important regulators of inflammatory mechanism in neurodegenerative diseases.[26,57]

12.4.4 NF-κB Signaling in Neuroinflammatory Injury from Manganese

Glial inflammatory activation is regulated by several different pathways, including mitogen-activated protein kinases (MAPK), activator protein-1 (AP-1), Janus Kinase (JAK)/signal transducer and activator of transcription (STAT), and interferon regulator factor families;[3,33] nevertheless, the nuclear factor kappa B (NFκB) appears to be the primary pathway involved in the activation of proinflammatory genes.[58] Deletion of NF-κB is detrimental to the ability of the immune system to initiate immunoprotective responses. Mice deficient in this pathway often succumb to opportunistic infections.[59] Deletion of this pathway in specific glial cells within the CNS has shown to be very neuroprotective with better recovery after spinal cord injury,[44] decreased pathology in mouse models of multiple sclerosis,[60] and decreased seizure-induced neuronal death in kainic acid model of seizure.[57]

NF-κB represents a family of transcription factors that is regulated by inhibitory κBs (IκB). Upon signal activation, IκBs are phosphorylated by IκB kinase complex (IKK) marking them for polyubiquination and ultimately degradation by the 26s proteasome, thus freeing the transcription factors, located as dimers within the cytosol, to translocate into the nucleus.[61] The IKK complex consists of three different proteins, including the two catalytic subunits IKKα/IKK1 and IKKβ/IKK2 and the regulatory subunit IKKγ. These two catalytic subunits mark the division of the two NF-κB activation pathways known as the classical pathway and the alternative pathway. The classical NF-κB pathway involves the heterodimers of p50 and p65/RelA and is activated by the action of IKKβ/IKK2 of the IKK complex. This pathway is primarily involved in immunoregulation, controlling innate immune responses and survival of immune cells. The alternative pathway is primarily involved in the development of secondary lymphoid organs and requires only IKKα/IKK1 and results in the processing of p100.[62–64] Deletion of IKKβ/IKK2 and not IKKα/IKK1 recapitulates similar mouse phenotypes as RelA knockout mice with almost complete inhibition of inflammatory responses and thus represents a major target in modulating glia neuroinflammatory activation.[59]

12.5 Neuroinflammation in Diseases of the CNS

The activation of microglia and astrocytes is one of the universal components of neuroinflammation and is implicated in the progression of neurodegeneration in ischemia, seizure, Alzheimer's disease, multiple sclerosis, amyotrophic lateral sclerosis, Parkinson's disease (PD) and manganism.[26,32,33,50,65] Since the first descriptions of activated glia in neurodegenerative disease almost 20 years ago, there has been an increasing number of nervous system pathologies describing activated glia. This research has revealed that although induction of inflammation may occur in a disease-specific manner, there is evidence of shared mechanisms for the sensing and transduction.[33] Research aimed at elucidating neuroinflammatory pathogenesis and developing targeting strategies is quickly expanding due to increasing indications of the importance of this mechanism in the progression, and possibly initiation, of many neuropathologies. It is useful to compare the role of neuroinflammation in Mn neurotoxicity to that of other better studied disorders to develop an appreciation for the mechanisms that are common to degenerative conditions of the CNS.

12.5.1 Seizure

Seizure is an event where there is spontaneous synchronization of neuronal activity whereby a set or subset of neurons undergoes continual and uncontrollable electrical burst of activity, and when this occurs chronically, the condition is termed epilepsy. Epilepsy, caused by both genetic and acquired factors, affects 50 million people worldwide; however, 60% of epileptic cases are idiopathic.[66] Mn is well described as an agent that causes damage to the CNS, in part, through increases in excitoxicity and altered glutamatergic neurotransmission,[67] and may therefore be a relevant environmental factor affecting the seizurogenic potential of other compounds, which is notoriously difficult to predict.[68] The events that lead to epileptogenesis are only partially defined and a large fraction of individuals with epilepsy are either refractory to current treatments or experience important side effects.[69] Postmortem examination of patients with intractable epilepsy has consistently identified the presence of astrogliosis and microgliosis,[70] stimulating intense research focused on the involvement of glial inflammation in the etiopathogenesis of this disease.[4]

The link between neuroinflammation and epilepsy is supported by both clinical and experimental evidence. Infection and gliomas have long been identified as an important trigger for secondary or symptomatic epilepsy and a polymorphism in IL-1β is an important cause of genetic-based seizures.[71,72] Activated astrocytes and microglia are commonly identified near or around seizure foci in many forms of epilepsy,[7] and in temporal lobe epilepsy (TLE), removal of astrogliotic tissue in the hippocampus is an important treatment in patients with intractable forms of the disease.[73] Additionally, studies examining the cerebral spinal fluid (CSF) in patients with both non-inflammatory and immune-based epilepsies often demonstrated increased levels of inflammatory cytokines such as IL-6, TNFα, and IL-1β.[74] Experimental seizure models have further supported glial involvement by revealing that neuroinflammatory activation of glia often proceeds seizure activity[65] and if specifically induced, can lead to alterations in synaptic physiology increasing hyperexcitability of local neurons.[75]

Proposed mechanisms for glia involvement in pathophysiology of seizures include both loss of function and inflammatory modulation of neuronal excitability. Non-inflammatory-based alterations include water dysregulation, alterations in potassium homeostasis, and perturbations in glutamate/GABA metabolism caused by changes in astrocytes. Animals deficient in proteins important in either of these pathways is sufficient enough to cause or potentiate seizures.[76,77] Inflammatory-mediated pathways include enhancement of glutamatergic activation and potentiation of NMDA-induced currents,[25] increased release of gliotransmitters, and inhibition of astrocyte functions.[35,65,72,78] Implicated molecules include pro-inflammatory cytokines, nitric oxide, and arachadonic acid.[25,79] Even with several implicated mechanisms, the most consistent alteration in both inflammatory and non-inflammatory pathways is the effects seen in glutamate/glutamine and GABA metabolism, with human and rodent models revealing high levels of glutamate and low levels of GABA stemming from loss of GS and glutamate transporters in astrocytes.[80] This results in both increased excitatory stimulation and decreased inhibitory transmission, and facilitates a pathological condition where astrocytes actually increase glutamate release thereby affecting thousands of synapses.[3,4] Heightened research into neuroinflammatory involvement in seizure has only begun to realize the extent and importance of glial mechanisms in this disease and understand the potential therapeutic implications of this research. The impact of Mn on these pathways will be an important area of investigation in future studies.

12.5.2 Parkinson's Disease

Manganese causes a degenerative condition of the cortex and basal ganglia that shares a number of neurological similarities to Parkinson's disease (PD), yet the etiology and pathology of the two disorders are distinct.[81,82] PD is a chronic and progressive neurodegenerative disease that affects 1.5% of the global population over the age of 65 and represents the most common movement disorder.[83] The most notable symptoms include bradykinesia, resting tremors, rigidity, and postural instability caused by loss of dopaminergic neurons in the nigrostriatal pathway, resulting in concomitant loss of dopamine in the striatum, which can often be alleviated by replacement dopamine therapy.[84] These symptoms are in contrast to manganism, which tends to present without resting tremor and with a distinctive gait disorder termed "cock walk" and is not usually responsive to dopamine therapy.[85] Other non-motor symptoms of PD include insomnia, depression, dementia, and autonomic failure, which are often unaddressed by current treatments.[86] Familial forms of PD are linked to mutations in the PARK family of genes, LRRK2, DJ-1, SNCA, and PINK-1;[87] however, 95% of all PD cases are idiopathic and are most likely caused by a complex interaction between genetic susceptibility and environmental exposure to pesticides, pollutants, and heavy metals.[33]

Several mechanisms are implicated in the etiology of sporadic PD, including glutamatergic toxicity, misfolding of α-synuclein and tau, decreased production of neurotrophic factors, exposure to environmental and infectious insults, mitochondrial dysfunction, oxidative stress, and neuroinflammation.[33,83] In particular, recent research has primarily focused on the involvement and interaction of protein misfolding, oxidative stress, and neuroinflammation due to evidence in experimental models linking these pathways in the induction, progression, and facilitation of neuronal loss.[32,50,88] Activated microglia and astrocytes are fundamental to the pathophysiology of these mechanisms because they are the primary sources of inflammatory and oxidative factors and can both be activated by and induce production of misfolded proteins.[89]

Inflammation is now considered a pathological hallmark of PD as astrogliosis and microgliosis is consistently revealed in examination of *post mortem* tissues as well as elevations in cytokines in patient CSFs.[26,90] Furthermore, infusion of the bacterial product lipopolysaccharide, a potent inducer of inflammation, in rodent brains is known to lead to specific loss of nigral dopamine neurons after initiation of glial activation in a manner that recapitulates the slow, progressive nature of PD.[26] Experimental models of PD using the toxin 1-methyl-4-phenyl-1,2,3,6-tetrahydropyridine (MPTP) have shown massive amounts of glial activation prior to and around neuron loss.[50,51,91,92] Release of glial inflammatory factors, such as NO and TNFα, are highly implicated in causing the progressive loss of neurons because suppression of NOS2 or deletion of TNFα receptors protect mice from MPTP induced neurodegeneration.[40,93,94] The products released from glial-damaged neurons, such as neuromelanin and α-synuclein, can further induce gliosis, creating a loop of interdependent neuronal loss and sustained glial activation.[30] The neuroinflammatory capacity of Mn may be a mechanism relevant to environmental influences in the etiology of a subset of PD cases.

12.5.3 Manganism

Manganese neurotoxicity, or manganism, is a Parkinson-like disease caused by excessive exposure to Mn and is marked by motor deficits such as gait disturbances, facial masking, hypo and dysphonia, dystonia, and action and postural tremor.[81,82] These Parkinson-like manifestations are due to the neuropathological changes, including neuronal loss, atrophy and gliosis within the globus pallidus (GP), substantia nigra pars reticulata (SNpr) and striatum (ST) of exposed individuals.[95,96] Traditional exposures to high levels of Mn occur occupationally in welders, miners, and steel workers;[97] however, the neurological consequences of environmental exposure to low levels of Mn through ingestion of crops with residues of the Mn-containing pesticide Maneb[98] and well water with high concentrations of Mn[99] is under scrutiny as an important route for non-occupational-based exposure. In particular, the there is increased concern with chronic Mn exposure in children due to their lower ability to clear Mn[100] and higher levels of iron deficiency, which has shown to elevate brain Mn levels[101] and more efficient oral uptake of Mn.[102] Recent epidemiological evidence has observed cognitive deficits in children exposed to high levels of Mn in drinking water,[103–105] highlighting the need for future studies addressing the long-term consequences of these exposures.

The mechanisms of how Mn exposure leads to specific neurodegenerative changes in the basal ganglia of exposed humans and animals are poorly understood. Elevated levels of Mn are routinely documented in the basal ganglia of exposed humans and animals[106] and experimental evidence has shown that Mn can be directly neurotoxic through inhibition of mitochondrial respiration leading to energy failure and oxidative stress[107] and through excitotoxicity.[108] Other mechanisms such as oxidative stress, glial toxicity, and neuroinflammation are implicated in the progression of Mn neurotoxicity because, despite cessation of exposure, the condition will continue to progress both clinically and in rodent models of the disease.[96,101,109,110]

The involvement of glia in Mn-induced neurotoxicity has only received attention within the last 20 years as a potential major mechanism in manganism.[109,110] Although activated astrocytes and microglia were often noted in *post mortem* evaluation of Mn patients,[82] few studies examined the importance of this glial activation. This was most likely due to the ability of Mn to be directly toxic to neurons through inhibition of mitochondrial respiration and induction of oxidative stress[107] and the historical focus on acute, high-level exposures. A study in 1998 by Spranger *et al.* radically altered perceptions of glia involvement in this disease showing that more modest levels of Mn exposures required the presence of glia to be neurotoxic. Other studies have now built upon these initial findings, revealing that Mn can exacerbate LPS and cytokine-induced activation of both microglia and astrocytes, resulting in increased levels of TNFα, IL-1β, ROS, and NOS2 expression.[111–115] Increased levels of the aforementioned inflammatory

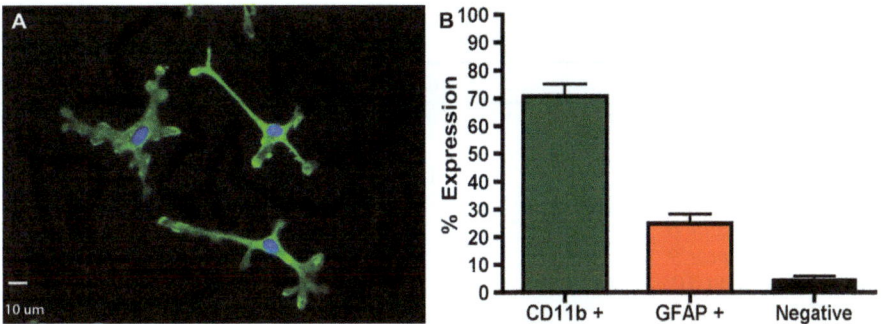

Figure 12.1 Isolation of microglia and astrocytes. Purity of glial cultures was assessed by immunofluorescence detection of GFAP- and IBA1-positive cells. (A) Representative images of IBA-1-positive (green) microglial cultures at high resolution with DAPI (blue) to visualize cell nuclei. (B) Quantification of the number of IBA-1 *versus* GFAP positive cells in microglia cultures. Cells are subject to further immunopurification to achieve monocultures of each cell type.

genes has also been measured in both rodent[29,116] and non-human primate[117] studies with deletion or inhibition of these pathways showing neuroprotection.[18,116,118] To decipher the complex signaling mechanisms likely to influence development of a neuroinflammatory phenotype in Mn neurotoxicity, primary astrocytes and microglia can be isolated from neonatal mouse and the relative contributions of each cell type defined by using immunopurification-based methods to establish pure astroglial and microglial cultures from mixed isolates (Figure 12.1).

Using pure cultures of microglia, treatment with 100 µM $MnCl_2$ resulted in expression of multiple NF-κB-regulated inflammatory genes, including those for chemokine ligands and receptors, interleukins and interleukin receptors, macrophage migration inhibitory factor, capsase-1, toll interacting protein, and TNFα (Table 12.1). Because microglia respond quickly to stress and injury in the CNS, it is highly likely that Mn-induced damage to neurons may quickly activate microglia, resulting in an inflammatory phenotype that increases the activation state of astrocytes and results in amplification of neuronal injury.

We tested this hypothesis using co-culture systems to examine cell–cell interactions between astrocytes and microglia using immunopurified cultures of each cell type (Figure 12.2).

The experimental design for such studies was constructed to enable the identification of factors derived from microglia that could enhance inflammatory activation of astrocytes. The first type of study employed in this regard uses microglial-condition medium (MCM) to treat naïve cultures of astrocytes. Microglia (Figure 12.2A) are treated with Mn independently from astrocytes and then Mn-containing MCM transferred to separate astrocyte cultures and the expression of inflammatory genes measured after 24 h. Levels of Mn in the microglial culture medium decrease following treatment

Table 12.1 Representative inflammatory genes induced in primary microglia following exposure to 100 μM MnCl$_2$.

Gene name	Control	+Manganese	Fold-change
Caspase 1	0	10.1	1097.5
Chemokine ligand 25	4.6	6	2.6
Chemokine ligand 5	10.4	12.1	3.2
Chemokine receptor 4	2.8	4.3	2.8
Chemokine receptor 6	0.2	2.6	5.3
Chemokine receptor 9	1.2	2.8	3
Chemokine ligand 10	7.5	10.2	6.5
Chemokine ligand 12	8.3	11.6	9.8
Chemokine ligand 9	2.8	5.1	4.9
Interleukin 11	4.7	6.2	2.8
Interleukin 13 receptor a1	8.8	10.9	4.3
Interleukin 15	9.7	11.5	3.5
Interleukin 1 receptor, type II	0	3	8
Interleukin 2 receptor, beta chain	1.4	3.1	3.2
Interleukin 2 receptor, gamma chain	9.3	10.6	2.5
Interleukin 4	1.9	3.2	2.5
Interleukin 8 receptor, beta	0	1.8	3.5
Macrophage migration inhibitory factor	13.3	16.8	11.3
Secreted phosphoprotein 1	16.3	18.2	3.7
Tumor necrosis factor	7.5	9	2.8
Toll interacting protein	8.8	10	2.3

Notes: Expression of representative inflammatory genes. Data are presented as fold-change from saline-treated controls, based upon the threshold cycle (Ct) values calculated by subtracting the raw Ct value from 35 (considered not expressed). Values are calculated according to $2^{(Ct, \text{ control} - Ct, \text{ treated})}$.

(Figure 12.2B), indicating active transport of the metal into culture microglial cells. The other paradigm for testing microglia–astrocyte interactions is to co-culture each cell type and treat them concurrently in the same medium. This is achieved through the use of micro-porous culture inserts, which allow diffusion of small molecules but not cell–cell contact (Figure 12.2C). Microglia are plated on the permeable culture inserts and astrocytes are plated on the bottom of a multi-well culture plate. Both cells are treated concurrently and mRNA and/or protein isolated for gene expression analysis. The results of each type of experiment are depicted in Figure 12.3.

Treatment with Mn in either MCM or co-culture paradigms caused upregulation of protypical inflammatory genes in astrocytes by qRT-PCR and revealed marked differences in astrocytes treated with Mn alone or in the presence of microglial factors (Figure 12.3A–C). Treatment of astrocytes with 100 μM MnCl$_2$ resulted in a large, albeit not significant, increase in the fold expression of Nos2, but did not alter levels of astrocyte expression of TNFα or Il-1β. However, treatment with MCM or co-culturing of astrocytes with microglia resulted in both potentiation of Nos2 expression and substantial increases in astrocyte expression of TNFα or Il-1β (Figure 12.3D–F). Two-way ANOVA analysis revealed that the extent of astrocyte activation in response to Mn was dependent on the presence of microglia or microglial-derived factors

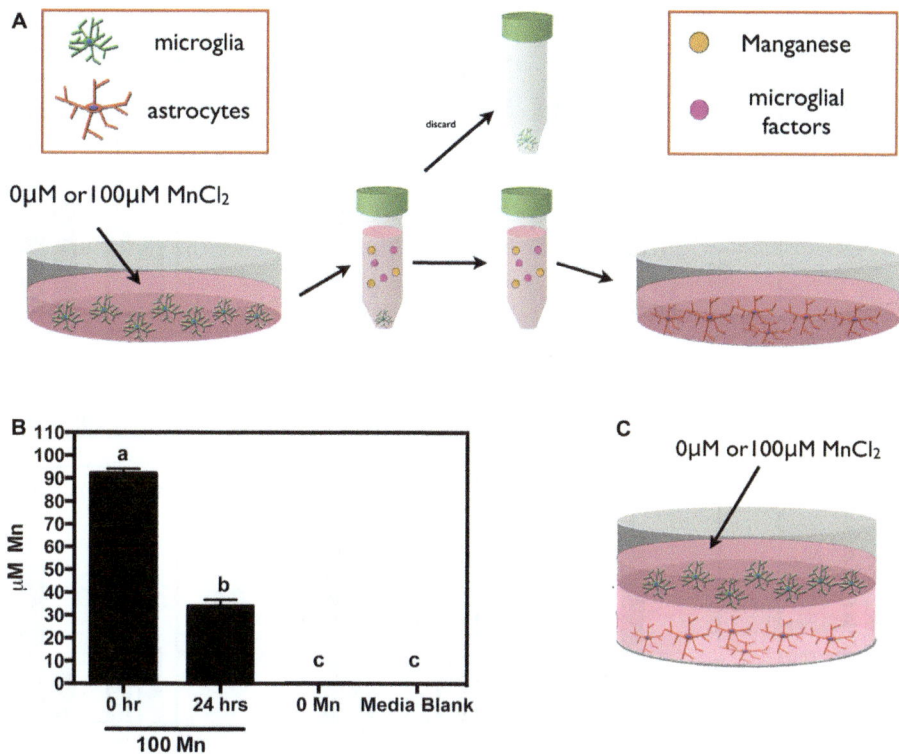

Figure 12.2 Schematics of microglia conditioned media (MCM) and co-culture experiments. (A) In MCM experiments, microglia are treated with either saline or 100 μM MnCl$_2$ for 24 h. After 24 h, media is removed, centrifuged to remove any unattached cells, and supernatants placed on astrocytes for 24 h. (B) The levels of Mn in microglial-conditioned media were measured *via* ICP-MS at time 0 and 24 h post treatment. Data are presented as the average micromolar concentration of Mn $\pm$ SEM (one-way ANOVA; different letters indicate $p < 0.05$). (C) In co-culture experiments, microglia seeded on permeable cell culture inserts were cultured with astrocytes plated on six-well tissue plates and simultaneously treated with 100 μM MnCl$_2$ for 24 h.

($p < 0.05$ for treatment and presence of microglial variables; $p < 0.05$ for interaction).

Despite the heightened focus on glial involvement in manganism, there are still many unanswered questions regarding mechanisms due to the limited number of *in vivo* studies and the inability of Mn to be a very potent glial activator in the absence of other inflammatory factors.[109,110] As with other disorders of the CNS with a neuroinflammatory component, most studies into glial involvement in manganism have used single or mixed cultures of microglia or astrocytes, with few studies examining cell–cell interactions. The studies described in this chapter suggest that complicated signaling mechanisms between microglia and astrocytes likely underlie

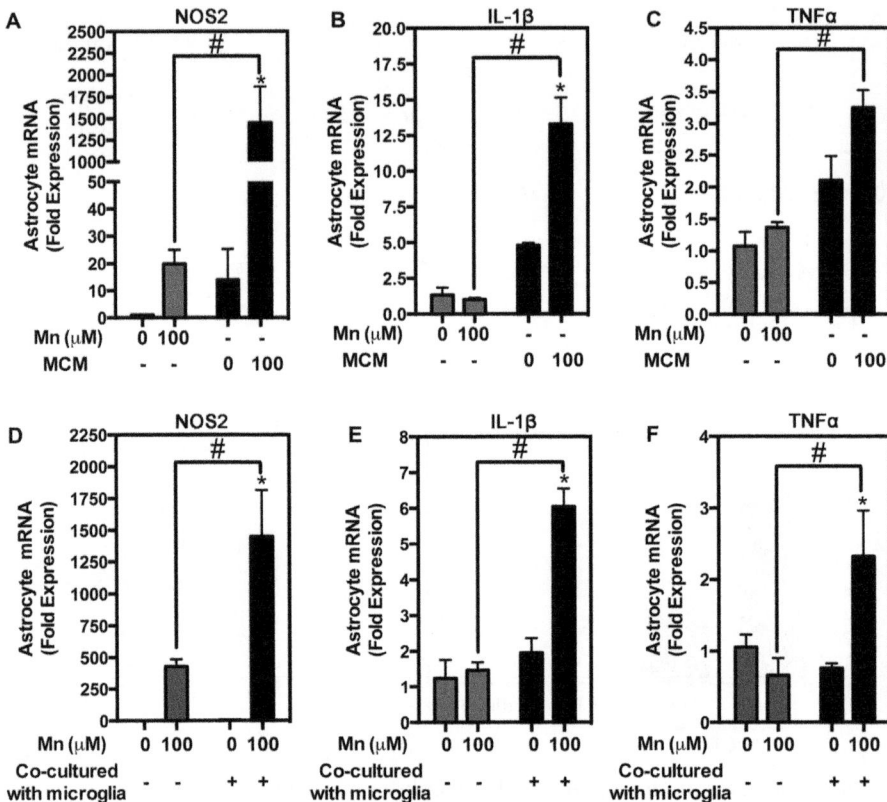

Figure 12.3 Presence of microglia or microglial derived factors are required for complete astrocyte activation. Levels of inflammatory gene expression in astrocytes treated with only saline or 100 μM MnCl₂ *versus* Mn with MCM (A–C) or when co-cultured with microglia (D–F) were determined *via* qRT-PCR. (A) Astrocyte expression of *Nos2* when treated with Mn alone *versus* MCM. (B) Astrocyte expression of *TNFα* when treated with Mn alone *versus* MCM. (C) Astrocyte expression of *Il-1β* when treated with Mn alone *versus* MCM. (D) Astrocyte expression of *Nos2* when treated with Mn alone *versus* when co-cultured with microglia. (E) Astrocyte expression of *TNFα* when treated with Mn alone *versus* when co-cultured with microglia. (F) Astrocyte expression of *Il-1β* when treated with Mn alone *versus* when co-cultured with microglia. Data are presented as the average fold change $\pm$ SEM. (Two-way ANOVA with Tukey *post hoc* analysis; $^*p < 0.05$ between saline and 100 μM MnCl₂; $^#p < 0.05$ *versus* 100 μM MnCl₂ alone).

development of a neuroinflammatory phenotype Mn neurotoxicity that ultimately results in the progression of neuronal injury leading to psychological and motor manifestations of manganism. Microglia produce a large number of pro-inflammatory factors that could amplify the activation state of astrocytes, including TNFα and IL1-β, as well as numerous cytokines and chemokines. It will be important in future studies to determine which of these factors are most relevant to the cell–cell interactions underlying the damaging effects of neuroinflammation following exposure to Mn. With limited or no treatment options for interdicting neuroinflammation in the CNS, it will be imperative to better identify underlying mechanisms in order to develop better therapies. This is particularly of concern, given the recent appreciation for how Mn exposures in children can lead to persistent adverse neurological affects. Thus, there is a need for a more systematic and comprehensive look at glial involvement and the potential importance of this response in chronic exposures.

12.6 Conclusion

Neuroinflammatory activation of glial cells is an important mechanism in Mn neurotoxicity and in other degenerative conditions of the CNS. Studies in the last several decades have redefined the importance of astrocytes and microglia to neuronal development, homeostasis, and survival, transforming our understanding of the role of these cells from inert structural components to important components of brain physiology and pathology. More specifically, the importance of astrocytes and microglia to neuronal survival has received increased attention because these two glial types are the most often altered during disease states and are now known to be fundamental components of the innate immune system of the brain. Inflammatory activation of glia, or neuroinflammation, is a classic and conserved marker of neuropathology and is implicated in the progression, and possibly initiation, of several CNS disorders including seizure, Parkinson's disease, and manganism. Yet, much of this information on neuroinflammation has been gleaned from rodent modeling, with few studies utilizing translational or environmental relevant models to examine these important mechanisms. Furthermore, because glial activation can also serve either neuroprotective or neurotoxic functions, there exists a need to better understand the timeline and pathways of glial activation with a more extensive focus on the relative contributions of different glial types and the dynamics of glial-to-glial signaling. By examining specific glial-derived mechanisms in several neurodegenerative diseases, we may better understand the implications of neuroinflammation for CNS pathology and discover new potential targets for therapeutic intervention.

References

1. J. Couper, On the effects of black oxide of manganese when inhaled into the lungs. *Br. Ann. Med. Pharm. Vital Stat. Gen. Sci.*, 1837, **1**, 41–42.
2. GribbinSimon. 5.2 Barres Q and A NS CNS, 2009, 1–3.
3. A. Verkhratsky and A. Butt, *Glial Neurobiology*, Wiley, New York, NY, 2007.
4. A. Araque, G. Carmignoto and P. G. Haydon, Dynamic signaling between astrocytes and neurons, *Annu. Rev. Physiol.*, 2001, **63**, 795–813.

5. V. Matyash and H. Kettenmann, Heterogeneity in astrocyte morphology and physiology, *Brain Res. Rev.*, 2010.
6. G. Perea and A. Araque, GLIA modulates synaptic transmission, *Brain Res. Rev.*, 2010, **63**, 93–102.
7. M. V. Sofroniew and H. V. Vinters, Astrocytes: biology and pathology, *Acta Neuropathol.*, 2010, **119**, 7–35.
8. H. K. Kimelberg, The problem of astrocyte identity, *Neurochem. Int.*, 2004, **45**, 191–202.
9. J. P. O'Callaghan and K. Sriram, Glial fibrillary acidic protein and related glial proteins as biomarkers of neurotoxicity, *Expert Opin. Drug Saf.*, 2005, **4**, 433–442.
10. A. Nimmerjahn, Astrocytes going live: advances and challenges, *J. Physiol.*, 2009, **587**, 1639–1647.
11. G. Carmignoto and M. Gómez-Gonzalo, The contribution of astrocyte signalling to neurovascular coupling, *Brain Res. Rev.*, 2010, **63**, 138–148.
12. V. Parpura, M. T. Heneka, V. Montana, S. H. R. Oliet, A. Schousboe, P. G. Haydon, R. F. Stout Jr, D. C. Spray, A. Reichenbach, T. Pannicke, M. Pekny, M. Pekna, R. Zorec and A. Verkhratsky, Glial cells in (patho)physiology, *J. Neurochem.*, 2012, **121**, 4–27.
13. M. Nedergaard and A. Verkhratsky, Artifact versus reality – how astrocytes contribute to synaptic events, *Glia*, 2012, **60**, 1013–1023.
14. G. Perea, M. Navarrete and A. Araque, Tripartite synapses: astrocytes process and control synaptic information, *Trends Neurosci.*, 2009, **32**, 421–431.
15. U. Lalo, Y. Pankratov, S. P. Wichert, M. J. Rossner, R. A. North, F. Kirchhoff and A. Verkhratsky, P2X1 and P2X5 subunits form the functional P2X receptor in mouse cortical astrocytes, *J. Neurosci.*, 2008, **28**, 5473–5480.
16. S. U. Kim and J. de Vellis, Microglia in health and disease, *J. Neurosci. Res.*, 2005, **81**, 302–313.
17. S. J. Mulligan and B. A. MacVicar, Calcium transients in astrocyte endfeet cause cerebrovascular constrictions, *Nature*, 2004, **431**, 195–199.
18. K. M. Streifel, J. A. Moreno, W. H. Hanneman, M. E. Legare and R. B. Tjalkens, Gene deletion of nos2 protects against manganese-induced neurological dysfunction in juvenile mice, *Toxicol. Sci.*, 2012, **126**, 183–192.
19. F. Doetsch, The glial identity of neural stem cells, *Nat. Neurosci.*, 2003, **6**, 1127–1134.
20. K. S. Christopherson, E. M. Ullian, C. C. A. Stokes, C. E. Mullowney, J. W. Hell, A. Agah, J. Lawler, D. F. Mosher, P. Bornstein and B. A. Barres, Thrombospondins are astrocyte-secreted proteins that promote CNS synaptogenesis, *Cell*, 2005, **120**, 421–433.
21. E. M. Powell and H. M. Geller, Dissection of astrocyte-mediated cues in neuronal guidance and process extension, *Glia*, 1999, **26**, 73–83.

22. Y. S. Kim, Matrix Metalloproteinase-3: a novel signaling proteinase from apoptotic neuronal cells that activates microglia, *J. Neurosci.*, 2005, **25**, 3701–3711.

23. M. K. Gehrmann, Microglia: intrinsic immuneffector cell of the brain, *Brain Res. Rev.*, 1995, **20**, 19.

24. L. J. L. Lawson, V. H. V. Perry, P. P. Dri and S. S. Gordon, Heterogeneity in the distribution and morphology of microglia in the normal adult mouse brain, *Neuroscience*, 1990, **39**, 151–170.

25. F. González-Scarano and G. Baltuch, Microglia as mediators of inflammatory and degenerative diseases, *Annu. Rev. Neurosci.*, 1999, **22**, 219–240.

26. M. L. Block and J.-S. Hong, Microglia and inflammation-mediated neurodegeneration: multiple triggers with a common mechanism, *Prog. Neurobiol.*, 2005, **76**, 77–98.

27. R. M. Ransohoff and V. H. Perry, Microglial physiology: unique stimuli, specialized responses, *Annu. Rev. Immunol.*, 2009, **27**, 119–145.

28. K. Nakajima and S. Kohsaka, Functional roles of microglia in the brain, *Neurosci. Res.*, 1993, **17**, 187–203.

29. J. A. Moreno, K. M. Streifel, K. A. Sullivan, M. E. Legare and R. B. Tjalkens, Developmental exposure to manganese increases adult susceptibility to inflammatory activation of glia and neuronal protein nitration, *Toxicol. Sci.*, 2009, **112**, 405–415.

30. T. Wyss-Coray and L. Mucke, Inflammation in neurodegenerative disease – a double-edged sword, *Neuron*, 2002, **35**, 419–432.

31. J. M. Craft, D. M. Watterson and L. J. Van Eldik, Neuroinflammation: a potential therapeutic target, *Expert Opin. Ther. Targets*, 2005, **9**, 887–900.

32. R. Lee Mosley, E. J. Benner, I. Kadiu, M. Thomas, M. D. Boska, K. Hasan, C. Laurie and H. E. Gendelman, Neuroinflammation, oxidative stress, and the pathogenesis of Parkinson's disease, *Clin. Neurosci. Res.*, 2006, **6**, 261–281.

33. C. K. Glass, K. Saijo, B. Winner, M. C. Marchetto and F. H. Gage, Mechanisms Underlying Inflammation in Neurodegeneration, *Cell*, 2010, **140**, 918–934.

34. M. G. Tansey, M. K. McCoy and T. C. Frank-Cannon, Neuroinflammatory mechanisms in Parkinson's disease: Potential environmental triggers, pathways, and targets for early therapeutic intervention, *Exp. Neurol.*, 2007, **208**, 1–25.

35. V. Kaushal and L. C. Schlichter, Mechanisms of microglia-mediated neurotoxicity in a new model of the stroke penumbra, *J. Neurosci.*, 2008, **28**, 2221–2230.

36. F. Di Virgilio, S. Ceruti, P. Bramanti and M. P. Abbracchio, Purinergic signalling in inflammation of the central nervous system, *Trends Neurosci.*, 2009, **32**, 79–87.

37. J. J. Husemann, J. D. J. Loike, R. R. Anankov, M. M. Febbraio and S. C. S. Silverstein, Scavenger receptors in neurobiology and

neuropathology: their role on microglia and other cells of the nervous system, *Glia*, 2002, **40**, 195–205.

38. J. C. Gensel, K. A. Kigerl, S. S. Mandrekar-Colucci, A. D. Gaudet and P. G. Popovich, Achieving CNS axon regeneration by manipulating convergent neuro-immune signaling, *Cell Tissue Res.*, 2012, **349**, 201–213.

39. W. Zhang, T. Wang, Z. Pei, D. S. Miller, X. Wu, M. L. Block, B. Wilson, W. Zhang, Y. Zhou, J.-S. Hong and J. Zhang, Aggregated alpha-synuclein activates microglia: a process leading to disease progression in Parkinson's disease, *FASEB J.*, 2005, **19**, 533–542.

40. M. F. McCarty, Down-regulation of microglial activation may represent a practical strategy for combating neurodegenerative disorders, *Med. Hypotheses*, 2006, **67**, 251–269.

41. T. C. Frank-Cannon, L. T. Alto, F. E. McAlpine and M. G. Tansey, Does neuroinflammation fan the flame in neurodegenerative diseases?, *Mol. Neurodegener.*, 2009, **4**, 47.

42. R. Kuno, Y. Yoshida, A. Nitta, T. Nabeshima, J. Wang, Y. Sonobe, J. Kawanokuchi, H. Takeuchi, T. Mizuno and A. Suzumura, The role of TNF-alpha and its receptors in the production of NGF and GDNF by astrocytes, *Brain Res.*, 2006, **1116**, 12–18.

43. J. Silver and J. H. Miller, Regeneration beyond the glial scar, *Nat. Rev. Neurosci.*, 2004, **5**, 146–156.

44. R. Brambilla, Inhibition of astroglial nuclear factor B reduces inflammation and improves functional recovery after spinal cord injury, *J. Exp. Med.*, 2005, **202**, 145–156.

45. R. Brambilla, T. Persaud, X. Hu, S. Karmally, V. I. Shestopalov, G. Dvoriantchikova, D. Ivanov, L. Nathanson, S. R. Barnum and J. R. Bethea, Transgenic Inhibition of Astroglial NF-κB Improves functional outcome in experimental autoimmune encephalomyelitis by suppressing chronic central nervous system inflammation, *J. Immunol.*, 2009, **182**, 2628–2640.

46. M. E. M. Hamby, J. A. J. Hewett and S. J. S. Hewett, TGF-beta1 potentiates astrocytic nitric oxide production by expanding the population of astrocytes that express NOS-2, *Glia*, 2006, **54**, 566–577.

47. X. Liu, K. A. Sullivan, J. E. Madl, M. Legare and R. B. Tjalkens, Manganese-induced neurotoxicity: the role of astroglial-derived nitric oxide in striatal interneuron degeneration, *Toxicol. Sci.*, 2006, **91**, 521–531.

48. D. L. Carbone, K. A. Popichak, J. A. Moreno, S. Safe and R. B. Tjalkens, Suppression of 1-methyl-4-phenyl-1,2,3,6-tetrahydropyridine-induced nitric-oxide synthase 2 expression in astrocytes by a novel diindolylmethane analog protects striatal neurons against apoptosis, *Mol. Pharmacol.*, 2008, **75**, 35–43.

49. T. Takano, N. Oberheim, M. L. Cotrina and M. Nedergaard, Astrocytes and Ischemic Injury, *Stroke*, 2009, **40**, S8–S12.

50. E. C. Hirsch and S. Hunot, Neuroinflammation in Parkinson's disease: a target for neuroprotection?, *Lancet Neurol.*, 2009, **8**, 382–397.

51. K. Saijo, B. Winner, C. T. Carso, J. G. Collier, L. Boyer, M. G. Rosenfeld, F. H. Gage and C. K. Glass, A Nurr1/CoREST pathway in microglia and astrocytes protects dopaminergic neurons from inflammation-induced death, *Cell*, 2009, **137**, 47–59.

52. S. C. Lee, W. Liu, D. W. Dickson, C. F. Brosnan and J. W. Berman, Cytokine production by human fetal microglia and astrocytes. Differential induction by lipopolysaccharide and IL-1 beta, *J. Immunol.*, 1993, **150**, 2659–2667.

53. W. J. W. Streit, Microglia as neuroprotective, immunocompetent cells of the CNS, *Glia*, 2002, **40**, 133–139.

54. D. Mastroeni, A. Grover, B. Leonard, J. N. Joyce, P. D. Coleman, B. Kozik, D. L. Bellinger and J. Rogers, Microglial responses to dopamine in a cell culture model of Parkinson's disease, *Neurobiol. Aging*, 2009, **30**, 1805–1817.

55. S. David and A. Kroner, Repertoire of microglial and macrophage responses after spinal cord injury, *Nat. Rev. Neurosci.*, 2011, **12**, 388–399.

56. K. A. Kigerl, J. C. Gensel, D. P. Ankeny, J. K. Alexander, D. J. Donnelly and P. G. Popovich, Identification of two distinct macrophage subsets with divergent effects causing either neurotoxicity or regeneration in the injured mouse spinal cord, *J. Neurosci.*, 2009, **29**, 13435–13444.

57. I. H. Cho, J. Hong, E. C. Suh, J. H. Kim, H. Lee, J. E. Lee, S. Lee, C.-H. Kim, D. W. Kim, E.-K. Jo, K. E. Lee, M. Karin and S. J. Lee, Role of microglial IKK in kainic acid-induced hippocampal neuronal cell death, *Brain*, 2008, **131**, 3019–3033.

58. M. Karin, Inflammation-activated Protein Kinases as Targets for Drug Development, *Proc. Am. Thorac. Soc.*, 2005, **2**, 386–390.

59. E. Alcamo, J. P. Mizgerd, B. H. Horwitz, R. Bronson, A. A. Beg, M. Scott, C. M. Doerschuk, R. O. Hynes and D. Baltimore, Targeted mutation of TNF receptor I rescues the RelA-deficient mouse and reveals a critical role for NF-kappa B in leukocyte recruitment, *J. Immunol.*, 2001, **167**, 1592–1600.

60. G. van Loo, R. de Lorenzi, H. Schmidt, M. Huth, A. Mildner, M. Schmidt-Supprian, H. Lassmann, M. R. Prinz and M. Pasparakis, Inhibition of transcription factor NF-κB in the central nervous system ameliorates autoimmune encephalomyelitis in mice, *Nat. Immunol.*, 2006, 7, 954–961.

61. J. A. DiDonato, M. Hayakawa, D. M. Rothwarf, E. Zandi and M. Karin, A cytokine-responsive IκB kinase that activates the transcription factor NF-κB, *Nature*, 1997, **388**, 548–554.

62. M. Karin, Inflammation-activated Protein Kinases as Targets for Drug Development, *Proceedings of the American Thoracic Society*, 1999, **2**, 386–390.

63. Z.-W. Li, S. A. Omori, T. Labuda, M. Karin and R. C. Rickert, IKK beta is required for peripheral B cell survival and proliferation, *J. Immunol.*, 2003, **170**, 4630–4637.

64. G. Bonizzi and M. Karin, The two NF-κB activation pathways and their role in innate and adaptive immunity, *Trends Immunol.*, 2004, **25**, 280–288.
65. A. Vezzani, E. Aronica, A. Mazarati and Q. J. Pittman, Epilepsy and brain inflammation, *Exp. Neurol.*, 2011, DOI: 10.1016/j.expneurol.2011.09.033.
66. A. K. Ngugi, S. M. Kariuki, C. Bottomley, I. Kleinschmidt, J. W. Sander and C. R. Newton, Incidence of epilepsy: a systematic review and meta-analysis, *Neurology*, 2011, **77**, 1005–1012.
67. B. Xu, Z.-F. Xu and Y. Deng, Manganese exposure alters the expression of N-methyl-d-aspartate receptor subunit mRNAs and proteins in rat striatum, *J. Biochem. Mol. Toxicol.*, 2010, **24**, 1–9.
68. A. Easter, B. M. Elizabeth, J. R. Damewood, W. S. Redfern, J.-P. Valentin, M. J. Winter, C. Fonck and R.A. Bialecki, Approaches to seizure risk assessment in preclinical drug discovery, *Drug Discovery Today*, 2009, **14**, 876–884.
69. M. A. Rogawski and W. Löscher, The neurobiology of antiepileptic drugs for the treatment of nonepileptic conditions, *Nat. Med.*, 2004, **10**, 685–692.
70. S. Najjar, D. Pearlman, D. C. Miller and O. Devinsky, Refractory epilepsy associated with microglial activation, *Neurologist*, 2011, **17**, 249–254.
71. G. Losi, M. Cammarota and G. Carmignoto, The role of astroglia in the epileptic brain, *Front. Pharmacol.*, 2012, **3**, 132.
72. M. L. Foresti, G. M. Arisi and L. A. Shapiro, Role of glia in epilepsy-associated neuropathology, neuroinflammation and neurogenesis, *Brain Res. Rev.*, 2011, **66**, 115–122.
73. G. Alarcón and A. Valentín, Mesial Temporal Lobe Epilepsy with Hippocampal Sclerosis. in *Atlas of Epilepsies*, Springer, London, 2010. pp. 1171–1175.
74. E. Aronica and P. B. Crino, Inflammation in epilepsy: clinical observations, *Epilepsia*, 2011, **52**, 26–32.
75. P. I. P. Ortinski, J. J. Dong, A. A. Mungenast, C. C. Yue, H. H. Takano, D. J. D. Watson, P. G. P. Haydon and D. A. D. Coulter, Selective induction of astrocytic gliosis generates deficits in neuronal inhibition, *Nat. Neurosci.*, 2010, **13**, 584–591.
76. D. J. Lee, M. S. Hsu, M. M. Seldin, J. L. Arellano and D. K. Binder, Decreased expression of the glial water channel aquaporin-4 in the intrahippocampal kainic acid model of epileptogenesis, *Exp. Neurol.*, 2012, **235**, 246–255.
77. O. Devinsky, A. Vezzani, S. Najjar, N. C. De Lanerolle and M. A. Rogawski, Glia and epilepsy: excitability and inflammation, *Trends Neurosci.*, 2013, **36**, 174–184.
78. G. C. Brown, Mechanisms of inflammatory neurodegeneration: iNOS and NADPH oxidase, *Biochem. Soc. Trans.*, 2007, **35**, 1119–1121.
79. R. Kovacs, A. Rabanus, J. Otahal, A. Patzak, J. Kardos, K. Albus, U. Heinemann and O. Kann, Endogenous nitric oxide is a key promoting factor for initiation of seizure-like events in hippocampal and entorhinal cortex slices, *J. Neurosci.*, 2009, **29**, 8565–8577.

80. D. A. Coulter and T. Eid, Astrocytic regulation of glutamate homeostasis in epilepsy, *Glia*, 2012, **60**, 1215–1226.

81. T. R. Guilarte, Manganese and Parkinson's disease: a critical review and new findings, *Environ. Health Perspect.*, 2010, **118**, 1071–1080.

82. D. P. Perl and C. W. Olanow, The neuropathology of manganese-induced Parkinsonism, *J. Neuropathol. Exp. Neurol.*, 2007, **66**, 675–682.

83. W. G. Meissner M. Frasier, T. Gasser, C. G. Goetz, A. Lozano, P. Piccini, J. A. Obeso, O. Rascol, A. Schapira, V. D. Voon, M. Weiner, F. Tison, E. Bezard, Priorities in Parkinson's disease research. *Nat. Rev. Drug. Discov.*, 2011, **10**, 377–393.

84. K. R. Chaudhuri and A. H. Schapira, Non-motor symptoms of Parkinson's disease: dopaminergic pathophysiology and treatment, *Lancet Neurol.*, 2009, **8**, 464–474.

85. C. S. Lu, C. C. Huang, N. S. Chu and D. B. Calne, Levodopa failure in chronic manganism, *Neurology*, 1994, **44**, 1600–1602.

86. J. Jankovic, Parkinson's disease: clinical features and diagnosis, *J. Neurol., Neurosurg. Psychiatry*, 2008, **79**, 368–376.

87. T. Gasser, Molecular pathogenesis of Parkinson disease: insights from genetic studies, *Expert Rev. Mol. Med.*, 2009, **11**, e22.

88. C. M. Lema Tomé, T. Tyson, N. L. Rey, S. Grathwohl, M. Britschgi and P. Brundin, Inflammation and α-synuclein's prion-like behavior in Parkinson's disease – is there a link?, *Mol. Neurobiol.*, 2012, **47**, 561–574.

89. S.-J. Lee, Origins and effects of extracellular alpha-synuclein: implications in Parkinson's disease, *J. Mol. Neurosci.*, 2008, **34**, 17–22.

90. T. Nagatsu and M. Sawada, Inflammatory process in Parkinson's disease: role for cytokines, *Curr. Pharm. Des.*, 2005, **11**, 999–1016.

91. S. Sugama, L. Yang, B. P. Cho, L. A. DeGiorgio, S. Lorenzl, D. S. Albers, M. F. Beal, B. T. Volpe and T. H. Joh, Age-related microglial activation in 1-methyl-4-phenyl-1, 2, 3, 6-tetrahydropyridine (MPTP)-induced dopaminergic neurodegeneration in C57BL/6 mice, *Brain Res.*, 2003, **964**, 288–294.

92. J. A. Miller, B. R. Trout, K. A. Sullivan, R. A. Bialecki, R. A. Roberts and R. B. Tjalkens, Low-dose 1-methyl-4-phenyl-1,2,3,6-tetrahydropyridine causes inflammatory activation of astrocytes in nuclear factor-κB reporter mice prior to loss of dopaminergic neurons, *J. Neurosci. Res.*, 2011, DOI: 10.1002/jnr.22549.

93. H.-M. Gao, H. Zhou, F. Zhang, B. C. Wilson, W. Kam and J.-S. Hong, HMGB1 Acts on microglia Mac1 to Mediate chronic neuroinflammation that drives progressive neurodegeneration, *J. Neurosci.*, 2011, **31**, 1081–1092.

94. K. Sriram, Deficiency of TNF receptors suppresses microglial activation and alters the susceptibility of brain regions to MPTP-induced neurotoxicity: role of TNF, *FASEB J.*, 2006, **20**, 670–682.

95. M. Aschner and J. L. Aschner, Manganese neurotoxicity: cellular effects and blood-brain barrier transport, *Neurosci. Biobehav. Rev.*, 1991, **15**, 333–340.

96. A. Sigel, H. Sigel and R. K. O. Sigel, *Metal ions in life sciences*, Wiley, New York, NY, 2007.
97. M. S. Hua and C. C. Huang, Chronic occupational exposure to manganese and neurobehavioral function, *J. Clin. Exp. Neuro.*, 1991, **13**, 495–507.
98. A. B. Santamaria, Manganese exposure, essentiality and toxicity, *Indian J. Med. Res.*, 2008, **128**, 484–500.
99. A. Woolf, R. Wright, C. Amarasiriwardena and D. Bellinger, A child with chronic manganese exposure from drinking water, *Environ. Health Perspect.*, 2002, **110**, 613–616.
100. P. J. Collipp, S. Y. Chen and S. Maitinsky, Manganese in infant formulas and learning disability, *Ann. Nutr. Metab.*, 1983, **27**, 488–494.
101. J. ASCHNER and M. ASCHNER, Nutritional aspects of manganese homeostasis, *Mol. Aspects Med.*, 2005, **26**, 353–362.
102. A. P. Neal and T. R. Guilarte, Mechanisms of heavy metal neurotoxicity: lead and manganese, *J. Drug Metab. Toxicol.*, 2012, **5**, 2.
103. J. A. Menezes-Filho, C. Novaes, O. de, J. C. Moreira, P. N. Sarcinelli and D. Mergler, Elevated manganese and cognitive performance in school-aged children and their mothers, *Environ. Res.*, 2011, **111**, 156–163.
104. H. Riojas-Rodríguez, R. Solís-Vivanco, A. Schilmann, S. Montes, S. Rodríguez, C. Ríos and Y. Rodríguez-Agudelo, Intellectual function in Mexican children living in a mining area and environmentally exposed to manganese, *Environ. Health Perspect.*, 2010, **118**, 1465–1470.
105. Y. Kim, B.-N. Kim, Y.-C. Hong, M.-S. Shin, H.-J. Yoo, J.-W. Kim, S.-Y. Bhang and S.-C. Cho, Co-exposure to environmental lead and manganese affects the intelligence of school-aged children, *Neurotoxicology*, 2013, **35**, 15–22.
106. C. W. Olanow, Manganese-induced parkinsonism and Parkinson's disease, *Ann. N. Y. Acad. Sci.*, 2004, **1012**, 209–223.
107. S. Zhang, Z. Zhou and J. Fu, Effect of manganese chloride exposure on liver and brain mitochondria function in rats, *Environ. Res.*, 2003, **93**, 149–157.
108. D. Centonze, Impaired excitatory transmission in the striatum of rats chronically intoxicated with manganese, *Exp. Neurol.*, 2001, **172**, 469–476.
109. N. M. Filipov and C. A. Dodd, Role of glial cells in manganese neurotoxicity, *J. Appl. Toxicol.*, 2011, **32**, 310–317.
110. M. Spranger, S. Schwab, S. Desiderato, E. Bonmann, D. Krieger and J. Fandrey, Manganese augments nitric oxide synthesis in murine astrocytes: a new pathogenetic mechanism in manganism?, *Exp. Neurol.*, 1998, **149**, 277–283.
111. R. Barhoumi, J. Faske, X. Liu and R. B. Tjalkens, Manganese potentiates lipopolysaccharide-induced expression of NOS2 in C6 glioma cells through mitochondrial-dependent activation of nuclear factor kappaB, *Mol. Brain Res.*, 2004, **122**, 167–179.
112. C.-J. Chen, Y.-C. Ou, S.-Y. Lin, S.-L. Liao, S.-Y. Chen and J.-H. Chen, Manganese modulates pro-inflammatory gene expression in activated glia, *Neurochem. Int.*, 2006, **49**, 62–71.

113. N. M. Filipov, R. F. Seegal and D. A. Lawrence, Manganese potentiates in vitro production of proinflammatory cytokines and nitric oxide by microglia through a nuclear factor kappa B-dependent mechanism, *Toxicol. Sci.*, 2005, **84**, 139–148.

114. J. A. Moreno, K. M. Streifel, K. A. Sullivan, W. H. Hanneman and R. B. Tjalkens, Manganese-induced NF-κB activation and nitrosative stress is decreased by estrogen in juvenile mice, *Toxicol Sci.*, 2011, **122**, 121–133.

115. J. A. Moreno, K. A. Sullivan, D. L. Carbone, W. H. Hanneman and R. B. Tjalkens, Manganese potentiates nuclear factor-kappaB-dependent expression of nitric oxide synthase 2 in astrocytes by activating soluble guanylate cyclase and extracellular responsive kinase signaling pathways, *J. Neurosci. Res.*, 2008, **86**, 2028–2038.

116. F. Zhao, T. Cai, M. Liu, G. Zheng, W. Luo and J. Chen, Manganese induces dopaminergic neurodegeneration via microglial activation in a rat model of manganism, *Toxicol. Sci.*, 2008, **107**, 156–164.

117. T. Verina, S. F. Kiihl, J. S. Schneider and T. R. Guilarte, Manganese exposure induces microglia activation and dystrophy in the substantia nigra of non-human primates, *Neurotoxicology*, 2011, **32**, 215–226.

118. P. Zhang, T. A. Wong, K. M. Lokuta, D. E. Turner, K. Vujisic and B. Liu, Microglia enhance manganese chloride-induced dopaminergic neurodegeneration: role of free radical generation, *Exp. Neurol.*, 2009, **217**, 219–230.

CHAPTER 13

Modeling Manganese Kinetics for Human Health Risk Assessment

MIYOUNG YOON,*[a] MICHAEL D. TAYLOR,[b]
HARVEY J. CLEWELL[a] AND MELVIN E. ANDERSEN[a]

[a] The Hamner Institutes for Health Sciences, Research Triangle Park, NC, USA; [b] Nickel Producers Environmental Research Association (NiPERA), Durham, NC, USA
*Email: myoon@thehamner.org

13.1 Introduction

Manganese (Mn) is an essential element, present in significant concentrations in the central nervous system (CNS) and all other tissues.[1-4] Mn is required for normal immune function, regulation of blood sugar and cellular energy, lipid and carbohydrate metabolism, skeletal cartilage development, reproduction, and brain functions. It participates in metabolic pathways and a co-factor metal for Mn superoxide dismutase and glutamine synthetase.[5,6] The body's requirements for Mn are met through food and drinking water, with adult human consumption ranging from 1 to 10 mg Mn per day, of which about 1–5% is absorbed in the gut.[2,5,7] Homeostatic mechanisms controlling uptake, storage, and elimination of dietary Mn maintain tissue Mn concentrations at a relatively constant level despite widely varying dietary levels of Mn.[8,9]

Issues in Toxicology No. 22
Manganese in Health and Disease
Edited by Lucio G. Costa and Michael Aschner
© The Royal Society of Chemistry 2015
Published by the Royal Society of Chemistry, www.rsc.org

As with other essential elements, Mn toxicity occurs with excessive exposure. The adverse effects associated with high exposure to Mn correlate with increases in Mn concentrations in affected tissues,[9] among which neurotoxicity is the major human health concerns.[10] The target tissues for Mn neurotoxicity are mid-brain structure such as striatum and globus pallidus that control motor functions.[1,8] A Parkinson's disease-like condition known as manganism has been observed after prolonged inhalation of high concentrations of Mn (>1 mg m^{-3}) containing dusts in occupational settings such as mining. More subtle neurological effects, such as subclinical deficits in fine motor control, occur in workers with chronic inhalation to lower concentrations (0.5–1 mg m^{-3}) of Mn in the workplace. High-dose oral ingestion through drinking water, parenteral exposure to Mn or conditions where hepatobiliary clearance of this element is impaired[6] also lead to similar neurological effects. Clearly, the CNS is the critical determinant for Mn toxicity, regardless of exposure route.

Pharmacokinetic (PK) models, especially physiologically based pharmacokinetic (PBPK) models, have value in human health assessment for many compounds by providing a means to predict the internal dose at the target for various exposure conditions. Use of internal exposure at the target tissue instead of external exposure metrics assists understanding the dose–response relationship for the effect of concern in the target tissue and aids extrapolation of the findings across species and/or exposure conditions. With Mn there is the challenge to consider both essentiality and toxicity. To support risk assessment of Mn, the focus of the modeling should be predicting exposure conditions that would lead to toxicologically significant increases in tissue Mn concentrations compared with those arising from normal dietary intake.[11] To this end, PK or PBPK models need to account for background amounts of Mn in tissues, the background uptake from normal daily ingestion; homeostatic regulation of body burden as an essential element; and other environmental or occupational exposures. This chapter describes how these considerations are incorporated into PBPK models that characterize brain Mn concentrations following Mn various routes of exposure. In addition, we review the Mn PK data used for model development and the PBPK modeling process, emphasizing their importance for understanding the control of tissue Mn for various exposure conditions and in different human subpopulations. We finish the chapter showing the application of the model to assist in a Mn risk assessment based on changes in target tissue concentrations of this essential element.

13.2 Key Findings from Mn Pharmacokinetic Studies

Since Mn is an essential element, the body has various biochemical mechanisms to ensure retention of Mn with low dietary intake and to limit uptake and enhance excretion at higher dietary intake. With inhalation, increasing intake can enhance biliary excretion, but the lung epithelium does not limit uptake of higher concentrations as does the gut. Findings from various PK studies on Mn over the past decade (Table 13.1)[12] supported the presence of dose-dependent transitions in Mn kinetics with increasing inhalation exposure: there are inhalation exposures below which tissue Mn levels are within normal bounds and above which tissue concentrations increase leading to toxicity. Key results supporting a dose-dependent transition leading to tissue overload include:

1. Mn arising from a defined level of dietary exposure is present in all tissues, *i.e.* basal steady state tissue levels.
2. Increase in brain (and other tissue) Mn concentrations do not occur following low to moderate level Mn inhalation.
3. High-concentration exposure, *e.g.* inhalation exposure at air concentration greater than 0.1 mg M m^{-3}, results in increased Mn concentrations in brain and other tissues. Tissues approach a new steady state rapidly even at high air concentrations, *i.e.* within 45 exposure days for 6 h per day exposure schedule.[13,14]

At low to moderate inhalation exposure, tissue concentrations do not increase because incremental uptake of Mn to the body by inhalation is simply much lower than that provided by dietary intake. At moderate inhalation levels, homeostatic mechanisms that increase biliary and pancreatic excretion of Mn serve to maintain basal (or background) tissue Mn concentrations.[13–18] At higher levels of inhaled Mn, tissue concentrations start to increase in closer proportion to the inhaled concentration with increases expected in all tissues. With cessation of the higher concentration inhalation, tissue Mn falls back to normal levels more quickly than expected from half-lives on Mn elimination observed in tracer studies.[13,14]

Other factors such as gender, age, and dietary Mn deficiency did not enhance Mn accumulation in the mid-brain.[13,16,19] The rat placenta effectively limited Mn delivery to the fetus including the brain.[19,20] At similar exposure concentrations, neonatal rats developed brain Mn concentrations comparable to levels seen in young male rats, non-pregnant female rats, and senescent rats.[13,19,21] Particle solubility influenced Mn pharmacokinetics, with inhalation of the more soluble sulfate form resulting in higher brain Mn concentrations than the less soluble phosphate or oxide forms.[13–15,22,23] Finally, rats and non-human primates exhibited qualitatively similar dose dependencies for Mn kinetics.[13,14] Both rats and monkeys maintained tissue Mn levels at relatively constant levels until high exposure concentrations, *e.g.* air concentrations of Mn greater than 0.1 mg m^{-3} led to marked increases of target tissue Mn. These PK similarities across species provide confidence that kinetic behaviors in rats are representative of general behaviors expected in other species, including humans.

Table 13.1 Description of various pharmacokinetics studies on Mn over the past 10 years.

Study	Mn form	Exposure concentrations	Exposure length	Description
With Spraque-Dawley or CD rats				
Vitarella *et al.*, 1998,[81] 2000[82]	Hurealite (inhalation), Chloride (tracer by iv)	0.03, 0.3, 3 mg Mn m^{-3} ($\sim$100 ppm Mn in diet)	6 h per day 5 days per week (10 exposures) or 7 days per week (14 exposures); single iv injection post-inhalation	Brain and tissue measurement
Brenneman *et al.*, 1999[83]	Chloride	25, 50 mg Mn kg^{-1} body weight in drinking water ($\sim$100 ppm Mn in diet)	7 days per week at PND 1 to 21 and 5 days per week at PND 22 to 49	Brain biochemistry and motor activity analysis of pups
Brenneman *et al.*, 2000[84]	Chloride (tracer)	0.5 mg Mn m^{-3} ($\sim$100 ppm Mn in diet)	90 min nose only (with/without occlusion)	Brain and tissue measurement and autoradiography imaging
Vitarella *et al.*, 2000[85]	Hurealite, sulfate, or oxide	0.04, 0.04, 0.16 ug Mn m^{-3} ($\sim$100 ppm Mn in diet)	1, 3, 14 days of intratracheal instillation	Brain and tissue measurement
Dorman *et al.*, 2000[86]	Chloride	25, 50 mg Mn kg^{-1} body weight in drinking water ($\sim$100 ppm Mn in diet)	21 consecutive days by oral gavage	Motor and behavioral analysis and brain measurement of pups and adult males
Dorman *et al.*, 2001[87]	Sulfate (inhalation), Chloride (tracer by iv)	0.03, 0.3, 3 mg Mn m^{-3} *vs.* 2, 10, 100 ppm Mn diets	6 h per day for 14 consecutive days; single iv injection post-inhalation	Brain and tissue measurement
Dorman *et al.*, 2001[88]	Sulfate or oxide	0.03, 0.3, 3 mg Mn m^{-3} ($\sim$100 ppm Mn in diet)	6 h per day for 14 consecutive days; single iv injection post-inhalation	Brain and tissue measurement
Dorman *et al.*, 2002[89]	Phosphate (tracer)	0.4 mg Mn m^{-3} ($\sim$100 ppm Mn in diet)	90 min nose only (with/without occlusion)	Brain and tissue measurement and autoradiography imaging

Table 13.1 (*Continued*)

Study	Mn form	Exposure concentrations	Exposure length	Description
Dorman *et al.*, 2002[90]	Oxide (inhalation), Chloride (tracer by iv)	0.03, 0.3, 3 mg Mn m^{-3} *vs.* 2, 10, 100 ppm Mn diets	6 h per day for 14 consecutive days; single iv injection post-inhalation	Brain and tissue measurement
Dorman *et al.*, 2004[91,92]	Hurealite or sulfate, Chloride (tracer by iv)	0.01, 0.1, 0.5 mg Mn m^{-3} (sulfate), 0.1 mg Mn m^{-3} (hurealite)(10 ppm Mn diet)	6 h per day 5 days per week for 13 consecutive weeks	Brain and tissue measurement
Dorman *et al.*, 2005[93,94]	Sulfate	0.05, 0.5, 1.0 mg Mn m^{-3} (10 ppm Mn diet)	6 h per day for 19 consecutive days (during and after pregnancy)	Brain and tissue measurement in pups and dams/lactating mother
Salehi *et al.*, 2001[95]	Hurealite	3 mg Mn m^{-3} (~100 ppm Mn in diet)	8 h per day 5 days per week for 5 consecutive weeks	Brain and tissue measurement, locomotor activity assessment, and neuropathology analysis
St. Pierre *et al.*, 2001[96]	Metallic (99%)	4 mg Mn m^{-3} (~100 ppm Mn in diet)	6 h per day 5 days per week for 13 consecutive weeks	Brain and tissue measurement along with locomotor activity assessment

Study	Form	Dose	Exposure	Endpoints
Normandin et al., 2002;[97] Salehi et al., 2003[98]	Hurealite	0.03, 0.3, 3 mg m^{-3} (95 ppm Mn in diet)	6 h per day 5 days per week for 13 consecutive weeks	Brain and tissue measurement, neurological assessment, and neuropathology analysis
Normandin et al., 2004;[99] Beaupré et al., 2004[100]	Metallic (99%), hurealite, or hurealite/sulfate mixture	4 mg metallic Mn m^{-3}, 3 mg m^{-3} hurealite and mixture (95 ppm Mn in diet)	6 h per day 5 days per week for 13 consecutive weeks	Brain and tissue measurement along with locomotor activity assessment
Tapin et al., 2005[101]	Sulfate	0.03, 0.3, 3 mg Mn m^{-3} ($\sim$100 ppm Mn in diet)	6 h per day 5 days per week for 13 consecutive weeks	Brain and tissue measurement, locomotor activity assessment, and neuropathology analysis
With Rhesus monkeys				
Dorman et al., 2005,[102] 2006[103]	Sulfate	0.06, 0.3, 1.5 mg Mn m^{-3} (133 ppm Mn in diet)	6 h per day 5 days per week for 13 consecutive weeks	Brain and tissue measurement and histopathology
Dorman et al., 2006[104]	Sulfate	0.06, 0.3, 1.5 mg Mn m^{-3} (133 ppm Mn in diet)	6 h per day 5 days per week for 13 consecutive weeks	Brain measurement and MRI assessment
Struve et al., 2007[105]	Sulfate	0.06, 0.3, 1.5 mg Mn m^{-3} (133 ppm Mn in diet)	6 h per day 5 days per week for 13 consecutive weeks	Brain neurochemistry analysis

Notes: hurealite: $Mn_5(PO_4)_2(PO_3(OH))_2 \cdot 4H_2O$; phosphate: $^{54}MnHPO_4$; sulfate: $MnSO_4$; chloride: $MnCl_2/^{54}MnCl_2$; oxide: Mn_3O_4; PND: postnatal days.
Source: Reprinted from Andersen et al.[12]

13.3   Pharmacokinetic Modeling of Mn

The development of PBPK models facilitated more rigorous quantitative analyses of the available PK data for Mn. Although PK or PBPK modeling for many exogenous compounds has become routine, there are significantly more challenges in understanding the full set of biological factors that control uptake, distribution, and clearance for essential trace metals, such as Mn. In addition, essential element biology concerning the mechanism of cellular fluxes and utilization in different tissues under various exposure conditions is still evolving.[6,24] Because of the limitations in understanding all the biological steps in control, the PBPK model developed does not contain a one-to-one correspondence for every biochemical process in Mn homeostasis. Our primary goal in developing the PBPK models was capturing the dose-dependent increases expected in rodents and then providing a confidence that a similar model structure was appropriate for humans. Although they are based on particular assumptions about processes controlling Mn within tissues, the PBPK models described in this chapter allowed coherent integration of the currently known Mn biology and recapitulated the observed tissue Mn in various tissues within the body.

13.3.1   Initial Development

Previously published essential element models for zinc[25,26] and copper[27,28] are based on a compartmental modeling approach. They captured the control of these essential elements under normal and deficient dietary conditions. Linear exchange rates captured distribution of the elements in different tissues and cellular components. Initially, this linear model structure was evaluated for Mn.[29,30] The model rate constants were fitted to reproduce the rate of [54]Mn tracer elimination from rats fed diets containing Mn concentrations ranging from what is considered to be marginally sufficient (1.5 ppm Mn) to standard, *i.e.* sufficient, level (125 ppm Mn) diets. The linear model structure described the decreased gastrointestinal Mn uptake and increased hepatobiliary excretion observed with increasing Mn levels in diet.

The first attempt to build a PBPK model for Mn was made with this linear PK model approach.[31] A physiological structure describing rat blood, brain, liver, lung, bone, kidneys, and remainder of the body was used. Storage of Mn in various tissues was described as an exchange between two separate compartments in the tissue. Tissue clearances were adjusted to fit measurements of [54]Mn elimination kinetics from various tissues.[32] Biliary clearance and dietary uptake of Mn in the gut were based on steady-state, *i.e.* basal or background, tissue Mn concentrations and [54]Mn elimination kinetics.[16] While adequate in depicting the slower Mn elimination kinetics in animals fed sufficient and deficient diets, this model with simple intercompartmental transfer rate constants was not able to capture the rapid increase and subsequent rapid decline in tissue Mn concentrations observed during and after Mn inhalation.[33] This observation suggested that Mn inhalation kinetics, *e.g.* kinetics in high dose conditions, do not reflect equilibrium between compartments but rather involves non-linear processes in storage and/or uptake and clearance of Mn.

13.3.2   Mn PBPK Models

13.3.2.1   Description of Tissue Mn

Nong and colleagues proposed an alternative model structure.[33] Here the model contained saturable tissue stores and asymmetrical transport of Mn in and out of the tissue (Figure 13.1). In this representation, transport and binding regulate tissue kinetics and tissue accumulation of Mn (Figure 13.1). The 'binding' does not correspond with specific binding sites, but represents the agglomeration of multiple Mn-binding proteins and tissue stores.[6,24,34] Mn binding is then described as a reversible, saturable process with tissue specific association (k_a) and dissociation (k_d) rate constants. As a result, Mn is present in either free or bound form in the tissue; the distribution between bound and free Mn in each tissue is determined by an equilibrium dissociation constant, equivalent to the ratio of rate constants (k_d/k_a). The maximum binding capacity (B_{max}) constrains the level of *bound* Mn in each tissue. Basal tissue Mn concentrations largely reflect *bound* Mn. Saturation of this binding capacity with increasing exposure concentrations leads to

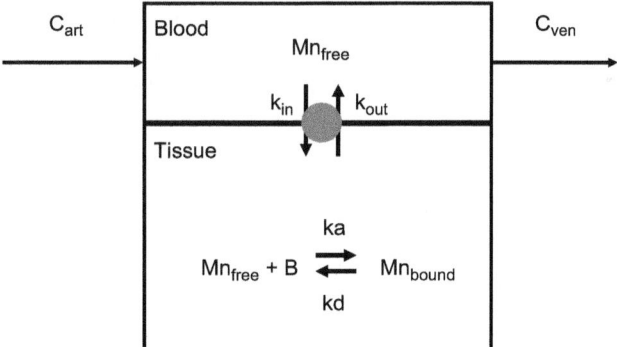

Figure 13.1   Description of tissue Mn kinetics with saturable tissue binding capacity and asymmetrical diffusional transport. Total Mn concentration (Mn_{total}) is the sum of free Mn (Mn_{free}) and bound Mn (Mn_{bound}). Maximal tissue capacity (B_{max}) is the sum of bound Mn and available free binding capacity (B). Equilibrium between free and bound Mn is determined by its dissociation constant K_D, the ratio between the dissociation (k_d) and association (k_a) rate constants. Preferential increase of free Mn in some brain regions are based on the greater ratio of diffusional influx (k_{in}) than efflux (k_{out}) occurring between Mn in the blood within the brain and brain tissue.
Reprinted from Andersen *et al.*[12]

more rapid increases as tissue Mn increases above the B_{max}. The rate of rise in *free* Mn in each tissue after exposure to high dose Mn depends on their respective binding rate constants and maximal binding capacity. The differential (or preferential) rise of *free* Mn in mid-brain regions is described by asymmetrical diffusion rate constants (k_{in} and k_{out}) that control transport of Mn across the blood–brain barrier. After cessation of exposure, tissue Mn concentrations rapidly return to their original level as *free* Mn is rapidly cleared in this description. As transport of a single form of Mn, *i.e.* Mn^{2+}, is likely to be dominant in determining the kinetic properties of this element in the body,[35] valence states of circulating Mn in blood were not described explicitly in the model.

This description of tissue compartmental kinetics adequately captured the observed Mn tissue kinetics over a wide range of daily uptake rates by both inhalation and oral exposure.[33] In essence, the concentration gradient maintained in brain tissues depends on the ratio of transport rate constants, k_{in}/k_{out}; the overall concentration at which the *free* Mn begins to dominate in all tissue depends on another ratio of rate constants, k_d/k_a, an effective dissociation constant (K_D) for Mn in the tissues and the B_{max}. This tissue description served as the basis for a series of Mn models subsequently covering different species and life stages.[36–40] These models, referred as "second generation" of Mn PMPK models in the review by Taylor and colleagues,[41] were developed in a stepwise fashion with increasing model complexity added sequentially to the basic description (Figure 13.2). The basic model structure was developed and evaluated in the rat first because the available PK data are most extensive in this species.[33,36] Extension of the basic model to other species and life stages maintained a similar physiological structure. In essence, the rat model allowed reparamaterization to other species. The ability to maintain good correspondence with kinetic data in the other species served as a test for overall model validity.

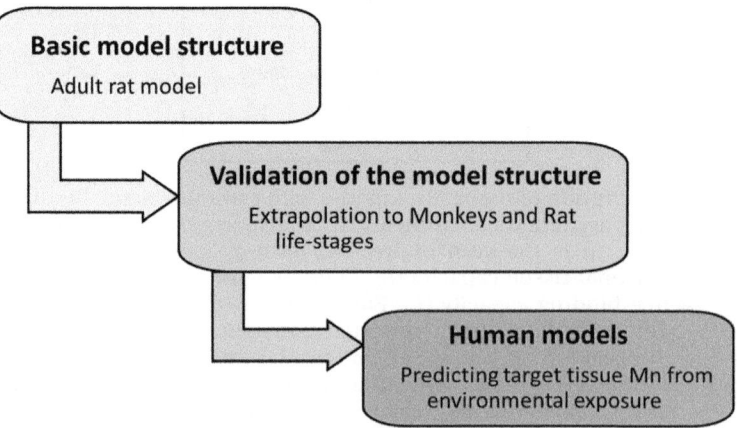

Figure 13.2   Steps to develop human PBPK models.

13.3.2.2   Homeostatic Control of Body Mn

Mn clearance from the body is slow at ingestion rates that maintain an essential (or normal) body burden of Mn. An exceptional feature of these PBPK models is the ability to reproduce the rapid rise of free Mn in the blood from high inhalation exposure and subsequent rapid clearance of Mn back to basal levels. The process of building and parameterizing a PBPK model for Mn as an essential element, which describes processes that maintain homeostasis by ingestion and still recapitulates the increase in tissue Mn that occurs with inhaled Mn above 0.1 mg m^{-3}, was first demonstrated in the rat.[33,36] Mn homeostasis with normal dietary exposure is maintained by controlling both biliary excretion and intestinal absorption.[29] Regulation of intestinal uptake by enterocytes occurs on a longer time frame, whereas changes in biliary excretion are rapid. Similar upregulation of biliary excretion was noted for increasing Mn intake, independent of dose routes.[15,16,42] To incorporate these homeostatic controls of body Mn at steady state with dietary exposure, time course data from a variety of dietary studies covering from slightly deficient to normal sufficient Mn intakes[15,16] were simulated with the PBPK model. Dietary uptake (F_{DietUp}) in the gut and biliary Mn excretion (k_{bile}) were varied in a dose-dependent pattern to be consistent with the observed tracer clearance kinetics with a range of dietary Mn levels.[36] The F_{DietUp} in the model is dietary Mn level specific. The model included two processes leading to dose-dependent increase in biliary excretion rate constants: one occurs at moderate Mn concentrations responding to slight increase in dietary Mn intake, while the other occurs to a greater extent at higher Mn intakes and the response is more rapid.[36] As a result, the biliary excretion increases and fractional dietary uptake decreases as the Mn dietary level increases. This feature is consistent with the findings from various dietary Mn balance studies and biliary elimination studies.[29] Since the dose-dependent control of Mn uptake and excretion with changes in dietary Mn levels is critical in maintaining steady state Mn levels in the body despite the fluctuation of dietary Mn levels, they were evaluated and incorporated in the model before estimating other tissue binding and transport parameters.

The model parameters regulating basal Mn levels at dietary steady state estimated from the dietary studies were subsequently applied for the inhalation kinetics. To describe dose-dependent transition of Mn kinetics, tissue capacity and diffusion parameters were calibrated to simulate tissue concentrations observed in rats exposed both through diet and *via* 6 h per day inhalation for 14 to 90 days to Mn concentrations from 0 to 3 mg m^{-3}.[15,43,36] After high-dose Mn inhalation exposure, tissue concentrations rapidly reached new steady state and rapidly returned to basal levels upon cessation of exposure.[13,14,16] These significantly shorter half-lives for the approach to steady state and return to basal levels after high-dose inhalation exposures compared to those of Mn elimination measured in tracer studies following diet only exposure, were key to designing a successful model structure. This dose-dependent behavior cannot be reproduced with linear PK models. The rapid increase in tissue Mn concentrations at the higher inhaled Mn concentrations were captured by the model by saturation of the tissue binding with increasing body burdens of Mn from inhalation and concomitant elevations in *free* Mn within the tissues (Figure 13.3). Tissue binding rate constants (k_a and k_d) were estimated to allow the basal Mn in a particular tissue to approximate the binding maximum. In this manner, most Mn in the tissue is in a *bound* form at steady state.

Different tissues show variable increases in Mn as daily intake by inhalation increases. Tissues observed with relatively higher accumulation of Mn include the mid-brain regions in rats and monkeys, the pituitary in the monkeys, and the heart in monkeys and rats.[14,16] This pattern of tissue- or brain region-specific uptake was captured in the model by incorporating asymmetrical transport between tissue intake and efflux (Figure 13.1). It is not clear why some tissues have increased capacity for Mn accumulation. Elevated Mn levels may be important for meeting energy production, neurotransmission, or antioxidant enzymatic functions in those tissues.[1] The mid-brain structures are capable of preferential Mn uptake and are also the major target organ for adverse response to excess Mn. In the rat model, influx and efflux rate constants (k_{in} and k_{out}) for different brain regions were adjusted along with the tissue binding constants (B_{max}, k_a, and k_d) to be consistent with the differential increase in tissue Mn concentrations. The binding maximum was estimated by curve fitting in a process that considered basal levels, the concentrations at which whole-body Mn begins to rise with increasing inhaled Mn, and the relative asymmetries in k_{in} and k_{out} for specific brain regions. Dose dependencies in the predicted tissue kinetics were consistent with the observed total Mn tissue levels in rats exposed at increasing inhaled Mn concentrations.[33,36]

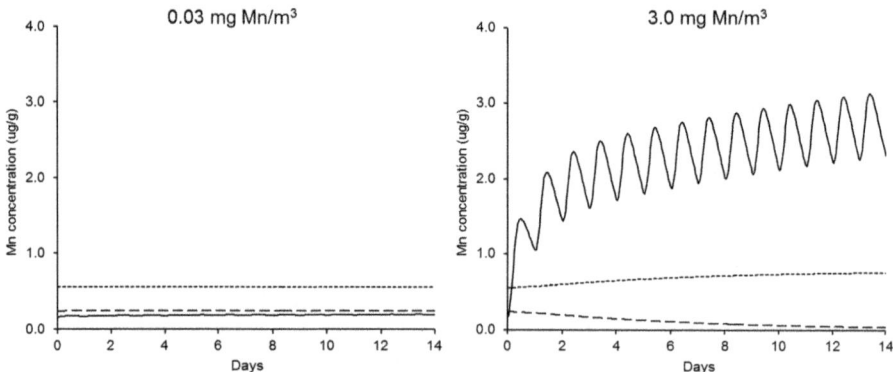

Figure 13.3 Prediction of free, bound, and free binding sites in the rat striatum following low (0.03 mg m^{-3}) and high (3 mg m^{-3}) inhalation exposure. Rats were exposed to Mn for 6 h per day for 14 consecutive days and were fed on a 125 ppm diet. The dashed lines represent free binding sites (tissue binding capacity), whereas solid and dotted lines represent free and bound Mn in the striatum, respectively.
Reprinted from Nong *et al.*[33]

13.3.2.3   Dose-Dependent Increases in Biliary Excretion

Initially, model simulations for the highest exposure concentrations (1.5 and 3.0 mg m^{-3}) used in the sub-chronic studies overestimated brain tissue Mn concentration in the rat; the agreement between the model predictions and the data was much better for shorter term inhalation, *e.g.* 14 day exposure duration (Figure 13.4).[36] A better fit to the sub-chronic studies was obtained by including a 2- to 2.5-fold increase in the rate constant for biliary excretion. This extent of induction of biliary excretion is much greater than the level of changes in this parameter in response to increase in dietary Mn within a normal fluctuation range. Enhanced biliary induction in the 90 day *versus* basal level in the 14 day study is with an adaptive response that required a more persistent increase in blood Mn to alter biochemical processes involved in excretion.

13.3.2.4   Nasal Uptake to the CNS

A deposition model based on particulate aerodynamics[44] was linked to the PBPK model to simulate pulmonary absorption of Mn. Fractional deposition of inhaled Mn particles into respiratory and nasal epithelium also had to be estimated. For this purpose, rat olfactory and respiratory regions were proportionally scaled to imaging representations of the rat nasal cavity.[45] In the deposition models, soluble MnSO$_4$ particles deposited in the respiratory region rapidly dissolved in mucus and tissue, and moved into the systemic

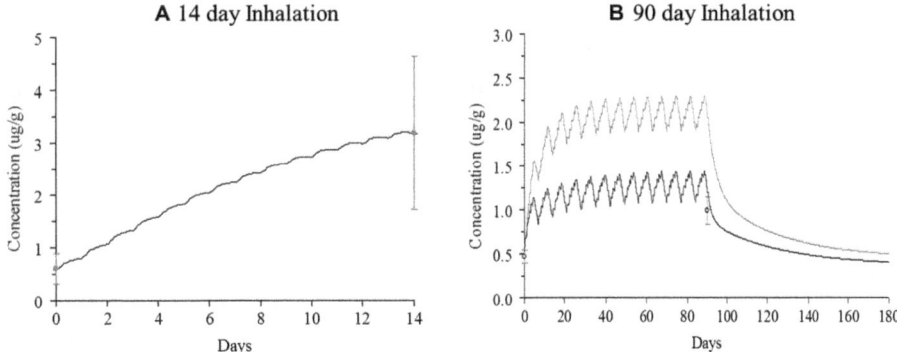

Figure 13.4 Comparison of model-simulated striatal Mn concentrations after 14 day or 90 day inhalation exposure to Mn: effect of biliary induction. Plot A shows the striatum Mn concentrations during 14 day inhalation studies[16] in rats exposed at 3 mg Mn m^{-3}; plot B shows the results from 90 day inhalation studies[13,43] at the same exposure concentration. The curve represents model simulations and symbols are data means and standard errors. The dark line in plot B represents the simulation with an increased biliary excretion rate constant, while the top line in plot B and the prediction in plot A represent simulations without biliary induction features.
Modified from Andersen *et al.*[12]

circulation. In describing the absorption of inhaled particles of Mn in PBPK models, there were also important contributions to understanding nasal transport. Leavens and colleagues[46] developed a PK model to describe the olfactory transport and systemic delivery of Mn in rats following acute inhalation exposure. This model evaluated data sets from a study of rats exposed over 90 min by inhalation to ^{54}MnCl$_2$ or ^{54}MnHPO$_4$ with either both nostrils patent or one nostril occluded.[47,48] The model provided estimates for the inter-compartmental transport of Mn from the olfactory mucosa, to olfactory bulb, to olfactory tract and tubercle, and into the striatum. In addition to influx from and efflux to the blood, each compartment contained free and bound Mn. After considering the relative solubility of the two forms of Mn, the model predictions substantiated that direct olfactory transport from the mucosa provided the majority of the Mn in the olfactory system. However, only a small amount of Mn tracer eliminated from the olfactory tract could be associated with neuronal transport: 3% for ^{54}MnHPO$_4$ and 0.1% for ^{54}MnCl$_2$. Olfactory delivery of Mn to the striatum appears minor compared to systemic uptake of Mn into other regions of the brain. This finding is consistent with the paucity of direct neural connections between the olfactory bulb and tract and the striatum.[19] In an inhalation study in non-human primates, utilizing magnetic resonance imaging (MRI) found that marked hyper-intensities were present in both the olfactory bulb and globus pallidus, but no tracts were visible in intervening tissues.[49] Collectively, these and other studies indicate that Mn does not undergo significant transport to the mid-brain structures directly from the nasal olfactory epithelium. Nonetheless, the more recent PBPK models included the possibility of some olfactory transport to account for pathway in subsequent dosimetry evaluations.

13.3.3 Inter-species Extrapolation to Non-human Primates

The PBPK model structure that successfully described Mn kinetics in rats for various exposure conditions was extended in a step-wise fashion to non-human primates and then to humans.[36,37] The monkey model was viewed as a critical step in the evolution of appropriate human models. Most structural features of the rat model were retained in order to maintain consistent descriptions of the controlling processes in Mn distribution, and physiological processes were scaled as appropriate between species. Successful extrapolation from rat to monkey supported the rationale behind the key determinants included in the model, thereby increasing confidence in the value of a human PBPK model for predicting Mn tissue dosimetry (Figure 13.2).

13.3.3.1 Adult Monkey PBPK Model

The adult rat model[36] has served as the basis for extrapolation to monkeys.[36,37] Tissue Mn concentrations were measured in rhesus monkeys exposed to 0.06, 0.3, or 1.5 mg m^{-3} of MnSO$_4$ (6 h per day and 5 day per week for 13 weeks). They had a diet containing about 133 ppm Mn.[14] Several steps were included in the process of adjusting and scaling basic model parameters in order to reproduce the kinetics seen in non-human primates.[33,36] Physiological parameters–body weight, tissue volumes, ventilation rates, olfactory and respiratory surface areas, and blood flows–were adjusted to describe species-specific physiology. After accounting for these physiological differences between rat and monkey, dietary uptake and basal biliary excretion rate were adjusted to fit background tissue Mn concentrations[14] with only the dietary intake. This adjustment from rat values was performed because dietary intake of Mn is one of the primary species differences that affect Mn kinetics. Laboratory animals are often maintained on diets that contain relatively high Mn concentrations. The human diet is typically much lower in Mn, ranging from around 1 to 10 mg per day; thus, the adjustment of the model parameters for dietary uptake and biliary excretion is critical for extrapolating animal models to humans. Using the Mn concentration data eliminated into bile, biliary excretion and its dose-dependent induction were adjusted (eqn 13.1) to have the biliary excretion rate increases by approximately three-fold in relation to the rise of monkey blood Mn concentration at the highest exposure concentration (1.5 mg m^{-3}).[36]

$$k_{bile} = k_{bile0} + \frac{k_{b\ max} \cdot C_{art}^n}{k_{b50}^n + C_{art}^n} \tag{13.1}$$

In this equation, k_{bile0} is the basal biliary excretion rate constant, k_{bmax} is the maximal excretion rate, k_{b50} is the arterial concentration at half the induced level of Mn in the arterial blood (C_{art}), and n is the slope factor. Blood Mn concentration (C_{art}) was used as a surrogate for free Mn concentration in the liver, the presumed driver for Mn excretion, because blood concentrations were directly measured in monkeys.[14,36] Consistent both with experimental rodent studies[16,42] and Mn experiments with humans,[50] the increased rate of biliary excretion of Mn rapidly regulates Mn levels in the body resulting from large inhaled contributions.

With these monkey-specific dietary intake and biliary excretion values, the monkey parameters for tissue Mn binding constants and brain diffusional flux rates were then estimated based on tissue concentrations observed in the 90 day inhalation PK studies.[49] These parameters were initially scaled allometrically to body weight or tissue weight, for example brain diffusional fluxes (k_{in} and k_{out}) were scaled to BW$^{-0.25}$, whereas tissue-specific binding capacity (B_{max}) was scaled to their respective tissue volumes. Tissue binding rate constants (k_a and k_d) were not allometrically scaled with the expectation that biochemical processes would be nearly constant across species. However, direct scaling of Mn kinetic parameters from the rat did not provide a good description of Mn tissue kinetics in the monkey.[37] Some of these parameters required further adjustment after generic allometric scaling approaches were attempted.[37] However, it bears emphasis that the description of underlying mechanisms governing Mn uptake and elimination in the model–inducible biliary excretion, tissue binding processes, and asymmetric diffusion fluxes–were preserved for all species, providing a consistent model structure. Further experiments would be helpful to determine the specific transporters involved in uptake and efflux from tissues and the nature of the binding sites. This information would refine understanding of the cellular processes of Mn utilization and trafficking, within tissues and the control mechanisms underlying intestinal uptake and biliary excretion/induction. Nonetheless, these non-human primates PBPK models have captured the main dose-dependent characteristics of Mn disposition in monkeys and provided a structure to organize and parameterize an equivalent PK description in humans.

13.3.4   Development of Human PBPK Models for Mn

The starting point of human model development was estimating basal uptake and excretion parameters that are consistent with normal dietary intake in humans. One of the primary species differences that affect Mn kinetics across species is differential dietary intake.[36,37] Dietary Mn uptake was calibrated to whole body elimination kinetics from intravenous (iv) tracer doses of ^{54}Mn in human volunteers[50,51] and the dose-dependent biliary induction (eqn 13.1) was allometrically scaled from the monkey model.[37] Typical human dietary intake was set at 3 mg Mn per day, a value considered to be around the adequate intake level.[5] Basal human tissue Mn levels were inferred from the tissue concentrations measured in cadaver studies and then tissue binding capacities (B_{max}) were estimated to be consistent with these basal concentrations.[37] The criterion for estimating tissue binding parameters (k_a and k_d) was to ensure that Mn in the basal condition was primarily in the bound form at steady-state. Even with this strategy, the range of the fraction bound varies depending on the dietary Mn intake compared to the Mn level required for normal body functions. For example, because the typical diet for monkeys contains more than adequate Mn, fraction of Mn bound in the tissue in monkeys would be lower than that in humans (*i.e.* more free available in the monkey tissue) with adequate dietary Mn intake. Brain diffusional flux parameters were estimated with the values from initial allometric scaling of monkey as starting points. As in the monkey, the resulting human model recapitulated the biphasic elimination of ^{54}Mn tracer studies in humans showing that the model accurately depicts the dose-dependence of Mn pharmacokinetics across exposure route and species.[37] The complete structure of human Mn PBPK model is shown in Figure 13.5.

As Mn exposure increases, *e.g.* through inhalation, homeostatic control associated with biliary elimination is overwhelmed by increased input rates of Mn. At higher exposure (*e.g.* inhaled concentrations greater than 0.1 mg m^{-3}), binding capacities and biliary excretion become saturated, leading to higher free Mn in the tissues (Figure 13.6). A comparison of rat,

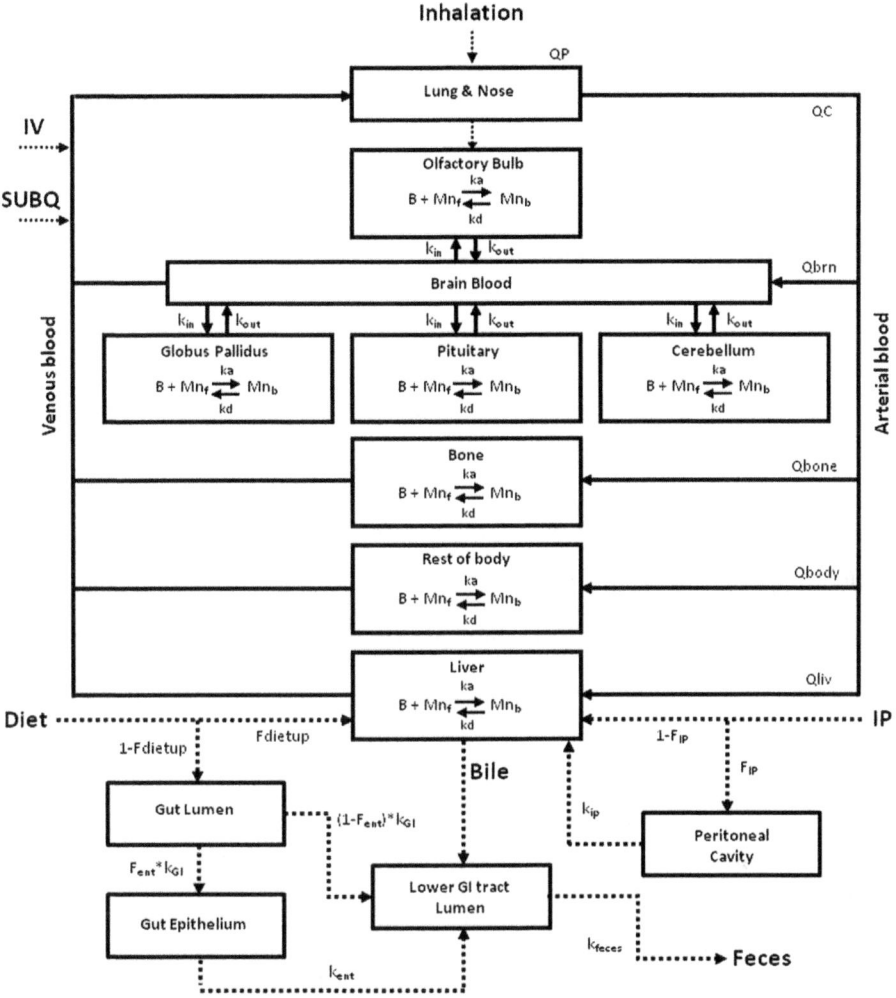

Figure 13.5 The PBPK model structure describing Mn tissue kinetics in adult monkeys and humans. In each tissue compartment, the amount of bound Mn is in equilibrium with the assumed binding capacity (B) and free Mn (Mn_f). Tissue-binding processes were controlled by association and dissociation rate constants (k_a and k_d, respectively). Free Mn moves in the blood throughout the body and is stored in each tissue as bound Mn (Mn_b). Influx and efflux diffusion rate constants (k_{in} and k_{out}, respectively) control preferential increases in free Mn in specific brain regions. QP, QC, and Q_{tissue} refer to alveolar ventilation, cardiac output, and tissue blood flows.
Reprinted from Schroeter *et al.*[37]

monkey, and human predictions of Mn concentrations in the mid-brain region reveals that all species exhibit similar behavior with increasing levels of exposure (Figure 13.7). The model simulations showed that mid-brain

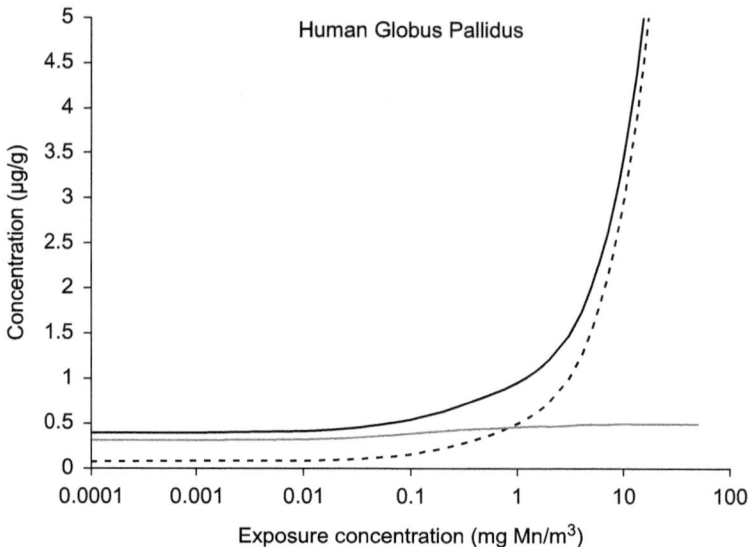

Figure 13.6 Simulated end-of-exposure (90 days) Mn tissue concentrations of total (black), free (dashed), and bound (gray) Mn in the monkey and human globus pallidus. The bound Mn fractions were assumed to be close to 60% and 80% in the monkey and human globus pallidus, respectively, under normal dietary exposure.
Modified from Schroeter *et al.*[37]

(globus pallidus) Mn concentrations are essentially unaffected by air concentrations lower than 10 µg Mn m^{-3}. The magnitude of increase differed among species. The larger proportionate changes seen in monkey simulations compared with those in humans at higher inhalation exposure concentrations are likely related to the high dietary Mn in the monkey, leading to saturation of tissue binding at lower inhaled concentrations. Consistency of our interspecies descriptions of Mn kinetics suggest that the human PBPK model should be adequate for predicting exposure conditions that will result in tissue concentrations appreciably greater than basal tissue levels in humans and can be used in a dosimetry-based risk assessment.[11]

13.3.5 Life Stage Extrapolation

Concerns for potential vulnerability to Mn neurotoxicity during fetal and neonatal development arise mainly due to increased needs for Mn during intrauterine and postnatal growth and pharmacokinetic differences between the young and adult. In addition, compared to the adult, sources of exposure to Mn differ during different stages of development: placental transfer, breast milk, and infant formula fortified with essential metals.[52] Since the body Mn homeostasis is mainly achieved at the level of absorption of Mn in

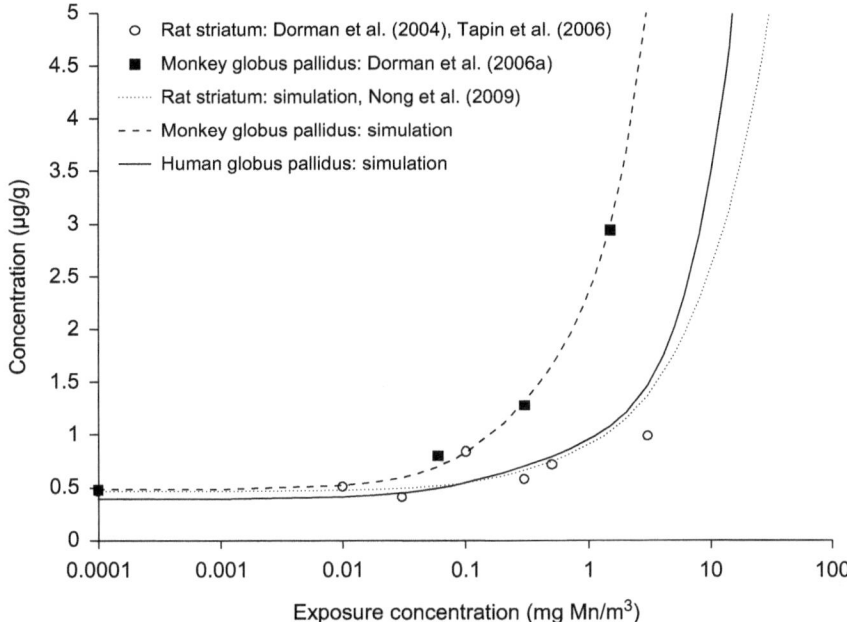

Figure 13.7 Simulated end-of-exposure tissue Mn levels in the rat striatum and monkey and human globus pallidus. The simulated rat striatal Mn levels are from Nong *et al.*[36] and are compared with data from Dorman *et al.*[13] and Tapin *et al.*[43] The simulated monkey globus pallidus Mn levels are compared with data from Dorman *et al.*[14] Rats and monkeys were exposed 6 h per day, 5 days per week for 90 days. Humans were exposed 8 h per day, 5 days per week for 90 days. Modified from Schroeter *et al.*[37]

the gut and biliary excretion of this element,[5] it is important to understand any age differences in these factors. Several factors may potentially predispose children to enhanced Mn uptake and subsequent toxic effects compared to adults. These factors include higher intestinal absorption of ingested Mn in children than in adults[53,54] and a lower basal excretion rate observed in young animals.[55] Along with the observation of enhanced delivery of Mn to the neonatal brain,[56] these pharmacokinetic differences are often regarded as a reflection of poorly developed homeostatic mechanisms in neonatal animals, all of which could result in greater uptake of Mn by the neonatal brain. However, differences in Mn pharmacokinetics observed in early ages should be evaluated with caution by taking into account the fact that Mn is essential for normal development. In fact, the avid retention of Mn seen in the neonates is likely a consequence of the scarcity of this essential element in diet.[57] The key question in addressing the potential risks of Mn in infants and children is to what extent these neonatal homeostatic controls are able to respond when challenged with Mn overexposure. In this context, some of the PBPK models integrated currently available information on Mn kinetics during fetal and neonatal development.[38–40] The models were first developed in the rat in order to identify differences in Mn homeostatic controls between young animals and adults and then appropriately extrapolated to consider expected behavior in human neonates and children.

A key question in addressing the potential sensitivity of the developing brain to Mn is the nature of dose-dependent changes in homeostatic control of Mn in early life. In other words, at what exposure level we expect to see a transition from normal steady state to over exposure conditions. The rat models for gestation and lactational periods in rats successfully addressed these questions by capturing underlying biology and physiology of Mn homeostasis during pre-and postnatal development in a PBPK model,[38,39] which was then adapted for the basis of a human developmental model.[40] For fetal and neonatal development, ensuring sufficient Mn supply from the mother through placenta and milk is a critical aspect in Mn homeostasis. On the other hand, keeping tissue Mn stable in response to the large daily fluctuation of Mn intake from a diet rich in this element is more important for maintaining homeostasis in the adult.

13.3.5.1   Describing Mn Homeostasis During Development

The essence of the feto–placental control of Mn during gestation was to ensure adequate supply of Mn to the developing fetus while maintaining maternal homeostasis.[38] The model included active/carrier-mediated transport of Mn from the placenta to the fetus that ensured adequate supply of Mn for growing fetus for maternal Mn intake ranging from marginally deficient to sufficient range; however, saturation of this active transfer of excessive Mn prevents fetal uptake for high maternal Mn loads. This description was supported by observation of no appreciable increase in fetal brain Mn after inhalation exposure in pregnant rats,[20] the accumulation of Mn in placenta in humans determined *in vitro*,[58] and studies in rhesus monkeys *in utero*.[59]

The most striking differences between the neonate and adult for Mn homeostasis are enhanced bioavailability in the gut and almost negligible biliary excretion of Mn observed in neonates. The enhanced gut uptake of Mn in the neonates[53] appears to be largely reflective of the requirement to absorb Mn to meet developmental requirements for Mn. This enhanced absorption occurs because of the immaturity of metal transport systems in the gut and limited supply of Mn from their major dietary Mn source, the mother's milk. These considerations are especially true for human infants due to even lower Mn content in human milk compared to other species.[5] The enhanced uptake of Mn in nursing infants–enhanced Mn bioavailability from mother's breast milk–was described by adjusting fractional uptake of Mn (F_{DietUp}) in the neonatal gut from that of the adult based on the receptor-mediated uptake of milk-protein-bound Mn, such as lactoferrin-bound Mn.[60–63] This process provides a mechanism for higher bioavailability of breast milk Mn minimizing the potential for Mn deficiency during nursing. In adults, divalent metal transporter-1 (DMT-1) plays an important role in absorption and transport of Mn in the gut mediated by regulation of its expression in response to the concentration/amount of Mn in food;[62,64] however, this regulation mechanism seems to be immature in neonates, as seen with iron.[65,66] The form of Mn binding-protein(s) in diet appears to serve as a cue for switching from the neonatal pattern of transporter expression (*e.g.* lactoferrin receptor) to the adult pattern (*e.g.* DMT-1) on weaning.[63] The ability to capture Mn source-specific regulation in the gut in nursing infants will be useful in evaluating the consequences of different levels of Mn in infant formula and/or infant food on tissue Mn.

13.3.5.2 Fetal and Neonatal Brain Mn

The apparent enhanced "permeability" of certain metals in the developing brain may not arise solely from immaturity of adult barrier mechanisms.[67,68] Mechanisms distinct from the adults appear to control uptake of compounds in the developing brain, thus allowing greater uptake of compounds required during development, including amino acids and growth factors.[68,69] Active transport of Mn across the blood–brain barrier is an example of specialized processes in place to meet the requirement for this element for normal brain development. The expression of transporter proteins known to be responsible for an active uptake of Mn and other essential elements appears comparable to or greater in fetuses/neonates than in adults.[52,70] Among other potential transporters, DMT-1 and transferrin receptor proteins are regarded as major contributors to Mn transport in the brain.[52,71] The elevated expression of these transporters likely results in the elevated influx of Mn into the developing brain compared to the adults. To capture this age-dependent change in brain transport characteristics, brain diffusion rate constants for influx and efflux, along with association rate constant in fetal and neonatal brains, were adjusted in the PBPK models.[38-40] For inter-species extrapolation, it is necessary to account for differences in the timing of neurodevelopment. The time profiles of age-dependent changes in brain diffusion and binding parameters were appropriately included. The success in rat PBPK modeling of developmental periods indicated that the brain uptake parameters, estimated based on the basal brain tissue Mn in the fetus and neonate, were applicable to the inhalation exposure conditions without requiring introduction of dose-dependencies for brain fluxes in early ages. The same strategy was applied in parameterizing the human developmental models.[40]

13.3.5.3 Control of Mn Biliary Excretion in Neonates

Neonatal biliary excretion at normal steady state is one of the key differences in Mn homeostasis during postnatal development compared to that in adults. The lack of basal biliary excretion during early postnatal period is, however, only "apparent" in response to the low level of Mn in mother's milk compared to the adult diet because the neonates are still able to excrete Mn when overdosed. This behavior is just as applicable for human neonates as for the young animals.[55,72] The biliary excretory pathway in the human infants seems comparable to the adult and is inducible when challenged with excess Mn by several different exposure routes, *e.g.* ingestion of high Mn-content food such as formula milk.[73–77] This age- and dose-dependent control of biliary excretion is described in the model (Figure 13.8). Since we had dose-dependent biliary excretion controlled by blood Mn in the model, it is critical to take into account the elevated basal blood concentration in infants and children compared to the adult to depict this key homeostatic mechanism. This process included applying a higher dissociation constant (k_{b50}) for biliary induction for infants and children. This alteration captured the increased steady-state blood levels in neonates. However, the responsiveness of biliary induction at much higher exposures may under-predict the actual induction capability.

These human developmental models included homeostatic controls of Mn during critical periods of development and, as with the adult model, could predict exposure conditions leading to increase in target tissue Mn beyond the normal steady-state in growing body (Figure 13.9).

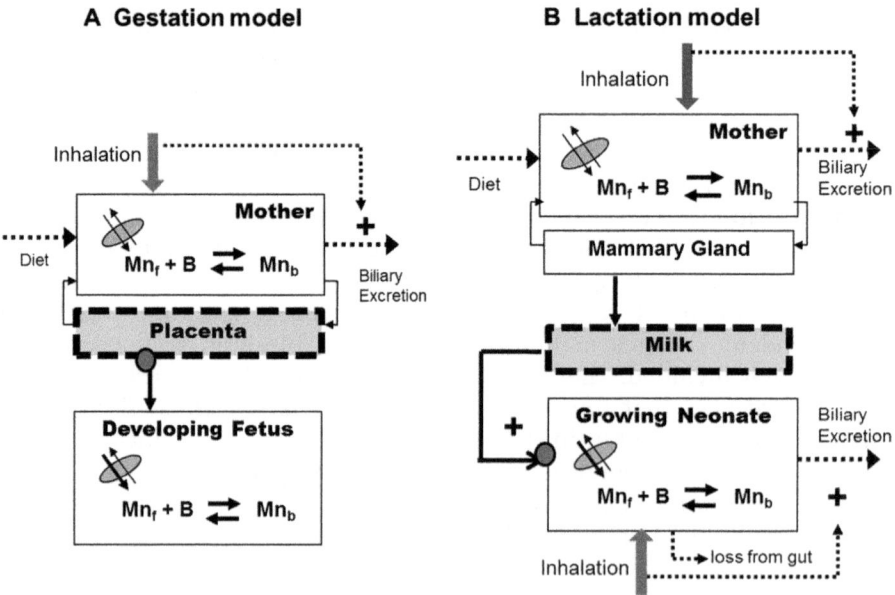

Figure 13.8 Depicting key characteristics of Mn PBPK models for gestation and lactation periods.

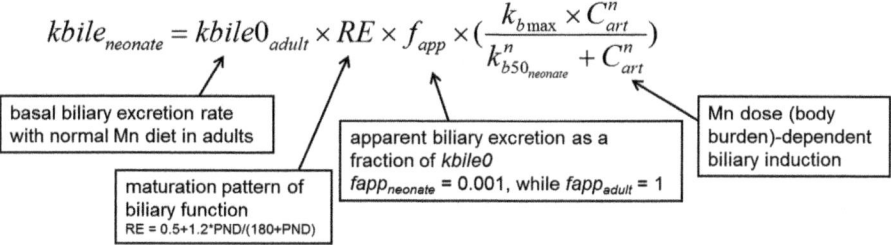

$$kbile_{neonate} = kbile0_{adult} \times RE \times f_{app} \times (\frac{k_{b\max} \times C_{art}^{n}}{k_{b50_{neonate}}^{n} + C_{art}^{n}})$$

| basal biliary excretion rate with normal Mn diet in adults |
| apparent biliary excretion as a fraction of *kbile0* $fapp_{neonate}$ = 0.001, while $fapp_{adult}$ = 1 |
| Mn dose (body burden)-dependent biliary induction |
| maturation pattern of biliary function RE = 0.5+1.2*PND/(180+PND) |

Figure 13.9 Control of Mn biliary excretion in neonates responding to both low and excess Mn.

13.4 Application of PBPK Models in Human Health Risk Assessment

Why does the tissue Mn matter for risk assessment of Mn? Mn is found in all mammalian tissues because it is essential in various bodily functions. As emphasized throughout this chapter, several homeostatic mechanisms have evolved to tightly regulate these tissue Mn concentrations within a normal range. For most tissues, this ranges from 0.15 to 4 µg Mn g^{-1}.[1] As described in the Introduction, Mn neurotoxicity occurs when Mn intake exceeds elimination, resulting in Mn accumulation in brain regions, including the globus pallidus, which is particularly sensitive to Mn accumulation during overexposure. Although Mn neurotoxicity is sensitive to exposure dose, it is relatively insensitive to route of exposure, because similar neurological responses have been linked to prolonged high-dose Mn inhalation, drinking water ingestion, long-term total parenteral nutrition (TPN), or impaired Mn clearance because of hepatobiliary impairment.[5] Because of the ubiquitous nature of Mn and the role of dietary Mn in establishing steady-state tissue concentrations, risk assessments of inhaled Mn should consider the essentiality of Mn from the diet to establish the tissue concentration that will be altered with increasing levels of inhaled or ingested Mn in addition to the normal dietary Mn. Therefore, to understand the risk to humans from excessive Mn exposure, it is important to determine the exposure conditions resulting in Mn concentrations in the brain that are increased significantly compared with brain Mn concentrations arising from normal dietary intake.[11]

Pharmacokinetic models may be used in an alternative risk assessment strategy for essential elements, such as Mn, based on dose to target tissue. PBPK models of Mn in animal species and humans introduced in this chapter showed that the Mn PBPK models can identify ranges of inhalation exposure (or other exposure routes of concern) concentrations where Mn levels in target tissues do not increase appreciably above background levels (Figure 13.7). Recently, the value of these Mn PBPK models in tissue dose-based risk assessment has been evaluated by accessing the large suite of monkey studies.[78] Experimental studies analyzed included various exposures such as inhalation, oral, iv, intra-peritoneal, and subcutaneous dose routes and spanned durations ranging from several weeks to over 2 years. This analysis demonstrated that model-predicted Mn tissue concentrations (or cumulative tissue Mn) in the critical brain regions in high dose studies in monkeys are better correlated with neurological effects of Mn than the administered doses.[79] This indicates that the Mn PBPK models can be used to integrate observations across studies performed in different animal species by different routes to investigate the dose-duration toxicity profile for Mn and suggest improved tissue dosimeters in conducting human risk assessment for Mn. The analysis provides confidence in using a human model to estimate air concentrations (or any other exposure route of concern) that could significantly increase brain Mn concentrations compared to the normal range resulting from dietary intake. Another example of application of Mn PBPK models towards risk assessment purposes was shown by Taylor and colleagues.[41] They applied the PBPK models to predict target tissue exposure to Mn under various exposure scenarios (short term *vs.* chronic) and in different subpopulations with altered physiology including young, aged, and diseased populations. These subpopulations were compared to adult males, which are the typical subjects evaluated in occupational Mn, to determine the pharmacokinetic adjustment factors necessary to extrapolate to these subpopulations in the context of risk assessment. It was discovered that no pharmacokinetic adjustment is needed when extrapolating from adult males to adult females, the aged, fetuses, neonates, and pregnant women (Figure 13.10), as well as those with high or low dietary Mn intake. Small adjustments could be made to adjust for duration of exposure or for those with moderate to severe hepatobiliary

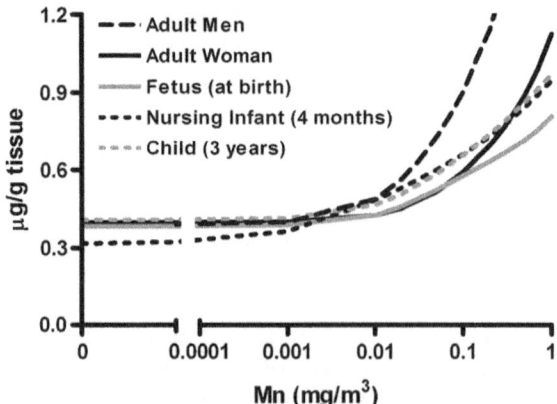

Figure 13.10   Comparison of dose-dependent changes in globus pallidus mn concentrations among different life stages.
Reprinted from Yoon *et al.*[40]

insufficiency. In total, the pharmacokinetic adjustment factors derived from utilizing the PBPK models are less than the pharmacokinetic portion of the typical uncertainty factor of 10 for human variability incorporated in current risk assessments of ambient Mn exposure. Future efforts can refine these comparisons and incorporate them into risk assessments as Chemical-Specific Adjustment Factors (CSAFs) for use *in lieu* of default uncertainty factors.

PBPK modeling of Mn is a good example showing the importance of using the tissue dose as a basis for risk assessment for an essential element. Similar PK model-based dosimetry approaches should also be considered for risk assessments with other essential metals, such as copper, zinc, *etc.*, where substantial background levels of which are always present in the body.

13.5 Suggested Research

The goal for a more complete PBPK model for Mn would be to describe the biology of the processes that control and regulate Mn concentrations throughout the body, especially in the mid-brain. The current PBPK model structure does not have direct biological correlation with all aspects of Mn biology. In its present form, the model might be regarded as a biologically motivated model, capturing key dose-dependencies and excursions of tissue Mn with increasing doses across species. What other information could better inform the current model structure?

The binding maximum represents a diverse group of tissue binding processes. What are these binding proteins? In addition, association and dissociation rate constants are small in the present model for processes that should occur on a faster time scale. What are the processes captured by these small rate constants? These dissociation rate constants capture two processes, the dissociation of Mn from binding sites and efflux of free Mn from tissues. During the past 15 years, which saw the evolution of our PBPK modeling strategies, several research groups proposed mechanisms that govern metal homeostasis. The majority of these suggested the control at a cellular transport level rather than cellular utilization. With Zn, efflux from bacteria is controlled by ligand-gated channels where a number of metal atoms (up to six) need to bind first before the binding of the last atom opens the channel.[80] Cellular transport and trafficking of Mn in yeast are likely under control of transporter expression in cellular and subcellular organelle membranes responding to the level of Mn present in extracellular environment.[24,34] Despite the lack of complete concordance with this newer biology, the current model structure with small k_d values provided quantitative understanding of the dose-dependent changes in tissue Mn from adequate to excess exposure conditions. The inventory of potential Mn binding sites (*e.g.* enzymes and metabolic processes that require Mn as cofactor and/or deposition of Mn in subcellular organelles during the normal cellular trafficking) could be further developed in relation to cellular/subcellular Mn transport processes. Similarly, the explicit description of forms of blood Mn could expand in a new model as the biochemical details of Mn transport become clear. This extension may be relevant for considering the type of transporters specific to different forms of Mn present in blood.[6] A very real advantage of additional biochemical detail for processes that control Mn fluxes and retention in the tissues will be the ability to assess individual differences in Mn uptake expected in mid-brain. Finally, introduction of emerging biochemical knowledge into the model will improve confidence in using a quantitative description of the dose-dependence of Mn distribution throughout body in risk assessments for this essential metal. Such a model is unlikely to alter conclusions about safe dose reached with current model structures, but enhance understanding of the integration of biological processes in achieving Mn homeostasis (Figure 13.10).

Acknowledgements

The invaluable contributions of Dr Andy Nong, Dr Jeffry Schroeter, and Dr David Dorman in the development of the manganese PBPK models are gratefully acknowledged. The authors would like to thank Dr Daniel Krewski, Dr Jeff Fisher, Dr Michele Medinsky, and Dr Lynne Haber for their helpful contributions and review of the manganese PBPK models. This publication and work are based on studies sponsored and funded by Afton Chemical Corporation in satisfaction of registration requirements arising under Section 211(a) and (b) of the Clean Air Act and corresponding regulations at 40 CFR Substance 79.50 et seq.

References

1. M. Aschner, K. M. Erikson and D. C. Dorman, Manganese dosimetry: species differences and implications for neurotoxicity, *Crit. Rev. Toxicol.*, 2005, **35**, 1–32.
2. ATSDR, *Toxicological profile for manganese*. Agency for Toxic Substances and Disease Registry, Atlanta, GA, US, 2000.
3. K. Sumino, K. Hayakawa, T. Shibata and S. Kitamura, Heavy metals in normal Japanese tissues. Amounts of 15 heavy metals in 30 subjects, *Arch. Environ. Health*, 1975, **30**, 487–494.
4. I. H. Tipton and M. J. Cook, Trace elements in human tissue. II. Adult subjects from the United States, *Health Phys.*, 1963, **9**, 103–145.
5. J. L. Aschner and M. Aschner, Nutritional aspects of manganese homeostasis, *Mol. Aspects Med.*, 2005, **26**, 353–362.
6. M. Aschner, T. R. Guilarte, J. S. Schneider and W. Zheng, Manganese: recent advances in understanding its transport and neurotoxicity, *Toxicol. Appl. Pharmacol.*, 2007, **221**, 131–147.
7. J. Freeland-Graves, Derivation of manganese estimated safe and adequate daily dietary intakes in *TITLE, Risk assessment of essential*

elements, ed. W. Mertz, C. O. Abernathy and S. S. Olin, ILSI Press, Washington, DC, 1994, pp. 237–252.

8. D. C. Dorman, M. F. Struve, D. Vitarella, F. L. Byerly, J. Goetz and R. Miller, Neurotoxicity of manganese chloride in neonatal and adult CD rats following subchronic (21-day) high-dose oral exposure, *J. Appl. Toxicol.*, 2000, **20**, 179–187.

9. H. A. Schroeder, J. J. Balassa and I. H. Tipton, Essential trace metals in man: Manganese: A study in homeostasis, *J. Chronic Dis.*, 1966, **19**, 545–571.

10. D. C. Dorman, M. E. Andersen, J. M. Roper and M. D. Taylor, Update on a Pharmacokinetic-Centric Alternative Tier II Program for MMT-Part I: Program Implementation and Lessons Learned, *J. Toxicol.*, 2012, **2012**, 946742.

11. M. E. Andersen, J. M. Gearhart and H. J. Clewell, 3rd, Pharmacokinetic data needs to support risk assessments for inhaled and ingested manganese, *Neurotoxicology*, 1999, **20**, 161–171.

12. M. E. Andersen, D. C. Dorman, H. J. Clewell III, M. D. Taylor and A. Nong, Multi-dose-route, multi-species pharmacokinetic models for manganese and their use in risk assessment, *J. Toxicol. Environ. Health, Part A*, 2010, **73**, 217–234.

13. D. C. Dorman, B. E. McManus, M. W. Marshall, R. A. James and M. F. Struve, Old age and gender influence the pharmacokinetics of inhaled manganese sulfate and manganese phosphate in rats, *Toxicol. Appl. Pharmacol.*, 2004, **197**, 113–124.

14. D. C. Dorman, M. F. Struve, M. W. Marshall, C. U. Parkinson, R. A. James and B. A. Wong, Tissue manganese concentrations in young male rhesus monkeys following subchronic manganese sulfate inhalation, *Toxicol. Sci.*, 2006, **92**, 201–210.

15. D. C. Dorman, M. F. Struve, R. A. James, M. W. Marshall, C. U. Parkinson and B. A. Wong, Influence of particle solubility on the delivery of inhaled manganese to the rat brain: manganese sulfate and manganese tetroxide pharmacokinetics following repeated (14-day) exposure, *Toxicol. Appl. Pharmacol.*, 2001, **170**, 79–87.

16. D. C. Dorman, M. F. Struve, R. A. James, B. E. McManus, M. W. Marshall and B. A. Wong, Influence of dietary manganese on the pharmacokinetics of inhaled manganese sulfate in male CD rats, *Toxicol. Sci.*, 2001, **60**, 242–251.

17. D. C. Dorman, M. F. Struve and B. A. Wong, Brain manganese concentrations in rats following manganese tetroxide inhalation are unaffected by dietary manganese intake, *Neurotoxicology*, 2002, **23**, 185–195.

18. D. Vitarella, B. A. Wong, O. R. Moss and D. C. Dorman, Pharmacokinetics of inhaled manganese phosphate in male Sprague-Dawley rats following subacute (14-day) exposure, *Toxicol. Appl. Pharmacol.*, 2000, **163**, 279–285.

19. D. C. Dorman, M. F. Struve, H. J. Clewell III and M. E. Andersen, Application of pharmacokinetic data to the risk assessment of inhaled manganese, *Neurotoxicology*, 2006, **27**, 752–764.

20. D. C. Dorman, A. M. McElveen, M. W. Marshall, C. U. Parkinson, R. Arden James, M. F. Struve and B. A. Wong, Maternal-fetal distribution of manganese in the rat following inhalation exposure to manganese sulfate, *Neurotoxicology*, 2005, **26**, 625–632.

21. D. C. Dorman, A. M. McElveen, M. W. Marshall, C. U. Parkinson, R. A. James, M. F. Struve and B. A. Wong, Tissue manganese concentrations in lactating rats and their offspring following combined in utero and lactation exposure to inhaled manganese sulfate, *Toxicol. Sci.*, 2005, **84**, 12–21.

22. L. Normandin, L. Ann Beaupré, F. Salehi, A. St-Pierre, G. Kennedy, D. Mergler, R. F. Butterworth, S. Philippe and J. Zayed, Manganese distribution in the brain and neurobehavioral changes following inhalation exposure of rats to three chemical forms of manganese, *Neurotoxicology*, 2004, **25**, 433–441.

23. F. Salehi, L. Normandin, D. Krewski, G. Kennedy, S. Philippe and J. Zayed, Neuropathology, tremor and electromyogram in rats exposed to manganese phosphate/sulfate mixture, *J. Appl. Toxicol.*, 2006, **26**, 419–426.

24. M. R. Bleackley and R. T. MacGillivray, Transition metal homeostasis: from yeast to human disease, *Biometals*, 2011, **24**, 785–809.

25. D. Foster, R. Aamodt, R. Henkin and M. Berman, Zinc metabolism in humans: a kinetic model, *Am. J. Phsyiol. Regul. Integr. Comp. Physiol.*, 1979, **237**, R340–R349.

26. L. V. Miller, N. F. Krebs and K. M. Hambidge, Development of a compartmental model of human zinc metabolism: identifiability and multiple studies analyses, *Am. J. Phsyiol. Regul. Integr. Comp. Physiol.*, 2000, **279**, R1671–R1684.

27. G. Cartwright and M. Wintrobe, Copper metabolism in normal subjects, *Am. J. Clin. Nutr.*, 1964, **14**, 224–232.

28. L. J. Harvey, J. R. Dainty, W. J. Hollands, V. J. Bull, J. H. Beattie, T. I. Venelinov, J. A. Hoogewerff, I. M. Davies and S. J. Fairweather-Tait, Use of mathematical modeling to study copper metabolism in humans, *Am. J. Clin. Nutr.*, 2005, **81**, 807–813.

29. J. G. Teeguarden, D. C. Dorman, T. R. Covington, H. J. Clewell III and M. E. Andersen, Pharmacokinetic modeling of manganese. I. Dose dependencies of uptake and elimination, *J. Toxicol. Environ. Health, Part A*, 2007, **70**, 1493–1504.

30. J. G. Teeguarden, D. C. Dorman, A. Nong, T. R. Covington, H. J. Clewell III and M. E. Andersen, Pharmacokinetic modeling of manganese. II. Hepatic processing after ingestion and inhalation, *J. Toxicol. Environ. Health, Part A*, 2007, **70**, 1505–1514.

31. J. G. Teeguarden, J. Gearhart, H. J. Clewell III, T. R. Covington, A. Nong and M. E. Andersen, Pharmacokinetic modeling of manganese. III. Physiological approaches accounting for background and tracer kinetics, *J. Toxicol. Environ. Health, Part A*, 2007, **70**, 1515–1526.

32. J. Furchner, C. Richmond and G. Drake, Comparative Metabolism of Radionuclides in Mammals-III Retention of Manganese-54 in the Mouse, *Health Phys.*, 1966, **12**, 1415–1424.

33. A. Nong, J. G. Teeguarden, H. J. Clewell III, D. C. Dorman and M. E. Andersen, Pharmacokinetic modeling of manganese in the rat IV: Assessing factors that contribute to brain accumulation during inhalation exposure, *J. Toxicol. Environ. Health, Part A*, 2008, **71**, 413–426.

34. A. R. Reddi, L. T. Jensen and V. C. Culotta, Manganese homeostasis in Saccharomyces cerevisiae, *Chem. Rev.*, 2009, **109**, 4722–4732.

35. J. A. Roth and M. D. Garrick, Iron interactions and other biological reactions mediating the physiological and toxic actions of manganese, *Biochem. Pharmacol.*, 2003, **66**, 1–13.

36. A. Nong, M. D. Taylor, H. J. Clewell, D. C. Dorman and M. E. Andersen, Manganese tissue dosimetry in rats and monkeys: Accounting for dietary and inhaled Mn with physiologically based pharmacokinetic modeling, *Toxicol. Sci.*, 2009, **108**, 22–34.

37. J. D. Schroeter, A. Nong, M. Yoon, M. D. Taylor, D. C. Dorman, M. E. Andersen and H. J. Clewell III, Analysis of Manganese Tracer Kinetics and Target Tissue Dosimetry in Monkeys and Humans with Multi-Route Physiologically-Based Pharmacokinetic Models, *Toxicol. Sci.*, 2011, kfq389, DOI: 10.1093/toxsci/kfq389.

38. M. Yoon, A. Nong, H. J. Clewell, M. D. Taylor, D. C. Dorman and M. E. Andersen, Evaluating placental transfer and tissue concentrations of manganese in the pregnant rat and fetuses after inhalation exposures with a PBPK model, *Toxicol. Sci.*, 2009, **112**, 44–58.

39. M. Yoon, A. Nong, H. J. Clewell, M. D. Taylor, D. C. Dorman and M. E. Andersen, Lactational transfer of manganese in rats: Predicting manganese tissue concentration in the dam and pups from inhalation exposure with a pharmacokinetic model, *Toxicol. Sci.*, 2009, **112**, 23–43.

40. M. Yoon, J. D. Schroeter, A. Nong, M. D. Taylor, D. C. Dorman, M. E. Andersen and H. J. Clewell, Physiologically based pharmacokinetic modeling of fetal and neonatal manganese exposure in humans: describing manganese homeostasis during development, *Toxicol. Sci.*, 2011, **122**, 297–316.

41. M. D. Taylor, H. J. Clewell, M. E. Andersen, J. D. Schroeter, M. Yoon, A. M. Keene and D. C. Dorman, Update on a Pharmacokinetic-Centric Alternative Tier II Program for MMT-Part II: Physiologically Based Pharmacokinetic Modeling and Manganese Risk Assessment, *J. Toxicol.*, 2012, **2012**, 791431.

42. E. A. Malecki, G. M. Radzanowski, T. J. Radzanowski, D. D. Gallaher and J. Greger, Biliary manganese excretion in conscious rats is affected by acute and chronic manganese intake but not by dietary fat, *J. Nutr.*, 1996, **126**, 489–498.

43. D. Tapin, G. Kennedy, J. Lambert and J. Zayed, Bioaccumulation and locomotor effects of manganese sulfate in Sprague-Dawley rats

following subchronic (90 days) inhalation exposure, *Toxicol. Appl. Pharmacol.*, 2006, **211**, 166–174.

44. S. Anjilvel and B. Asgharian, A multiple-path model of particle deposition in the rat lung, *Fundam. Appl. Toxicol.*, 1995, **28**, 41–50.

45. J. D. Schroeter, J. S. Kimbell, E. A. Gross, G. A. Willson, D. C. Dorman, Y.-M. Tan and H. J. Clewell III, Application of physiological computational fluid dynamics models to predict interspecies nasal dosimetry of inhaled acrolein, *Inhalation Toxicol.*, 2008, **20**, 227–243.

46. T. L. Leavens, D. Rao, M. E. Andersen and D. C. Dorman, Evaluating transport of manganese from olfactory mucosa to striatum by pharmacokinetic modeling, *Toxicol. Sci.*, 2007, **97**, 265–278.

47. K. A. Brenneman, B. A. Wong, M. A. Buccellato, E. R. Costa, E. A. Gross and D. C. Dorman, Direct Olfactory Transport of Inhaled Manganese (^{54}MnCl$_2$) to the Rat Brain: Toxicokinetic Investigations in a Unilateral Nasal Occlusion Model, *Toxicol. Appl. Pharmacol.*, 2000, **169**, 238–248.

48. D. C. Dorman, K. A. Brenneman, A. M. McElveen, S. E. Lynch, K. C. Roberts and B. A. Wong, Olfactory transport: a direct route of delivery of inhaled manganese phosphate to the rat brain, *J. Toxicol. Environ. Health, Part A*, 2002, **65**, 1493–1511.

49. D. C. Dorman, M. F. Struve, B. A. Wong, J. A. Dye and I. D. Robertson, Correlation of brain magnetic resonance imaging changes with pallidal manganese concentrations in rhesus monkeys following subchronic manganese inhalation, *Toxicol. Sci.*, 2006, **92**, 219–227.

50. J. P. Mahoney and W. J. Small, Studies on manganese: III. The biological half-life of radiomanganese in man and factors which affect this half-life, *J. Clin. Invest.*, 1968, **47**, 643.

51. I. Mena, O. Marin, S. Fuenzalida and G. C. Cotzias, Chronic manganese poisoning Clinical picture and manganese turnover, *Neurology*, 1967, **17**, 128.

52. K. M. Erikson, K. Thompson, J. Aschner and M. Aschner, Manganese neurotoxicity: a focus on the neonate, *Pharmacol. Ther.*, 2007, **113**, 369–377.

53. K. Dorner, S. Dziadzka, A. Hohn, E. Sievers, H. D. Oldigs, G. Schulzlell and J. Schaub, Longitudinal Manganese and Copper Balances in Young Infants and Preterm Infants Fed on Breast-Milk and Adapted Cows Milk Formulas, *Br. J. Nutr.*, 1989, **61**, 559–572.

54. C. L. Keen, J. G. Bell and B. Lonnerdal, The Effect of Age on Manganese Uptake and Retention from Milk and Infant Formulas in Rats, *J. Nutr.*, 1986, **116**, 395–402.

55. N. Ballatori, E. Miles and T. W. Clarkson, Homeostatic control of manganese excretion in the neonatal rat, *Am. J. Physiol.*, 1987, **252**, R842–R847.

56. A. Takeda, S. Ishiwatari and S. Okada, Manganese uptake into rat brain during development and aging, *J. Neurosci. Res.*, 1999, **56**, 93–98.

57. M. Aschner, Manganese: brain transport and emerging research needs, *Environ. Health Perspect.*, 2000, **108**(Suppl 3), 429–432.

58. R. K. Miller, D. R. Mattison, M. Panigel, T. Ceckler, R. Bryant and P. Thomford, Kinetic assessment of manganese using magnetic resonance imaging in the dually perfused human placenta in vitro, *Environ. Health Perspect.*, 1987, **74**, 81–91.

59. H. H. Kay and D. R. Mattison, Magnetic resonance imaging in non-human primates, in *Animal models in fetal medicine*, ed. P. W. Nathanielsz, Perinatology Press, Ithaca, NY, 1986, pp. 269–323.

60. L. A. Davidson and B. Lonnerdal, Specific binding of lactoferrin to brush-border membrane: ontogeny and effect of glycan chain, *Am. J. Physiol.*, 1988, **254**, G580–G585.

61. L. A. Davidson and B. Lonnerdal, Fe-saturation and proteolysis of human lactoferrin: effect on brush-border receptor-mediated uptake of Fe and Mn, *Am. J. Physiol.*, 1989, **257**, G930–G934.

62. B. Lonnerdal, C. L. Keen and L. S. Hurley, Manganese binding proteins in human and cow's milk, *Am. J. Clin. Nutr.*, 1985, **41**, 550–559.

63. J. Pacha, Development of intestinal transport function in mammals, *Physiol. Rev.*, 2000, **80**, 1633–1667.

64. W. Y. Chan, J. M. Bates, Jr. and O. M. Rennert, Comparative studies of manganese binding in human breast milk, bovine milk and infant formula, *J. Nutr.*, 1982, **112**, 642–651.

65. K. J. Collard, Iron homeostasis in the neonate, *Pediatrics*, 2009, **123**, 1208–1216.

66. L. K. Shannon, Effects of age and mineral intake on the regulation of iron absorption in infants, *J. Pediatr.*, 2006, **149**, S69–S73.

67. N. Ballatori, Transport of toxic metals by molecular mimicry, *Environ. Health Perspect.*, 2002, **110**, 689–694.

68. U.S. EPA, *Exploration of perinatal pharmacokinetic issues*, Report EPA/630/R-01/004, 2001.

69. K. M. Dziegielewska, Y. Daikuhara, T. Ohnishi, M. Waite, J. Ek, M. D. Habgood, M. A. Lane, A. Potter and N. R. Saunders, Fetuin in the developing neocortex of the rat: distribution and origin, *J. Comp. Neurol.*, 2000, **423**, 373–388.

70. A. J. M. Siddappa, R. B. Rao, J. D. Wobken, E. A. Leibold, J. R. Connor and M. K. Georgieff, Developmental changes in the expression of iron regulatory proteins and iron transport proteins in the perinatal rat brain, *J. Neurosci. Res.*, 2002, **68**, 761–775.

71. S. J. Garcia, K. Gellein, T. Syversen and M. Aschner, A manganese-enhanced diet alters brain metals and transporters in the developing rat, *Toxicol. Sci.*, 2006, **92**, 516–525.

72. K. Kostial, M. Blanuša and M. Piasek, Regulation of manganese accumulation in perinatally exposed rat pups, *J. Appl. Toxicol.*, 2005, **25**, 89–93.

73. K. M. Hambidge, R. J. Sokol, S. J. Fidanza and M. A. Goodall, Plasma manganese concentrations in infants and children receiving parenteral nutrition, *JPEN, J. Parenter. Enteral Nutr.*, 1989, **13**, 168–171.

74. S. Hatano, K. Aihara, Y. Nishi and T. Usui, Trace-Elements (Copper, Zinc, Manganese, and Selenium) in Plasma and Erythrocytes in

Relation to Dietary-Intake during Infancy, *J. Pediatr. Gastroenterol. Nutr.*, 1985, **4**, 87–92.

75. B. Sampson, G. B. Barlow and A. W. Wilkinson, Manganese balance studies in infants after operations on the heart, *Pediatr. Res.*, 1983, **17**, 263–266.

76. D. Stastny, R. S. Vogel and M. Picciano, Manganese intake and serum manganese concentration of human milk-fed and formula-fed infants, *Am. J. Clin. Nutr.*, 1984, **39**, 872–878.

77. S. Zlotkin and B. Buchanan, Manganese intakes in intravenously fed infants, *Biol. Trace Elem. Res.*, 1986, **9**, 271–280.

78. R. Gwiazda, R. Lucchini and D. Smith, Adequacy and consistency of animal studies to evaluate the neurotoxicity of chronic low-level manganese exposure in humans, *J. Toxicol. Environ. Health, Part A*, 2007, **70**, 594–605.

79. J. D. Schroeter, D. C. Dorman, M. Yoon, A. Nong, M. D. Taylor, M. E. Andersen and H. J. Clewell, Application of a multi-route physiologically based pharmacokinetic model for manganese to evaluate dose-dependent neurological effects in monkeys, *Toxicol. Sci.*, 2012, **129**, 432–446.

80. D. H. Nies, How cells control zinc homeostasis, *Science*, 2007, **317**, 1695–1696.

81. D. Vitarella, Development of an inhalation system for the simultaneous exposure of rat dams and pups during developmental neurotoxicity studies, *Inhal. Toxicol.*, 1998, **10**, 1095–1117.

82. D. Vitarella, B. A. Wong, O. R. Moss and D. C. Dorman, Pharmacokinetics of inhaled manganese phosphate in male Sprague–Dawley rats following subacute (14-day) exposure, *Toxicol. Appl. Pharmacol.*, 2000, **163**, 279–285.

83. K. A. Brenneman, R. C. Cattley, S. F. Ali and D. C. Dorman, Manganese-induced developmental neurotoxicity in the CD rat: is oxidative damage a mechanism of action?, *Neurotoxicology*, 1999, **20**, 477–487.

84. K. A. Brenneman, B. A. Wong, M. A. Buccellato, E. R. Costa, E. A. Gross and D. C. Dorman, Direct olfactory transport of inhaled manganese ((54)MnCl(2)) to the rat brain: toxicokinetic investigations in a unilateral nasal occlusion model, *Toxicol. Appl. Pharmacol.*, 2000, **169**, 238–248.

85. D. Vitarella, O. Moss and D. C. Dorman, Pulmonary clearance of manganese phosphate, manganese sulfate, and manganese tetraoxide by CD rats following intratracheal instillation, *Inhal. Toxicol.*, 2000, **12**, 941–957.

86. D. C. Dorman, M. F. Struve, D. Vitarella, F. L. Byerly, J. Goetz and R. Miller, Neurotoxicity of manganese chloride in neonatal and adult CD rats following subchronic (21-day) high-dose oral exposure, *J. Appl. Toxicol.*, 2000, **20**, 179–187.

87. D. C. Dorman, M. F. Struve, R. A. James, B. E. McManus, M. W. Marshall and B. A. Wong, Influence of dietary manganese on the

pharmacokinetics of inhaled manganese sulfate in male CD rats, *Toxicol. Sci.*, 2001, **60**, 242–251.

88. D. C. Dorman, M. F. Struve, R. A. James, M. W. Marshall, C. U. Parkinson and B. A. Wong, Influence of particle solubility on the delivery of inhaled manganese to the rat brain: manganese sulfate and manganese tetroxide pharmacokinetics following repeated (14-day) exposure, *Toxicol. Appl. Pharmacol.*, 2001, **170**, 79–87.

89. D. C. Dorman, K. A. Brenneman, A. M. McElveen, S. E. Lynch, K. C. Roberts and B. A. Wong, Olfactory transport: a direct route of delivery of inhaled manganese phosphate to the rat brain, *J. Toxicol. Environ. Health, Part A*, 2002, **65**, 1493–1511.

90. D. C. Dorman, M. F. Struve and B. A. Wong, Brain manganese concentrations in rats following manganese tetroxide inhalation are unaffected by dietary manganese intake, *Neurotoxicology*, 2002, **23**, 185–195.

91. D. C. Dorman, B. E. McManus, M. W. Marshall, R. A. James and M. F. Struve, Old age and gender influence the pharmacokinetics of inhaled manganese sulfate and manganese phosphate in rats, *Toxicol. Appl. Pharmacol.*, 2004, **197**, 113–124.

92. D. C. Dorman, B. E. McManus, C. U. Parkinson, C. A. Manuel, A. M. McElveen and J. I. Everitt, Nasal toxicity of manganese sulfate and manganese phosphate in young male rats following subchronic (13-week) inhalation exposure, *Inhal. Toxicol.*, 2004, **16**, 481–488.

93. D. C. Dorman, A. M. McElveen, M. W. Marshall, C. U. Parkinson, R. Arden James, M. F. Struve and B. A. Wong, Maternal-fetal distribution of manganese in the rat following inhalation exposure to manganese sulfate, *Neurotoxicology*, 2005, **26**, 625–632.

94. D. C. Dorman, A. M. McElveen, M. W. Marshall, C. U. Parkinson, R. A. James, M. F. Struve and B. A. Wong, Tissue manganese concentrations in lactating rats and their offspring following combined in utero and lactation exposure to inhaled manganese sulfate, *Toxicol. Sci.*, 2005, **84**, 12–21.

95. F. Salehi, G. Carrier, L. Normandin, G. Kennedy, R. F. Butterworth, A. Hazell, G. Therrien, D. Mergler, S. Philippe and J. Zayed, Assessment of bioaccumulation and neurotoxicity in rats with portacaval anastomosis and exposed to manganese phosphate: a pilot study, *Inhal. Toxicol.*, 2001, **13**, 1151–1163.

96. A. St-Pierre, L. Normandin, G. Carrier, G. Kennedy, R. Butterworth and J. Zayed, Bioaccumulation and locomotor effect of manganese dust in rats, *Inhal. Toxicol.*, 2001, **13**, 623–632.

97. L. Normandin, G. Carrier, P. F. Gardiner, G. Kennedy, A. S. Hazell, D. Mergler, R. F. Butterworth, S. Philippe and J. Zayed, Assessment of bioaccumulation, neuropathology, and neurobehavior following subchronic (90 days) inhalation in Sprague-Dawley rats exposed to manganese phosphate, *Toxicol. Appl. Pharmacol.*, 2002, **183**, 135–145.

98. F. Salehi, D. Krewski, D. Mergler, L. Normandin, G. Kennedy, S. Philippe and J. Zayed, Bioaccumulation and locomotor effects of

manganese phosphate/sulfate mixture in Sprague-Dawley rats following subchronic (90 days) inhalation exposure, *Toxicol. Appl. Pharmacol.*, 2003, **191**, 264–271.

99. L. Normandin, L. Ann Beaupre and F. Salehi, A. St -Pierre, G. Kennedy, D. Mergler, R. F. Butterworth, S. Philippe and J. Zayed, Manganese distribution in the brain and neurobehavioral changes following inhalation exposure of rats to three chemical forms of manganese, *Neurotoxicology*, 2004, **25**, 433–441.

100. L. A. Beaupre, F. Salehi, J. Zayed, P. Plamondon and G. L'Esperance, Physical and chemical characterization of Mn phosphate/sulfate mixture used in an inhalation toxicology study, *Inhal. Toxicol.*, 2004, **16**, 231–244.

101. D. Tapin, G. Kennedy, J. Lambert and J. Zayed, Bioaccumulation and locomotor effects of manganese sulfate in Sprague-Dawley rats following subchronic (90 days) inhalation exposure, *Toxicol. Appl. Pharmacol.*, 2006, **211**, 166–174.

102. D. C. Dorman, M. F. Struve, E. A. Gross, B. A. Wong and P. C. Howroyd, Sub-chronic inhalation of high concentrations of manganese sulfate induces lower airway pathology in rhesus monkeys, *Respir. Res.*, 2005, **6**, 121.

103. D. C. Dorman, M. F. Struve, M. W. Marshall, C. U. Parkinson, R. A. James and B. A. Wong, Tissue manganese concentrations in young male rhesus monkeys following subchronic manganese sulfate inhalation, *Toxicol. Sci.*, 2006, **92**, 201–210.

104. D. C. Dorman, M. F. Struve, B. A. Wong, J. A. Dye and I. D. Robertson, Correlation of brain magnetic resonance imaging changes with pallidal manganese concentrations in rhesus monkeys following subchronic manganese inhalation, *Toxicol. Sci.*, 2006, **92**, 219–227.

105. M. F. Struve, B. E. McManus, B. A. Wong and D. C. Dorman, Basal ganglia neurotransmitter concentrations in rhesus monkeys following subchronic manganese sulfate inhalation, *Am. J. Ind. Med.*, 2007, **50**, 772–778.

CHAPTER 14

Significance and Usefulness of Biomarkers of Exposure to Manganese

PERRINE HOET* AND HARRY A. ROELS*

Université catholique de Louvain (UCL), Institut de Recherche Expérimentale et Clinique (IREC), Louvain Centre for Toxicology and Applied Pharmacology (LTAP), Bruxelles, Belgium
*Email: perrine.hoet@uclouvain.be; roelsharry@telenet.be

14.1 Introduction

Since the evidence that the central nervous system can be adversely affected at low level of exposure to manganese (Mn), accurate estimation of exposure to this trace element is highly needed. By integrating uptake from all sources and exposure pathways, biological monitoring of exposure is generally acknowledged to be a useful tool to assess exposure to chemical agents at both the occupational and the environmental levels. Measuring the internal dose is also more likely related to the possible systemic health effects than the "external" doses as measured by ambient monitoring (air, water, food, *etc.*). Being an essential element, Mn is, in healthy adults, tightly controlled by a homeostatic mechanism regulating systemic absorption and excretion of Mn. Hence, at the individual level this will affect the association between indicators of external and internal Mn exposure.

To be validated as a biomarker (BM) of Mn exposure, to be implemented in routine biomonitoring surveillance, and to be correctly

Issues in Toxicology No. 22
Manganese in Health and Disease
Edited by Lucio G. Costa and Michael Aschner
© The Royal Society of Chemistry 2015
Published by the Royal Society of Chemistry, www.rsc.org

interpreted, the biological parameter should ideally exhibit the following characteristics:

- sufficiently specific and sensitive to distinguish exposed from non-exposed subjects. This implies, among others, not only the availability of an analytical method meeting the criteria of reliability, including sensitivity, accuracy, intra- and interday repeatability, and reproducibility between laboratories, but also accurate reference values;
- displaying a dose-related association with the external level of exposure; the higher the level of exposure, the higher the level of the biomarker;
- demonstrating a reasonable dose–effect/response relationship with the development of adverse effect(s) as to the central nervous system being considered as the critical target of Mn exposure. The higher the level of the biomarker, the more severe the damage or the more subjects affected.

14.2 Biological Matrices and Reference Values

The measurement of Mn has been explored in several biological media (blood, urine, hair, nail, saliva, placenta, faeces, cerebrospinal fluid, exhaled breath condensate) for its usefulness as biomarker to estimate the level of occupational or non-occupational exposure to Mn. The most investigated fluid is blood. Since the early 1990s, there is growing interest in measuring Mn *in vivo* in the brain by a non-invasive method, *i.e.* magnetic resonance imaging (MRI).

14.2.1 Mn in Blood

In healthy reference adults on a normal diet most of the Mn content of whole blood is in the cellular components with the erythrocytes containing about 66% of the total amount of Mn in blood, the leukocytes and platelets 30%, and plasma less than 5%.[1] Because of the much higher Mn levels in red blood cells, whole blood manganese (MnB) might be preferred over serum or plasma manganese as biomarker of exposure. Several investigations have illustrated good correlations between MnB concentrations and MRI scores.[2–5] There is indication that MnB and/or Mn in red blood cells (RBC) are better associated with the relative signal intensities of a T1-weighted MRI pallidal index than plasma or urine Mn in smelter workers,[6] in welders,[7,8] as well as in liver cirrhotics.[9,10] Hence, it is not unreasonable to infer that MnB or MnRBC might reflect to some extent Mn concentration in the target tissue.

On the other hand, based on the hypothesis that Mn in plasma (MnP) reflects the diffusible fraction of Mn in blood (*i.e.* the most readily biologically available fraction of Mn in the blood pool), and that it is in steady state with Mn in the brain, MnP could also be of interest as biomarker of exposure. Given the partitioning of Mn in blood, a rise of MnP without a rise in total cellular fraction (TCF) would cause a negligible rise in whole blood Mn concentration. Still, this rise in MnP would increase the amount of Mn in the blood circulation available to cross cell membranes of the blood–brain barrier and eventually reach the basal ganglia.

Investigations in welders also demonstrated that the majority (about 94%) of Mn is partitioned in the cellular fraction of the blood.[11] In spite of the lack of association between whole blood and plasma Mn levels, a highly significant inverse association between MnB and the relative fraction (%) of Mn contained in the plasma ($r = 0.699$, $p < 0.001$) was observed in these welders of whom 50% were currently exposed. *Ex vivo* analyses have shown that the relatively small fraction of Mn in plasma can be explained by the rapid clearance of Mn from plasma.[11]

Interestingly, on the basis that the changes of Mn and Fe concentrations in biological media may go in opposite directions,[12,13] suggests that combining these two critical measurements into one parameter may widen the differences between exposed and control individuals and, therefore, increase the sensitivity of Mn exposure assessment. They introduced the concept of Mn/Fe ratio in biological matrices and particularly for RBC and plasma. It is important to note that, because of the much higher concentration of Mn in RBC, even a slight hemolysis can result in accidental increase in MnP concentration, which may compromise the outcome of the exposure assessment.

It is likely that levels of Mn in whole blood, in RBC or total cellular fraction, and in plasma or serum are to be interpreted differently (see sections 14.3.1.1–14.3.1.3).

14.2.2 Mn in Urine

Mn is excreted in very little proportion in urine (<0.02% per day of the body burden); its main pathway for excretion is *via* the hepatobiliary system. Moreover, based on 24 h urine collection in six portions, urinary Mn (MnU) showed clear diurnal variation in welders, although there were some individual differences in the timing of the highest and lowest concentrations.[14]

Urinary levels of a parameter are often adjusted to the urinary creatinine concentration to integrate the effects of fluid balance on spot samples. Carrying out this adjustment systematically for all urinary biomarkers is questionable, as creatinine excretion rate is influenced by many factors (race, gender, age, muscle mass, physical activity, *etc.*).[15] Moreover, the validity of creatinine adjustment is, among others, dependent on whether the parameter is excreted in urine *via* the same pathway as creatinine, which for Mn is not established.

14.2.3 Mn in Hair and Nail

Hair and nail have been investigated as possible biomarkers of integrated exposure to Mn over the duration of their growth. The rate of growth is generally considered to be about 10 mm per month for hair and 1 mm per month for nail. Growth rates of 12.5 mm per month for hair and 1.4 mm per month for toenail were calculated in a Japanese US community.[16] The average fingernail growth rate was reported to be faster than that of toenails (3.47 *versus* 1.62 mm per month) in American young adults,[17] and big toenail growth rate was significantly greater than that of the other toenails.[18]

Interpreting the levels of trace elements in hair is tricky. The usefulness of hair is limited to only a few months after exposure because of the cycle of hair growth and loss. It is not yet clearly established to what extent reported Mn hair (MnH) levels reflect the Mn body burden (absorbed dose) *versus* external environmental Mn exposure. Moreover, the risk of exogenous contamination (*e.g.* fine dust) is high, while differences in reported MnH levels of occupationally or environmentally exposed subjects may also reflect the fact that studies have often used different, not always rigorous, methods for cleaning hair prior to analysis. Elimination of such extraneous increase of hair Mn appears highly challenging. Some authors claimed to have developed an appropriate hair cleaning methodology for removing exogenous Mn contamination, which would improve the potential utility of hair Mn as an exposure biomarker.[19] In addition, inherent differences in hair Mn concentrations between individuals can result from differences in the pigmentation and chemical composition of hair (*e.g.* amino acid and melanin content) (higher levels being reported in darker colored hair) and personal habits (*e.g.* dye, bleaching).[20–22]

Nails, particularly toenails, are considered to be somewhat better sheltered from environmental contaminants and their cycle of growth and loss is slower;[23] however, removing all contamination can be an issue.[24]

14.2.4   Mn in Brain

This clinical neurological manifestation results essentially from episodes during which the capacity of the Mn homeostatic control is overwhelmed and Mn is accumulating preferentially in basal ganglia and to a lesser extent in the cortex. The paramagnetic properties of the Mn^{2+}-ion, the most stable valence of manganese in tissues, enables detection of Mn deposition in the brain by using MRI. The bulk of Mn that reaches brain tissues is transferred from the blood circulation by crossing the blood–brain barrier. Through bypassing the bloodstream, direct exposure of some brain parts *via* the olfactory nerve is likely to occur at a minor level, but nevertheless this pathway of exposure escapes detection by the measurement of exposure biomarkers such as MnB.

Mn^{2+} shortens $T1$ relaxation times of protons (*e.g.* water) in tissues, causing a positive contrast enhancement or a "bright" signal in $T1$-weighted MRI images of tissues where it has accumulated. These hyperintensities on $T1$-weighted images are bilateral and symmetrical. Results are expressed as MRI intensity, $T1$ value or pallidal index (PI). The PI compares the intensity of the $T1$ signal in the globus pallidus (GP), the main site of Mn accumulation in humans, to that of the subcortical frontal white matter [(GP intensity/frontal white matter intensity)×100] to assess the extent of Mn deposition.[25] However, because Mn also accumulates in caudate and putamen, diffusion weighted imaging measuring $T1$ signal intensity in extra-pallidal basal ganglia has been recently suggested.[26] In some cases, inter-ference is possible with the presence of other paramagnetic metals like Cu and Fe that, in some conditions, may also accumulate in basal ganglia. Despite MRI scanning of the brain being hardly suitable for exposure monitoring, an exposure metric associated directly with the site of toxic action is usually a better biomarker than a peripheral biomarker of exposure.

14.2.5   Reference Values

The knowledge of accurate reference values (RVs) in the appropriate bio-logical matrices is crucial for correctly interpreting biomonitoring data and to assess whether particular exposure levels are higher than would normally be expected. Manganese is an essential element, a normal component of the diet and is present in all human tissues and fluids. The reference values in biological matrices are influenced by environmental, physiological, and lifestyle factors, and may differ between countries/regions, hence they should be established at a national/regional level. Secular trends are also possible.

References values and limits (RL) are statistically derived from measure-ments obtained from a representative sample of a characterized population, and are *per se* of no health relevance, *i.e.* observed values above RLs do not necessarily indicate a health risk. Table 14.1 lists "control" values of Mn in different biological media as reported in the literature for healthy non-occupationally exposed population groups ($n \geq 50$). Reported blood, serum/plasma, and urinary Mn concentrations appear to be highly variable. It should be noted that most "control" values included in this table are from epidemiological studies, which were not necessarily designed to establish reference values/limits. Moreover, successful participation in external qual-ity assessment schemes is far from systematically confirmed, so that ac-curate reference values/limits are scarce.

A significant gender difference in blood Mn levels is observed, in that women generally show higher values.[27–32] The finding that men generally absorb less Mn was associated to the fact that men usually have higher Fe stores than women. Lower Fe stores (ferritin levels) are known to be con-nected with increased Mn absorption.[33] MnB levels are known to increase throughout pregnancy and higher levels have been observed after delivery in mother and newborn.[34–37] Because of increased needs for Fe, maternal MnB at full term can be twice as high as in non-pregnant women.[38–40] In a study involving 225 children (aged 1–6 years) from Liverpool, UK, only 59 had a normal iron status while the others were diagnosed with either a borderline iron status ($n = 66$), iron deficiency ($n = 80$), or iron deficiency anemia ($n = 20$). The median MnB level was 16.4 µg L^{-1} (range 11.4–42.4 µg L^{-1}) in

Table 14.1 Main studies reporting control and/or reference values for biomarkers of Mn in healthy populations not occupationally exposed to Mn. Only studies reporting results for at least 50 subjects are included. Other studies reporting reference (control) values are shown in Table 14.2.

Location	Ref. M: male, F: female	Method		Blood (µg/L)	Serum/Plasma (µg/L)	Urine (µg/L or µg/g creat)	Scalp hair (µg/g)	Fingernail/toenails (µg/g)
EUROPE								
Belgium	Hoet et al. [15]	ICP-MS	P50			<0.043		
	460M, 541F (18–80 yrs)		P5–95			0.043–0.36		
	Nationwide survey		AM (range)			<0.043 (<0.043–1.72)		
Denmark	Kristiansen et al. [128]	ET-AAS	P50	8.64				
	85M, 93F (40–70 yrs)		P5–95	5.50–14.9				
			AM (SD; range)	9.08 (2.8; 4.1–40.3)				
France	Takser et al. [129]	F-AAS	P5–95	11.1–40.4			0.16–0.87	
	222F (29.7 ± 4.3 yrs)		AM (SD; range)	23 (15; 6.3–151.2)				
	Northeastern Paris		GM (range)	20.4			0.36 (0.10–3.24)	
	Goullé et al. [130]	ICP-MS	P50	7.6	1.12	0.31	0.067	
	100 subjects (Hair: 45)		P5–95	5.0–12.8	0.63–2.26	0.11–1.32	0.016–0.57	
Germany	Rükgauer et al. [45]	E-AAS	P50		1.32			
	78M, 59F (1 mo–18 yrs)		AM (SD; range)		1.4 (0.63; 0.17–2.9)			
	46M, 23F (22–75 yrs)		P50		0.71			
			AM (SD; range)		0.79 (0.31; 0.17–1.51)			
	Heitland et al. [131]	ICP-MS	P5–95	5.7–14.6				
	50M, 80F (18–70 yrs)		AM (range)	9.0 (4.8–18)				
	North		GM	8.6				
	Goën et al. [132]							
	BAR		P95	**15**				
	Biologische Arbeitsstoff–Toleranz–Werte*							
Ireland	Afridi et al. [133]	ICP-AES	AM (SD)				M 3.93 (0.21)	
	31M, 25F healthy adults						F 4.35 (0.15)	
	Urban area							
Italy	Alimonti et al. [134]	ICP-MS	P50	7.85	0.60			
	110 adults (20–61 yrs)		P5–95	1.53–13.2	0.31–1.02			
	Urban area, Rome		AM (SD)	7.70 (3.13)	0.62 (0.23)			
	Bocca et al. [135]	SF-ICP-MS	P50		0.88			
	289 adults		P5–95		0.55–1.50			
	Umbria		AM (range)		0.92 (0.55–1.40)			
			GM		0.88			
	218 adults		P50		0.80			
	Calabria		P5–95		0.50–1.22			
			AM (range)		0.83 (0.37–1.53)			
			GM		0.81			

Country	Reference / population	Method	Statistic	Blood	Urine	Hair / Nails
	Bocca et al.[27] 11M, 104F (18–89 yrs) Sardinia	SF-ICP-MS	P50 P5–95 GM	8.79 4.73–17 8.91		
	Lucchini et al.[111] 125–151 (11–14 yrs) Valcamonica Vincinity previous ferro-Mn alloy plants	GF-AAS ICP-MS:Hair	P50 P25–75 AM (range)	10.9 8.70–13.2 11.0 (4.00–21.6)	0.10 0.10–0.10 0.22 (0.10–7.60)	0.11 0.07–0.19 0.16 (0.02–1.27)
	133–150 (11–14 yrs) Garda Lake Reference zone		P50 P25–75 AM (range)	11.0 8.9–12.8 11.2 (6.00–24.1)	0.10 0.10–0.10 0.16 (0.10–1.50)	0.12 0.06–0.20 0.18 (0.02–3.45)
	Eastman et al.[19] 121 children (11–14 yrs) Vincinity of ferro-Mn alloy plant		P50 AM (range)			0.073 0.121 (0.011–0.736)
Finland	Jarvisalo et al.[14] 58M, 96F (18–58 yrs) Helsinki	GF-AAS	P97.5 AM	20.9 10.5 (blood, N=65)	2.1 0.313	1.5
Portugal	Lourenço et al.[136] 14M, 16F (17–75 yrs) Non–mining rural area	ICP-MS	AM (SD)	9.1 (2.6)		
	19M, 35F (13–91 yrs) Abandoned U mining site		AM (SD)	15.2 (3.3)		
Spain	Torra et al.[137] 122M, 128F (15–90 yrs) Barcelona	GF-AAS	AM (SD; range) GM	0.99 (0.31; 0.3–2.5) 1.1		
Sweden	Rodushkin et al.[138] 114 (hair)–96 (nails) (1–76 yrs) Northeast	ICP-MS	P50 AM (SD; range)			hair 0.35 0.56 (0.55; 0.08–2.41) nails 0.65 0.90 (0.75;0.19–3.30)
UK	White and Sabbioni[139] 188 urine – 206 blood samples South Scotland, North Midlands North & West Yorkshire	ET-AAS	P50 P5–95 (range)	7.40 4.2–12.00 (1.5–22.0)	0.3 0.09–1.89 (<0.09–7.8)	
	Sieniawska et al.[140] 77M, 34F (21–85 yrs) Southampton		P50 P97.5 GM		*0.47* *1.96* *0.48* *24h collection*	
NORTH AMERICA						
USA	Paschal et al.[141] 496 (6–88 yrs) Rural, urban areas	ICP-MS	P50 P5–95 AM GM		**1.00** <0.2–3.33 1.19 0.53	

Table 14.1 (*Continued*)

Location	Ref. M: male, F: female	Method		Blood (µg/L)	Serum/Plasma (µg/L)	Urine (µg/L or µg/g creat)	Scalp hair (µg/g)	Fingernail/toenails (µg/g)
	Haynes et al. [99] 135 subjects (hair: 73) (2–81 yrs) Vincinity ferro–Mn refinery Southeast Ohio	GF–AAS	P50	8.7			3.57	
			AM (SD; range)	9.2 (3.89; 1.8–22)			5.80 (6.42; 0.64–41)	
			GM	8.34			3.94	
	Mordukhovich et al. [142] 639M (72 ± 6.6 yrs) (Normative Aging Study)		P50					0.28
			IQR					0.4
Canada	Takser et al. [39] 149 pregnant women (T2) (15–39 yrs) Rural, urban areas	GF–AAS	P50	5.9–15.3				
			P5–95					
			AM (range)	9.9 (3.7–25.3)				
			GM	9.5				
	Clark et al. [28] 33M, 28F (30–65 yrs) British Columbia	ICP–MS	P50	10.7	0.671			
			P95	15.0	4.46			
			AM (range)	11.0 (6.9–16.4)	1.84 (0.32–48.3)			
			GM	10.8	0.81			
	Canadian Health measures [143] Survey Cycle II 2687M, 2888F (6–79 yrs)		P50	9.5		<0.2		
			P95	15		0.36		
			GM	9.22				
	Bouchard et al. [95] 168M, 194F (6–13 yrs)	ICP–MS	P50				0.7	
			P5–95				0.2–4.7	
			(range)				(0.1–21.0)	
SOUTH AMERICA								
Brazil	Batista et al. [144] 169M, 243F (15–60 yrs) Sao Paulo, Para	ICP–MS	P5–95			0.5–4.4		
	Nunes et al. [145] 506M, 619F (18–60 yrs) Sao Paulo, Para, Minas Gerais Rio Grande do Sul, Goias	ICP–MS	P5–95	6.9–18.5				
			AM	9.6				
	Menezes-Filho et al. [101] 53M, 56F (1–10 yrs) Vicinity ferro–Mn alloy plant 4 areas	GF–AAS	P50				6.90–31.30[a] (1.10–95.5)	
			(range)					
			GM				6.36–27.4[a]	
	19M, 24F (1–10 yrs) Reference area Bahia		P50				1.19 (0.39–8.58)	
			(range)					
			GM				1.37	
	Menezes-Filho et al. [102] 44M, 39F (6–12 yrs)	GF–AAS	AM (SD; range)	8.2 (3.6; 2.7–23.4)				

			GM (SD; range)	
Mexico	Vicinity ferro–Mn alloy plant Bahia		GM (SD; range)	5.83 (11.5; 0.10–86.7)
	Solís-Vivanco et al.[32] 120M, 168W (20–87 yrs) Mn–mining district of Molango Hidalgo State	GF-AAS	P50 AM (SD; range)	9.50 10.2 (4.14; 5.00–31.0)
	Riojas-Rodríguez et al.[103] 42M, 37F (7–11 yrs) Mn–mining district of Molango	GF-AAS	P50 P95 (range) GM	9.50 14.00 (5.50–18.0) 9.71
	45M, 49F (7–11 yrs) Reference community Hidalgo state, Central Mexico		P50 P95 (range) GM	8.00 13.0 (5.00–14.0) 8.22
	Henn et al.[43] 139M, 131F (1 yr)	ICP-MS	P50 AM (SD)	23.7 24.3 (4.5)
	219M, 211F (2 yrs) Mexico City		P50 AM (SD)	20.3 21.1 (6.2)
ASIA				
Bangladesh	Berglund et al.[146] 111M, 127F (8–12 yrs)	ICP-MS	P50 P5–95	0.54 0.13–141
	53M, 54F (14–15 yrs)		P50 P5–95	0.51 0.10–29
	86M, 154F (30–40 yrs)		P50 P5–95	0.58 0.12–31
	88M, 101F (41–50 yrs)		P50 P5–95	0.55 0.14–26
	129M, 152F (51–88 yrs) Rural area, Matlab		P50 P5–95	0.60 0.11–34
Japan	Ohashi et al.[147] 1000F (20–81 yrs)	GF-AAS	P50 (range) GM	0.16 (<0.05–5.7) 0.14
	Ikeda et al.[148] 1420F (20–81 yrs)	SF-ICP-MS	P50 max GM	13.1 33.4 13.2
Korea	Lee et al.[29] 1556M, 3531F (>20 yrs) Costal, rural, urban areas	GF-AAS	P50 P95 AM GM	10.7 19.8 11.6 10.8

Table 14.1 (*Continued*)

Location	Ref. M: male, F: female	Method	Blood (µg/L)	Serum/Plasma (µg/L)	Urine (µg/L or µg/g creat)	Scalp hair (µg/g)	Fingernail/toenails (µg/g)
India	Sukumar & Subramanian[149]	Flame–AAS					
	40M (31–75 yrs)	AM (SE)				3.8 (0.4)	
	Rural, costal area, Pondicherry						
	35M (7–30 yrs)	AM (SE)				2.7 (0.5)	
	Urban residential area, Madras						
	37M (16–72 yrs)	AM (SE)				3.2 (0.7)	
	Urban residential area, Madras						
	50M (16–60 yrs)	AM (SE)				1.2 (0.2)	
	Vincinity of lignite open mine						
	Mehra and Juneja[150]	GF–AAS	AM (SD)			M 5.91 (3.89) F 6.21 (2.03)	M 12.1 (4.06) F 22.4 (6.75) M 11.5 (2.68) F 11.8 (5.99)
	32M, 32F (20–24 yrs)						
	Rajasthan						
Pakistan	Khalique et al.[151]	ICP–AES	P50			M 3.23 F 4.76	
	58M, 30F (3–100 yrs)						
	Rural area, Chakwal		AM (SD; range)			M 4.02 (3.55; 0.01–14.3) F 6.52 (6.00; 0.36–28.0)	
	Shah et al.[152]	AAS					
	62M (3–54 yrs)		P50			1.68	
	Islamabad, Pakistan		AM (SD; range)			1.93 (0.94; 0.66–5.08)	
	62M (3–54 yrs)		P50			1.41	
	Tripoli, Libya		AM (SD; range)			1.73 (1.10; 0.54–5.98)	
	Afridi et al.[133]	GF–AAS					
	59M, 62F		AM (SD)			M 3.8 (0.63) F 4.3 (0.49)	
	Urban area, Hyderabad						

AFRICA

South Africa	Rollin et al. [106]	ICP–AES	range	1.5–32.8
381 schoolchildren			P50	6.2
Johannesburg			IQR	4.7–7.8
			AM (SD)	6.74 (3.47)
355 schoolchildren			P50	9.5
Kimberley			IQR	7.5–11.2
			AM (SD)	9.72 (3.12)
427 schoolchildren			P50	6.2
Cape Town			IQR	4.7–7.8
			AM (SD; range)	6.75 (3.47; 1.6–32.8)
119 schoolchildren			P50	7.4–9.2
Northern Cape (3 towns)			IQR	6.0–11.9
(sub)urban, rural areas			AM (SD; range)	7.75–9.86[a] (3.3–17.4)
	Batterman et al. [153]	GF–AAS	P50	9.6
149M, 193F (8–16 yrs)			P5–95	5.8–16.6
Durban			AM (SD; range)	10.1 (3.4; 3.0–25.0)

OCEANIA

Australia	Gulson et al. [105]	HR–ICP–MS	P50	11.0
56M, 57F (0.3–4 yrs)			AM (SD; range)	M 12.2 (6.0; 1.8–45.0)
Sydney				F 12.3 (4.8; 4.1–31.0)
			GM	M 11.0
				F 11.5

In bold, values considered (by the authors of the study) as reference values

*Manganese concentrations in the general population (P95) not occupationally exposed to Mn, but in the working age.

[a] According to the area

the latter group and 11.1 µg L^{-1} (range 5.9–20.9 µg L^{-1}) in the group with adequate iron status. Across the whole group, a strong association ($R^2 = 34.3\%$) was demonstrated between MnB and iron deficiency with hematological indices and soluble transferrin receptors as the primary contributing factors.[41] Whether iron deficient young children would be at higher risk of Mn-induced neurotoxicity remains to be elucidated. Newborns and infants have higher values of MnB than children or adults.[42–45] MnB values are probably rapidly decreasing in infants once the liver/bile function is fully maturated a few weeks after birth.

Since manganese is mainly excreted in the bile, it is not surprising that increased MnB levels have been observed in patients with liver failure;[3,9,46–49] however, not systematically.[50]

The blood manganese concentrations do not appear to show a strong association with smoking habits.[29]

14.3   Occupationally Exposed Subjects

14.3.1   External–Internal Exposure Relationship: Ambient Air-Biomarker

To be implemented in a routine biological monitoring scheme and for correct interpretation of its results, the level of a biomarker of exposure should, by definition, increase with the level of exposure.

Most biological occupational guideline values aiming at protecting workers are established on the basis of the knowledge of the relationship between external exposure (ambient air concentration) and internal exposure (biomarker level). They generally represent the levels of the biomarker that are most likely observed in biological samples collected from healthy workers who have been exposed to the chemical to the same extent as workers with inhalation exposure at the occupational exposure limit value.

As shown in Table 14.2, numerous studies have examined the suitability of potential biomarkers of occupational Mn exposure on the basis of the evaluation of their relationship with external Mn exposure; MnB and MnU being the most widely investigated. Most studies reported significantly higher mean levels in occupationally exposed workers than in their controls; however, values are often overlapping and are mostly found to be inconsistently related to Mn in the air (MnA), although some studies reported a significant dose–response between current or cumulative MnA exposure and MnB. Overall, the relationships between external Mn exposure indices and the biomarkers of this Mn exposure are poor or not existing.

Different methodological approaches, however, may render direct comparison and interpretation of results complex.

Table 14.2 Main studies investigating internal exposure levels in workers occupationally exposed to Mn in different industrial settings.

Authors, location	Air Sampling (AS)	AM (SD)	GM (GSD) [median]	Range	Biological Sampling (BS)		AM (SD)	GM (GSD) [median]	Range	Main results			
E: number of exposed subjects	Personal/stationary, sampling duration				Time of the day					Correlation between biomarkers (BM)			
length of exposure:	Time window for AS & BS				Method for Mn determination					BM level: exposed vs controls			
AM (SD; min–max) yrs	MnA: Mn in air (T: total; I: inhalable; R: respirable) (mg/m³)				Mn-urine: MnU (µg/L; µg/g creat)					Correlation: BM vs exposure			
exE: former exposed subjects	CEI: cumulative exposure index (mg/m³ × yrs)				Mn-blood: MnB (µg/L)					Correlation: BM vs effect			
C: number of controls					Mn-serum: MnSe (µg/L); Mn-plasma: MnP (µg/L)								
					Mn-erythrocytes: MnRBC (µg/L)								
					Mn-saliva: MnSa (µg/L)								
					Mn-hair: MnH (µg/g)								
					Mn-nail (µg/g)								
					Pallidal Index: PI								

Mining

Asia

Chia et al.[85] Singapore

	Air Sampling	AM (SD)	GM (GSD)[median]	Range	Biological Sampling		AM (SD)	GM (GSD)[median]	Range	Main results			
Mn ore milling (2 plants)	Personal, 6 h				nd					BM level exposed vs controls			
E: 17 baggers	nd				GF-AAS					MnB E	>	MnB C	p = nd
7.4 (4.3) yrs	MnA-T	1.59			MnU	E	6.1		1.7–17.9	MnU E	>	MnU C	p = nd
C: 17						C	3.9		0.7–9.6	MnSe E	>	MnSe C	p = nd
					MnB	E	25.3		15.0–92.6	Correlation BM – CNS effect			
						C	23.3		17.3–30.1	MnB	&	NBT/NPT	ns
					MnSe	E	4.5		2.0–32.8	MnU	&	NBT/NPT	ns
						C	3.9		1.5–6.4	MnSe	&	NBT/NPT	ns

Chia et al.[86] Singapore

	Air Sampling	AM (SD)	GM (GSD)[median]	Range	Biological Sampling		AM (SD)	GM (GSD)[median]	Range	Main results			
Mn ore milling (2 plants)	Personal, 6 h				nd					Correlation BM – CNS effect			
E: 32	nd				GF-AAS					MnU	&	postural stability	ns
6.6 (1.1–15.7) yrs	MnA-T	1.59			MnU	E	6.0		0.6–53.3				

Boojar and Goodarzi[88] Iran

	Air Sampling		GM (GSD)[median]	Range	Biological Sampling		AM (SD)	GM (GSD)[median]	Range	Main results			
Mn mine: tunnel digging, drilling, mixing, transport	Personal, half workshift				nd					BM level Exposed vs Controls			
E: 145	nd				GF-AAS					MnB Eb	>	MnB C	p < 0.05
Ea at time of exposure	MnA-T	Ea 62 (41)		29–206	MnU*	Ea	1.72 (0.43)			Ec	>	MnB C	p < 0.05
Eb 4.3 (0.5) yrs after recruitment		Eb 94 (52)				Eb	8.55 (1.53)			MnU Eb	>	MnU C	p < 0.05
Ec 3.2 (0.3) yrs after Eb		Ec 114 (66)				Ec	13.2 (3.17)			Ec	>	MnU C	p < 0.05
C: 65	MnA-R	Ea 27.6 (21.4)		9.2–83.6		C	1.52 (0.32)			MnH Eb	>	MnH C	p < 0.05
		Eb 38.1 (28.5)			MnB*	Ea	17.3 (4.2)			Ec	>	MnH C	p < 0.05
		Ec 43.3 (31.1)				Eb	137 (22.1)						
	CEI-R	Ea 1,365 (645)				Ec	167 (34.6)						
	µg/m³ x yr	Eb 1,738 (786)				C	18.9 (4.1)						
		Ec 1,935 (892)			MnH*	Ea	1.16 (0.41)						
						Eb	14.2 (2.85)						
	Respirable fraction: 40–45%					Ec	19.8 (3.81)						
	Main Mn species: silicate, oxides					C	1.78 (0.42)						

*results for non-smokers (also available for smokers)

Table 14.2 (Continued)

Ferro/silico-Mn alloy production, foundry, steel mill

North America

Mergler et al.,[87]; Bouchard et al.[154] Canada

FeMn & SiMn alloy plant
(crushing as well as smelting)

				Before last day of work cycle Flameless-AAS					BM level exposed vs controls			
E: 56–69				MnU	E	1.07	0.225	0.014–11.5	MnB E	>	MnB C	p = 0.0001
16.7 (3.2) yrs					C	1.05	0.035	0.001–1.27	MnU E	=	MnU C	
C: 56–69				MnB	E	11.2 (4.8)	1,186	1.6–25.9	*Correlation BM vs exposure*			
					C	7.2	0.122		MnB	vs	MnA–R (same day)	r = 0.38, p = 0.006
Personal & stationary, fullshift AS same day as BS										vs	MnA–R over 45d	r = 0.44, p = 0.002
				CEI calculated, not reported						vs	MnA–R over 182d	r = 0.48, p = 0.0005
										vs	CEI–T	ns
										vs	CEI–R	ns
									Correlation BM - CNS effect (POMS)			
									MnB <10 vs MnB >10 µg/L			ns

Europe

Lucchini et al.[67] Italy (Brescia)

FeMn, SiMn production (1 plant)

				nd* ET-AAS					*Correlation between BM*			
E: 58				MnU	E1	1.7 (1.2)	0.177 (0.213)	0.001–0.943	MnB	vs	MnU	r = 0.48, p = 0.0002
13 (1–28) yrs	E1				E2	2.3 (1.3)	0.199 (0.260)	0.009–1.023	*Correlation BM vs exposure*			
E1 20 (clerks, foremen, lab)	E2				E3	2.8 (0.8)	0.668 (0.590)	0.015–2.130	MnB	vs	CEI	r = 0.6, p = 0.0001
E2 19 (maintenance)	E3								MnU	vs	CEI	r = 0.4, p = 0.002
E3 19 (furnace workers)				MnB	E1	6.0 (0.7)		4–7.4	*Correlation BM - CNS effect (NBT)*			
					E2	8.6 (0.6)		7.9–9.5	MnB	vs	symbol digit	r = 0.45, p = 0.004
95% MnO₂					E3	11.9 (1.8)		9.6–18		vs	additions	r = 0.47, p = 0.001
Respirable fraction: 50–60%										vs	digit span	r = 0.35, p = 0.05
Personal & stationary, nd										vs	finger tapping	r = 0.35, p = 0.05
Before lay-off period (not same day as BS)				*Subjects examined during temporary lay-off period from work. Latency time after the cessation of exposure: 1 to 42 d.*					MnU	vs	additions	r = 0.36, p = 0.02
CEI–T									*Other results: ns*			

Lucchini et al.[68] Italy (Brescia)

FeMn, SiMn production (1 plant)
(crushing as well as smelting)

				"In the morning" ET-AAS					*Correlation between BM*			
E: 57				MnU	E	1.81	0.176	0.005–1.50	MnB	vs	MnU	ns
15.2 yrs					C	0.67	0.054	0.06–5	*BM level Exposed vs Controls*			
C: 87				MnB	E	9.71	0.067	4–19	MnB E	>	MnB C	p < 0.0001
					C	6.00	0.017	2–9.5	MnU E	>	MnU C	p < 0.0001
				nd					*Correlation BM vs exposure*			
MnA–T			1,205			1.53	0.3–5		MnB	vs	MnA–T	r = 0.36, p = 0.007
CEI–T						0.4	0.06–5			vs	CEI	ns
MnA–R						9.18			MnU	vs	MnA–T	ns
						5.74				vs	CEI	ns
MnO₂, Mn₃O₄									*correlations not reported for MnA–R*			
Respirable fraction: 40–60%									*Correlation BM - CNS effect (NBT, NPT)*			
Same plant as Lucchini et al.[68] but not on a lay-off period									MnB			ns
									MnU			ns

Apostoli et al. [155] Italy (Brescia)
FeMn, SiMn production (2 plants) (crushing as well as smelting)
E: 94
15.8 (7; 3–38) yrs
C: 87

Personal, halfshift
Same day as BM

MnA-T	0.20 (0.19)	0.097	0.005–0.740
CEI-T	calculated, not reported		

MnO$_2$, Mn$_3$O$_4$
Ratio total/respirable fraction: 2.8–2.0
Masks worn regularly, not constantly

End of shift
ET-AAS

MnU	E	4.9 (3.7)	3.8	0.4–21
	C	1.2 (1.4)	0.7	0.1–7
MnB	E	10.3 (3.8)	9.7	4–27
	C	5.9 (1.7)	5.7	2–10

Correlation between BM

MnB	vs	MnU C	r = 0.47, p < 0.0001
BM level Exposed vs Controls			
MnB E	>	MnB C	p < 0.0001
MnU E	>	MnU C	p < 0.0001
Correlation BM vs exposure			
MnB	vs	MnA	r = 0.34, p = 0.001
	vs	CEI	ns
	vs	length of exposure	ns
MnU	vs	MnA	r = 0.35, p = 0.0009
	vs	CEI	ns
	vs	length of exposure	ns

Smith et al. [11] Italy (Brescia)
FeMn, SiMn production (2 plants)
E: 147 smelters

Personal, nd
AS same day as BS

MnA	E1	[0.42]	≤0.7
	E2	[4.2]	≤1.2–≥13.5
	E3	[292]	≥17.5

End of shift
GF-AAS

MnB	E1	7.90 (2.81)	[7.35]	0.05–61.9
	E2	7.61 (2.52)	[7.20]	0.05–6.4
	E3	9.77 (3.08)	[9.00]	

Correlation BM vs exposure

MnB E2	=	MnB E1	
E3	>	MnB E2, E1	p ≤ 0.012
E1	vs	MnA E1	r = 0.523, p = 0.009
E2	vs	MnA E2	r = 0.275, p = 0.007
E3	vs	MnA E3	ns

Ellingsen et al. [79]; Bast-Petersen et al. [88] Norway
FeMn, SiMn production (3 plants)
E: 100
20.2 (8.4; 2.1–41) yrs
C: 100

Personal, fullshift
AS 3 days before &/or after BS

MnA-I	0.753	0.301	0.009–11.5
MnA-R	0.064	0.036	0.003–0.356
MnA-I-soluble	0.570	0.197	0.009–9.00
MnA-R-soluble	0.049	0.025	0.002–0.320

Respirable fraction: 10.6%
9% (raw material dpt) - 39% (crane operator furnace)
Correlation between respirable & inhalable: r = 0.7; p<0.001
Inhalable fraction: 28% insoluble, 8.2% water soluble
Respirable fraction: 23.4% insoluble, 10.8% water soluble

Before shift (U: first void in the morning)
ET-AAS

MnU	E	1.9	0.4	0.05–61.9
	C	0.4	0.19	0.05–6.4
MnB	E	10.4	9.94	4.6–23.4
	C	9.1	8.8	3.9–20.5

BM level Exposed vs Controls

MnB E	>	MnB C	p = 0.002
MnU E	>	MnU C	p = < 0.001
Correlation BM vs exposure			
MnB	vs	MnA-I	ns
	vs	MnA-I-soluble	ns
	vs	MnA-R	ns
	vs	MnA-R-soluble	ns
MnU	vs	MnA-I	r = 0.23, p < 0.05
	vs	MnA-I-soluble	r = 0.21, p < 0.05
	vs	MnA-R	r = 0.34, p < 0.01
	vs	MnA-R-soluble	r = 0.38, p < 0.01

Correlation BM - CNS effect (tremor dominant hand)

MnB <8.5	µg/L	ns from C
8.5–11	µg/L	ns from C
>11	µg/L	p < 0.001

Table 14.2 (*Continued*)

Africa

Myers et al. [81;] Young et al. [156] South Africa
Ferro-Mn alloy smelting (8 plants)
E: 509
17.2 (8.1; 0.4–42) yrs
C: 67

Personal, fullshift
nd
GF-AS

BM	Group	Value	Value	Value	Value	Range	Range
MnA-I	E	0.3 (5.5)	0.8 (1.1)	9.2 (19.1)	3.3 (3.9)	0.5–170	
	E1	<0.1		[1.4]		0.5–51	0–5.1
	E2	0.1–<0.2		[3.0]		0.5–98.7	
	E3	0.2–<1.0		[4.1]		0.5–124	
	E4	1.0–<2.0		[6.2]		0.6–48.4	
	E5	>2.0		[5.7]		0.7–170	
	C			[0.7]		0.5–35	
MnA-R		0.058					0.003–0.21
MnB	E	11.7 (5.6)	10.6 (1.6)			3.3–44	
	E1			[7.6]		3.3–19.3	
	E2			[9.4]		4.6–41.2	
	E3			[11.7]		5.1–44	
	E4			[13.1]		8.2–43.3	
	E5			[14.8]		5.5–38.7	
	C			[6.2]		3.3–10.9	
CEI-H		16.0 (22.4)	5.1 (6.7)			0–138	
CEI-R		0.92					0.015–13.26

Correlation between BM

MnB	vs	MnU C	r = 0.43, p < 0.0001

BM level Exposed vs Controls

MnB E	>	MnB C	p = nd
MnB E5 > MnB E4 > MnB E3 > MnB E2 > MnB E1 > MnB C			
MnU E	>	MnU C	p = nd
no difference between levels of exposure			

Correlation BM vs exposure

Individual basis

MnB	vs	MnA-I	r = 0.57, p = nd
	vs	CEI	r = 0.53, p = nd
	vs	length of exposure	r = 0.27, p = nd
MnU		10 µg/l discriminates individuals exposed to 0.2 mg/m³	
	vs	MnA-I	r = 0.26, p = nd
	vs	CEI	r = 0.25, p = nd
	vs	length of exposure	r = 0.16, p = nd

Correlation BM – effect

MnB	vs	prolactin serum	ns
MnU	vs	prolactin serum	ns

Asia

Jiang et al. [6] China (Guangxi Province)
Ferro-Mn alloy smelter
E: 18
14.1 (10.5; 5–33) years
E1 5 (power distribution/control room)
E2 13 (furnace smelting)
C: 9

Stationary (breathing zone), 5 h
nd
ICP-MS
"In the morning"

BM	Group	Value	Range
MnA-T	E1	0.66	0.36–0.96*
	E2	1.26	0.31–2.93*
			* = 95%CI
MnB µg/100 mL	E1	50 (10)	40–60*
	E2	50 (30)	30–70*
	C	40 (20)	20–50*
MnP	E1	40 (10)	30–50*
	E2	50 (20)	40–70*
	C	50 (30)	40–80*
MnRBC	E1	150 (20)	120–180*
	E2	160 (50)	120–180*
	C	140 (10)	140–160*
			* = 95%CI

Correlation between BM

MnB	vs	MnP	ns
	vs	MnRBC	r = 0.56, p < 0.05
	vs	PI	ns
MnP	vs	MnRBC	ns
	vs	PI	ns
MnRBC	vs	PI	r = 0.55, p = 0.02

BM level Exposed vs Controls

MnB E	=	MnB C
MnP E	=	MnP C
MnRBC E	=	MnRBC C

Cowan et al. [12,13] China (Guizhou Province)
Ferro-Mn alloy smelter
E: 217
~5 (0.1–12) yrs
E1 122 (supervisors, support staff)
E2 95 (smelters + maintenance)
C: 106

Personal, 2 × 4 h
BS within 24 h of AS
GF-AS
First urine in the morning (blood, saliva: nd)

BM	Group	Value	Range
MnA-T	E1	0.026 (0.028)	0.01–0.11
	E2	0.177 (0.103)	0.098–0.374
MnU	E1	2.73 (1.46)	
	E2	2.17 (1.35)	
	C	0.64 (0.31)	
MnP	E1	23.3 (19.2)	
	E2	30.4 (19.5)	
	C	9.97 (7.97)	
MnRBC	E1	7.45 (6.15)	
	E2	15.9 (9.01)	
	C	4.68 (3.59)	
MnH	E1	32.1 (29.2)	
	E2	37.6 (22.2)	
	C	1.21 (2.00*)	
MnSa	E1	22.3 (11.3)	

BM level Exposed vs Controls

MnP E	>	MnP C	p < 0.01
MnRBC E	>	MnRBC C	p < 0.01
MnH E	>	MnH C	p < 0.01
MnU E	>	MnU C	p < 0.01

Correlation BM vs exposure

MnP	vs	MnA	r = 0.66, p < 0.01
	vs	length of exposure	ns
MnRBC	vs	MnA	r = 0.69, p < 0.01
	vs	length of exposure	ns
MnH	vs	MnA	ns
	vs	length of exposure	ns
MnU	vs	MnA	ns
	vs	length of exposure	ns
MnSa	vs	MnA	r = 0.77, p < 0.01

Afridi et al. [82] Pakistan

Steel mill
E: 91
nd
E1 56 steel production workers
E2 35 quality control workers
C: 75

U: in the morning; B: nd
ET-AAS

	Group	Value mean (SD)	Range
	E2	31.3 (13.6)	
	C	9.98 (6.10)	
MnU	E1	2.49 (0.7)	1.78–3.09
	E2	1.87 (0.6)	1.08–2.56
	C	1.3 (0.3)	0.95–1.69
MnB	E1	65.6 (4.8)	61.3–70.5
	E2	58.6 (5.2)	53.2–64.5
	C	48.5 (6.9)	41.9–54.2
MnH	E1	7.2 (1.5)	6.14–8.9
	E2	5.9 (1.3)	4.57–7.45
	C	4.7 (1.02)	3.55–5.83

Correlation BM - effect

MIR-P	vs	length of exposure	ns
MIR-P	vs	MnA	r = 0.70, p < 0.01
MIR-RBC	vs	MnA	r = 0.77, p < 0.01
MnRBC	vs	PP-SCI	ns
MIR-P	vs	PP-SCI	r = –0.261, p = 0.002
MIR-RBC	vs	PP-SCI	ns

BM level Exposed vs Controls

MnB E	>	MnB C	p = 0.001–0.009
MnU E	>	MnU C	
MnH E	>	MnH C	

Welding

North America

Barrington et al. [84] USA

Mn alloy welders (railway track frogs) & machinists
E: 8
(subjects with irritability, headaches, poor memory)
(3–30) yrs

Personal, stationary; nd
nd

MnA	0.015–4.3

29% of 21 personal samplers > 1 mg/m³
(absence of adequate air sampling)

nd
nd

	Value mean (SD)	Range
MnU post-EDTA µg/24 h	29 (7.1)	17–37
MnB*	1.2 (0.4)	0.7–2.0

*= serum or blood ?

Correlation BM vs exposure

MnB	vs	length of exposure	ns
MnU	vs	length of exposure	ns

Correlation BM - CNS effect

MnB	vs	NBT	ns
MnB	vs	symptoms	ns
MnU	vs	NBT	ns
MnU	vs	symptoms	ns

Bowler et al. [70], Smith et al. [11] USA (San Francisco Bay Bridge)

E: 43
14.2 yrs
(Bay Bridge: 16.5 mo (6; 6–28))
E1 21 active welders
E2 16 stopped 1 mo ago

Work in confined space
>50% of the welders were not currently exposed
to the measured MnA levels at the time of BS

Personal & stationary, fullshift
nd

MnA-T	0.21 (0.08)	0.01–0.38
CEI (mg/m³ × mo)	2.56 (1.2)	0.07–4.72

nd
HR-ICP-MS

	Group	Value mean (SD)	[ ]	Range
MnU	E	0.28 (0.46)	[9.07]	0.00–1.93
	active E1	9.58 (2.53)		5.13–15.3
	stop 1 mo E2	10.3 (2.82)		
		8.7 (1.74)		
MnP	E	0.58 (0.13)	[0.59]	0.22–0.85

Correlation between BM

MnB	vs	MnP	ns
MnB	vs	MnU	ns

Correlation BM vs exposure

MnB	vs	CEI	r = 0.407; p = 0.01
MnP	vs	CEI	ns
MnU	vs	CEI	ns

Correlation BM - CNS effect

MnB	vs	some NBT	r = 0.245–0.469; p = 0.095–0.004

FU after 3.5 yrs
E 26 (among which 13 still active welders)

MnB	8.4 (2.6)	5.1–13.4
still active	9.9 (2.38)	

MnB: significant decrease
MnB still welders > MnB non welders

Table 14.2 (*Continued*)

Laohaudomchok et al.[23] USA
Boilermakers (welding school)
E: 40

Personal (shoulder), 5.83 (1) h
AS same day as BM
U & B before and after shift (N = 26); toenail clipping (N = 49)
DRC-ICP-MS

MnA	[0.013]	0.002–0.137	
CEI*	[1.23]	0.21.9	
mg/m³ × h 1–6 mo	[0.52]	0–6.88	
7–9 mo	[0.44]	0–0.05	
10–12 mo	[1.15]	0–10.1	
7–12 mo	[3.41]	0.24.1	
1–12 mo			
MnU before shift		1.5	0.45–3.02
MnB before shift		18.9	15.3–28.6
MnToenail*		[0.80]	0.05–10.4

* averaged over clippings from all 10 toes

MnB & MnU after shift are not cited here

* exposure window before BS

Correlation between BM

MnB before	vs	MnToenail	ns
MnU before	vs	MnToenail	ns
MnB	vs	MnU	ns

Correlation BM vs exposure

MnB after	vs	MnA	ns
MnU after	vs	MnA	ns
MnToenail	vs	MnA	ns
		CEI 1–6 mo	rs = 0.35; p = 0.006
		CEI 7–9 mo	
		CEI 10–12 mo	rs = 0.32, p = 0.031
		CEI 7–12 mo	rs = 0.32; p = 0.027
		CEI 1–12 mo	ns

Europe

Järvisalo et al.[34] Finland
Shipyard welders
E: 15
"several years"
C: 65 (blood) ~154 (urine)

Personal (under helmet for 5 welders) & stationary, fullshift
AS same day as BS (for some welders)
At different time, for several days (for 5 welders)
GF-AAS

MnA-T			0.05–1.47
MnU	E	1.07 (0.79)	0.33–5.75
	C	0.313 (0.62)	
MnB	E	16.5 (4.95)	
	C	10.5 (3.85)	
MnSe	E	1.21 (0.16)	

Correlation between BM

MnB	vs	MnU	ns

BM level Exposed vs Controls

MnB E	>	MnB C	p < 0.001
MnU E	>	MnU C	p < 0.05

Correlation BM vs exposure

MnU	vs	MnA	ns

MnU E after 4-week vacation — no decrease
MnSe E after 4-week vacation — ns decrease

Sjögren et al.[157] Sweden
Railway track welders
E: 12 Mn welders
welding hours: 270 (100–1760)
C1: 38 Al welders
C2: 39

nd
ET-AAS

MnB	E	[8.5]	5–14
	C1	[8.0]	2–17
	C2	[7.0]	2–16

BM level Exposed vs Controls

MnB E	=	MnB C	ns

Hoet et al.[58] Belgium
Looms, industrial washing machines, manufacture of gas stations
E: 28
9 (1–14) yrs
C: 5

Personal (under helmet), fullshift
AS on same days as BS (Monday, Tuesday)
Before & after shift, Monday & Tuesday
ICP-MS

MnU	E			<0.2–2.85
	C			65% <0.2
MnP	E			<0.2–0.34
	C			80% <0.2
MnP**	E	2.08	2.04	1.4–4.2
	C	1.51	1.52	1.0–1.9
MnA-T	0.078	0.028	0.001–0.729	

Correlation between BM

MnP	vs	MnU	ns

BM level Exposed vs Controls

MnP E	>	MnP C	p < 0.05
MnU E	=	MnU C	

Correlation BM vs exposure

MnP**	vs	MnA (MnA > 10 µg/m³)	ns
Monday*	vs	MnA (Monday)	r = 0.685, p < 0.001 (n = 20)
Tuesday*	vs	MnA (Tuesday)	r = 0.519, p = 0.027 (n = 18)
MnU*	vs	MnA	ns

* = after shift

Pesch et al. [80] Germany

Weldox study (23 companies)
Shipyards, containers & vessels, machine & tool building
E: 241

Personal (underneath welding helmet), 3.5 (1.9–5) h
AS same day as BS

After shift
GF-AAS

MnA-I	[0.073]	0.010–0.340*	MnB	[10.3]	8.33–13.15* *IQR
MnA-R	[0.062]	0.008–0.320* *IQR			

correlation MnA-I & MnA-R: r = 1.00

Correlation BM vs exposure
MnB vs MnA-R r = 0.44; p < 0.0001

Lehnert et al. [158] Germany

Weldox study, FU after intervention in high exposure setting (Improvements of exhaust ventilation & respiratory protection)
E: 17

MnA-R before intervention	0.399	0.100–2.17	MnB	12.8	14.1	6.2–21.9
MnA-R after intervention	0.007	0.0001–0.140		8.9	10.0	5.6–14.0

MnB decrease after intervention: p = 0.008

Ellingsen et al. [68,89] Russia (St Petersburg)

Heavy machinery & shipyard
E: 96
13.5 (1–40) yrs
exE* 27
C: 96

Personal (underneath welding helmet), fullshift
AS during 2 days before BS
MnA-T 0.238 0.121 0.007–2.322

U: first void in the morning; B: same morning (8.30–9.30)
ICP-MS

MnU	E	0.37	0.17	0.03–5.5
	exE*		0.07	0.03–0.17
	C		0.12	0.02–10.2
MnB	E	8.6		4.7–21.7
	E0 5.5			3.7–6.5
	E1 7.5			6.6–8.6
	E2 12.6			8.7–23.5
(MnB stratified in 3 equally large groups)				
	exE*	8.7		5.2–19.1
	C	6.9		3.8–14.3

BM level Exposed vs Controls
MnB E > MnB C p < 0.001
MnB exE* > MnB C p < 0.05
MnU E = MnU C ns increase

Correlation BM vs exposure
MnB vs MnA 2 days before r = 0.22, p < 0.05
     vs MnA 1 day before r = 0.31, p < 0.01
MnU vs MnA ns

Correlation BM - CNS effect (NBT)
MnB E2 poorer Digit Symbol test performance than age-matched referents p = 0.01

*27 former welders with *manganism* who stopped exposure 5.8 yrs prior to the study

Asia

Chandra et al. [159] India

E: 60
14–21 (2–32) yrs
E1 20 (heavy engineering)
E2 20 (railway)
E3 20 (ship repair)
C: 20

? "breathing zone", nd
nd

nd
AAS

MnA-T	0.31		0.44–0.99
	0.57		0.50–0.80
	1.74		0.88–2.6
MnSe	E1		13–52
	E2		13–59
	E3		13–44
	C		10–26
MnU	E1		3.0–57.0
	E2		5.0–240
	E3		3.0–45
	C		1.8–9.4

Li et al. [160] China (Beijing)

Vehicle factory
E: 37
(2–36) yrs
C: 50

Stationary (breathing zone), 5 h
nd
MnA-T 1.45

U: 24 h; Ser: nd
HR-ICP-MS

MnU	E	~1.5
	C	~0.95
MnSe	E	2.86 (2.64)
	C	0.66 (0.38)

BM level Exposed vs Controls
MnSe E > MnSe C p < 0.01
MnU E = MnU C

Table 14.2　(*Continued*)

Lu et al. [71] China (Beijing)
Vehicle factory — Stationary (breathing zone), 5 h

E: 97	nd			
15.9 (8.0; 1–33) yrs	MnA-T	0.56	0.47–1.47	
C: 91				

"In the morning"; GF-AAS

			BM level Exposed vs Controls	
MnSe	E	3.40 (1.68)	MnSe E > MnSe C	p < 0.001
	C	1.14 (2.43)	*Correlation BM vs exposure*	
			MnSe E　vs　length of exposure	ns

Yuan et al. [161] China
Machine building factory — Stationary, nd

E: 56	nd			
nd	MnA	0.138	0.038–0.198	
C: 32	CEI	2.23		

"In the morning"; GF-AAS

			BM level Exposed vs Controls	
MnB	E	48.4	16.2–75.3	MnB E > MnB C　　p < 0.05
	C	19.2	8.1–25.4	

Wang et al. [72] China (Beijing & Shandong Province)
Stationary, 4 h

E: 53	nd			
8.1 (0.5–35) yrs	MnA-T	E1	0.24	0.01–1.0
E1 28 (steel frames manufacturers)		E2	2.21	0.02–11.1
E2 25 (oil cylinder tank manufacturers)				
C: 33				

"During the morning"; ICP-MS

			Correlation between BM	
MnSe	E1	3.85 (0.9)	MnSe　vs　MnSa	r = 0.575; p < 0.05
	E2	5.68 (0.91)	*BM level Exposed vs Controls*	
	C	2.70 (1.50)	MnSe E > MnSe C	p < 0.01
MnSa	E1	3.47 (1.42)	MnSa E > MnSa C	p < 0.01
	E2	5.55 (2.31)	*Correlation BM vs exposure*	
	C	3.04 (1.40)	MnSe　vs　length of exposure	ns

Wongwit et al. [162] Thailand
Thermal power plant
E: 135

"In the morning"; GF-AAS; µg/mg

			Correlation between BM	
MnB	12.0 (3.9)	3.8–23.7	MnB　vs　MnToenail	ns
MnToenail	4.86 (3.13)	0.02–14.68		

Kim et al. [163] South Korea
Nationwide survey: smelters, steel mill, automobile, shipyards, welding rod manufactures — Personal, nd

E: 111–121	nd			
9.95 (7.88; 0–29) yrs	MnA-T	0.50 (3.65)	<0.04–4.8	

GF-AAS

			Correlation between BM	
MnB	E	12.4 (13.9)	4.9–34.5	MnB　vs　PI　　r = 0.358, p < 0.001
PI	E	107 (1.1)	98–132	*Correlation BM vs exposure*
				MnB　vs　MnA　ns
				MnB　vs　length of exposure　r = 0.446, p < 0.001
				PI　vs　MnA　ns
				PI　vs　length of exposure　ns

Kim et al. [164] Choi et al. [165] South Korea
Shipyard. — Personal, nd

E: 20	nd			
21 (2.1; 17–23) yrs	MnA-T	0.399 (0.205)	0.076–0.679	
C: 10	*(recent exposure level within last 3 months)*			
	CEI	6.85 (2.01)	2.86–9.14	

After shift; GF-AAS

			Correlation between BM	
MnB	E	14.2 (6.5)	7.3–37.1	MnB　vs　PI　　r = 0.35, p = 0.066
	C	10.5 (2.1)	7.5–14.0	*BM level Exposed vs Controls*
PI	E	1.12 (0.04)	1.06–1.12	MnB E = MnB C
	C	1.06 (0.03)	1.01–1.11	PI E > PI C　　p = 0.0006
				Correlation BM vs exposure
				MnB　vs　MnA　ns
				PI　vs　MnA　r = 0.68, p < 0.0001
				PI　vs　CEI　r = 0.54, p = 0.002

Chang et al. [90,93] South Korea
Shipyard.
E: 43
20.8 (7.0; 7–35) yrs
C: 29

Personal, at least 6h
nd

MnA-T	0.14 (0.08)	(0.002–0.254)	

At least 12h after previous workshift
GF-AAS

MnB	E	15.3 (4.6)	6.8–29.2
	C	11.4 (3.2)	6.1–20.5
PI	E	124 (11)	109–157
	C	117 (3)	112–125

Correlation between BM
MnB vs PI   r = 0.51, p < 0.01
BM level Exposed vs Controls
MnB E > MnB C
PI E > PI C
Correlation BM vs exposure
PI vs MnA   r = 0.34, p < 0.01
Correlation BM - CNS effect (NBT)
MnB vs grooved pegboard   p = 0.016
    vs other NBT   ns
PI vs grooved pegboard,   p < 0.05
    vs digit symbol,
    vs digit span backward,
    vs Stroop Word, Stroop error index
    vs other NBT   ns

Hassani et al. [166] Iran
Gas transmission pipeline welders
E: 118
3.78 (1.51) yrs
C: 37

Personal, nd
AS on the same day as BM

MnA-T	0.125 (0.150)		

"During the entire workshift (8h)"
GF-AAS

MnU	E	4.38 (3.29)	0.77–7.58
	C	0.77 (1.05)	nd – 4.10

Correlation BM vs exposure
MnU vs MnA   r = 0.598, p < 0.001
Correlation BM - respiratory effect
MnU vs FVC   r = –0.283, p < 0.05
    vs FEV1   r = –0.286, p < 0.05
    vs FEF 25-75   ns

Other occupational exposures

Chemical production

Roels et al. [76,187] Belgium
Mn oxides & salts production
E: 141
7.1 (5.5; 1–19) yrs
C: 104

Personal, 3–8 h (median 7.5 h)
AS same day as BM for 24 (B)–34 (U) workers

MnA-T	1.33 (0.14)	0.94	0.07–8.61

MnO, MnO$_2$, Mn$_3$O$_4$
MnSO$_4$, MnCO$_3$, Mn(NO$_3$)$_2$, Mn acetate

past integrated exposure:
subjective estimation by chief foreman

morning & afternoon
GF-AAS

MnU	E	4.76	1.59 (3.73)	0.06–140.6
	C	0.30	0.15 (3.09)	0.01–5.04
MnB	E	13.6 (6.4)	12.2	1.0–35.9
	C	5.7 (2.7)	4.9	0.4–13.1

Correlation between BM
MnB vs MnU   ns
BM level Exposed vs Controls
MnB E > MnB C   p < 0.001
MnU E > MnU C   p < 0.001
Correlation BM vs exposure
individual basis
MnB vs MnA   ns
    vs length of exposure   ns
aftershift vs preshift   ns
MnU vs MnA   ns
    vs length of exposure   ns
group basis
MnB vs MnA   ns
    vs past integrated exp   r = 0.83, p < 0.05
MnU vs MnA   r = 0.62, p < 0.05
    vs past integrated exp   ns
Correlation BM - CNS effect (NBT)
MnB vs eye–hand coord.   p < 0.02
    vs hand steadiness   p < 0.03
    vs short term memory   ns
    vs simple reaction time   ns

Table 14.2 (Continued)

Dry cell battery production

Roels et al. [77,188] Belgium

E: 92
5.3 (0.2–17.7) yrs
C: 101

Personal, >4–5 h (80%)
AS at different day as BM
morning & afternoon
GF-AAS

		Value	Value	Range
MnA-T		1,780	0.948	0.046–10.84
MnA-R		0.301	0.215	0.021–1.317
CEI-T			3.505	0.191–27.47
CEI-R			0.793	0.040–4.43
MnU	E		0.84	0.15–7.33
	C		0.09	0.01–0.49
MnB	E		8.1	2.1–21.0
	C		6.8	2.5–13.1

Respirable fraction: ~25%
(proportion decreasing exponentially with increasing MnA-T)
Correlation MnA-T & MnA-R: $R^2 = 0.81$
MnO_2

BM level Exposed vs Controls

MnB E	>	MnB C	$p < 0.001$
MnU E	>	MnU C	$p < 0.001$

Correlation BM vs exposure

individual basis

MnB	vs	MnA-T	ns
	vs	MnA-R	ns
	vs	CEI-T	ns
	vs	CEI-R	ns
MnU	vs	MnA-T	ns
	vs	MnA-R	ns
	vs	CEI-T	ns
	vs	CEI-R	ns

group basis

MnU	vs	MnA-T	rs = 0.83, $p < 0.05$
	vs	MnA-R	rs = 0.83, $p < 0.05$

Bader et al. [169] Germany

E: 100
mean: 11–14 yrs
E1 39 (quality control)
E2 22 (assembly lines)
E3 39 (pellets preparation from MnO_2 dust)
C: 17

Stationary, 2 h
nd
GF-AAS

	Group	Mean (SD) [median]	Range
MnA	E1	0.004 (0.004)	0.001–0.012
	E2	0.037 (0.028)	0.012–0.064
	E3	0.387 (0.283)	0.137–0.794
MnU	E1	0.26 (0.35) [0.10]	0.10–1.80
	E2	0.33 (0.35) [0.10]	0.10–1.30
	E3	0.49 (0.49) [0.40]	0.10–2.20
	C	0.39 (0.37) [0.30]	0.10–1.20
MnB	E1	10.7 (4.3) [9.4]	3.9–25.8
	E2	11.7 (5.6) [10.4]	3.2–23.0
	E3	13.8 (4.4) [13.8]	6.1–23.3
	C	7.5 (2.7) [7.7]	2.6–15.1
MnHa* µg/g	E1	4.6 (5.8) [2.6]	0.4–29.8
	E2	5.2 (4.5) [3.4]	0.5–17.2
	E3	8.2 (6.7) [5.7]	0.9–27.5
	C	2.2 (1.8) [1.7]	0.4–6.2

* proximal axillary hair

Correlation between BM

MnB	vs	MnHa	ns

BM level Exposed vs Controls

MnB	E	>	MnB C	$p < 0.05$
	E3	>	MnB E1	$p < 0.05$
MnHa	E	>	MnHa C	$p < 0.05$
	E3	>	MnHa E1	$p < 0.05$
MnU	E	≈	MnU C	

Glaze, enamels, pigments

Deschamps et al. [170] France

1 enamels factory
E: 138
19.9 (9) yrs
C: 45

sampling: personal (P) & stationary (S), nd
nd
GF-AAS

		Mean (SD)	Range
MnA-T	P	3.24 (2.91)	0.5–10.2
	S	0.86 (1.17)	0.5–5.2
MnA-R	P	0.06 (0.84)	0.01–0.29
	S	0.13 (0.09)	0.01–0.45
MnB	E	9.35 (3.57)	0.50–19.5
	C	9.13 (3.35)	3.13–18.2

BM level Exposed vs Controls

MnB E	≈	MnB C

Arai et al. [171] Japan

Cloisonne ware workers
E: 70 cloisonne ware workers
E1 49 glaze workers
E2 16 silver plating workers
E3 5 plant office workers
C: 62

U: during working hours; B: nd
GF-AAS

				Correlation between BM		
MnU	E1	2.48 (0.82)	0.45–5.06	MnB	vs	MnU	r = 0.575, p < 0.01
	E2	3.05 (0.93)	1.49–4.94	BM level Exposed vs Controls		
	E3	2.55 (1.39)	0.59–4.02	MnB E1	>	MnB C	p < 0.01
	C	1.00 (0.69)	0.11–3.33	E2	>	MnB C	p < 0.01
MnB	E1	33.7 (9.3)	22.2–69.4	MnU E1	>	MnU C	p < 0.01
	E2	36.1 (7.8)	24.3–50.0	E2	>	MnU C	p < 0.01
	E3	28.7 (6.6)	23.6–39.2			
	C	14.5 (8.7)	3.5–58.3			

Other

Lander et al. [172] Denmark (Island of Funen)

Cast iron foundries
E: 46
12 (1–26) yrs
Stationary, 2 h
nd
nd
GF-AAS

				BM level Exposed vs Controls		
MnA-T	E1	0.039	0.007–0.064	MnB E1	>	MnB C	p < 0.05
	E2	0.005	0.002–0.008	E2	=	MnB C	
	E3	nd	nd	E3	>	MnB C	p < 0.05
	E4	0.006	0.004–0.008	E4	=	MnB C	
MnA-R	E1	0.027	0.005–0.022	MnB	vs	length of exposure	ns

MnB			
	E1	13.1	8.1–20.4
	E1*	10.9	7.3–14.8
	E2	9.2	7.5–11.2
	E2*	8.2	5.5–10.6
	E3	16.8	13.9–25.1
	E3*	10.3	9.4–11.3
	E4	8.8	5.0–19.5
	C	8.7	4.1–19.0

*3–4 weeks after decreasing or stopping exposure

Mari et al. [173] Spain

Hazardous waste incinerator workers
E: 29
E1 15 plant workers
E2 7 laboratory workers
E3 7 administrative workers
nd
ICP-MS

MnB	E1	16.7 (4.4)	12.3–30.5	after 4-week cessation: decrease in MnB E1 & MnB E2
	E2	17.8 (4.4)	13.8–26.2	
	E3	17.7 (6.2)	12.7–29.9	

nd: no data; ns: not significant; r: Pearson coefficient; rs: Spearman coefficient
CNS: central nervous system
NBT: neurobehavioral testing
NPT: neurophysiological testing
MIR: Manganese–iron ratio
PP-SCI: Purdue Pegboard test - standardized composite index
POMS: Profile of Mood State
FU: follow-up

- The quality of the inhalation exposure characterization varies greatly between studies. An accurate description of the type and duration of air sampling used is not always available.
- In case of personal air sampling, the location of the sampler is often not fully clarified. For instance in welder studies, the air sampler can be fixed on the welder's chest or collar, and sometimes it is fixed inside the welding shield, which makes comparison of exposure levels difficult.
- Another problem is that the correctness of the use of respiratory protection is generally difficult to assess.
- In many studies, the information concerning the time lapse between the air sampling and the biological sampling is not provided. Whether the air sample is collected on the same day as the biological sample or not may yield different results.
- As to the sampling of biological media and the measurement of the biomarker, important information is also often not provided; for instance, the moment of the sample collection (before, during or after work) or the participation of the laboratory in external quality control programs.
- It is not always specified whether statistical analyses were done on an individual or on a group basis.

The lack of these specifications and details precludes from properly concluding about the potential usefulness of an association between Mn air exposure and internal Mn exposure. Nevertheless, some features inherent of the Mn metabolism can offer an explanation, at least partly, for the lack of or poor association between air and Mn biomarker levels, including (a) the existence of a homeostatic regulation, (b) the issue of the half-life of the biomarker and the exposure windows, and (c) the variable nature of the profile of external Mn exposure. These three aspects are discussed next.

14.3.1.1   Homeostatic Regulation

The Mn absorption rate for intake of Mn *via* ingestion is low (3 to 5%) and subjected to a homeostatic control as well as its excretion rate. Because excess of Mn in blood is rapidly removed by the liver and Mn excretion *via* the kidneys is little compared to the bile, the levels of Mn in whole blood and urine, the most commonly used biological media, are not expected to be sensitive indicators of external Mn exposure. Such a tight regulation of Mn absorption and excretion, which is limiting MnB levels, is not clearly demonstrated in the case of Mn inhalation exposure, the main route of exposure in workers. Inhalation exposure results in a time-window during which blood Mn circulates first to the brain and thus bypasses the liver. However, some arguments could be advanced for the existence of a certain degree of homeostasis. It is well established that the biliary elimination rate constant and the proportion of tracer dose [54]Mn eliminated in the rapid phase increase with increasing dose of Mn ingestion.[51–53] There is some experimental indication that a similar response, probably smaller in extent, may exist for Mn inhalation exposure.[54–56] In rats exposed to Mn_3O_4 for 14 days, increased whole-body clearance of [54]Mn was only detected at the highest level of exposure (3 mg Mn m^{-3}),[57] which suggests that overwhelming of the homeostatic control may occur at high inhalation exposure.

Two studies on a limited number of Mn-exposed welders reported intriguing findings that offer evidence of the existence of a homeostatic control in humans exposed to increased levels of MnA. The first study in welders described an inverse correlation between pre-work shift MnB and change in MnB over the work shift ($r = -0.66$, $p < 0.01$). This might be explained by a homeostatic pressure that limits the rise in MnB; those who had a high MnB concentration before the work shift did not have an increase in MnB concentration as much as those whose MnB was lower.[23] The second study on a group of 28 welders showed that, in spite of similar MnA exposure on Monday (after 64 h without occupational exposure) and Tuesday, the relationships between MnA and after-shift MnP strikingly differed, in that on Tuesday the relationship was less obvious and the slope of the regression line was 2.3 times lower than that on Monday. Furthermore, on Monday the increment in MnP (after-shift MnP minus pre-shift MnP) correlated well with log MnA of Monday; on Tuesday, however, this correlation disappeared. A plausible interpretation of this finding is that change(s) in homeostatic control influenced the Mn distribution on Tuesday, in that the rate of hepatobiliary clearance of Mn from plasma would be stimulated after the exposure on Monday leading to a lower MnP on Tuesday for a similar exposure range.[58] The fact that Mn in welder fume (more than 90% of the aerosol particulate is in the respirable fraction) is rapidly bio-available explains the good correlation ($r = 0.69$, $p < 0.001$) between MnA and MnP on Monday. This study showed, on an individual basis, a direct effect of current MnA exposure on MnP in humans at exposure levels >10 µg Mn m^{-3}.

14.3.1.2 Biomarker Half-life and the Exposure Windows

Since their different half-lives, the diverse biomarkers of Mn exposure presumably represent different exposure windows, *e.g.* current, recent, or time-integrated exposure. For a biomarker with a long half-life, one assumes that its level reflects the body burden, which will increase over the years of exposure. Therefore, one would expect to see a correlation between the biomarker and a MnA cumulative exposure index (CEI) or a higher biomarker level in Mn-exposed workers with a longer employment history. For a more rapidly disappearing biomarker, whose concentration primarily reflects exposures during the previous days or hours, the timing of the sample collection can be critical.

Whether Mn in blood reflects current or more long-term Mn exposure remains a topic of discussion. Some authors suggest that MnB (or Mn in total cellular fraction) is fairly stable over relatively long periods of time in occupationally exposed workers and thus can be used as a surrogate of the Mn body burden, while others point out that because of its rather short blood half-life, MnB may at best reflect recent exposure, *i.e.* the last 2 or 3 months.

A biomarker of exposure should actually be assessed in a timeframe that reflects its half-life in the fluid/tissue of interest (*e.g.* blood) while simultaneously taking into account the half-life in the whole body and whether or not the exposure is still ongoing. The short terminal-phase elimination half-time following intravenous injection of $MnCl_2$ in rats is only about 2 h, which would suggest no important risk of accumulation of Mn in the body;[59] however, brain tissue Mn analyses described a biological half-life ranging between 51 and 74 days.[60,61] In humans, the biological half-life of Mn in blood ranged from 12 h in healthy miners to 40 h in healthy volunteers injected intravenously with [54]Mn,[62] but the clearance half-times for the whole body were 15 and 37 days, respectively.[63] The half-life in brain appeared to be even longer, with 54 days in humans given a Mn-compound intravenously.[64]

In addition to homeostatic regulation, the discrepancy between blood and tissue/body half-life may render MnB levels difficult to interpret. If one considers a relatively short half-life for MnB and that long-term exposure to Mn leads to accumulation of Mn in the body, then whole blood Mn levels may be influenced by both current exposure and release of Mn into the blood circulation from internal stores constituted by past exposures (*e.g.* bone marrow, which would explain increased levels of Mn in RBC). Therefore, at some time-lapse away from active exposure, MnB levels would better reflect the body burden. Besides the existence of an endogenous compartment with a long elimination half-life, it could also be hypothesized that following prolonged inhalation of poorly soluble Mn particles, a certain amount is retained in the lungs, providing a depot that slowly releases Mn into the bloodstream. Because of the intricate relation between Mn exposure, Mn homeostasis, different half-lives of Mn in blood and brain, and accumulation of Mn in tissues, the recently developed PBPK model for oral and inhalation exposure to Mn in monkeys and humans[65,66] would help to clarify the significance of Mn in the two most important blood compartments, *i.e.* plasma and RBC, in view of risk assessment of Mn exposure.

Lucchini *et al.*[67] assessed ferro–Mn alloy production workers 1–2 weeks following cessation of exposure during a plant closedown and observed, on an individual basis, a good logarithmic correlation between MnB and MnA CEI. The correlation coefficient improved when considering the subjects who were tested a longer period after cessation of exposure. As postulated by the authors, it is possible that in the weeks following cessation of exposure, the effect of recent exposure on the MnB levels has decreased resulting in a better reflection of the cumulative Mn exposure or the bioavailability of inhaled Mn. In a further study, they reported a modest correlation between current Mn exposure and MnB.[68] Ellingsen *et al.*[69] observed that MnB was higher in former welders than in controls of similar age, even several years after the cessation of exposure, a finding which provides suggestive evidence for the existence of an endogenous Mn compartment. On the other hand, MnB was significantly associated with MnA concentration measured 1 day prior to blood sampling, but not with MnA levels measured 2 days before blood collection, suggesting that MnB may at least partly reflect recent exposure. Similarly, Bowler *et al.*[70] found a modest but significant correlation between the MnA CEI and MnB in welders, 50% of which were not exposed to the air Mn levels measured at the time the blood samples were collected. This observation is also in agreement with the existence of at least one compartment with a long elimination half-life. It should be pointed out that in the latter study, serum Mn was not correlated with Mn in air, which is in accordance with the very short half-life of Mn in the blood plasma compartment. Mn in plasma is effectively considered to undergo a particularly rapid clearance.[11] In other studies on welders, the concentration of Mn in serum was not associated with the duration of exposure as a welder.[71,72] It may be concluded that serum/plasma Mn levels could serve as an indicator for the very recent Mn exposure, but are not suitable for the estimation of the historical accumulation in the body following long-term exposure.

14.3.1.3 Exposure Scenarios and Biomarkers of Exposure

The variable profile of the exposure scenario may have a considerable impact on the toxicokinetics of Mn and hence on the relationship between MnA and biomarker levels.

For a similar ambient air exposure level of Mn, the rate and amount of Mn taken up by the body depends on the main determinant of the particles, *i.e.* the size, which may have a significant influence on the bioavailability of Mn. This can explain, at least partly, that results regarding the correlation between the external and internal exposure to Mn can vary depending on the occupational sector evaluated. The degree of respiratory uptake of Mn by inhalation depends primarily on particle size. Mn particles in the respirable fraction (<10 μm) are small enough to reach and deposit in the alveoli where the process of Mn absorption into the bloodstream occurs. In contrast to the coarse airborne Mn particulate as found in mining, crushing/milling of Mn ores, dry-battery assembly, and ferro/silico–Mn alloy production, a particular feature of welding fumes is the better bioavailability of Mn as more than 90% of the airborne particles is respirable.[73,74] Apart from welding operations, welders may carry out other tasks involving Mn exposure, such as metal grinding or weld finishing operations entailing coarse dust exposure. The likelihood is high that the coarse particles in the latter exposure scenario will have a different bioavailability of Mn (most likely lower) than the fine particles in welding fume aerosol.

The solubility and speciation/oxidation state of the Mn particles can also affect the time course of respiratory tract absorption. Mn of readily soluble salts ($MnSO_4$, $MnCl_2$) is more rapidly absorbed and thus taken up into the bloodstream with a higher rate compared to equimolar exposures to relatively insoluble species, such as Mn-oxides.[56,57,75–78] In the case of inhalation exposure to sparingly soluble Mn species, a significant amount of Mn particulate matter can be retained in the lungs, providing a depot that slowly releases Mn into the bloodstream. In line with this, dose–response correlations, if any, are generally better for the respirable than the inhalable fraction. For instance, a study in Mn-alloy production workers showed higher correlation coefficients between biological exposure indices and current exposure of Mn in the respirable aerosol fraction than in the inhalable fraction. Pearson's correlation coefficient was 0.38 between "soluble" respirable Mn and MnU, while no association was found between chemically "insoluble" Mn in the respirable aerosol fraction and MnU; the best fit was found for the association between MnU and Mn species with the $3^+/4^+$ oxidation state in the respirable aerosol fraction ($r = 0.48; p < 0.001$).[79] A non-linear association between respirable Mn and MnB (Spearman $r = 0.44$) was described in welders.[80] However, in dry cell battery workers exposed to a poorly soluble MnO_2 compound, no correlation was found on an individual basis between either MnA-Total or MnA-Respirable (current or CEI) and MnB or MnU. On a group basis, only MnU correlated to both MnA-T and MnA-R (Spearman $r = 0.83$), highlighting the complexity of the relationships.[77]

In ferro/Mn-alloy workers, Myers *et al.*[81] found only modest correlations at the individual level between MnA and MnB (33% of the variance in log atmospheric Mn intensity in the current job was explained by log MnB). However, a receiver operating characteristic (ROC) analysis showed that it is possible to use a MnB cut-off of 10 µg L^{-1} (P95 in the unexposed group) as a screening tool to discriminate (sensitivity 80%, specificity 81%) between individual exposures exceeding or falling below a relatively strict atmospheric Mn exposure threshold at the 1995-ACGIH threshold limit value of 0.2 mg m^{-3}. The welder study by Hoet *et al.*[58] showed that after-shift MnP of Monday (the first workday of the week) could discriminate welders exposed to more than 20 µg m^{-3} from those exposed to less than 20 µg m^{-3} using a MnP cut-off value of 2 µg L^{-1} (sensitivity 69%, specificity 82%). These MnB and MnP cut-off values[58,81] are to be interpreted with caution because they fall within "normal" values in some populations. In smelter workers, the MnRBC–iron ratio (MIR–RBC) and MIR-P, which more accurately distinguished exposed from controls than Mn in RBC and Mn in plasma, were shown to exhibit strong correlations with airborne Mn levels. An MIR–RBC cut-off value of 8.8, established using the ROC analysis, indicated that more than 88% of the smelters in the high-exposure group were above the cut-off value, whereas a similar percentage (87%) of the control workers was below this value. Similarly, with a MIR-P cut-off value of 2.75, 75% of smelters in the high-exposure group were above the cut-off value and 84% of controls were below this value.[12,13]

In a similar Mn exposure scenario, different biomarkers of Mn exposure may lead to different conclusions. In a large welder study, Pesch *et al.*[80] found no obvious correlation between MnB and Mn in the welding fume up to 50–100 µg m^{-3}, but beyond that threshold, an increase of MnB with rising respirable MnA was detected. When using MnP as biomarker of current Mn exposure in welders on the first workday of the week, Hoet *et al.*[58] observed that after-shift MnP started to increase from a MnA concentration between 10 and 20 µg m^{-3}.

The interest of Mn concentration in hair and nail as biomarker of occupational exposure to Mn has been less explored, but a number of studies have demonstrated increased levels in hair.[12,82,83] In a study among welders at a union welding school, the average Mn concentration of clippings from all 10 toenails was modestly but significantly correlated with cumulative exposure windows encompassing months 7–9, 10–12, and 7–12 before the toenail clipping date, but not with the earlier exposure window of months 1–6.[23]

14.3.2   Internal Exposure – Effect Relationship

Ideally, the level of a biomarker of exposure should be related to the occurrence of adverse effect(s). Health risk can only be estimated when a quantitative relationship between a biomarker level and the occurrence of a health effect has been demonstrated. The knowledge of such relationships makes it possible to establish true health-based guideline values and hence to directly evaluate the health risk.

Some investigators have examined the dose–effect (response) relationships between biomarkers of exposure to Mn and neurobehavioral or motor performances, sometimes with contradictory results (see Table 14.2). Several researchers concluded that there is no good relationship between MnB or MnU and the occurrence or severity of health effects.[84-87]

In a manganese oxide and salt production plant, however, Roels *et al.*[76] observed a relationship between abnormal eye–hand coordination and hand steadiness in workers with increasing MnB levels. Significant dose–effect relationships between MnB and the performance of several neurofunctional tests (*i.e.* additions, symbol digit, digit span, and finger tapping) were also found by Lucchini *et al.*[67] in ferro/Mn-alloy workers temporarily laid-off from work. However, these correlations totally disappeared in a further study when workers were currently exposed to Mn at the time of the neurofunctional testing.[68] Smelter workers with MnB >11 µg L^{-1} showed increased postural tremor of the dominant hand when executing a visually guided task in comparison with the referents after adjustment for age and tobacco consumption.[88]

Dose–effect relationships between neuropsychological variables and MnB, but not MnP, were also reported in a group of 43 welders, for 50% of which the exposure stopped about 1 month before the examination.[70] In a study by Ellingsen *et al.*,[89] welders with the highest mean MnB concentration (AM 12.6 µg L^{-1}, range 8.7–23.5 µg L^{-1}) scored significantly poorer ($p < 0.01$) on the Digit Symbol test when compared to age-matched controls (AM MnB 7.5 µg L^{-1}, range 3.7–14.3 µg L^{-1}). In asymptomatic welders, Chang *et al.*[90,91] demonstrated that MnB levels were associated significantly with grooved pegboard (dominant hand) neurobehavioral motor test performance (assessment of manual dexterity and fine motor coordination). However, the pallidal index appeared a better predictor of neurobehavioral performances than MnB.

Cowan *et al.*[13] observed in a group of smelter workers that none of the measured biomarkers (MIR–RBC, MIR-P, and MnRBC) correlated with groove hand steadiness, nine-hole hand steadiness, and the Benton memory test results. When adjusted for age, years of education, sex, income, and years of employment, only the MIR-P values were significantly associated with individual Purdue pegboard scores, as well as standardized composite index of Purdue pegboard scores, suggesting that the MIR-P is a more appropriate biomarker of exposure for subtle changes in neurobehavioral function.

Importantly, adverse health effects due to previous high Mn exposure can be present in the absence of positive biological exposure measurements when the patient is diagnosed.[92] Even the hyperintensity on the $T1$ MRI scan disappears almost completely within 6 months after cessation of Mn exposure, despite permanent neurological damage.[8,93]

14.4   Non-Occupational Exposure

14.4.1   Environmental Exposure

Miscellaneous studies have investigated environmental Mn levels and biomarkers of Mn exposure in adults and children environmentally exposed to Mn, either by ingestion of contaminated food and/or drinking water,[20,94–98] or by inhalation of increased MnA while living in the vicinity of mining areas or ferro/Mn-alloy production plants,[19,32,99–103] or by inhalation of motor exhaust from vehicles using the anti-knock gasoline additive methylcyclopentadienyl manganese tricarbonyl (MMT).[104–107]

Although many of these studies show that environmental exposure to Mn may result in a tendency to increased levels of Mn biomarkers, the relationships between external and internal Mn exposure are currently not clearly established.

While some authors reported that neither MnU nor MnP were correlated with the subjects' dietary intakes of manganese,[96] others have shown a positive correlation between the total daily intake of Mn and MnU for different populations.[94]

Even at very high Mn content in drinking water, no relation was observed between water manganese (MnW) and MnB in children[98] or adults.[97] In contrast, Mn intake from water ingestion, but not from the diet, was significantly associated with elevated Mn in hair (MnH), suggesting that the homeostatic regulation of Mn does not prevent overload upon exposure from water.[95,97] No increase in toenail Mn was observed with supra-nutritional consumption of this element.[24]

Blood and/or hair Mn levels have been associated with neurotoxic outcomes in several studies. For example, results from a community-based study ($n = 273$ adults) in Quebec revealed that MnB levels above 7.5 µg L^{-1} were significantly associated with worse coordinated upper limb movements and poorer learning and recall.[108] An association between MnB and poor performance on general cognition using the Mini Mental State (MMS) examination was reported in a pilot study ($n = 73$ subjects; MnB 7.5–88 µg L^{-1}, median 15) carried out in two communities of the mining district of Hidalgo, Mexico. Lower MMS performance and poor motor function tests were identified at MnB levels above 15 µg L^{-1}, obvious clinical tremor, and numbness at MnB above 25 µg L^{-1}.[109] However, in a further and larger study, MnB levels did not associate to any of the measured outcomes ($n = 288$, median MnB 9.5 (5.0–31.0).[31,32] A significant relationship between MnH (mean 4.4 µg g^{-1}; range 1.2–12.4 µg g^{-1}) but not with MnB (9.4 µg L^{-1}; range 4.2–21.7 µg L^{-1}) and postural balance was observed in selected adults residing near a ferro/Mn refinery in Ohio.[110]

MnH (mean 5.83 µg g^{-1}; range: 0.1–87), but not MnB (mean 8.2 µg L^{-1}, range 2.7–23.4 µg L^{-1}), was found to be associated with poorer cognitive performance, especially in the verbal domain in school-aged children living near a ferro/Mn-alloy plant.[102]. Bouchard *et al.*[95] carried out a cross-sectional study of 362 children, 6–13 years of age, living in communities supplied by groundwater. Although MnW was on average rather low (34 µg L^{-1}; range 0.1–2700 µg L^{-1}), MnH (median 0.7 µg g^{-1}, range: 0.1–21 µg g^{-1}) increased with Mn intake from water consumption and was significantly associated with lower IQ scores. In children aged 11 − 14 years, exposed to environmental Mn from historical ferro/Mn-alloy plant emissions, tremor intensity was positively associated with MnB ($p = 0.005$) and MnH ($p = 0.01$), but not with MnU.[111]

14.4.2 Other Mn Exposures

Hypermanganesemia is an issue in patients receiving long-term parenteral nutrition. High whole blood or serum/plasma Mn levels, as well as increased signal intensity on *T*1-weighted MRI have been measured in neonates, children, and adults, particularly in case of hepatobiliary dysfunction.[112–119] Some patients developed overt clinical evidence of Mn toxicity, and routine whole blood measurement of Mn in combination with MRI has been recommended by some authors to monitor potential accumulation.[120,121] However, no threshold level can be established because in view of the Mn toxicokinetics, "normal" levels of Mn in blood and/or serum do not necessarily preclude high Mn levels in the brain. Autopsy examinations have confirmed the persistence of elevated brain Mn levels in the presence of normal serum values in a fatal case of intoxication.[122] In a Mn on–off clinical study, the presence of high signal intensity on *T*1-weighted MRI scans of the brain was reversibly and reproducibly related to the administration or withdrawal of manganese. MnB concentration showed strong correlations with both the MRI intensity and the *T*1 value ($r = 0.769$, -0.701) – correlations being somewhat lower with MnP ($r = 0.543$, -0.635). The time-course change of MnB concentration was more similar to the profile of the MRI intensity than the profile of MnP[121]. Mn-caused neurotoxicity by long-term parenteral nutrition can be irreversible while the hyperintensity on *T*1-weighted images had disappeared about 5–6 months after cessation of parenteral nutrition. On the other hand, no neurological abnormalities were detected in patients despite obvious MRI changes.[113,117,123]

Subjects suffering from severe liver dysfunction are reported to exhibit increased MnB levels. In series of cirrhotic patients with MnB levels (*e.g.* AM (SD), 20.6 (10.2) µg L^{-1},[3] 18.2 (10.5),[5] 26.01 (12.82),[48] and 23.4 [range: 7.6–21][9]) higher than in controls, investigations revealed correlations between MnB and scores of T1-weighted signal hyperintensity on MRI. However, there was no significant correlation between MnB and extrapyramidal symptoms.[5] MnP or MnU, which were not different from the control values, did not correlate with the signal intensities of T1-weighted MRI.[9] In another series, the MnB levels, which were not elevated (0.46 to 2.34 µg L^{-1}), did not reflect the MRI signal or the neurological damages. The pallidal MRI signal was not predictive of clinical manifestations of chronic acquired hepatocerebral degeneration; a hyperintense pallidal MRI signal was found to be a necessary but not a sufficient prerequisite for such neurological degeneration.[50]

Cases of high blood/serum or hair Mn levels and positive MRI T1 signals have been described in methcathinone abusers who developed Parkinsonism following intravenous use of a solution consisting of ephedrine, acetylsalicylic acid and potassium permanganate as a psycho-stimulant, popularly known as "Russian Cocktail".[124,125] Even extremely elevated MnB levels (2100 µg L^{-1} and 3176 µg L^{-1}) with MRIs showing bilateral symmetric hyperintensities on T1-weighted images (in the dentate nucleus, subcortical white substance of cerebellar hemisphere, GP, and putamen) were found to decrease after abstinence and treatment with EDTA without change in objective neurological findings.[126]

In a case of acute Mn intoxication after ingesting a pharmaceutical preparation containing hydrated manganese sulfate instead of hydrated magnesium sulfate, the *post mortem* MnB level reached 6650 µg L^{-1}.[127]

14.5   Conclusion

There is a growing concern about the potential central nervous system effects at low levels of Mn exposure and correct assessment of the internal Mn dose is urgently needed. Several potential biomarkers have been evaluated for use in Mn exposure assessment with limited success and the utility of implementing biological monitoring in routine surveillance programs is questionable.

There is little evidence to support the view of MnU as a biomarker of interest. Urinary Mn excretion is low, appears to show diurnal variation, exhibits wide variations among individuals, is poorly correlated to the exposure level, and is a poor biomarker for health risk assessment.

Mn concentrations in blood have been found to be increased in Mn-exposed subjects by comparison with control subjects; however, inter-individual variability is high, MnB concentrations are often overlapping between exposed and control subjects, and accurate reference vales are scarce. MnB levels that are increased in some settings are within the range measured in some non-exposed populations. No clear relationship could be demonstrated between the level of external Mn exposure and MnB, and associations between MnB and changes in neuropsychological or motor functions are not always straightforward. Because nearly 95% of Mn in blood is found in the cellular fraction, it has been postulated that whole blood or the RBC, which accounts for 66% of blood Mn, might be a good indicator of tissue accumulation/body burden. However, although some correlations between MnB and/or RBC and MRI have been described, it is important to keep in mind that the half-life of Mn in the brain is considerably longer than in blood. Plasma Mn, which is rapidly cleared from the blood circulation, can be assumed to reflect very recent exposure and the bioavailable fraction of Mn in blood.

Despite growing interest in the use of MRI parameters as a biomarker of Mn exposure directly linked to the target organ dose and as a promising tool to predict neurobehavioral disturbances, it has to be confirmed that increases in hyperintensities indicate actual brain Mn concentrations and reflect Mn neurotoxicity. Nonetheless, the implementation of such a technique in a routine heath surveillance program is unlikely in the near future.

Although some results are promising, hair and nail Mn concentrations have yet to be explored as biomarkers of Mn exposure. Data are insufficient to determine dose–response and/or dose–effect relationships.

Overall, available studies suggest that the measurement of Mn in whole blood, RBC, plasma (serum), hair and nail, and *via* MRI may, in some circumstances, be useful. Data are sometimes insufficient and tricky to interpret because each of these biomarkers may represent a different exposure window (current, recent, or time-integrated exposure). Regarding the relationship with the occurrence of neurological effect, Mn in plasma/serum is likely to be less interesting because of its very short half-life. High Mn blood/serum values and MRI *T*1 hyperintensity have been used to support a diagnosis of Mn neurotoxicity in workers, in individuals receiving total parenteral nutrition, as well as in patients with hepatobiliary insufficiency, or in drug abusers. Although high levels of Mn biomarkers can be used to confirm Mn intoxication, within-reference-range values cannot be used to exclude sequelae from past Mn exposure. MRI findings also disappear a few months after cessation of exposure.

Currently available evidence does not allow recommending a reliable well-validated marker of exposure to Mn, and certainly not a threshold level establishing excessive exposure. The appropriate method(s) for accurately assessing internal dose of Mn are still debated and further research is needed.

References

1. D. B. Milne, R. L. Sims and N. V. Ralston, Manganese content of the cellular components of blood, *Clin. Chem.*, 1990, **36**, 450.
2. R. F. Butterworth, L. Spahr, S. Fontaine and G. P. Layrargues, Manganese toxicity, dopaminergic dysfunction and hepatic encephalopathy, *Metab. Brain Dis.*, 1995, **10**, 259.

3. R. A. Hauser, T. A. Zesiewicz, C. Martinez, A. S. Rosemurgy and C. W. Olanow, Blood manganese correlates with brain magnetic resonance imaging changes in patients with liver disease, *Can. J. Neurol. Sci.*, 1996, **23**, 95.

4. H. Saito and A. Ejima, Liver dysfunction and probable manganese accumulation in the brainstem and basal ganglia, *J. Neurol. Psychiatry*, 1995, **58**, 760.

5. L. Spahr, R. F. Butterworth, S. Fontaine, L. Bui, G. Therrien, P. C. Milette, L. H. Lebrun, J. Zayed, A. Leblanc and G. Pomier-Layrargues, Increased blood manganese in cirrhotic patients: relationship to pallidal magnetic resonance signal hyperintensity and neurological symptoms, *Hepatology*, 1996, **24**, 1116.

6. Y. Jiang, W. Zheng, L. Long, W. Zhao, X. Li, X. Mo, J. Lu, X. Fu, W. Li, S. Liu, Q. Long, J. Huang and E. Pira, Brain magnetic resonance imaging and manganese concentrations in red blood cells of smelting workers: search for biomarkers of manganese exposure, *Neurotoxicology*, 2007, **28**, 126.

7. Y. Chang, Y. Kim, S. T. Woo, H. J. Song, S. H. Kim, H. Lee, Y. J. Kwon, J. H. Ahn, S. J. Park, I. S. Chung and K. S. Jeong, High signal intensity on magnetic resonance imaging is a better predictor of neurobehavioral performances than blood manganese in asymptomatic welders, *Neurotoxicology*, 2009, **30**, 555.

8. Y. Kim, K. S. Kim, J. S. Yang, I. J. Park, E. Kim, Y. Jin, K. R. Kwon, K. H. Chang, J. W. Kim, S. H. Park, H. S. Lim, H. K. Cheong, Y. C. Shin, J. Park and Y. Moon, Increase in signal intensities on T1-weighted magnetic resonance images in asymptomatic manganese-exposed workers, *Neurotoxicology*, 1999, **20**, 901.

9. Y. Choi, J. K. Park, N. H. Park, J. W. Shin, C. I. Yoo, C. R. Lee, H. Lee, H. K. Kim, S. R. Kim, T. H. Jung, J. Park, C. S. Yoon and Y. Kim, Whole blood and red blood cell manganese reflected signal intensities of T1-weighted magnetic resonance images better than plasma manganese in liver cirrhotics, *J. Occup. Health*, 2005, **47**, 68.

10. N. H. Park, J. K. Park, Y. Choi, C. I. Yoo, C. R. Lee, H. Lee, H. K. Kim, S. R. Kim, T. H. Jeong, J. Park, C. S. Yoon and Y. Kim, Whole blood manganese correlates with high signal intensities on T1-weighted MRI in patients with liver cirrhosis, *Neurotoxicology*, 2003, **24**, 909.

11. D. Smith, R. Gwiazda, R. Bowler, H. Roels, R. Park, C. Taicher and R. Lucchini, Biomarkers of Mn exposure in humans, *Am. J. Ind. Med.*, 2007, **50**, 801.

12. D. M. Cowan, Q. Fan, Y. Zou, X. Shi, J. Chen, M. Aschner, F. S. Rosenthal and W. Zheng, Manganese exposure among smelting workers: blood manganese–iron ratio as a novel tool for manganese exposure assessment, *Biomarkers*, 2009, **14**, 3.

13. D. M. Cowan, W. Zheng, Y. Zou, X. Shi, J. Chen, F. S. Rosenthal and Q. Fan, Manganese exposure among smelting workers: relationship

between blood manganese–iron ratio and early onset neurobehavioral alterations, *Neurotoxicology*, 2009, **30**, 1214.

14. J. Järvisalo, M. Olkinuora, M. Kiilunen, H. Kivisto, P. Ristola, A. Tossavainen and A. Aitio, Urinary and blood manganese in occupationally nonexposed populations and in manual metal arc welders of mild steel, *Int. Arch. Occup. Environ. Health*, 1992, **63**, 495.

15. P. Hoet, Ch. Jacquerye, Gl. Deumer, D. Lison and V. Haufroid, Reference values and upper reference limits for 26 trace elements in the urine of adults living in Belgium, *Clin. Chem.*, 2013, **51**, 839.

16. T. Hinners, A. Tsuchiya, A. H. Stern, T. M. Burbacher, E. M. Faustman and K. Marien, Chronologically matched toenail-Hg to hair-Hg ratio: temporal analysis within the Japanese community (US), *Environ. Health*, 2012, **11**, 81.

17. S. Yaemsiri, N. Hou, M. M. Slining and K. He, Growth rate of human fingernails and toenails in healthy American young adults, *J. Eur. Acad. Dermatol. Venereol.*, 2010, **24**, 420.

18. M. A. Buzalaf, J. P. Pessan and K. M. Alves, Influence of growth rate and length on fluoride detection in human nails, *Caries Res.*, 2006, **40**, 231.

19. R. R. Eastman, T. P. Jursa, C. Benedetti, R. G. Lucchini and D. R. Smith, Hair as a biomarker of environmental manganese exposure, *Environ. Sci. Technol.*, 2013, **47**, 1629.

20. M. Bouchard, F. Laforest, L. Vandelac, D. Bellinger and D. Mergler, Hair manganese and hyperactive behaviors: pilot study of school-age children exposed through tap water, *Environ. Health Perspect.*, 2007, **115**, 122.

21. A. Lyden, B. S. Larsson and N. G. Lindquist, Melanin affinity of manganese, *Acta Pharmacol. Toxicol.*, 1984, **55**, 133.

22. A. Sturaro, G. Parvoli, L. Doretti, G. Allegri and C. Costa, The influence of color, age, and sex on the content of zinc, copper, nickel, manganese, and lead in human hair, *Biol. Trace Elem. Res.*, 1994, **40**, 1.

23. W. Laohaudomchok, X. Lin, R. F. Herrick, S. C. Fang, J. M. Cavallari, D. C. Christiani and M. G. Weisskopf, Toenail, blood, and urine as biomarkers of manganese exposure, *J. Occup. Environ. Med.*, 2011, **53**, 506.

24. J. M. Guthrie, J. D. Brockman, J. S. Morris and J. D. Robertson, The "one source" cohort – evaluating the suitability of the human toenail as a manganese biomonitor, *J. Radioanal. Nucl. Chem.*, 2008, **276**, 41.

25. V. A. Fitsanakis, N. Zhang, M. J. Avison, J. C. Gore and J. L. Aschner, The use of magnetic resonance imaging (MRI) in the study of manganese neurotoxicity, *Neurotoxicology*, 2006, **27**, 798.

26. S. R. Criswell, J. S. Perlmutter, J. L. Huang, N. Golchin, H. P. Flores, A. Hobson, M. Aschner, K. M. Erikson, H. Checkoway and B. A. Racette, Basal ganglia intensity indices and diffusion weighted imaging in manganese-exposed welders, *Occup. Environ. Med.*, 2012, **69**, 437.

27. B. Bocca, R. Madeddu, Y. Asara, P. Tolu, J. A. Marchal and G. Forte, Assessment of reference ranges for blood Cu, Mn, Se and Zn in a selected Italian population, *J. Trace Elem. Med. Biol.*, 2011, **25**, 19.

28. N. A. Clark, K. Teschke, K. Rideout and R. Copes, Trace element levels in adults from the west coast of Canada and associations with age, gender, diet, activities, and levels of other trace elements, *Chemosphere*, 2007, **70**, 155.

29. J. W. Lee, C. K. Lee, C. S. Moon, I. J. Choi, K. J. Lee, S. M. Yi, B. K. Jang, B. J. Yoon, D. S. Kim, D. Peak, D. Sul, E. Oh, H. Im, H. S. Kang, J. Kim, J. T. Lee, K. Kim, K. L. Park, R. Ahn, S. H. Park, S. C. Kim, C. H. Park and J. H. Lee, Korea National Survey for Environmental Pollutants in the Human Body 2008: heavy metals in the blood or urine of the Korean population, *Int. J. Hyg. Environ. Health*, 2012, **215**, 449.

30. S. Montes, H. Riojas-Rodriguez, E. Sabido-Pedraza and C. Rios, Biomarkers of manganese exposure in a population living close to a mine and mineral processing plant in Mexico, *Environ. Res.*, 2008, **106**, 89.

31. Y. Rodriguez-Agudelo, H. Riojas-Rodriguez, C. Rios, I. Rosas, E. Sabido,Pedraza, J. Miranda, C. Siebe, J. L. Texcalac and C. Santos-Burgoa, Motor alterations associated with exposure to manganese in the environment in Mexico, *Sci. Total Environ.*, 2006, **368**, 542.

32. R. Solis-Vivanco, Y. Rodriguez-Agudelo, H. Riojas-Rodriguez, C. Rios, I. Rosas and S. Montes, Cognitive impairment in an adult Mexican population non-occupationally exposed to manganese, *Environ. Toxicol. Pharmacol.*, 2009, **28**, 172.

33. H. M. Meltzer, A. L. Brantsaeter, B. Borch-Iohnsen, D. G. Ellingsen, J. Alexander, Y. Thomassen, H. Stigum and T. A. Ydersbond, Low iron stores are related to higher blood concentrations of manganese, cobalt and cadmium in non-smoking, Norwegian women in the HUNT 2 study, *Environ. Res.*, 2010, **110**, 497.

34. M. Baldwin, D. Mergler, F. Larribe, S. Belanger, R. Tardif, L. Bilodeau and K. Hudnell, Bioindicator and exposure data for a population based study of manganese, *Neurotoxicology*, 1999, **20**, 343.

35. A. Spencer, Whole blood manganese levels in pregnancy and the neonate, *Nutrition*, 1999, **15**, 731.

36. K. Tholin, B. Sandstrom, R. Palm and G. Hallmans, Changes in blood manganese levels during pregnancy in iron supplemented and non supplemented women, *J. Trace Elem. Med. Biol.*, 1995, **9**, 13.

37. H. Tsuchiya, K. Mitani, K. Kodama and T. Nakata, Placental transfer of heavy metals in normal pregnant Japanese women, *Arch. Environ. Health*, 1984, **39**, 11.

38. A. Smargiassi, L. Takser, A. Masse, M. Sergerie, D. Mergler, G. St-Amour, P. Blot, G. Hellier and G. Huel, A comparative study of manganese and lead levels in human umbilical cords and maternal blood from two urban centers exposed to different gasoline additives, *Sci. Total Environ.*, 2002, **290**, 157.

39. L. Takser, J. Lafond, M. Bouchard, G. St-Amour and D. Mergler, Manganese levels during pregnancy and at birth: relation to environmental factors and smoking in a Southwest Quebec population, *Environ. Res.*, 2004, **95**, 119.

40. A. R. Zota, A. S. Ettinger, M. Bouchard, C. J. Amarasiriwardena, J. Schwartz, H. Hu and R. O. Wright, Maternal blood manganese levels and infant birth weight, *Epidemiology*, 2009, **20**, 367.

41. E. A. Smith, P. Newland, K. G. Bestwick and N. Ahmed, Increased whole blood manganese concentrations observed in children with iron deficiency anaemia, *J. Trace Elem. Med. Biol.*, 2013, **27**, 65.

42. O. M. Alarcon, J. A. Reinosa-Fuller, T. Silva, F. M. Ramirez, de and J. Gamboa, Manganese levels in serum of healthy Venezuelan infants living in Merida, *J. Trace Elem. Med. Biol.*, 1996, **10**, 210.

43. B. K. Henn, A. S. Ettinger, J. Schwartz, M. M. Téllez-Rojo, H. Lamadrid-Figueroa, M. Henrnandez-Avila, L. Schnaas, Ch. Amarasiriwardena, D. C. Bellinger, H. Hu and R. O. Wright, Early postnatal blood manganese levels and children's neurodevelopment, *Epidemiology*, 2010, **21**, 433.

44. D. C. Rice, R. Lincoln, J. Martha, L. Parker, K. Pote, S. Xing and A. E. Smith, Concentration of metals in blood of Maine children 1-6 years old, *J. Exposure Sci. Environ. Epidemiol.*, 2010, **20**, 634.

45. M. Rükgauer, J. Klein and J. D. Kruse-Jarres, Reference values for the trace elements copper, manganese, selenium, and zinc in the serum/plasma of children, adolescents, and adults, *J. Trace Elem. Med. Biol.*, 1997, **11**, 92.

46. D. Krieger, S. Krieger, O. Jansen, P. Gass, L. Theilmann and H. Lichtnecker, Manganese and chronic hepatic encephalopathy, *Lancet*, 1995, **346**, 270.

47. R. Mehta and J. J. Reilly, Manganese levels in a jaundiced long-term total parenteral nutrition patient: potentiation of haloperidol toxicity? Case report and literature review, *JPEN, J. Parenter. Enteral Nutr.*, 1990, **14**, 428.

48. R. B. Pinto, P. E. Froehlich, E. H. Pitrez, J. A. Bragatti, J. Becker, A. F. Cornely, A. C. Schneider and T. R. da Silveira, MR findings of the brain in children and adolescents with portal hypertension and the relationship with blood manganese levels, *Neuropediatrics*, 2010, **41**, 12.

49. H. M. Zeron, M. R. Rodriguez, S. Montes and C. R. Castaneda, Blood manganese levels in patients with hepatic encephalopathy, *J. Trace Elem. Med. Biol.*, 2011, **25**, 225.

50. E. Maffeo, A. Montuschi, G. Stura and M. T. Giordana, Chronic acquired hepatocerebral degeneration, pallidal T1 MRI hyperintensity and manganese in a series of cirrhotic patients, *Neurol. Sci.*, 2014, **35**, 523.

51. A. A. Britton and G. C. Cotzias, Dependence of manganese turnover on intake, *Am. J. Physiol.*, 1966, **211**, 203.

52. A. Nong, M. D. Taylor, H. J. Clewell III, D. C. Dorman and M. E. Andersen, Manganese tissue dosimetry in rats and monkeys:

accounting for dietary and inhaled Mn with physiologically based pharmacokinetic modeling, *Toxicol. Sci.*, 2009, **108**, 22.

53. M. Suzuki and W. E. Wacker, Determination of manganese in biological materials by atomic absorption spectroscopy, *Anal. Biochem.*, 1974, **57**, 605.

54. D. C. Dorman, M. F. Struve, R. A. James, B. E. McManus, M. W. Marshall and B. A. Wong, Influence of dietary manganese on the pharmacokinetics of inhaled manganese sulfate in male CD rats, *Toxicol. Sci.*, 2001, **60**, 242.

55. D. C. Dorman, M. F. Struve, M. W. Marshall, C. U. Parkinson, R. A. James and B. A. Wong, Tissue manganese concentrations in young male rhesus monkeys following subchronic manganese sulfate inhalation, *Toxicol. Sci.*, 2006, **92**, 201.

56. D. Vitarella, O. Moss and D. C. Dorman, Pulmonary clearance of manganese phosphate, manganese sulfate, and manganese tetraoxide by CD rats following intratracheal instillation, *Inhalation Toxicol.*, 2000, **12**, 941.

57. D. C. Dorman, M. F. Struve, R. A. James, M. W. Marshall, C. U. Parkinson and B. A. Wong, Influence of particle solubility on the delivery of inhaled manganese to the rat brain: manganese sulfate and manganese tetroxide pharmacokinetics following repeated (14-day) exposure, *Toxicol. Appl. Pharmacol.*, 2001, **170**, 79.

58. P. Hoet, E. Vanmarcke, T. Geens, G. Deumer, V. Haufroid and H. A. Roels, Manganese in plasma: a promising biomarker of exposure to Mn in welders. A pilot study, *Toxicol. Lett.*, 2012, **213**, 69.

59. W. Zheng, H. Kim and Q. Zhao, Comparative toxicokinetics of manganese chloride and methylcyclopentadienyl manganese tricarbonyl (MMT) in Sprague-Dawley rats, *Toxicol. Sci.*, 2000, **54**, 295.

60. M. C. Newland, C. Cox, R. Hamada, G. Oberdorster and B. Weiss, The clearance of manganese chloride in the primate, *Fundam. Appl. Toxicol.*, 1987, **9**, 314.

61. A. Takeda, J. Sawashita and S. Okada, Biological half-lives of zinc and manganese in rat brain, *Brain Res.*, 1995, **695**, 53.

62. J. P. Mahoney and W. J. Small, Studies on manganese. 3. The biological half-life of radiomanganese in man and factors which affect this half-life, *J. Clin. Invest.*, 1968, **47**, 643.

63. I. Mena, O. Marin, S. Fuenzalida and G. C. Cotzias, Chronic manganese poisoning. Clinical picture and manganese turnover, *Neurology*, 1967, **17**, 128.

64. G. C. Cotzias, K. Horiuchi, S. Fuenzalida and I. Mena, Chronic manganese poisoning. Clearance of tissue manganese concentrations with persistance of the neurological picture, *Neurology*, 1968, **18**, 376.

65. J. D. Schroeter, A. Nong, M. Yoon, M. D. Taylor, D. C. Dorman, M. E. Andersen and H. J. Clewell III, Analysis of manganese tracer kinetics and target tissue dosimetry in monkeys and humans with multi-route physiologically based pharmacokinetic models, *Toxicol. Sci.*, 2011, **120**, 481.

66. J. D. Schroeter, D. C. Dorman, M. Yoon, A. Nong, M. D. Taylor, M. E. Andersen and H. J. Clewell III, Application of a multi-route physiologically based pharmacokinetic model for manganese to evaluate dose-dependent neurological effects in monkeys, *Toxicol. Sci.*, 2012, **192**, 432.

67. R. Lucchini, L. Selis, D. Folli, P. Apostoli, A. Mutti, O. Vanoni, A. Iregren and L. Alessio, Neurobehavioral effects of manganese in workers from a ferroalloy plant after temporary cessation of exposure, *Scand. J. Work, Environ. Health*, 1995, **21**, 143.

68. R. Lucchini, P. Apostoli, C. Perrone, D. Placidi, E. Albini, P. Migliorati, D. Mergler, M.-P. Sassine, S. Palmi and L. Alessio, Long-term exposure to "low levels" of manganese oxides and neurofunctional changes in ferroalloy workers, *Neurotoxicology*, 1999, **20**, 287.

69. D. G. Ellingsen, L. Dubeikovskaya, K. Dahl, M. Chashchin, V. Chashchin, E. Zibarev and Y. Thomassen, Air exposure assessment and biological monitoring of manganese and other major welding fume components in welders, *J. Environ. Monit.*, 2006, **8**, 1078.

70. R. M. Bowler, H. A. Roels, S. Nakagawa, M. Drezgic, E. Diamond, R. Park, W. Koller, R. P. Bowler, D. Mergler, M. Bouchard, D. Smith, R. Gwiazda and R. L. Doty, Dose-effect relationships between manganese exposure and neurological, neuropsychological and pulmonary function in confined space bridge welders, *Occup. Environ. Med.*, 2007, **64**, 167.

71. L. Lu, L. L. Zhang, G. J. Li, W. Guo, W. Liang and W. Zheng, Alteration of serum concentrations of manganese, iron, ferritin, and transferrin receptor following exposure to welding fumes among career welders, *Neurotoxicology*, 2005, **26**, 257.

72. D. Wang, X. Du and W. Zheng, Alteration of saliva and serum concentrations of manganese, copper, zinc, cadmium and lead among career welders, *Toxicol. Lett.*, 2008, **176**, 40.

73. M. Harris, *Welding Health and Safety: A Field Guide for OEHS Professionals*, American Industrial Hygiene Association Press, Fairfax, VA, 2002, pp. 214–224.

74. M. K. Harris, W. M. Ewing, W. Longo, C. DePasquale, M. D. Mount, R. Hatfield and R. Stapleton, Manganese exposures during shielded metal arc welding (SMAW) in an enclosed space, *J. Occup. Environ. Hyg.*, 2005, **2**, 375.

75. L. Normandin, B. L. Ann, F. Salehi, A. St -Pierre, G. Kennedy, D. Mergler, R. F. Butterworth, S. Philippe and J. Zayed, Manganese distribution in the brain and neurobehavioral changes following inhalation exposure of rats to three chemical forms of manganese, *Neurotoxicology*, 2004, **25**, 433.

76. H. Roels, R. Lauwerys, J. P. Buchet, P. Genet, M. J. Sarhan, I. Hanotiau, I. M. de Fays, A. Bernard and D. Stanescu, Epidemiological survey among workers exposed to manganese: effects on lung, central nervous system, and some biological indices, *Am. J. Ind. Med.*, 1987, **11**, 307.

77. H. A. Roels, P. Ghyselen, J. P. Buchet, E. Ceulemans and R. R. Lauwerys, Assessment of the permissible exposure level to manganese in workers exposed to manganese dioxide dust, *Br. J. Ind. Med.*, 1992, **49**, 25.

78. H. Roels, G. Meiers, M. Delos, I. Ortega, R. Lauwerys, J.-P. Buchet and D. Lison, Influence of the route of administration and the chemical form (MnCl2, MnO2) on the absorption and cerebral distribution of manganese in rats, *Arch. Toxicol.*, 1997, **71**, 223.

79. D. G. Ellingsen, S. M. Hetland and Y. Thomassen, Manganese air exposure assessment and biological monitoring in the manganese alloy production industry, *J. Environ. Monit.*, 2003, **5**, 84.

80. B. Pesch, T. Weiss, B. Kendzia, J. Henry, M. Lehnert, A. Lotz, E. Heinze, H. U. Käfferlein, R. Van Gelder, M. Berges, J. U. Hahn, M. Mattenklott, E. Punkenburg, A. Hartwig and T. Brüning, Levels and predictors of airborne and internal exposure to manganese and iron among welders, *J. Exposure Sci. Environ. Epidemiol.*, 2012, **22**, 291.

81. J. E. Myers, M. L. Thompson, I. Naik, P. Theodorou, E. Esswein, H. Tassel, A. Daya, K. Renton, A. Spies, J. Paicker, T. Young, M. Jeebhay, S. Ramushu, L. London and D. J. Rees, The utility of biological monitoring for manganese in ferroalloy smelter workers in South Africa, *Neurotoxicology*, 2003, **24**, 875.

82. H. I. Afridi, T. G. Kazi, N. G. Kazi, M. K. Jamali, M. B. Arain, Sirajuddin, G. A. Kandhro, A. Q. Shah and J. A. Baig, Evaluation of arsenic, cobalt, copper and manganese in biological Samples of Steel mill workers by electrothermal atomic absorption Spectrometry, *Toxicol. Ind. Health*, 2009, **25**, 59.

83. M. M. Boojar and F. Goodarzi, A longitudinal follow-up of pulmonary function and respiratory symptoms in workers exposed to manganese, *J. Occup. Environ. Med.*, 2002, **44**, 282.

84. W. W. Barrington, C. R. Angle, N. K. Willcockson, M. A. Padula and T. Korn, Autonomic function in manganese alloy workers, *Environ. Res.*, 1998, **78**, 50.

85. S. E. Chia, S. C. Foo, S. L. Gan, J. Jeyaratnam and C. S. Tian, Neurobehavioral functions among workers exposed to manganese ore, *Scand. J. Work, Environ. Health*, 1993, **19**, 264.

86. S. E. Chia, S. L. Gan, L. H. Chua, S. C. Foo and J. Jeyaratnam, Postural stability among manganese exposed workers, *Neurotoxicology*, 1995, **16**, 519.

87. D. Mergler, G. Huel, R. Bowler, A. Iregren, S. Bélanger, M. Baldwin, R. Tardif, A. Smargiassi and L. Martin, Nervous system dysfunction among workers with long-term exposure to manganese, *Environ. Res.*, 1994, **64**, 151.

88. R. Bast-Pettersen, D. G. Ellingsen, S. M. Hetland and Y. Thomassen, Neuropsychological function in manganese alloy plant workers, *Int. Arch. Occup. Environ. Health*, 2004, 77, 277.

89. D. G. Ellingsen, R. Konstantinov, R. Bast-Pettersen, L. Merkurjeva, M. Chashchin, Y. Thomassen and V. Chashchin, A neurobehavioral

study of current and former welders exposed to manganese, *Neurotoxicology*, 2008, **29**, 48.

90. Y. Chang, Y. Kim, S. T. Woo, H. J. Song, S. H. Kim, H. Lee, Y. J. Kwon, J. H. Ahn, S. J. Park, I. S. Chung and K. S. Jeong, High signal intensity on magnetic resonance imaging is a better predictor of neurobehavioral performances than blood manganese in asymptomatic welders, *Neurotoxicology*, 2009, **30**, 555.

91. Y. Chang, S. T. Woo, Y. Kim, J. J. Lee, H. J. Song, H. J. Lee, S. H. Kim, H. Lee, Y. J. Kwon, J. H. Ahn, S. J. Park, I. S. Chung and K. S. Jeong, Pallidal index measured with three-dimensional T1-weighted gradient echo sequence is a good predictor of manganese exposure in welders, *J. Magn. Reson. Imaging*, 2010, **31**, 1020.

92. C. C. Huang, N. S. Chu, C. S. Lu, R. S. Chen and D. B. Calne, Long-term progression in chronic manganism: ten years of follow-up, *Neurology*, 1998, **50**, 698.

93. K. Nelson, J. Golnick, T. Korn and C. Angle, Manganese encephalopathy: utility of early magnetic resonance imaging, *Br. J. Ind. Med.*, 1993, **50**, 510.

94. S. W. Al-Rmalli, R. O. Jenkins and P. I. Haris, Betel quid chewing as a source of manganese exposure: total daily intake of manganese in a Bangladeshi population, *BMC Public Health*, 2011, **11**, 85.

95. M. F. Bouchard, S. Sauve, B. Barbeau, M. Legrand, M. E. Brodeur, T. Bouffard, E. Limoges, D. C. Bellinger and D. Mergler, Intellectual impairment in school-age children exposed to manganese from drinking water, *Environ. Health Perspect.*, 2011, **119**, 138.

96. J. L. Greger, C. D. Davis, J. W. Suttie and B. J. Lyle, Intake, serum concentrations, and urinary excretion of manganese by adult males, *Am. J. Clin. Nutr.*, 1990, **51**, 457.

97. X. G. Kondakis, N. Makris, M. Leotsinidis, M. Prinou and T. Papapetropoulos, Possible health effects of high manganese concentration in drinking water, *Arch. Environ. Health*, 1989, **44**, 175.

98. G. A. Wasserman, X. Liu, F. Parvez, H. Ahsan, D. Levy, P. Factor-Litvak, J. Kline, A. van Geen, V. Slavkovich, N. J. LoIacono, Z. Cheng, Y. Zheng and J. H. Graziano, Water manganese exposure and children's intellectual function in Araihazar, Bangladesh, *Environ. Health Perspect.*, 2006, **114**, 124.

99. E. N. Haynes, P. Heckel, P. Ryan, S. Roda, Y. K. Leung, K. Sebastian and P. Succop, Environmental manganese exposure in residents living near a ferromanganese refinery in Southeast Ohio: a pilot study, *Neurotoxicology*, 2010, **31**, 468.

100. D. Hernandez-Bonilla, A. Schilmann, S. Montes, Y. Rodríguez-Agudelo, S. Rodríguez-Dozal, R. Solís-Vivanco, C. Ríos and H. Riojas-Rodríguez, Environmental exposure to manganese and motor function of children in Mexico, *Neurotoxicology*, 2011, **32**, 615.

101. J. A. Menezes-Filho, C. R. Paes, A. M. Pontes, J. C. Moreira, P. N. Sarcinelli and D. Mergler, High levels of hair manganese in

children living in the vicinity of a ferro-manganese alloy production plant, *Neurotoxicology*, 2009, **30**, 1207.

102. J. A. Menezes-Filho, C. O. Novaes, J. C. Moreira, P. N. Sarcinelli and D. Mergler, Elevated manganese and cognitive performance in school-aged children and their mothers, *Environ. Res.*, 2011, **111**, 156.

103. H. Riojas-Rodriguez, R. Solis-Vivanco, A. Schilmann, S. Montes, S. Rodríguez, C. Ríos and Y. Rodríguez-Agudelo, Intellectual function in Mexican children living in a mining area and environmentally exposed to manganese, *Environ. Health Perspect.*, 2010, **118**, 1465.

104. S. Bolte, L. Normandin, G. Kennedy and J. Zayed, Human exposure to respirable manganese in outdoor and indoor air in urban and rural areas, *J. Toxicol. Environ. Health, Part A*, 2004, **67**, 459.

105. B. Gulson, K. Mizon, A. Taylor, M. Korsch, J. Stauber, J. M. Davis, H. Louie, M. Wu and H. Swan, Changes in manganese and lead in the environment and young children associated with the introduction of methylcyclopentadienyl manganese tricarbonyl in gasoline–preliminary results, *Environ. Res.*, 2006, **100**, 100.

106. H. B. Rollin, A. Mathee, J. Levin, P. Theodorou, H. Tassell, I. Naik and F. Wewers, Examining the association between blood manganese and lead levels in schoolchildren in four selected regions of South Africa, *Environ. Res.*, 2007, **103**, 160.

107. H. B. Rollin, C. V. Rudge, Y. Thomassen, A. Mathee and J. O. Odland, Levels of toxic and essential metals in maternal and umbilical cord blood from selected areas of South Africa – results of a pilot study, *J. Environ. Monit.*, 2009, **11**, 618.

108. D. Mergler, M. Baldwin, S. Belanger, F. Larribe, A. Beuter, R. Bowler, M. Panisset, R. Edwards, A. de Geoffroy, M. P. Sassine and K. Hudnell, Manganese neurotoxicity, a continuum of dysfunction: results from a community based study, *Neurotoxicology*, 1999, **20**, 327.

109. C. Santos-Burgoa, C. Rios, L. A. Mercado, R. Arechiga-Serrano, F. Cano-Valle, R. A. Eden-Wynter, J. L. Texcalac-Sangrador, J. P. Villa-Barragan, Y. Rodriguez-Agudelo and S. Montes, Exposure to manganese: health effects on the general population, a pilot study in central Mexico, *Environ. Res.*, 2001, **85**, 90.

110. J. S. Standridge, A. Bhattacharya, P. Succop, C. Cox and E. Haynes, Effect of chronic low level manganese exposure on postural balance: a pilot study of residents in southern Ohio, *J. Occup. Environ. Med.*, 2008, **50**, 1421.

111. R. G. Lucchini, S. Guazzetti, S. Zoni, F. Donna, S. Peter, A. Zacco, M. Salmistraro, E. Bontempi, N. J. Zimmerman and D. R. Smith, Tremor, olfactory and motor changes in Italian adolescents exposed to historical ferro-manganese emission, *Neurotoxicology*, 2012, **33**, 687.

112. R. Abdalian, O. Saqui, G. Fernandes and J. P. Allard, Effects of manganese from a commercial multi-trace element supplement in a population sample of canadian patients on long-term parenteral nutrition, *JPEN, J. Parenter. Enteral Nutr.*, 2013, **37**, 538.

113. D. B. Bertinet, M. Tinivella, F. A. Balzola, A. de Francesco, O. Davini, L. Rizzo, P. Massarendi, M. A. Leonardi and F. Balzola, Brain manganese deposition and blood levels in patients undergoing home parenteral nutrition, *JPEN, J. Parenter. Enteral Nutr.*, 2000, **24**, 223.
114. I. F. Btaiche, P. L. Carver and K. B. Welch, Dosing and monitoring of trace elements in long-term home parenteral nutrition patients, *JPEN, J. Parenter. Enteral Nutr.*, 2011, **35**, 736.
115. J. A. Chalela, L. Bonillha, R. Neyens and A. Hays, Manganese encephalopathy: an under-recognized condition in the intensive care unit, *Neurocrit. Care*, 2011, **14**, 456.
116. J. M. Fell, A. P. Reynolds, N. Meadows, K. Khan, S. G. Long, G. Quaghebeur, W. J. Taylor and P. J. Milla, Manganese toxicity in children receiving long-term parenteral nutrition, *Lancet*, 1996, **347**, 1218.
117. Y. Iinuma, M. Kubota, M. Uchiyama, M. Yagi, S. Kanada, S. Yamazaki, H. Murata, K. Okamoto, M. Suzuki and K. Nitta, Whole-blood manganese levels and brain manganese accumulation in children receiving long-term home parenteral nutrition, *Pediatr. Surg. Int.*, 2003, **19**, 268.
118. S. A. Mirowitz and T. J. Westrich, Basal ganglial signal intensity alterations: reversal after discontinuation of parenteral manganese administration, *Radiology*, 1992, **185**, 535.
119. S. Nagatomo, F. Umehara, K. Hanada, Y. Nobuhara, S. Takenaga, K. Arimura and M. Osame, Manganese intoxication during total parenteral nutrition: report of two cases and review of the literature, *J. Neurol. Sci.*, 1999, **162**, 102.
120. G. Hardy, Manganese in parenteral nutrition: who, when, and why should we supplement?, *Gastroenterology*, 2009, **137**, S29.
121. Y. Takagi, A. Okada, K. Sando, M. Wasa, H. Yoshida and N. Hirabuki, On-off study of manganese administration to adult patients undergoing home parenteral nutrition: new indices of in vivo manganese level, *JPEN, J. Parenter. Enteral Nutr.*, 2001, **25**, 87.
122. G. Alves, J. Thiebot, A. Tracqui, T. Delangre, C. Guedon and E. Lerebours, Neurologic disorders due to brain manganese deposition in a jaundiced patient receiving long-term parenteral nutrition, *JPEN, J. Parenter. Enteral Nutr.*, 1997, **21**, 41.
123. Y. Kafritsa, J. Fell, S. Long, M. Bynevelt, W. Taylor and P. Milla, Long-term outcome of brain manganese deposition in patients on home parenteral nutrition, *Arch. Dis. Child.*, 1998, **79**, 263.
124. A. Koksal, S. Baybas, V. Sozmen, N. S. Koksal, Y. Altunkaynak, A. Dirican, B. Mutluay, H. Kucukoglu and C. Keskinkilic, Chronic manganese toxicity due to substance abuse in Turkish patients, *Neurol. India*, 2012, **60**, 224.
125. K. Sikk, S. Haldre, S. M. Aquilonius, A. Asser, M. Paris, A. Roose, J. Petterson, S. L. Eriksson, J. Bergquist and P. Taba, Manganese-induced parkinsonism in methcathinone abusers: bio-markers of exposure and follow-up, *Eur. J. Neurol.*, 2013, **20**, 915.

126. F. Varlibas, I. Delipoyraz, G. Yuksel, G. Filiz, H. Tirelis and N. O. Gecim, Neurotoxicity following chronic intravenous use of "Russian cocktail, *Clin. Toxicol.*, 2009, **47**, 157.

127. B. Sanchez, J. Casalots-Casado, S. Quintana, A. Arroyo, C. Martin-Fumado and I. Galtes, Fatal manganese intoxication due to an error in the elaboration of Epsom salts for a liver cleansing diet, *Forensic Sci. Int.*, 2012, **223**, e1.

128. J. Kristiansen, J. M. Christensen, B. S. Iversen and E. Sabbioni, Toxic trace element reference levels in blood and urine: influence of gender and lifestyle factors, *Sci. Total Environ.*, 1997, **204**, 147.

129. L. Takser, D. Mergler, G. Hellier, J. Sahuquillo and G. Huel, Manganese, monoamine metabolite levels at birth, and child psychomotor development, *Neurotoxicology*, 2003, **24**, 667.

130. J. P. Goullé, L. Mahieu, J. Castermant, N. Neveu, L. Bonneau, G. Lainé, D. Bouige and C. Lacroix, Metal and metalloid multi-elementary ICP-MS validation in whole blood, plasma, urine and hair reference values, *Forensic Sci. Int.*, 2005, **153**, 39.

131. P. Heitland and H. D. Koster, Biomonitoring of 37 trace elements in blood samples from inhabitants of northern Germany by ICP-MS, *J. Trace Elem. Med. Biol.*, 2006, **20**, 253.

132. Th. Goën, K. H. Schaller and H. Drexler, Biological reference values for chemical compounds in the work area (BARs): an approach for evaluating biomonitoring data, *Int. Arch. Occup. Environ. Health*, 2012, **85**, 571.

133. H. I. Afridi, T. G. Kazi, D. Brabazon, S. Naher and F. N. Talpur, Comparative metal distribution in scalp hair of Pakistani and Irish referents and diabetes mellitus patients, *Clin. Chim. Acta*, 2013, **415**, 207.

134. A. Alimonti, B. Bocca, E. Mannella, F. Petrucci, F. Zennaro, R. Cotichini, C. D'Ippolito, A. Agresti, S. Caimi and G. Forte, Assessment of reference values for selected elements in a healthy urban population, *Ann. Ist. Super. Sanita*, 2005, **41**, 181.

135. B. Bocca, D. Mattei, A. Pino and A. Alimonti, Italian network for human biomonitoring of metals: preliminary results from two Regions, *Ann. Ist. Super. Sanita*, 2010, **46**, 259.

136. J. Lourenço, R. Pereira, F. Pinto, T. Caetano, A. Silva, T. Carvalheiro, A. Guimaraes, F. Gonçalves, A. Paiva and S. Mendo, Biomonitoring a human population inhabiting nearby a deactivated uranium mine, *Toxicology*, 2013, **305**, 89.

137. M. Torra, M. Rodamilans and J. Corbella, Biological monitoring of environmental exposure to manganese in blood samples from residents of the city of Barcelona, Spain, *Sci. Total Environ.*, 2002, **289**, 237.

138. I. Rodushkin and M. D. Axelsson, Application of double focusing sector field ICP-MS for multielemental characterization of human hair and nails. Part II. A study of the inhabitants of northern Sweden, *Sci. Total Environ.*, 2000, **262**, 21.

139. M. A. White and E. Sabbioni, Trace element reference values in tissues from inhabitants of the European Union. X. A study of 13 elements in blood and urine of a United Kingdom population, *Sci. Total Environ.*, 1998, **216**, 253.

140. C. E. Sieniawska, L. C. Jung, R. Olufadi and V. Walker, Twenty-four-hour urinary trace element excretion: reference intervals and interpretive issues, *Ann. Clin. Biochem.*, 2012, **49**, 341.

141. D. C. Paschal, B. G. Ting, J. C. Morrow, J. L. Pirkle, R. J. Jackson, E. J. Sampson, D. T. Miller and K. L. Caldwell, Trace metals in urine of United States residents: reference range concentrations, *Environ. Res.*, 1998, **76**, 53.

142. I. Mordukhovich, R. O. Wright, H. Hu, C. Amarasiriwardena, A. Baccarelli, A. Litonjua, D. Sparrow, P. Vokonas and J. Schwartz, Associations of toenail arsenic, cadmium, mercury, manganese, and lead with blood pressure in the normative aging study, *Environ. Health Perspect.*, 2012, **120**, 98.

143. Health Canada. Second Report on Human Biomonitoring of Environmental Chemicals in Canada. Results of the Canadian Health Measures Survey Cycle 2 (2009–2011), April 2013.

144. B. L. Batista, J. L. Rodrigues, L. Tormen, A. J. Curtius and F. Barbosa, Jr, Reference concentrations for trace elements in urine for the Brazilian population based on q-ICP-MS with a simple dilute and shoot procedure, *J. Braz. Chem. Soc.*, 2009, **8**, 1406.

145. J. A. Nunes, B. L. Batista, J. L. Rodrigues, N. M. Caldas, J. A. Neto and F. Barbosa Jr, A simple method based on ICP-MS for estimation of background levels of arsenic, cadmium, copper, manganese, nickel, lead, and selenium in blood of the Brazilian population, *J. Toxicol. Environ. Health, Part A*, 2010, **73**, 878.

146. M. Berglund, A. L. Lindberg, M. Rahman, M. Yunus, M. Grandér, B. Lönnerdal and V. Vahter, Gender and age differences in mixed metal exposure and urinary excretion, *Environ. Res.*, 2011, **111**, 1271.

147. F. Ohashi, Y. Fukui, S. Takada, J. Moriguchi, T. Ezaki and M. Ikeda, Reference values for cobalt, copper, manganese, and nickel in urine among women of the general population in Japan, *Int. Arch. Occup. Environ. Health*, 2006, **80**, 117.

148. M. Ikeda, F. Ohashi, Y. Fukui, S. Sakuragi and J. Moriguchi, Cadmium, chromium, lead, manganese and nickel concentrations in blood of women in non-polluted areas in Japan, as determined by inductively coupled plasma-sector field-mass spectrometry, *Int. Arch. Occup. Environ. Health*, 2011, **84**, 139.

149. A. Sukumar and R. Subramanian, Elements in the hair of non-mining workers of a lignite open mine in Neyveli, *Ind. Health*, 2003, **41**, 63.

150. R. Mehra and M. Juneja, Variation of concentration of heavy metals, calcium and magnesium with sex as determined by atomic absorption spectrophotometry, *Indian J. Environ. Health*, 2003, **45**, 317.

151. A. Khalique, S. Ahmad, T. Anjum, M. Jaffar, M. H. Shah, N. Shaheen, S. R. Tariq and S. Manzoor, A comparative study based on gender and age dependence of selected metals in scalp hair, *Environ. Monit. Assess.*, 2005, **104**, 45.

152. M. H. Shah, N. Shaheen, A. Khalique, A. A. Alrabti and M. Jaffar, Comparative metal distribution in hair of Pakistani and Libyan population and source identification by multivariate analysis, *Environ. Monit. Assess.*, 2006, **114**, 505.

153. S. Batterman, F. C. Su, C. Jia, R. N. Naidoo, T. Robins and I. Naik, Manganese and lead in children's blood and airborne particulate matter in Durban, South Africa, *Sci. Total Environ.*, 2011, **409**, 1058.

154. M. Bouchard, D. Mergler, M. Baldwin, M. P. Sassine, R. Bowler and B. MacGibbon, Blood manganese and alcohol consumption interact on mood states among manganese alloy production workers, *Neurotoxicology*, 2003, **24**, 641.

155. P. Apostoli, R. Lucchini and L. Alessio, Are current biomarkers suitable for the assessment of manganese exposure in individual workers?, *Am. J. Ind. Med.*, 2000, **37**, 283.

156. T. Young, J. E. Myers and M. L. Thompson, The nervous system effects of occupational exposure to manganese – measured as respirable dust – in a South African manganese smelter, *Neurotoxicology*, 2005, **26**, 993.

157. B. Sjögren, A. Iregren, W. Frech, M. Hagman, L. Johansson, M. Tesarz and A. Wennberg, Effects on the nervous system among welders exposed to aluminium and manganese, *Occup. Environ. Med.*, 1996, **53**, 32.

158. M. Lehnert, T. Weiss, B. Pesch, A. Lotz, S. Zilch-Schöneweis, E. Heinze, R. Van Gelder, J. U. Hahn, T. Brüning and the WELDOX Study Group, Reduction in welding fume and metal exposure of stainless steel welders: an example from the WELDOX study, *Int. Arch. Occup. Environ. Health*, 2014, **87**, 483.

159. S. V. Chandra, G. S. Shukla, R. S. Srivastava, H. Singh and V. P. Gupta, An exploratory study of manganese exposure to welders, *Clin. Toxicol.*, 1981, **18**, 407.

160. G. J. Li, L. L. Zhang, L. Lu, P. Wu and W. Zheng, Occupational exposure to welding fume among welders: alterations of manganese, iron, zinc, copper, and lead in body fluids and the oxidative stress status, *J. Occup. Environ. Med.*, 2004, **46**, 241.

161. H. Yuan, S. He, M. He, Q. Niu, L. Wang and S. Wang, A comprehensive study on neurobehavior, neurotransmitters and lymphocyte subsets alteration of Chinese manganese welding workers, *Life Sci.*, 2006, **78**, 1324.

162. W. Wongwit, J. Kaewkungwal, Y. Chantachum and V. Visesmanee, Comparison of biological specimens for manganese determination among highly exposed welders, *Southeast Asian J. Trop. Med. Public Health*, 2004, **35**, 764.

163. E. Kim, Y. Kim, H. K. Cheong, S. Cho, Y. C. Shin, J. Sakong, K. S. Kim, J. S. Yang, Y. W. Jin and S. K. Kang, Pallidal index on MRI as a target organ dose of manganese: structural equation model analysis, *Neurotoxicology*, 2005, **26**, 351.

164. E. A. Kim, H. K. Cheong, D. S. Choi, J. Sakong, J. W. Ryoo, I. Park and D. M. Kang, Effect of occupational manganese exposure on the central nervous system of welders: 1H magnetic resonance spectroscopy and MRI findings, *Neurotoxicology*, 2007, **28**, 276.

165. D. S. Choi, E. A. Kim, H. K. Cheong, H. S. Khang, J. W. Ryoo, J. M. Cho, J. Sakong and I. Park, Evaluation of MR signal index for the assessment of occupational manganese exposure of welders by measurement of local proton T1 relaxation time, *Neurotoxicology*, 2007, **28**, 284.

166. H. Hassani, F. Golbabaei and A. Ghahri, Occupational exposure to manganese-containing welding fumes and pulmonary function indices among natural gas transmission pipeline welders, *J. Occup. Health*, 2012, **54**, 316.

167. H. Roels, R. Lauwerys, P. Genet, M. J. Sarhan, M. de Fays, I. Hanotiau and J.-P. Buchet, Relationship between external and internal parameters of exposure to manganese in workers from a manganese oxide and salt producing plant, *Am. J. Ind. Med.*, 1987, **11**, 297.

168. H. A. Roels, M. I. Ortega Eslava, E. Ceulemans, A. Robert and D. Lison, Prospective study on the reversibility of neurobehavioral effects in workers exposed to manganese dioxide, *Neurotoxicology*, 1999, **20**, 255.

169. M. Bader, M. C. Dietz, A. Ihrig and G. Triebig, Biomonitoring of manganese in blood, urine and axillary hair following low-dose exposure during the manufacture of dry cell batteries, *Int. Arch. Occup. Environ. Health*, 1999, **72**, 521.

170. F. J. Deschamps, M. Guillaumot and S. Raux, Neurological effects in workers exposed to manganese, *J. Occup. Environ. Med.*, 2001, **43**, 127.

171. F. Arai, Y. Yamamura, M. Yoshida and T. Kishimoto, Blood and urinary levels of metals (Pb, Cr, Cd, Mn, Sb, Co and Cu) in cloisonne workers, *Ind. Health*, 1994, **32**, 67.

172. F. Lander, J. Kristiansen and J. M. Lauritsen, Manganese exposure in foundry furnacemen and scrap recycling workers, *Int. Arch. Occup. Environ. Health*, 1999, **72**, 546.

173. M. Mari, M. Schuhmacher and J. L. Domingo, Levels of metals and organic substances in workers at a hazardous waste incinerator: a follow-up study, *Int. Arch. Occup. Environ. Health*, 2009, **82**, 519.

Manganese: Health Effects

CHAPTER 15

Manganese and Parenteral Nutrition

JUDY L. ASCHNER*[a] AND NATHALIE L. MAITRE[b]

[a] Department of Pediatrics, The Children's Hospital at Montefiore and The Albert Einstein College of Medicine of Yeshiva University, 1300 Morris Park Avenue – Belfer 706, Bronx, NY 10461, USA; [b] Department of Pediatrics, Vanderbilt Kennedy Center for Human Development, Vanderbilt Center for Molecular Toxicology, Vanderbilt University Medical Center, 11111 Doctor's Office Tower, 2200 Children's Way, Nashville, TN 37232-9544, USA
*Email: judy.aschner@einstein.yu.edu

15.1 Parenteral Nutrition and Manganese (Mn) Supplementation

Parenteral nutritional therapy is a mainstay of modern medicine. The goal of parenteral nutrition therapy is to provide both the macronutrient building blocks and the micronutrients required for optimal human nutrition. Parenteral nutrition is designed to address a critical need in patients unable to tolerate sufficient enteral nutrition to meet energy requirements or support growth. Prior to the development of modern parenteral nutrition formulations in the late 1960s, patients with functional gastrointestinal diseases or short bowel syndrome died of malnutrition. Today, more than 30 000 patients are permanently dependent on parenteral nutrition to survive. Tens of thousands more are treated with short-term parenteral nutrition therapy during an acute illness, or perioperatively. Despite the unquestionable value

Issues in Toxicology No. 22
Manganese in Health and Disease
Edited by Lucio G. Costa and Michael Aschner
© The Royal Society of Chemistry 2015
Published by the Royal Society of Chemistry, www.rsc.org

of parenteral nutrition for patients with structural or functional intestinal diseases, complications associated with its use are common. These include acute, life-threatening conditions caused by microbial and particulate contaminants as well as long-term complications such as parenteral nutrition-associated liver disease (PNALD). Unintended adverse effects of parenteral nutrition can also occur as a result of supplementation with potentially toxic trace metals, such as Mn.[1]

All nutrients have an optimal range of intake, with either excess or deficiency being potentially harmful. Until 40 years ago, trace elements were thought to be of little significance in human nutrition. Our current state of knowledge points to many factors that render individuals and populations susceptible to diseases resulting from trace element excess or deficiencies. Therefore, the goals for trace metal supplementation in parenteral nutrition include prevention of trace metal deficiencies and avoidance of toxicity from excess intake. However, neither the nutritional requirements and the causes and effects of micronutrient deficiencies, nor their toxicities, are fully understood. This is particularly true for parenteral Mn.

Manganese is an essential metal required for normal growth, development and cellular homeostasis. The principal route for Mn intake is *via* food consumption. Thus, it seemed prudent to include Mn in the trace element supplements of patients dependent on parenteral solutions for their nutrition. However, exposure to excess Mn can result in an irreversible psychological and neurologic syndrome, known as manganism. Manganism is associated with destructive symmetric lesions in the basal ganglia (globus pallidus, substantia nigra, subthalamic nucleus) and, to a lesser extent, in the caudate and putamen. Parkinsonian-like symptoms are the clinical manifestation of manganism. Mn is paramagnetic and can be detected as a hyperintense T1 signal in the basal ganglia by magnetic resonance imaging (MRI).

The trace metals that are typically added to parenteral nutrition solutions include zinc, copper, chromium, manganese and selenium. The most common supplementation strategy includes a cocktail with fixed ratios of zinc, copper, chromium and manganese, with selenium added separately. Recognized and unrecognized risks to human health stem from insufficient knowledge of the optimal dosing of parenteral nutrition additives. In the case of Mn, infants and children are particularly vulnerable to inappropriate supplementation. The potential toxicities associated with Mn in parenteral nutrition, either from excessive amounts or when alterations to normal metabolism and excretion occur, have received insufficient attention and are the focus of this chapter.

15.2 Existing Guidelines for Parenteral Manganese Supplementation

15.2.1 Adult Guidelines

In adults, the dietary concentration of Mn influences its gastrointestinal absorption and biliary excretion. High dietary Mn levels induce adaptive changes that reduce gastrointestinal absorption, enhance hepatic metabolism, and increase biliary excretion of Mn.[2–10] Thus, in healthy adults, Mn levels in the brain and other tissues remain relatively constant, despite large fluctuations in dietary Mn intake. No formal Recommended Dietary Allowance for Mn exists, but the US National Research Council has established an estimated safe and adequate dietary intake (ESADDI) of 2–5 mg per day for adults.[11] The National Academy of Science defines adequate Mn intake as 2.3 and 1.8 mg per day for adult men and women, respectively,[12] reflecting lower gastrointestinal Mn absorption in men *vs.* women.[6] Pregnancy and lactation likely increase dietary Mn requirements.[12] The impact of disease states, such as liver disease, and nutritional factors, such as iron (Fe) status, is covered later in this chapter.

When long-term parenteral nutrition was first introduced for patients with intestinal failure, it was accompanied by morbidities resulting from inadequate replacement of micronutrients, especially zinc. Notably, there were no reported cases of Mn deficiency in patients given long-term parenteral nutrition.[13,14] To this date, Mn deficiency has yet to be identified in this population. Despite limited evidence that Mn is an essential additive for humans who are dependent on parenteral nutrition, the American Medical Association developed guidelines for its inclusion in parenteral trace element supplements in 1979.[15] The published literature cites a broad, 200-fold range for the recommended adult daily parenteral dose of Mn, from a low of 0.18–0.91 μmol (10–50 μg) to a high of 40 μmol (2.2 mg),[16] underscoring the lack of an evidence base to inform guidelines. The 2004 American Society for Parenteral and Enteral Nutrition (A.S.P.E.N.) Task Force for the Revision of Safe Practices for Parenteral Nutrition recommended the daily addition of 60–100 μg Mn supplement to adult parenteral nutrition formulations.[17] However, currently available multiple trace element products provide considerably more Mn when following the recommended daily dosing of other trace elements. A 2012 A.S.P.E.N. position paper recommended modifications to decrease Mn to 55 μg per day in commercially available adult multiple trace element products.[18] In 2013 and 2014, shortages of trace elements for parenteral use led A.S.P.E.N. to suggest there was no need for Mn supplements (during shortage), unless signs and symptoms of clinical deficiency were noted.[19]

15.2.2 Pediatric Guidelines

Age and development influence dietary Mn requirements. In infants under 6 months of age, adequate Mn intake is defined as 3 μg per day; by 7 to 12 months of age adequate Mn intake increases to 600 μg per day. At 1–3 and 4–8 years of age, adequate daily manganese recommendations are 1.2 and 1.5 mg per day, respectively.[12]

Infant diets contain a wide range of Mn concentrations, from 3–10 μg l^{-1} in human milk to 200–300 μg l^{-1} in soy formula,[20] as illustrated in Table 15.1. Davidsson *et al.* reported that adults absorb 8% of the Mn from

Table 15.1　Mn content in neonatal enteral and parenteral nutrition.

	Human milk	Cow milk formula	Soy formula	PN (no TES)	Multitrace®-4 Neonatal
Mn content (μg l^{-1})	3–10	30–50	200–300	7 ± 1	2500
Mn intake (μg kg^{-1} per day)^a	0.45–1.5	4.5–7.5	30–45	~ 1.0	7.5^b
Absorption (%)	8	2	<1	100	100

^{a}Based on intake of 150 ml kg^{-1} per day.
^{b}Based on adding 0.2 ml dl^{-1}. Infant diets contain a wide range of Mn levels. Parenteral nutrition (PN) contains Mn as a contaminant, even in the absence of a TES (trace element supplement). It is standard clinical practice to add 0.2 ml dl^{-1} of the TES, Multitrace®-4 Neonatal or Multitrace®-4 Pediatric (American Regent, Inc. Shirley, NY) containing 25 μg ml^{-1} Mn. For an infant receiving 150 ml kg^{-1} per day of PN, TES provides an additional 7.5 μg kg^{-1} per day of Mn.

human milk, but <1% from soy formula.[21] Wilson et al.[22] reported no significant differences in mean plasma Mn concentrations between preterm infants fed maternal milk containing 4.1 μg Mn l^{-1} or preterm infant formula containing 303 μg Mn l^{-1}. Therefore, even preterm infants have effective homeostatic mechanisms that ensure steady-state circulating Mn over a wide range of Mn intake. These homeostatic mechanisms exert themselves at the level of intestinal absorption and Mn excretion *via* the biliary tract.[23]

In 1985, The American Academy of Pediatrics Committee on Nutrition recommended that preterm infants requiring exclusive parenteral nutrition receive daily Mn supplementation at a dose of 2–10 μg kg^{-1} per day.[24] This recommendation was derived from the concentration of Mn in term human milk and quickly became the standard of care for all parenterally fed infants in the US and Europe. It also directed the composition of currently available commercial multiple trace element products used for neonates and children receiving parenteral nutrition therapy today. However, the 1985 recommendation failed to account for the gastrointestinal control of Mn absorption. Even accounting for the higher retention of ingested Mn in neonates compared to adults,[25,26] Mn absorption represents 8–20% of enteral dietary Mn content. Since then, the recommendations for Mn supplementation in parenteral nutrition for pediatric and neonatal patients have been revised downward but the content of Mn in commercially available multiple trace element solutions has remained unchanged.

It is standard clinical practice to supplement pediatric (and adult) patients receiving parenteral nutrition with a multiple trace elements solution (TES) containing fixed ratios of zinc, copper, chromium, manganese ± selenium. The use of TES with fixed ratios of multiple components limits any flexibility to adjust Mn dosing (or that of other trace elements) in response to clinical conditions or updated recommendations. Thus, in 2003, when the *Pediatric Nutrition Handbook* (5th edition) recommended a reduction in daily parenteral Mn from 2–10 μg kg^{-1} per day to 1 μg kg^{-1} per day, clinicians were unable to respond. Reducing Mn dosing posed a risk of providing too little zinc or other trace metals included in the trace metal cocktail for preterm infants. Then, the 2004 A.S.P.E.N. Special Report and the 2012 A.S.P.E.N. position paper also recommended 1 μg kg^{-1} per day for preterm neonates <3 kg, and for term neonates of 3–10 kg, further compounding the problem of adjusting TES.[17,18] However, the evidence base for these newer recommendations was not clear and, as highlighted below, this may not represent the optimal approach to Mn supplementation for all infants and children. In particular, the new recommendations may not be applicable to those with liver disease or those who are dependent on life-sustaining parenteral nutrition for long periods of time.

Mn is a contaminant of parenteral nutrition solutions, even in the absence of intentional Mn supplementation.[22,23,27] Wilson *et al.* reported that the Mn content of parenteral nutrition solutions in the absence of TES was 7.3 μg l^{-1} (range 5.6–8.9 μg l^{-1}; see Table 15.1), more than sufficient to meet daily recommended amounts of Mn.[22] When this is combined with the standard clinical practice of adding 0.2 ml dl^{-1} of a TES containing 25 μg ml^{-1} Mn, the total amount of Mn delivered far exceeds the amount that would be absorbed by an infant receiving a diet of human milk (Table 15.1 and Figure 15.1).

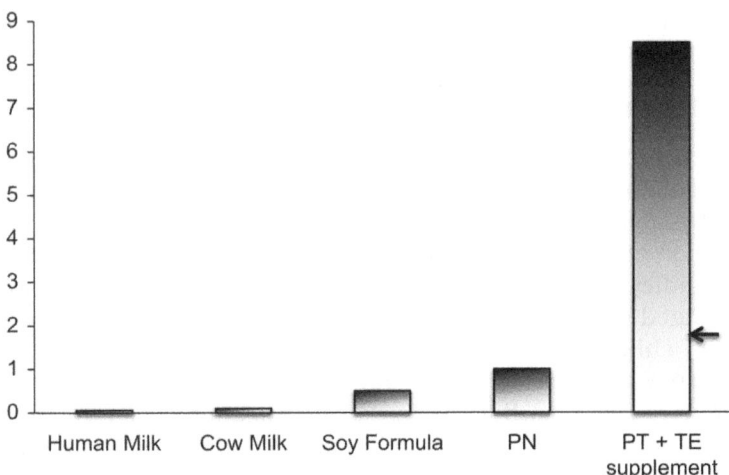

Figure 15.1 Calculated Mn absorption as a function of enteral or parenteral content of Mn. The figure shows calculated dietary Mn absorption in an infant consuming 150 ml kg^{-1} per day. Trace element-supplemented parenteral nutrition provides >100 times more Mn than would be absorbed by an infant fed human milk. The arrow points to the toxicity level in adults: Mn intoxication has been reported in association with PN solutions providing ≥0.1 mg Mn per day, approximately 1.5–2.0 μg kg^{-1} per day for an average adult.

15.3   Risk Factors for Manganese Excess and Toxicity

Since the routine inclusion of TES containing Mn in parenteral nutrition solutions, recognition of hypermanganesemia has been growing, along with its associated complications, in patients receiving long-term parenteral nutrition. Parenteral exposure poses a risk of Mn toxicity, because parenteral administration bypasses the normal regulatory mechanisms of the gastrointestinal track. Furthermore, the bioavailability of Mn in parenteral fluids is near 100%, *vs.* about 5% for enteral dietary Mn.[28]

Premature and sick neonates may be at increased risk of brain Mn deposition when exposed to parenteral Mn, relative to older children and adults. Neonates have an immature and more permeable blood-brain barrier, hence more Mn may diffuse into the central nervous system during critical stages of brain development. In addition, biliary excretion of Mn in the neonate may be impaired, due to either immaturity or to cholestasis (a common complication of prolonged parenteral nutrition in infants).

Critically ill infants and children, such as those with necrotizing enterocolitis and short bowel syndrome, may be dependent on parenteral nutrition for weeks, months or sometimes years. Yet, a longer duration of parenteral nutrition therapy is associated with higher risk of Mn toxicity in this already vulnerable population. It has been reported that over 50% of patients on long-term parenteral nutrition can have elevated blood Mn levels.[29] To further complicate matters, overt clinical signs and symptoms are not observed in all patients with hypermanganesemia, making clinical monitoring a challenging prospect.

15.3.1   Disease States

Patients with cholestatic liver disease are at particularly high risk for hypermanganesemia, because more than 90% of Mn is excreted in the bile.[8] Hepatic dysfunction and cholestasis are known risk factors for increased accumulation of Mn in the brain in both humans and animal models.[8,30–32] Autopsy studies in adult patients with chronic hepatic encephalopathy revealed elevated Mn levels in the basal ganglia.[31] Patients with portosystemic shunts and children with biliary atresia display hypermanganesemia, even in the absence of increased dietary Mn.[33–36]

In parenterally fed newborns, Mn excretion is limited as a consequence of poor defecation patterns, immature biliary excretion and the common development of cholestasis and hepatic dysfunction, further increasing their risk of toxic Mn levels. As early as 1989, Hambidge and colleagues recommended that Mn supplements be omitted from parenteral nutrition solutions for patients with evidence of cholestasis. They concluded that the amount of Mn present as a contaminant in intravenous solutions provided adequate Mn for these patients.[23]

Additional disease states hypothesized to play a role in Mn metabolism include renal dysfunction, diabetes, and metabolic syndrome.[37,38] However, these associations are less well documented and their modifying effects on parenteral nutrition Mn requirements or on the potential adverse effects of Mn are unclear.

15.3.2 Nutritional Iron Status

Iron (Fe) deficiency anemia and low serum ferritin are associated with increased blood and brain Mn levels in animal models and human adults.[39,40] Recent studies indicate that this holds true for children as well.[41,42] In women, the effect of low ferritin is thought to be further modulated by hormonal influences: in premenopausal women, increased ferritin results in lower blood Mn levels, in contrast to postmenopausal women in whom ferritin levels no longer influence Mn levels.[43]

The mechanisms and interactions between Mn and Fe are well studied. In the plasma, both Mn and Fe are bound to albumin and transferrin. Both metals also compete for the same carrier transport systems *via* transferrin-mediated endocytosis and the divalent metal transporter 1 (DMT-1).[44–50] In the presence of low Fe, both DMT-1 and transferrin receptors (TfR) are upregulated, resulting in increased Mn transport. Conversely, DMT-1 and TfR levels are decreased in the presence of high levels of Fe, resulting in decreased Mn transport.[51–54] The roles of many other transporters involved in Mn uptake and efflux are discussed at length in Chapter 3 of this book.

In vivo, iron deficiency results in more Mn uptake into the brain, both in adults[44,45,52,55] and in neonates.[49,50] Fe stores in premature newborns are low, as preterm birth interrupts the accumulation of fetal Fe stores that predominantly occurs in the third trimester of human gestation.[56] Fe deficiency in this population is exacerbated by the lack of a safe preparation for parenteral Fe supplementation, and by frequent phlebotomy losses. As with other Fe deficient populations, it is hypothesized that they are more susceptible to Mn intoxication, even when exposed to low or normal levels of Mn.

15.4 Consequences of Excessive Parenteral Mn

15.4.1 Adults: Manganism and other Neuropsychiatric Disorders

Exposure to excess Mn can result in an irreversible psychological and neurologic syndrome known as manganism. Manganism is associated with destructive symmetric lesions in the basal ganglia (globus pallidus, substantia nigra, subthalamic nucleus) and, to a lesser extent, in the caudate and putamen. Early symptoms of Mn toxicity include compulsive or violent behavior, emotional instability, hallucinations and other psychiatric symptoms.[57] Systemic symptoms include headache, fatigue, muscle cramps, and insomnia. Severe Mn intoxication presents with neurologic symptoms resembling Parkinson's disease, including dystonia, hypokinesia, rigidity and muscle tremors.[58]

Manganese intoxication in adults has been reported in association with parenteral nutrition solutions providing ≥0.1 mg Mn per day, or approximately 1.5–2.0 µg kg^{-1} per day for an average adult.[59–63] These patients exhibited elevated serum Mn levels and, similar to other forms of Mn poisoning, demonstrated symmetrical high intensity T1-weighted MR signals in the globus pallidus, consistent with Mn accumulation. Some patients also displayed characteristic psychiatric symptoms and extrapyramidal signs of Mn-induced parkinsonism-like syndrome. Withdrawal of the Mn supplement from parenteral nutrition significantly decreased Mn levels in both the blood and brain of these patients.[61]

In a population-based sample of Canadian patients on prolonged parenteral nutrition containing a commercial TES, the mean daily Mn supplementation exceeded the A.S.P.E.N. 2002 recommendation[64] of 60–100 µg per day (1.09–1.82 µmol per day) and mean whole blood Mn levels exceeded the upper limit of normal. More than 80% of patients had high signals on T1-weighted MRI and two patients with positive MRI had a clinical diagnosis of Parkinson's disease.[65] The recent literature is replete with reports of Mn intoxication in adults, associated with long-term exposure to Mn in parenteral nutrition.[23,34,59–62,66–74] Most of the case reports of Mn toxicity in adults reflect exposures to daily doses of >500 µg per day of parenteral Mn.[14,34,59,66,71,75]

15.4.2 Infants and Children: Cognition and Neurodevelopment

An immature and more permeable blood–brain barrier confers an enhanced risk of Mn-induced central nervous system (CNS) toxicity in neonates. Rodent studies suggest that, compared to adult animals, neonatal animals are at increased risk of manganese-induced neurotoxicity, based on their propensity to achieve higher brain Mn levels and altered brain dopamine concentrations following similar oral exposures.[4,76–78] In monkeys and rats, Mn brain concentrations correlate with the severity of neurologic symptoms, with both the rate and extent of Mn transport into the CNS influencing the clinical outcome.[79,80] Tran et al. demonstrated that a high dietary Mn intake in neonatal rats resulted in developmental deficits manifested, for example, by a significant increase in passive avoidance errors.[81]

Cohort studies in children have contributed to the mounting concern that high level Mn exposure may contribute to poor neurodevelopment, with decreased cognitive and verbal performance in early childhood.[82,83] Associations with Mn exposure and school age neurobehavioral outcomes point to an increase in externalizing behaviors and attention problems.[84,85] The effect of Mn on neurodevelopment may be magnified by teratogenic prenatal exposures or by neurocognitive and behavioral deficits of Mn-exposed caregivers.[82,86]

The specific effects of parenteral Mn on neurologic function in infants and children are difficult to assess. There have been sporadic cases of children with excessive exposure to Mn who have developed overt signs of neurotoxicity,[74,87–89] typically children receiving >40 µg kg per day.[74,88] Fell *et al.* reported 11 children given prolonged parenteral nutrition who had hypermanganesemia and cholestasis.[88] The T1-weighted MRI of one of these children, who presented with a movement disorder, showed bilaterally symmetric increased signal intensity in the globus pallidus and subthalamic nuclei. Similar changes were seen in five other children, one with hypermagnesemia and cholestasis, and four on long-term parenteral nutrition without liver disease but with elevated blood Mn levels.[88] Among six children receiving long-term parenteral nutrition, MRI showed T1-hyperintensity in the globus pallidus in all patients and in the anterior pituitary in one patient; most MRI findings were reversed 12 months after removal of Mn from the parenteral nutrition. The neurodevelopmental consequences of this exposure were unfortunately not reported.[68] Our group has similarly documented increased T1-weighted signal intensity in the globus pallidus of children on prolonged Mn-supplemented parenteral nutrition. Figure 15.2 shows a representative T1-weighted MR image obtained from a 4 year-old child, who had been dependent since birth on parenteral nutrition. After early development of cholestasis and hepatic dysfunction, TES supplementation was limited to twice weekly. Nonetheless, the image clearly demonstrates enhanced signal intensity in the basal ganglia (Figure 15.2).

Multiple case reports also link parenteral Mn excess in infants and young children with seizure disorders, which resolve when Mn levels

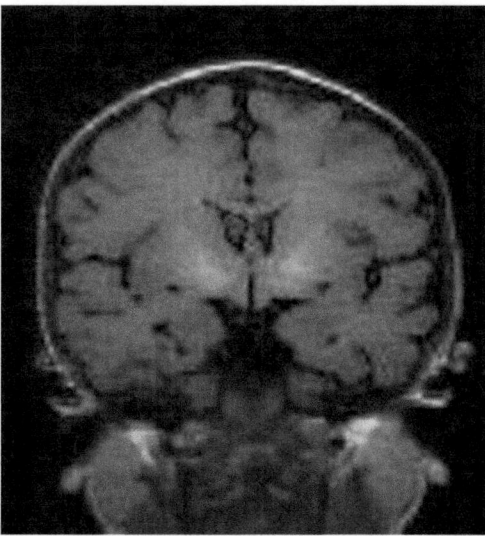

Figure 15.2 T1-weighted MR images of a 4 year-old child who had been receiving Mn-supplemented home parenteral nutrition therapy from birth. The globus pallidus and basal ganglia appear bright (hyperintense).

decrease.[67,69,90] Mechanisms for the epileptogenic disorders in these infants were not pursued.

15.5   Detection of Manganese Body Burden

15.5.1   Mn Measurements in Blood

Lack of a practical method for assessing Mn status has hampered progress in the field of Mn nutrition.[21] Monitoring of Mn levels is recommended for patients requiring more than 30 days of parenteral nutrition; however, there is no consensus on the optimal biomarker for body burden of Mn, or how frequently to monitor for toxicity. Mn levels are usually assessed by measurement of blood or serum levels in patients receiving parenteral nutrition. However, whether circulating levels actually reflect nutritional status or excessive accumulation is unclear. In particular, the utility of serum Mn has been questioned as a marker for Mn body burden, because it was reported that intra-cerebral Mn levels were elevated in the presence of normal serum values.[66]

The most common approach to monitoring Mn is the analysis of whole blood Mn levels. Whole blood Mn concentrations are thought to be more accurate than serum levels and to correlate with signal intensity on MRI.[57] However, there is large individual variation in whole blood Mn levels, and neither signal intensity on MRI nor whole blood Mn correlate with neurologic symptoms of Mn neurotoxicity.[91]

In a review by Iyengar and Woittiez, median Mn values in the general population were 13.6 (8.0–18.7) µg l^{-1} in whole blood, 0.63 (0.54–1.76) µg l^{-1} in serum and 0.6 (0.5–9.8) µg l^{-1} in urine (values in parentheses represent the range). These median values were obtained from a population of 100 000 individuals in 55 countries.[92] A publication focused on patients in North America suggested a normal range of 7–16 µg l^{-1} of Mn in whole blood.[57] Despite the consistently large individual variability in whole blood Mn levels, a community-based study in Québec showed an association between higher whole blood Mn (>7.5 µg l^{-1}) levels and motor deficits.[93]

Some experts consider neither serum nor whole blood Mn levels to be suitable indicators, and have proposed that the Mn content of red blood cells (RBCs) may better reflect tissue accumulation.[61] Whole blood Mn levels are affected by concurrent intravenous (IV) delivery of Mn,[94] and RBCs account for approximately 60–80% of Mn in whole blood.[72,95,96] RBCs also have a slower turnover than other types of blood cell,[97] and, importantly, RBC Mn levels are not directly dependent on concurrent IV administration. Taken together, these characteristics make them potentially better indicators of Mn body burden.

Urinary Mn levels have not been shown to serve as a useful biomarker of exposure. Urinary Mn excretion is low, representing only about 0.01 ± 1% of the absorbed dose,[98] and about 6% of the total excreted amount.[99] The Mn concentration in urine is poorly correlated with recent and cumulative doses.[100] Biliary excretion is the main pathway by which Mn is excreted, with most ultimately being excreted in the feces.[3,16]

New evidence from piglet models suggests that, in neonates, IV Mn may result in much higher accumulation in liver, kidney and bone than administration of an enteral dose.[101] Mn levels from hair are more useful in reflecting long-term exposures than in biomonitoring and titrating Mn dosing in parenteral nutrition. Although they are costly and difficult to perform in some populations, neuroimaging, neurologic and behavioral assessments provide the most relevant and accurate measures of parenteral Mn exposure.

15.5.2 Magnetic Resonance Imaging (MRI)

When present in large quantities, Mn, which has paramagnetic properties, can be detected in the tissues by MRI. In the brain, Mn deposition appears as a hyperintense signal (shortened T1) in the basal ganglia. Specifically the globus pallidus displays hyperintense signals on T1 (short echo time (TE)/ short repetition time (TR)) but not on T2-weighted (long TE/long TR) MRI.[102] Other metals, such as copper and iron, do not manifest this appearance. The T1 MRI intensity in the globus pallidus is therefore a useful means of detecting brain Mn accumulation.[62,103] Most patients receiving parenteral nutrition supplemented with Mn are asymptomatic and are diagnosed with hypermanganesemia only on the basis of abnormal T1 signal intensity in the globus pallidus.[72,104] Because routine cerebral MRI scans may not always be clinically feasible, it has been recommended that whole blood Mn determination in combination with less frequently scheduled MRIs may provide a practical approach to monitoring Mn accumulation in patients given long-term parenteral nutrition.[57]

A recent meta-analysis demonstrated that the pallidal index (the ratio of signal intensity in the globus pallidus compared to the frontal white matter on T1-weighted imaging) may prove to be a consistent marker of Mn accumulation in the basal ganglia.[105] In a study of asymptomatic Mn-exposed adults, the pallidal index appears to be a better predictor of neurobehavioral function than blood Mn. The pallidal index explained approximately 20–30% of the variance in standardized test scores in this study, while blood Mn rarely exceeds 20%.[106] While the pallidal index is still the best-studied marker of Mn accumulation, recent studies point to the frontal cortex[107,108] and the hippocampus and thalamus[109,110] as structures also evidencing Mn deposition. More global Mn deposition, including in the frontal cortex, would reduce the sensitivity of the pallidal index. Furthermore, these neuroanatomic structures are essential to complex cognitive processes and executive function. Thus T1 indices and metabolite utilization changes in the frontal cortex and limbic structures may be even more relevant markers of Mn excess than the pallidal index. Despite limitations in adults, the T1 indices in the globus pallidus remain a crucial marker for neonates exposed to Mn through parenteral nutrition because globus pallidus indices are relatively stable in the face of brain immaturity.[111] In contrast, white matter myelination is still evolving and hippocampal indices are profoundly affected by immaturity and prematurity.[112-114]

15.6 Future Directions for Optimizing Mn in Parenteral Nutrition

15.6.1 Knowledge Gaps and Research Priorities

Given numerous gaps in our knowledge, the study of micronutrient requirements for Mn in children and adults receiving parenteral nutrition is a fertile area for translational research. Areas of research priority include the following:

- Identification of factors that increase the risk of neurotoxicity in individuals and specific patient populations, such as preterm infants and patients on parenteral nutrition. The urgency of this research cannot be understated.
- Improved definitions of safe and effective doses of parenteral Mn.
- Studies on Mn contamination in the various components used in parenteral nutrition components used in neonatal and pediatric formulations.
- Development of practical methods for assessing tissue stores in patients requiring long-term parenteral nutrition.[14,115]
- Research and development of appropriate monitoring strategies for other trace element and vitamin deficiencies and toxicities in patients given parenteral nutrition.
- Commercial product development to address the needs of individualized patient-based therapy and trace element supplementation.
- Investigations of nutritional modifiers of Mn neurotoxicity and, specifically, the role of Fe status in the distribution patterns and accumulation of brain Mn in premature and term infants receiving parenteral nutrition.
- Prospective studies and randomized controlled trials of Mn supplementation in parenterally fed adults and children.
- Short- and long-term neurodevelopmental studies of the consequences of the full range of Mn exposures.

15.6.2 Recommendations for Clinical Practice Modifications

The parenteral TES preparations that are commercially available in the US require significant modifications. The use of TES with fixed ratios of multiple components limits the flexibility that may be required to regulate Mn intake (or that of other trace elements) under various clinical conditions, such as cholestatic liver disease. The commercially available TES cocktails for neonates, pediatric patients and adults deliver an excessive dose of Mn if recommended doses of zinc are given. Based on current knowledge of trace element homeostasis in health and disease, it seems likely that an individualized approach to trace metal supplementation in patients given parenteral nutrition is required. Single-entity trace element products can be used to meet individual patient needs when the multiple-element products are inappropriate.[18] Hambidge and colleagues recommended that Mn supplements be omitted from parenteral nutrition solutions for patients with evidence of cholestasis, concluding that the amount of Mn present as a contaminant in intravenous solutions appears to provide adequate Mn for these patients.[23] Indeed, there is little evidence that any patient on parenteral nutrition requires Mn supplementation beyond that which currently is delivered as a contaminant of other components of parenteral nutrition.

Increased awareness of the risks, manifestations and approach to diagnosis of Mn excess and neurotoxicity on the part of clinicians prescribing inpatient or home-based parenteral nutrition is needed. Routine screening for Mn deposition in the brain by T1-weighted MRI should be incorporated into the monitoring of patients of all ages requiring parenteral nutrition for prolonged periods of time.

Acknowledgements

Supported by a research grant from the Gerber Foundation.

References

1. M. Aschner, K. M. Erikson, E. H. Hernández and R. Tjalkens, Manganese and Its Role in Parkinson's Disease: From Transport to Neuropathology, *Neuromol. Med.*, 2009, **11**, 252–266.
2. Dependence of Manganese Turnover on Intake. 1966.
3. C. D. Davis, L. Zech and J. L. Greger, Manganese Metabolism in Rats: an Improved Methodology for Assessing Gut Endogenous Losses, *Exp. Biol. Med.*, 1993, **202**, 103–108.
4. D. C. Dorman, M. F. Struve and B. A. Wong, Brain Manganese Concentrations in Rats Following Manganese Tetroxide Inhalation Are Unaffected by Dietary Manganese Intake, *Neurotoxicology*, 2002, **23**, 185–195.
5. D. C. Dorman, M. F. Struve, R. A. James, B. E. McManus, M. W. Marshall and B. A. Wong, Influence of Dietary Manganese on the Pharmacokinetics of Inhaled Manganese Sulfate in Male CD Rats, *Toxicol. Sci.*, 2001, **60**, 242–251.
6. J. W. Finley, Manganese Absorption and Retention by Young Women Is Associated with Serum Ferritin Concentration, *Am. J. Clin. Nutr.*, 1999.
7. J. P. Mahoney and W. J. Small, Studies on Manganese: III. the Biological Half-Life of Radiomanganese in Man and Factors Which Affect This Half-Life, . *J. Clin. Invest.*, 1968.

8. E. A. Malecki, G. M. Radzanowski, T. J. Radzanowski, D. D. Gallaher and J. L. Greger, Biliary Manganese Excretion in Conscious Rats Is Affected by Acute and Chronic Manganese Intake but Not by Dietary Fat, *J. Nutr.*, 1996, **126**, 489–498.

9. P. S. Papavasiliou, S. T. Miller and G. C Cotzias, Role of liver in regulating distribution and excretion of manganese, *Am. J. Physiol.*, 1966, **211**, 211–216.

10. P. S. Papavasiliou, S. T. Miller and G. C. Cotzias, Role of Liver in Regulating Distribution and Excretion of Manganese, *Am. J. Physiol.*, 1966, **211**, 211–216.

11. J. L. Greger, Dietary Standards for Manganese: Overlap Between Nutritional and Toxicological Studies, *J. Nutr.*, 1998, **128**, 368S–371S.

12. P. Trumbo, A. A. Yates, S. Schlicker and M. Poos, Food and Nutrition Board, Institute of Medicine, The National Academies. Dietary reference intakes: vitamin A, vitamin K, arsenic, boron, chromium, copper, iodine, iron, manganese, molybdenum, nickel, silicon, vanadium, and zinc, *J. Am. Diet. Assoc.*, 2001, **101**, 294–301.

13. D. A. Frankel, Supplementation of Trace Elements in Parenteral Nutrition: Rationale and Recommendations, *Nutr. Res.*, 1993, **13**, 583–596.

14. R. N. Dickerson, Manganese Intoxication and Parenteral Nutrition, *Nutrition*, 2001, **17**, 689–693.

15. H. L. Greene; D. Van Der Vorm; G. L. Helinek; G. Nichoalds, Trace Elements in Parenteral Feeding of Infants, in *Nutrition and Metabolism of the Fetus and Infant*, Springer, Netherlands, Dordrecht, 1979, pp. 377–389.

16. J. L. Aschner and M. Aschner, Nutritional Aspects of Manganese Homeostasis, *Mol. Aspects Med.*, 2005, **26**, 353–362.

17. P. J. Charney, A.S.P.E.N. Statement on aluminum in parenteral nutrition solutions, *Nutr. Clin. Pract.*, 2004, **19**, 416–417.

18. V. W. Vanek, P. Borum, A. Buchman, T. A. Fessler, L. Howard, K. Jeejeebhoy, M. Kochevar, A. Shenkin and C. J. Valentine, A.S.P.E.N. Position Paper: Recommendations for Changes in Commercially Available Parenteral Multivitamin and Multi-trace Element Products, *Nutr. Clin. Pract.*, 2012, **27**, 440–491.

19. S. Plogsted, G. Brooks, J. DiBaise, T. Fuhrman, J. Ybarra, B. Holcombe, D. A. Andris, D. R. Houston and S. W. Plogsted, A.S.P.E.N. parenteral nutrition trace element product shortage considerations, *Nutr. Clin. Pract.*, 2014, **29**, 249–251.

20. B. Lönnerdal, Nutritional Aspects of Soy Formula, *Acta Paediatr.*, 1994, **83**, 105–108.

21. L. Davidsson, B. Lönnerdal, B. Sandström, C. Kunz and C. L. Keen, Identification of Transferrin as the Major Plasma Carrier Protein for Manganese Introduced Orally or Intravenously or After in Vitro Addition in the Rat, *J. Nutr.*, 1989, **119**, 1461–1464.

22. D. C. Wilson, T. R. J. Tubman, H. L. Halliday and D. McMaster, Plasma Manganese Levels in the Very Low Birth Weight Infant Are High in Early Life, *Neonatology*, 1992, **61**, 42–46.

23. K. M. Hambidge, R. J. Sokol, S. J. Fidanza and M. A. Goodall, Plasma Manganese Concentrations in Infants and Children Receiving Parenteral Nutrition, *JPEN, J. Parenter. Enteral Nutr.*, 1989, **13**, 168–171.

24. American Academy of Pediatrics, Committee on Nutrition. Nutritional needs of low-birth-weight infants, *Pediatrics*, 1985, **75**, 976–986.

25. S. H. Zlotkin, S. Atkinson and G. Lockitch, Trace Elements in Nutrition for Premature Infants, *Clin. Perinatol.*, 1995, **22**, 223–240.

26. K. Dörner, S. Dziadzka, A. Höhn, E. Sievers, H.-D. Oldigs, G. Schulz-Lell and J. Schaub, Longitudinal Manganese and Copper Balances in Young Infants and Preterm Infants Fed on Breast-Milk and Adapted Cow's Milkformulas, *Br. J. Nutr.*, 1989, **61**, 559–572.

27. J. Kurkus, N. W. Alcock and M. E. Shils, Manganese Content of Large-Volume Parenteral Solutions and of Nutrient Additives, *JPEN, J. Parenter. Enteral Nutr.*, 1984, **8**, 254–257.

28. A. W. Dobson, K. M. Erikson and M. Aschner, Manganese Neurotoxicity, *Ann. N. Y. Acad. Sci.*, 2004, **1012**, 115–128.

29. J. K. Siepler, R. A. Nishikawa, T. Diamantidis and R. Okamoto, Asymptomatic Hypermanganesemia in Long-Term Home Parenteral Nutrition Patients, *Nutr. Clin. Prac.*, 2003, **18**, 370–373.

30. N. Ballatori, E. Miles and T. W. Clarkson, Homeostatic Control of Manganese Excretion in the Neonatal Rat, *Am. J. Physiol.*, 1987, **252**, R842–R847.

31. D. Krieger, S. Krieger, O. Jansen, P. Gass, L. Theilmann and H. Lichtnecker, Manganese and Chronic Hepatic Encephalopathy, *Lancet*, 1995, **346**, 270–274.

32. S. Montes, M. Alcaraz-Zubeldia, P. Muriel and C. Ríos, Striatal Manganese Accumulation Induces Changes in Dopamine Metabolism in the Cirrhotic Rat, *Brain Res.*, 2001, **891**, 123–129.

33. C. Rose, R. F. Butterworth, J. Zayed, L. Normandin, K. Todd, A. Michalak, L. Spahr, P. M. Huet and G. Pomier-Layrargues, Manganese Deposition in Basal Ganglia Structures Results From Both Portal-Systemic Shunting and Liver Dysfunction, *Gastroenterology*, 1999, **117**, 640–644.

34. J. M. Reimund, J. L. Dietemann, J. M. Warter, R. Baumann and B. Duclos, Factors Associated to Hypermanganesemia in Patients Receiving Home Parenteral Nutrition, *Clin. Nutr.*, 2000, **19**, 343–348.

35. S. Ikeda, Y. Yamaguchi, Y. Sera, H. Ohshiro, S. Uchino, Y. Yamashita and M. Ogawa, Manganese Deposition in the Globus Pallidus in Patients with Biliary Atresia, *Transplantation*, 2000, **69**, 2339–2343.

36. S. Ikeda, Y. Sera, M. Yoshida, H. Ohshiro, S. Uchino, Y. Oka, K. J. Lee and A. Kotera, Manganese Deposits in Patients with Biliary Atresia After Hepatic Porto-Enterostomy, *J. Pediatr. Surg.*, 2000, **35**, 450–453.

37. E. S. Koh, S. J. Kim, H. E. Yoon, J. H. Chung, S. Chung, C. W. Park, Y. S. Chang and S. J. Shin, Association of Blood Manganese Level with

Diabetes and Renal Dysfunction: a Cross-Sectional Study of the Korean General Population, *BMC Endocr. Disord.*, 2014, **14**, 24.

38. M.-K. Choi and Y.-J. Bae, Relationship Between Dietary Magnesium, Manganese, and Copper and Metabolic Syndrome Risk in Korean Adults: the Korea National Health and Nutrition Examination Survey (2007-2008), *Biol. Trace Elem. Res.*, 2013, **156**, 56–66.

39. Y. Kim and B.-K. Lee, Iron Deficiency Increases Blood Manganese Level in the Korean General Population According to KNHANES 2008, *Neurotoxicology*, 2011, **32**, 247–254.

40. H. M. Meltzer, A. L. Brantsaeter, B. Borch-Iohnsen, D. G. Ellingsen, J. Alexander, Y. Thomassen, H. Stigum and T. A. Ydersbond, Low Iron Stores Are Related to Higher Blood Concentrations of Manganese, Cobalt and Cadmium in Non-Smoking, Norwegian Women in the HUNT 2 Study, *Environ. Res.*, 2010, **110**, 497–504.

41. M. A. Rahman, B. Rahman and N. Ahmed, High Blood Manganese in Iron-Deficient Children in Karachi, *Public Health Nutr.*, 2013, **16**, 1677–1683.

42. E. A. Smith, P. Newland, K. G. Bestwick and N. Ahmed, Increased Whole Blood Manganese Concentrations Observed in Children with Iron Deficiency Anaemia, *J. Trace Elem. Med. Biol.*, 2013, **27**, 65–69.

43. B.-K. Lee and Y. Kim, Effects of Menopause on Blood Manganese Levels in Women: Analysis of 2008-2009 Korean National Health and Nutrition Examination Survey Data, *Neurotoxicology*, 2012, **33**, 401–405.

44. I. Mena, The Role of Manganese in Human Disease, *Ann. Clin. Lab. Sci.*, 1974, **4**, 487–491.

45. M. Aschner and J. L. Aschner, Manganese Transport Across the Blood-Brain Barrier: Relationship to Iron Homeostasis, *Brain Res. Bull.*, 1990, **24**, 857–860.

46. M. Aschner and J. L. Aschner, Manganese Neurotoxicity: Cellular Effects and Blood-Brain Barrier Transport, *Neurosci. Biobehav. Rev.*, 1991, **15**, 333–340.

47. K. Erikson and M. Aschner, Manganese Causes Differential Regulation of Glutamate Transporter (GLAST) Taurine Transporter and Metallothionein in Cultured Rat Astrocytes, *Neurotoxicology*, 2002, **23**, 595–602.

48. A. C. Chua and E. H. Morgan, Manganese Metabolism Is Impaired in the Belgrade Laboratory Rat, *J. Comp. Physiol., B*, 1997, **167**, 361–369.

49. S. J. Garcia, K. Gellein, T. Syversen and M. A. Aschner, Manganese-Enhanced Diet Alters Brain Metals and Transporters in the Developing Rat, *Toxicol. Sci.*, 2006, **92**, 516–525.

50. S. J. Garcia, K. Gellein, T. Syversen and M. Aschner, Iron Deficient and Manganese Supplemented Diets Alter Metals and Transporters in the Developing Rat Brain, *Toxicol. Sci.*, 2007, **95**, 205–214.

51. K. M. Erikson, D. J. Pinero, J. R. Connor and J. L. Beard, Regional Brain Iron, Ferritin and Transferrin Concentrations During Iron Deficiency and Iron Repletion in Developing Rats, *J. Nutr.*, 1997, **127**, 2030–2038.

52. K. M. Erikson, Z. K. Shihabi, J. L. Aschner and M. Aschner, Manganese Accumulates in Iron-Deficient Rat Brain Regions in a Heterogeneous Fashion and Is Associated with Neurochemical Alterations, *Biol. Trace Elem. Res.*, 2002, **87**, 143–156.

53. J. A. Roth and M. D. Garrick, Iron Interactions and Other Biological Reactions Mediating the Physiological and Toxic Actions of Manganese, *Biochem. Pharmacol.*, 2003, **66**, 1–13.

54. N. C. Andrews, M. D. Fleming and J. E. Levy, Molecular Insights Into Mechanisms of Iron Transport, *Curr. Opin. Hematol.*, 1999, **6**, 61.

55. K. M. Erikson, T. Syversen, E. Steinnes and M. Aschner, Globus Pallidus: a Target Brain Region for Divalent Metal Accumulation Associated with Dietary Iron Deficiency, *J. Nutr. Biochem.*, 2004, **15**, 335–341.

56. R. Rao and M. K. Georgieff, Perinatal Aspects of Iron Metabolism, *Acta. Paediatr. Suppl.*, 2002, **91**, 124–129.

57. I. J. Hardy, L. Gillanders and G. Hardy, Is Manganese an Essential Supplement for Parenteral Nutrition?, *Curr. Opin. Clin. Nutr. Metab. Care*, 2008, **11**, 289–296.

58. V. A. Fitsanakis, C. Au, K. M. Erikson and M. Aschner, The Effects of Manganese on Glutamate, Dopamine and Gamma-Aminobutyric Acid Regulation, *Neurochem. Int.*, 2006, **48**, 426–433.

59. J. Ono, K. Harada, R. Kodaka, K. Sakurai, H. Tajiri, Y. Takagi, T. Nagai, T. Harada, A. Nihei and A. Okada, Manganese Deposition in the Brain During Long-Term Total Parenteral Nutrition, *JPEN, J. Parenter. Enteral Nutr.*, 1995, **19**, 310–312.

60. S. Nagatomo, F. Umehara, K. Hanada, Y. Nobuhara, S. Takenaga, K. Arimura and M. Osame, Manganese Intoxication During Total Parenteral Nutrition: Report of Two Cases and Review of the Literature, *J. Neurol. Sci.*, 1999, **162**, 102–105.

61. D. B. Bertinet, M. Tinivella, F. A. Balzola, A. de Francesco, O. Davini, L. Rizzo, P. Massarenti, M. A. Leonardi and F. Balzola, Brain Manganese Deposition and Blood Levels in Patients Undergoing Home Parenteral Nutrition, *JPEN, J. Parenter. Enteral Nutr.*, 2000, **24**, 223–227.

62. Y. Takagi, A. Okada, K. Sando, M. Wasa, H. Yoshida and N. Hirabuki, Evaluation of Indexes of in Vivo Manganese Status and the Optimal Intravenous Dose for Adult Patients Undergoing Home Parenteral Nutrition, *Am. J. Clin. Nutr.*, 2002, **75**, 112–118.

63. Y. Takagi, A. Okada, K. Sando, M. Wasa, H. Yoshida and N. Hirabuki, On-Off Study of Manganese Administration to Adult Patients Undergoing Home Parenteral Nutrition: New Indices of in Vivo Manganese Level, *JPEN, J. Parenter. Enteral Nutr.*, 2001, **25**, 87–92.

64. D. G. Kelly, Guidelines and Available Products for Parenteral Vitamins and Trace Elements, *JPEN, J. Parenter. Enteral Nutr.*, 2002, **26**, S34–S36.

65. R. Abdalian, O. Saqui, G. Fernandes and J. P. Allard, Effects of Manganese From a Commercial Multi-Trace Element Supplement in a Population Sample of Canadian Patients on Long-Term Parenteral Nutrition, *JPEN, J. Parenter. Enteral Nutr.*, 2013, **37**, 538–543.

66. G. Alves, J. Thiebot, A. Tracqui, T. Delangre, C. Guedon and E. Lerebours, Neurologic Disorders Due to Brain Manganese Deposition in a Jaundiced Patient Receiving Long-Term Parenteral Nutrition, *JPEN, J. Parenter. Enteral Nutr.*, 1997, **21**, 41–5.

67. C.-T. Hsieh, J.-S. Liang, S. S.-F. Peng and W.-T. Lee, Seizure Associated with Total Parenteral Nutrition–Related Hypermanganesemia, *Pediatr. Neurol.*, 2007, **36**, 181–183.

68. Y. Iinuma, M. Kubota, M. Yagi, S. Kanada, S. Yamazaki, H. Murata, K. Okamoto, M. Suzuki, K. Nitta and M. Uchiyama, Whole-Blood Manganese Levels and Brain Manganese Accumulation in Children Receiving Long-Term Home Parenteral Nutrition, *Pediatr. Surg. Int.*, 2003, **19**, 268–272.

69. H. Komaki, S. Maisawa, K. Sugai and Y. Kobayashi, Tremor and Seizures Associated with Chronic Manganese Intoxication, *Brain Dev.*, 1999, **21**, 122–124.

70. I. F. Btaiche, P. L. Carver and K. B. Welch, Dosing and Monitoring of Trace Elements in Long-Term Home Parenteral Nutrition Patients, *JPEN, J. Parenter. Enteral Nutr.*, 2011, **35**, 736–747.

71. A. Ejima, T. Imamura, S. Nakamura and H. Saito, Manganese Intoxication During Total Parenteral Nutrition, *Lancet*, 1992, **339**, 426.

72. S. A. Mirowitz, T. J. Westrich and J. D. Hirsch, Hyperintense Basal Ganglia on T1-Weighted MR Images in Patients Receiving Parenteral Nutrition, *Radiology*, 1991, **181**, 117–120.

73. K. Fitzgerald, V. Mikalunas, H. Rubin, R. McCarthy, A. Vanagunas and R. M. Craig, Hypermanganesemia in Patients Receiving Total Parenteral Nutrition, *JPEN, J. Parenter. Enteral Nutr.*, 1999, **23**, 333–336.

74. A. P. Reynolds, E. Kiely and N. Meadows, Manganese in Long Term Paediatric Parenteral Nutrition, *Arch. Dis. Child.*, 1994, **71**, 527–528.

75. S. Taylor and A. R. Manara, Manganese Toxicity in a Patient with Cholestasis Receiving Total Parenteral Nutrition, *Anaesthesia*, 1994, **49**, 1013.

76. S. V. Chandra and G. S. Shukla, Manganese Encephalopathy in Growing Rats, *Environ. Res.*, 1978, **15**, 28–37.

77. P. J. Kontur and L. D. Fechter, Brain Regional Manganese Levels and Monoamine Metabolism in Manganese-Treated Neonatal Rats, *Neurotoxicol. Teratol.*, 1988, **10**, 295–303.

78. B. A. Pappas, D. Zhang, C. M. Davidson, T. Crowder, G. A. Park and T. Fortin, Perinatal Manganese Exposure: Behavioral, Neurochemical, and Histopathological Effects in the Rat, *Neurotoxicol. Teratol.*, 1997, **19**, 17–25.

79. Y. Suzuki, T. Mouri, K. Nishiyama and N. Fujii, Study of Subacute Toxicity of Manganese Dioxide in Monkeys, *Tokushima J. Exp. Med.*, 1975, **22**, 5–10.

80. H. Roels, G. Meiers, M. Delos, I. Ortega, R. Lauwerys, J. P. Buchet and D. Lison, Influence of the Route of Administration and the Chemical

Form (MnCl2, MnO2) on the Absorption and Cerebral Distribution of Manganese in Rats, *Arch. Toxicol.*, 1997, **71**, 223–230.

81. T. T. Tran, W. Chowanadisai, F. M. Crinella, A. Chicz-Demet and B. Lönnerdal, Effect of High Dietary Manganese Intake of Neonatal Rats on Tissue Mineral Accumulation, Striatal Dopamine Levels, and Neurodevelopmental Status, *Neurotoxicology*, 2002, **23**, 635–643.

82. J. A. Menezes-Filho, C. Novaes, O. de, J. C. Moreira, P. N. Sarcinelli and D. Mergler, Elevated Manganese and Cognitive Performance in School-Aged Children and Their Mothers, *Environ. Res.*, 2011, **111**, 156–163.

83. K. Khan, G. A. Wasserman, X. Liu, E. Ahmed, F. Parvez, V. Slavkovich, D. Levy, J. Mey, A. van Geen, J. H. Graziano and P. Factor-Litvak, Manganese Exposure From Drinking Water and Children's Academic Achievement, *Neurotoxicology*, 2012, **33**, 91–97.

84. J. A. Menezes-Filho, C. F. de Carvalho-Vivas, G. F. S. Viana, J. R. D. Ferreira, L. S. Nunes, D. Mergler and N. Abreu, Elevated Manganese Exposure and School-Aged Children's Behavior: a Gender-Stratified Analysis, *Neurotoxicology*, 2013, 1–8.

85. K. Khan, G. A. Wasserman, X. Liu, E. Ahmed, F. Parvez, V. Slavkovich, D. Levy, J. Mey, A. van Geen, J. H. Graziano and P. Factor-Litvak, Manganese Exposure From Drinking Water and Children's Classroom Behavior in Bangladesh, *Environ. Health Perspect.*, 2011, **119**, 1501–1506.

86. J. Liu, L. Jin, Li, Z. Le Zhang, L. Wang, R. Ye, Y. Zhang and A. Ren, Placental Concentrations of Manganese and the Risk of Fetal Neural Tube Defects, *J. Trace Elem. Med. Biol.*, 2013, **27**, 322–325.

87. J. Cawte, Psychiatric Sequelae of Manganese Exposure in the Adult, Foetal and Neonatal Nervous Systems, *Aust. N. Z. J. Psychiatry*, 1985, **19**, 211–217.

88. J. M. Fell, A. P. Reynolds, N. Meadows, K. Khan, S. G. Long, G. Quaghebeur, W. J. Taylor and P. J. Milla, Manganese Toxicity in Children Receiving Long-Term Parenteral Nutrition, *Lancet*, 1996, **347**, 1218–1221.

89. A. Woolf, R. Wright, C. Amarasiriwardena and D. Bellinger, A Child with Chronic Manganese Exposure From Drinking Water, *Environ. Health Perspect.*, 2002, **110**, 613–616.

90. E. Herrero Hernandez, G. Discalzi, P. Dassi, L. Jarre and E. Pira, Manganese Intoxication: the Cause of an Inexplicable Epileptic Syndrome in a 3 Year Old Child, *Neurotoxicology*, 2003, **24**, 633–639.

91. D. Santos, C. Batoreu, L. Mateus, A P. Marreilha Dos Santos and M. Aschner, Manganese in Human Parenteral Nutrition: Considerations for Toxicity and Biomonitoring, *Neurotoxicology*, 2013, **43**, 1–10.

92. V. Iyengar and J. Woittiez, Trace Elements in Human Clinical Specimens: Evaluation of Literature Data to Identify Reference Values, *Clin. Chem.*, 1988, **34**, 474–481.

93. A. Beuter, R. Edwards, A. deGeoffroy, D. Mergler and K. Hundnell, Quantification of Neuromotor Function for Detection of the Effects of Manganese, *Neurotoxicology*, 1998, **20**, 355–366.

94. C. L. Keen, M. S. Clegg, B. Lönnerdal and L. S. Hurley, Whole-Blood Manganese as an Indicator of Body Manganese, *N. Engl. J. Med.*, 1983, **308**, 1230.

95. J. P. Buchet, R. Lauwerys, H. Roels and C. De Vos, Determination of Manganese in Blood and Urine by Flameless Atomic Absorption Spectrophotometry, *Clin. Chim. Acta*, 1976, **73**, 481–486.

96. K. Hagenfeldt, L. O. Plantin and E. Diczfalusy, Trace Elements in the Human Endometrium. 2. Zinc, Copper and Manganese Levels in the Endometrium, Cervical Mucus and Plasma, *Acta Endocrinol.*, 1973, **72**, 115–126.

97. D. B. Milne, R. L. Sims and N. V. Ralston, Manganese Content of the Cellular Components of Blood, *Clin. Chem.*, 1990, **36**, 450–452.

98. P. Apostoli, R. Lucchini and L. Alessio, Are Current Biomarkers Suitable for the Assessment of Manganese Exposure in Individual Workers?, *Am. J. Ind. Med.*, 2000, **37**, 283–290.

99. A. Smargiassi and A. Mutti, Peripheral Biomarkers and Exposure to Manganese, *Neurotoxicology*, 1999, **20**, 401–406.

100. R. Lucchini, L. Selis, D. Folli, P. Apostoli, A. Mutti, O. Vanoni, A. Iregren and L. Alessio, Neurobehavioral Effects of Manganese in Workers From a Ferroalloy Plant After Temporary Cessation of Exposure, *Scand. J. Work, Environ. Health*, 1995, **21**, 143–149.

101. R. F. Bertolo, P. B. Pencharz and R. O. Ball, Tissue Mineral Concentrations Are Profoundly Altered in Neonatal Piglets Fed Identical Diets via Gastric, Central Venous, or Portal Venous Routes, *JPEN, J. Parenter. Enteral Nutr.*, 2014, **38**, 227–235.

102. Y. Kim, High Signal Intensities on T1-Weighted MRI as a Biomarker of Exposure to Manganese, *Ind. Health*, 2004, **42**, 111–115.

103. Y. Finkelstein, N. Zhang, V. A. Fitsanakis, M. J. Avison, J. C. Gore and M. Aschner, Differential Deposition of Manganese in the Rat Brain Following Subchronic Exposure to Manganese: a T1-Weighted Magnetic Resonance Imaging Study, *Isr. Med. Assoc. J.*, 2008, **10**, 793–798.

104. Y. Saitoh, S. Kimura, A. Nezu, N. Ohtsuki, T. Kobayashi, H. Osaka and S. Uehara, Hyperintense Brain Lesions on T1-Weighted MRI After Parenteral Nutrition, *No to Hattatsu*, 1996, **28**, 39–43.

105. S.-J. Li, L. Jiang, X. Fu, S. Huang, Y.-N. Huang, X.-R. Li, J.-W. Chen, Y. Li, H.-L. Luo, F. Wang, S.-Y. Ou and Y.-M. Jiang, Pallidal Index as Biomarker of Manganese Brain Accumulation and Associated with Manganese Levels in Blood: a Meta-Analysis, *PLoS One*, 2014, **9**, e93900.

106. Y. Chang, Y. Kim, S.-T. Woo, H.-J. Song, S. H. Kim, H. Lee, Y. J. Kwon, J.-H. Ahn, S.-J. Park, I.-S. Chung and K. S. Jeong, High Signal Intensity on Magnetic Resonance Imaging Is a Better Predictor of Neurobehavioral Performances Than Blood Manganese in Asymptomatic Welders, *Neurotoxicology*, 2009, **30**, 555–563.

107. T. Verina, J. S. Schneider and T. R. Guilarte, Manganese Exposure Induces A-Synuclein Aggregation in the Frontal Cortex of Non-Human Primates, *Toxicol. Lett.*, 2013, **217**, 177–183.

108. T. R. Guilarte, Manganese Neurotoxicity: New Perspectives From Behavioral, *Neuroimaging, and Neuropathological Studies in Humans and Non-Human Primates*, 2013, 1–10.

109. Z. Long, X.-R. Li, J. Xu, R. A. E. Edden, W.-P. Qin, L.-L. Long, J. B. Murdoch, W. Zheng, Y.-M. Jiang and U. Dydak, Thalamic GABA Predicts Fine Motor Performance in Manganese-Exposed Smelter Workers, *PLoS One*, 2014, **9**, e88220.

110. Z. Long, Y.-M. Jiang, X.-R. Li, W. Fadel, J. Xu, C.-L. Yeh, L.-L. Long, H.-L. Luo, J. Harezlak, J. B. Murdoch, W. Zheng and U. Dydak, Vulnerability of Welders to Manganese Exposure – a Neuroimaging Study, *Neurotoxicology*, 2014, 10.1016/j.neuro.2014.03.007.

111. N. L. Maitre, J. C. Slaughter, A. R. Stark, J. L. Aschner and A. W. Anderson, Validation of a Brain MRI Relaxometry Protocol to Measure Effects of Preterm Birth at a Flexible Postnatal Age, *BMC Pediatr.*, 2014, **14**, 84.

112. D. K. Thompson, C. Adamson, G. Roberts, N. Faggian, S. J. Wood, S. K. Warfield, L. W. Doyle, P. J. Anderson, G. F. Egan and T. E. Inder, Hippocampal Shape Variations at Term Equivalent Age in Very Preterm Infants Compared with Term Controls: Perinatal Predictors and Functional Significance at Age 7, *Neuroimage*, 2013, **70**, 278–287.

113. D. K. Thompson, C. Omizzolo, C. Adamson, K. J. Lee, R. Stargatt, G. F. Egan, L. W. Doyle, T. E. Inder and P. J. Anderson, Longitudinal Growth and Morphology of the Hippocampus Through Childhood: Impact of Prematurity and Implications for Memory and Learning, *Hum. Brain Mapp.*, 2014, **35**, 4129–4139.

114. L. J. Woodward, C. A. C. Clark, S. Bora and T. E. Inder, Neonatal White Matter Abnormalities an Important Predictor of Neurocognitive Outcome for Very Preterm Children, *PLoS One*, 2012, 7, e51879.

115. K. M. Erikson, K. Thompson, J. Aschner and M. Aschner, Manganese Neurotoxicity: a Focus on the Neonate, *Pharmacol. Ther.*, 2007, **113**, 369–377.

CHAPTER 16

Developmental Effects of Manganese

SCOTT M. LANGEVIN AND ERIN N. HAYNES*

Department of Environmental Health, University of Cincinnati College of
Medicine, Cincinnati, OH, USA
*Email: haynesen@ucmail.uc.edu

16.1 Introduction

Manganese is an abundant, naturally occurring transition (metal) element
that is ubiquitously found in food and water supplies, and, therefore, is
readily available through dietary consumption. It is also used in a wide
variety of industrial applications, including manufacturing of steel, production of dry-cell batteries, and fuel oil and gasoline additives, and is often
a constituent of industrial and automotive air pollution. Manganese from
these airborne sources can be inhaled and absorbed through the pulmonary
epithelium or the olfactory cells in the nasal epithelium, bypassing the usual
digestive absorption route of entry.[1]

The impact of manganese on human development can seem somewhat
paradoxical: it is an essential nutrient that is involved in critical physiologic
processes relating to growth, development, oxidation defense, and enzymatic function, yet manganese insufficiency and/or excessive exposure can
have negative consequences. As is the case with many other essential metals,
this gives rise to an inverted U-shaped risk-curve for adverse health effects
and developmental anomalies. The remainder of this chapter will focus on
describing the impact of both extremes (that is, insufficient and excess
uptake) in the context of human growth and development.

Issues in Toxicology No. 22
Manganese in Health and Disease
Edited by Lucio G. Costa and Michael Aschner

16.2 Manganese Deficiency and Development

Manganese is an essential nutrient involved in the synthesis of the mucopolysaccharide matrix of cartilage, bone growth, immune function, carbohydrate, lipid, and protein metabolism, regulation of cellular energy, reproductive function, and blood clotting.[2,3] Thus, insufficient intake can adversely impact growth and development, and can lead to skeletal and neurologic abnormalities,[4] as well as an increased susceptibility to seizures.[1] Since, as previously discussed, manganese is a common, naturally occurring element that is readily available through food and water, dietary deficiency is relatively rare; however, manganese deficiency can also arise from a number of health conditions, including epilepsy, pancreatic exocrine insufficiency, Perthes disease, or phenylketonuria (PKU).[5] Therefore, special attention should be paid to children with these disorders with respect to their manganese levels because they may be at higher risk for skeletal and neurologic development issues.

16.3 Manganese Toxicity and Development

While manganese deficiency is relatively rare, manganese toxicity may be on the rise owing to environmental pollution.[6] The United States Environmental Protection Agency has previously voiced concerns that chronic inhalation of low levels of atmospheric manganese compounds stemming from air pollution, including from the gasoline additive methylcyclopentadienyl manganese tricabonyl (MMT), may result in an accumulation of manganese in human tissues.[7]

The adverse impact of excess manganese exposure is presently much better established for adults, although young, developing children may in fact be more susceptible to its health effects. Although manganese homeostasis is tightly regulated in adults through physiologic control of intestinal absorption and biliary excretion of any excess, making it difficult to take in high levels of manganese through ingestion, this regulatory mechanism can be circumvented by absorption through the airway epithelium and direct entry into the bloodstream. In children, the rate of gastrointestinal absorption is comparatively higher during early development, potentially making them susceptible to excess dietary intake in addition to airborne exposure. In normal healthy adults, the typical dietary intake ranges from 0.9 to 10 mg of manganese per day, and it is absorbed at a rate of about 1–5%,[8] with women absorbing at a slightly higher rate than men. In contrast with adults, for whom the estimated safe and adequate dietary intake (ESADDI) is 2–5 mg of manganese per day, the ESADDI is 1.5 mg per day for children aged 4–8 years, 1.2 mg per day for children aged 1–3 years, 0.6 mg per day for children aged 7–12 months, and 0.003 mg per day for newborns.[8] Moreover, neonates have an immature (incompletely formed) blood–brain barrier, and there is a lack of biliary excretion prior to weaning, all of which can contribute to increased rates of manganese uptake and deposition,[9] and higher rates of delivery and deposition in the developing neonatal brain and fetal tissue.[10,11] It has been estimated[12] that the retention rate in formula-fed infants is around 20%, which may be of particular concern for infants fed with soy-based infant formulas because these can contain relatively high levels of manganese.[13]

Manganese crosses the placenta *via* active transport mechanisms[14] that are in part facilitated by the divalent metal transporter-1 (DMT-1) transport protein, which is ubiquitously expressed in placental tissue.[15] Initially, blood manganese levels tend to be higher in neonates but regress to adult levels at around 1 year of age.[16] Pregnant women also tend to have higher levels of blood manganese,[16] which is likely due to elevated expression of DMT-1, enhancing manganese uptake.[17] Pharmacokinetic models suggest that placental manganese levels increase with higher maternal inhalation exposure,[10] although it is important to note that these estimated increases remain within "normal" range.

16.3.1 Brain Development

At excessive levels, manganese is an established neurotoxin,[18] as demonstrated in adult workers who experience chronic levels of occupational exposure.[19] Manganese neurotoxicity, in part, stems from oxidative stress,[8,20] which again is somewhat paradoxical because manganese is an important cofactor involved in the reduction of free radicals. The potential neurologic effects are of particular concern in young children with developing brains, who, as previously discussed, are less well equipped to maintain homeostasis. Manganese crosses the blood–brain barrier 4 times more efficiently in young children than in adults[21] and accumulates in the brain in a dose-dependent manner.[22] Excess deposition and accumulation in the hippocampus of developing children, in particular the dentate gyrus, may be especially problematic, as this region is crucial for higher brain functions, such as memory and learning, and continues to develop after birth into early adulthood.[23] In fact, this susceptibility has been reflected in a number of study publications that reported an inverse association between cumulative manganese levels and neurologic or behavioral indices.[24–26] This relationship is depicted in Figure 16.1.

In a prospective study of 247 healthy pregnant women and their eventual offspring, Takser and colleagues found that higher levels of manganese exposure *in utero* (as measured by maternal and cord blood concentrations) were inversely associated with early childhood psychomotor development.[27] Lin and colleagues found that *in utero* exposure to high levels of both lead and manganese (each ≥75th percentile), determined by cord blood concentrations, adversely impacted the neurodevelopmental progress in 2 year olds, as demonstrated by lower overall, cognitive, and language components of the Comprehensive Developmental Inventory for Infants and Toddlers (CDIIT).[28] Similarly, in a study of 448 children born in Mexico City, Claus Henn and colleagues observed a U-shaped association with blood manganese levels in 12 month olds: decreased concurrent mental development

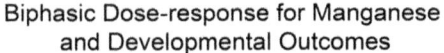

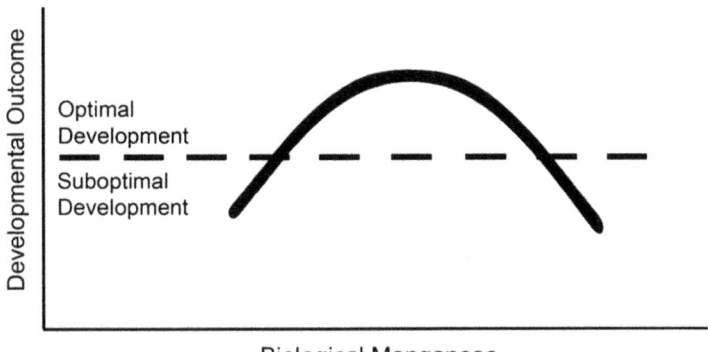

Figure 16.1 Illustration of biphasic dose–response relationship between biological manganese concentration and neurodevelopment outcomes.

scores (as determined by Baley Scales of Infant Development) were associated with the highest and lowest manganese quintiles (relative to the middle three quintiles).[29] The findings of the human studies are further corroborated by experimental evidence that demonstrated that neonatal rodents had a relatively higher risk of manganese neurotoxicity than adults after oral exposure.[30–33] Collectively, these results suggest a relationship between manganese and cognitive development, in which prenates/neonates at either extreme of manganese exposure (*i.e.* high or low intake) are at an elevated risk for impairment.

There is some evidence that selenium is protective against manganese toxicity in neonates. This is biologically plausible, because selenium is a cofactor for seleno-protein antioxidants[34] and, as stated earlier in this section, part of the neurotoxicity from manganese can be attributed to oxidative stress. Yang and colleagues observed an inverse association between cord blood manganese concentrations and Neonatal Neurological Assessment score, which was mitigated by increased levels of selenium in cord blood.[35] This is supported by experimental evidence that showed that ebselen (an organoselenium compound) caused a decrease in the concentration of manganese in the blood and brain of adult rats, as well as a reduction in neuroinflammation, oxidative stress, and locomotor impairment.[36]

Mechanistically, an experimental study has suggested that exposure of pregnant mice to manganese chloride through the drinking water induces apoptosis and neuronal mismigration of immature granule cells, resulting in sustained increases in immature-reelin synthesizing gamma aminobutyric acid (GABA)-ergic interneurons in the dentate gyrus.[23] Reelin is an extracellular matrix that plays a critical role in neuronal migration and positioning during brain development.[37] The authors speculate that this may in fact indicate an aberration of the postnatal neurogenic process.

Similarly, maternal exposure of pregnant rats to manganese chloride impacts late-stage differentiation in the dentate gyrus of the offspring.[38]

DNA methylation patterns, which are tightly coupled to cellular differentiation,[39] were disrupted in the offspring of mice exposed to manganese chloride during pregnancy[40] through aberrant promoter methylation of genes involved in neurogenic development, which include *Pvalb* and *Atp1a3*. Interestingly, the observed methylation differences for *Pvalb* eroded in older pups, while the increases in *Atp1a3* promoter methylation were sustained into adulthood. While these findings introduce the possibility of epigenetic involvement in manganese-associated slowing of neurogenic development, it is important to note that it remains unclear whether these changes are a direct result of the manganese exposure or rather an artifact stemming from changes in the neural cell population in the dentate gyrus. This is an important distinction because there is mounting evidence that environmental or dietary exposures that are sustained during early development can result in epigenetic remodeling of somatic cells.[41,42] While such remodeling may help the developing fetus or infant cope with the present stimulus, it can potentially predispose him or her to adverse health effects later on in adult life (often referred to as fetal origin of adult disease). To this point, although far from conclusive, Moreno and colleagues found that developmental exposure to manganese in juvenile mice at postnatal day 20–34 (where day 20 is the approximate murine equivalent of a human 2 year old) resulted in increased susceptibility to neuroinflammation as an adult.[43] Therefore, from a mechanistic standpoint, additional studies are warranted to clarify the biological significance of the observed association between manganese and methylation patterns and the potential impact, if any, on downstream disease susceptibilities.

Of additional interest, neural tube defects have been associated with low concentrations of manganese in maternal serum[44] and hair,[45,46] but higher concentrations in placenta, with a dose response,[47] after adjusting for neural tube defect risk factors and mineral cofactors. It is presently unclear why these measures act in opposite directions. More studies are needed before any definitive conclusions can be reached.

16.3.2 Birth Outcomes

In addition to neurotoxic effects, manganese intake and blood levels may also impact fetal and infant growth. This can have important ramifications on fetal or neonatal development and the risk of adverse birth outcomes, such as mortality or health problems in both the short- and long-term.[48]

Vigeh and colleagues found that low concentrations of manganese in maternal whole blood, but higher levels in fetal cord blood, were associated with intrauterine growth restriction.[49] It is presently unclear why cord and maternal blood tend in opposite directions, as the literature on this matter is sparse, although it is notable that similar findings have been reported for neurodevelopment, as previously discussed. Zota and colleagues found no association between birth weight and manganese concentration in cord blood but reported an inverted U-shaped relationship with maternal blood, where birth weight was significantly decreased at both extremes (high and low manganese concentrations).[50] Yu and colleagues observed an association between high cord blood manganese concentration and the pondoral index,[51] which is an anthropometric measure commonly used in pediatric populations because it better accounts for height variations than does body mass index, may predict adult obesity and cardiovascular disease risk, and may be indicative of glucose intolerance.

There is some experimental evidence of an impact of excess manganese exposure on fetal/neonatal growth restriction. Intravenous or subcutaneous manganese dosing of pregnant mice has been reported to result in decreased fetal size.[52,53] Further, Hiney and colleagues found that low-dose manganese exposure (*via* manganese chloride supplementation in the drinking water) induced expression of transforming growth factor alpha (TGFα) and insulin-like growth factor 1 (IGF-1), which are important factors in fetal growth, in the hypothalamus of prepubertal rats.[54] However, it is important to keep in perspective that these are also inflammatory markers, which is a significant consideration because manganese stimulates the release of proinflammatory cytokines (including TGFα).[55]

There is also limited evidence of an adverse impact of high paternal levels of manganese exposure and the rate of live births. In a cross-sectional study of male workers exposed to manganese dust, Lauwerys and colleagues observed a decreased rate of live births.[56] A second similar study found no such effect,[57] although it is important to note that the blood and urinary manganese levels were about 50% lower in the latter study. In a prospective study of 1875 women, Rahman and colleagues reported that elevated manganese concentrations ($>599~\mu g~l^{-1}$) in drinking water were associated with *decreased* risk of spontaneous abortion after adjusting for gravidity and iron and arsenic levels in the water source.[58] However, the estimates were no longer significant after additional adjustment for body mass index, education, and socioeconomic status, although the point estimates remained similar. These findings are in sharp contrast with the results from a cross-sectional study, which suggested almost doubling of the risk of spontaneous abortion for women with manganese levels $>0.4~mg~l^{-1}$ in their drinking water,[59] and an ecological study conducted in the US state of North Carolina showing a higher rate of infant mortality in counties with higher manganese concentrations in their drinking water;[60] however, another ecological study conducted in Bangladesh found no association.[61] Therefore, more studies are needed before any definitive conclusions can be made regarding the relationship between fetal or infant mortality and manganese exposure.

16.3.3 Onset of Puberty

The onset of puberty in girls, as measured by breast development and menarche, shows a secular trend towards earlier ages in the US and other parts of the world.[62] There is some limited experimental evidence stemming from a series of studies by Les Dees' group that suggests a possible role of low-dose manganese exposure in this observed phenomenon.

Environmental manganese acts on the hypothalamus to stimulate pre-pubertal luteinizing hormone-releasing hormone (LHRH) in female rats,[63] and the time of puberty onset is accelerated at low-dose exposures for female[63] and male[64] rats, with females being more sensitive. Low-level manganese exposure leads to increased accumulation of manganese in the hypothalamus of prepubertal rats and is associated with local activation of genes that regulate LHRH.[65] Although these findings are provocative, additional epidemiologic and experimental evidence is needed before any definitive conclusions can be drawn.

16.4   Conclusions

The impact of manganese on human development is paradoxical: it is both an essential nutrient and a toxicant. Emerging research provides evidence of manganese toxicity as an inverted U-shaped curve where both low and high biological levels are associated with negative consequences, which are summarized in Figure 16.2. Given that manganese is essential for normal growth and development, excessive exposure to manganese is cause for

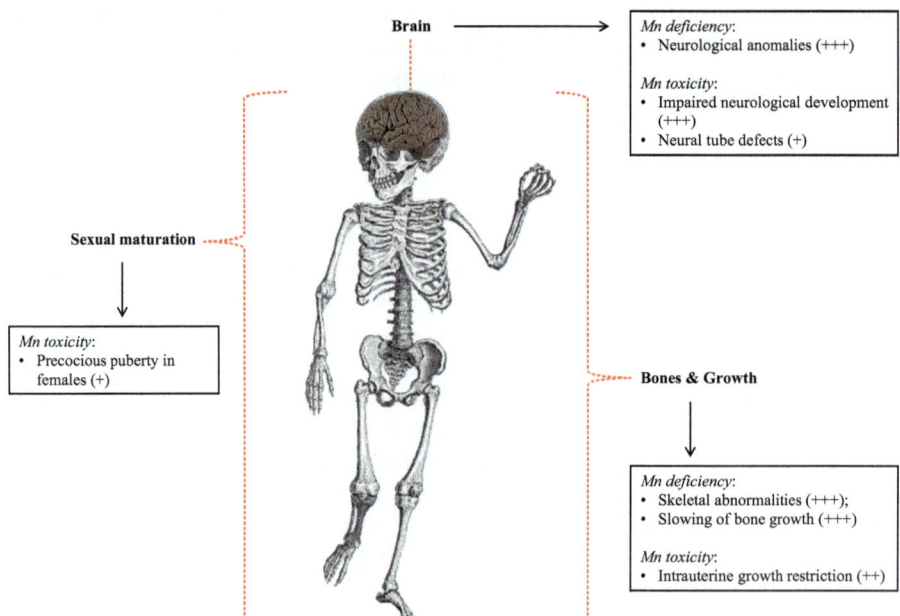

Figure 16.2   Summary of the developmental effects of manganese deficiency and toxicity. The overall strength of evidence is denoted after each effect in parentheses, where (+) denotes weak evidence, (++) denotes moderate evidence, and (+++) denotes strong evidence.

concern because the body has developed mechanisms to absorb and retain manganese. Developing brains are of particular concern as manganese can cross the blood–brain barrier in infants 4 times more efficiently than in adults. Both human and animal data are providing evidence of the impacts of manganese on fetal and infant growth, as measured by intrauterine growth restriction concentrations, pondoral index, and birth weight. Manganese may also impact pubertal development as manganese accumulates in the hypothalamus and stimulates puberty-related hormones. Although this has been demonstrated in rats, these findings have not been replicated in humans. Additionally, as evidence increases surrounding epigenetic changes due to environmental and dietary exposures during critical developmental windows, research is needed to examine epigenetic modifications resulting from insufficient and excessive exposure to manganese during development. Given the necessity for manganese in normal growth and development, it is important to evaluate the potential negative impact of both insufficient and excessive manganese on these similar developmental pathways.

References

1. C. Au, A. Benedetto and M. Aschner, *Neurotoxicology*, 2008, **29**, 569–576.
2. C. Castillo-Duran and F. Cassorla, *J. Pediatr. Endocrinol. Metab.*, 1999, **12**, 589–601.
3. K. M. Erikson, K. Thompson, J. Aschner and M. Aschner, *Pharmacol. Ther.*, 2007, **113**, 369–377.
4. L. S. Hurley, *Philos. Trans. R. Soc., B*, 1981, **294**, 145–152.
5. O. P. Soldin and M. Aschner, *Neurotoxicology*, 2007, **28**, 951–956.
6. M. F. Bouchard, S. Sauve, B. Barbeau, M. Legrand, M. E. Brodeur, T. Bouffard, E. Limoges, D. C. Bellinger and D. Mergler, *Environ. Health Perspect.*, 2011, **119**, 138–143.
7. U. S. E. P. Agency, *Reevaluation of inhalation risks associated with methylcyclopentadienyl manganese tricarbonyl (MMT) in gasoline*, EPA Office of Research and Development, 1994.
8. J. L. Aschner and M. Aschner, *Mol. Aspects Med.*, 2005, **26**, 353–362.
9. N. Ballatori, E. Miles and T. W. Clarkson, *Am. J. Physiol.*, 1987, **252**, R842–847.
10. M. Yoon, J. D. Schroeter, A. Nong, M. D. Taylor, D. C. Dorman, M. E. Andersen and H. J. Clewell, 3rd, *Toxicol. Sci.*, 2011, **122**, 297–316.
11. M. Aschner, K. M. Erikson and D. C. Dorman, *Crit. Rev. Toxicol.*, 2005, **35**, 1–32.
12. K. Dorner, S. Dziadzka, A. Hohn, E. Sievers, H. D. Oldigs, G. Schulz-Lell and J. Schaub, *Br. J. Nutr.*, 1989, **61**, 559–572.
13. M. Krachler, W. Domej and K. J. Irgolic, *Biol. Trace Elem. Res.*, 2000, **75**, 253–263.
14. M. Krachler, E. Rossipal and D. Micetic-Turk, *Eur. J. Clin. Nutr.*, 1999, **53**, 486–494.

15. W. S. Chong, P. C. Kwan, L. Y. Chan, P. Y. Chiu, T. K. Cheung and T. K. Lau, *Hum. Reprod.*, 2005, **20**, 3532–3538.
16. R. J. Wood, *Nutr. Rev.*, 2009, **67**, 416–420.
17. K. N. Millard, D. M. Frazer, S. J. Wilkins and G. J. Anderson, *Gut*, 2004, **53**, 655–660.
18. A. B. Bowman, G. F. Kwakye, E. Herrero Hernandez and M. Aschner, *J. Trace Elem. Med. Biol.*, 2011, **25**, 191–203.
19. R. M. Park, *Saf. Health Work*, 2013, **4**, 123–135.
20. E. J. Martinez-Finley, C. E. Gavin, M. Aschner and T. E. Gunter, *Free Radical Biol. Med.*, 2013, **62**, 65–75.
21. I. Mena, *Ann. Clin. Lab.Sci.*, 1974, **4**, 487–491.
22. J. C. Lai, M. J. Minski, A. W. Chan, T. K. Leung and L. Lim, *Neurotoxicology*, 1999, **20**, 433–444.
23. L. Wang, T. Ohishi, A. Shiraki, R. Morita, H. Akane, Y. Ikarashi, K. Mitsumori and M. Shibutani, *Toxicol. Sci.*, 2012, **127**, 508–521.
24. M. Rodriguez-Barranco, M. Lacasana, C. Aguilar-Garduno, J. Alguacil, F. Gil, B. Gonzalez-Alzaga and A. Rojas-Garcia, *Sci. Total Environ.*, 2013, **454–455**, 562–577.
25. J. A. Menezes-Filho, M. Bouchard, N. Sarcinelli Pde and J. C. Moreira, *Revista panamericana de salud publica/Pan American journal of public health*, 2009, **26**, 541–548.
26. A. Bhattacharya, F. Rugless, P. Succop, K. Dietrich, C. Cox, J. Alden, K. Kuhnell, M. Barnas, R. Wright, P. J. Parsons, M. L. Praamsma, C. D. Palmer, C. Beidler, R. Wittberg and E. N. Haynes, *Neurotoxicol. Teratol.*, 2013, Responding to Reviewer Comments.
27. L. Takser, D. Mergler, G. Hellier, J. Sahuquillo and G. Huel, *Neurotoxicology*, 2003, **24**, 667–674.
28. C. C. Lin, Y. C. Chen, F. C. Su, C. M. Lin, H. F. Liao, Y. H. Hwang, W. S. Hsieh, S. F. Jeng, Y. N. Su and P. C. Chen, *Environ. Res.*, 2013, **123**, 52–57.
29. B. Claus Henn, A. S. Ettinger, J. Schwartz, M. M. Tellez-Rojo, H. Lamadrid-Figueroa, M. Hernandez-Avila, L. Schnaas, C. Amarasiriwardena, D. C. Bellinger, H. Hu and R. O. Wright, *Epidemiology*, 2010, **21**, 433–439.
30. S. V. Chandra and G. S. Shukla, *Environ. Res.*, 1978, **15**, 28–37.
31. D. C. Dorman, M. F. Struve, D. Vitarella, F. L. Byerly, J. Goetz and R. Miller, *J. Appl. Toxicol.*, 2000, **20**, 179–187.
32. P. J. Kontur and L. D. Fechter, *Neurotoxicol. Teratol.*, 1988, **10**, 295–303.
33. B. A. Pappas, D. Zhang, C. M. Davidson, T. Crowder, G. A. Park and T. Fortin, *Neurotoxicol. Teratol.*, 1997, **19**, 17–25.
34. L. V. Papp, J. Lu, A. Holmgren and K. K. Khanna, *Antioxid. Redox Signaling*, 2007, **9**, 775–806.
35. X. Yang, Y. Bao, H. Fu, L. Li, T. Ren and X. Yu, *PLoS One*, 2014, **9**, e86611.
36. A. P. Santos, R. L. Lucas, V. Andrade, M. L. Mateus, D. Milatovic, M. Aschner and M. C. Batoreu, *Toxicol. Appl. Pharmacol.*, 2012, **258**, 394–402.

37. G. D'Arcangelo, K. Nakajima, T. Miyata, M. Ogawa, K. Mikoshiba and T. Curran, *J. Neurosci.*, 1997, **17**, 23–31.
38. T. Ohishi, L. Wang, H. Akane, A. Shiraki, K. Goto, Y. Ikarashi, K. Suzuki, K. Mitsumori and M. Shibutani, *Reprod. Toxicol.*, 2012, **34**, 408–419.
39. S. M. Langevin and K. T. Kelsey, *Environ. Mol. Mutagen.*, 2013, **54**, 533–541.
40. L. Wang, A. Shiraki, M. Itahashi, H. Akane, H. Abe, K. Mitsumori and M. Shibutani, *Toxicol. Sci.*, 2013, **136**, 154–165.
41. K. Calkins and S. U. Devaskar, *Curr. Probl. Pediatr. Adolesc. Health Care*, 2011, **41**, 158–176.
42. P. D. Gluckman, M. A. Hanson and F. M. Low, *Birth Defects Res., Part C*, 2011, **93**, 12–18.
43. J. A. Moreno, K. M. Streifel, K. A. Sullivan, M. E. Legare and R. B. Tjalkens, *Toxicol. Sci.*, 2009, **112**, 405–415.
44. H. M. Jiang, *Zhonghua Yufang Yixue Zazhi*, 1991, **25**, 102–104.
45. W. Zhang, A. Ren, L. Pei, L. Hao and L. Ouyang, *Chin. J. Public Health*, 2005, **21**, 1153–1155.
46. G. Saner, T. Dagoglu and T. Ozden, *Am. J. Clin. Nutr.*, 1985, **41**, 1042–1044.
47. J. Liu, L. Jin, L. Zhang, Z. Li, L. Wang, R. Ye, Y. Zhang and A. Ren, *J. Trace Elem. Med. Biol.*, 2013, **27**, 322–325.
48. E. Cosmi, T. Fanelli, S. Visentin, D. Trevisanuto and V. Zanardo, *J. Pregnancy*, 2011, **2011**, 364381.
49. M. Vigeh, K. Yokoyama, F. Ramezanzadeh, M. Dahaghin, E. Fakhriazad, Z. Seyedaghamiri and S. Araki, *Reprod. Toxicol.*, 2008, **25**, 219–223.
50. A. R. Zota, A. S. Ettinger, M. Bouchard, C. J. Amarasiriwardena, J. Schwartz, H. Hu and R. O. Wright, *Epidemiology*, 2009, **20**, 367–373.
51. X. Yu, L. Cao and X. Yu, *Environ. Res.*, 2013, **121**, 79–83.
52. M. T. Colomina, J. L. Domingo, J. M. Llobet and J. Corbella, *Vet. Hum. Toxicol.*, 1996, **38**, 7–9.
53. D. J. Sanchez, J. L. Domingo, J. M. Llobet and C. L. Keen, *Toxicol. Lett.*, 1993, **69**, 45–52.
54. J. K. Hiney, V. K. Srivastava and W. L. Dees, *Toxicol. Sci.*, 2011, **121**, 389–396.
55. J. M. Antonini, M. D. Taylor, A. T. Zimmer and J. R. Roberts, *J. Toxicol. Environ. Health, Part A*, 2004, **67**, 233–249.
56. R. Lauwerys, H. Roels, P. Genet, G. Toussaint, A. Bouckaert and S. De Cooman, *Am. J. Ind. Med.*, 1985, **7**, 171–176.
57. J. P. Gennart, J. P. Buchet, H. Roels, P. Ghyselen, E. Ceulemans and R. Lauwerys, *Am. J. Epidemiol.*, 1992, **135**, 1208–1219.
58. S. M. Rahman, A. Akesson, M. Kippler, M. Grander, J. D. Hamadani, P. K. Streatfield, L. A. Persson, S. El Arifeen and M. Vahter, *PLoS One*, 2013, **8**, e74119.
59. D. Hafeman, P. Factor-Litvak, Z. Cheng, A. van Geen and H. Ahsan, *Environ. Health Perspect.*, 2007, **115**, 1107–1112.
60. A. H. Spangler and J. G. Spangler, *Ecohealth*, 2009, **6**, 596–600.

61. N. Cherry, K. Shaik, C. McDonald and Z. Chowdhury, *Arch. Environ. Occup. Health*, 2010, **65**, 148–153.
62. F. M. Biro, L. C. Greenspan and M. P. Galvez, *J. Pediatr. Adolesc. Gynecol.*, 2012, **25**, 289–294.
63. M. Pine, B. Lee, R. Dearth, J. K. Hiney and W. L. Dees, *Toxicol. Sci.*, 2005, **85**, 880–885.
64. B. Lee, M. Pine, L. Johnson, V. Rettori, J. K. Hiney and W. L. Dees, *Reprod. Toxicol.*, 2006, **22**, 580–585.
65. V. K. Srivastava, J. K. Hiney and W. L. Dees, *Toxicol. Sci.*, 2013, **136**, 373–381.

CHAPTER 17

The Effects of Manganese on Female Pubertal Development

WILLIAM L. DEES,* JILL K. HINEY AND VINOD K. SRIVASTAVA

Department of Veterinary Integrative Biosciences, College of Veterinary Medicine, Texas A&M University, College Station, Texas, USA
*Email: ldees@cvm.tamu.edu

17.1 Introduction

Manganese (Mn) is an essential element that is required for normal mammalian physiological processes including growth and development of bone and cartilage[1] as well as connective tissue and the reproductive system.[2,3] With regard to reproduction, it is known that Mn deficiencies in laboratory animals are associated with impaired growth and reproduction,[4,5] thus suggesting a role in reproductive function. Furthermore, while exposure to high levels of Mn is toxic and also causes reproductive dysfunction,[6,7] the effects of exposure to lower, but still elevated, levels of the metal on reproductive function have not been adequately assessed. This is especially true with regard to our understanding as to whether exposure to low levels of Mn facilitates or inhibits neuroendocrine development at the time of puberty.

In order for a substance to be actively involved in the onset of puberty it must act within the reproductive hypothalamus to stimulate the synthesis and secretion of gonadotropin-releasing hormone (GnRH). The hypothalamus is adjacent to the third ventricle and, thus, Mn can enter this brain region through the cerebral vasculature and the cerebral spinal fluid (CSF). This element crosses the blood–brain barrier by binding to transport systems such as transferrin.[8,9] As its levels in blood increase there is an influx

Issues in Toxicology No. 22
Manganese in Health and Disease
Edited by Lucio G. Costa and Michael Aschner
© The Royal Society of Chemistry 2015
Published by the Royal Society of Chemistry, www.rsc.org

into the CSF and entry across the choroid plexus becomes more important.[10] Mn crosses the blood–brain barrier over 4 times more efficiently in the young than in adults.[11] Furthermore, the young do not yet have the full capability to eliminate Mn,[12] and it is known to accumulate in the hypothalamus.[13] Additionally, infants and children have been classified as being potentially more sensitive to excess Mn exposure,[14] largely because the optimum levels of exposure for these ages have not been well defined.[2] Because of these collective actions of Mn, we have investigated whether prepubertal exposure to low but elevated levels of the element would alter events associated with the normal acquisition of puberty. Thus, this review will describe results demonstrating the hypothalamic effects and mechanisms of action of low dose $MnCl_2$ exposure on puberty-related genes and hormones that are critical for regulating pubertal development.

17.2 Critical Events Associated with the Normal Onset of Female Puberty

Before discussing the actions of Mn on the pubertal process, it is necessary to discuss some of the most critical events associated with the initiation of puberty. Briefly, the central nervous system (CNS) plays the critical role of synchronizing events leading to the onset of puberty by controlling pituitary function *via* hypothalamic neurohormones and gonadal function *via* pituitary hormones and direct neural inputs. Pituitary luteinizing hormone (LH) and follicle stimulating hormone (FSH) secretions are periodic or pulsatile in nature. The initiation of puberty is characterized by an increase in the frequency and amplitude of the LH pulses. This basic secretory pattern is similar in rats and primates, including humans,[15–17] is centrally mediated, and is reflective of the pulsatile release of hypothalamic GnRH.[18] In the rat, this peptide is synthesized in neurons mainly located in the preoptic area (POA) of the brain and then transported caudally to the median eminence (ME) region of the medial basal hypothalamus (MBH), from where it is released. Activation of GnRH secretion at puberty is due to the removal of an inhibitory tone, such as that provided by gamma aminobutyric acid and opioid peptides,[19,20] as well as due to the developmental responsiveness to excitatory inputs, such as those provided by insulin like growth factor-1 (IGF-1), excitatory amino acids (EAAs), kisspeptin (Kp) and Mn, the latter of which is the focus of this review. Importantly, all of these excitatory influences have been shown to drive the pubertal process in rats[21–26] and monkeys.[27–30] Following activation of the GnRH releasing system there is an increase in the serum levels of gonadotropins and estradiol (E_2), resulting in enhanced steroid positive feedback on the hypothalamus. The rising levels of E_2 eventually produce preovulatory GnRH and gonadotropin surges, resulting in first ovulation and reproductive maturity.

The pituitary and gonad are capable of functioning at any age after a short period of priming;[31] therefore, the age of pubertal onset is variable and depends on the timing of GnRH activation. In recent years evidence has been presented suggesting that the increased release of GnRH at puberty is a result of interactive communications between neuronal circuits and glial cells.[32–34] Additionally, it is widely accepted that these neuronal and glial functions can be influenced by peripheral metabolic signals, genetic and environmental influences, as well as drugs of abuse. Thus, any endogenous or exogenous substance that is capable of inhibiting or stimulating pre-pubertal GnRH secretion could impact the central activation of this peptide and, therefore, alter developmental events leading to puberty.

17.3 Acute Effects of Mn on Puberty-Related Hormones: A Hypothalamic Site of Action

A series of *in vivo* and *in vitro* experiments were used to determine whether manganese was capable of acting at the hypothalamic level to activate the reproductive hormone axis in immature female rats.[25] The initial experiment investigated the potential for Mn, when administered centrally, to elicit the pituitary release of LH. In this regard, third ventricular injections of $MnCl_2$ caused a marked, dose-dependent increase in LH release compared to basal levels (Figure 17.1), hence suggesting a hypothalamic action. To confirm that Mn acts at the hypothalamus and not the pituitary, both *in vitro* and *in vivo* assessments were conducted. Importantly, Mn dose-dependently stimulated

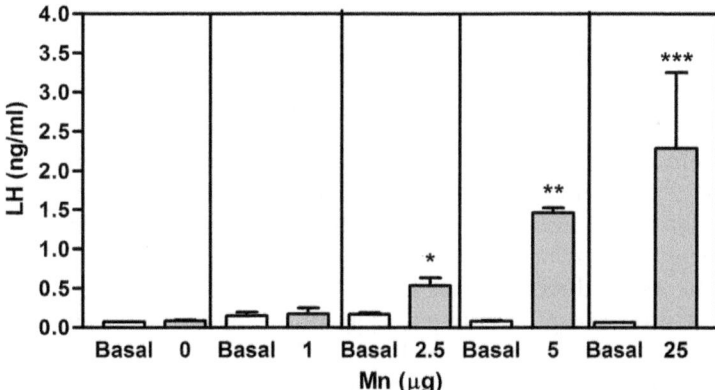

Figure 17.1 The effect of third ventricular administration of $MnCl_2$ on LH release during the late juvenile phase of developing female rats. Each concentration point indicates basal LH levels *vs.* stimulated levels. The animals that received the saline and the 1.0 µg dose of Mn showed no significant changes in LH secretion when compared to basal levels. However, animals that received 2.5 µg, 5.0 µg, and 25 µg doses of Mn showed marked increases in LH secretion when compared to their respective basal levels. Values represent mean ± standard error of the mean (SEM). $N = 7$–16 per panel. $^*p < 0.05$; $^{**}p < 0.01$; $^{***}p < 0.001$. (Reprinted with permission from Oxford University Press in Pine *et al.* (2005).[25])

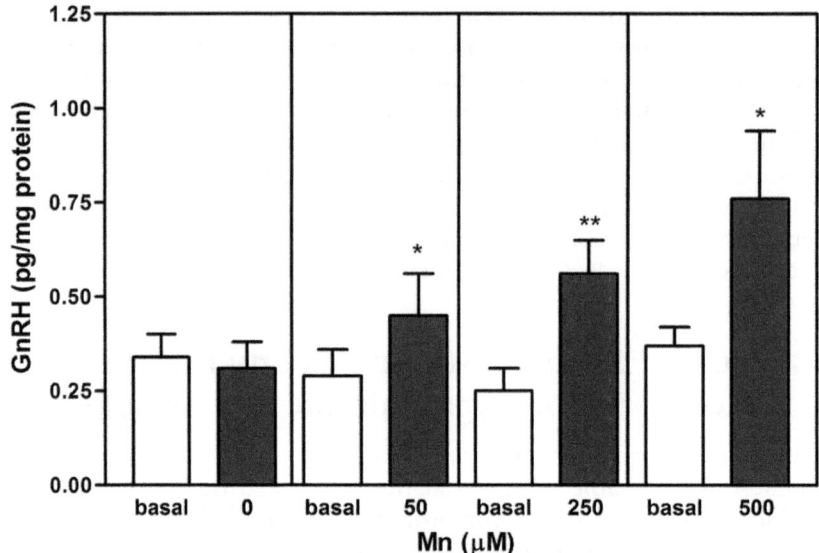

Figure 17.2 The effect of $MnCl_2$ on GnRH release from the medial basal hypothalamus *in vitro*. Each concentration point indicates basal GnRH levels
vs. stimulated levels. Tissue incubated in Locke's buffer only showed no
significant changes in GnRH secretion when compared to basal levels.
Tissues incubated in 50 μM, 250 μM, and 500 μM concentrations of Mn
showed significant increases in GnRH secretion when compared to
their respective basal levels. Values represent mean ± SEM. $N = 11$–15
per panel. $^{*}p < 0.05$; $^{**}p < 0.02$.
(Reprinted with permission from Oxford University Press in Pine *et al.*
(2005).[25])

GnRH release from basal hypothalamic explants incubated *in vitro*
(Figure 17.2), but did not induce the *in vitro* release of LH from the pituitary.
Further demonstration of a central action of Mn was shown by blocking
GnRH receptors on the pituitary prior to the administration of Mn. In this
regard, the Mn-induced release of LH was blocked in the animals treated
with the GnRH receptor antagonist acyline (Figure 17.3). Taken together,
these studies demonstrated conclusively that Mn acts at the hypothalamic
level to induce prepubertal LH secretion.

17.4   Chronic Effects of Mn on Puberty-Related Hormones and the Timing of Puberty

Prepubertal female rats were fed a rodent laboratory chow and supplemented daily with a previously determined minimal effective dose of
$MnCl_2$ (10 mg kg^{-1}), or an equal volume of saline, by a single gastric gavage
injection from day 12 until day 29. At this time, the animals were killed for
Mn and hormone analysis. Exposure to Mn during juvenile development

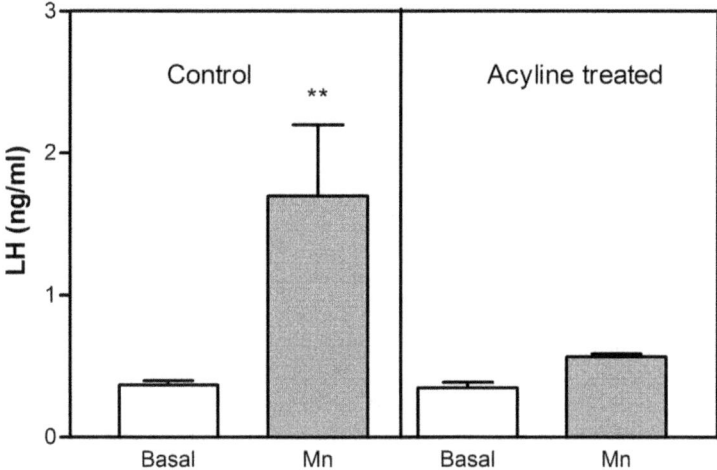

Figure 17.3 The effect of pretreatment with the GnRH receptor antagonist acyline on $MnCl_2$ stimulated LH release. Acyline treated animals showed no significant change in LH release after third ventricular (3V) administration of $MnCl_2$ as compared to basal levels. Animals that were not pretreated with acyline exhibited a four-fold increase in LH compared to basal levels. Values represent the mean ± SEM. $N = 4$–6 per panel. $^{**}p < 0.01$.
(Reprinted with permission from Oxford University Press in Pine *et al.* (2005)[25].)

caused increases in the serum levels of LH, FSH and E_2 by 29 days of age in the animals exposed to Mn.[25] Importantly, this regimen of Mn exposure caused the accumulation of the element in both the POA and the MBH.[25] In order to assess the timing of puberty, another study was conducted exactly as above except that the dosing of the animals with Mn continued until the day of vaginal opening (VO). Thus, when the dosing regimen was continued for several more days we noted that there were no differences between the mean (± SEM) daily weight gain of the animals (controls: 3.8 ± 0.07 g *vs.* Mn treated: 3.7 ± 0.08 g per day), but that the age at VO was advanced in the rats that were exposed to Mn (Figure 17.4).

It is important to note that the 10 mg kg^{-1} dose described here was low compared to previous studies using rats, although it is comparable on a mg kg^{-1} basis to doses causing effects on movement-responsiveness tests in adult primates.[35] While different endpoints were assessed, the low-level effects of Mn associated with increased LH, FSH and E_2 in prepubertal animals are important, because many of the detrimental, neurotoxicological endpoint effects described previously as a result of Mn exposure in adult rats and primates occurred with much higher doses.[36] As a result of the effect of Mn on the onset of puberty, we suggest the possibility that a premature Mn-induced activation of the GnRH releasing system too early in life may place an individual at risk of precocious pubertal development.

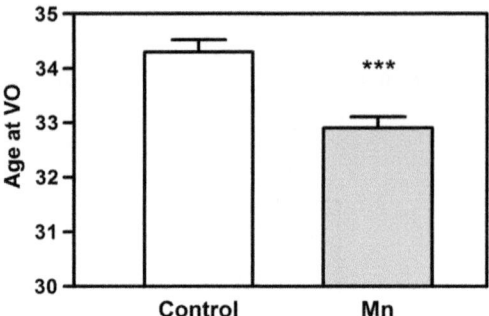

Figure 17.4   Effect of chronic oral administration of $MnCl_2$ on the age at vaginal opening (VO). Note that low level Mn exposure during the juvenile period moderately advanced the age at VO in terms of days, yet this was a highly significant trend. Values represent the mean $\pm$ SEM. $N = 31$ controls and 47 Mn-treated animals. $^{***}p < 0.001$.
(Reprinted with permission from Oxford University Press in Pine et al. (2005).[25])

17.5   Downstream Mechanism(s) of Mn Action on GnRH Release

While we have described above that the site of Mn action to facilitate GnRH secretion is at the hypothalamic level, the physiological mechanism behind this action needed further investigation. Nitric oxide (NO) is a known stimulator of hypothalamic GnRH release.[37] Because NO activates soluble guanylyl cyclase (sGC), and since Mn is the preferred cofactor for the activation of this enzyme,[38] we determined whether Mn-induced GnRH release involves an action within the NO–guanylyl cyclase (GC)–cyclic guanylyl monophosphate (cGMP)–protein kinase G (PKG) signaling pathway. In this regard, a static in vitro system for assessing GnRH release from the MBH of prepubertal female rats was used.[39] Results indicated that the addition of 50–500 µM $MnCl_2$ caused increases in GnRH release from tissues in medium alone and from tissues in medium containing the nitric oxide synthase (NOS) inhibitor N-monomethyl-L-arginine (NMMA), thus demonstrating that NOS inhibition was ineffective in blocking GnRH release. This is demonstrated in Figure 17.5 using the 50 µM dose of $MnCl_2$. To assess more directly a potential action of Mn on tissue NO formation, we assessed total nitrite, an indicator of NO generation, as well as GnRH secretion from the same MBH incubates. The 50 µM dose of Mn did not cause an increase in total nitrite (Figure 17.6A); however, it was capable of stimulating the release of GnRH (Figure 17.6B). This same effect was also observed with a 100 µM dose of Mn (not shown). Importantly, only the much higher 250 µM dose of Mn caused increases in nitrite production (Figure 17.6C) and GnRH secretion (Figure 17.6D). Therefore, taken together, these results suggest that low doses of Mn do not induce GnRH release by first inducing NOS/NO activity.

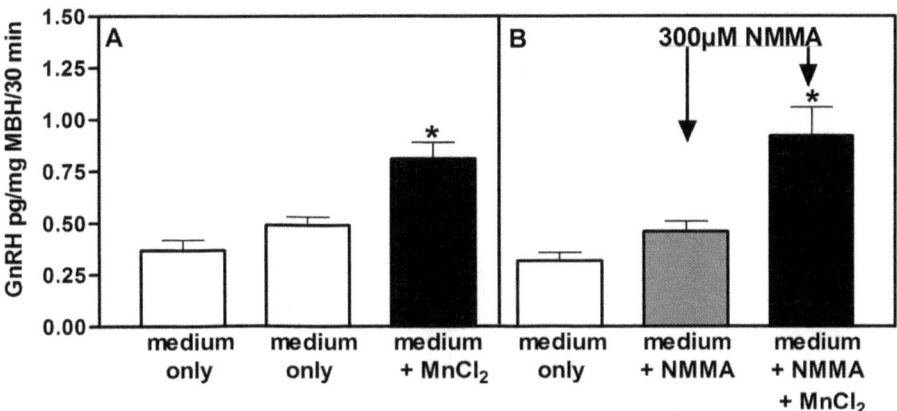

Figure 17.5 Effect of nitric oxide synthase (NOS) inhibition using *N*-monomethyl-L-arginine (NMMA) on $MnCl_2$-induced GnRH release. Open bars represent basal GnRH release. Gray bar represents GnRH release in the presence of NMMA only. Black bars represent GnRH released following 50 µM $MnCl_2$ stimulation in the absence (panel A) or presence (panel B) of NMMA. Note that Mn stimulated GnRH release in control tissues (panel A), and that this action was not altered by the presence of the NOS blocker (panel B). $^*p < 0.05$ *vs.* medium only or medium plus NMMA. Each bar represents the mean ± SEM. $N = 8$–9 per panel. (Reprinted with permission from John Wiley and Sons in Lee *et al.* (2007).[39])

Further experimentation determined that $1H$-[$1'2'4$]oxadiazolo[4,3-*a*]quinoxalin-1one (ODQ), a specific inhibitor of sGC,[40] dose-dependently blocked GnRH release induced by Mn, hence indicating that sGC is, at least in part, a site of action for this element to facilitate GnRH secretion (Figure 17.7). Additionally, Mn elicited the release of both cGMP and GnRH from the same tissue block (Figure 17.8) and, furthermore, KT5823, a downstream inhibitor of PKG, blocked the Mn-stimulated release of GnRH (Figure 17.9). Thus, these results indicate that Mn is involved in the activation of sGC, which subsequently induces the stimulation of the cGMP–PKG pathway, facilitating the release of GnRH. This ability of Mn to stimulate sGC in prepubertal animals supports an earlier study showing that Mn is an important element during development that is capable of activating over 50 enzymes, including GC and protein kinases.[41] Although it is well known that Mn is the preferred cofactor for GC and that it induces the activity of GC either directly or as a cofactor with NO,[38,42] it has now been shown that this element can stimulate GC directly in a physiological setting relevant to neuropeptide hormone secretion.[39]

17.6 Effect of Mn on GnRH Gene Expression

Thus far we have discussed the importance of the increased release of GnRH from the MBH at the time of puberty, as well as the effect of Mn in this

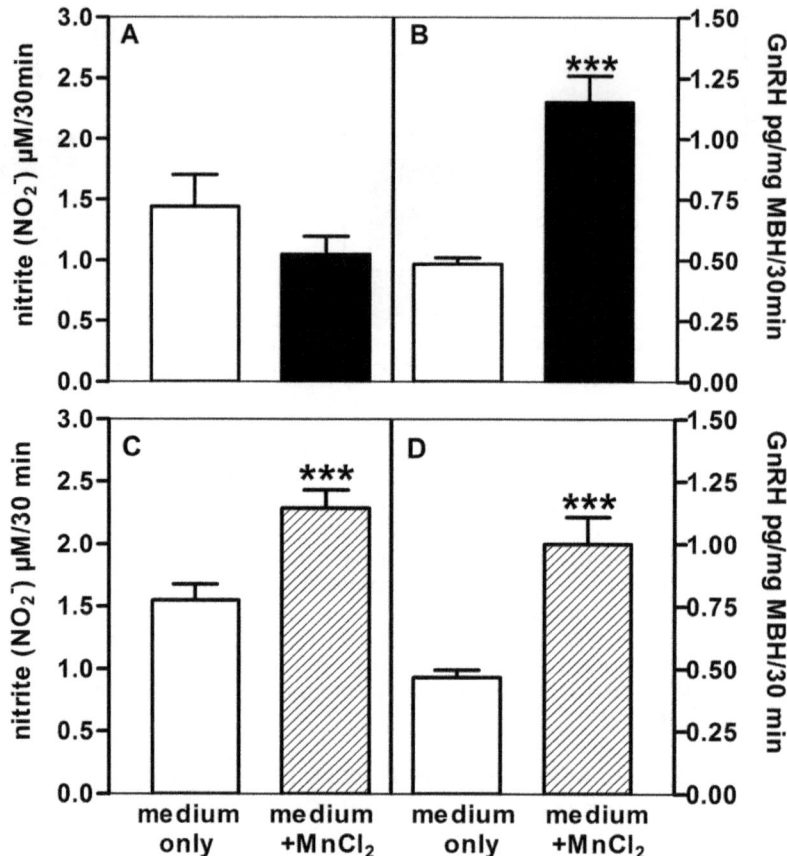

Figure 17.6 Effect of MnCl$_2$ on nitrite (NO$_2^-$) production and GnRH release from the same tissues. Open bars represent basal NO$_2^-$ and GnRH released into the medium. Black or hatched bars represent NO$_2^-$ and GnRH released following MnCl$_2$ stimulation at doses of 50 μM and 250 μM, respectively. Note that the 50 μM MnCl$_2$ dose did not increase NO$_2^-$ (panel A), but markedly induced GnRH release (panel B). Similar effects were observed with the 100 μM dose of MnCl$_2$ (not shown). The 250 μM MnCl$_2$ dose caused increases in both NO$_2^-$ (panel C) and GnRH (panel D). Together, this suggests that MnCl$_2$ can only stimulate NO production at higher concentrations. Each bar represents the mean $\pm$ SEM. $^{***}p < 0.001$ *vs.* respective medium only. $N = 9$–14 per panel. (Reprinted with permission from John Wiley and Sons in Lee *et al.* (2007).[39])

regard. It is equally important to note that the increased synthesis of the peptide is necessary to drive the pubertal process. Identifying factors capable of regulating GnRH gene expression is critical for understanding the mechanisms that control or alter the onset of puberty. With regard to Mn, we have recently shown that chronic exposure to this element upregulated the GnRH gene.[43] Figure 17.10 shows that, compared with controls, prepubertal

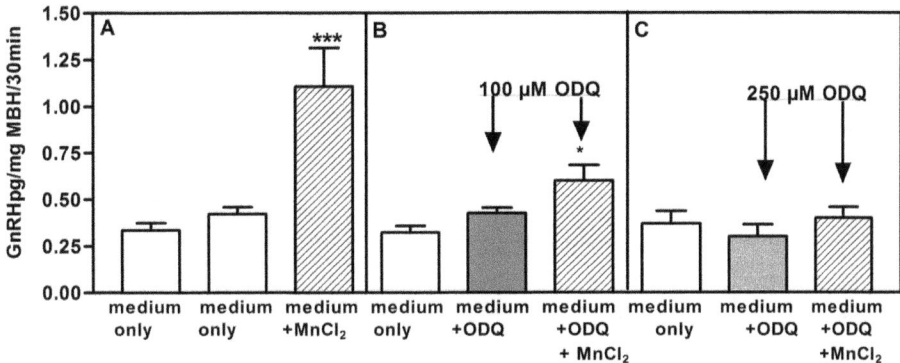

Figure 17.7 Effect of GC inhibition using ODQ on MnCl₂-induced GnRH release. Open bars represent basal GnRH release. Gray bars represent GnRH released in the presence of ODQ. Hatched bars represent GnRH released following 250 μM MnCl₂ stimulation in the absence (panel A) or presence (panels B and C) of the inhibitor. Note that the MnCl₂ stimulated GnRH release from control tissues (panel A), but this stimulatory effect was dose-dependently blocked (panels B and C) by the presence of the sGC inhibitor in the medium. $^{***}p < 0.01$ *vs.* medium only. $^{*}p < 0.05$ *vs.* medium only and medium plus ODQ. Each bar represents the mean ± SEM. $N = 20$, 11 and 15 for panels A, B and C, respectively.
(Reprinted with permission from John Wiley and Sons in Lee *et al.* (2007).[39])

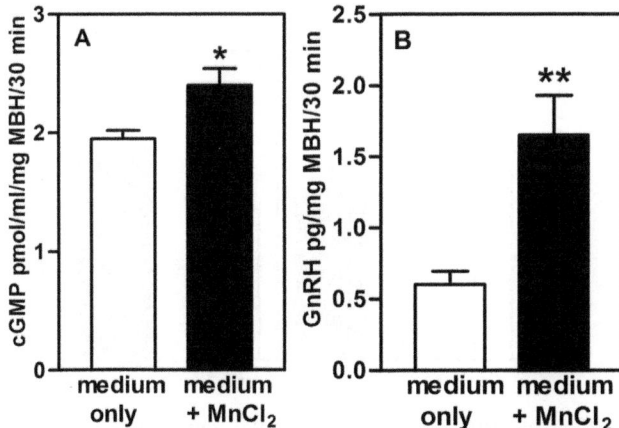

Figure 17.8 MnCl₂ stimulates cGMP and GnRH release. Open bar represents basal cGMP and GnRH release. Solid bars represent cGMP and GnRH released following addition of 50 μM MnCl₂ into the medium. Note that MnCl₂ stimulated both cGMP (panel A) and GnRH (panel B) secretion from the same tissue incubates. $^{*}p < 0.05$ and $^{**}p < 0.01$ *vs.* basal levels. Each bar represents the mean ± SEM. $N = 8$ for panels (A) and (B), respectively.
(Reprinted with permission from John Wiley and Sons in Lee *et al.* (2007).[39])

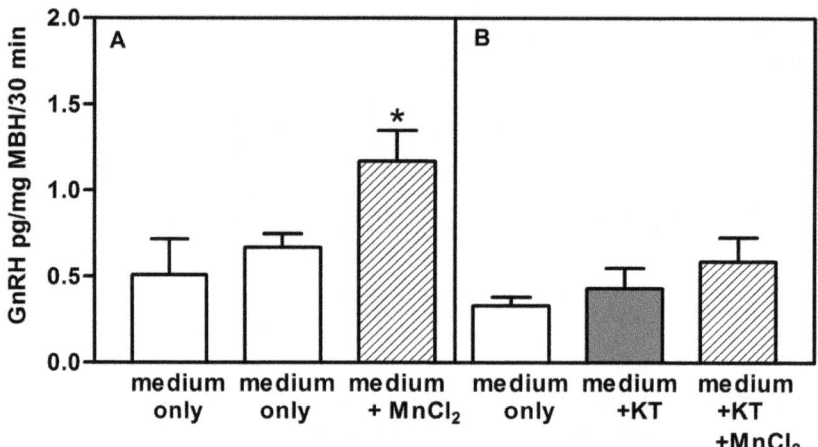

Figure 17.9   Effect of protein kinase G (PKG) inhibition using KT5823 on MnCl$_2$-induced GnRH release. Open bars represent basal GnRH release. Gray bar represents GnRH released in the presence of 10 μM KT5823. Hatched bars represent GnRH released following 250 μM MnCl$_2$ stimulation in the absence (panel A) or presence (panel B) of the inhibitor. Note that MnCl$_2$ stimulated GnRH release in control tissues (panel A), but was unable to stimulate GnRH release when the PKG blocker was present in the medium. $^*p < 0.05$ *vs.* medium only. Each bar represents the mean ± SEM. $N = 8$ for panels (A) and (B), respectively. (Reprinted with permission from John Wiley and Sons in Lee *et al.* (2007).[39])

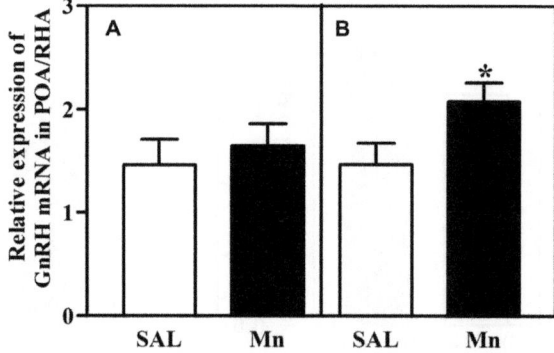

Figure 17.10   Effect of MnCl$_2$ exposure on GnRH gene expression in the POA/RHA. Note that MnCl$_2$ did not alter GnRH mRNA at 22 days (A), but induced increased expression at 29 days (B) of age. Bars represent an N of 5 at 22 days and 7 at 29 days of age in each group. $^*p < 0.05$ *vs.* saline. (Reprinted with permission from Oxford University Press in Srivastava *et al.* (2013).[43])

GnRH gene expression was increased at 29 days of age in a tissue block containing both the POA and the rostral hypothalamic area (POA/RHA); an action that was associated with increased ($p < 0.01$) serum levels of E$_2$

(controls: 12.9 ± 0.085 pg ml^{-1}; $N = 13$; *vs.* Mn-treated: 16.4 ± 0.59 pg ml^{-1}; $N = 13$). As a result of this early increase in GnRH expression, we aimed to determine whether Mn could affect other genes which are known or have the potential to influence GnRH neuronal activity and puberty.

17.7 Upstream Mechanisms of Mn Action in the Control of GnRH Neuronal Activity

17.7.1 Mn Action on Kiss-1 Gene Expression

In recent years, specific genes associated with tumor growth or suppression, referred to as tumor associated genes (TAGs), have been linked to events leading to puberty. Thus far, the most important of these genes with regard to puberty is the KiSS-1 gene. This gene encodes the kisspeptin (Kp) family of peptides which were first identified as tumor suppressors,[44,45] but more recently have been shown to act through specific G protein receptors (GPR54) on GnRH neurons,[46] resulting in the stimulation of GnRH neuronal activity.[47,48] KiSS-1 gene expression increases in the hypothalamus as puberty approaches in both primates[29] and rats.[23] Because of its relationship with GnRH, Kp is considered critical for attaining puberty in every species studied, including humans.[24,49–51] It is recognized that identifying upstream regulators of KiSS-1 gene expression will be important for better understanding events leading to the onset of puberty. In this regard, only limited progress has emerged in identifying substances associated with KiSS-1 upregulation.[52,53]

There have been several other TAGs recently shown to have increased expression in the hypothalamus at puberty,[54] although their definitive relationships to KiSS-1 and the pubertal process are not known. Some evidence, however, has been presented to suggest that several genes appear to be organized in an intricate network, and that a select few of them may be upstream to KiSS-1.[54] Determining factors influencing activation of KiSS-1 and these other TAGs at puberty, as well as assessing potential interrelationships among them, has been the focus of some of our recent investigations. Specifically, given that Mn is an environmental substance with the ability to stimulate prepubertal GnRH gene expression and peptide release resulting in precocious pubertal development, we assessed, as detailed below, whether prepubertal exposure to this element is capable of upregulating the expression of a select group of puberty-related TAGs.[43]

Because of the relationship between KiSS-1 and GnRH, initial assessments were made to determine whether Mn could upregulate the expression of the genes for KiSS-1 and the Kp receptor, GPR54, in a similar manner to that observed with the expression of GnRH at 29 days of age. In this regard, Mn-treated rats showed increases in both genes over control levels within the POA/RHA (Figure 17.11). It is important to note that the POA/RHA brain region includes the anteroventral periventricular (AVPV) nucleus, in which the KiSS-1 expressing neurons are localized. Importantly, in this brain region

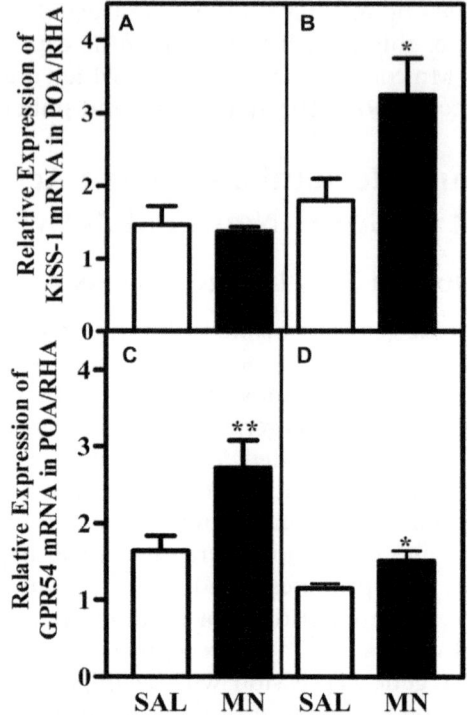

Figure 17.11 Effects of MnCl$_2$ exposure on KiSS-1 and GPR54 gene expression in the POA/RHA. Note that MnCl$_2$ did not alter KiSS-1 mRNA at 22 days (A), but induced increased expression at 29 days (B) of age. GPR54 expression was increased at both 22 and 29 days of age (C and D, respectively). Bars represent an N of 5 at 22 days and 7 at 29 days in each group. $^*p < 0.05$ *vs.* saline; $^{**}p < 0.01$ *vs.* saline. (Reprinted with permission from Oxford University Press in Srivastava *et al.* (2013).[43])

KiSS-1 is under the positive influence of E$_2$,[51,55,56] which, along with GnRH gene expression, was also increased in Mn-treated rats at 29 days. The fact that Mn exposure is capable of inducing KiSS-1 gene expression is important, and opened the question as to whether this element may be acting on a specific pathway of genes upstream to KiSS-1.

Given that Mn is an essential nutrient, specific genes were identified to determine whether it contributes to expression of the KiSS-1 and GnRH genes, at least in part, through a nutrient signaling action. Therefore, because the ras homologue expressed in brain (Rheb) is a TAG[57] that acts as a mediator of the nutrient signaling input to mammalian target of rapamycin (mTOR),[58] and because central mTOR signaling is a key modulator of puberty and reproduction *via* its regulation of KiSS-1,[59] the effect of Mn on the expression of these two genes was assessed. Figure 17.12 shows that Mn induced the expression of both Rheb and mTOR in the POA/RHA at 29 days. While mTOR has not been designated a TAG, it was assessed because it is

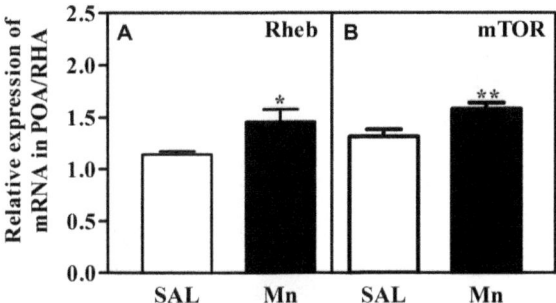

Figure 17.12 Effects of MnCl$_2$ exposure on Rheb and mTOR gene expression in the POA/RHA. Panels (A) and (B) show that MnCl$_2$ exposure caused increased mRNA expression levels of both genes at 29 days of age. Bars represent an N of 5 for Rheb and 11 for mTOR in each group. $^*p < 0.05$ *vs.* saline; $^{**}p < 0.01$ *vs.* saline.
(Reprinted with permission from Oxford University Press in Srivastava *et al.* (2013).[43])

activated by Rheb.[60] More work is still needed to define whether Mn is acting as a nutrient signal or by some other biochemical mechanism to activate Rheb. Once activated, however, Rheb binds directly to mTOR,[60] and this interaction is essential for activation of the mTOR complex 1. This complex is the nutrient-responsive mediator of cell growth regulation,[61–63] and is known to be involved in the central activation of puberty.[59]

It is now clear that expression of the Rheb, mTOR and KiSS-1 genes is increased precociously by 29 days in the POA/RHA of Mn-treated animals and, importantly, that these increases are associated with the increased expression of the prepubertal GnRH gene. An earlier study showed that the central blockade of mTOR caused decreased KiSS-1 and GnRH gene expression.[59] Taken together, these two studies show that either the activation or inactivation of this pathway results in opposite actions with regard to KiSS-1 and GnRH gene expression. These studies also further demonstrate the role of KiSS-1/Kp in the regulation of GnRH gene expression, an action supported by the fact that GnRH neurons express the Kp receptor.[46] We now suggest that this pathway in prepubertal animals can be activated by early life exposure to low amounts of Mn.

17.7.2 A Potential Role for Divalent Metal Transporter-1

In order for Mn to initiate these puberty-related events, it must enter and accumulate in the hypothalamus at an early prepubertal age. The optimal levels and timing of Mn entry into the brain are important, because deficiencies and excesses can result in altered brain functions. Divalent metal transporter-1 (DMT1) is a TAG protein that transports metals, including Mn, across the cell membrane.[64] Brain DMT1 is localized in neurons, capillary endothelial cells associated with the blood–brain barrier, and epithelial cells of the choroid plexus that are associated with the blood and cerebral spinal

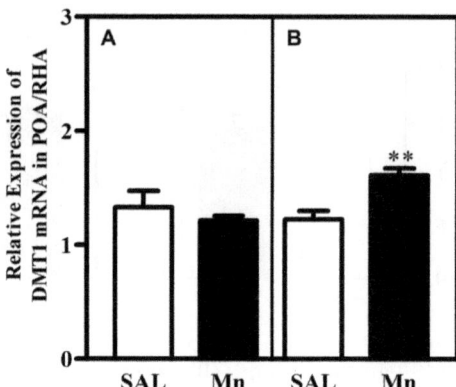

Figure 17.13   Effect of MnCl$_2$ exposure on DMT1 gene expression in the POA/RHA. Note that MnCl$_2$ did not alter DMT1 mRNA at 22 days (A), but induced increased expression at 29 days (B) of age. Bars represent an N of 5 at 22 days and 6 at 29 days. $^{**}p < 0.01$ *vs.* saline.
(Reprinted with permission from Oxford University Press in Srivastava *et al.* (2013).[43])

fluid (CSF) barrier.[65] Mn enters the hypothalamus either through the blood vasculature or the CSF. This element readily accumulates in the choroid plexus,[66] a structure involved in brain maturation, homeostasis and neuroendocrine functions.[67,68] As Mn levels increase in the blood, it moves into the CSF and its entry across the choroid plexus becomes more important.[69] Of potential relevance is our observation that Mn upregulates the DMT1 gene in the POA/RHA by 29 days of age (Figure 17.13),[43] hence demonstrating the same precocious increase as the proposed pathway of puberty-related genes described above. While this is interesting, more research is needed to discern the specific actions and interactions between Mn exposure and DMT1 with regard to Mn-induced precocious pubertal development.

17.8   Low Level Mn Exposure and Precocious Puberty

Despite the normal variation in the time of pubertal onset, evidence has come forth in recent years indicating that puberty may be occurring at an earlier age, especially in females.[70,71] While the cause of this trend is known, it has been suggested that assessing the onset of puberty may be a marker of interactions between environmental conditions and genetic susceptibility that influence the pubertal process.[71] The onset of puberty before the age of 8 years in girls and 9.5 years in boys is considered precocious.[31] Central or "true" precocious puberty is due to a premature activation of the GnRH secretory system; thus, it is GnRH dependent and is characterized by hormonal changes similar to those changes that occur at the normal time of puberty. This activation in boys is usually due to hypothalamic harmatomas, other CNS lesions or familial disease, with less than 10% of the cases being idiopathic. Conversely, in girls over 95% of true precocious puberty cases are idiopathic.[72] Therefore, there must be some underlying cause, and any substance that can act centrally to induce the release of GnRH could possibly be involved. Results presented herein suggest that Mn is a likely candidate for such an action. While we have focused on the effects of Mn in pre-pubertal female rats, evidence indicates that this element can cause early signs of puberty in male rats as well;[73] however, the dose of Mn required to elevate puberty-related hormones and cause precocious puberty in males was 2.5 times greater than that required for females. This suggests that females are more sensitive to the element. This sensitivity issue in females may be important with regard to identifying environmental factors contributing to the onset of puberty, as well as gaining a better understanding of potential causes of precocious puberty in females.

17.9 Conclusions

Evidence has been provided indicating that Mn may be a peripherally derived metabolic signal capable of contributing to the activation of the hypothalamus at the time of puberty. We have described the effects of low level Mn exposure to precociously activate an upstream pathway leading to increased neuronal synthesis of GnRH in the POA/RHA brain region of prepubertal rats (Figure 17.14). The most important gene with regard to regulation of GnRH at puberty is KiSS-1, which is responsible for the production of the Kp family of peptides. These neuropeptides bind to the GPR54 receptor expressed on GnRH neurons, resulting in the stimulation of GnRH neuronal activity. It has previously been shown that mTOR can regulate KiSS-1. Importantly, we have described that Mn is an upstream regulator of mTOR because it upregulates Rheb, a mediator of the nutrient signaling input to mTOR. More research will be required to determine whether this essential element acts as a nutrient signal or by some other biochemical mechanism to activate the Rheb to GnRH neuronal pathway. The fact that GnRH synthesis must keep up with release to drive puberty is important because Mn also precociously activates a pathway within the basal hypothalamus causing increased prepubertal GnRH release.

While activation of the GnRH system is critical at the normal time of puberty, and thus is beneficial, it could also be harmful if its activation occurs too early in life. In this regard, we have clearly shown, at least in laboratory animals, that prepubertal exposure to low levels of Mn is associated with the activation of this pathway, resulting in increased GnRH neuronal function. We have discussed that this Mn-induced GnRH activation results in elevated puberty-related hormones and, thus, suggest that it may be an environmental factor capable of working in concert with other metabolic signals, as well as with genetic factors, to influence normal pubertal development. In addition, we have provided evidence that prepubertal exposure to the element is associated with signs of precocious puberty in both sexes, with females appearing to be more sensitive. The potential for Mn to influence the GnRH system and pubertal development is supported by

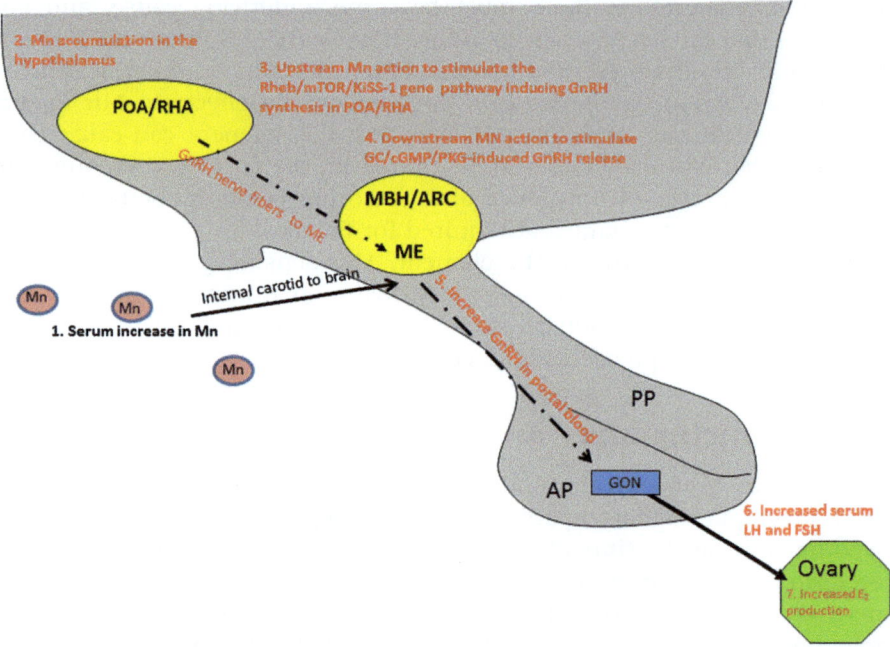

Figure 17.14 Effects of MnCl$_2$ exposure on GnRH neuronal activity. Numbers indicate the proposed order of upstream actions controlling GnRH synthesis in the preoptic area/rostral hypothalamic area (POA/RHA) and downstream actions controlling GnRH release from the medial basal hypothalamus (MBH)/median eminence (ME) area. AP, anterior pituitary; E$_2$, estradiol; FSH, follicle stimulating hormone; cGMP, cyclic guanylyl monophosphate; GC, guanylyl cyclase; GON, gonadotropin cell; KiSS-1, kisspeptin gene; LH, luteinizing hormone; mTOR, mammalian target of rapamycin; PKG, protein kinase G; PP, posterior pituitary; Rheb, ras homologue expressed in brain.
(Reproduced with permission from John Wiley and Sons in Dees *et al.* (2009).[74])

several facts. It accumulates in the reproductive hypothalamus, and infants and children are more sensitive to excess amounts of the element as a result of its minimum exposure levels not being well defined and because they do not yet have the full capacity to excrete the element. Therefore, should exposure to low levels of Mn occur prior to puberty, resulting in its modest accumulation in the hypothalamus too early in life and reaching levels not normally attained until later, then a potential risk of developing precocious puberty could occur. Epidemiological research in children and experimental studies in primates will likely be able further to address this potentially important issue concerning child health.

Acknowledgement

This work was supported by NIH/NIEHS grant ES013143 to WLD.

References

1. L. Hurley, Teratogenic aspects of manganese, zink and copper nutrition, *Physiol. Rev.*, 1981, **61**, 249.
2. J. L. Greger, Nutrition versus toxicology of manganese in humans: evaluation of potential biomarkers, *Neurotoxicology*, 1999, **20**, 205.
3. C. L. Keen, J. L. Ensunsa, M. H. Watson, D. L. Baly, S. M. Donavan and M. H. Monaco, Nutritional aspects of manganese from experimental studies, *Neurotoxicology*, 1999, **20**, 213.
4. P. H. Boyer, J. H. Shaw and P. H. Philips, Studies on manganese deficiency in the rat, *J. Biol. Chem.*, 1941, **143**, 417.
5. S. E. Smith, M. Medlicott and G. H. Ellis, Manganese deficiency in the rabbit, *Arch. Biochem. Biophys.*, 1944, **4**, 281.
6. L. E. Gray and L. W. Laskey, Multivariate analysis of the effects of manganese on the reproductive physiology and behavior of the male house mouse, *J. Toxicol. Environ. Health*, 1980, **6**, 861.
7. L. W. Laskey, J. F. Reinberg, J. F. Hein and S. D. Carter, Effects of chronic manganese (Mn_3O_4) exposure on selected reproductive parameters in rats, *J. Toxicol. Environ. Health*, 1982, **9**, 677.
8. M. Aschner, Manganese: brain transport and emerging research needs, *Environ. Health Perspect.*, 2000, **108**, 429.
9. M. Aschner and J. L. Aschner, Manganese transport across the blood brain barrier: relationship to iron homeostasis, *Brain Res. Bull.*, 1990, **24**, 857.
10. V. A. Murphy, K. C. Wadhwami, Q. R. Smith and S. I. Rapoport, Saturable transport of manganese across the rat blood brain barrier, *J. Neurochem.*, 1991, **57**, 948.
11. I. Mena, The role of manganese in human disease, *Ann. Clin. Lab. Sci.*, 1974, **4**, 487.
12. L. D. Fechter, Distribution of manganese in development, *Neurotoxicology*, 1999, **20**, 197.
13. R. Deskin, S. J. Bursain and F. W. Edens, Neurochemical alterations induced by manganese chloride in neonatal rats, *Neurotoxicology*, 1980, **2**, 65.
14. Environmental Protection Agency. (EPA Report 02-029). Health effects support document for manganese, 2002, US-EPA, Washington, DC.
15. H. F. Urbanski and S. R. Ojeda, The juvenile-peripubertal transition period in the female rat: establishment of a diurnal pattern of pulsatile luteinizing hormone secretion, *Endocrinology*, 1985, **117**, 644.
16. T. M. Plant, Puberty in primates, *The Physiology of Reproduction*, ed. E. Knobil and J. D. Neill, Raven Press, New York, 1994, ch. 41, vol. 2, p. 453.
17. H. A. Delemarre-Van De Waal, Gonadotropin and growth hormone secretion throughout puberty, *Acta Paediatr. Scand.*, 1991, **372**, 26.
18. E. Knobil, The neuroendocrine control of the menstrual cycle, *Recent Prog. Horm. Res.*, 1980, **36**, 53.

19. E. Terasawa, Hypothalamic control of the onset of puberty, *Curr. Opin. Endocrinol. Diabetes*, 1999, **6**, 44.

20. E. Terasawa and D. L. Fernandez, Neurobiological mechanisms of the onset of puberty in primates, *Endocr. Rev.*, 2001, **22**, 111.

21. J. K. Hiney, S. R. Ojeda and W. L. Dees, Insulin-like growth factor-1 (IGF-1) stimulates LHRH release from the prepubertal female median eminence in vitro, *Neuroendocrinology*, 1991, **54**, 420.

22. J. K. Hiney, V. K. Srivastava, C. L. Nyberg, S. R. Ojeda and W. L. Dees, Insulin-like growth factor-1 (IGF-1) of peripheral origin acts centrally to accelerate the initiation of female puberty, *Endocrinology*, 1996, **137**, 3717.

23. V. M. Navarro, J. M. Castellano, R. Fernandez-Fernandez, M. L. Barriero, J. Roa, J. E. Sanchez-Criado, E. Aguilar, C. Dieguez, L. Pinilla and M. Tena-Sempere, Developmental and hormonally regulated mRNA expression of KiSS-1 and its putative receptor, GPR54, in rat hypothalamus and potent LH-releasing activity of KiSS-1 peptide, *Endocrinology*, 2004a, **145**, 4565.

24. V. M. Navarro, R. Fernandez-Fernandez, J. M. Castellano, J. Roa, A. Mayen, M. L. Barreiro, F. Gaytan, E. Aguilar, L. Pinilla, C. Dieguez and M. Tena-Sempere, Advanced vaginal opening and precocious activation of the reproductive axis by KiSS-1 peptide, the endogenous ligand of GRP54, *J. Physiol.*, 2004b, **561**, 379.

25. M. D. Pine, B. Lee, R. K. Dearth, J. K. Hiney and W. L. Dees, Manganese acts centrally to stimulate LH secretion in immature female rats. A potential influence on female pubertal development, *Toxicol. Sci.*, 2005, **85**, 880.

26. H. F. Urbanski and S. R. Ojeda, Activation of LHRH release advances the onset of female puberty, *Neuroendocrinology*, 1987, **46**, 273.

27. L. E. Claypool, E. Ksuya, Y. Saitoh, F. Marzban and E. Terasawa, N-methyl D, L aspartate induces the release of luteinizing hormone-releasing hormone in the prepubertal and pubertal rhesus monkey as measured by in vivo push-pull perfusion in the stalk-median eminence, *Endocrinology*, 2000, **141**, 219.

28. V. L. Gay and T. M. Plant, EEAs stimulate LHRH release in prepubertal male rhesus monkeys, *Endocrinology*, 1987, **120**, 2289.

29. M. Shahab, C. Mastronardi, S. Seminara, W. F. Crowley, S. R. Ojeda and T. M. Plant, Increased hypothalamic GPR54 signaling: A potential mechanism for initiation of puberty in primates, *Proc. Natl. Acad. Sci. U. S. A.*, 2005, **102**, 2129.

30. M. E. Wilson, Premature elevation of insulin-like growth factor-1 advances first ovulation in rhesus monkeys, *J. Endocrinol.*, 1998, **158**, 247.

31. P. A. Lee, Disorders of puberty, *Pediatric Endocrinology*, ed. F. Lifshitz, Marcel Dekker, Inc., New York, 1996, ch. 13, pp. 175–195.

32. S. R. Ojeda, A. Lomniczi, C. Mastronardi, S. Heger, C. Roth, A. Parent, V. Matagne and A. E. Mungenast, The neuroendocrine regulation of

puberty: is the time ripe for a systems biology approach?, *Endocrinology*, 2006, **147**, 1166.

33. V. K. Srivastava, J. K. Hiney and W. L. Dees, Hypothalamic actions and interactions of alcohol and IGF-1 on the expression of glial receptor protein tyrosine phosphatase-β during female pubertal development, *Alcohol.: Clin. Exp. Res.*, 2011, **35**, 1812.

34. J. K. Hiney, V. K. Srivastava and W. L. Dees, Manganese induces IGF-1 and cyclooxygenase-2 gene expressions in the basal hypothalamus during prepubertal female development, *Toxicol. Sci.*, 2011, **121**, 389.

35. M. C. Newland and B. Weiss, Persistent effects of manganese on effortful responding and their relationship to manganese accumulation in the primate globus pallidus, *Toxicol. Appl. Pharmacol.*, 1991, **113**, 87.

36. M. C. Newland, Animal models of manganese neurotoxicity, *Neurotoxicology*, 1999, **20**, 415.

37. V. Rettori, N. Belova, W. L. Dees, C. L. Nyberg, M. Gimeno and S. M. McCann, Role of nitric oxide in the control of LHRH release in vivo and in vitro, *Proc. Natl. Acad. Sci. U. S. A.*, 1993, **90**, 10130.

38. F. Murad, Regulation of cytosolic guanylyl cyclase by nitric oxide: the NO cyclic-GMP signal transduction system, *Adv. Pharmacol.*, 1994, **26**, 19.

39. B. Lee, J. K. Hiney, M. D. Pine, V. K. Srivastava and W. L. Dees, Manganese stimulates luteinizing hormone releasing hormone secretion in prepubertal female rats: hypothalamic site and mechanism of action, *J. Physiol.*, 2007, **578**, 765.

40. Y. Zhao, P. E. Brandish, M. DiValentin, J. P. M. Schelvis, G. T. Babcock and M. A. Marletta, Inhibition of soluable guanylate cyclase by ODQ, *Biochemistry*, 2000, **39**, 10848.

41. F. C. Welder, Biological significance in mammalian systems, *Progress in Medicinal Chemistry*, ed. G. P. Ellis and D. K. Luscombe, Elsevier Science Publishers B. V., The Netherlands, 1993, ch. 3, pp. 89–116.

42. D. L. Garbers, Purification of soluble guanylyl cyclase from rat lung, *J. Biol. Chem.*, 1979, **254**, 240.

43. V. K. Srivastava, J. K. Hiney and W. L. Dees, Early-Life Manganese Exposure Up-regulates Tumor Associated Genes in the Hypothalamus of Female Rats: Relationship to Manganese-Induced Precocious Puberty, *Toxicol. Sci.*, 2013, **136**, 373.

44. J. H. Lee, M. E. Miele, D. J. Hicks, K. K. Phillips, J. M. Trent, B. E. Weissman and D. R. Welch, KiSS-1, a novel human malignant melanoma metastasis-suppressor gene, *J. Natl. Cancer Inst.*, 1996, **88**, 1731.

45. T. Ohtaki, Y. Shintani, S. Honda, A. Matsumoto, A. Hori, K. Kanehashi, Y. Terao, S. Kumano, Y. Takatsu, Y. Masuda, Y. Ishibashi, T. Watanabe, M. Asada, T. Yamada, M. Suenaga, C. Kitada, S. Usuki, T. Kurokawa, H. Onda, O. Nishimura and M. Fujino, Metastasis suppressor gene KiSS-1 encodes peptide ligand of a G-protein-coupled receptor, *Nature*, 2001, **411**, 613.

46. S. Messager, E. E. Chatzidaki, D. Ma, A. G. Hendrick, D. Zahn, J. Dixon, R. R. Thresher, I. Malinge, D. Lomet, M. B. Carlton, W. H. Colledge, A. Caraty and S. A. Aparicio, Kisspeptin directly stimulates gonadotropin-releasing hormone release via G protein-coupled receptor 54, *Proc. Natl. Acad. Sci. U. S. A.*, 2005, **102**, 1761.

47. K. L. Keen, F. H. Wegner, S. R. Bloom, M. A. Ghatei and E. Terasawa, An increase in kisspeptin-54 release occurs with the pubertal increase in luteinizing hormone-releasing hormone-1 release in the stalk-median eminence of female rhesus monkeys in vivo, *Endocrinology*, 2008, **149**, 4151.

48. E. L. Thompson, M. Patterson, K. G. Murphy, K. L. Smith, W. S. Dhillo, J. F. Todd, M. A. Ghatei and S. R. Bloom, Central and peripheral administration of kisspeptin-10 stimulates the hypothalamic-pituitary-gonadal axis, *J. Neuroendocrinol.*, 2004, **16**, 850.

49. N. de Roux, E. Genin, J. Carel, F. Matsuda, J. Chaussin and E. Milgrom, Hypogonadotropic hypogonadism due to loss of function of the KiSS1-derived peptide receptor GRP54, *Proc. Natl. Acad. Sci. U. S. A.*, 2003, **100**, 10972.

50. S. B. Seminara, S. Messager, E. E. Chatzidaki, R. R. Thresher, J. S. Acierno, J. K. Shagoury, Y. Bo-Abbas, W. Kuohung, K. M. Schwinof, A. G. Hendrick, D. Zahn, J. Dixon, U. B. Kaiser, S. A. Slaugenhaupt, J. F. Gusella, S. O'Rahilly, M. B. Carlton, W. F. Crowley Jr, S. A. Aparicio and W. H. Colledge, The GPR54 gene as a regulator of puberty, *N. Engl. J. Med.*, 2003, **349**, 1614.

51. J. T. Smith, C. M. Clay, A. Caraty and I. J. Clark, KiSS-1 messenger ribonucleic acid expression in the hypothalamus of the ewe is regulated by sex steroids and season, *Endocrinology*, 2007, **148**, 1150.

52. J. K. Hiney, V. K. Srivastava, M. D. Pine and W. L. Dees, Insulin-like growth factor-1 activates KiSS-1 gene expression in the brain of the prepubertal female rat, *Endocrinology*, 2009, **50**, 376.

53. J. K. Mueller, A. Dietzel, A. Lomniczi, A. Loche, K. Tefs, W. Kiess, T. Danne, S. R. Ojeda and S. Heger, Transcriptional regulation of the human *KiSS1* gene, *Mol. Cell. Endocrinol.*, 2011, **342**, 8.

54. C. L. Roth, C. Mastronardi, A. Lomniczi, H. Weight, R. Cabrera, A. E. Mungenast, S. Heger, H. Jung, C. Dubay and S. R. Ojeda, Expression of a tumor-related gene network increases in the mammalian hypothalamus at the time of female puberty, *Endocrinology*, 2007, **148**, 5147.

55. C. Mayer, M. Acosta-Martinez, S. L. Dubois, A. Wolfe, S. Radovick, U. Boehm and J. E. Levine, Timing and completion of puberty in female mice depend on estrogen receptor α-signaling in kisspeptin neurons, *Proc. Natl. Acad. Sci. U. S. A.*, 2010, **107**, 22693.

56. J. T. Smith, M. J. Cunningham, E. F. Rissman, D. K. Clifton and R. A. Steiner, Regulation of KiSS-1 gene expression in the brain of the female mouse, *Endocrinology*, 2005, **146**, 3686.

57. X. Jiang, R. L. Elliott and J. F. Head, Manipulation of ion transporter genes results in suppression of human and mouse mammary adenocarcinomas, *Anticancer Res.*, 2010, **30**, 759.

58. A. R. Tee, B. D. Manning, P. P. Roux, L. C. Cantley and J. Blenis, Tuberous Sclerosis complex gene products, tuberin and hamartin, control mTOR signaling by acting as a GTPase-activating protein complex toward Rheb, *Curr. Biol.*, 2003, **13**, 1259.

59. J. Roa, D. Garcia-Giliano, L. Varela, M. A. Sanchez-Garrido, R. Pineda, J. M. Castellano, F. Ruiz-Pino, M. Romero, E. Aguilar, M. Lopez, F. Gaytan, C. Dieguez, L. Pinilla and M. Tena-Sempere, The mammalian target of rapamycin as novel central regulator of puberty onset via modulation of hypothalamic Kiss1 system, *Endocrinology*, 2009, **150**, 5016.

60. J. Avruch, X. Long, S. Ortiz-Vega, J. Rapley, A. Papageorgiou and N. Dai, Amino acid regulation of TOR complex 1, *Am. J. Physiol.: Endocrinol. Metab.*, 2009, **296**, 592.

61. D. H. Kim, D. D. Sarbassov, S. M. Ali, J. E. King, R. R. Latek, H. Erdjument-Bromage, P. Tempst and D. M. Sabatini, mTOR interacts with raptor to form a nutrient-sensitive complex that signals to the cell growth machinery, *Cell*, 2002, **110**, 163.

62. D. H. Kim, D. D. Sarbassov, S. M. Ali, R. R. Latek, K. V. Guntur, H. Erdjument-Bromage, P. Tempst and D. M. Sabatini, GbetaL, a positive regulator of rapamycin-sensitive pathway required for the nutrient-sensitive interaction between raptor and mTOR, *Mol. Cell*, 2003, **11**, 895.

63. R. Loewith, E. Jacinto, S. Wullschleger, A. Lorberg, J. L. Crespo, D. Bonenfant, W. Oppliger, P. Jenoe and M. N. Hall, Two TOR complexes, only one of which is rapamycin sensitive, have distinct roles in cell growth control, *Mol. Cell*, 2002, **10**, 457.

64. H. Gunshin, B. Mackenzie, U. V. Berger, Y. Gunshin, M. F. Romero, W. F. Boron, S. Nussberger, J. L. Gollan and M. A. Hediger, Cloning and characterization of mammalian proton-coupled metal ion transporter, *Nature*, 1997, **388**, 482.

65. J. R. Burdo, S. L. Menzies, I. A. Simpson, L. M. Garrick, M. D. Garrick, K. G. Dolan, D. J. Haile, J. L. Beard and J. R. Connor, Distribution of divalent metal transporter 1 and metal transport protein 1 in the normal and Belgrade rat, *J. Neurosci. Res.*, 2001, **66**, 1198.

66. Y. Michotte, D. L. Massart, A. Lowenthal, L. Knaepen, J. Pelsmaekers and M. Collard, A morphological and chemical study of calcification of the choroid plexus, *J. Neurol.*, 1977, **216**, 127.

67. N. Strazielle and J. Ghersi-Egea, Choroid plexus in the central nervous system: biology and physiopathology, *J. Neuropathol. Exp. Neurol.*, 2000, **59**, 561.

68. W. Zheng, H. Kim and Q. Zhao, Comparative toxicokinetics of manganese chloride and methylcyclopentadienyl manganese tricarbonyl (MMT) in Sprague-Dawley rats, *Toxicol. Sci.*, 2000, **54**, 295.

69. V. A. Murphy, K. C. Wadhwami, Q. R. Smith and S. I. Rapoport, Saturable transport of manganese across the rat blood brain barrier, *J. Neurochem.*, 1991, **57**, 948.

70. P. A. Herman-Giddings, E. Slora, R. C. Wasserman, C. J. Bourdony, M. V. Bhapkar and G. G. Koch, Secondary sexual characteristics and

menses in young girls seen in office practice: a study from the pediatric research office setting network, *Pediatrics*, 1997, **88**, 505.

71. A. S. Parent, G. Teilmann, A. Fuul, N. E. Skakkebaek, J. Toppaari and J. P. Bourguignon, The timing of normal puberty and the age limits of sexual precocity: Variations around the world, secular trends and changes after migration, *Endocr. Rev.*, 2003, **24**, 668.

72. R., L. Rosenfield, Puberty in the female and its disorders, *Pediatric Endocrinology*, ed. M. A. Sperling, Saunders, Philadelphia, PA, 2nd edn, 2002, ch. 16, pp. 455–518.

73. B. Lee, M. D. Pine, L. Johnson, V. Rettori, J. K. Hiney and W. L. Dees, Manganese acts centrally to activate reproductive hormone secretion and pubertal development in male rats, *Reprod. Toxicol.*, 2006, **22**, 580.

74. W. L. Dees, V. Srivastava and J. K. Hiney, Actions and interactions of alcohol and insulin-like growth factor-1 on female pubertal development, *Alcohol.: Clin. Exp. Res.*, 2009, **33**, 1847.

CHAPTER 18

A Decade of Studies on Manganese Neurotoxicity in Non-Human Primates: Novel Findings and Future Directions

TOMÁS R. GUILARTE

Department of Environmental Health Sciences, Mailman School of Public Health, Columbia University, 722 West 168th St., Room 1105-E, New York, NY 10032, USA
Email: trguilarte@columbia.edu

18.1 The Early Studies on Manganese Neurotoxicity in Non-Human Primates

James Parkinson provided the first description of idiopathic Parkinson's disease in 1817.[1] Twenty years later, John Couper described a similar but yet different parkinsonian syndrome resulting from excess exposure to manganese (Mn).[2] In the early part of the 20th century, there was an emergence of reports related to Mn-induced parkinsonism in miners, which is when the term manganism was coined.[3,4] Importantly, miners exposed to high concentrations of Mn exhibited not only parkinsonian features but a syndrome with the early presentation of what was called *locura manganica* or "manganese madness", which was the expression of psychiatric symptoms typically observed in the early stages of Mn-induced neurological dysfunction. This early stage was followed by the expression of movement abnormalities,

Issues in Toxicology No. 22
Manganese in Health and Disease
Edited by Lucio G. Costa and Michael Aschner
© The Royal Society of Chemistry 2015
Published by the Royal Society of Chemistry, www.rsc.org

including parkinsonism.[5] Therefore, Mn-induced psychiatric symptoms and parkinsonism were prevalent at high levels of exposure and, presumably, at least the parkinsonism was likely to involve dysfunction of the dopaminergic system, because degeneration of nigrostriatal dopaminergic neurons and decreased dopamine concentrations in the striatum have been shown to be the cause of the clinical expression of idiopathic Parkinson's disease.[6-8]

In the early 1990s and thereafter, new studies from different parts of the world appeared, suggesting effects of Mn exposure on neuropsychiatric symptoms and cognitive function.[9-16] Further, at that time Mn had been approved as a gasoline additive in the form of methylcyclopentadienyl manganese tricarbonyl (MMT) to increase the fuel's octane rating and to replace tetraethyl lead from gasoline.[17] The scientific community and government regulatory agencies were concerned that the widespread and long-term use of MMT in gasoline would increase the levels of Mn exposure in the general population and cause neurological effects.[18,19] One of the shortcomings in regards to the Mn neurotoxicity literature was the limited number of studies in animals and humans demonstrating the neurological consequences of chronic low-level exposure to Mn. In 1999, I initiated conversations with Dr Jay Schneider in the Department of Pathology, Anatomy and Cell Biology at Thomas Jefferson University and Drs Dean Wong and Peter Barker in the Department of Radiology at Johns Hopkins Hospital and School of Medicine to initiate studies on the neurological consequences of chronic Mn exposure. It was clear that, in order to provide scientific data that would be useful to the human condition, we needed to select an animal model that would express neurological symptoms similar to humans, and the non-human primate was the best choice. Dr Schneider is a renowned expert in behavioral assessment of cognitive and motor function deficits in non-human primates resulting from MPTP treatment and he would provide the behavioral expertise needed to perform the studies. The goal was to assess the effect of chronic exposure to Mn and follow, in a prospective fashion, the behavior and brain chemistry changes using positron emission tomography (PET), T1-weighted magnetic resonance imaging (T1w-MRI) and magnetic resonance spectroscopy (MRS). At the termination of the Mn exposure period, there would be confirmation of the *in vivo* PET studies and extensive neuropathology in the brain of the same animals. Therefore, the study was a multidisciplinary effort to understand the cognitive and motor function effects of chronic exposure to Mn and assess in parallel the *in vivo* brain chemistry changes that may be responsible for the behavioral deficits.

We recognized early on that an important aspect of the study was the route of Mn administration. The most relevant route of exposure to the human condition would be *via* inhalation. However, this route of administration was prohibitively expensive at a research university. It should be noted that, at approximately the same time, Dr David Dorman, then at CIIT Centers for Health Research in Research Triangle Park, North Carolina, had received funding by Afton Chemical Corporation for a multimillion dollar effort to expose non-human primates to Mn *via* inhalation. The studies carried out by

Dr Dorman provided the basis for important information on the pharmaco-kinetics and biological effects of inhaled Mn.[20–22] However, the design and construction of inhalation chambers for non-human primates and the exposure and long-term monitoring of Mn exposure *via* inhalation were not feasible for our studies because of the cost. After significant deliberation it was decided that the best possible exposure route besides inhalation would be *via* intravenous (i.v.) infusion. Using this approach we would know the exact amount of Mn that each animal had received and the amount of time over which it had received it. Thus, we would have highly controlled exposure conditions and Mn dose.

18.2 Early Behavioral and Neuroimaging Findings

It is important to emphasize the complexity and time-consuming nature of these studies because they involved teaching the animals a variety of cognitive tasks and tracking their motor function. Dr Jay Schneider's team at Thomas Jefferson University in Philadelphia, PA was in charge of the be-havioral studies. The teaching of the behavioral tasks to criteria takes several months for the animals to learn. Once the animals are performing the cognitive tasks at criteria, they are transferred to Johns Hopkins Medical Institutions in Baltimore, MD where they receive a series of "baseline" neuroimaging studies in order to measure specific aspects of the nigro-striatal dopaminergic system in the basal ganglia (caudate/putamen) using PET, brain metabolites using MRS, and the deposition of Mn in the brain using T1w-MRI. The latter is possible because Mn is paramagnetic and it shortens the T1-relaxation time, resulting in a hyperintense signal in the MRI image. Once the "baseline" imaging is performed, animals are returned to Thomas Jefferson University, they begin to receive Mn and their behavior is followed prospectively. The same neuroimaging studies as in the "base-line" neuroimaging set are then repeated at two additional time points during the course of Mn administration, for a total of three neuroimaging sets. Once the entire protocol is completed, the animals are euthanized and the brain harvested for extensive neuropathological assessment and con-firmation of the PET studies.

18.2.1 Behavioral Findings

In 2006, we published the first findings from a subset of animals in the first cohort, in which we describe behavioral abnormalities, dopamine system changes and the first findings of Mn effects in the cerebral cortex.[23–25] From a behavioral perspective, using cognitive and motor function assessment as well as behavioral ratings from videotape analysis, we found that by the end of the Mn exposure period, the animals developed subtle deficits in spatial working memory and decreased spontaneous activity and impaired manual dexterity.[25] We also observed that Mn-exposed animals expressed stereotypic and compulsive-like behaviors such as compulsive grooming that increased in frequency by the end of the Mn-exposure period.[25] Importantly, these behavioral changes induced by chronic Mn exposure occurred at whole blood Mn levels that were in the upper range of those reported in non-occupationally exposed human populations.[25] Three aspects of these results should be noted. First, this was the first demonstration in non-human primates that Mn exposure may be affecting working memory. Secondly, from a motor function perspective, even though Mn exposure resulted in a significant increase in the parkinsonian rating scale for non-human primates, the effect was very subtle; that is, there was a small change in the rating scale.[25] Lastly, we found that, at this level of exposure, Mn affected fine motor control, an effect that was later confirmed in humans occupationally exposed to Mn.[26]

18.2.2 Positron Emission Tomography Findings

The PET studies of the nigrostriatal dopaminergic system were a very important aspect of the experimental design because they were aimed at assessing three different components of the dopaminergic synapse in the caudate/putamen. Given that one of the most prominent effects of high exposure to Mn in humans is the expression of parkinsonism, the studies were focused on understanding the putative mechanism by which Mn exposure results in parkinsonism. One of the advantages of performing the PET studies at the Johns Hopkins Medical Institutions is that the PET Center is one of the leading PET centers in the world, with a wide variety of radioligands to interrogate a number of neuronal systems including the dopaminergic system. In fact, it was at the Johns Hopkins Medical Institutions that the first human PET study was performed to image the D2-dopamine receptor (D2R) using [^{11}C]-*N*-methylspiperone.[27] The PET studies in the Mn-exposed and control animals used [^{11}C]-methylphenidate to measure dopamine transporter (DAT) levels and [^{11}C]-raclopride with amphetamine administration. The later PET study uses a bolus plus continuous infusion of the D2R ligand [^{11}C]-raclopride and, at the end of the first 40 minutes of the PET scan, at a time when the [^{11}C]-raclopride reaches equilibrium in the brain, one can assess D2R levels.[28] Further, at exactly 40 minutes from the beginning of the [^{11}C]-raclopride infusion, a 2.0 mg kg^{-1} dose of amphetamine is slowly injected i.v. to cause the release of dopamine (DA) from dopaminergic terminals. The amphetamine-induced release of dopamine into the synaptic space displaces the binding of [^{11}C]-raclopride to D2R in the caudate/putamen. The degree of the displacement of [^{11}C]-raclopride binding to D2R as a result of amphetamine-induced DA release is a function of the DA that is released from DA terminals.[29] These studies, with a limited number of Mn-exposed animals from the entire cohort, provided the first evidence that the most significant effect of Mn exposure on the dopaminergic system was a marked decrease of *in vivo* DA release in the absence of a change in DA terminals as measured by DAT levels.[23] In this study we also showed that there were no significant differences in the levels of DAT, D2R, tyrosine hydroxylase (TH) protein (immunohistochemistry), DA concentrations or the DA metabolite homovallinic acid (HVA) in the caudate or putamen of Mn-exposed animals, relative to controls.[23] Finally, analysis of metals in various parts of the non-human primate brain showed that Mn was significantly increased not only in the caudate/putamen and globus pallidus, but also in white matter.[23] We also found that in these brain regions of Mn-exposed non-human primates there was a small but significant increase in copper concentrations with no change in zinc or iron. Consistent with the multidisciplinary nature of the studies, the metal analysis in brain tissue and blood was performed by Dr Tore Syversen in the Department of Neuroscience at the Norwegian University of Science and Technology in Trondheim, Norway.

18.2.3   Magnetic Resonance Spectroscopy and T1-Weighted Magnetic Resonance Imaging Findings

The last of the three studies published in 2006 was related to our preliminary findings in the deposition of Mn in the monkey brain using T1w-MRI and brain metabolite changes using MRS.[24] Several prominent findings were described in this study. First, prior to our Mn studies in non-human primates, there was a consensus in the literature that Mn accumulation was selective to the basal ganglia. However, our studies as well as non-human primate studies by David Dorman's group at CIIT showed that Mn accumulated not only in the basal ganglia, but, in fact, it also accumulates throughout the brain, including white matter.[20–22,24] The latter observation was important because at the time the most common way to express the distribution of Mn in the brain using T1w-MRI was the "pallidal index", defined as the ratio of the T1-signal intensity in the globus pallidus divided by the frontal white matter signal.[30] This approach assumed that Mn does not accumulate in the white matter, but in fact the new studies showed that exposure to Mn results in a significant increase in Mn concentrations in white matter.[20–22,24] Thus, the pallidal index, while originally valuable, is not as sensitive to increases in brain Mn because white matter concentrations also increase. The T1-relaxation time is a more appropriate measure for evaluating *in vivo* Mn deposition in the brain using T1w-MRI.

The analysis of brain metabolites using MRS provided the first evidence in non-human primates that cerebral cortex structures were affected by Mn exposure.[24] We found that the *N*-acetyl aspartate (NAA) to creatine (Cr) ratio (NAA:Cr) was significantly decreased in the parietal cortex of Mn-exposed animals, with a nearly significant decrease in frontal white matter.[24] These findings indicated that, at least in the parietal cortex, chronic Mn exposure produces brain metabolite changes that are indicative of neuronal death or dysfunction. NAA is a brain metabolite that is localized in neurons,[31] and studies have shown that the NAA:Cr ratio is decreased in diseases that involve neuronal death and/or dysfunction.[31–33] Therefore, these studies provided important preliminary evidence on the effects of chronic Mn exposure on cognitive and motor function, the nigrostriatal dopaminergic system and novel effects on the cerebral cortex, in particular the parietal cortex.

18.3 Chronic Mn Exposure Impairs Dopamine Neuron Function in the Striatum and Produces Extensive Degeneration in the Frontal Cortex

The preliminary studies published in 2006 and described above only used a subset of the full cohort of Mn-exposed animals. Further, while having "naïve" control animals (defined as not experiencing the behavioral and neuroimaging studies and receiving no Mn) was important, it was also essential to include another control group that we called "imaged controls". This control group experienced the same behavioral and neuroimaging studies as the Mn-exposed animals, but they were not exposed to Mn. The addition of the "imaged control" group was needed in order to correct for any potential effects that the behavioral and/or neuroimaging studies may have on the various endpoints being measured. For example, in the PET study in which we assess *in vivo* DA release, *i.e.*, the $[^{11}C]$-raclopride with amphetamine PET, amphetamine is used to release DA from synaptic terminals. Given that amphetamine is known to affect various aspects of the dopaminergic synapse,[34] this potential effect of amphetamine needed to be accounted for in our studies because it was not associated with the Mn exposure. An additional aspect of the design of our studies that should be noted is that each animal served as its own control because we have "baseline" measures of behavioral performance and baseline neuroimaging in each animal prior to Mn administration.

18.3.1 Effects of Chronic Mn Exposure on Dopaminergic Neuron Terminals in the Striatum Measured by PET and Confirmation by *Ex Vivo* Methods

As noted above, preliminary findings from a subgroup of Mn-exposed animals from the first cohort of animals provided initial evidence that chonic Mn exposure did not alter DAT levels or D2R in the caudate/putamen, but it produced a marked inhibitory effect on *in vivo* DA release.[23] In 2008, we published a follow-up study in the larger cohort of control and Mn-exposed animals that had completed the behavioral and neuroimaging protocol and *ex vivo* studies of the dopaminergic system was also performed.[35] The result from the full cohort of animals confirmed the preliminary findings that the most significant effect of chronic Mn exposure on dopaminergic terminals in the caudate/putamen was a marked inhibitory effect on *in vivo* DA release. This effect of Mn exposure was present in the absence of a change in DAT levels, a marker of dopamine terminals. Therefore, the PET studies provide clear evidence that, at the dose and length of Mn exposure used in our studies, there were significant effects of Mn on the ability of DA neurons to release DA in the absence of DA terminal degeneration.[35]

An important aspect of the studies described above is that the brains of the same animals used for the behavioral and neuroimaging studies were harvested once animals completed the entire behavioral/neuroimaging experimental protocol. This was done in order to confirm the PET results and to perform additional neurochemical and neuropathological studies. We performed quantitative receptor autoradiography at the level of the caudate/putamen for DAT, vesicular monoamine transporter type-2 (VMAT-2), D2R and D1-dopamine receptors (D1R).[35] We also performed quantitative receptor autoradiography for cannabinoid-1 receptors at the level of the caudate/putamen that included the internal and external globus pallidus.[35] Further, we performed immunohistochemistry for TH and DAT at the level of the caudate/putamen and analyzed DA and metabolite levels. The results of the *ex vivo* studies indicate that, from all possible comparisons, the only differences detected were when the Mn-exposed animals were compared to the "naïve controls". The endpoint measures that differed significantly between Mn-exposed animals and "naïve control" animals were: (1) DA and the DA metabolite DOPAC in the putamen; (2) DAT and VMAT-2 in the caudate and putamen; (3) and D2R receptors in the putamen.[35] However, although in the study we compared Mn-exposed animals to "naïve controls", this was not the most appropriate comparison because the Mn-exposed animals received amphetamine and the naïve control animals did not. Thus, a more appropriate experimental comparison was to combine the naïve controls and imaged controls as an "all controls" group or to compare the Mn-exposed animals to the "imaged control" group. The results indicate that when the Mn-exposed animals were compared to "all controls", which included the "naïve" animals and the "imaged controls" which received the amphetamine in the PET studies and no Mn, none of the comparisons were significantly different. When one examines the effect of amphetamine on the various dopaminergic endpoints in the "imaged control" group, it is clear that the levels of these endpoints are much lower in the "imaged control" animals than in the Mn-exposed animals (see Figures 3, 4 and 6 in ref. 35). This finding suggests that chronic Mn exposure seems to be "protective" against the amphetamine effect on dopaminergic markers. Although initially this may seem counter-intuitive, these findings suggest that Mn interferes with the amphetamine effect on dopaminergic markers. Relevant to this point, previous studies have shown that cocaine, a DAT antagonist, can block the uptake of Mn into dopaminergic terminals, suggesting that Mn can enter DA terminals *via* DAT.[36,37] Given that one of the effects of amphetamine is to reverse the DAT flux,[34] it is likely that Mn interferes with the DAT-mediated effects of amphetamine or, as suggested previously, Mn may interfere with DA release dynamics *via* an interaction with α-synuclein, a presynaptic protein involved in neurotransmitter release.[35] Consistent with the later, we have shown α-synuclein aggregation in brain tissue from Mn-exposed animals.[38]

The results from the PET studies and the *ex vivo* studies on dopaminergic terminals confirmed the results from the subset of animals published in 2006 that chronic Mn exposure produces marked inhibition of *in vivo* DA release in the absence of a change in DA levels, DAT, VMAT-2, and TH levels.[35] The mechanism by which Mn impairs DA release is yet to be defined, but the inhibitory effect of Mn on DA release has recently been confirmed in rodent studies.[39,40] Two recent rodent studies examining the effects of Mn exposure on the dopaminergic system using stereological cell counting have shown that the main effect of Mn is to inhibit potassium-stimulated DA release, with no change in DA tissue concentrations in the striatum or the number of DA neurons in the substantia nigra pars compacta.[39,40]

18.3.2 Effects of Chronic Mn Exposure on the Glutamatergic and GABAergic Systems

In 2006, there were studies in rodents and limited studies in non-human primates in which measures of glutamatergic and gamma aminobutyric acid (GABA)ergic neurotransmission had been examined (see ref. 41). However, much of the data was contradictory and it was an opportunity for us to assess both of these neuronal systems in our animals under highly controlled conditions and with precise knowledge of the Mn doses administered. In these studies, we measured total tissue glutamate, glycine and GABA concentrations in several basal ganglia regions, the thalamus, cerebellum and several areas in the cerebral cortex.[42] The results showed that there were no significant differences in the regional brain tissue concentrations of glutamate, glycine or GABA in Mn-exposed animals relative to controls.[42] We also performed quantitative receptor autoradiography for GABAa receptors using [3H]-muscimol, *N*-methyl-D-aspartate (NMDA) receptors using [3H]-MK-801 and glutamate transporters using [3H]-D-aspartate. Out of 23 brain regions examined, comprising the cerebral cortex, basal ganglia, hippocampus and cerebellum, the only measure found to be significant was an increase in [3H]-MK-801 binding in one area of the parietal cortex of Mn-exposed animals, suggesting an effect on the glutamatergic system. Importantly, the parietal cortex was a brain region in which we had found a significant decrease in the NAA:Cr ratio in the same animals, which was suggestive of degenerating or dysfunctional neurons.[24] Lastly, in this same study we performed glutamine synthase (GS) immunohistochemistry in the basal ganglia because GS is a Mn-containing enzyme responsible for the conversion of glutamate to glutamine. That is, GS is present in astrocytes and is involved in the glutamate–glutamine cycle responsible for the recycling of "synaptic" glutamate. We found significant decreases in GS levels in the external and internal globus pallidus of Mn-exposed animals relative to controls, with no significant effect in the caudate and putamen.[42] Further, the change in GS immunostaining was inversely proportional to the level of tissue Mn in the internal and external aspects of the globus pallidus. In summary, for the most part, we did not find significant effects of Mn exposure on glutamatergic and GABAergic markers, with the exception of NMDA receptor levels in the parietal cortex measured by [³H]-MK-801 binding and GS protein levels in the globus pallidus based on immunohistochemistry, suggesting a direct effect on astrocyte function related to glutamate metabolism.

18.3.3 The Frontal Cortex in Mn-Exposed Non-Human Primates: Alzheimer's Disease-like Pathology and Neurodegeneration

One of our initial goals for the Mn studies in non-human primates was to maximize the use of these valuable animals in order to obtain the maximal amount of information possible from a behavioral, neuroimaging and neuropathological perspective. We also wanted to cast a broad net in regards to neurological effects and not only concentrate on the basal ganglia as a target for Mn neurotoxicity but also examine other brain regions and in particular the cerebral cortex. Prior to our studies, there were limited studies that examined the cerebral cortex in the context of Mn neurological effects. To this aim, we performed gene array studies in frontal cortex tissue from Mn-exposed and control animals. The gene array work was led by Dr Kevin Becker of the Gene Expression and Genomics Unit of the National Institute on Aging.

The gene array results from the frontal cortex indicated that, from a total of 6766 unique genes represented in the array, 61 genes were significantly upregulated and four genes were significantly downregulated.[43] The genes found to be significantly changed in the frontal cortex of Mn-exposed animals relative to controls were categorized under the following biological functions: (1) amyloid precursor protein regulation; (2) apoptosis; (3) cholesterol metabolism/transport; (4) axonal/vesicular transport; (5) inflammatory/immune response; (6) cell cycle/transcription/DNA repair and biosynthesis; (7) proteasome/protein folding/protein turnover; and (8) Mn-responsive genes with no known function.[43] We found that the majority of the genes that were changed by Mn exposure were either known to be activated by p53 or their gene product interacts with p53 to modify its function. Consistent with p53 playing a central role in the Mn-induced gene changes in the frontal cortex, we found that the p53 protein was increased in the frontal cortex of Mn-exposed animals relative to controls, based on immunohistochemical staining.

From the array results we found that the most highly upregulated gene was amyloid precursor-like protein 1 (APLP1), a member of the amyloid precursor protein (APP) family, a gene whose protein proteolytic cleavage products are involved in Alzheimer's disease (AD). Immunohistochemistry confirmed that the APLP1 protein was significantly increased in frontal cortex gray and white matter from Mn-exposed animals relative to controls. Based on these findings, it became clear that we needed to investigate the possibility of β-amyloid plaques in the Mn-exposed animals, and we performed immunostaining with 6E10, a commonly used antibody for the detection of β-amyloid plaques in human tissue. Our results showed that the Mn-exposed animals expressed diffuse β-amyloid aggregates in the frontal cortex.[43,44] This was an important finding because these animals were research-naïve young adults (7–8 years of age) and they were not supposed to have any β-amyloid plaques at this age. It should be noted that diffuse β-amyloid plaques are part of the normal aging process of non-human primates and diffuse β-amyloid is observed in aged monkeys, 20 years of age or older.[45] However, our Mn-exposed animals were young, and the findings of diffuse β-amyloid plaques in the frontal cortex suggested that the Mn exposure was in a sense "accelerating" the aging process.

From a neuropathological perspective, the gene array findings and their biological functions were confirmed in Mn-exposed frontal cortex tissue. That is, we found that besides the increased APLP1 protein and diffuse β-amyloid plaques (relevant to the APP regulation gene changes in the gene array results) and increased p53 protein (relevant to the cell cycle/transcription/DNA repair and biosynthesis gene changes in the gene array results), there was a significant degree of cellular degeneration in the Mn-exposed frontal cortex that expressed apoptotic stigmata (consistent with apoptosis gene changes in the gene array results). We also found expression of axonal processes in the gray and white matter undergoing degeneration (consistent with axonal/vesicular transport gene changes in the gene array results), and β-amyloid aggregation with a later finding of α-synuclein aggregation in the frontal cortex[38] (consistent with gene changes associated with the proteasome/protein folding/protein turnover biological function in the gene array results). Finally, we also found microglia activation in the substantia nigra pars reticulata of the same Mn-exposed monkeys,[46] and although this was not assessed in the frontal cortex, it is consistent with the gene changes associated with the inflammatory/immune response category in the gene array results. It should be noted that glial cell activation was found in the frontal cortex of Mn-exposed animals in the form of astrocytosis. Therefore, the neuropathological changes documented in the frontal cortex of Mn-exposed animals are consistent with the gene expression changes in the gene array results.[43]

We also found that there were unique morphological changes in frontal cortex neurons from Mn-exposed animals, including single or multiple intracytoplasmic vacuoles, and in some cases we observed neurons with hypertrophic nuclei in which the nucleus almost completely filled the soma with little or no visible cytoplasm.[43] This was an important observation because this type of nuclear morphological transformation has been described in cortical tissue from patients with mild cognitive impairment and in the AD brain.[47] In summary, these findings provided strong evidence of significant degeneration in cellular elements in the white and gray matter in the frontal cortex from Mn-exposed animals and indicated AD-like changes as a result of Mn exposure. To our knowledge these findings were the first to implicate significant pathology in the frontal cortex and AD-like changes as a result of Mn exposure.

18.4 Behavioral Studies Reveal Significant Impairment in Working Memory and Visuospatial Paired Associative Learning in Mn-Exposed Non-Human Primates

The early cognitive function studies in a subgroup of Mn-exposed animals provided preliminary evidence that working memory was affected.[25] The results from the entire cohort of Mn-exposed animals and controls confirmed the preliminary findings, and we concluded that chronic Mn exposure has detrimental effects on working memory and that there was an inverse relationship between working memory performance and increased brain Mn concentrations.[48] The data showed that non-spatial working memory, assessed by the delay matching to sample test (DMST), was more affected than spatial working memory, assessed by variable-delay response (DVR) performance. These behavioral findings were consistent with the neuropathological studies described above in that significant cellular changes and degeneration were present in the frontal cortex of Mn-exposed animals, a brain region that plays a critical role in working memory.[49]

More recently we have described impairment of paired-associative learning (PAL), a visuospatial task using Cambridge Neuropsychologial Test Automated Battery (CANTAB) for non-human primates.[50] While this finding is based on a new cohort of animals exposed to similar levels as in the previous cohort of animals, it is important to note because of its potential significance to AD-like pathology and future directions of our work. PAL is a complex fronto-executive task that requires both visual pattern recognition and visuospatial memory, both of which are known to be dependent on the functional integrity of the frontal cortex.[51] The PAL task involves learning to associate visual stimuli with distinct spatial locations on a trial-by trial basis. This task has attentional, working memory, and executive function components. At the easiest level, a single stimulus is presented in different locations on the screen and, in the response phase, the animal must touch the stimulus in the same location in which it was originally shown. On more difficult trials (containing two or three different stimuli), each stimulus is presented consecutively with a one-second delay between presentations. After all stimuli have been presented, one of the sample stimuli is presented again in two, three, or four different locations on the screen. The animal must touch the target location in which the stimulus was originally presented in order to receive a reward. If an animal fails to complete a trial successfully, the same trial is presented again, up to six times. If it is still performed incorrectly, the trial is aborted and the system moves to the next trial (also referred to as a sequence). Each testing session consisted of 10 trials (or sequences), each at the different levels of task difficulty. The measures analyzed were the number of trials (sequences) completed on the first attempt and the average number of attempts needed to complete a trial (sequence) successfully at each level of difficulty. Importantly, this task has been shown to be sensitive for detecting symptoms of cognitive decline associated with mild cognitive impairment and the conversion to an AD phenotype.[52–55] This task was selected on the basis of our previous findings of diffuse β-amyloid plaques in the frontal cortex of Mn-exposed animals.[43,44] Thus, we could test the hypothesis that Mn exposure disrupts a complex cognitive task that is sensitive and similarly impaired in the early stages of AD.

We found that performance in the PAL task was disrupted very early in time following Mn administration to the animals.[50] Two main aspects of PAL performance were significantly affected by Mn exposure: (1) the number of attempts needed to complete a sequence successfully and (2) the percentage of sequences that were successfully completed on the first attempt. These findings indicate that Mn interferes with incremental learning of the task, in that the animals needed more attempts to complete a sequence at the intermediate and difficult levels of the task. This study showed that a task that has been previously suggested as a marker for preclinical AD[56] is disrupted in non-human primates early after the initiation of Mn administration. In general, this preliminary study which needs to be confirmed in the larger cohort of ongoing Mn-exposed animals and controls, suggests that chronic Mn exposure may initiate or accelerate disruption of molecular pathways and cellular functions that could predispose animals to AD-like cognitive dysfunction and pathology.

18.5   Novel Findings and Future Directions

The studies that we have been doing for the last decade have provided novel and significant discoveries ranging from effects of Mn exposure on *in vivo* DA release in the absence of DA terminal degeneration to neurodegenerative changes in the frontal and parietal cortex and indication of AD-like pathology with relevant cognitive deficits. These studies bring to light new questions about the neurotoxicity of Mn that have important and broad implications for human health. For example, while there are now significant data from both human occupational studies and non-human primate studies on the effects of chronic Mn exposure on neurological, neurochemical and neuropathological endpoints, there are virtually no data on how reversible these effects are if subjects are removed from the exposure. There has been one study of the San Francisco Bay Bridge welders that attempted to determine the reversibility of effects from occupational exposure to Mn-containing welding fumes.[57] However, the study was limited in the number of subjects that remained from the initial cohort. Therefore, it would be important to examine in non-human primates under highly controlled conditions which of the cognitive, motor, brain chemistry and neuropathological changes induced by chronic Mn exposure are potentially reversible once animals are removed from the Mn exposure.

Second, the findings of impaired working memory and PAL in conjunction with the neurodegenerative changes and AD-like pathology in the frontal cortex suggest that chronic Mn exposure may be a risk factor for dementia or AD later in life. Therefore, it would be important for future studies to focus on this aspect of the work by using state-of-the-art neuroimaging and behavioral methods that may be predictive of mild cognitive impairment and AD-like cognitive decline and pathology in Mn-exposed non-human primates. Further, with the advent of new enzyme-linked immunosorbent assay (ELISA) methods for the analysis of Aβ peptide fragments and oligomers in cerebrospinal fluid (CSF), one could measure in a prospective fashion the CSF levels of Aβ peptides and oligomers before and after Mn administration as a putative marker of β-amyloid deposition and aggregation in the brain. The neurological effects of Mn related to working memory function and AD-like pathology and cognitive impairment seem to occur at much lower cumulative Mn doses than are needed to produce Mn-induced parkinsonism. Therefore, it seems obvious that focusing on Mn-induced impairments in cognitive domains related to executive function, and underlying neurochemical and pathological correlates, would provide valuable information to further our understanding of the neurological consequences of chronic environmental and occupational Mn exposure in humans.

Acknowledgements

This work would not have been possible without the collaboration, support, and dedication of many colleagues and friends. Dr Jay Schneider has been an outstanding collaborator who was immediately interested in getting involved with the behavioral studies from our very first conversation. Drs Dean Wong, Peter Barker, and Susumu Mori have not only been long-time friends and colleagues but have provided valuable insights into the analysis of brain chemistry changes using different neuroimaging platforms. Dr Tore Syversen was instrumental in the analysis of different metals in blood and brain tissue. Ms Jennifer Dziedzic, MS, has been my right hand in the coordination of these studies and has spent endless hours with a great deal of dedication to organize the transport of animals, anesthesia, and performance of the neuroimaging studies and harvesting of brain tissue, data collection and analysis. A great deal of thanks and gratitude goes to the many other colleagues, doctoral students and post-doctoral fellows that have contributed significantly to the analysis of neurochemical and neuropathological endpoints during the last 10 years that we have dedicated to this project. I am immensely grateful to Dr Annette Kirshner, Program Administrator at the National Institute of Environmental Health Sciences, for the continued support of our quest to obtain funding under NIEHS grant number ES010975 and for her dedication and persistence in supporting studies that will advance our understanding of the role of the environment on the human brain in health and disease.

Dedication

This book chapter is dedicated to the memory of my mentor, colleague and friend, Dr Henry N. Wagner, Jr., MD who passed away on September 25, 2012. A preeminent scientist and founding father of nuclear medicine, he taught me about the scientific process and the dedication needed to tackle the important questions in science. It was due to his guidance and his ability to provide the freedom to take my "own fork on the road" that I understood early on in my career the value and importance of the new and powerful neuroimaging modalities that were being developed at the time. As a result of this early experience, my current research uses many of these imaging modalities to study the toxic effects of environmental pollutants, including manganese, on the brain. He was tenacious about the scientific discovery and it was on his own brain where the first PET image of a neuroreceptor was performed in 1983. This event alone opened up a new world of molecular imaging that has allowed the advancement of many fields, including neuroscience, psychiatry, neurology, and neurotoxicology, for the benefit of humankind.

References

1. J. Parkinson, *An Essay of the Shaking Palsy*, Sherwood, Neely, and Jones, London, 1817.
2. J. Couper, On the effects of black oxide of manganese when inhaled in the lungs, *Br. Ann. Med. Pharm. Vital Stat. Gen. Sci.*, 1837, **1**, 41.
3. L. Casamajor, An unusual form of mineral poisoning affecting the nervous system: manganese?, *JAMA, J. Am. Med. Assoc.*, 1913, **60**, 646.
4. M. Canavan and S. Cobb, Chronic manganese poisoning: report of a case, with autopsy, *Arch. Neurol. Psychiatry*, 1934, **32**, 501.
5. J. Rodier, Manganese poisoning in Moroccan miners, *Br. J. Ind. Med.*, 1955, **12**, 21.
6. O. Hornykiewicz, The discovery of dopamine deficiency in the parkinsonian brain, *J. Neural Transm.*, 2006, **70**, 9.
7. J. M. Savitt, V. L. Dawson and T. M. Dawson, Diagnosis and treatment of Parkinson Disease: molecules to medicine, *J. Clin. Invest.*, 2006, **116**, 1744.
8. W. Dauer and S. Przedborski, Parkinson's Disease: mechanisms and models, *Neuron*, 2003, **39**, 889.
9. D. Mergler, G. Huel, R. Bowler, A. Iregen, S. Belanger, M. Baldwin, R. Tardif, A. Smargiassi and L. Martin, Nervous system dysfunction among workers with long-term exposure to manganese, *Environ. Res.*, 1994, **64**, 151.

10. D. Mergler, M. Baldwin, S. Belanger, F. Larribe, A. Beuter, R. Bowler, M. Panisset, R. Edwards, A. de Geoffroy, M. P. Sassine and K. Hudnell, Bioindicator and exposure data for a population based study of manganese, *Neurotoxicology*, 1999, **20**, 327.

11. R. M. Bowler, D. Mergler, M. P. Sassine, Larribe and K. Hudnell, Neuropsychiatric effects of manganese on mood, *Neurotoxicology*, 1999, **20**, 367.

12. R. Lucchini, L. Sellis, D. Folli, P. Apostoli, A. Mutti, O. Vanoni, A. Iregren and L. Alessio, Neurobehavioral effects of manganese in workers from a ferroalloy plant after temporary cessation of exposure, *Scand. J. Work, Environ. Health*, 1995, **21**, 143.

13. R. Lucchini, P. Apostoli, C. Perrone, D. Placidi, E. Albini, P. Migliorati, D. Mergler, M. P. Sassine, S. Palmi and L. Alessio, Long-term exposure to "low levels" of manganese oxides and neurofunctional changes in ferroalloy workers, *Neurotoxicology*, 1999, **20**, 287.

14. K. A. Josephs, J. E. Ahlskog, K. J. Klos, N. Kumar, R. D. Fealey, M. R. Trenerry and C. T. Cowl, Neurologic manifestations in welders with pallidal MRI T1 hyperintensity, *Neurology*, 2005, **64**, 2033.

15. R. M. Bowler, S. Gysens, E. Diamond, S. Nakagawa, M. Drezgic and H. A. Roels, Manganese exposure: neuropsychological and neurological symptoms and effects in welders, *Neurotoxicology*, 2006, **27**, 315.

16. K. J. Klos, K. Chandler, N. Kumar, J. E. Ahiskog and K. A. Josephs, Neuropsychological profiles of manganese neurotoxicity, *Eur. J. Neurol.*, 2006, **13**, 1139.

17. M. P. Walsh, The global experience with lead in gasoline and the lessons we should apply to the use of MMT, *Am. J. Ind. Med.*, 2007, **50**, 853.

18. W. C. Cooper, The health implications of increased manganese in the environment resulting from the combustion of fuel additives: a review of the literature, *J. Toxicol. Environ. Health*, 1984, **14**, 23.

19. J. M. Davis, A. M. Jarabek, D. T. Mage and J. A. Graham, The EPA health risk assessment of methylcyclopentadienyl manganese tricarbonyl (MMT), *Risk Anal.*, 1998, **18**, 57.

20. D. C. Dorman, M. F. Struve, M. W. Marshall, C. U. Parkinson, R. A. James and B. A. Wong, Tissue manganese concentrations in young male rhesus monkeys following subchronic manganese sulfate inhalation, *Toxicol. Sci.*, 2006, **92**, 201.

21. D. C. Dorman, M. F. Struve, B. A. Wong, J. A. Dye and I. D. Robertson, Correlation of brain magnetic resonance imaging changes with pallidal manganese concentrations in rhesus monkeys following subchronic manganese inhalation, *Toxicol. Sci.*, 2006, **92**, 219.

22. M. F. Struve, B. E. McManus, B. A. Wong and D. C. Dorman, Basal ganglia neurotransmitter concentrations in rhesus monkeys following subchronic manganese sulfate inhalation, *Am. J. Ind. Med.*, 2007, **50**, 772.

23. T. R. Guilarte, M. K. Chen, J. L. McGlothan, T. Verina, D. F. Wong, Y. Zhou, M. Alexander, C. A. Rohde, T. Syversen, E. Decamp, A. J. Koser, S. Fritz, H. Gonczi, D. W. Anderson and J. S. Schneider, Nigrostriatal

dopamine system dysfunction and subtle motor deficits in manganese-exposed non-human primates, *Exp. Neurol.*, 2006, **202**, 381.

24. T. R. Guilarte, J. L. McGlothan, M. Degaonkar, M. K. Chen, P. B. Barker, T. Syversen and J. S. Schneider, Evidence for cortical dysfunction and widespread manganese accumulation in the nonhuman primate brain following chronic manganese exposure: a 1H-MRS and MRI study, *Toxicol. Sci.*, 2006, **94**, 351.

25. J. S. Schneider, E. Decamp, A. J. Koser, S. Fritz, H. Gonczi, T. Syversen and T. R. Guilarte, Effects of chronic manganese exposure on cognitive and motor functioning in non-human primates, *Brain Res.*, 2006, **1118**, 222.

26. D. M. Cowan, W. Zheng, Y. Zou, X. Shi, J. Chen, F. S. Rosenthal and Q. Fan, Manganese exposure among smelting workers: relationship between blood manganese-iron ratio and early onset neurobehavioral alterations, *Neurotoxicology*, 2009, **30**, 1214.

27. H. N. Wagner Jr., H. D. Burns, R. F. Dannals, D. F. Wong, B. Langstrom, T. Duelfer, J. J. Frost, H. T. Ravert, J. M. Links, S. B. Rosenbloom, S. E. Lukas, A. V. Kramer and M. J. Kuhar, Imaging dopamine receptors in the human brain by positron tomography, *Science*, 1983, **221**, 1264.

28. Y. Zhou, M.-K. Chen, C. J. Endres, W. Ye, J. R. Brasic, M. Alexander, A. H. Crabb, T. R. Guilarte and D. F. Wong, An extended simplified reference tissue model for the quantification of dynamic PET with amphetamine challenge, *NeuroImage*, 2006, **33**, 550.

29. M. Laruelle, Imaging synaptic neurotransmission with in vivo binding competition techniques: a critical review, *J. Cereb. Blood Flow Metab.*, 2000, **20**, 423.

30. D. Krieger, S. Krieger, O. Jansen, P. Gass, L. Theilmann and H. Lichtnecker, Manganese and chronic hepatic encephalopathy, *Lancet*, 1995, **346**, 270.

31. C. Demougeot, P. Garnier, C. Mossiat, N. Bertrand, M. Giroud, A. Beley and C. Marie, N-Acetylaspartate, a marker of both cellular dysfunction and neuronal loss: its relevance to studies of acute brain injury, *J. Neurochem.*, 2001, **77**, 408.

32. W. Block, F. Traber, S. Flacke, F. Jessen, C. Pohl and H. Schild, In vivo proton MR-spectroscopy of the human brain: assessment of N-acetylaspartate (NAA) reduction as a marker for neurodegeneration, *Amino Acids*, 2002, **23**, 317.

33. J. B. Clark, N-acetylaspartate: a marker for neuronal loss or mitochondrial dysfunction, *Dev. Neurosci.*, 1998, **20**, 271.

34. D. Sulzer, How addictive drugs disrupt presynaptic dopamine neurotransmission, *Neuron*, 2011, **69**, 628.

35. T. R. Guilarte, N. C. Burton, J. L. McGlothan, T. Verina, Y. Zhou, M. Alexander, L. Pham, M. Griswold, D. F. Wong, T. Syversen and J. S. Schneider, Impairment of nigrostriatal dopamine neurotransmission by manganese is mediated by pre-synaptic mechanism(s):

implications to manganese-induced parkinsonism, *J. Neurochem.*, 2008, **107**, 1236.

36. J. G. Anderson, P. T. Cooney and K. M. Erikson, Inhibition of DAT function attenuates manganese accumulation in the globus pallidus, *Environ. Toxicol. Pharmacol.*, 2007, **23**, 179.

37. R. T. Ingersoll, E. B. Montgomery Jr. and H. V. Aposhian, Central nervous system toxicity of manganese. II: Cocaine or reserpine inhibit manganese concentration in the rat brain, *Neurotoxicology*, 1999, **20**, 467.

38. T. Verina, J. S. Schneider and T. R. Guilarte, Manganese exposure induces α-synuclein aggregation in the frontal cortex of non-human primates, *Toxicol. Lett.*, 2013, **217**, 177.

39. T. M. Peneder, P. Scholze, M. L. Berger, H. Riether, G. Heinze, J. Bertl, J. Bauer, E. K. Richfield, O. Hornykiewicz and C. Pifl, Chronic exposure to manganese decreases striatal dopamine turnover in human alpha-synuclein transgenic mice, *Neuroscience*, 2011, **180**, 280.

40. M. Khalid, R. A. Aoun and T. A. Mathews, Altered striatal dopamine release following a sub-acute exposure to manganese, *J. Neurosci. Methods*, 2011, **202**, 182.

41. N. C. Burton and T. R. Guilarte, Manganese neurotoxicity: lessons learned from longitudinal studies in nonhuman primates, *Environ. Health Perspect.*, 2009, **117**, 325.

42. N. C. Burton, J. A. Schneider, T. Syversen and T. R. Guilarte, Effects of chronic manganese exposure on glutamatergic and GABAergic neurotransmitter markers in the nonhuman primate brain, *Toxicol. Sci.*, 2009, **111**, 131.

43. T. R. Guilarte, N. C. Burton, T. Verina, V. V. Prabhu, K. G. Becker, T. Syversen and J. S. Schneider, Increased APLP1 expression and neurodegeneration in the frontal cortex of manganese-exposed non-human primates, *J. Neurochem.*, 2008, **105**, 1948.

44. T. R. Guilarte, APLP1, Alzheimer's-like pathology and neurodegeneration in the frontal cortex of manganese-exposed non-human primates, *Neurotoxicology*, 2010, **31**, 572.

45. N. Kimura, K. Yanagisawa, K. Terao, F. Ono, I. Sakakibara, Y. Ishii, S. Kyuwa and Y. Yoshikawa, Age-related changes of intracellular Abeta in cynomolgus monkey brains, *Neuropathol. Appl. Neurobiol.*, 2005, **31**, 170.

46. T. Verina, S. F. Kiihl, J. S. Schneider and T. R. Guilarte, Manganese exposure induces microglia activation and dystrophy in the substantia nigra of non-human primates, *Neurotoxicology*, 2011, **32**, 215.

47. M. A. Riudavets, D. Iacono, S. M. Resnick, R. O'Brien, A. B. Zonderman, L. J. Martin, G. Rudow, O. Pletnikova and J. C. Troncoso, Resistance to Alzheimer's pathology is associated with nuclear hypertrophy in neurons, *Neurobiol. Aging*, 2007, **28**, 1484.

48. J. S. Schneider, E. Decamp, K. Clark, C. Bouquio, T. Syversen and T. R. Guilarte, Effects of chronic manganese exposure on working memory in non-human primates, *Brain Res.*, 2009, **1258**, 86.

49. T. R. Guilarte, Manganese neurotoxicity: new perspectives from be-
    havioral, neuroimaging, and neuropathological studies in humans and
    non-human primates, *Front. Aging Neurosci.*, 2013, **5**, 23.
50. J. S. Schneider, C. Williams, M. Ault and T. R. Guilarte, Chronic man-
    ganese exposure impairs visuospatial associative learning in non-human
    primates, *Toxicol. Lett.*, 2013, **221**, 146.
51. A. M. Owen, B. J. Sahakian, J. Semple, C. E. Polkey and T. W. Robbins,
    Visuo-spatial short-term recognition memory and learning after tem-
    poral lobe excisions, frontal lobe excisions or amygdalo-hippo-
    campectomy in man, *Neuropsychologia*, 1995, **33**, 1.
52. P. Wang, J. Li, H. Li and S. Zhang, Differences in learning rates for item
    and associative memories between amnestic mild cognitive impairment
    and healthy controls, *Behav. Brain Funct.*, 2013, **9**, 29.
53. K. S. Fowler, M. M. Saling, E. L. Conway, J. M. Semple and W. J. Louis,
    Paired associate performance in the early detection of DAT, *J. Int.
    Neuropsychol. Soc.*, 2002, **8**, 58.
54. R. Swainson, J. R. Hodges, C. J. Galton, J. Semple, A. Michael,
    B. D. Dunn, J. L. Iddon, T. W. Robbins and B. J. Sahakian, Early detection
    and differential diagnosis of Alzheimer's disease and depression with
    neuropsychological tasks, *Dementia Geriatr. Cognit. Disord.*, 2001,
    12, 265.
55. A. C. Lee, S. Rahman, J. R. Hodges, B. J. Sahakian and K. S. Graham,
    Associative and recognition memory for novel objects in dementia:
    implications for diagnosis, *Eur. J. Neurosci.*, 2003, **18**, 1660.
56. A. D. Blackwell, B. J. Sahakian, R. Vesey, J. M. Semple, T. W. Robbins and
    J. R. Hodges, Detecting dementia: novel neuropsychological markers of
    preclinical Alzheimer's disease, *Dementia Geriatr. Cognit. Disord.*, 2004,
    17, 42.
57. R. M. Bowler, V. Gocheva, M. Harris, L. Ngo, N. Abdelouahab, J. Wilkinson,
    R. L. Doty, R. Park and H. A. Roels, Prospective study on neurotoxic effects
    in manganese-exposed bridge construction welders, *Neurotoxicology*, 2011,
    32, 596.

CHAPTER 19

Imaging Modalities for Manganese Toxicity

ULRIKE DYDAK*[a,b] AND SUSAN R. CRISWELL[c,d]

[a] School of Health Sciences, Purdue University, 550 Stadium Mall Drive, West Lafayette, Indiana 47907, USA; [b] Department of Radiology and Imaging Sciences, Indiana University School of Medicine, Indianapolis, Indiana, USA; [c] Department of Neurology, Washington University School of Medicine, St. Louis, Missouri, USA; [d] American Parkinson Disease Association Advanced Center for Parkinson Research, St. Louis, Missouri, USA
*Email: udydak@purdue.edu

19.1 Introduction

Over the past decades, the novel possibilities and available technology of *in vivo* imaging have significantly advanced the field of neuroscience, and in particular they have contributed to the study of manganese (Mn) neurotoxicity. This chapter describes several of the imaging modalities that have made an impact in Mn neurotoxicity research. Some of these imaging modalities are targeted at measuring the body burden of Mn, either indirectly as with magnetic resonance imaging (MRI), or directly with X-ray fluorescence imaging. Others are imaging morphological and biochemical changes due to Mn exposure using positron emission tomography (PET), single photon emission computed tomography (SPECT), magnetic resonance (MR) volumetry, MR diffusion weighted imaging or MR spectroscopy, thereby measuring the effect of Mn exposure.

Issues in Toxicology No. 22
Manganese in Health and Disease
Edited by Lucio G. Costa and Michael Aschner
© The Royal Society of Chemistry 2015
Published by the Royal Society of Chemistry, www.rsc.org

For each modality, the basic principle of the imaging technique will be briefly described to facilitate proper interpretation and understanding of the limitations with regard to imaging Mn neurotoxicity. This will be followed by a discussion of the main findings using that modality, and how they have shaped our understanding of Mn neurotoxicity.

19.2   Magnetic Resonance Imaging

19.2.1   Basics of MRI

MRI is a non-invasive medical imaging technique that enables the *in vivo* study of tissue, with high (~ 1 mm) resolution and good soft tissue contrast. It also provides a variety of functional assessments, such as the measurement of blood oxygenation, blood flow, metabolism, and diffusion properties. MRI makes use of the nuclear magnetic properties of hydrogen nuclei (protons), which possess an intrinsic angular momentum (spin). Given that living tissue contains a large fraction of water, the MRI signal stems primarily from the hydrogen nuclei in water molecules. Placed into an external magnetic field, the proton spins will align either with or against the magnetic field and precess about the field at a specific frequency called the Larmor frequency. The Larmor frequency depends both on the type of nucleus (in MRI mostly hydrogen) and on the strength of the magnetic field at the location of the nucleus. By irradiating the tissue with electromagnetic radiation at this particular Larmor frequency, a resonance effect can be obtained. Absorption of the radiation results in "flipping" the spins into the higher energy state, also called excitation of the spins. When the irradiation stops, the spins stochastically flip back to their ground state, emitting an exponentially decaying electromagnetic signal of the same frequency – the MRI signal. At clinical magnetic field strengths of MRI scanners, 1.5 and 3 tesla (T), the resonance frequency lies in the radiofrequency (RF) domain and is thus considered harmless to the human body.

The image contrast of different tissue types results from the different decay constants of the emitted MRI signal, which are also called relaxation times. Relaxation can be achieved through two different mechanisms, which occur simultaneously but at different rates. A spin can shed excessive energy by interaction between the spin and its atomic environment, causing it to return to its longitudinal alignment parallel to the magnetic field. The time associated with this relaxation mechanism is the spin–lattice relaxation time, $T1$. Another way for the observed signal to decay is through random spin–spin interaction, which causes the coherently precessing individual spins to dephase, thereby cancelling out their sum. The associated relaxation time is termed the spin–spin relaxation time, $T2$. External field inhomogeneities, introduced by tissue type variations, blood flow, or even air in the tissue to be imaged, cause an even faster dephasing and thus signal decay, characterized by the $T2^*$ relaxation time. In general the image intensity (contrast) of a particular tissue depends on its proton density, and on its relaxation times, $T1$, $T2$ and $T2^*$. By varying several imaging parameters, in particular the timing between repeated RF excitations, or between excitation and recording of the signal, the image contrast can be changed to depend mostly on the $T1$ relaxation time ($T1$-weighted imaging), or mostly on the $T2$ relaxation time ($T2$-weighted imaging).

19.2.2 Manganese as MRI Contrast Agent

Manganese, in the form of Mn^{2+}, is strongly paramagnetic and thus highly influences the relaxation properties of neighboring protons. It is important to note that MRI images depicting Mn deposition in tissue do so not by directly imaging the Mn nuclei, but they instead reflect its role as a contrast agent. The presence of Mn changes the magnetic relaxation properties of the hydrogen nuclei that are imaged (Figure 19.1).

As a paramagnetic contrast agent, Mn shortens both the $T1$ and the $T2$ relaxation times of the protons in water molecules.[1-6] Tissue with short $T1$ will appear bright in $T1$-weighted imaging, and thus Mn accumulation in the brain results in higher signal intensity in $T1$-weighted images (positive contrast). In $T2$-weighted images, tissue with longer $T2$ will contribute more

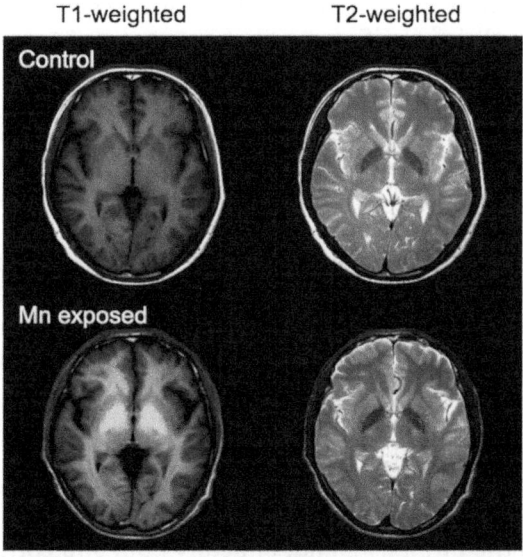

Figure 19.1 $T1$-weighted (left) and $T2$-weighted (right) MRI images of the same brain slice in an unexposed control subject (top row) and a Mn-exposed metal worker (bottom row). The bright signal in the globus pallidus and surrounding areas in the $T1$-weighted image of the exposed subject reflects increased Mn concentration. The dark signal intensity of the globus pallidus in the $T2$-weighted images reflects the effects of a large iron content in the globus pallidus (seen in both subjects) as well as the increased Mn deposition (a combined effect in the exposed subject).

to the signal and appear brighter. By shortening $T2$, Mn decreases the signal intensity in $T2$-weighted images (negative contrast). However, $T1$-weighted imaging is used more commonly to detect Mn deposition, because for most tissues $T1$-weighted imaging is somewhat more sensitive than $T2$-weighted imaging.

The relaxation times depend on other properties besides the Mn concentration alone, such as the molecular configuration in which Mn resides.[2] In addition, other paramagnetic metals, such as iron, have similar effects on MRI relaxation properties. The combined influence with Mn on MRI signal contrast is complicated, especially because Mn and iron (Fe) concentrations are not fully independent of each other in physiological systems. A model for the analysis of competitive relaxation effects of Mn and Fe *in vivo* is presented in Zhang *et al.*[7] Absolute quantification of manganese by means of MRI is therefore anything but straightforward.

19.2.2.1   *T1-Weighted MRI in Mn Toxicity Studies*

Newland *et al.* were among the first to use MRI to assess Mn deposition in a toxicity study in non-human primates,[8] reporting Mn uptake in the caudate nucleus, lenticular nuclei, substantia nigra, pituitary gland and a region corresponding to the subthalamic nucleus and ventromedial hypothalamus. Since then, a wealth of human studies have shown that increased Mn exposure can result in significant signal changes in $T1$-weighted images of Mn-exposed workers.[9–21] Similar hyperintensities are found in patients with reduced hepatobiliary excretion of Mn,[22–34] in patients receiving total parenteral nutrition (TPN),[35–41] as well as in subjects addicted to the drug methcathinone (ephedrone).[42–49]

In non-human primates, as well as in humans, Mn-induced signal changes are highest in the globus pallidus, adjacent basal ganglia regions, and the pituitary gland (see Figure 19.2), intermediate in the caudate and putamen, and lowest in other gray matter and white matter regions. However, many brain structures have by now been shown to exhibit $T1$-weighted hyperintensities, or a reduction in measured $T1$ relaxation times after Mn exposure, including the medial cerebral peduncle[50] and the olfactory bulb.[21] Moreover, quantitative evaluation of the signal intensities shows that white matter areas also accumulate Mn.[21,51,52]

In contrast to human and non-human primates, Mn-induced $T1$ hyperintensities in rodents first appear in the choroid plexus and ventricles,[53] and subsequently in the pituitary gland, olfactory bulb and cortical regions such as the hippocampus.[50,54]

$T1$-weighted MRI can also be used to monitor the efficacy of treatment or to assess the effect of cessation of Mn exposure. For example, in one report, chelation therapy resulted in a reversal of the $T1$ signal increase,[55] while other studies have reported vanishing hyperintensities within approximately six months after workers were no longer exposed in an occupational setting.[14,18] In contrast, the average half-life of Mn in the rat brain has been reported as

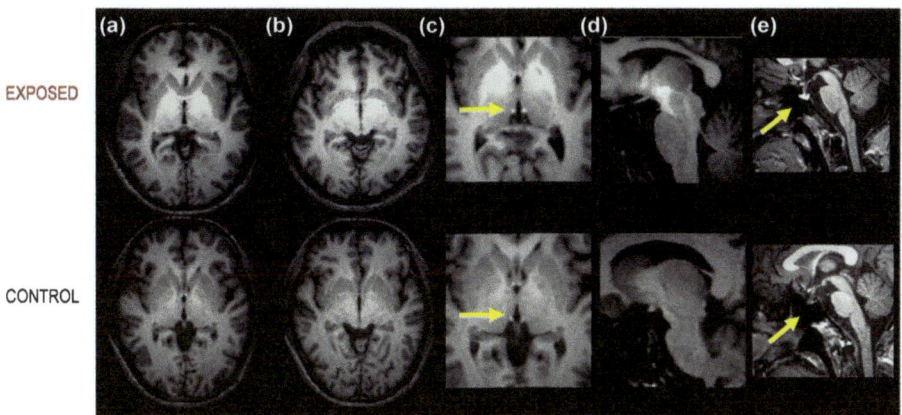

Figure 19.2 *T*1-weighted MRI images from a Mn-exposed smelter (top row) and a non-exposed control subject, showing hyperintensities in (a) the globus pallidus and part of the thalamus, (b) subthalamic nucleus, (c) pineal stalk, (d) medial cerebral peduncle and (e) pituitary gland.[50]
(Reproduced with permission from Dydak *et al.*, *Proc. Intl. Soc. Mag. Reson. Med.*, 2011, **19**, 1428).

51–74 days.[56] In non-human primates the elimination rate is suggested to be brain region specific, with an average half-life ranging from 33 days[57] to 53 days[58] after inhalational and subcutaneous Mn exposure, respectively.

19.2.2.2 The Pallidal Index

The pallidal index (PI) is a popular approach to quantifying the hyper-intensities in *T*1-weighted MRI images that correspond to increased Mn concentrations. First introduced by Krieger *et al.* in 1995,[27] the PI reflects the relative signal intensity in a *T*1-weighted image of the globus pallidus *vs.* the adjacent subcortical frontal white matter. This measure is relatively easy to obtain from MRI data, usually by placing a region of interest (ROI) in the globus pallidus and one in frontal white matter (Figure 19.3), then taking the ratio of the mean signal intensities of each ROI and multiplying by 100.

$$PI = \frac{Signal_{Global\ Pallidus}}{Signal_{Frontal\ WM}} \times 100 \qquad (19.1)$$

The PI has been used in many MRI studies on Mn neurotoxicity, both in animal models and in humans. In rats, significant correlations between the concentration of Mn in the globus pallidus, as measured by *ex vivo* techniques, and the PI measured by MRI were reported as early as 2001.[59] In 2006 Dorman *et al.*[52] presented the first comprehensive study comparing the PI, the relaxation rate *R*1 (=1/*T*1), and a histological analysis of brain Mn concentrations. This paper established linear relationships between the PI in rhesus monkeys and manganese concentrations in both the globus pallidus and in whole blood. They also reported linear relationships between the *R*1

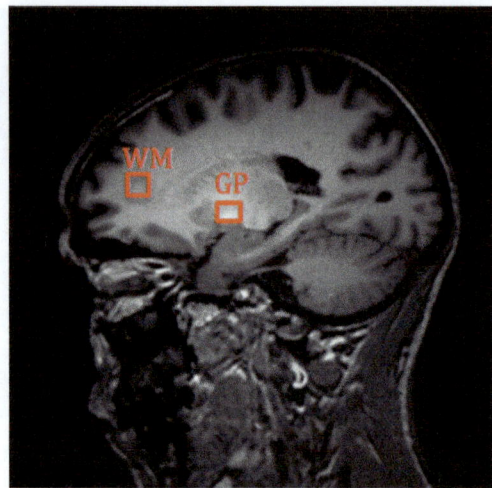

Figure 19.3 Sagittal image of a Mn-exposed worker, showing placement of ROIs in the globus pallidus and in frontal white matter for the calculation of the pallidal index.
(Reproduced with permission from Dydak *et al.*, *Environmental Health Perspectives*, 2011, **119**, 219–224).

relaxation rate and white matter Mn, pituitary Mn and globus pallidus Mn concentrations. Finding a decrease in $T1$ in white matter as well, Dorman *et al.* were the first to note that the PI may be less sensitive than direct measurement of $T1$ relaxation, because the calculation of the PI assumes that white matter is unaffected. For this reason Guilarte *et al.*[51] suggested using a "PI equivalent", using signal from pericranial muscle – which is not assumed to take up any Mn – as the reference region, and expanding the idea of the PI to many other brain regions in addition to the globus pallidus. Using the "PI equivalent", this group found higher signal intensity ratios in Mn-exposed non-human primates, not only for the globus pallidus, but also for the pituitary gland, caudate, putamen and substantia nigra.[51] Their findings confirmed that the $T1$-weighted signal in white matter also increases during Mn administration.

Early human studies already suggested that an increase in PI could document increases in brain Mn levels prior to the onset of clinical symptoms of manganism, showing correlations between the PI and blood Mn levels[15] and cumulative exposure.[10] These dependencies were confirmed in an imaging study on 111 Mn exposed workers (welders, smelters and welding rod manufacturers) conducted by Kim *et al.*,[60] whose regression model reveals significant contributions of both airborne and blood manganese to the PI. They further describe a correlation between the PI and decreased performance on neurological tests. Another study looking at neurobehavioral performance also found the PI to be a predictor for several neurobehavioral test scores such as the digit symbol, digit span backward, Stroop word, Stroop error index, and grooved pegboard scores.[61]

19.2.2.3 Relaxometry to Quantify Brain Mn

The exact value of the pallidal index is dependent on image parameters, such as resolution, slice thickness, and contrast; on the choice and reproducibility of the ROI used to calculate the signal intensity; and finally on the assumption that the reference region is not affected by Mn. A multitude of studies have compared the PI to the direct measurement of the physical MR property of $T1$ relaxation. The value of $T1$ (or the relaxation rate $R1$) is dependent on the main magnetic field strength (*e.g.* using a 1.5 T or 3 T scanner), but should be fully user independent. The MRI sequences necessary for this measurement are more complex and take longer than the scans that yield the PI, and they require sophisticated calculations to extract the $T1$ value. Nonetheless, most clinical MRI scanners today offer the necessary sequences and postprocessing tools.

Early animal studies established correlations between relaxation rates and tissue Mn concentrations.[59,62] In 2006 Dorman *et al.* pointed out that the direct measurement of the $R1$ rate is more exact than using the PI,[52] as noted above. In 2007, Choi *et al.* found in human MRI studies that the blood Mn level correlated only with the $T1$ relaxation time, but not with the PI, and that the $T1$ correlation with PI was only present at higher levels of the PI.[63] Similarly, Sen *et al.* reported group differences in $T1$ relaxation times and normalized $T1$w signal intensities (not using a ratio) in several brain regions between welders and non-exposed workers in the US, yet saw no differences in the PI.[21] Moreover, they found no correlations of the PI with fine motor measures, yet these correlations were present for the normalized signal intensities. While the PI may prove robust for higher Mn exposures that lead to clearly visible $T1$ hyperintensities in MRI images, as were observed for example in many of the early human studies, it appears that measurement of the $T1$ relaxation time is more reliable at the lower exposure settings found in newer studies, where hyperintensities are seldom seen with the naked eye due to recent regulatory limitations on Mn exposure in work settings.

Measurement of $T2$ and $T2^*$ in the substantia nigra as a means of assessing the iron concentration in this brain area has been suggested to provide a good marker of disease progression in Parkinson disease.[64] Yet to date only one study has looked at $T2^*$ in Mn exposed workers: Long *et al.* reported reduced $T2^*$ levels in the frontal cortex in welders, another indication for the involvement of the frontal cortex in Mn neurotoxicity.[16] Whether such a reduction of $T2^*$ is caused by iron deposition as a side effect of Mn exposure, or as direct effect of Mn deposition, or as a combination of both, is subject of ongoing research.

19.2.2.4 MEMRI – Mn-Enhanced MRI

In parallel to the first use of MRI to study Mn accumulation in the brains of exposed animals and humans, the research field of Mn-enhanced MRI (MEMRI) has evolved. In MEMRI, a technique first described in 1998, an injection of Mn^{2+} is used as a powerful MRI contrast agent. MEMRI has revolutionized the field of neuroscience by enabling the *in vivo* tracing of neuronal connections, monitoring neuronal activity and visualizing axonal regeneration.[65] Extensive reviews on the field of MEMRI research can be found by Silva,[66] Koretsky[67] and Inoue,[65] describing three main groups of applications. The first is neuronal tract tracing *in vivo*, first described by Pautler *et al.*,[68] which makes use of the fact that Mn^{2+} enters neurons *via* voltage gated calcium (Ca^{2+}) channels, gets released at the synapse, and is then taken up by postsynaptic neurons. This property enables MRI-detectable *in vivo* trans-synaptic tract tracing, with the paramagnetic Mn^{2+} ion providing localized enhancement of $T1$ relaxation. The second group of applications uses activity-induced Mn-enhanced MRI (AIM-MRI) as a functional MRI technique to monitor neuronal activity.[69] Given that Mn^{2+} serves as an analog for Ca^{2+}, increases in Ca^{2+} influx in response to increased neuronal activity lead to increased local concentrations of Mn^{2+} and thus can be monitored by MRI.[70] The third group of applications uses Mn^{2+} as a contrast agent to enhance anatomical detail in the visualization of neural architecture.[71,72]

Owing to the cellular toxicity of larger concentrations of Mn^{2+}, MEMRI remains confined to research in animal models, where it has been successfully applied to study small amphibians, such as frogs, as well as songbirds, rodents, and non-human primates. Since MEMRI is based on single injections of high concentrations of Mn^{2+} (up to 180 mg kg^{-1} in rodents with systemic administration), it does not serve the purpose of studying the effects of *chronic* exposure to Mn, as is of interest in environmental and occupational health sciences. However, the very same characteristics of Mn applied in MEMRI are also useful for the study of Mn deposition in the brain due to chronic, low-concentration Mn exposure, both in animals and in humans. For example MEMRI studies confirm that temporal changes of the relaxation times $T1$ and $T2$, measured in the olfactory bulb and cortex for 35 days after systemic administration of Mn, are inversely proportional to the underlying tissue Mn concentration and reflect the total amount of Mn present in the tissue.[73] The same authors also used MEMRI to detect Mn^{2+} transport from the rat olfactory bulb through appropriate brain structures to the amygdala in individual animals.[74]

In recent years, researchers using MEMRI have become more alert to the toxic effects of Mn to the brain and therefore the number of MEMRI studies investigating these toxic effects is increasing. Several MEMRI studies have looked into ways to reduce the toxicity and increase tolerance to MEMRI by studying small-dose ($\sim$30 mg kg^{-1} $MnCl_2$) fractionated injection schemes,[75] the biological half-life of Mn,[56] and the spatial distribution and time course of Mn uptake and washout in different animal models (Figure 19.4).[66] These studies in turn are of high interest for the interpretation of Mn-induced MRI signal hyperintensities in human studies with chronic exposure settings.

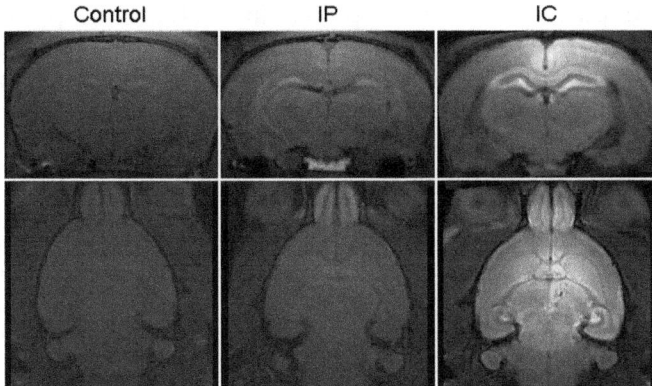

Figure 19.4 MEMRI images. Coronal (top) and axial (bottom) MEMRI images of rat hippocampus 24 h after Mn injection: control, intraperitoneal (IP) injection (MnCl$_2$, 30 mg kg^{-1}) and intracerebral (IC) injection (MnCl$_2$, 10 µl, 50 mM). Increased Mn accumulation in the hippocampus is clearly visible following both types of injection, with the IC injection of Mn resulting in increased contrast.
(Reprinted from NeuroImage 64, A. Daoust, E. L. Barbier, S. Bohic, Manganese enhanced MRI in rat hippocampus: A correlative study with synchrotron X-ray microprobe, Pages 10–18, Copyright 2013, with permission from Elsevier.)

19.2.3 Morphological Changes Assessed by MRI

In addition to the indirect measure of tissue Mn enabled by the Mn-induced contrast change in MRI images, MRI offers a variety of morphological and functional measures of physiological states. Amongst these are changes in brain region volumes (atrophy) as measured by MR volumetry, changes in metabolism as measured by magnetic resonance spectroscopy (MRS), changes in tissue diffusion properties as measured by diffusion weighted imaging (DWI), and changes in functional activity of particular brain areas as measured by functional MRI (fMRI). While these measures are not specific to Mn neurotoxicity, they have been used extensively to assess functional outcomes or to search for biomarkers of Mn neurotoxicity. These MRI methods and their findings relevant to Mn neurotoxicity are briefly discussed in the following subsections.

MR volumetry techniques, which include voxel-based morphometry (VBM), use whole-brain high resolution MRI images (usually $T1$-weighted) with good contrast between gray and white matter. The image dataset is segmented into different types of brain matter (white matter, gray matter and cerebrospinal fluid), as well as into individual brain areas. This enables the determination of total brain volume as well as volumes of individual brain regions. According to its contrast, each image pixel gets assigned a probability of belonging to gray matter, white matter or cerebrospinal fluid. In addition each pixel may also be attributed to a particular brain region, *e.g.*

the hippocampus, with a certain probability. Adding the probabilities of belonging to gray matter over all pixels within a particular brain region gives a measure of "gray matter density" for the respective brain region. Usually images are normalized to a brain atlas and undergo several smoothing and filtering steps, which then allows for the assessment of group differences in brain volumes from particular brain areas between exposed and non-exposed individuals.

While currently ongoing research is making more and more use of MR volumetry, relatively few reports of volume changes associated with Mn exposure are found in the literature. In 2011 a MEMRI study investigated long-term consequences of using Mn as contrast agent in rats. Comparing animals studied with MEMRI, *i.e.* with Mn administered on a regular basis, to animals receiving MRI scans without contrast agents over a period of six months, the authors found that the MEMRI animals showed progressive signs of cerebral toxicity, including progressive brain volume decrease as measured by MR volumetry.[76] Atrophy was observed in whole brain volume differences, as well as in regional differences between MEMRI animals and untreated animals in amygdala, hippocampus, thalamus and cortex. In Mn-exposed welders, Chang *et al.* found significantly diminished brain volumes in the globus pallidus and cerebellar regions that were associated with cognitive and fine motor performance.[77] In combination with the hyper-intensities seen in Mn-exposed workers, this measure may indicate subtle structural abnormalities in the exposed group.

A potential confounder of this method is the fact that image contrast is used to classify the tissue type of each pixel – yet Mn accumulation affects the contrast by changing the relaxation times. This may result in incorrect tissue classifications and thus wrongly calculated volumes. For example, the obvious hyperintensity in the $T1$-weighted image of the Mn-exposed worker displayed in Figure 19.1 clearly does not allow for correct segmentation of the globus pallidus using those images. In more subtle cases of Mn accumulation, thorough testing of the segmentation algorithm is necessary before volumetric results may be interpreted.

19.2.4 Magnetic Resonance Spectroscopy

Magnetic resonance spectroscopy (MRS) is based on the same physical principles as MRI. It complements MRI by providing biochemical infor-mation based on the chemical shift effect: the same nuclei can absorb and emit electromagnetic energy at slightly different resonance frequencies depending on their chemical environment in molecules. Thanks to this frequency separation (expressed as a chemical shift difference), concen-trations of different metabolites and chemical compounds can be measured in living tissue. MRS can be used with a variety of nuclei, but the proton (^{1}H) is the most common choice *in vivo*, owing to its abundance in the human body, its high intrinsic sensitivity, and the fact that MR scanner hardware is tuned by default for protons.

A typical short-echo-time ^{1}H spectrum acquired from a human brain region at 3 T features a number of important metabolites, such as *N*-acetyl aspartate (NAA, a marker for neuronal integrity), total creatine (tCr, an energy buffer and energy shuttle), choline (Cho, involved in phospholipid synthesis and degradation), myo-inositol (mI, a glial cell marker), the main excitatory neurotransmitter glutamate (Glu), and the closely related compound glutamine (Gln) (Figure 19.5). Since the MRS signals of Glu and Gln are hard to differentiate at lower field strength, the sum of Glu and Gln, known as Glx, is often reported. The spectrum also contains macromolecules and lipids, which give rise to the broad baseline underneath the prominent peaks. A variety of other neurochemical compounds, such as the main inhibitory neurotransmitter gamma aminobutyric acid (GABA) and the antioxidant glutathione (GSH), also contribute to a ^{1}H MRS spectrum, but cannot readily be measured by acquiring a standard brain spectroscopy scan at clinical magnetic field strengths. This inability is due to the low *in vivo* concentration of these compounds and the fact that their resonances are concealed by much larger peaks of other brain metabolites. To detect them, an indirect intramolecular spin–spin interaction known as J-coupling (which splits peaks into multiplets) can be exploited to select species of interest while cancelling out unwanted signal. In particular, J-difference editing

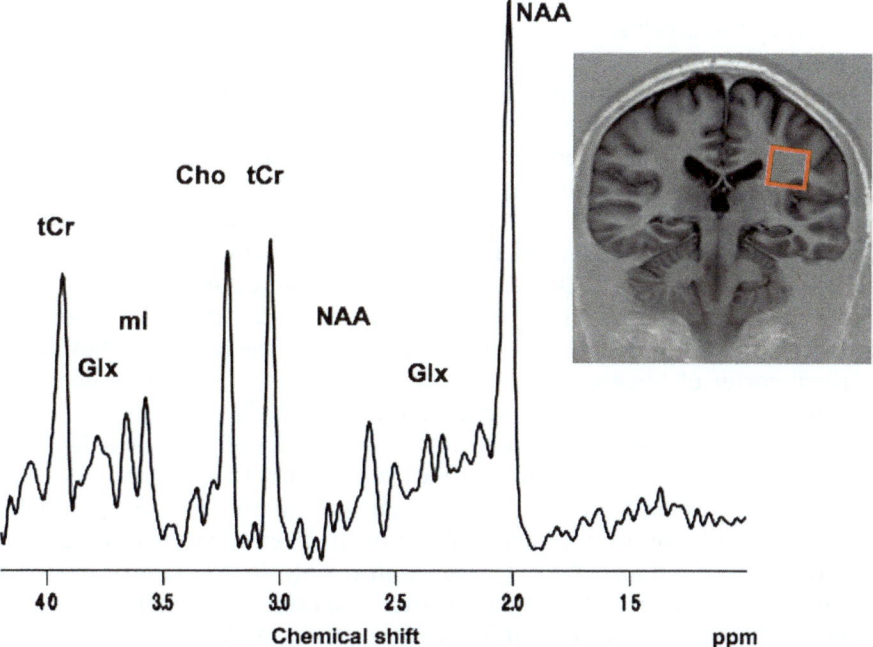

Figure 19.5 Typical spectrum acquired in parietal white matter of a human brain at 3 T, showing the signal peaks of *N*-acetyl aspartate (NAA), Glx (=Glutamate + Glutamine), total creatine (tCr), choline-containing compounds (Cho) and myo-inositol (mI).

using point resolved spectroscopy with Mescher–Garwood (MEGA) suppression (MEGA-PRESS) is a commonly used sequence for spectral editing of GABA,[78–80] revealing a GABA resonance at a chemical shift of 3.0 ppm. *In vivo* dopamine cannot be measured by MRS owing to its very low concentration, but it can be assessed by PET and SPECT as discussed later in this chapter. The higher field strength (typically 7–14 T) and better magnetic field homogeneity of animal MR scanners or nuclear magnetic resonance (NMR) spectrometers allow for quantification and identification of several additional brain metabolites in rodent brains or tissue extracts. The resulting full set of MRS-derived metabolite information is often called the neurochemical profile.[81]

The ability of MRS to make an early, non-invasive diagnosis is highly valued by clinicians, and MRS is increasingly incorporated into clinical protocols for brain examinations in selected patients.[82] However, the lack of standardized protocols, analysis techniques and reporting of results has led in many cases to a wide range of reported changes for the same disorder. Nevertheless, it is generally accepted that the characteristic MRS feature of neurodegenerative diseases such as Alzheimer disease and Parkinson disease is primarily a decrease in total NAA, reflecting the degeneration of the neurons. In addition, decreased Glu levels, an elevated Cho level and elevated mI levels have been associated with neurodegeneration in various studies.[82]

To study the mechanism of manganese neurotoxicity, MRS has been used to obtain metabolic information from Mn-exposed cell cultures and animal models, as well as occupationally exposed human workers. In an MRS study of Mn-treated cultured cells, Glu was found to decrease in neurons and neuron–astrocyte co-cultures, and decreases in mI were also observed in the co-cultures.[83] A high resolution NMR spectroscopy study on tissue extracts found a selective decrease of NAA in the globus pallidus of rats after Mn exposure, as well as decreased Gln, Cho and Glu, which was paralleled by accumulation of GABA.[84] Decreased NAA and Glu levels were reported in the hypothalamus of overnight food-suppressed rats after Mn dosing.[85] Furthermore, Guilarte *et al.* found a significant decrease of NAA in the parietal cortex of Mn-exposed non-human primates,[51] indicating decreased neuronal integrity in these Mn-treated animal models.

In humans, Kim *et al.*[86] studied the basal ganglia of welders using MRS, but did not see any significant changes in NAA:Cr, Cho:Cr and NAA:Cho ratios. Chang *et al.*[87] investigated frontal gray matter and parietal white matter and only found decreased mI:Cr in the frontal cortex of welders. In another study, reduced NAA:Cr was found only in the frontal cortex of Mn-exposed smelters, but not in the thalamus, putamen, or globus pallidus.[11] Additionally, a significant increase in GABA level, similar to the report of elevated GABA in globus pallidus tissue by Zwingmann *et al.*,[84] was observed in a larger thalamus-centered volume of interest in the same study.[11] This finding of increased thalamic GABA has been reproduced in additional populations of smelters and welders in different countries and settings;[88–90] moreover, thalamic GABA levels were recently found to predict fine motor performance in Mn-exposed workers.[89] Finally, Long *et al.* also reported reduced Glu in the frontal cortex of Mn-exposed smelters and welders *vs.* matched controls, as well as reduced mI in the thalamus and posterior cingulate cortex.[16]

The wide range of metabolite changes reported in these studies most likely can be explained by differences in Mn-exposure settings and brain regions explored, as well as different scan and analysis protocols. However, a general trend is a decrease in NAA, especially at higher exposure settings such as in animal models or the smelter population reported by Dydak *et al.*,[11] in line with dysfunction or even degeneration of neurons. Another interesting aspect is the fact that many MRS studies report metabolic changes in the frontal cortex, supporting the notion that cortical areas, in particular the frontal cortex, are involved in and vulnerable to Mn neurotoxicity.

19.2.5 Diffusion Weighted Imaging

Diffusion weighted imaging (DWI) and diffusion tensor imaging (DTI) are forms of MRI based upon the Brownian motion of water molecules (diffusion).[91] Tissue cellularity and the presence of intact cell membranes limit the diffusion of water in tissue. The overall magnitude of diffusion is measured in terms of the mean diffusivity (MD) or apparent diffusion co-efficient (ADC) while the fractional or relative anisotropy (FA or RA) represents the main vector or direction of water molecule diffusion within a voxel of tissue.[92] Changes in the MD, ADC and FA represent disruption in the movement of water secondary to changes in cellular architecture,[91] and all have been reported in association with Mn neurotoxicity. McKinney *et al.*[93] first reported restricted diffusion in the globus pallidus of a patient with Mn neurotoxicity secondary to liver failure and long-term TPN administration. Similarly, Criswell *et al.*[9] reported lower ADC values in the globus pallidus ($p = 0.04$) and anterior putamen ($p = 0.005$) of 18 Mn-exposed welders when compared to age- and sex-matched non-exposed controls. Both studies found the restricted diffusion in areas of gray matter with elevated *T*1-weighted intensity indices. Interestingly, in the study by Criswell *et al.* these changes were present in welders without symptomatic complaints and only minimal parkinsonian signs on examination.

Kim *et al.* reported a reduction in the FA of the corpus callosum and frontal white matter in Mn-exposed welders that correlated with elevated blood Mn levels, pallidal indices, and impaired neurobehavioral performance, suggesting that Mn also affects the microstructural abnormalities in white matter.[94] Specifically, digit span (backward), verbal fluency, Stroop's, and motor test outcomes were significantly associated with FA changes in the corpus callosum and frontal white matter. Finally, Stepens *et al.* reported both a diffuse decrease in white matter FA and an increase of 7% in average mean diffusivity within the bilateral globus pallidus in patients with symptomatic Mn neurotoxicity secondary to intravenous methcathinone (ephedrone) abuse.[95] This differs from the restricted diffusion reported in the basal ganglia by McKinney *et al.*[93] and Criswell *et al.*[9] It is possible the different patterns of diffusion are secondary to differences in the underlying mechanisms of Mn neurotoxicity associated with occupational exposure, liver dysfunction, and ephedrone abuse. However, Favrole *et al.*[96] provide an interesting alternative hypothesis in their study of Wilson's disease. Wilson's disease is a genetic disorder of copper metabolism resulting in a similar clinical phenotype of parkinsonism and dystonia secondary to copper deposition within the basal ganglia. They found that subjects with pre-symptomatic Wilson's disease demonstrated restricted diffusion in the putamen, whereas symptomatic Wilson's disease was associated with increased putamen ADC values. Favrole *et al.*[96] hypothesized that this may represent an inflammatory processes or gliosis with increased cellularity preceding the typical degenerative lesions usually seen at autopsy.[96] Similarly, the difference in gray matter diffusivity may represent a spectrum of evolving inflammatory and destructive lesions in Mn neurotoxicity. Further studies in asymptomatic and symptomatic Mn neurotoxicity would be required to test this hypothesis.

19.2.6   Functional MRI

Functional MRI (fMRI) is an MRI technique that images brain activity by assessing changes in blood oxygenation and flow in response to neural activity – also called the hemodynamic response.[97–99] It is based on the fact that active brain areas will consume more oxygen, and that the regional blood flow is increased to meet this increased oxygen need. Given that oxygenated (Hb) and deoxygenated hemoglobin (dHb) have different magnetic properties, being paramagnetic and diamagnetic, respectively, the replacement of dHb by Hb will create a short (~ 5 s) increase in signal upon a stimulus, which can be detected by specialized MRI sequences and data processing. The fMRI images are usually presented in form of brain activation maps that show the brain areas that are active in response to a particular task or mental process. These maps are quantitative and thus can show whether certain brain regions are recruited less or more in a particular population.

   Two studies have been conducted to date on Mn-exposed human subjects using fMRI to evaluate functional correlates of Mn-induced brain dysfunction. A first study investigated motor-related brain activation in Mn-exposed welders. Using a finger-tapping paradigm, the authors found increased activation of the primary sensorimotor cortex, bilateral supplementary motor area (SMA), bilateral premotor cortex, bilateral superior parietal cortex and ipsilateral dentate nucleus in the exposed group.[100] Furthermore, motor behavior, as measured by the Grooved Pegboard test of the right hand, correlated with the bilateral activation signal in the SMA obtained during finger tapping of the right hand.[100] The second study used the 2-back verbal memory task to look at neural correlates of working memory alterations due to Mn exposure in the same cohort of welders.[101] While task performance showed no difference between exposed and non-exposed groups, the working memory networks of welders were significantly more activated and welders recruited additional brain regions for this task, such as the inferior frontal cortex, basal ganglia (including the putamen) and the cerebellum. Again, correlations between brain activity and cognitive testing outside the MRI scanner were found.[101] In summary these two studies not only confirm subclinical deficits in motor and cognitive function due to Mn exposure, but also suggest that Mn-exposed subjects activate common networks to a higher degree, and engage additional brain networks to perform the same task when compared with non-exposed subjects, possibly as a compensatory mechanism.

19.3 PET and SPECT Imaging

Positron emission tomography (PET) is a non-invasive functional imaging technique used to create three-dimensional images of the biochemical processes in the living brain. The PET system detects pairs of gamma rays indirectly emitted by a tracer labeled with a positron-emitting radionuclide; common examples include carbon-11 (^{11}C) and fluorine-18 (^{18}F). Multiple tracer molecules have been developed that target both general and specific biochemical sites including those within the pre- and postsynaptic dopaminergic nerve terminal. Fluorodeoxyglycose (^{18}F-FDG) is a general marker of tissue glucose uptake used in cancer screening and cerebral metabolisms studies.[102,103] Commonly used dopamine specific presynaptic radiotracers include tagged enzyme substrates such as 6-[^{18}F]fluoro-L-dopa and [β-^{11}C]-L-dopa (reflects aromatic L-amino acid decarboxylase activity[104]) and molecules with an affinity for dopamine presynaptic reuptake sites (also called dopamine transporter or DAT sites) including [^{11}C]-nomifensin, [^{11}C]-methylphenidate, and [^{11}C]-WIN.[105] Postsynaptic studies of the dopaminergic system commonly utilize tagged dopamine receptor ligands to measure dopamine receptor density. Studies in Mn neurotoxicity have primarily used the radiotracer [^{11}C]-raclopride which has an affinity for D2 post-synaptic dopamine receptors.[105] By combining radiotracer studies with specific biochemical targets, PET imaging can be used to pinpoint functional areas within pathways affected by neurodegenerative diseases and neurotoxic agents such as Mn to elucidate the underlying pathological mechanisms.

Just as with PET, single photon emission computed tomography (SPECT) may be used to assess dopamine neuron terminal markers. SPECT is similar to PET in its use of radioligands, which are chosen to bind to particular targets, and its detection of gamma rays emitted by the radioligand. In contrast to PET, SPECT tracers emit gamma radiation, which is measured directly with gamma cameras, resulting in lower spatial resolution ($\sim$1 cm resolution). However, SPECT scans utilize longer-lived and more easily obtained radioisotopes, making them less expensive than PET scans.

Dopamine transporter (DAT) SPECT has been utilized to study the integrity of the nigrostriatal dopaminergic system in both idiopathic Parkinson disease and manganese-induced parkinsonism.[106–109]

19.3.1 PET Studies in Non-Human Primates

Eriksson *et al.*[110] first used PET imaging in 1992 to study two monkeys exposed to manganese oxide by subcutaneous injections over a 16 month period. Both monkeys developed increased $T1$ intensities in the caudate, putamen, globus pallidus and internal capsule on MRI imaging and clinical signs of toxicity including unsteady gait, minor clumsiness of the hands, and hypoactivity. In both monkeys, uptake of striatal [^{11}C]-nomifensin (dopamine reuptake transporters) progressively declined to reach a 60% reduction from baseline. The first animal was also scanned with [β^{11}C]-L-dopa (pre-synaptic decarboxylase activity) and expressed normal uptake throughout the striatum. The second monkey underwent [^{11}C]-raclopride PET (D2 postsynaptic receptors) with the levels transiently decreased in the early stages of Mn intoxication, but normalized by the end of the study. The combination of the [β-^{11}C]-L-dopa and [^{11}C]-nomifensin data suggested that Mn toxicity caused either a loss of presynaptic dopamine nerves with a compensatory upregulation in decarboxylase activity and dopamine turnover or no change in the number of nerve terminals with a reduction in functional reuptake transporters. In either case, no definitive conclusions could be made from this small, initial study.

Next, Shinotoh *et al.*[111] performed the combination of [^{18}F]-fluoro-L-dopa and [^{11}C]-raclopride PET with [^{18}F]-fluoro-deoxyglucose PET (cerebral metabolic rate of glucose) before and after serial administration of 10–14 mg kg^{-1} of MnCl$_2$ to three adult male rhesus monkeys. After Mn administration one monkey demonstrated elevated $T1$ signal intensity in the caudate and putamen with hypoactivity and dystonic posturing. A second monkey demonstrated similar clinical findings but normal MRI imaging. The last monkey demonstrated no clinical or imaging findings of Mn neurotoxicity. However all monkeys were reported to have cell loss and gliosis in the globus pallidus and substantial nigra pars reticularis. There was no significant alteration in the cerebral metabolic rate of glucose as measured by [^{18}F]-fluoro-deoxyglucose PET. Similarly there was no change in [^{11}C]-raclopride or [^{18}F]-fluoro-deoxyglucose, suggesting preservation of the nigrostriatal dopaminergic pathway, despite clinical deficits and neuropathic changes in the globus pallidus.

More recently, Guilarte *et al.*[112,113] performed a series of studies in 13 adult male macaques administered serial intravenous (i.v.) injections of manganese sulfate at 3.3–5.0 mg Mn kg^{-1}, 5.0–6.7 mg Mn kg^{-1}, or 8.3–10.0 mg Mn kg^{-1} for 7–59 weeks and two imaging control monkeys. PET studies including [^{11}C]-methylphenidate (dopamine reuptake transporters), [^{11}C]-raclopride (D2 postsynaptic receptors) and [^{11}C]-raclopride with amphetamine challenge (*in vivo* dopamine release) were performed at baseline and during chronic Mn exposure. Mn exposures did not affect dopamine transporters as measured by [^{11}C]-methylphenidate binding potential. The D2 postsynaptic dopamine receptors expressed a small but significant decrease (14.5%) in [^{11}C]-raclopride binding potential at 285 days post Mn administration. However, the primary finding of this study was a 51% decrease of amphetamine-induced dopamine release at 285 days post Mn administration relative to baseline.[112,113] The marked decrease in amphetamine-induced dopamine release was associated with a subtle decrease in fine motor skills and the general activity levels of the animals.[112,114] These studies also included post-mortem tissue analysis demonstrating no effect of Mn exposure on dopamine transport levels, D2 receptors, tyrosine hydroxylase levels, and dopamine or HVA (a dopamine metabolite) concentrations, indicating that the dopaminergic presynaptic terminals in the striatum were intact. Mn exposure in these animals appeared to cause changes in amphetamine-induced dopamine release without obvious nigrostriatal terminal degeneration.

Concurrently, Chen *et al.*[115] found that administration of high dose MnSO$_4$ caused an acute but transient increase in striatal dopamine transport levels measured by [^{11}C]-WIN 35,428 PET in two adult baboons. As part of this study they also demonstrated that acute Mn administration inhibits *in vitro* binding of [^{3}H]-WIN 35,428 to dopamine transporters in rat striatal membranes and uptake of [^{3}H]-DA by dopamine transporters into rat striatal synaptosomes, suggesting that the transient increase of [^{11}C]-WIN 35,428 (dopamine transporters) may be a compensatory response to its inhibitory effect on the transporter. This acute effect of Mn on dopamine transporter levels is different from the decreased DAT levels found after 16 months of repeat Mn exposure by Eriksson *et al.*[110] and the normal dopamine transporter levels described by Guilarte *et al.*[112] It seems likely the differences in these effects may be due to the duration of Mn exposure.

The PET studies from non-human primate models suggest that Mn neurotoxicity is mediated by presynaptic mechanisms that inhibit dopamine release, causing dysfunction but not degeneration of the nigrostriatal pathway. Differences between individual PET studies may be related to the variation in duration and magnitude of Mn exposure. However, the duration of exposure in all the non-human primate studies is limited (days *vs.* years) relative to typical occupational Mn exposures in humans. Therefore these PET findings may still represent early/transient changes in dopamine synapses, and the possibility of progressive nerve terminal degeneration or neuronal injury after chronic Mn exposure cannot be excluded. Further, these studies report striatal findings as a whole; individual differences between the caudate nucleus and putamen cannot be elucidated.

19.3.2   PET Studies in Human Subjects

The first PET study in Mn-exposed humans was completed by Wolters *et al.*[116] in four Taiwanese smelter workers with high levels of Mn exposure and clinical parkinsonism. All four patients exhibited masked facies, bradykinesia, and gait abnormalities with two patients demonstrating rest tremor. Initial [18F]-fluorodopa PET (presynaptic decarboxylase activity) in these subjects demonstrated normal uptake in the caudate and putamen. Eight years later, the same subjects underwent repeat [18F]-fluoro-L-dopa studies with [11C]-raclopride PET added to examine postsynaptic D2-dopamine receptors.[117] [18F]-fluoro-L-dopa uptake remained normal in all subjects despite progression of their parkinsonian symptoms. [11C]-raclopride binding was mildly reduced in the caudate and normal in the putamen, suggesting that the nigrostriatal pathway is not affected by Mn exposure.

In contrast, Kim *et al.*[118] performed [18F]-fluoro-L-dopa PET on a parkinsonian welder with a 10 year history of welding exposure and elevated T1 MR signal in the globus pallidus. [18F]-fluoro-L-dopa uptake was reduced in the left putamen. In addition, Racette *et al.*[119] describe two additional parkinsonian welders with Mn exposure. They found reduced [18F]-fluoro-L-dopa uptake across the striatum with the greatest reduction in the posterior putamen, similar to the ranges and anatomical uptake patterns of the comparison group of early idiopathic Parkinson (IPD) subjects. Together these studies argue that Mn does affect function within the presynaptic dopaminergic neurons of exposed subjects.

Two studies have reported on [18F]-fluoro-L-dopa PET in Mn neurotoxicity secondary to liver dysfunction and impaired biliary clearance. Racette *et al.*[120] reported reduced [18F]-fluoro-L-dopa PET uptake in the caudate and putamen of a patient with severe parkinsonism in the setting of alcoholic cirrhosis, elevated blood Mn levels, and increased T1-weighted MR signal in the globus pallidum. [18F]-fluoro-L-dopa uptake was reduced across the caudate, anterior, and posterior putamen within the range demonstrated by comparison subjects with IPD. However [18F]-fluoro-L-dopa uptake in the caudate nucleus appeared to be relatively more affected in the cirrhotic patient with a caudate : posterior putamen uptake ratio of 1.43 compared to the mean IPD uptake ratio of 2.86. Criswell *et al.*[121] reported on another individual with Mn neurotoxicity secondary to alcoholic cirrhosis but mild parkinsonism in the setting of elevated blood Mn levels and MR pallidal signal. [18F]-fluoro-L-dopa uptake was reduced across the striatum by more than two standard deviations (caudate, 24.7%; anterior putamen, 28.0%; posterior putamen, 29.3%) compared to healthy controls. [18F]-fluoro-L-dopa again appeared to be more evenly affected across the striatum with a caudate : posterior putamen ratio of 0.99. Both studies indicate presynaptic dysfunction in Mn neurotoxicity secondary to liver dysfunction.

Sikk *et al.*[122] performed FDG PET in four former ephedrone addicts with extrapyramidal symptoms after long-term intravenous use of ephedrone (methcathinone) contaminated with high concentrations of Mn. All patients demonstrated a widespread, but heterogeneous, pattern of reduced FDG uptake within the basal ganglia and the surrounding white matter. However, it is unclear whether these changes are specific to Mn or related to long-term ephedrone abuse.

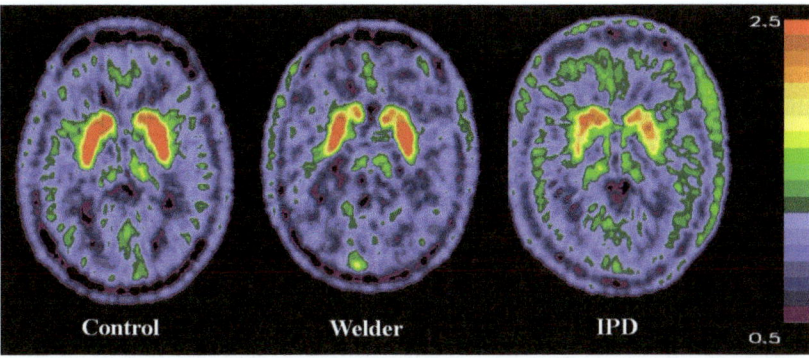

Figure 19.6 FDOPA PET composite images of decay-corrected counts from 24–94 minutes from a representative control, welder, and subject with idiopathic Parkinson disease (IPD) normalized to the reference region. FDOPA uptake is reduced in the caudate region of the welder in comparison to the control subject, while the posterior putamen is the most affected region in the subject with IPD.
(Reprinted from Neurology 76, S. Criswell, J. Perlmutter, T. Videen, S.Moerlein, H. Flores, A. Birke, B. Racette. Reduced uptake of [18F]FDOPA PET in asymptomatic welders with occupational manganese exposure, pp. 1296–1301, Copyright 2011, with permission from Wolters Kluwer Health.)

One major confound common to all these initial human PET studies on symptomatic Mn neurotoxicity is the inability to differentiate subjects with IPD or, more recently, ephedrone use and superimposed Mn exposure from those with only Mn neurotoxicity. A recent study by Criswell *et al.*[123] eliminated this confound by performing [18F]-fluoro-L-dopa PET in 20 asymptomatic welders exposed to Mn fumes. They found that caudate [18F]-fluoro-L-dopa PET uptake was reduced by 11.71% in asymptomatic welders compared to control subjects ($p \leq 0.002$); this was slightly higher than the 17% reduction seen in the symptomatic IPD subjects (Figure 19.6). This finding has significance for two reasons. First, the presence of reduced [18F]-fluoro-L-dopa caudate uptake in asymptomatic welders may represent a useful, early marker of Mn neurotoxicity in humans. Second, the preferential effect on caudate [18F]-fluoro-L-dopa uptake was anatomically reversed from the pattern in IPD subjects (Figure 19.6). In combination with the previous studies by Racette *et al.*[120] and Criswell *et al.*,[121] this suggests that Mn in humans has a unique pattern of neurotoxicity. A previous case series collected by Bhatia and Marsden[124] report that lesions in the putamen are more likely to cause motor symptoms, while caudate lesions are more often associated with psychiatric and cognitive changes. This could explain the clinical phenotype associated with Mn neurotoxicity in which neuropsychiatric symptoms, including cognitive impairment, depression, and hallucinations, are often present before or concurrent with motor symptomatology, which would be unusual in early IPD.[125–127]

The differences between early PET studies by Wolters *et al.*[116] and subsequent studies may be related to differences in the method of Mn exposure, upgrades in scanner resolution, or more likely sample size, because these studies were not powered to detect the approximately 10% difference in [18F]-fluoro-L-dopa uptake identified by Criswell *et al.*[123] While these [18F]-fluoro-L-dopa PET studies demonstrate evidence of presynaptic dopaminergic dysfunction, [18F]-fluoro-L-dopa measurements alone cannot distinguish a neurotoxic effect on nigrostriatal neurons from a dysfunctional process causing a regulatory effect on presynaptic decarboxylase activity. In combination with the non-human primate studies by Guilarte *et al.*,[112] these findings support the hypothesis that chronic Mn exposure produces clinical neurotoxicity through presynaptic dopamine terminal dysfunction. However, determining the specific presynaptic mechanism responsible for Mn neurotoxicity uptake in humans will require further research.

19.3.3   SPECT Studies

Various ligands binding to DAT, such as Tc-TRODAT-1, [123I]-β-CIT, [123I]-FP-CIT and [123I]-ioflupane, have been used to study presynaptic dopaminergic function in Mn-exposed workers,[106–108] Mn neurotoxicity in the context of severe liver dysfunction,[33] and patients exposed to Mn through their ephedrone addiction.[46,49,122] While SPECT has shown a progressive loss of DAT density in PD patients,[109] DAT studies in Mn exposure have been mixed. An early SPECT study demonstrated significantly reduced DAT levels in the striatum of two Mn-exposed workers by [123I]-β-CIT.[108] However, a subsequent report in four patients with chronic Mn intoxication from a ferromanganese smelting plant identified only mildly reduced DAT in the putamen.[106,107] SPECT scanning in one subject with Mn accumulation secondary to severe liver disease demonstrated normal overall DAT uptake in the striatum with scattered small hypodense regions within the putamen.[33] SPECT 123I-Ioflupane (DaTscan GE) scans of ephedrone addicts with clinical symptoms of manganism report normal DAT uptake.[46,49,122] Sikk *et al.* further performed iodine-123 iobenzamine (IBZM) SPECT to measure the density of dopamine D2 receptors in their cohort of ephedrone addicts.[45] They found normal tracer uptake in the striatum, indicating preserved postsynaptic D2 receptors, which suggests that manganism in ephedrone users may be related to dopaminergic dysfunction rather than degeneration, as suggested by Guilarte.[128]

There are several possible causes for the range of SPECT results in Mn neurotoxicity. First, the differences may be secondary to the underlying etiologies of Mn neurotoxicity (occupational exposure, liver failure, and contaminated ephedrone). Second, SPECT studies to date have been limited to symptomatic individuals, therefore confounding with early IPD, liver failure, or long-term ephedrone abuse cannot be eliminated. Lastly, detecting small or subtle differences may be difficult owing to the inherent limitations in the resolution of SPECT imaging. Future SPECT studies in larger cohorts may help to further interpret SPECT imaging in these Mn-exposed cohorts and elucidate the underlying mechanisms of Mn neurotoxicity.

19.4   X-Ray Fluorescence

X-ray fluorescence (XRF) may be used for elemental imaging, which means it is a direct measure of tissue quantities of a particular chemical element, *e.g.* Mn. This type of direct measurement of tissue Mn makes it unique compared to all other imaging methods discussed in this chapter, which image Mn indirectly, as in MRI, or which image effects of Mn exposure on the body's biochemistry, as in PET and SPECT. To date XRF has found its applications in the study of Mn neurotoxicity in single cell imaging,[129–132] as well as in high-resolution (order of microns) elemental imaging of *ex vivo* brain slices of rats exposed to Mn.[133–135]

The generation of an XRF signal begins with the expulsion of a core electron (*e.g.* 1s) from the atomic shell by an incoming X-ray. The core hole produced in this process is rapidly filled by an electron from a higher orbital (*e.g.* 2p, 3p) while emitting a photon with an energy specific to the difference in orbital energy levels. Since every element has a unique orbital structure, peak locations of the resulting energy spectrum indicate the presence of a given element in the sample and peak height is proportional to the concentration of said element. To perform XRF imaging, a sample is raster scanned across a focused X-ray beam while an energy sensitive detector records the fluorescence spectrum on a pixel-by-pixel basis. Recorded data can then be fitted and elemental maps constructed. X-ray beams can be focused to several microns to provide imaging of large areas such as coronal sections of rodent or human brain. Alternately, focusing on a nanometer scale (currently down to 30 nm) allows for single cell imaging. While in principle Mn should be detectable at 30 nm resolution, practical considerations, such as image acquisition time, influence the choice of resolution. In addition to requiring a synchrotron source to perform measurements, XRF imaging of tissue is performed *ex vivo* and therefore cannot be used to study kinetics of metal uptake. Furthermore, sample drying and/or radiation damage prevents imaging of the same sample by a secondary method, such as immuno-histochemical staining.

Measuring Mn in single cells by XRF was reported as early as 2003,[129] followed by studies on Mn accumulation in dopaminergic cells of the Golgi apparatus.[131] The same authors, Carmona *et al.*,[136] recently used single cell micro-XRF to compare the cytotoxicity towards dopamine-producing cells of several environmental Mn sources: inorganic compounds of different oxidation state and solubility ($MnCl_2$, $MnSO_4$, and Mn_2O_3) and organic compounds (MMT, a gasoline additive, and maneb, a diththiocarbamate fungicide). The authors found that maneb exhibited the highest toxicity, followed by $MnCl_2$, $MnSO_4$ and MMT with intermediate toxicity, whereas the insoluble Mn_2O_3 was the least toxic compound.[136] The micro-XRF

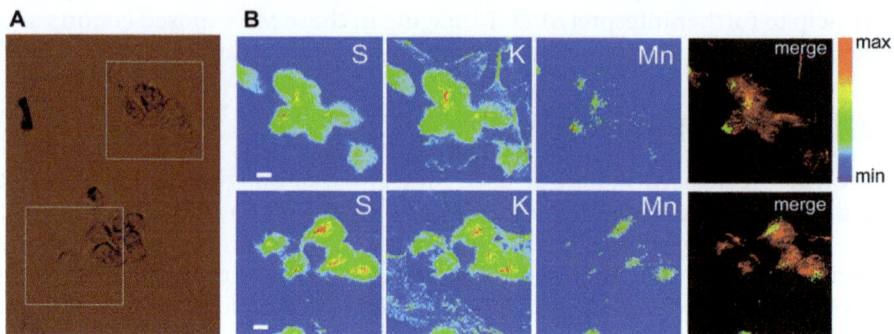

Figure 19.7   (A) Optical image of PC12 cells in culture exposed to 500 µM MnCl$_2$
during 24 h. White squares (100 µm×100 µm) indicate the two groups
of cells analyzed by micro-XRF. (B) Maps of chemical element distributions (S, K and Mn) obtained by micro-SXRF and merged image of K
(red) and Mn (green) distributions showing the perinuclear localization
of Mn. Scale bar (white bar in sulfur images): 10 µm.
(Reproduced from Carmona, A. *et al.*, *Metallomics*, 2014, **6**, 822, with
permission from the Royal Society of Chemistry.)

imaging technique used in this study enabled the creation of Mn, sulfur (S)
and potassium (K) distribution maps within single cells with a resolution
of 1 µm×1 µm, revealing a perinuclear localization of Mn in cells exposed
to MnCl$_2$ (Figure 19.7).

Recently XRF has also been used to image whole brain slices from rodents
exposed to Mn (Figure 19.8).[133–135] The penetrating nature of X-rays allows
for relatively thick sections (*e.g.* 30 µm), and the technique is sensitive to the
total metal content regardless of the binding environment. Robison *et al.*
used XRF imaging to investigate the spatial and quantitative distribution of
Mn relative to other biologically relevant metal ions, such as Fe, copper or
zinc, in brains of rats chronically exposed to Mn. They found that Mn did not
follow the distribution of any of these metals in the brain, with highest Mn
concentrations in the globus pallidus, substantia nigra compacta and the
thalamus.[135] Subsequently they investigated spatial correlations between Mn
and Zn, as well as Mn and Fe, in exposed brain slices of the hippocampal
formation both at the tissue level (40 µm×40 µm) and the cellular level
(300 nm×300 nm).[134]

Most noteworthy is the comparison of some of the high-resolution imaging techniques described in this chapter. Daoust *et al.*[133] studied the Mn
distribution within the rat hippocampus by XRF, comparing an intracranial
to an intraperitoneal route of Mn, and correlated the XRF results to the
MEMRI signal. His results are important in validating MEMRI as quantitative measure by finding a clear correlation between the Mn-enhanced MRI
contrast in MEMRI and the total amount of Mn measured by XRF, demonstrating in addition that MEMRI is sensitive to low Mn concentrations
(Figure 19.9).

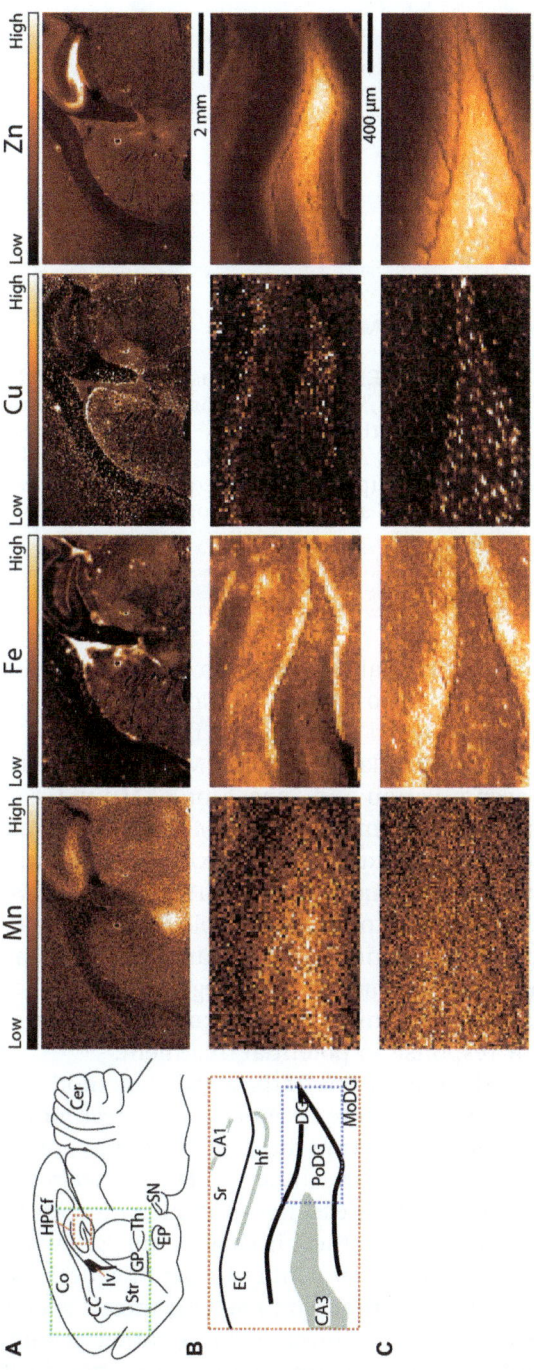

Figure 19.8 XRF images of Mn, Fe, Cu and Zn in a 30 μm thick sagittal section of the brain of a rodent chronically exposed to Mn, showing different resolutions. (A) Diagram of the sagittal section (lateral ~1.90 mm) imaged by XRF at "tissue level" resolution (40 μm×40 μm pixel size) displayed to the right. The scale bar on the right represents a length of 2 mm, and the green dashed box in the diagram represents the approximate location of tissue level XRF imaging. CC, corpus callosum; Cer, cerebellum; Co, cortex; EP, entopeduncular nucleus; HPCf, hippocampal formation; lv, lateral ventricle; SN, substantia nigra; Str, striatum; Th, thalamus. (B) Higher resolution image (20 μm×20 μm pixel size) taken of the HPCf which demonstrates heterogeneous distribution of transition metals. The red dashed box in schematic (A) indicates the approximate location of the scan. Scale bar represents a length of 400 μm. CA1 & 3, cornus ammonis 1 & 3; DG, granular layer of the dentate gyrus; EC, entorhinal cortex; hf, hippocampal fissure; MoDG/PoDG, molecular/polymorphic layer of the dentate gyrus; Sr, striatum radiatum. (C) Highest resolution image (5 μm×5 μm pixel size) of the crest of the DG, indicated by the blue dashed box in the schematic displayed in (B). The scale bar represents a length of 150 μm. (Data courtesy of Pushkar Y, Sullivan B, and Robison G, Purdue University.)

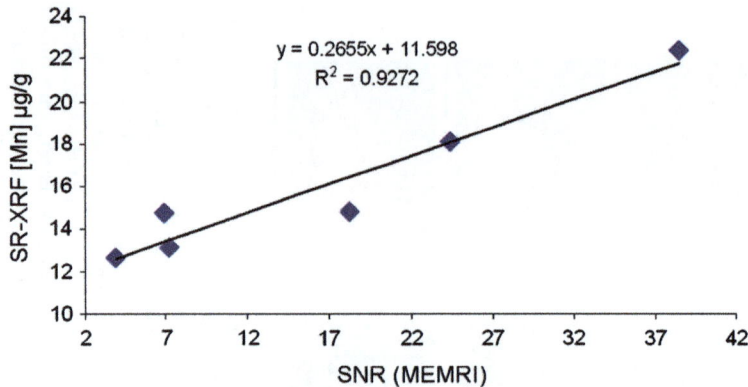

Figure 19.9   Signal to noise ratio (SNR) in MEMRI imaging as a function of Mn
concentration determined by XRF. Each point represents one animal.
The line represents a linear fit to the data.
(Reprinted from NeuroImage 64, A. Daoust, E. L. Barbier, S. Bohic,
Manganese enhanced MRI in rat hippocampus: A correlative study with
synchrotron X-ray microprobe, Pages 10–18, Copyright 2013, with
permission from Elsevier.)

19.5   Conclusions

In summary, a wide range of imaging modalities has been used successfully
to measure and study the effects of Mn exposure *in vivo*. Notably, all of these
imaging studies have focused on the brain and thus on the neurotoxic aspects of Mn. MRI (*T*1-weighted imaging, relaxometry and MEMRI) as well as
XRF may be used to study the accumulation and deposition of Mn in tissue,
ranging from millimeter resolution in humans using MRI to subcellular
micrometer resolution in *ex vivo* tissue extracts using XRF. On the other
hand, "functional" imaging techniques such as PET, SPECT, MRS, DWI,
fMRI and MR volumetry have been shown to be useful tools for studying the
toxic effects of Mn exposure on neurochemistry, neurotransmitter systems,
neuronal and axonal integrity, and ultimately neurodegeneration. These
methods allow investigation of the mechanism of neurotoxicity, and monitoring of disease progression or response to potential treatment.

   In all of these imaging studies, the limitations of the applied methods, *e.g.*
in spatial resolution or sensitivity, as well as the specificity of the techniques,
need to be taken into account when interpreting the data. Note that none of
the techniques discussed in this chapter that can be applied *in vivo* gives a
direct measurement of the quantity of interest. For example, *T*1-weighted
MRI does not measure Mn itself, but rather hydrogen nuclei, whose
relaxation properties are changed by close proximity to paramagnetic Mn
ions. To complicate matters, proximity to other paramagnetic metals such as
Fe may also contribute to the same contrast mechanism. Another example
is PET imaging, which does not directly measure neurodegeneration itself,
but rather binding at presynaptic dopamine transporter sites or postsynaptic

dopamine receptors. Finally, the choice of imaging parameters may greatly influence the variability in results from different studies, as they influence sensitivity and specificity.

Over the past few years, several very good and detailed review articles have been written about neuroimaging of Mn toxicity by Kim *et al.*[137] and Fitsanakis *et al.*[138] (who focused on MRI), as well as portions of more general reviews by Aschner[139] and Guilarte.[128] Technological developments in neuroimaging are advancing rapidly, and more sophisticated imaging methodology is becoming more widely available to study the effects of Mn neurotoxicity. This trend is likely to increase in coming years, yielding higher resolution both *in vivo* and *in vitro* or the ability to assess a wider range of functional and biochemical measures of toxicity.

Acknowledgements

The authors want to thank James B. Murdoch for careful proof reading and discussions, and Zaiyang Long, Ruoyun Ma and Chien-Lin Yeh for assistance in putting together this chapter. We further acknowledge funding from NIH R01 ES020529 (Dydak) and NIH K23ES021444-01 (Criswell).

References

1. M. H. Mendonca-Dias, E. Gaggelli and P. C. Lauterbur, Paramagnetic contrast agents in nuclear magnetic resonance medical imaging, *Semin. Nucl. Med.*, 1983, **13**, 364–376.
2. Y. S. Kang and J. C. Gore, Studies of tissue NMR relaxation enhancement by manganese. Dose and time dependences, *Invest. Radiol.*, 1984, **19**, 399–407.
3. C. F. Geraldes, A. D. Sherry, R. D. Brown, 3rd and S. H. Koenig, Magnetic field dependence of solvent proton relaxation rates induced by Gd^{3+} and Mn^{2+} complexes of various polyaza macrocyclic ligands: implications for NMR imaging, *Magn. Reson. Med.*, 1986, **3**, 242–250.
4. D. A. Cory, D. J. Schwartzentruber and B. H. Mock, Ingested manganese chloride as a contrast agent for magnetic resonance imaging, *Magn. Reson. Imaging*, 1987, **5**, 65–70.
5. D. Fornasiero, J. C. Bellen, R. J. Baker and B. E. Chatterton, Paramagnetic complexes of manganese(II), iron(III), and gadolinium(III) as contrast agents for magnetic resonance imaging. The influence of stability constants on the biodistribution of radioactive aminopolycarboxylate complexes, *Invest. Radiol.*, 1987, **22**, 322–327.
6. A. C. Silva, J. H. Lee, I. Aoki and A. P. Koretsky, Manganese-enhanced magnetic resonance imaging (MEMRI): methodological and practical considerations, *NMR Biomed.*, 2004, **17**, 532–543.
7. W. J. Zhang, Z. L. Liu and H. Shao, [Biomarkers of workers exposed to manganese], *Zhonghua Laodong Weisheng Zhiyebing Zazhi*, 2010, **28**, 926–928.

8. M. C. Newland, T. L. Ceckler, J. H. Kordower and B. Weiss, Visualizing manganese in the primate basal ganglia with magnetic resonance imaging, *Exp. Neurol.*, 1989, **106**, 251–258.

9. S. R. Criswell, J. S. Perlmutter, J. L. Huang, N. Golchin, H. P. Flores, A. Hobson, M. Aschner, K. M. Erikson, H. Checkoway and B. A. Racette, Basal ganglia intensity indices and diffusion weighted imaging in manganese-exposed welders, *Occup. Environ. Med.*, 2012, **69**, 437–443.

10. M. C. Dietz, W. Wrazidlo, A. Ihrig, M. Bader and G. Triebig, [Magnetic resonance tomography of the brain in workers with chronic occupational manganese dioxide exposure], *Rofo*, 2000, **172**, 514–520.

11. U. Dydak, Y. M. Jiang, L. L. Long, H. Zhu, J. Chen, W. M. Li, R. A. Edden, S. Hu, X. Fu, Z. Long, X. A. Mo, D. Meier, J. Harezlak, M. Aschner, J. B. Murdoch and W. Zheng, In vivo measurement of brain GABA concentrations by magnetic resonance spectroscopy in smelters occupationally exposed to manganese, *Environ. Health Perspect.*, 2011, **119**, 219–224.

12. Y. Jiang, W. Zheng, L. Long, W. Zhao, X. Li, X. Mo, J. Lu, X. Fu, W. Li, S. Liu, Q. Long, J. Huang and E. Pira, Brain magnetic resonance imaging and manganese concentrations in red blood cells of smelting workers: search for biomarkers of manganese exposure, *Neurotoxicology*, 2007, **28**, 126–135.

13. K. A. Josephs, J. E. Ahlskog, K. J. Klos, N. Kumar, R. D. Fealey, M. R. Trenerry and C. T. Cowl, Neurologic manifestations in welders with pallidal MRI T1 hyperintensity, *Neurology*, 2005, **64**, 2033–2039.

14. Y. Kim, High signal intensities on T1-weighted MRI as a biomarker of exposure to manganese, *Ind. Health*, 2004, **42**, 111–115.

15. Y. Kim, K. S. Kim, J. S. Yang, I. J. Park, E. Kim, Y. Jin, K. R. Kwon, K. H. Chang, J. W. Kim, S. H. Park, H. S. Lim, H. K. Cheong, Y. C. Shin, J. Park and Y. Moon, Increase in signal intensities on T1-weighted magnetic resonance images in asymptomatic manganese-exposed workers, *Neurotoxicology*, 1999, **20**, 901–907.

16. Z. Long, Y. M. Jiang, X. R. Li, W. Fadel, J. Xu, C. L. Yeh, L. L. Long, H. L. Luo, J. Harezlak, J. B. Murdoch, W. Zheng and U. Dydak, Vulnerability of welders to manganese exposure – A neuroimaging study, *Neurotoxicology*, 2014.

17. R. Lucchini, E. Albini, D. Placidi, R. Gasparotti, M. G. Pigozzi, G. Montani and L. Alessio, Brain magnetic resonance imaging and manganese exposure, *Neurotoxicology*, 2000, **21**, 769–775.

18. K. Nelson, J. Golnick, T. Korn and C. Angle, Manganese encephalopathy: utility of early magnetic resonance imaging, *Br. J. Ind. Med.*, 1993, **50**, 510–513.

19. A. H. Sadek, R. Rauch and P. E. Schulz, Parkinsonism due to manganism in a welder, *Int. J. Toxicol.*, 2003, **22**, 393–401.

20. K. Sato, H. Ueyama, R. Arakawa, T. Kumamoto and T. Tsuda, [A case of welder presenting with parkinsonism after chronic manganese exposure], *Rinsho shinkeigaku/Clin. Neurol.*, 2000, **40**, 1110–1115.

21. S. Sen, M. R. Flynn, G. Du, A. I. Troster, H. An and X. Huang, Manganese accumulation in the olfactory bulbs and other brain regions of "asymptomatic" welders, *Toxicol. Sci.*, 2011, **121**, 160–167.

22. R. F. Butterworth, L. Spahr, S. Fontaine and G. P. Layrargues, Manganese toxicity, dopaminergic dysfunction and hepatic encephalopathy, *Metab. Brain Dis.*, 1995, **10**, 259–267.

23. Y. Choi, J. K. Park, N. H. Park, J. W. Shin, C. I. Yoo, C. R. Lee, H. Lee, H. K. Kim, S. R. Kim, T. H. Jung, J. Park, C. S. Yoon and Y. Kim, Whole blood and red blood cell manganese reflected signal intensities of T1-weighted magnetic resonance images better than plasma manganese in liver cirrhotics, *J. Occup. Health*, 2005, **47**, 68–73.

24. R. A. Hauser, T. A. Zesiewicz, C. Martinez, A. S. Rosemurgy and C. W. Olanow, Blood manganese correlates with brain magnetic resonance imaging changes in patients with liver disease, *Can. J. Neurol. Sci.*, 1996, **23**, 95–98.

25. R. A. Hauser, T. A. Zesiewicz, A. S. Rosemurgy, C. Martinez and C. W. Olanow, Manganese intoxication and chronic liver failure, *Ann. Neurol.*, 1994, **36**, 871–875.

26. K. J. Klos, J. E. Ahlskog, K. A. Josephs, R. D. Fealey, C. T. Cowl and N. Kumar, Neurologic spectrum of chronic liver failure and basal ganglia T1 hyperintensity on magnetic resonance imaging: probable manganese neurotoxicity, *Arch. Neurol.*, 2005, **62**, 1385–1390.

27. D. Krieger, S. Krieger, O. Jansen, P. Gass, L. Theilmann and H. Lichtnecker, Manganese and chronic hepatic encephalopathy, *Lancet*, 1995, **346**, 270–274.

28. E. A. Malecki, A. G. Devenyi, T. F. Barron, T. J. Mosher, P. Eslinger, C. V. Flaherty-Craig and L. Rossaro, Iron and manganese homeostasis in chronic liver disease: relationship to pallidal T1-weighted magnetic resonance signal hyperintensity, *Neurotoxicology*, 1999, **20**, 647–652.

29. N. H. Park, J. K. Park, Y. Choi, C. I. Yoo, C. R. Lee, H. Lee, H. K. Kim, S. R. Kim, T. H. Jeong, J. Park, C. S. Yoon and Y. Kim, Whole blood manganese correlates with high signal intensities on T1-weighted MRI in patients with liver cirrhosis, *Neurotoxicology*, 2003, **24**, 909–915.

30. L. Spahr, R. F. Butterworth, S. Fontaine, L. Bui, G. Therrien, P. C. Milette, L. H. Lebrun, J. Zayed, A. Leblanc and G. Pomier-Layrargues, Increased blood manganese in cirrhotic patients: relationship to pallidal magnetic resonance signal hyperintensity and neurological symptoms, *Hepatology*, 1996, **24**, 1116–1120.

31. A. Rovira, J. Alonso and J. Cordoba, MR imaging findings in hepatic encephalopathy, *AJNR Am. J. Neuroradiol.*, 2008, **29**, 1612–1621.

32. J. Alonso, J. Cordoba and A. Rovira, Brain magnetic resonance in hepatic encephalopathy, *Seminars in ultrasound, CT, and MRI*, 2014, **35**, 136–152.

33. J. Kim, J. M. Kim, Y. K. Kim, J. W. Shin, S. H. Choi, S. E. Kim and Y. Kim, Dopamine transporter SPECT of a liver cirrhotic with atypical parkinsonism, *Ind. Health*, 2007, **45**, 497–500.

34. K. J. Klos, J. E. Ahlskog, N. Kumar, S. Cambern, J. Butz, M. Burritt, R. D. Fealey, C. T. Cowl, J. E. Parisi and K. A. Josephs, Brain metal concentrations in chronic liver failure patients with pallidal T1 MRI hyperintensity, *Neurology*, 2006, **67**, 1984–1989.
35. S. A. Mirowitz, T. J. Westrich and J. D. Hirsch, Hyperintense basal ganglia on T1-weighted MR images in patients receiving parenteral nutrition, *Radiology*, 1991, **181**, 117–120.
36. J. Ono, K. Harada, R. Kodaka, K. Sakurai, H. Tajiri, Y. Takagi, T. Nagai, T. Harada, A. Nihei, A. Okada, *et al.*, Manganese deposition in the brain during long-term total parenteral nutrition, *JPEN, J. Parenter. Enteral Nutr.*, 1995, **19**, 310–312.
37. Y. Kafritsa, J. Fell, S. Long, M. Bynevelt, W. Taylor and P. Milla, Long-term outcome of brain manganese deposition in patients on home parenteral nutrition, *Arch. Dis. Child.*, 1998, **79**, 263–265.
38. Y. Iinuma, M. Kubota, M. Uchiyama, M. Yagi, S. Kanada, S. Yamazaki, H. Murata, K. Okamoto, M. Suzuki and K. Nitta, Whole-blood manganese levels and brain manganese accumulation in children receiving long-term home parenteral nutrition, *Pediatr. Surg. Int.*, 2003, **19**, 268–272.
39. H. Suzuki, J. Takanashi, N. Saeki and Y. Kohno, Temporal parenteral nutrition in children causing t1 shortening in the anterior pituitary gland and globus pallidus, *Neuropediatrics*, 2003, **34**, 200–204.
40. R. Abdalian, O. Saqui, G. Fernandes and J. P. Allard, Effects of manganese from a commercial multi-trace element supplement in a population sample of Canadian patients on long-term parenteral nutrition, *JPEN, J. Parenter. Enteral Nutr.*, 2013, **37**, 538–543.
41. S. A. Mirowitz and T. J. Westrich, Basal ganglial signal intensity alterations: reversal after discontinuation of parenteral manganese administration, *Radiology*, 1992, **185**, 535–536.
42. R. M. de Bie, R. M. Gladstone, A. P. Strafella, J. H. Ko and A. E. Lang, Manganese-induced Parkinsonism associated with methcathinone (Ephedrone) abuse, *Arch. Neurol.*, 2007, **64**, 886–889.
43. Y. Sanotsky, R. Lesyk, L. Fedoryshyn, I. Komnatska, Y. Matviyenko and S. Fahn, Manganic encephalopathy due to "ephedrone" abuse, *Mov. Disord.*, 2007, **22**, 1337–1343.
44. K. Sikk, P. Taba, S. Haldre, J. Bergquist, D. Nyholm, G. Zjablov, T. Asser and S. M. Aquilonius, Irreversible motor impairment in young addicts–ephedrone, manganism or both?, *Acta Neurol. Scand.*, 2007, **115**, 385–389.
45. K. Sikk, S. Haldre, S. M. Aquilonius, A. Asser, M. Paris, A. Roose, J. Petterson, S. L. Eriksson, J. Bergquist and P. Taba, Manganese-induced parkinsonism in methcathinone abusers: bio-markers of exposure and follow-up, *Eur. J. Neurol.*, 2013, **20**, 915–920.
46. C. Colosimo and M. Guidi, Parkinsonism due to ephedrone neurotoxicity: a case report, *Eur. J. Neurol.*, 2009, **16**, e114–115.
47. A. Stepens, I. Logina, V. Liguts, P. Aldins, I. Eksteina, A. Platkajis, I. Martinsone, E. Terauds, B. Rozentale and M. Donaghy,

A Parkinsonian syndrome in methcathinone users and the role of manganese, *N. Engl. J. Med.*, 2008, **358**, 1009–1017.

48. F. Varlibas, I. Delipoyraz, G. Yuksel, G. Filiz, H. Tireli and N. O. Gecim, Neurotoxicity following chronic intravenous use of "Russian cocktail", *Clin. Toxicol.*, 2009, **47**, 157–160.

49. M. Selikhova, L. Fedoryshyn, Y. Matviyenko, I. Komnatska, M. Kyrylchuk, L. Krolicki, A. Friedman, A. Taylor, H. R. Jager, A. Lees and Y. Sanotsky, Parkinsonism and dystonia caused by the illicit use of ephedrone–a longitudinal study, *Mov. Disord.*, 2008, **23**, 2224–2231.

50. U. Dydak, J. Xu, A. Epur, X. Li, S. Streitmatter, L. L. Long, W. Zheng and Y. M. Jiang, Brain Regions showing Manganese Accumulation in the Human versus the Rat Brain, *Proc. Intl. Soc. Mag. Reson. Med.*, 2011, **19**, 1428.

51. T. R. Guilarte, J. L. McGlothan, M. Degaonkar, M. K. Chen, P. B. Barker, T. Syversen and J. S. Schneider, Evidence for cortical dysfunction and widespread manganese accumulation in the nonhuman primate brain following chronic manganese exposure: a 1H-MRS and MRI study, *Toxicol. Sci.*, 2006, **94**, 351–358.

52. D. C. Dorman, M. F. Struve, B. A. Wong, J. A. Dye and I. D. Robertson, Correlation of brain magnetic resonance imaging changes with pallidal manganese concentrations in rhesus monkeys following subchronic manganese inhalation, *Toxicol. Sci.*, 2006, **92**, 219–227.

53. R. E. London, G. Toney, S. A. Gabel and A. Funk, Magnetic resonance imaging studies of the brains of anesthetized rats treated with manganese chloride, *Brain Res. Bull.*, 1989, **23**, 229–235.

54. Y. T. Kuo, A. H. Herlihy, P. W. So, K. K. Bhakoo and J. D. Bell, In vivo measurements of T1 relaxation times in mouse brain associated with different modes of systemic administration of manganese chloride, *J. Magn. Reson. Imaging*, 2005, **21**, 334–339.

55. G. Discalzi, E. Pira, E. Herrero Hernandez, C. Valentini, M. Turbiglio and F. Meliga, Occupational Mn parkinsonism: magnetic resonance imaging and clinical patterns following CaNa2-EDTA chelation, *Neurotoxicology*, 2000, **21**, 863–866.

56. A. Takeda, J. Sawashita and S. Okada, Biological half-lives of zinc and manganese in rat brain, *Brain Res.*, 1995, **695**, 53–58.

57. D. C. Dorman, M. F. Struve, M. W. Marshall, C. U. Parkinson, R. A. James and B. A. Wong, Tissue manganese concentrations in young male rhesus monkeys following subchronic manganese sulfate inhalation, *Toxicol. Sci.*, 2006, **92**, 201–210.

58. M. C. Newland, C. Cox, R. Hamada, G. Oberdorster and B. Weiss, The clearance of manganese chloride in the primate, *Fundam. Appl. Toxicol.*, 1987, **9**, 314–328.

59. B. Gallez, R. Demeure, C. Baudelet, N. Abdelouahab, N. Beghein, B. Jordan, M. Geurts and H. A. Roels, Non invasive quantification of manganese deposits in the rat brain by local measurement of NMR proton T1 relaxation times, *Neurotoxicology*, 2001, **22**, 387–392.

60. E. Kim, Y. Kim, H. K. Cheong, S. Cho, Y. C. Shin, J. Sakong, K. S. Kim, J. S. Yang, Y. W. Jin and S. K. Kang, Pallidal index on MRI as a target organ dose of manganese: structural equation model analysis, *Neurotoxicology*, 2005, **26**, 351–359.

61. Y. Chang, S. T. Woo, Y. Kim, J. J. Lee, H. J. Song, H. J. Lee, S. H. Kim, H. Lee, Y. J. Kwon, J. H. Ahn, S. J. Park, I. S. Chung and K. S. Jeong, Pallidal index measured with three-dimensional T1-weighted gradient echo sequence is a good predictor of manganese exposure in welders, *J. Magn. Reson. Imaging*, 2010, **31**, 1020–1026.

62. S. H. Kim, K. H. Chang, J. G. Chi, H. K. Cheong, J. Y. Kim, Y. M. Kim and M. H. Han, Sequential change of MR signal intensity of the brain after manganese administration in rabbits. Correlation with manganese concentration and histopathologic findings, *Invest. Radiol.*, 1999, **34**, 383–393.

63. D. S. Choi, E. A. Kim, H. K. Cheong, H. S. Khang, J. W. Ryoo, J. M. Cho, J. Sakong and I. Park, Evaluation of MR signal index for the assessment of occupational manganese exposure of welders by measurement of local proton T1 relaxation time, *Neurotoxicology*, 2007, **28**, 284–289.

64. M. Ulla, J. M. Bonny, L. Ouchchane, I. Rieu, B. Claise and F. Durif, Is R2* a new MRI biomarker for the progression of Parkinson's disease? A longitudinal follow-up, *PLoS One*, 2013, **8**, e57904.

65. T. Inoue, T. Majid and R. G. Pautler, Manganese enhanced MRI (MEMRI): neurophysiological applications, *Rev. Neurosci.*, 2011, **22**, 675–694.

66. A. C. Silva and N. A. Bock, Manganese-enhanced MRI: an exceptional tool in translational neuroimaging, *Schizophr. Bull.*, 2008, **34**, 595–604.

67. A. P. Koretsky and A. C. Silva, Manganese-enhanced magnetic resonance imaging (MEMRI), *NMR Biomed.*, 2004, **17**, 527–531.

68. R. G. Pautler, A. C. Silva and A. P. Koretsky, In vivo neuronal tract tracing using manganese-enhanced magnetic resonance imaging, *Magn. Reson. Med.*, 1998, **40**, 740–748.

69. I. Aoki, C. Tanaka, T. Takegami, T. Ebisu, M. Umeda, M. Fukunaga, K. Fukuda, A. C. Silva, A. P. Koretsky and S. Naruse, Dynamic activity-induced manganese-dependent contrast magnetic resonance imaging (DAIM MRI), *Magn. Reson. Med.*, 2002, **48**, 927–933.

70. Y. J. Lin and A. P. Koretsky, Manganese ion enhances T1-weighted MRI during brain activation: an approach to direct imaging of brain function, *Magn. Reson. Med.*, 1997, **38**, 378–388.

71. T. Watanabe, O. Natt, S. Boretius, J. Frahm and T. Michaelis, In vivo 3D MRI staining of mouse brain after subcutaneous application of $MnCl_2$, *Magn. Reson. Med.*, 2002, **48**, 852–859.

72. I. Aoki, Y. J. Wu, A. C. Silva, R. M. Lynch and A. P. Koretsky, In vivo detection of neuroarchitecture in the rodent brain using manganese-enhanced MRI, *Neuroimage*, 2004, **22**, 1046–1059.

73. K. H. Chuang, A. P. Koretsky and C. H. Sotak, Temporal changes in the T1 and T2 relaxation rates (DeltaR1 and DeltaR2) in the rat brain are

consistent with the tissue-clearance rates of elemental manganese, *Magn. Reson. Med.*, 2009, **61**, 1528–1532.

74. K. H. Chuang and A. Koretsky, Improved neuronal tract tracing using manganese enhanced magnetic resonance imaging with fast T(1) mapping, *Magn. Reson. Med.*, 2006, **55**, 604–611.

75. N. A. Bock, F. F. Paiva and A. C. Silva, Fractionated manganese-enhanced MRI, *NMR Biomed.*, 2008, **21**, 473–478.

76. V. Bouilleret, L. Cardamone, C. Liu, A. S. Koe, K. Fang, J. P. Williams, D. E. Myers, T. J. O'Brien and N. C. Jones, Confounding neurodegenerative effects of manganese for in vivo MR imaging in rat models of brain insults, *J. Magn. Reson. Imaging*, 2011, **34**, 774–784.

77. Y. Chang, S. U. Jin, Y. Kim, K. M. Shin, H. J. Lee, S. H. Kim, J. H. Ahn, S. J. Park, K. S. Jeong, Y. C. Weon and H. Lee, Decreased brain volumes in manganese-exposed welders, *Neurotoxicology*, 2013, **37**, 182–189.

78. R. A. Edden and P. B. Barker, Spatial effects in the detection of gamma-aminobutyric acid: improved sensitivity at high fields using inner volume saturation, *Magn. Reson. Med.*, 2007, **58**, 1276–1282.

79. M. Mescher, H. Merkle, J. Kirsch, M. Garwood and R. Gruetter, Simultaneous in vivo spectral editing and water suppression, *NMR Biomed.*, 1998, **11**, 266–272.

80. P. G. Mullins, D. J. McGonigle, R. L. O'Gorman, N. A. Puts, R. Vidyasagar, C. J. Evans, Cardiff Symposium on MRS of GABA and R. A. Edden, Current practice in the use of MEGA-PRESS spectroscopy for the detection of GABA, *NeuroImage*, 2014, **86**, 43–52.

81. J. Pfeuffer, I. Tkac, S. W. Provencher and R. Gruetter, Toward an in vivo neurochemical profile: quantification of 18 metabolites in short-echo-time (1)H NMR spectra of the rat brain, *J. Magn. Reson.*, 1999, **141**, 104–120.

82. G. Oz, J. R. Alger, P. B. Barker, R. Bartha, A. Bizzi, C. Boesch, P. J. Bolan, K. M. Brindle, C. Cudalbu, A. Dincer, U. Dydak, U. E. Emir, J. Frahm, R. G. Gonzalez, S. Gruber, R. Gruetter, R. K. Gupta, A. Heerschap, A. Henning, H. P. Hetherington, F. A. Howe, P. S. Huppi, R. E. Hurd, K. Kantarci, D. W. Klomp, R. Kreis, M. J. Kruiskamp, M. O. Leach, A. P. Lin, P. R. Luijten, M. Marjanska, A. A. Maudsley, D. J. Meyerhoff, C. E. Mountford, S. J. Nelson, M. N. Pamir, J. W. Pan, A. C. Peet, H. Poptani, S. Posse, P. J. Pouwels, E. M. Ratai, B. D. Ross, T. W. Scheenen, C. Schuster, I. C. Smith, B. J. Soher, I. Tkac, D. B. Vigneron, R. A. Kauppinen and M. R. S. C. Group, Clinical proton MR spectroscopy in central nervous system disorders, *Radiology*, 2014, **270**, 658–679.

83. C. Zwingmann, D. Leibfritz and A. S. Hazell, Energy metabolism in astrocytes and neurons treated with manganese: relation among cell-specific energy failure, glucose metabolism, and intercellular trafficking using multinuclear NMR-spectroscopic analysis, *J. Cereb. Blood Flow Metab.*, 2003, **23**, 756–771.

84. C. Zwingmann, D. Leibfritz and A. S. Hazell, Nmr spectroscopic analysis of regional brain energy metabolism in manganese neurotoxicity, *Glia*, 2007, **55**, 1610–1617.

85. N. Just, C. Cudalbu, H. Lei and R. Gruetter, Effect of manganese chloride on the neurochemical profile of the rat hypothalamus, *J. Cereb. Blood Flow Metab.*, 2011, **31**, 2324–2333.

86. E. A. Kim, H. K. Cheong, D. S. Choi, J. Sakong, J. W. Ryoo, I. Park and D. M. Kang, Effect of occupational manganese exposure on the central nervous system of welders: 1H magnetic resonance spectroscopy and MRI findings, *Neurotoxicology*, 2007, **28**, 276–283.

87. Y. Chang, S. T. Woo, J. J. Lee, H. J. Song, H. J. Lee, D. S. Yoo, S. H. Kim, H. Lee, Y. J. Kwon, H. J. Ahn, J. H. Ahn, S. J. Park, Y. C. Weon, I. S. Chung, K. S. Jeong and Y. Kim, Neurochemical changes in welders revealed by proton magnetic resonance spectroscopy, *Neurotoxicology*, 2009, **30**, 950–957.

88. U. Dydak, E. J. Ward, R. Ma, S. Snyder, S. E. Zauber, J. B. Murdoch, Z. Long and F. Rosenthal, Occupational Manganese Exposure Levels Correlate with Brain GABA Levels, *Proc. Intl. Soc. Mag. Reson. Med.*, 2014, **22**, 1895.

89. Z. Long, X. R. Li, J. Xu, R. A. Edden, W. P. Qin, L. L. Long, J. B. Murdoch, W. Zheng, Y. M. Jiang and U. Dydak, Thalamic GABA predicts fine motor performance in manganese-exposed smelter workers, *PLoS One*, 2014, **9**, e88220.

90. L. Long, Y. M. Jiang, X. R. Li, J. Xu, C. L. Yeh, L. L. Long, W. Zheng, J. B. Murdoch and U. Dydak, Increased Thalamic GABA and Decreased Glutamate-Glutamine in Chronic Manganese-exposed Metal Workers and Manganism Patients., *Proc. Intl. Soc. Mag. Reson. Med.*, 2014, **22**, 3777.

91. D. Le Bihan, J. F. Mangin, C. Poupon, C. A. Clark, S. Pappata, N. Molko and H. Chabriat, Diffusion tensor imaging: concepts and applications, *J. Magn. Reson. Imaging*, 2001, **13**, 534–546.

92. C. Beaulieu, The basis of anisotropic water diffusion in the nervous system – a technical review, *NMR Biomed.*, 2002, **15**, 435–455.

93. A. M. McKinney, R. W. Filice, M. Teksam, S. Casey, C. Truwit, H. B. Clark, C. Woon and H. Y. Liu, Diffusion abnormalities of the globi pallidi in manganese neurotoxicity, *Neuroradiology*, 2004, **46**, 291–295.

94. Y. Kim, K. S. Jeong, H. J. Song, J. J. Lee, J. H. Seo, G. C. Kim, H. J. Lee, H. J. Kim, J. H. Ahn, S. J. Park, S. H. Kim, Y. J. Kwon and Y. Chang, Altered white matter microstructural integrity revealed by voxel-wise analysis of diffusion tensor imaging in welders with manganese exposure, *Neurotoxicology*, 2011, **32**, 100–109.

95. A. Stepens, C. J. Stagg, A. Platkajis, M. H. Boudrias, H. Johansen-Berg and M. Donaghy, White matter abnormalities in methcathinone abusers with an extrapyramidal syndrome, *Brain*, 2010, **133**, 3676–3684.

96. P. Favrole, H. Chabriat, J. P. Guichard and F. Woimant, Clinical correlates of cerebral water diffusion in Wilson disease, *Neurology*, 2006, **66**, 384–389.

97. S. Ogawa, T. M. Lee, A. R. Kay and D. W. Tank, Brain magnetic resonance imaging with contrast dependent on blood oxygenation, *Proc. Natl. Acad. Sci. U. S. A.*, 1990, **87**, 9868–9872.

98. S. A. S. Huettel, A. W. Song and G. McCarthy, *Functional Magnetic Resonance Imaging*, Sinauer Associates, Inc., Sunderland, MA, USA, 2nd edn, 2009.

99. R. B. Buxton, *Introduction to functional magnetic resonance imaging: Principles and techniques*, Cambridge University Press, 2002.

100. Y. Chang, H. J. Song, J. J. Lee, J. H. Seo, J. H. Kim, H. J. Lee, H. J. Kim, Y. Kim, J. H. Ahn, S. J. Park, J. H. Kwon, K. S. Jeong and D. K. Jung, Neuroplastic changes within the brains of manganese-exposed welders: recruiting additional neural resources for successful motor performance, *Occup. Environ. Med.*, 2010, **67**, 809–815.

101. Y. Chang, J. J. Lee, J. H. Seo, H. J. Song, J. H. Kim, S. J. Bae, J. H. Ahn, S. J. Park, K. S. Jeong, Y. J. Kwon, S. H. Kim and Y. Kim, Altered working memory process in the manganese-exposed brain, *Neuroimage*, 2010, **53**, 1279–1285.

102. J. B. Bomanji, D. C. Costa and P. J. Ell, Clinical role of positron emission tomography in oncology, *Lancet Oncol.*, 2001, **2**, 157–164.

103. L. Mosconi, Brain glucose metabolism in the early and specific diagnosis of Alzheimer's disease. FDG-PET studies in MCI and AD, *Eur. J. Nucl. Med. Mol. Imaging*, 2005, **32**, 486–510.

104. R. E. Yee, I. Irwin, C. Milonas, D. B. Stout, S. C. Huang, K. Shoghi-Jadid, N. Satyamurthy, L. E. Delanney, D. M. Togasaki, K. F. Farahani, K. Delfani, A. M. Janson, M. E. Phelps, J. W. Langston and J. R. Barrio, Novel observations with FDOPA-PET imaging after early nigrostriatal damage, *Mov. Disord.*, 2001, **16**, 838–848.

105. S. Thobois, S. Guillouet and E. Broussolle, Contributions of PET and SPECT to the understanding of the pathophysiology of Parkinson's disease, *Neurophysiologie clinique/Clin. Neurophysiol.*, 2001, **31**, 321–340.

106. C. C. Huang, Parkinsonism induced by chronic manganese intoxication–an experience in Taiwan, *Chang Gung Med. J.*, 2007, **30**, 385–395.

107. C. C. Huang, Y. H. Weng, C. S. Lu, N. S. Chu and T. C. Yen, Dopamine transporter binding in chronic manganese intoxication, *J. Neurol.*, 2003, **250**, 1335–1339.

108. Y. Kim, J. M. Kim, J. W. Kim, C. I. Yoo, C. R. Lee, J. H. Lee, H. K. Kim, S. O. Yang, H. K. Chung, D. S. Lee and B. Jeon, Dopamine transporter density is decreased in parkinsonian patients with a history of manganese exposure: what does it mean?, *Mov. Disord.*, 2002, **17**, 568–575.

109. K. Tatsch and G. Poepperl, Nigrostriatal dopamine terminal imaging with dopamine transporter SPECT: an update, *J. Nucl. Med.*, 2013, **54**, 1331–1338.

110. H. Eriksson, J. Tedroff, K. A. Thuomas, S. M. Aquilonius, P. Hartvig, K. J. Fasth, P. Bjurling, B. Langstrom, K. G. Hedstrom and E. Heilbronn, Manganese induced brain lesions in Macaca fascicularis

as revealed by positron emission tomography and magnetic resonance imaging, *Arch. Toxicol.*, 1992, **66**, 403–407.

111. H. Shinotoh, B. J. Snow, K. A. Hewitt, B. D. Pate, D. Doudet, R. Nugent, D. P. Perl, W. Olanow and D. B. Calne, MRI and PET studies of manganese-intoxicated monkeys, *Neurology*, 1995, **45**, 1199–1204.

112. T. R. Guilarte, N. C. Burton, J. L. McGlothan, T. Verina, Y. Zhou, M. Alexander, L. Pham, M. Griswold, D. F. Wong, T. Syversen and J. S. Schneider, Impairment of nigrostriatal dopamine neurotransmission by manganese is mediated by pre-synaptic mechanism(s): implications to manganese-induced parkinsonism, *J. Neurochem.*, 2008, **107**, 1236–1247.

113. T. R. Guilarte, M. K. Chen, J. L. McGlothan, T. Verina, D. F. Wong, Y. Zhou, M. Alexander, C. A. Rohde, T. Syversen, E. Decamp, A. J. Koser, S. Fritz, H. Gonczi, D. W. Anderson and J. S. Schneider, Nigrostriatal dopamine system dysfunction and subtle motor deficits in manganese-exposed non-human primates, *Exp. Neurol.*, 2006, **202**, 381–390.

114. J. S. Schneider, E. Decamp, K. Clark, C. Bouquio, T. Syversen and T. R. Guilarte, Effects of chronic manganese exposure on working memory in non-human primates, *Brain Res.*, 2009, **1258**, 86–95.

115. M. K. Chen, J. S. Lee, J. L. McGlothan, E. Furukawa, R. J. Adams, M. Alexander, D. F. Wong and T. R. Guilarte, Acute manganese administration alters dopamine transporter levels in the non-human primate striatum, *Neurotoxicology*, 2006, **27**, 229–236.

116. E. C. Wolters, C. C. Huang, C. Clark, R. F. Peppard, J. Okada, N. S. Chu, M. J. Adam, T. J. Ruth, D. Li and D. B. Calne, Positron emission tomography in manganese intoxication, *Ann. Neurol.*, 1989, **26**, 647–651.

117. H. Shinotoh, B. J. Snow, N. S. Chu, C. C. Huang, C. S. Lu, C. Lee, H. Takahashi and D. B. Calne, Presynaptic and postsynaptic striatal dopaminergic function in patients with manganese intoxication: a positron emission tomography study, *Neurology*, 1997, **48**, 1053–1056.

118. Y. Kim, J. W. Kim, K. Ito, H. S. Lim, H. K. Cheong, J. Y. Kim, Y. C. Shin, K. S. Kim and Y. Moon, Idiopathic parkinsonism with superimposed manganese exposure: utility of positron emission tomography, *Neurotoxicology*, 1999, **20**, 249–252.

119. B. A. Racette, L. McGee-Minnich, S. M. Moerlein, J. W. Mink, T. O. Videen and J. S. Perlmutter, Welding-related parkinsonism: clinical features, treatment, and pathophysiology, *Neurology*, 2001, **56**, 8–13.

120. B. A. Racette, J. A. Antenor, L. McGee-Minnich, S. M. Moerlein, T. O. Videen, V. Kotagal and J. S. Perlmutter, [18F]FDOPA PET and clinical features in parkinsonism due to manganism, *Mov. Disord.*, 2005, **20**, 492–496.

121. S. R. Criswell, J. S. Perlmutter, J. S. Crippin, T. O. Videen, S. M. Moerlein, H. P. Flores, A. M. Birke and B. A. Racette, Reduced

uptake of FDOPA PET in end-stage liver disease with elevated manganese levels, *Arch. Neurol.*, 2012, **69**, 394–397.

122. K. Sikk, P. Taba, S. Haldre, J. Bergquist, D. Nyholm, H. Askmark, T. Danfors, J. Sorensen, L. Thurfjell, R. Raininko, R. Eriksson, R. Flink, C. Farnstrand and S. M. Aquilonius, Clinical, neuroimaging and neurophysiological features in addicts with manganese-ephedrone exposure, *Acta Neurol. Scand.*, 2010, **121**, 237–243.

123. S. R. Criswell, J. S. Perlmutter, T. O. Videen, S. M. Moerlein, H. P. Flores, A. M. Birke and B. A. Racette, Reduced uptake of [(1)(8)F]FDOPA PET in asymptomatic welders with occupational manganese exposure, *Neurology*, 2011, **76**, 1296–1301.

124. K. P. Bhatia and C. D. Marsden, The behavioural and motor consequences of focal lesions of the basal ganglia in man, *Brain*, 1994, **117**(Pt 4), 859–876.

125. J. Rodier, Manganese poisoning in Moroccan miners, *Br. J. Ind. Med.*, 1955, **12**, 21–35.

126. R. M. Park, R. M. Bowler and H. A. Roels, Exposure-response relationship and risk assessment for cognitive deficits in early welding-induced manganism, *J. Occup. Environ. Med./Am. Coll. Occup. Environ. Med.*, 2009, **51**, 1125–1136.

127. R. M. Bowler, S. Nakagawa, M. Drezgic, H. A. Roels, R. M. Park, E. Diamond, D. Mergler, M. Bouchard, R. P. Bowler and W. Koller, Sequelae of fume exposure in confined space welding: a neurological and neuropsychological case series, *Neurotoxicology*, 2007, **28**, 298–311.

128. T. R. Guilarte, Manganese and Parkinson's disease: a critical review and new findings, *Environ. Health Perspect.*, 2010, **118**, 1071–1080.

129. B. S. Twining, S. B. Baines, N. S. Fisher, J. Maser, S. Vogt, C. Jacobsen, A. Tovar-Sanchez and S. A. Sanudo-Wilhelmy, Quantifying trace elements in individual aquatic protist cells with a synchrotron X-ray fluorescence microprobe, *Anal. Chem.*, 2003, 75, 3806–3816.

130. M. D. de Jonge, C. Holzner, S. B. Baines, B. S. Twining, K. Ignatyev, J. Diaz, D. L. Howard, D. Legnini, A. Miceli, I. McNulty, C. J. Jacobsen and S. Vogt, Quantitative 3D elemental microtomography of Cyclotella meneghiniana at 400-nm resolution, *Proc. Natl. Acad. Sci. U. S. A.*, 2010, **107**, 15676–15680.

131. A. Carmona, G. Deves, S. Roudeau, P. Cloetens, S. Bohic and R. Ortega, Manganese accumulates within golgi apparatus in dopaminergic cells as revealed by synchrotron X-ray fluorescence nanoimaging, *ACS Chem. Neurosci.*, 2010, **1**, 194–203.

132. R. Ortega, G. Deves and A. Carmona, Bio-metals imaging and speciation in cells using proton and synchrotron radiation X-ray microspectroscopy, *J. R .Soc., Interface*, 2009, **6**(Suppl 5), S649–658.

133. A. Daoust, E. L. Barbier and S. Bohic, Manganese enhanced MRI in rat hippocampus: a correlative study with synchrotron X-ray microprobe, *Neuroimage*, 2013, **64**, 10–18.

134. G. Robison, T. Zakharova, S. Fu, W. Jiang, R. Fulper, R. Barrea, W. Zheng and Y. Pushkar, X-ray fluorescence imaging of the hippocampal formation after manganese exposure, *Metallomics*, 2013, 5, 1554–1565.

135. G. Robison, T. Zakharova, S. Fu, W. Jiang, R. Fulper, R. Barrea, M. A. Marcus, W. Zheng and Y. Pushkar, X-ray fluorescence imaging: a new tool for studying manganese neurotoxicity, *PLoS One*, 2012, 7, e48899.

136. A. Carmona, S. Roudeau, L. Perrin, G. Veronesi and R. Ortega, Environmental manganese compounds accumulate as Mn(ii) within the Golgi apparatus of dopamine cells: relationship between speciation, subcellular distribution, and cytotoxicity, *Metallomics*, 2014, **6**, 822–832.

137. Y. Kim, Neuroimaging in manganism, *Neurotoxicology*, 2006, **27**, 369–372.

138. V. A. Fitsanakis, N. Zhang, M. J. Avison, J. C. Gore, J. L. Aschner and M. Aschner, The use of magnetic resonance imaging (MRI) in the study of manganese neurotoxicity, *Neurotoxicology*, 2006, **27**, 798–806.

139. M. Aschner, T. R. Guilarte, J. S. Schneider and W. Zheng, Manganese: recent advances in understanding its transport and neurotoxicity, *Toxicol. Appl. Pharmacol.*, 2007, **221**, 131–147.

CHAPTER 20

Epidemiological Studies of Parkinsonism in Welders

HARVEY CHECKOWAY,*[a] SUSAN SEARLES NIELSEN[b] AND
BRAD A. RACETTE[c]

[a] Department of Family and Preventive Medicine, University of California,
San Diego, La Jolla, California, USA; [b] Department of Environmental and
Occupational Health Sciences, University of Washington, Seattle,
Washington, USA; [c] Department of Neurology, Washington University,
St. Louis, Missouri, USA
*Email: hcheckoway@ucsd.edu

20.1 Parkinsonism: Clinical, Pathological, and Epidemiological Features

Parkinsonism (PS), of which Parkinson disease (PD) is the most common cause, is a progressive neurological motor system disorder. The British physician Parkinson first described the syndrome among six patients in 1817, considering "severe fright" to have been the cause of their condition, as evidenced by a masked-like facial expression.[1] The cardinal clinical signs of PD have since been defined as: bradykinesia, rest tremor, rigidity, and postural instability. Non-motor features such as cognitive and autonomic dysfunction are also common.[2] The underlying pathology of PD is selective destruction of dopaminergic neurons, especially in mid-brain substantia nigra.[3] The prevalence of PD in persons aged 60 years and older is approximately 2%,[4] and there are estimates that PS signs occur in

Issues in Toxicology No. 22
Manganese in Health and Disease
Edited by Lucio G. Costa and Michael Aschner
© The Royal Society of Chemistry 2015
Published by the Royal Society of Chemistry, www.rsc.org

approximately one-third of older adults.[5] Men are consistently observed to have approximately 30–50% greater rates of PD relative to women.[6]

In this review of epidemiological evidence of associations among welders, we will summarize evidence for PD and PS because of the substantial phenotypic overlap of both clinical designations. The causes of PD remain poorly understood, despite considerable toxicological and epidemiological research during the past 50 years. Mendelian inheritance accounts for a relatively small fraction ($\sim$5–10%) of PD.[7] Although additional genetic variants may also contribute to risk,[8] environmental, lifestyle, and medical factors play predominant etiologic roles.[9] Apart from an increasing risk with age, and a male preponderance, the most consistent epidemiological finding in PD has been the observed reduced risk among ever smokers of cigarettes, whose risks are approximately half those of never smokers.[10] Environmental factors that have received most attention are pesticides, solvents, and metals,[9,11–13] including manganese (Mn), which is the predominant neurotoxicant exposure experienced by welders.

20.2 Manganese and Parkinsonism: Historical Background

Couper's 1837 report of parkinsonian-like symptoms–tremor, gait disturbance, whispering speech–in five Scottish Mn ore crushers[14] is generally regarded as the first documentation of occupation-related Mn-induced neurotoxicity. More than a century had passed when Rodier[15] described a spectrum of physical, behavioral, and psychological symptoms from his survey of 3849 Moroccan miners. Prominent motor symptoms that occurred frequently and with varying degrees of severity were tremor, a dystonic "cock-like walk", and retropulsion. Noteworthy findings were that the large majority of cases held jobs with the most intense Mn exposures (although not quantified), and that roughly half of the cases occurred within two years of initial exposure, suggesting that manganism may often be an acute sequela of high dose exposure. Rodier's case description remains the benchmark for "manganism". In fact, clinical differences in case presentation and symptom onset between manganism and idiopathic PD have been invoked as arguments against a causal link of Mn and PD.[16] Pathological and neuroimaging differences between manganism and parkinsonism have also been noted as evidence supporting distinct disease entities.[17] Inhalation of Mn is the predominant exposure route relevant to neurotoxicity,[18] although there may also be transport through the olfactory system.[19] As evidenced by imaging studies, Mn appears to accumulate in the nigrostriatal system and other brain regions following divalent cation transport[20] from the lungs or olfactory system.[21] Once in brain tissue, Mn can induce neuronal damage that potentially can culminate in PS/PD.[22]

In the next section of this chapter we will summarize evidence derived from epidemiological studies of PS/PD among welders, who represent a segment of the working population with particularly high Mn exposures. Our review is not intended to be comprehensive, but rather is restricted to literature that we regard as most informative.

20.3   Epidemiological Studies of PS/PD among Welders

Possible associations between welding and PS/PD have been investigated in numerous epidemiological investigations of defined cohorts of professional welders,[23–27] and from general population surveys of occupations and disease risks.[28–33] Associations with PD have also been examined in the context of some population-based case-control studies in which welding occupation was considered as one of many potential occupational and environmental risk factors.[34–37] We regard the studies of defined welder cohorts as most informative because of the direct focus on welding exposures. Accordingly, we will devote most attention to investigations of defined cohorts of welders.

Table 20.1 presents findings from epidemiological studies of PD or PS in defined welder cohorts. The results vary considerably among studies, possibility owing to differences in case definitions and case ascertainment methods. Sources of case data were based mainly on cross-sectional assessments of PS prevalence or mortality follow-up for PD. The strongest associations were observed in cohorts of US professional welders enumerated by Racette *et al.*,[24,27] in whom clinical parkinsonism was determined by standardized neurological examinations performed by neurologists specializing in movement disorders. In these studies, risk was related to specific occupational type[24] or an estimate of cumulative welding hours weighted for potential intensity of Mn exposure.[27] With the exception of a South Korean study that indicated elevated risks associated with job titles classified by Mn exposure levels,[26] all other studies reported null findings.

No excess risks for PD among welders were reported from the general population surveys of hospitalization claims or mortality statistics (Table 20.2). Similarly, as shown in Table 20.3, findings for welder occupation observed in population- or clinic-based case-control studies, based on subjects' self-report or from review of medical records, do not support etiologic relations. In fact, relative risk estimates are lower than the null value in some instances, although not statistically significantly below 1.0.

20.4   Discussion

Mn remains a widespread exposure in many occupational settings, including welding, steel smelting, and mining. Environmental exposures from industrial sources also occur from air, soil, and water pollution. Given the virtually ubiquitous nature of Mn exposure, combined with clinical and toxicological evidence supporting the neurotoxicity of Mn, a possible role of Mn in PD etiology has been a topic of considerable scientific and public health interest since the 1970s.

Table 20.1 Risks of parkinsonism (PS) and Parkinson disease (PD) in studies of defined cohorts of welders.

Study	Location	Cohort size	Outcome, Source	Exposure assessment, Reference group	Main findings (RR, 95% CI)[a]
Fryzek et al.[23] (2005)	Denmark	6163	PD, Hospitalization	Welder occupation; National rates	Overall: 0.9 (0.4–1.5) Employed >20 years: 0.8 (0.2–2.0)
Racette et al.[24] (2005a)	US	1423	PS, Neurological exam	Welder occupation type; Community rates in Alabama	Overall: 7.6 (3.3–17.7) Boilermakers: 10.3 (2.6–40.5)
Marsh and Gula[25] (2006)	US	12 595	PD, Medical insurance claims	Welding job title; Controls (n = 2230) without PD	*Incident cases* Overall: 0.8 (0.3–2.2) ≥30 years welding: 0.8 (0.3–2.0) *Prevalent cases* Overall: 0.9 (0.4–2.1) ≥30 years welding: 1.1 (0.5–2.2)
Park et al.[26] (2005)	S. Korea	38 560	PD, Hospitalization billing claims		Low exposure: 3.6 (0.7–18.6)
Racette et al.[27] (2012)	US	716	PS, Neurological exam	Weighted welding hrs; Non-welders (n = 57)	Overall: 15.6% welders, 0% non-welders Highest exposed welders: 1.0 (0.6–1.6)

[a]Relative risk estimate (95% confidence interval).

Table 20.2 Risks for parkinsonism (PS) and Parkinson disease (PD) in general population surveys of welding occupation.

Study	Location	Source	Exposure assessment; Reference group	Main findings (RR, 95% CI)[a]
Park et al.[28] (2005)	US	Mortality	Welder occupation; Non-neurological deaths	Overall: 0.9 (0.8–1.0) Age < 65: 1.8 (1.1–2.8)
Fored et al.[29] (2006)	Sweden	Hospitalization, mortality	Welder occupation; National rates	0.9 (0.8–1.0)
Stampfer et al.[30] (2009)	US	Mortality	Welder occupation on death certificate; National rates	0.9 (0.8–0.9)
Li et al.[31] (2009)	Sweden	Hospitalization	Welding occupation on census; National rates	0.9 (0.7–1.3)
Feldman et al.[32] (2011)	Sweden	Hospitalization, Mortality	Welding "smoke" exposure from job/ exposure matrix; Reference group	0.7 (0.2–2.6)[b]
Kenborg et al.[33] (2012)	Denmark	Hospitalization	Welding and welding fume exposed occupations; Reference group	Overall: 1.1 (0.8–1.5) RR/ 10 years welding: 0.8 (0.6–1.2)

[a]Relative risk estimate (95% confidence interval).
[b]Occupations with >2/3 exposed.

Evidence from the epidemiological literature on the welding occupation in particular as a risk factor for PS/PD, taken together, does not indicate consistent or strong etiologic associations. With a few notable exceptions, most of the findings reported are null. In their meta-analysis of many of the studies reviewed here, Mortimer et al.[38] reported a summary relative risk of 0.86 (95% confidence interval 0.80–0.92), which is consistent with the literature findings we review here. One might therefore conclude that welders are not at increased risk for PS/PD. However, certain important pathophysiological and epidemiological considerations deserve attention.

Whether manganism and parkinsonism are distinct clinical entities remains a controversial question. There is considerable clinical overlap in signs and symptoms of manganism and idiopathic PD. In the US welders cohort study,[27] PS cases, defined as welders with Unified Parkinson Disease Rating Scale Motor Subsection 3 (UPDRS3)[39] scores ≥15, exhibited similar motor sub-scores to patients with untreated idiopathic PD, suggesting substantial phenotypic overlap. The largest molecular imaging study of the presynaptic dopaminergic system (the pathway primarily affected in PD) demonstrated dopaminergic dysfunction in Mn-exposed welders.[40,41] Although the pattern of dopaminergic dysfunction was different from that seen in PD, findings from molecular imaging of the dopaminergic system in patients with Mn toxicity secondary to liver failure suggest that the pattern of dopaminergic dysfunction may be more typical of PD.[42,43]

Table 20.3 Population-based and clinic-based case-control studies of Parkinson disease (PD) reporting associations with welding occupation.

Study	Location	No. cases, controls	Source of subjects	Exposure assessment	Main findings (RR, 95%CI)[a]
Frigerio et al.[34] (2005)	US	202, 202	Clinic	Welding occupation from medical record	0.1 (0.01–2.7)
Dick et al.[35] (2007)	Five European countries	767, 1989	Clinics	Self-reported occupation, classified as having high Mn exposure from job/exposure matrix	0.9 (0.6–1.3)[b]
Tanner et al.[36] (2009)	US	519, 511	Clinics	Self-reported welding occupation	1.0 (0.6–1.7)
Firestone et al.[37] (2010)	US	252, 326	Health maintenance organization	Self-reported occupation	0.6 (0.4–1.1)[b]

[a]Relative risk estimate (95% confidence interval).
[b]Males only.

Much of the epidemiological research conducted to date suffers from severe limitations. PD mortality is an incomplete and possibly biased indication of PD incidence, which is the fundamental epidemiological disease frequency measure most suitable for understanding disease etiology. Prevalence surveys of PS signs and symptoms offer valuable clinical data, but likewise may not provide an unbiased reflection of disease incidence. Notably, studies in which disease prevalence, rather than incidence, is ascertained may be prone to a "healthy worker survivor" bias due to affected workers leaving employment or transferring to lower-exposure jobs. Moreover, the clinical expertise of the investigator may substantially influence the results. Population-based case-control studies have the advantages of accruing relatively large numbers of PD cases, and often obtaining data to confirm diagnoses, together with data on potential confounding factors such as cigarette smoking and co-morbid conditions. However, case-control studies are typically limited by small numbers of welders, because welding is not a common occupation in the population at large.

Another critical consideration is that exposure assessment in most studies has been very limited, relying on relatively crude indicators of exposure–job title as a welder or years in welding occupations. Even most attempts to quantify exposures more precisely have not been based on actual environmental measurements of welding fumes or its components. It should be appreciated that welding fume composition varies by type of welding, and typically includes multiple metals, in addition to Mn. Importantly, iron, a common welding fume metal, and manganese compete for biological uptake in the brain;[44] thus, data on multiple metal concentrations that span presumably etiologically relevant time periods of exposure would be needed to provide physiologically meaningful dose estimates. At most, biological measurements of Mn, such as blood, hair, or toenail concentrations, may possibly indicate short-term Mn exposure.[45–48] These may not provide reliable dose estimates, and it is debatable whether short-term exposure is as relevant as cumulative (lifetime) exposure in studies focused on PS.

Insofar as a possible etiologic relation between welding and PS remains an important public health and scientific concern, it is worthwhile considering what might be an optimal epidemiological approach, recognizing that certain aspects may be difficult or impossible to achieve. An idealized study would be a longitudinal assessment of clinical signs and symptoms of PS, administered by movement disorders specialists, among a large cohort of professional welders. Inclusion of a non-welder comparison group, similar to welders in terms of gender, age, socioeconomic status, and lifestyle would be valuable. A thorough exposure assessment that permits quantitative dose estimation for Mn and other common welding fume metals would be necessary for dose–response estimation, and ultimately for revising occupational and environmental exposure standards. Future studies with these design considerations should help reconcile the inconsistencies observed in the literature to date.

References

1. J. Parkinson, An Essay on the Shaking Palsy, *J. Neuropsychiatry Clin. Neurosci.*, 2002, **14**, 223–236, discussion 222.
2. A. Samii, J. G. Nutt and B. R. Ransom, Parkinson's disease, *Lancet*, 2004, **363**, 1783–1793.
3. Y. Agid, Parkinson's disease: pathophysiology, *Lancet*, 1991, **337**, 1321–1324.
4. A. Elbaz, J. H. Bower, D. M. Maraganore, S. K. McDonnell, B. J. Peterson, J. E. Ahlskog, D. J. Schaid and W. A. Rocca, Risk tables for parkinsonism and Parkinson's disease, *J. Clin. Epidemiol.*, 2002, **55**, 25–31.
5. E. D. Louis, J. A. Luchsinger, M. X. Tang and R. Mayeux, Parkinsonian signs in older people - prevalence and associations with smoking and coffee, *Neurology*, 2003, **61**, 24–28.
6. A. Wright Willis, B. A. Evanoff, M. Lian, S. R. Criswell and B. A. Racette, Geographic and ethnic variation in Parkinson disease: a population-based study of US Medicare beneficiaries, *Neuroepidemiology*, 2010, **34**, 143–151.
7. I. Martin, V. L. Dawson and T. M. Dawson, Recent advances in the genetics of Parkinson's disease, *Annu. Rev. Genomics Hum. Genet.*, 2011, **12**, 301–325.
8. C. M. Lill, J. T. Roehr, M. B. McQueen, F. K. Kavvoura, S. Bagade, B. M. Schjeide, L. M. Schjeide, E. Meissner, U. Zauft, N. C. Allen, T. Liu, M. Schilling, K. J. Anderson, G. Beecham, D. Berg, J. M. Biernacka, A. Brice, A. L. DeStefano, C. B. Do, N. Eriksson, S. A. Factor, M. J. Farrer, T. Foroud, T. Gasser, T. Hamza, J. A. Hardy, P. Heutink, E. M. Hill-Burns, C. Klein, J. C. Latourelle, D. M. Maraganore, E. R. Martin, M. Martinez, R. H. Myers, M. A. Nalls, N. Pankratz, H. Payami, W. Satake, W. K. Scott, M. Sharma, A. B. Singleton, K. Stefansson, T. Toda, J. Y. Tung, J. Vance, N. W. Wood, C. P. Zabetian, P. Young, R. E. Tanzi, M. J. Khoury, F. Zipp, H. Lehrach, J. P. Ioannidis and L. Bertram, Comprehensive research synopsis and systematic meta-analyses in Parkinson's disease genetics: The PDGene database, *PLoS Genet.*, 2012, **8**, e1002548.
9. K. Wirdefeldt, H. O. Adami, P. Cole, D. Trichopoulos and J. Mandel, Epidemiology and etiology of Parkinson's disease: a review of the evidence, *Eur. J. Epidemiol.*, 2011, **26**(Suppl 1), S1–58.
10. B. Ritz, A. Ascherio, H. Checkoway, K. S. Marder, L. M. Nelson, W. A. Rocca, G. W. Ross, D. Strickland, S. K. Van Den Eeden and J. Gorell, Pooled analysis of tobacco use and risk of Parkinson disease, *Arch. Neurol.*, 2007, **64**, 990–997.
11. M. T. Allen and L. S. Levy, Parkinson's disease and pesticide exposure–a new assessment, *Crit. Rev. Toxicol.*, 2013, **43**, 515–534.
12. S. M. Goldman, P. J. Quinlan, G. W. Ross, C. Marras, C. Meng, G. S. Bhudhikanok, K. Comyns, M. Korell, A. R. Chade, M. Kasten, B. Priestley, K. L. Chou, H. H. Fernandez, F. Cambi, J. W. Langston and C. M. Tanner, Solvent exposures and Parkinson disease risk in twins, *Ann. Neurol.*, 2012, **71**, 776–784.

13. A. W. Willis, B. A. Evanoff, M. Lian, A. Galarza, A. Wegrzyn, M. Schootman and B. A. Racette, Metal emissions and urban incident Parkinson disease: a community health study of Medicare beneficiaries by using geographic information systems, *Am. J. Epidemiol.*, 2010, **172**, 1357–1363.

14. J. Couper, On the effects of black oxide of manganese when inhaled into the lungs, *Br. Ann. Med. Pharmacol.*, 1837, **1**, 41–42.

15. J. Rodier, Manganese Poisoning in Moroccan Miners, *Br. J. Ind. Med.*, 1955, **12**, 21–35.

16. J. Jankovic, Searching for a relationship between manganese and welding and Parkinson's disease, *Neurology*, 2005, **64**, 2021–2028.

17. R. G. Lucchini, C. J. Martin and B. C. Doney, From Manganism to Manganese-Induced Parkinsonism: A Conceptual Model Based on the Evolution of Exposure, *NeuroMol. Med.*, 2009, **11**, 311–321.

18. E. J. Martinez-Finley, C. E. Gavin, M. Aschner and T. E. Gunter, Manganese neurotoxicity and the role of reactive oxygen species, *Free Radical Biol. Med.*, 2013, **62**, 65–75.

19. K. Thompson, R. M. Molina, T. Donaghey, J. E. Schwob, J. D. Brain and M. Wessling-Resnick, Olfactory uptake of manganese requires DMT1 and is enhanced by anemia, *FASEB J.*, 2007, **21**, 223–230.

20. J. A. Roth, Are there common biochemical and molecular mechanisms controlling manganism and parkisonism, *NeuroMol. Med.*, 2009, **11**, 281–296.

21. A. Elder, R. Gelein, V. Silva, T. Feikert, L. Opanashuk, J. Carter, R. Potter, A. Maynard, Y. Ito, J. Finkelstein and G. Oberdörster, Translocation of inhaled ultrafine manganese oxide particles to the central nervous system, *Environ. Health Perspect.*, 2006, **114**, 1172–1178.

22. W. M. Caudle, T. S. Guillot, C. R. Lazo and G. W. Miller, Industrial toxicants and Parkinson's disease, *Neurotoxicology*, 2012, **33**, 178–188.

23. J. P. Fryzek, J. Hansen, S. Cohen, J. P. Bonde, M. T. Llambias, H. A. Kolstad, A. Skytthe, L. Lipworth, W. J. Blot and J. H. Olsen, A cohort study of Parkinson's disease and other neurodegenerative disorders in Danish welders, *J Occup Environ Med*, 2005, **47**, 466–472.

24. B. A. Racette, S. D. Tabbal, D. Jennings, L. Good, J. S. Perlmutter and B. Evanoff, Prevalence of parkinsonism and relationship to exposure in a large sample of Alabama welders, *Neurology*, 2005, **64**, 230–235.

25. G. M. Marsh and M. J. Gula, Employment as a welder and Parkinson disease among heavy equipment manufacturing workers, *J. Occup. Environ. Med.*, 2006, **48**, 1031–1046.

26. J. Park, C. I. Yoo, C. S. Sim, H. K. Kim, J. W. Kim, B. S. Jeon, K. R. Kim, O. Y. Bang, W. Y. Lee, Y. Yi, K. Y. Jung, S. E. Chung and Y. Kim, Occupations and Parkinson's disease: a multi-center case-control study in South Korea, *Neurotoxicology*, 2005, **26**, 99–105.

27. B. A. Racette, S. R. Criswell, J. I. Lundin, A. Hobson, N. Seixas, P. T. Kotzbauer, B. A. Evanoff, J. S. Perlmutter, J. Zhang, L. Sheppard and

H. Checkoway, Increased risk of parkinsonism associated with welding exposure, *Neurotoxicology*, 2012, **33**, 1356–1361.

28. R. M. Park, P. A. Schulte, J. D. Bowman, J. T. Walker, S. C. Bondy, M. G. Yost, J. A. Touchstone and M. Dosemeci, Potential occupational risks for neurodegenerative diseases, *Am. J. Ind. Med.*, 2005, **48**, 63–77.
29. C. M. Fored, J. P. Fryzek, L. Brandt, G. Nise, B. Sjögren, J. K. McLaughlin, W. J. Blot and A. Ekbom, Parkinson's disease and other basal ganglia or movement disorders in a large nationwide cohort of Swedish welders, *Occup. Environ. Med.*, 2006, **63**, 135–140.
30. M. J. Stampfer, Welding occupations and mortality from Parkinson's disease and other neurodegenerative diseases among United States men, 1985–1999, *J. Occup. Environ. Hyg.*, 2009, **6**, 267–272.
31. X. Li, J. Sundquist and K. Sundquist, Socioeconomic and occupational groups and Parkinson's disease: a nationwide study based on hospitalizations in Sweden, *Int. Arch. Occup. Environ. Health*, 2009, **82**, 235–241.
32. A. L. Feldman, A. L. Johansson, G. Nise, M. Gatz, N. L. Pedersen and K. Wirdefeldt, Occupational exposure in parkinsonian disorders: a 43-year prospective cohort study in men, *Parkinsonism Relat. Disord.*, 2011, **17**, 677–682.
33. L. Kenborg, C. F. Lassen, J. Hansen and J. H. Olsen, Parkinson's disease and other neurodegenerative disorders among welders: a Danish cohort study, *Mov. Disord.*, 2012, **27**, 1283–1289.
34. R. Frigerio, A. Elbaz, K. R. Sanft, B. J. Peterson, J. H. Bower, J. E. Ahlskog, B. R. Grossardt, M. de Andrade, D. M. Maraganore and W. A. Rocca, Education and occupations preceding Parkinson disease: a population-based case-control study, *Neurology*, 2005, **65**, 1575–1583.
35. E. F. D. Dick, G. De Palma, A. Ahmadi, A. Osborne, N. W. Scott, G. J. Prescott, J. Bennett, S. Semple, S. Dick, C. Counsell, P. Mozzoni, N. Haites, S. B. Wettinger, A. Mutti, M. Otelea, A. Seaton, P. Söderkvist and A. Felice, Gene-environment interactions in parkinsonism and Parkinson's disease: the Geoparkinson study, *Occup. Environ. Med.*, 2007, **64**, 666–672.
36. C. M. Tanner, G. W. Ross, S. A. Jewell, R. A. Hauser, J. Jankovic, S. A. Factor, S. Bressman, A. Deligtisch, C. Marras, K. E. Lyons, G. S. Bhudhikanok, D. F. Roucoux, C. Meng, R. D. Abbott and J. W. Langston, Occupation and risk of parkinsonism: a multicenter case-control study, *Arch. Neurol.*, 2009, **66**, 1106–1113.
37. J. A. Firestone, J. I. Lundin, K. M. Powers, T. Smith-Weller, G. M. Franklin, P. D. Swanson, W. T. Longstreth and H. Checkoway, Occupational factors and risk of Parkinson's disease: A population-based case-control study, *Am. J. Ind. Med.*, 2010, **53**, 217–223.
38. J. A. Mortimer, A. R. Borenstein and L. M. Nelson, Associations of welding and manganese exposure with Parkinson disease: review and meta-analysis, *Neurology*, 2012, **79**, 1174–1180.
39. S. Fahn and R. L. Elton, Members of the UPDRS Development Committee, *Recent developments in Parkinson's disease*. ed. S. Fahn, C.D.

Marsden, M. Goldstein and D.B. Calne, Macmillan, Florham Park, NJ, 1987, pp. 153–163.

40. S. R. Criswell, J. S. Perlmutter, T. O. Videen, S. M. Moerlein, H. P. Flores, A. M. Birke and B. A. Racette, Reduced uptake of [18F]FDOPA PET in asymptomatic welders with occupational manganese exposure, *Neurology*, 2011, **76**, 1296–1301.

41. S. R. Criswell, J. S. Perlmutter, J. L. Huang, N. Golchin, H. P. Flores, A. Hobson, M. Aschner, K. M. Erikson, H. Checkoway and B. A. Racette, Basal ganglia intensity indices and diffusion weighted imaging in manganese-exposed welders, *Occup. Environ. Med.*, 2012, **69**, 437–443.

42. S. R. Criswell, J. S. Perlmutter, J. S. Crippin, T. O. Videen, S. M. Moerlein, H. P. Flores, A. M. Birke and B. A. Racette, Reduced uptake of FDOPA PET in end-stage liver disease with elevated manganese levels, *Arch. Neurol.*, 2012, **69**, 394–397.

43. B. A. Racette, J. A. Antenor, L. McGee-Minnich, S. M. Moerlein, T. O. Videen, V. Kotagal and J. S. Perlmutter, [18F]FDOPA PET and clinical features in parkinsonism due to manganism, *Mov. Disord.*, 2005, **20**, 492–496.

44. J. D. Park, K. Y. Kim, D. W. Kim, S. J. Choi, B. S. Choi, Y. H. Chung, J. H. Han, J. H. Sung, I. H. Kwon, J. H. Mun and I. J. Yu, Tissue distribution of manganese in iron-sufficient or iron-deficient rats after stainless steel welding-fume exposure, *Inhalation Toxicol.*, 2007, **19**, 563–572.

45. M. Bader, M. C. Dietz, A. Ihrig and G. Triebig, Biomonitoring of manganese in blood, urine and axillary hair following low-dose exposure during the manufacture of dry cell batteries, *Int. Arch. Occup. Environ. Health*, 1999, **72**, 521–527.

46. M. G. Baker, C. D. Simpson, B. Stover, L. Sheppard, H. Checkoway, B. A. Racette and N. S. Seixas, Blood manganese as an exposure biomarker: state of the evidence, *J. Occup. Environ. Hyg.*, 2014, **11**, 210–217.

47. D. M. Cowan, Q. Fan, Y. Zou, X. Shi, J. Chen, M. Aschner, F. S. Rosenthal and W. Zheng, Manganese exposure among smelting workers: blood manganese-iron ratio as a novel tool for manganese exposure assessment, *Biomarkers*, 2009, **14**, 3–16.

48. W. Laohaudomchok, X. Lin, R. F. Herrick, S. C. Fang, J. M. Cavallari, D. C. Christiani and M. G. Weisskopf, Toenail, blood, and urine as biomarkers of manganese exposure, *J. Occup. Environ. Med.*, 2011, **53**, 506–510.

CHAPTER 21

Cognitive Effects of Manganese in Children and Adults

ROBERTO LUCCHINI*[a,b] AND SILVIA ZONI[b]

[a] Department of Preventive Medicine, Icahn School of Medicine at Mount Sinai, New York; [b] Occupational Medicine and Industrial Hygiene, University of Brescia, Italy
*Email: roberto.lucchini@mssm.edu

21.1 Introduction

Manganese (Mn) is an essential metal for humans, but high concentrations of manganese may be neurotoxic and can cause damage to motor and cognitive functions. Several reviews[1-5] have shown that a significant effect of manganese on the central nervous system can result from chronic and excessive exposure. Manganese neurotoxicity has been considered as a continuum of dysfunction ranging from early neurofunctional changes to subclinical neurologic signs and finally to a clinically well-defined illness.[2]

The exposure routes for Mn are gastrointestinal, most commonly for the general population, and inhalation, typical of occupational and environmental exposure (the fastest route to transfer Mn to the brain). Inhalation is highly relevant for occupational and environmental exposure also because Mn may access the brain directly through olfactory uptake,[6] bypassing the homeostatic mechanisms that regulate absorption and excretion to keep Mn levels in the desired range.

Exposure to Mn by inhalation or ingestion may impact adults and children differently, being based on different mechanisms.[7] The intestinal absorption

Issues in Toxicology No. 22
Manganese in Health and Disease
Edited by Lucio G. Costa and Michael Aschner
© The Royal Society of Chemistry 2015
Published by the Royal Society of Chemistry, www.rsc.org

rate of ingested Mn in children is higher than in adults,[8] and the demand for Fe related to growth may enhance the absorption of ingested Mn.[9]

The excretion pathway, mostly through the bile for Mn, is also not fully developed until months after birth. Therefore, especially with inhaled Mn, the ratio of inhaled air : body weight is much higher in children, leading to a higher absorption dose.

Manganese can accumulate in the central nervous system, and the basal ganglia are the target sites for Mn accumulation.[10-13] Given that the basal ganglia are parts of different corticostriatal loops,[14,15] they are of importance for both cognitive and motor performance. Moreover, they are closely related to executive functions such as the selection and adjustment of behavior[16] and working memory.[17]

Accumulation of Mn causes alterations in a neurotransmitter system, namely the dopaminergic system, in brain areas responsible for attention, cognition, and motor coordination.[18,19] Mn is a potent dopamine oxidant: excessive exposure can cause loss or inactivation of dopamine receptors.[20]

Mn neurotoxicity has been widely studied and the interest in toxic effects has grown constantly in the last 20 years. The average number of publications per year, ascertained in PubMed for the search term "Mn AND neurotox*", increased from 77 between 1981 and 1990 to 452 between 2001 and 2010.[21]

The first studies focused especially on motor and movement impairments resulting from Mn-induced disruptions in the basal ganglia. After 2000, cognitive deficits were reported in various studies on occupationally exposed workers,[22-26] in children and adults environmentally exposed to Mn from a variety of sources, including airborne particulates from industrial emissions, for example, steel and ferroalloy foundries, dry-alkaline production, welding and mining, in addition to drinking water and the use of Mn-based pesticides such as Maneb and Mancozeb.

A variety of tests and test batteries have been used to detect and quantify cognitive effects of Mn exposure.[27] Significant results have been obtained, for adults, with the measurement of perceptual and motor speed using Digit Symbol and auditory recall and attention with the Digit Span. In children one of the most commonly used tests is the Wechsler Intelligence Scale for Children (WISC). Emerging evidence of the cognitive effects of Mn has focused attention on brain regions such as the frontal cortex, and other cortical and subcortical structures related to the cognitive domains.[28]

21.2 Cognitive Effects in Children

Several studies on children residing in high-risk areas have revealed effects on cognitive functions, especially on the reduction of intelligence quotient (IQ),[29-34] executive functions,[35,36] memory,[30,36-38] academic achievement,[37,39,40] mental development[41] and other functions such as perceptual reasoning[38] and attention.[40,42] The main features and characteristics of these studies are summarized in Table 21.1.

Table 21.1 Summary of studies investigating cognitive functions in early life manganese exposure.

Study	Country	Design	n.	Age	Exposure biomarker	Main observed effects
He et al. 1994	China	Cross-sectional	92	11–13 years	MnH	Neurobehavioral effects
Takser et al. 2003	France	Prospective	247	neonates	Mother's blood	Psychomotor development
Wasserman et al. 2006	Bangladesh	Cross-sectional	142	10.0 ± 0.4 years	WMn	Cognitive deficit (Intelligent Quotient-IQ)
Wright et al. 2006	USA	Cross-sectional	32	11–13 years	Mn, As, Cd in hair	IQ inversely related to hair Mn
Ericson et al. 2007	United States	Cross sectional in a follow-up cohort	27	From 20th week to preschoo lage	Mn in tooth enamel, between 20th and 62–64th gestational week	Behavioral disinibition, errors in the Stroop Test, externalizing and attention problem
Kim et al. 2009	Korea	Cross-sectional	261	8–11 years	PbB, MnB	Cognitive deficit (IQ)
Clauss-Henn et al. 2010	Mexico City	Longitudinal	448	12–36 months	MnB, PbB	Mixed exposure M n-Pb related with deficits in mental and psychomotor development
Riojas-Rodríguez et al. 2010	Mexico	Cross-sectional	172	7–11 years	MnB, MnH	Cognitive deficit (IQ)
Bouchard et al. 2011	Quebec	Cross-sectional	362	6–13 years	WMn, HMn	Cognitive deficit (IQ)
Menezes-Filho et al. 2011	Brazil	Cross-sectional	83	6–12 years	MnH, MnB, PbB	Cognitive deficit (IQ)
Wasserman et al. 2011	Bangladesh	Cross-sectional	299	8–11 years	WMn, WAs	Cognitive deficit (IQ)
Henn et al. 2012	Mexico	Longitudinal	455	12–36 months	MnB, PbB	Mental and psychomotor development score
Khan et al. 2012	Bangladesh	Cross-sectional	840	8–11 years	WMn, WAs	loss in mathematics test scores
Lucchini et al. 2012	Italy	Cross-sectional	299	11–14 years	MnB, PbB, MnH	Reduction of IQ points resulted from an increase of BPb, not Mn
Bhang et al. 2013	South Korea	Cross-sectional	1089	8–11 years	MnB, PbB	Attention and academic function deficit; lower scores of thinking; reading, calculation, and learning quotient
Carvalho et al. 2013	Brazil	Cross-sectional	70	7–12 years	MnH	Lower IQ and neuropsychological performance in tests of executive function of inhibition responses, strategic visual formation and verbal working memory

Most studies indicate that hair Mn (MnH), a biomarker of occupational and environmental exposure, is inversely associated with intellectual function in young school-age children. Although these findings were quite consistent across studies and in different countries, potential problems with external contamination of hair samples have been shown by one study,[43] which demonstrated the effectiveness of a more rigorous method for cleaning hair samples before the chemical analysis. This pre-analytical aspect is worthy of consideration for future studies using levels in hair as a biomarker of Mn recent exposure.

The birth cohort studies considered biomarkers including maternal and cord blood.[44] Another exposure variable measured in these studies on early exposure was Mn in drinking water in areas such as Bangladesh and Quebec with high natural Mn levels.

Neurodevelopmental toxicity is determined by co-exposure to multiple neurotoxicants,[45,46] and co-exposure to neurotoxicants may have an additive effect.[47] This is more evident for economically disadvantaged children who are exposed to various combinations of toxic materials.[48,49] The interaction created by mixed exposure is particularly relevant to neurodevelopmental effects but has been rarely addressed.[30] Concurrent exposure to lead (Pb) has not been consistently considered, regardless of the ubiquitous presence of lead in the environment due to the previous use in gasoline, paint, and other exposure sources. Most of the studies on early life exposure to manganese have not considered the potential separate effect of lead on cognition. When considered, lead has been treated as a potential confounder by adjusting the Mn *vs.* IQ association for blood lead.[32,34] More appropriately, blood lead has been considered as an independent variable in regression models, allowing a better understanding of potential interaction with Mn in blood.[31,41,50]

The separate influence of lead exposure has also been observed as the predominant predictor of IQ reduction.[51] The study analyzed the IQ of 299 Italian adolescents (11–14 years old), environmentally exposed to Mn and Pb caused by emissions from the ferroalloy industry. Results of multivariate regression modeling demonstrated significant adverse effects of lead on cognitive functions, with a reduction of about 2.4 IQ points in the IQ score for a two-fold increase in Pb. Neither a manganese effect nor a lead interaction was observed in this study. In the same population, Mn exposure was associated with impairment of fine motor coordination and odor discrimination.[52]

Another important consideration, besides the exposure and co-exposure metrics, is represented by the critical exposure windows during early life. Manganese is an essential element and therefore there is high physiological demand in the first developmental period. After birth, Mn levels exceeding the required range can become toxic. Factors such as iron metabolism and liver excretion may play an important role in susceptible individuals. A U-shaped relation showing improved mental development at low manganese blood level and deterioration at high levels was shown[41] at 12 months of age.

21.2.1   Reduction of IQ

A cross-sectional study[29] reported the results on intellectual function in a group of 142 10 year-old children exposed to Mn from wells with drinking water averaging 793 µg l^{-1} Mn in Bangladesh. Water Mn (MnW) was associated with reduced Total, Verbal, and Performance IQ raw scores in a dose–response manner.

The associations between hair levels of As, Mn and Cd and neuropsychological function and behavior in school-aged children was assessed in 32 children,[30] 11–13 years old. The authors observed an inverse relationship between Total and Verbal IQ and MnH levels averaging 471.5 parts per billion.

Another study[31] investigated the association of blood Pb and Mn with the intelligence of 261 school-aged Korean children. The results showed a significant linear relationship between both blood Pb (mean 1.73 µg dl^{-1}, median = 1.55) and blood Mn (14.3 µg l^{-1}, median = 14.0) and the Full-Scale IQ and Verbal IQ, suggesting an additive interaction and effect modification between Pb and Mn.

A total of 172 Mexican children divided into 79 exposed to Mn from mining operations and 93 from a reference area, aged 7–11 years, showed an inverse association between MnH (median in the exposed area 12.6 µg g^{-1}) and Verbal, Performance and Total Scale IQ. Blood Mn (median in the exposed area 9.5 µg l^{-1}) was inversely but not significantly associated with Total and Verbal IQ scores.[32]

A group of 362 children aged 6–13 years was investigated in Quebec to assess the impact of Mn exposure through drinking water on the IQ.[33] Manganese was measured in residential tap water (median 34 µg l^{-1}), and in children's hair (median 0.7 µg g^{-1}). Adjusted results showed that higher MnW was significantly associated with lower Performance, Verbal and Full Scale IQ scores and that higher MnH was associated with lower Full Scale IQ scores. In particular, the authors found that MnW was more strongly associated with Performance IQ than with Verbal IQ.

A study in Brazil[34] investigated the relation of MnH (mean 5.83 µg g^{-1}) and MnB (8.2 µg l^{-1}) with the IQ score among 86 children aged 6–12 years, living near a ferro-manganese plant. The results showed that the children's MnH was negatively associated with Verbal and Total IQ. The β coefficients were respectively −5.78 (95%CI −10.71 to −0.21) and −6.72 (−11.81 to −0.63), adjusted for maternal education and nutritional status.

21.2.2   Executive Functions

The ability to generate new and coherent responses to the context and the voluntary control of actions has been identified in the domain of executive functions (EF)[53] and has been related to the fronto-striatal circuit, which may be disrupted by Mn accumulation in the brain. EFs integrate skills such as self-monitoring, self-regulation, inhibition and flexibility, and abandon or replace ineffective strategies in favor of others that are more effective for problem solving.[36]

Components of EFs were assessed[35] in 27 preschool children; the authors found that high Mn levels in tooth enamel dating to prenatal life were significantly correlated with impulsive errors on the Continuous Performance Test and a children's Stroop test at 4 years old.

Another recent study[36] aimed to assess the association between elevated Mn exposure and performance on EF and attention neuropsychological tests in 70 children, aged 7–12 years, residing in two communities near a ferromanganese alloy plant in Brazil. MnH levels were used as a biomarker of exposure and averaged 1.48 (range 0.52–55.74) µg g^{-1}. The results suggested that airborne Mn exposure may be associated with lower IQ and neuropsychological performance in tests of EF of inhibition responses, strategic visual formation and verbal working memory.

21.2.3 Memory

A sample of 92 children, aged 11–13 years, living in a rural community in China, was evaluated.[37] The Mn level in contaminated drinking water ranged from 0.24 to 0.35 mg l^{-1} in an area following the use of high-Mn sewage water for irrigation. The authors found lower performance in exposed children, with hair Mn averaging 1.25 µg g^{-1}, compared with controls, on 5 of 12 neurobehavioral tests administered: one of the tests was the Digit Span that assesses working memory.

Arsenic (As) and manganese exposure and children's intellectual functions were evaluated[38] in children (8–11 years old), living in Bangladesh, stratified on As and Mn concentrations in domestic well-water. When adjusted only for each other, MnB and AsB were significantly and negatively related to most WISC-IV subscales. After adjustment for sociodemographic variables and ferritin, the authors observed a negative association between MnB and Percepetual Reasoning and Working Memory scores.

Another study[36] found a negative association between verbal immediate and working memory scores and MnH levels. These observations are also consistent with previous studies showing impairment associated with Mn exposure in immediate memory[54] and memory for histories and a word learning list.[30]

21.2.4 Academic Achievement

Memory deficit and cognitive effects related to Mn exposure suggest consequences for children's academic achievement. Lowered academic achievement in languages, math, science and other disciplines has been related to lead exposure, while associations with Mn have less often been reported. However, Mn-exposed Chinese children exhibited significantly lower school performance in mathematics and language compared to not-exposed children.[37]

The potential interaction between Mn and As and their effect on children's academic achievement have been considered.[39] The sample was composed of a group of 840 children, aged 8–11 years, from Bangladesh, with data on well-water samples from each house, urinary As (UAs) measurement, and academic achievement in three disciplines (Bangla, English, Mathematics). The authors observed a significant negative association between MnW and mathematics scores, also in adjusted models. The relation between MnW and language scores was not significant.

Another study[40] examined 1089 children, 8–11 years old, living in five areas in South Korea. The results showed that attention and academic function in children were affected by the Mn level in the blood, which averaged 14.14 µg L^{-1}. After adjusting for urine ecotinine, blood lead, children's IQ, and other potential confounders, the high Mn group showed lower scores in thinking, reading, calculations and learning quotient. Excess Mn was associated with lower scores for thinking, reading, calculation, and learning quotient (LQ), a global measure of learning disabilities.

21.2.5   Mental Development

The interaction between manganese and lead in early childhood and its possible association with neurodevelopment deficiencies was investigated.[41] The authors considered a sample of 455 children from Mexico City, assessing child neurodevelopment through Bayley Scales (BSID-IIS). Biological markers of exposure were Mn and Pb in blood. The results indicated that a mixed exposure to manganese and lead was related to greater deficits in mental and psychomotor development than exposure to one of the two metals alone. Mixed-effects models showed a significant interaction over time.

21.3   Cognitive Effects in Adults

The neurological effects following massive exposure to manganese (Mn) were described as early as 1837.[55] Symptoms of manganism include loss of equilibrium, trembling of fingers, rigidity, decreased memory, and headache.

Early evidence of preclinical neuropsychological alteration includes reduced performance on neuropsychological testing, poor eye–hand coordination and hand steadiness, reduced reaction time, reduced cognitive flexibility and poor postural stability.[5] Other symptoms commonly reported include headache, weakness, memory loss, sleep disturbance, irritability, anxiety disorders, and gait disturbance. These effects have been associated with Mn deposition in the brain as measured with magnetic resonance imaging.[56]

21.3.1   Occupational Studies

Occupational studies were predominant until the 2000s. A large body of evidence shows that occupational exposure to manganese (Mn) can cause neurotoxicity,[57] which persists in retired workers.[58] Although symptoms and deficits varied, workers with elevated exposures commonly reported headache, weakness, memory loss, sleep disturbance, irritability, anxiety disorders and gait disturbance. These effects have been associated with Mn deposition in the brain as measured with magnetic resonance imaging (MRI) in otherwise normal industrial populations.

A recent review[59] of the epidemiological literature on Mn identified: (1) studies focused on idiopathic Parkinson disease without considering manganism; (2) studies with a healthy workers effect bias; (3) studies with problematic statistical modeling; and (4) studies arising from case series derived from litigation. Investigations with adequate study design and exposure assessment revealed consistent neurobehavioral effects and subclinical and clinical signs and symptoms of impairment attributable to Mn exposure. Twenty-eight studies showed an exposure–response relationship between Mn and neurobehavioral effects. The review reported a summary of a study that observed statistically significant associations between Mn exposure status and cognitive function, and between motor symptom outcomes and mean airborne exposure concentrations.

A meta-analysis[21] of eight studies on occupationally exposed workers found deficits in attention and short-term memory. This meta-analysis provided evidence for lower motor and cognitive performance scores in workers occupationally exposed to Mn when compared with reference workers. In terms of cognitive functioning, the speed of information processing and short-term memory were lower among exposed groups. The deficits were found in workers exposed to mean concentrations of inhalable Mn ranging from 0.05 to 0.30 mg m^{-3}.

In contrast to the significant effects on motor function, results on memory, attention and cognitive functions were less clearly related to Mn. However, there is agreement in the occupational literature about the reduction of cognitive function in exposed workers, especially for short-term memory.[60–63]

Functional MRI (fMRI) was used[63] to assess the neural correlates of Mn-induced memory impairment in 23 welders with subclinical dysfunction in working memory networks as a result of chronic Mn exposure averaging 16.5 ± 4.5 μg l^{-1} of blood Mn and 90 ± 80 μg m^{-3} of airborne Mn. Within-group and between-group analyses revealed that brain activity in working memory networks was increased in welders with chronic Mn exposure during the 2-back verbal working memory task compared to healthy control individuals. Welders showed impairment in various memory functions (*e.g.*, verbal/visual, short-term/long-term, and working memory) compared with healthy controls. In addition, attention and executive functions were lower in welders than in controls. These results are consistent with previous reports[22–26,62] showing impaired cognitive functions in welders examined with neuropsychological testing. The same group has shown reduced volume in the basal ganglia and cerebellum associated with the cognitive and motor changes.[64]

Reduction in perceptual and motor speed, assessed with the Digit Symbol test, was observed in several studies.[24,25,60,65–69] Visual conceptualization and visuomotor tracking were also found to be impaired in exposed workers.[24,25,65,70] Attention/concentration, visuospatial scanning abilities and visuospatial dysfunction such as spatial neglect were found to be impaired in studies on workers.[23,70] These last studies also found an impairment in verbal fluency among Mn-exposed welders examined with the Vocabulary Test. To evaluate general intelligence, the authors administered the WAIS-III: in the first study, a significant impairment in mental arithmetic (Arithmetic Subtest) and processing speed involving a visuomotor component (Processing Speed Index) was observed. In the second study, the WAIS-III Processing Speed Index, the WAIS-III Verbal Comprehension Index and the Vocabulary scores were impaired among the welders.

21.3.2 Environmental Studies

While there is clear evidence of manganese neurotoxicity in pediatric and adult occupational populations, little is known about the effects in the general population who may exhibit enhanced susceptibility as a result of compromised physiology, when compared to young adult workers. Therefore, results from studies of exposures to environmental Mn are less conclusive and the existing community studies of Mn exposure have focused more on motor disorders, with less frequent comprehensive neuropsychological assessment.

Non-occupational studies on adult populations showed both neuromotor and cognitive abnormalities,[56,71] and an increased frequency of parkinsonism associated with Mn in airborne particles[72] and deposited dust.[73]

In a study[74] in Marietta, Ohio, USA, a community with elevated air-Mn from industrial emissions, a comprehensive neuropsychological test battery was administered, including tests of category fluency (Animal Naming), processing speed (WAIS-III Digit Symbol Coding), visuospatial learning (NAB Shape Learning), and visuospatial memory (Rey–Osterrieth Complex Figure Test). Although no differences in neuropsychological test scores were found between the groups, the observed relationships of CEI (cumulative exposure index) and HQ (hazard quotient) with generalized anxiety, category fluency, processing speed, visuospatial learning and memory suggested the existence of some association of anxiety states and neuropsychological function with environmental air-Mn exposure at the Marietta site.

Another study[75] investigated the relationship between environmental Mn contamination and neurological health outcomes in 250 elderly subjects (65–75 years old) residing in Valcamonica, an area in the province of Brescia, Italy, that was affected by ferromanganese plant emissions for about a century until 2001. The neuropsychological battery included the Mini-Mental State Examination (MMSE) for general cognitive assessment, the Story Recall Test to assess long-term verbal memory, the Raven's Colored Progressive Matrices (CPM) to measure clear-thinking ability, the Trail

Making test to assess the ability for spatial planning in a visual-motor track, and the Digit Span (from WAIS) to evaluate short-term memory. The multivariate analysis showed association of Mn in blood and urine and distance from the exposure point sources with the CPM and the Trail Making tests.

21.4 Conclusions

Several studies support the concept that exposure to Mn over different temporal windows throughout the life span, even at relatively low levels of exposure, may lead to similar long-lasting neurotoxic endpoints.[76] Not only motor functions but also cognition seem to be impaired by Mn. Studies on children indicate a negative impact of excess environmental Mn on cognitive development, although there are still many questions to answer with respect to adequate biomarkers of Mn exposure, and other sources of Mn exposure, such as from spraying of Mn-based pesticides.[71] The basic knowledge on mechanism(s) by which Mn produces cognitive impairment is lacking. Future research should target the effects of Mn on cognitive domains associated with the frontal cortex and other cortical and subcortical structures.

References

1. *Manganese*, Geneva, World Health Organization, 1981 (Environmental Health Criteria, No. 17).
2. D. Mergler and M. Balwin, Early manifestations of manganese neurotoxicity in humans: an update, *Environ. Res.*, 1997, **73**, 92–100.
3. A. Iregren, Manganese neurotoxicity in industrial exposures: proof of effects, critical exposure level, and sensitive tests, *Neurotoxicology*, 1999, **20**, 315–23.
4. H. K. Hudnell, Effects from environmental manganese exposure; a review of the evidence from non-occupational studies, *Neurotoxicology*, 1999, **20**, 379–398.
5. B. S. Levy and W. J. Nassetta, Neurologic effects of manganese in humans: a review, *Int. J. Occup. Environ. Health*, 2003, **9**, 153–63.
6. H. Tjälve and J. Henriksson, Uptake of metals in the brain via olfactory pathways, *Neurotoxicology*, 1999, **20**, 181–95.
7. J. A. Menezes-Filho, C. R. Paes, A. M. Pontes, J. C. Moreira, P. N. Sarcinelli and D. Mergler, High levels of hair manganese in children living in the vicinity of a ferro-manganese alloy production plant, *Neurotoxicology*, 2009, **30**, 1207–13.
8. K. Dorner, S. Dziadzka, A. Hohn, E. Sievers, H. D. Oldigs, G. Schulz-Lell, *et al.*, Longitudinal manganese and copper balances in young infants and preterm infants fed on breast-milk and adapted cow's milk formulas, *Br. J. Nutr.*, 1989, **61**, 559–72.

9. I. Mena, K. Horiuchi, K. Burke and G. C. Cotzias, Chronic manganese poisoning: individual susceptibility and absorption of iron, *Neurology*, 1969, **19**, 1000–6.

10. M. C. Newland, T. L. Ceckler, J. H. Kordower and B. Weiss, Visualizing manganese in the primate basal ganglia with magnetic resonance imaging, *Exp. Neurol.*, 1989, **106**(3), 251–8.

11. H. Eriksson, J. Tedroff, K. A. Thuomas, S. M. Aquilonius, P. Hartvig, K. J. Fasth, *et al.*, Manganese induced brain lesions in Macaca fascicularis as revealed by positron emission tomography and magnetic resonance imaging, *Arch. Toxicol.*, 1992, **66**(6), 403–7.

12. Y. Kim, K. S. Kim, J. S. Yang, I. J. Park, E. Kim, Y. Jin, *et al.*, Increase in signal intensities on T1-weighted magnetic resonance images in asymptomatic manganese-exposed workers, *Neurotoxicology*, 1999, **20**(6), 901–7.

13. E. A. Kim, H. K. Cheong, D. S. Choi, J. Sakong, J. W. Ryoo, I. Park, *et al.*, Effect of occupational manganese exposure on the central nervous system of welders: 1H magnetic resonance spectroscopy and MRI findings, *Neurotoxicology*, 2007, **28**(2), 276–83.

14. G. E. Alexander, M. R. DeLong and P. L. Strick, Parallel organization of functionally segregated circuits linking basal ganglia and cortex, *Annu. Rev. Neurosci.*, 1986, **9**, 357–81.

15. J. A. Saint-Cyr, Frontal–striatal circuit functions: context, sequence, and consequence, *J. Int. Neuropsychol. Soc.*, 2003, **9**(1), 103–27.

16. Y. Chudasama and T. W. Robbins, Functions of frontostriatal systems in cognition: comparative neuropsychopharmacological studies in rats, monkeys and humans, *Biol. Psychol.*, 2006, **73**(1), 19–38.

17. T. E. Hazy, M. J. Frank and R. C. O'Reilly, Towards an executive without a homunculus: computational models of the prefrontal cortex/basal ganglia system, *Philos. Trans. R. Soc. London, Ser. B*, 2007, **362**(1485), 1601–13.

18. A. W. Dobson, K. M. Erikson and M. Aschner, Manganese neurotoxicity, *Ann. N. Y. Acad. Sci.*, 2004, **1012**, 115–28.

19. C. H. Kern, G. D. Stanwood and D. R. Smith, Preweaning manganese exposure causes hyperactivity, disinhibition, and spatial learning and memory deficits associated with altered dopamine receptor and transporter levels, *Synapse*, 2010, **64**(5), 363–78.

20. V. A. Fitsanakis, C. Au, K. M. Erikson and M. Aschner, The effects of manganese on glutamate, dopamine and gamma-aminobutyric acid regulation, *Neurochem. Int.*, 2006, **48**(6-7), 426–33.

21. M. Meyer-Baron, M. Schäper, G. Knapp, R. Lucchini, S. Zoni, R. Bast-Pettersen, D. G. Ellingsen, Y. Thomassen, S. He, H. Yuan, Q. Niu, X. L. Wang, Y. J. Yang, A. Iregren, B. Sjögren, M. Blond, P. Laursen, B. Netterstrom, D. Mergler, R. Bowler and C. van Thriel, The neurobehavioral impact of manganese: results and challenges obtained by a meta-analysis of individual participant data, *Neurotoxicology*, 2013, **36**, 1–9.

22. R. M. Bowler, S. Gysens, E. Diamond, A. Booty, C. Hartney and H. A. Roels, Neuropsychological sequelae of exposure to welding fumes in a group of occupationally exposed men, *Int. J. Hyg. Environ. Health*, 2003, **206**, 517–529.
23. R. M. Bowler, S. Gysens, E. Diamond, S. Nakagawa, M. Drezgic and H. A. Roels, Manganese exposure: neuropsychological and neurological symptoms and effects in welders, *Neurotoxicology*, 2006, **27**, 315–326.
24. R. M. Bowler, H. A. Roels, S. Nakagawa, M. Drezgic, E. Diamond, R. Park, W. Koller, R. P. Bowler, D. Mergler, M. Bouchard, D. Smith, R. Gwiazda and R. L. Doty, Dose-Effect Relations Between Manganese Exposure And Neurological, Neuropsychological And Pulmonary Function In Confined Space Bridge Welders, *Occup. Environ. Med.*, 2007a, **64**(3), 167–77.
25. R. M. Bowler, S. Nakagawa, M. Drezgic, H. A. Roels, R. M. Park, E. Diamond, D. Mergler, M. Bouchard, R. P. Bowler and W. Koller, Sequelae of fume exposure in confined space welding: a neurological and neuropsychological case series, *Neurotoxicology*, 2007b, **28**(2), 298–311.
26. D. G. Ellingsen, R. Konstantinov, R. Bast-Pettersen, L. Merkurjeva, M. Chashchin, Y. Thomassen and V. Chashchin, A neurobehavioral study of current and former welders exposed to manganese, *Neurotoxicology*, 2008, **29**, 48–59.
27. S. Zoni, E. Albini and R. Lucchini, Neuropsychological testing for the assessment of manganese neurotoxicity: a review and a proposal, *Am. J. Ind. Med.*, 2007, **50**(11), 812–30.
28. T. R. Guilarte, Manganese neurotoxicity: new perspectives from behavioral, neuroimaging, and neuropathological studies in humans and non-human primates, *Front. Aging Neurosci.*, 2013, **24**(5), 23.
29. G. A. Wasserman, X. Liu, F. Parvez, H. Ahsan, D. Levy, P. Factor-Litvak, *et al.*, Water manganese exposure and children's intellectual function in Araihazar, Bangladesh, *Environ. Health Perspect.*, 2006, **114**(1), 124–9.
30. R. O. Wright, C. Amarasiriwardena, A. D. Woolf, R. Jim and D. C. Bellinger, Neuropsychological correlates of hair arsenic, manganese, and cadmium levels in school-age children residing near a hazardous waste site, *Neurotoxicology*, 2006, 27(2), 210–6.
31. Y. Kim, B. N. Kim, Y. C. Hong, M. S. Shin, H. J. Yoo, J. W. Kim, S. Y. Bhang and S. C. Cho, Co-exposure to environmental lead and manganese affects the intelligence of school-aged children, *Neurotoxicology*, 2009, **30**(4), 564–71.
32. H. Riojas-Rodriguez, R. Solis-Vivanco, A. Schilmann, S. Montes, S. Rodriguez, C. Rios, *et al.*, Intellectual function in Mexican children living in a mining area and environmentally exposed to manganese, *Environ. Health Perspect.*, 2010, **118**(10), 1465–70.
33. M. F. Bouchard, S. Sauvé, B. Barbeau, *et al.*, Intellectual impairment in school-age children exposed to manganese from drinking water, *Environ. Health Perspect.*, 2011, **119**(1), 138–43.

34. J. A. Menezes-Filho, O. Novaes Cde, J. C. Moreira, P. N. Sarcinelli and D. Mergler, Elevated manganese and cognitive performance in school-aged children and their mothers, *Environ. Res.*, 2011, **111**(1), 156–63.

35. J. E. Ericson, F. M. Crinella, K. A. Clarke-Stewart, V. D. Allhusen, T. Chan and R. T. Robertson, Prenatal manganese levels linked to childhood behavioral disinhibition, *Neurotoxicol. Teratol.*, 2007, **29**(2), 181–7.

36. C. F. Carvalho, J. A. Menezes-Filho, V. P. Matos, J. R. Bessa, J. Coelho-Santos, G. F. Viana, N. Argollo and N. Abreu, Elevated airborne manganese and low executive function in school-aged children in Brazil, *Neurotoxicology*, 2013, DOI: 10.1016/j.neuro.2013.11.006.

37. P. He, D. H. Liu and G. Q. Zhang, Effects of high-level-manganese sewage irrigation on children's neurobehavior, *Zhonghua Yufang Yixue Zazhi*, 1994, **28**, 216–8.

38. G. A. Wasserman, X. Liu, F. Parvez, P. Factor-Litvak, H. Ahsan, D. Levy, *et al.*, Arsenic and manganese exposure and children's intellectual function, *Neurotoxicology*, 2011, **32**, 450–7.

39. K. Khan, G. A. Wasserman, X. Liu, *et al.*, Manganese exposure from drinking water and children's academic achievement, *Neurotoxicology*, 2012, **33**(1), 91–7.

40. S. Y. Bhang, S. C. Cho, J. W. Kim, Y. C. Hong, M. S. Shin, H. J. Yoo, I. H. Cho, Y. Kim and B. N. Kim, Relationship between blood manganese levels and children's attention, cognition, behavior, and academic performance--a nationwide cross-sectional study, *Environ. Res.*, 2013, **126**, 9–16.

41. B. Claus Henn, A. S. Ettinger, J. Schwartz, M. M. Téllez-Rojo, H. Lamadrid-Figueroa, M. Hernández-Avila, L. Schnaas, C. Amarasiriwardena, D. C. Bellinger, H. Hu and R. O. Wright, Early postnatal blood manganese levels and children's neurodevelopment, *Epidemiology*, 2010, **21**(4), 433–439.

42. A. C. Farias, A. Cunha, C. R. Benko, J. T. McCracken, M. T. Costa, L. G. Farias and M. L. Cordeiro, Manganese in children with attention-deficit/hyperactivity disorder: relationship with methylphenidate exposure, *J. Child Adolesc. Psychopharmacol.*, 2010, **20**(2), 113–8.

43. R. R. Eastman, T. P. Jursa, C. Benedetti, R. G. Lucchini and D. R. Smith, Hair as a biomarker of environmental manganese exposure, *Environ. Sci. Technol.*, 2013, **47**(3), 1629–37.

44. L. Takser, D. Mergler, G. Hellier, J. Sahuquillo and G. Huel, Manganese, monoamine metabolite levels at birth, and child psychomotor development, *Neurotoxicology*, 2003, **24**(4–5), 667–674.

45. D. C. Bellinger, Very low lead exposures and children's neurodevelopment, *Curr. Opin. Pediatr.*, 2008a, **20**, 172–7.

46. D. A. Cory-Slechta, B. Weiss and J. Cranmer, The environmental etiologies of neurobehavioral deficits and disorders: weaving complex outcomes and risk modifiers into the equation, *Neurotoxicology*, 2008, **29**, 759–60.

47. B. Weiss and D. C. Bellinger, Social ecology of children's vulnerability to environmental pollutants, *Environ. Health Perspect.*, 2006, **114**, 1479–85.

48. O. Naess, F. N. Piro, P. Nafstad, G. D. Smith and A. H. Leyland, Air pollution, social deprivation, and mortality: a multilevel cohort study, *Epidemiology*, 2007, **18**, 686–94.

49. D. C. Bellinger, Lead neurotoxicity and socioeconomic status: conceptual and analytical issues, *Neurotoxicology*, 2008b, **29**, 828–32.

50. B. Claus Henn, L. Schnaas, A. S. Ettinger, *et al.*, Associations of early childhood manganese and lead coexposure with neurodevelopment, *Environ. Health Perspect.*, 2012, **120**(1), 126–31.

51. R. G. Lucchini, S. Zoni, S. Guazzetti, E. Bontempi, S. Micheletti, K. Broberg, G. Parrinello and D. R. Smith, Inverse association of intellectual function with very low blood lead but not with manganese exposure in Italian adolescents, *Environ. Res.*, 2012a, **118**, 65–71.

52. R. G. Lucchini, S. Guazzetti, S. Zoni, F. Donna, S. A. Peter, A. Zacco, E. Bontempi, M. Salmistraro, N. J. Zimmerman and D. R. Smith, Tremor, olfactory and motor changes in Italian adolescents exposed to historical ferro-manganese emission, *Neurotoxicology*, 2012b, **33**(4), 687–96.

53. R. Elliott, Executive functions and their disorders, *Br. Med. Bull.*, 2003, **65**, 49–59.

54. R. Torres-Agustín, Y. Rodríguez-Agudelo, A. Schilmann, R. Solís-Vivanco, S. Montes, H. Riojas- Rodríguez, *et al.*, Effect of environmental manganese exposure on verbal learning and memory in Mexican children, *Environ. Res.*, 2012, **121**(2013), 39–44.

55. J. Couper, On the effects of black oxide of manganese when inhaled into the lungs, *Br. Ann. Med. Pharmacol.*, 1837, **1**, 41–2.

56. Y. Kim, K. S. Jeong, H. J. Song, J. J. Lee, J. H. Seo, G. C. Kim, *et al.*, Altered white matter microstructural integrity revealed by voxel-wise analysis of diffusion tensor imaging in welders with manganese exposure, *Neurotoxicology*, 2011, **32**(1), 100–9.

57. M. Meyer-Baron, G. Knapp, M. Schaper and C. van Thriel, Performance alterations associated with occupational exposure to manganese – a meta-analysis, *Neurotoxicology*, 2009, **30**(4), 487–96.

58. M. Bouchard, D. Mergler, M. Baldwin, M. Panisset, R. Bowler and H. Roels, Neurobehavioral functioning after cessation of manganese exposure: a follow-up after 14 years, *Am. J. Ind. Med.*, 2007, **50**, 83.

59. R. Park, Neurobehavioral Deficits and Parkinsonism in Occupations with Manganese Exposure: A Review of Methodological Issues in the Epidemiological Literature, *Saf. Health Work*, 2013, **4**(3), 123–135.

60. J. E. Myers, J. teWaterNaude, M. Fourie, H. B. Zogoe, I. Naik, P. Theodorou, H. Tassel, A. Daya and M. L. Thompson, Nervous system effects of occupational manganese exposure on South African manganese mineworkers, *Neurotoxicology*, 2003a, **24**, 649–56.

61. D. Mergler, M. Baldwin, S. Belanger, F. Larribe, A. Beuter, R. Bowler, *et al.*, Manganese neurotoxicity, a continuum of dysfunction: results from a community based study, *Neurotoxicology*, 1999, **20**(2–3), 327–42.

62. Y. Chang, Y. Kim, S.-T. Woo, H.-J. Song, S. H. Kim, H. Lee, Y. J. Kwon, J. H. Ahn, S.-J. Park, I.-S. Chung and K. S. Jeong, High signal intensity on magnetic resonance imaging is a better predictor of neurobehavioral performances than blood manganese in asymptomatic welders, *Neurotoxicology*, 2009, **30**, 555–563.

63. Y. Chang, J. J. Lee, J. H. Seo, H. J. Song, J. H. Kim, S. J. Bae, J. H. Ahn, S. J. Park, K. S. Jeong, Y. J. Kwon, S. H. Kim and Y. Kim, Altered working memory process in the manganese-exposed brain, *Neuroimage*, 2010, **53**(4), 1279–85.

64. Y. Chang, S. U. Jin, Y. Kim, K. M. Shin, H. J. Lee, S. H. Kim, J. H. Ahn, S. J. Park, K. S. Jeong, Y. C. Weon and H. Lee, Decreased brain volumes in manganese-exposed welders, *Neurotoxicology*, 2013, **37**, 182–9.

65. S. E. Chia, S. C. Foo, S. L. Gan, J. Jeyaratnam and C. S. Tian, Neurobehavioral functions among workers exposed to manganese ore, *Scand. J. Work, Environ. Health*, 1993, **19**, 264–270.

66. R. Lucchini, L. Selis, D. Folli, P. Apostoli, A. Mutti, O. Vanoni, A. Iregren and L. Alessio, Neurobehavioral effects of manganese in workers from a ferroalloy plant after temporary cessation of exposure, *Scand. J. Work, Environ. Health*, 1995, **21**, 143–149.

67. R. Lucchini, P. Apostoli, C. Perrone, D. Placidi, E. Albini, P. Migliorati, D. Mergler, M. P. Sassine, S. Palmi and L. Alessio, Long-term exposure to "low levels" of manganese oxides and neurofunctional changes in ferroalloy workers, *Neurotoxicology*, 1999, **20**(2–3), 287–97.

68. J. E. Myers, M. l. Thompson, S. Ramushu, T. Young, M. F. Jeebhay, L. London, E. Esswein, K. Renton, A. Spies, A. Boulle, I. Naik, A. Iregren and D. J. Rees, The nervous system effects of occupational exposure on workers in a South African manganese smelter, *Neurotoxicology*, 2003b, **24**, 885–94.

69. H. Yuan, S. He, M. He, Q. Niu, L. Wang and S. Wang, A comprehensive study on neurobehavior, neurotransmitters and lymphocyte subtest alteration of Chinese manganese welding workers, *Life Sci.*, 2006, **78**, 1324–8.

70. D. Mergler, G. Huel, R. Bowler, A. Iregren, S. Belanger, M. Baldwin, R. Tardif, A. Smargiassi and L. Martin, Nervous system dysfunction among workers with long-term exposure to manganese, *Environ. Res.*, 1994, **64**, 151–80.

71. H. A. Roels, R. M. Bowler, Y. Kim, B. Claus Henn, D. Mergler, P. Hoet, V. V. Gocheva, D. C. Bellinger, R. O. Wright, M. G. Harris, Y. Chang, M. F. Bouchard, H. Riojas-Rodriguez, J. A. Menezes-Filho and M. M. Téllez-Rojo, Manganese exposure and cognitive deficits: A growing concern for manganese neurotoxicity, *Neurotoxicology*, 2012, **33**(4), 872–80.

72. M. M. Finkelstein and M. Jerrett, A study of the relationship between Parkinson's disease and markers of traffic derived and environmental manganese air pollution in two Canadian cities, *Environ. Res.*, 2007, **104**(3), 420–32.

73. R. G. Lucchini, E. Albini, L. Benedetti, S. Borghesi, R. Coccaglio, E. Malara, G. Parrinello, S. Garattini, S. Resola and L. Alessio, High prevalence of parkinsonian disorders associated to manganese exposure in the vicinities of ferroalloy industries, *Am. J. Ind. Med.*, 2007, **50**(11), 788–800.

74. R. M. Bowler, M. Harris, V. Gocheva, K. Wilson, Y. Kim, S. I. Davis, *et al.*, Anxiety affecting parkinsonian outcome and motor efficiency in adults of an Ohio community with environmental airborne manganese exposure, *Int. J. Hyg. Environ. Health*, 2012, **215**(3), 393–405.

75. R. G. Lucchini, S. Guazzetti, S. Zoni, F. Donna, S. Micheletti, E. Bontempi, L. Borgese, R. Ferri, S. Marchetti, C. Benedetti, C. Fedrighi, M. Peli and D. R. Smith, Neurofunctional dopaminergic impairment in elderly after lifetime exposure to manganese, *Neurotoxicology*, DOI: 10.1016/j.neuro.2014.05.006.

76. R. G. Lucchini and N. Zimmerman, Lifetime cumulative exposure as a threat for neurodegeneration: need for prevention strategies on a global scale, *Neurotoxicology*, 2009, **30**, 1144–1148.

CHAPTER 22

Manganese and Huntington Disease

ANDREW M. TIDBALL, TERRY JO BICHELL AND
AARON B. BOWMAN*

Vanderbilt University Medical Center, 465 21st Avenue South, 6140 MRB3,
Nashville, TN 37232-8552, USA
*Email: aaron.bowman@vanderbilt.edu

22.1 Huntington Disease Pathobiology and Environmental Influence

Huntington disease (HD) is a devastating neurodegenerative disease
presenting with impaired movement, psychological and behavioral disturbances, and cognitive decline. The most pronounced symptoms are motor
impairments including chorea (a dance-like involuntary movement), motor
impersistence, and deterioration of coordination and motor skills.[1] In late
stage HD, or earlier in the juvenile onset form, patients can present as
hypokinetic and rigid, similar to patients with Parkinson's disease.[2] HD
typically results in death in the second decade after the time of clinical
diagnosis, often from cardiovascular disease or complications such as falls,
difficulty swallowing, aspiration leading to pneumonia, and an increased
suicide rate.[3] The motor symptoms are thought to result primarily from
neuronal degeneration in the caudate and putamen regions of the basal
ganglia. The gamma aminobutyric acid (GABA)ergic medium spiny neurons in
these regions are the most susceptible to HD-related degeneration;[4] however,
neurons in the substantia nigra, globus pallidus, thalamus, subthalamic

Issues in Toxicology No. 22
Manganese in Health and Disease
Edited by Lucio G. Costa and Michael Aschner
Published by the Royal Society of Chemistry, www.rsc.org

nucleus, subregions of the hypothalamus, and cortical layers 3, 5, and 6 are also vulnerable to cell death in HD.[1,2]

HD is caused by a *CAG* repeat expansion within the coding region of the Huntingtin gene (*HTT*).[5] Individuals with more than 39 repeats have complete disease penetrance.[6] The length of the repeat also has a strong inverse relationship to the age of onset; however, additional genetic and environmental factors play a critical role.[7,8] Although the length of the repeat region contributes over half of the variability in age of onset, one landmark epidemiological study attributed the majority of the remaining age of onset variability to environmental factors.[9] This contribution of an environmental modifier to age of onset is thought to increase with shorter repeat lengths. In fact, patients with a repeat length of 40 can have ages of onset spanning four decades.[9] Even monozygotic twins with HD have shown differences in symptomatic manifestations and age of onset of up to 7 years, despite identical genetics.[10–13] These data taken together suggest that environmental exposures play a role in HD disease progression though few specific environmental modifiers have been identified.[14–16] Thus far, non-genetic influences such as environmental enrichment, exercise, and diet interventions have been shown to delay disease progression.[14–21] Factors that may accelerate the pace of HD are those that cause oxidative stress and mitochondrial damage, such as heavy metals and pollutants which are known to cause degeneration of basal ganglia subregions.[22–24] Although it has been more than 20 years since the *HTT* gene was identified, it has been difficult to identify environmental disease modifiers because the function of the ubiquitous Htt protein throughout development is still only partially explained.

Knockout of the *Htt* gene is embryonically lethal in mouse models.[25–27] Additionally, partial knockdown of *Htt* leads to neurodevelopmental abnormalities in both zebrafish and mice.[28,29] Mutant *HTT* is able to rescue both of these phenotypes, suggesting that neurodegeneration in HD results from a toxic gain of function rather than a of loss of wild-type *HTT* function.[29] Interestingly, *HTT* is understood to be expressed in all tissues of the body at all developmental time points and, considering the late onset of neurodegeneration (typically between 30 and 50 years of age), this may indicate that mutant *HTT* expression is not sufficient to cause pathology but requires normal age-related environmental stress.[30] This potential environmental role in HD opens up the possibility of delaying onset by avoiding potential toxic exposures or by enriching environments and increasing nutritional protection and thereby altering the impact of age-related environmental stress. Furthermore, the identification of specific pollutants or heavy metals, which may serve as environmental modifiers, could lead to potential targets for pharmaceutical intervention to delay the onset of HD symptoms.

22.2 A History of HD and Metal Ions

Heavy metals are closely linked with both function and dysfunction in the basal ganglia, and are, therefore, likely candidates to be the environmental modifiers for age of onset in HD.[23] The high metabolic requirements of the brain as well as the need for tight regulation and detoxification of reactive oxygen species necessitate relatively high concentrations of the biologically essential heavy metal ions (*i.e.* Fe^{2+}, Mn^{2+}, Cu^{2+}, and Zn^{2+}) which are important cofactors for enzymes that regulate these processes. On the other hand, excessive metal ion concentrations result in increased oxidative stress, mitochondrial dysfunction, protein aggregation, and apoptosis. For this reason, heavy metal toxicity and neurodegeneration have many shared mechanisms and symptomatic features.[22,31,32] Excess accumulation of heavy metals in the brain, either from exposure (*e.g.* manganism) or *via* genetic disturbance of metal ion homeostasis (*e.g.* Wilson's disease), can lead to degeneration of brain regions, particularly in the basal ganglia.[33,34] Abnormally low concentrations of micronutritive metals in the brain are also detrimental to brain development and function.[35-40] The similarities in pathobiology between neurodegeneration and metal toxicity or deficiency warrant further investigation into shared mechanisms. Because manganism and HD overlap in symptomology, a derangement in manganese exposure or handling may explain some of the variation in timing of disease onset, as well as some of the function of the Htt protein in the brain.

22.2.1   Manganese and HD

Few studies have measured the level of manganese in the brains of patients with HD. Dexter *et al.* (1991) did not find any differences in manganese levels, but a more recent study, by Rosas *et al.* (2012), found a significant decrease in manganese concentration in parts of HD cortex.[41,42] Additionally, increased iron accumulation in the striatum and pallidum of patients with HD has been seen in multiple studies,[41-43] and manganese has a strong inverse correlation to iron concentration in the basal ganglia.[44] Therefore, increased iron accumulation could result in a slight reduction in manganese levels in HD striatum beyond the detection of the previous studies. Manganese is known to accumulate in the globus pallidus and caudate nucleus, two areas highly susceptible to HD neurodegeneration.[45,46] This accumulation may highlight a specific necessity of manganese for proper function, and may also make these brain areas more susceptible to fluctuations in manganese levels.

Recently, an immortalized striatal murine HD model cell line (ST*Hdh*$^{Q7/Q7}$ and ST*Hdh*$^{Q111/Q111}$) showed differential toxicological sensitivity to manganese (Mn^{2+}) and cadmium (Cd^{2+}) but no other metal ions tested (Fe^{3+}, Cu^{2+}, Pb^{2+}, Co^{2+}, Zn^{2+}, Ni^{2+}).[47] Mutant Huntingtin expression conferred a survival advantage under manganese exposure conditions in these cells, due to a dramatic decrease in manganese uptake.[47] This differential manganese level was also observable under basal conditions despite the extremely low concentration of manganese in normal cell culture conditions.[47] Expression of human mutant Huntingtin also caused a decrease in manganese levels in the brains of manganese-exposed mice.[47] Interestingly, this difference was seen only in the mouse striatum, the specific region that degenerates in the brains of patients with HD, and not the cerebellum, cortex, or hippocampus, which are largely spared early in disease progression. The proximal cause for this alteration in manganese homeostasis is currently under investigation; however, the possibility of a manganese-handling defect in HD will be further explored in this chapter.

Glia make up 50% of total brain tissue and >75% of the cerebral cortex.[48] These cells are reported to contain ~80% of total brain manganese, and gliosis is a known manifestation of HD.[49–51] A more dramatic deficit in the amount of manganese in the neurons of patients with HD could, therefore, exist despite relatively small differences in overall cortical manganese and no observable differences in manganese in the basal ganglia, when measured without regard to cell type.[41,42] The *in vivo* striatal differences revealed in the HD mice were seen using a manganese over-exposure paradigm but not under basal conditions.[47] Therefore, the human HD caudate and putamen may have a cell-type specific manganese defect that would only become apparent with over-exposure. To address a cell-type specific defect, *ex vivo* primary cell culture of both medium spiny neurons and astrocytes from the HD mouse striatum are warranted.

In addition to the differences in manganese accumulation, manganese exposure exacerbated abnormalities in dendritic complexity and arborization seen in the medium spiny neurons of YAC128 HD model mice; these mice contain a mutant human HTT gene containing 128 CAG repeats.[52] In this same *in vivo* model, curiously, the HD genotype suppressed the Mn-dependent decrease in striatal dopamine levels.[52] Therefore, HD brains may have manganese deficits that can endanger processes requiring manganese as a cofactor, while also being more susceptible to some aspects of manganese toxicity. These studies into the interactions between manganese and HD have produced useful insights into the potential altered regulation of manganese homeostasis; however, the full range of potential reasons for the underlying disease–toxicant interaction has not yet been explored. The breadth of HD-related cellular phenotypes is diverse and multifactorial.[2,53] For this reason, a clear hypothesis connecting the similarities in manganese and HD pathology is needed. Section 22.3 will outline numerous HD-related phenotypes that may have relevance to manganese biology. Within each section, HD-associated manganese-containing enzymes and similarities in manganese deficiency and/or toxicity to HD pathology will be highlighted because of the critical balance of this metal.

22.2.2 Iron Homeostasis and HD

In addition to abnormalities in manganese homeostasis, HD has also been associated with abnormal iron homeostasis. Patients with HD have increased accumulation of iron in the caudate, putamen, and pallidum.[41–43] In the same studies, ferritin levels were unchanged between patients with HD and control subjects in all of the brain regions examined; however, other authors have found increased ferritin levels in striatum and cortex of HD brains, with staining primarily in microglia.[54,55] Neuroferritinopathy, a condition caused by mutation in the ferritin light chain gene (*FTL1*), results in accumulation of iron and ferritin in the basal ganglia.[56] Neuroferritinopathy is associated with low serum ferritin levels, accumulation of iron and ferritin in the basal ganglia, oxidative stress, and signs of mitochondrial dysfunction, which are all seen in HD.[57,58] Neuroferritinopathy results in neurodegeneration in the basal ganglia as well as other symptomology similar to HD, and patients are often misdiagnosed as having HD.[57,59] Several studies have also linked wild-type Huntingtin protein function directly to iron homeostasis. Partial knock-down of Huntingtin in a zebrafish embryo model resulted in a lack of bioavailable iron for hemoglobin in red blood cells despite adequate levels of total iron in the cells. Therefore, Huntingtin may be needed for iron release from endocytic vesicles.[28] Increasing the morpholino concentration to knock-down more Huntingtin in this fish model resulted in a dramatic effect on overall central nervous system (CNS) development, with reduced size of brain and eye, and brain necrosis, which may or may not be linked to iron availability.[28]

This close link between iron homeostasis and HD suggested that an iron trafficking defect could have caused the altered manganese concentration in striatal neurons seen by Williams *et al.* 2010.[47] A shared defect seems likely because many iron transport processes (*e.g.* transferrin, DMT1) also traffic manganese ions.[60] Iron transporters have been assessed in cultured striatal lines, and the transferrin receptor was found to be significantly lower (<50%) in the mutant cells compared with wild type.[61] High concentrations of iron were then used to saturate iron-dependent transporters, which partially blocked manganese uptake. This paradigm diminished Mn-dependent cell death in both cell lines equally, suggesting that iron transporters, such as the transferrin receptor, do account for a portion of manganese uptake in the mouse striatal cells but do not account for altered manganese accumulation in HD cells. Accumulation of iron (Fe^{3+}) in the mutant cells was also lower in HD cells; however, the fold change was much less than that of manganese and seemed to correlate with the difference in transferrin receptor expression. Unlike manganese levels, iron levels were the same in both cell lines when they were not exposed to excess iron.[61]

In addition to the need for bioavailable iron for many key cellular processes, excess iron and/or copper can increase oxygen free radicals *via* Fenton chemistry. In this process, a metal ion cycles between redox states II and III by disproportionate cleavage of a hydrogen peroxide molecule, resulting in the formation of water and a highly reactive free radical (either $HO^{\bullet}$ or $HOO^{\bullet}$).[62] These reactive oxygen species (ROS) can then cause DNA damage and both protein and lipid peroxidation, processes that cause cellular dysfunction and activate necrotic and apoptotic events. HD models demonstrate higher levels of ROS and increased sensitivity to ROS mediated damage.[63]

Derangement of metal ion uptake and transport may partially contribute to the neurodegeneration seen in HD through this iron-related formation of ROS, along with damage caused by imbalances in other metal ions such as manganese and copper.

22.2.3 Copper and HD

In addition to iron, copper was also found to be increased in the brains of patients with HD in some studies, but not others.[41,42] Accumulations were seen in the striatum and cortex of R6/2 HD model mice.[64] Expression of mutant exon1 of the *Huntingtin* gene in yeast led to increased expression of eight copper binding proteins including metallothioneins 1 and 2 (MT1 and MT2).[65] These two genes also showed increased mRNA expression in the brains of patients with HD.[66] Copper has been shown to facilitate mutant Huntingtin aggregation.[64] Mutant Huntingtin is known to cause inclusions *via* aggregation in the nucleus and cytoplasm.[67] These aggregates were thought to be important mediators of HD pathology; however, recent studies have shown a lack of direct connection between cell death and aggregates, indicating a potentially protective effect of aggregation.[68]

Despite this potentially positive role of copper in inducing mutant aggregation, reduction of copper *via* chelation or *via* genetic knock-down of copper transporters has mitigated HD pathology in both mouse and fly models of HD while simultaneously reducing aggregation.[69,70] Copper treatment also further decreased the life span of HD flies.[70] These models, however, express only exon1 of the mutant gene, and aggregation may be a necessary step in pathology with these imperfect model systems. Therefore, copper may play a different (or no) role in patients and experimental animals expressing the full-length mutant *Huntingtin* gene. Alternatively, Boll *et al.* (2008) showed decreased activity of Cu/Zn superoxide dismutase (SOD) in patients with HD, with a trend for increased free copper in the patients' cerebrospinal fluid (CSF).[71] Given that free copper is a Fenton reagent leading to increased ROS production, copper may add to HD pathology in this way. The alterations in iron, copper, and manganese seen in patients with HD and in HD model organisms suggest that metal-handling may be one of the functions of the Huntingtin protein that are affected by expansion of the pathogenic polyglutamine domain.

22.3 Manganese Essentiality and Toxicity in HD Related Phenotypes

As discussed above, patients with HD and mouse *in vivo* and *in vitro* models exhibit decreased manganese levels either upon basal and/or elevated exposure to manganese (Section 22.2.1). Manganese deficiency results in many dysfunctional manifestations similar to HD, including urea cycle dysfunction, altered glutamate regulation, increased oxidative stress, and metabolic disturbances including altered IGF–AKT signaling. The following subsections will investigate the interrelationship among these manganese-dependent processes and HD-related pathological phenotypes, highlighting enzymes in each process that use manganese as a cofactor.

22.3.1 Regulation of Amines and Nitric Oxide

Nitrogen is one of the most important elements in biological systems. Amino acids, protein monomers, all contain one amine group containing nitrogen, and many contain an additional amine on their R-subunit (*i.e.* asparigine, arginine, glutamine, and lysine). During normal cellular biochemical processes, deamination releases ammonia into the cytoplasm and eventually the bloodstream, but elevated ammonia levels are toxic and can lead to encephalopathy and death. To regulate ammonia levels, the liver produces urea to be excreted into urine to deplete the human body of excess amines. Arginine is an important precursor in the urea cycle as well as in an additional process called the nitric oxide cycle. Nitric oxide (NO) is a compound produced from arginine *via* nitric oxide synthase (NOS). NO regulates many important functions and in the CNS it acts as a neurotransmitter and neuromodulator, particularly regulating glutamate release in the cortex, striatum, and hippocampus. Although its production is vital to many biological functions, excess NO can be neurotoxic *via* glutamate excitotoxicity and oxidative stress.[72,73] Increased arginine levels can activate NOS activity increasing NO signaling.[74] Therefore, although the liver is the primary producer of excreted urea, the brain also converts arginine to ornithine and urea, to regulate the production of NO.

22.3.1.1 Arginase, a Manganese-Dependent Enzyme

Arginase is a manganese-dependent enzyme that catalyzes the conversion of arginine to ornithine. Arginase contains a binuclear manganese cluster of six Mn ions that is vital for its function and cannot be substituted with magnesium like many other manganese-containing enzymes.[75] Dietary manganese deficiency decreases arginase activity and increases NOS activity.[76,77] Two genes (*ARG1* and *ARG2*) encode the two isozymes, ARG1 and ARG2. These two enzymes have identical enzymatic function, but ARG1 is primarily in the cytoplasm whereas ARG2 is located in the mitochondria. ARG1 has been found at higher levels in the brain than ARG2.[78] ARG2 expression is especially high in particular areas of the brain affected by HD, including the putamen and ventral striatum.[79]

Evidence of altered arginase activity has been reported in both mouse models of HD and peripheral measures from patients with HD. At least two different HD model mice (the "R6/2" and a Hdh knock-in model) have shown increased ammonia and citrulline levels, both indicative of decreased arginase activity.[80,81] Similarly, patients with HD were shown to have increased blood citrulline levels, indicating increased NOS activity.[80] Arginase transcript levels were also found to be reduced in HD mouse liver.[82] Other groups using a non-neuronal cell model showed increased *ARG1* expression, NOS, and arginase activity in HD mutant cells that correlated with an increase in arginine uptake.[83] Dysregulation of arginase by altered manganese homeostasis/bioavailability could result in localized ammonia and NO-related toxic processes as well as altered production of glutamate and GABA, which require the conversion of ornithine from arginine as a precursor for their synthesis.[84] These HD phenotypes could be the result of altered bioavailability of manganese, as has been seen in manganese exposed mouse striatal cells and striatal tissues.[47]

Knock-down of neuronal NOS in the R6/1 mouse model of HD delayed disease symptoms and increased dietary arginine accelerated symptom onset by increasing NO synthesis.[85,86] This interaction between NO level and HD pathology may be due to the necessity of NO in a process called glutamate excitotoxicity, a well-studied HD pathological phenotype.[72] Excessive NO can also interact with superoxide radicals to form more ROS as well as toxic reactive nitrogen species (RNS) that can cause severe cellular dysfunction. Increased oxidative stress due to reactive oxygen and nitrogen species (ROS and RNS) has been reported for several HD model systems.[63,87] Increased nitrates and nitrites, which are downstream products of NO signaling, are found in the CSF of patients with HD.[71] Arginase 1 regulation of nitric oxide production has been shown to be key to survival of trophic factor-deprived motor neurons.[88] Thus, altered arginase activity *via* altered manganese homeostasis in HD could potentially contribute to a wide range of HD-related phenotypes.

22.3.1.2 Arginase, Polyamines, and HD

Arginase activity increases polyamine synthesis downstream of the arginase reaction, causing beneficial effects. The reduced arginase activity seen in HD models could also lead to dysfunction resulting from reduced production of these polyamines. The production of spermidine in particular has been found to be neuroprotective and it actually promotes axonal regeneration.[89] Spermine is a scavenger of ROS and may therefore be protective against HD pathological processes.[90] Spermine can rescue HD-related memory deficits in a rodent model.[91] One study found that spermine increased the rate of aggregate formation in a poly-Q cell model with increased cell death;[83] however, as previously mentioned, aggregation is no longer thought to be a part of the disease process and may actually be protective against mutant Huntingtin pathogenesis.[68,92] In fact, this may be part of the protective nature of spermine in HD.

22.3.2 Glutamate Excitotoxicity

Glutamate excitotoxicity is a well-supported pathophysiological process in HD.[93,94] This neurotoxic mechanism involves excessive glutamate-induced intracellular calcium release, resulting in mitochondrial dysfunction. Figure 22.1 illustrates potential steps of glutamate cycling and signaling, the first steps of excitotoxicity, which manganese may impinge on. There is no

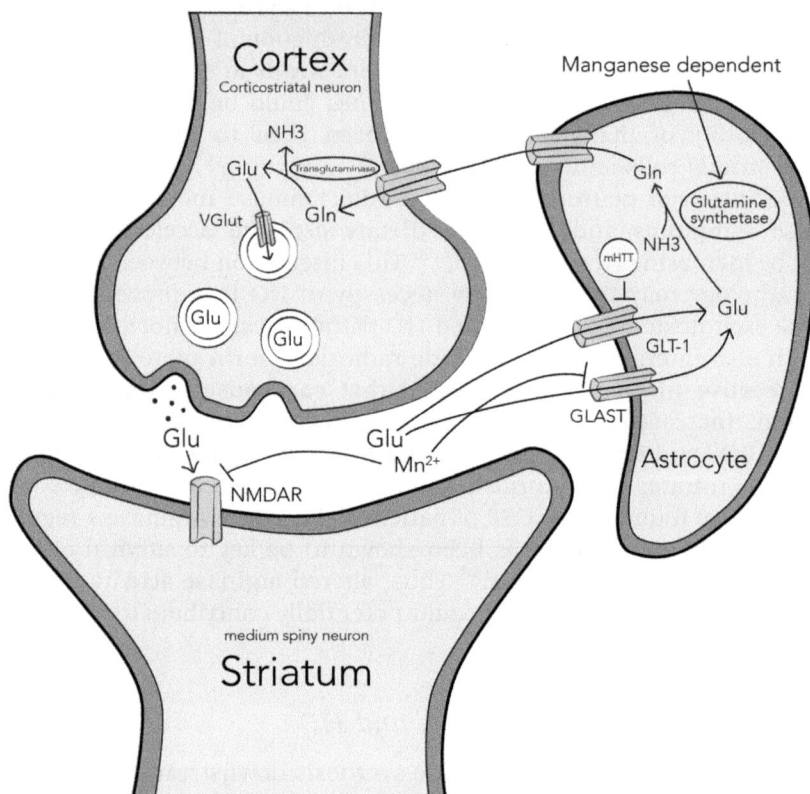

Figure 22.1   Effects of manganese and mutant Huntingtin on glutamate signaling and cycling at the corticostriatal synapse. A cortical projection neuron (presynaptic) is synaptically joined to the dendrite of a medium spiny neuron in the striatum (postsynaptic). Upon depolarization, glutamate is released from the presynaptic terminal where it interacts with glutamate receptors such as the NDMA receptor. This receptor is calcium permeable and is important for calcium signaling, but excess glutamate can cause excess calcium influx leading to excitotoxicity. Manganese is a known inhibitor of the NMDA receptor. Glutamate is cleared from the synapse by supporting astrocytes through glutamate transporters, GLAST and GLT-1, whose expression is known to be inhibited by manganese and mutant Huntingtin (mHTT) respectively. Glutamate is converted to glutamine in the astrocyte by glutamine synthetase, a manganese-dependent enzyme. Glutamine is released from the astrocyte to be taken back up in the presynaptic terminal. It is then converted back into glutamate *via* transglutaminase, releasing ammonia. The glutamate is then repackaged into vesicles by a VGlut transporter.
(Figure illustrated by Angela Tidball.)

consensus yet about how mutant Huntingtin leads to elevated exictotoxicity. This section will outline several steps in the process from glutamate release to mitochondrial dysfunction and identify points in the process where either manganese and/or mutant Huntingtin may play a role in this pathophysiological process.

22.3.2.1 Glutamate Release and Clearance

Glutamate is the primary excitatory neurotransmitter of the CNS. In the caudate and putamen, a large component of excitatory tone comes from glutamatergic cortical presynaptic axon terminals synapsing with dendrites of the medium spiny neurons. Glutamate then binds to AMPA and N-methyl-D-aspartate (NMDA) glutamate receptors to cause membrane depolarization. NMDA receptors are permeable to Ca^{2+} in addition to Na^+ and K^+. This influx of calcium during depolarization is important for long-term potentiation and memory formation. Excessive glutamate released at the synapse can result in abnormal Ca^{2+} influx.

To maintain efficient neurotransmission, glutamate must be efficiently cleared from the synapse after release. Astrocytes typically account for 80% of synaptic glutamate uptake.[95] Manganese decreases glutamate uptake and decreases GLAST (astrocytic glutamate transporter) expression in astrocytes.[96,97] Manganese can also increase synaptic glutamate release and inhibit NMDA receptor conductance as a channel blocker.[98,99] HD models have shown reductions in protein and mRNA of another astrocytic glutamate transporter, GLT1, as well as reductions in astrocytic glutamate uptake.[94,100] As already discussed, the manganese-dependent enzyme arginase can also play a role in glutamate release by regulating production of ornithine, a precursor of glutamate. In addition to contributing to glutamate release and clearance, manganese also regulates glutamate levels through the process of conversion of glutamate to glutamine.

22.3.2.2 Glutamine Synthetase, a Mn-Dependent Enzyme

Glutamate that is taken up by astrocytes is converted into glutamine by glutamine synthetase within the glial cell. Glutamine can then be exported from the astrocyte for reuptake by the presynaptic neuron for conversion back into glutamate, to help maintain neurotransmitter levels. Glutamine synthetase is the most abundant manganese-containing enzyme in the brain and is primarily expressed in glia.[101] Like arginase, glutamine synthetase has a high specificity for manganese over magnesium.[102]

Loss of glutamine synthetase function by genetic mutation or inhibition of gene expression can lead to epileptic seizures.[103,104] These seizures are the result of glutamate–glutamine cycle dysregulation, leading to excess extracellular glutamate.[105] Elevated risk of seizures occurs in HD, and post-mortem brains of patients with HD show a significant decrease in glutamine synthetase activity in the caudate and putamen compared with matched controls.[106,107] Mouse models have shown reductions in glutamine synthetase mRNA levels as well as activity.[94,100] Additionally, patients with juvenile onset HD have even further increased risk of seizures, with ~25% of patients presenting in this way.[108,109] A manganese-handling defect in isolated brain regions could potentially be reflected in reduced glutamine synthetase activity in HD although this direct relationship has not yet been explored.

Manganese deficiencies alone are related to an increased seizure rate in both rats and humans.[110–112] Given that glutamate is the most abundant compound in the brain, cycling between glutamate and glutamine requires a high enzymatic output *via* glutamine synthetase and transglutaminase.[113] Manganese deficiency is proposed to decrease glutamine synthetase activity, resulting in increased extracellular glutamate, which increases the propensity for seizures.

The production of glutamine from glutamate also uses a free ammonia molecule. Increased ammonia levels in hyperammonemia have been known to increase the rate of glutamine production, n potentially buffering the brain from toxic levels of ammonia.[114] Reductions in hepatic glutamine synthetase have also been associated with hyperammonemia.[115] Therefore, altered action of glutamine synthetase in HD and manganese deficiency could result in localized dysregulation of ammonia levels. Manganese is clearly a crucial player in neuronal energetics as well as intercellular detoxification processes.

22.3.2.3 Calcium Dysregulation

Calcium is tightly regulated in neurons. As mentioned above, excess glutamate increases calcium influx through NMDA receptors. NMDA and voltage-gated calcium channels allow calcium into the postsynaptic neuron during glutamate-mediated membrane depolarization. In addition, metabotropic glutamate receptors can cause calcium release from the endoplasmic reticulum *via* the inositol 1,4,5-triphosphate (IP3) pathway. These processes are vital for long-term potentiation and memory formation but can cause damage in excess. Medium spiny neurons containing mutant Huntingtin have been found to have abnormally high calcium release when exposed to the same concentration of glutamate.[116] In addition to alterations in glutamate processing in HD, which may lead to excess Ca^{2+} influx, mutant Huntingtin has been found to cause calcium dysregulation by binding to the IP3 receptor (IP3R), resulting in constitutive release of calcium.[116] Cytoplasmic calcium is buffered by mitochondria; however, when calcium concentrations become excessively high, overwhelmed mitochondria can swell, depolarize, and open the mitochondrial membrane transition pore, leading to apoptosis.[117] Interestingly, along with excess calcium, excess manganese also accumulates in the mitochondria where it can lead to dysfunction and depolarization.

22.3.3 Mitochondrial Dysfunction: Oxidative Stress and Energetics

22.3.3.1 Mitochondrial Calcium and Manganese Buffering

Mitochondrial stress can be caused by excess calcium as well as excess manganese (Figure 22.2). Manganese and calcium are taken up by the mitochondrial calcium uniporter, and manganese is released very slowly by the $Na^{(+)}$-independent efflux process.[118] These slow release kinetics help to buffer the cytosol from potentially toxic effects of these divalent cations; however, when the mitochondria become overloaded with either manganese

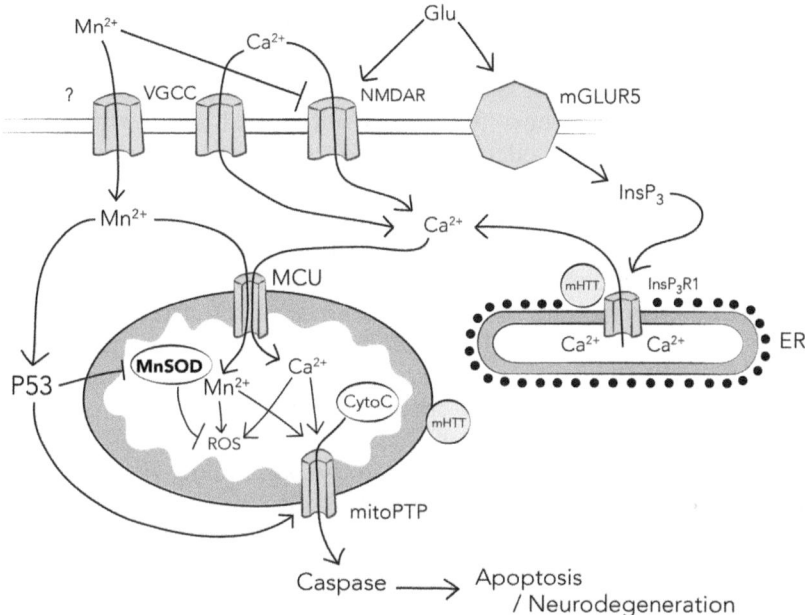

Figure 22.2 Effects of manganese and mutant Huntingtin on calcium signaling and mitochondrial function. Models of HD have excess glutamate release at the synapse, which can cause excessive influx of calcium into the cell from NMDA receptors. Mutant huntingtin (mHTT) has also been shown to cause leaking of the IP3 (InsP₃) receptor, causing dysregulated release of calcium from the endoplasmic reticulum (ER). Excess calcium is buffered by the mitochondria where it is taken up by the mitochondrial calcium uniporter (MCU); however, in excess, calcium can cause mitochondrial swelling, depolarization, and opening of the mitochondrial permeability transition pore (mitoPTP), leading to apoptosis. In the same way, excess manganese is buffered in the mitochondria and can lead to cell death in a similar mechanism. Additionally, manganese exposure has recently been shown to activate p53, which accelerates apoptosis by inhibiting manganese superoxide dismutase, a manganese-dependent free radical scavenger, and by potentiating the opening of the mitoPTP.
(Figure illustrated by Angela Tidball.)

or calcium, the imbalance can lead to impaired adenosine triphosphate (ATP) synthesis, membrane depolarization, uncoupling of the electron transport chain, and production of ROS.[119] Manganese toxicity is known to inhibit all four mitochondrial complexes of the electron transport chain *via* oxidative stress.[120] In isolated mitochondria, significant complex inhibition was achieved by 50 µM manganese that was blocked by several antioxidants, indicating that oxidative stress is a major pathway of manganese inhibition of mitochondrial complexes.[120] Additionally, manganese toxicity inhibits the release of calcium from the mitochondria, leading to a potentially toxic buildup.[118] In addition to mitochondrial toxicity due to excess manganese, reduced manganese levels are also harmful, altering the activity of the protective manganese super oxide dismutase (discussed in Section 22.3.3.2), which also leaves mitochondria vulnerable to damage (Figure 22.2). Striatal neurons are particularly sensitive to glutamate-mediated excitotoxic cell death. This apoptotic event is dependent on elevated mitochondrial calcium levels.[121] Several groups have found that mitochondria are particularly susceptible to calcium-induced membrane depolarization in the presence of mutant Huntingtin.[122]

In summary, this HD-related mitochondrial vulnerability could stem from increased cytosolic calcium levels due to altered glutamate release and clearance (see Sections 22.3.2.1–22.3.2.2), from increased calcium release from the endoplasmic reticulum (see Section 22.3.2.3), from slower recovery kinetics of mitochondrial calcium buffering,[122] or from increased mitochondrial sensitivity. In any case, depolarization of the membrane results in opening of the mitochondrial membrane permeability transition pore (MPTP), which releases cytochrome *c* and initiates apoptotic caspase signaling.[117] Striatal neurons have proven to be particularly susceptible to this process of cell death, which may explain the selective degeneration in HD.[123] Blockade of MPTP formation by either cyclosporine A or bongkrekic acid was able to block NMDA-induced and HD-related striatal apoptosis, highlighting the importance of this process in HD pathology.[93,122]

22.3.3.2   Manganese Superoxide Dismutase

Mitochondria may be especially dependent on proper manganese handling because manganese superoxide dismutase (MnSOD) is critical for detoxifying reactive oxygen species in the mitochondria, and manganese is a necessary cofactor for this activity. Manganese supplementation is known to increase MnSOD activity in lymphocytes, while iron supplementation decreases MnSOD activity, presumably due to the inverse relationship between iron and manganese accumulation.[124] Neuronal concentration of manganese is very low (<10 µM); however, the demand for high mitochondrial activity in neurons may be the reason for increased MnSOD activity compared with that found in glia.[125] Manganese deficiency results in decreased MnSOD activity and increased mitochondrial lipid peroxidation in rats,[35] and patients with HD have demonstrated increased lipid peroxidation in their CSF.[71] Manganese deficiency also causes reduced oxygen uptake and dysmorphic elongation of mitochondria, likely due to lack of ROS detoxification.[126]

Partial loss of MnSOD in heterozygous knockout mice results in increased sensitivity to 3-nitroproprionic acid.[127] 3-Nitroproprionic acid (3NPA) causes oxidative stress in neurons by inhibiting succinate dehydrogenase, complex II of the electron transport chain.[128] This uncoupling of the electron transport chain results in increased production of superoxide, a potent ROS that can result in mitochondrial dysfunction as well as oxidative damage to proteins and DNA. 3NPA is also known to cause a preferential lesion to the striatum and was commonly used to model HD in rodents and primates prior to the cloning of *HTT*.[129] Furthermore, cells expressing mutant Huntingtin are more susceptible to this toxicant.[47,130] Together, these data argue for a mechanistic link between MnSOD and HD *via* mitochondrial biology. Clearance of the toxic byproducts of cellular respiration is clearly inefficient in neurons bearing the HD gene mutation, but other functions, such as cellular energetics, are also impacted by mutant HTT, the delicate balance of intracellular metal ions, and possibly by the interaction between the two.

22.3.3.3 Energetics

Given that mitochondria are the primary site of ATP production in the cell, mitochondrial function is tightly linked to ATP levels. Many disease models of HD demonstrate reduced ATP levels, hinting at underlying mitochondrial dysfunction.[130–133] Furthermore, a striking inverse relationship between ATP/ADP (adenosine diphosphate) levels and CAG repeat lengths, both pathogenic and non-pathogenic, was observed in human lymphoblastoid lines.[133] Defects in extra-mitochondrial energy metabolism have been observed in mouse striatal models of HD.[134] In addition to the potential mitochondrial dysfunction *via* decreased MnSOD discussed in the previous section, reductions in important metabolites necessary for the tricarboxylic acid (TCA) cycle can also lead to reduced mitochondrial respiration and ATP production. One important enzyme that produces metabolites for the TCA cycle is pyruvate carboxylase.

Pyruvate carboxylase (PC) is a manganese-dependent enzyme that converts pyruvate to oxaloacetate, which is a metabolite for the TCA cycle.[135] Deficiencies in PC can lead to increased lactate in the blood and lactic acidosis, a serious medical condition.[136] The activity of pyruvate carboxylase was found to be highly variable in post-mortem brains of patients with HD;[106] however, many studies have found alterations that indicate reduced activity of this enzyme in patients with HD. Lactate has reduced clearance in HD.[137] Increased lactate was also seen in the frontal lobe, occipital lobe, and striata of patients with HD.[138,139] In addition to increasing lactate levels, reduced PC activity decreases the availability of metabolite precursors to the TCA cycle, which is the major source of ATP production. Reduced ATP levels have been seen in many disease models of HD.[130–133]

Additionally, *N*-acetyl aspartate (NAA), for which oxaloacetate (OAA) is a precursor, is decreased in the same areas that see an increase in lactate in HD, and both of these potential biomarkers for HD correlate in a concentration-dependent manner with duration of the illness.[138,139] There is a high concentration of NAA in brain, and production of NAA from OAA is an alternate route for the removal of free cellular ammonia *via* the addition of an amine.[140] In fact, NAA is the second most abundant molecule in the brain after glutamate.[113] The urea cycle alterations seen in HD may be interconnected with HD-related reduction in NAA. Additionally, the amino acid asparagine, for which NAA is a precursor, is found to be reduced in the brains of patients with HD, with high predictive ability.[141] Interestingly, creatine treatment can reduce the levels of lactic acid in patients with lactic acidosis.[142] Creatine has been shown in multiple HD mouse models to ameliorate many of the pathological phenotypes, including motor deficits, weight loss, hyperglycemia, and brain atrophy.[17,18] Although there may be other derangements of the TCA cycle stemming from the HD gene mutation, some of alterations that lead to increased lactate may be explained by the reduced regional uptake of manganese, which is necessary for crucial cofactors in this cycle.

In theory, manganese deficiency should lead to a reduction in PC activity; however, manganese-deficient rat pups have elevated PC activity by postnatal day 3. Despite this finding, plasma glucose levels are reduced on postnatal days 1 and 2, suggesting that glucose metabolism is compromised by manganese deficiency.[143] PC expression may increase to compensate for diminished activity from a lack of available manganese, with this increase in PC production becoming evident once manganese is replaced in the diet. Of note, gene expression profiling data from a striatal model of HD implicated several genes in glucose metabolism as being altered in HD; in particular, expression of *PCX* (the gene which encodes PC) is elevated in an HD mutant line.[134] Thus, models of both HD and manganese deficiency in rodents result in increased PC expression or activity.

22.3.4   IGF/PI3K/AKT Signaling in HD and Manganese Exposure

Metabolic disturbances in HD may also be due to altered cell signaling *via* the insulin and insulin-like growth factor (IGF)-1 pathways that regulate important rate-limiting steps in energy production including glucose transport, glycogen synthesis, and production of precursors for the TCA cycle. As described in Section 22.1, HD is a neurodegenerative disorder, with select regions showing severe neuronal death, indicating a cell-type specific response to the pathogenic mutation of the Huntingtin protein. Growth factors such as IGF-1 and brain-derived neurotrophic factor (BDNF) have been shown be essential for the survival of striatal neurons, and lack of these growth factor signals may contribute to HD.[2] IGF signaling, and insulin to a lesser extent, leads to activation of the phosphoinositide 3-kinase(PI3K)–protein kinase B (AKT) pathway, which is an important signaling cascade that promotes proliferation and cell survival. IGF-1 can stimulate this pathway *via* the IGF-1 receptor (IGF-1R) and can also activate the insulin receptor.

22.3.4.1 AKT Signaling and HD

Several studies have found reduced activation of the AKT pathway as measured by the ratio of phospho-AKT(S473) over total AKT in HD cells from patients and mice.[47,144,145] The increased risk of developing diabetes and insulin resistance in HD may result from the altered ability of IGF-1 and insulin to activate AKT and its downstream targets.[146,147] Blood glucose levels have also been shown to be increased in mice expressing mutant Huntingtin, and patients with HD have decreased cerebral glucose consumption.[148,149]

Increased activation of the AKT pathway by IGF-1 supplementation has been shown to protect striatal neurons from mutant Huntingtin induced toxicity, indicating that this phenomenon may be important in HD neurodegeneration.[150] An alternative route to ameliorate HD-related AKT dysfunction has been administration of ganglioside GM1, which is reduced in HD mice and patient fibroblasts.[145] Surprisingly, the deficit in AKT phosphorylation can also be ameliorated in a striatal model of HD by manganese supplementation, and the underlying deficiency in manganese accumulation in these cells may provide a potential mechanism for AKT deficiency.[47]

22.3.4.2 AKT Pathway Signaling and Manganese

Several groups have reported alterations in AKT signaling with manganese deficiency and exposure, although not in the context of HD. Manganese deficiency has been shown to decrease expression of both insulin and IGF-1.[151] Alternatively, exposure to manganese has been shown to increase IGF-1 expression,[152] increase AKT phosphorylation at serine473 in the striatum,[153] and activate both AKT and inducible (i)NOS activity in microglia.[154] These increases in AKT activity could be due to stimulation of IGF-1 and insulin production or *via* a more direct route resulting from an observed insulin-like mimetic activity of manganese.[155,156]

In any case, changes in manganese levels, perhaps *via* effects on insulin and IGF-1, result in altered glucose regulation and metabolism. Manganese deficiency has been shown to decrease glucose metabolism, insulin secretion, and insulin-induced glucose uptake, while manganese exposure can cause hypoglycemia.[157,158] In fact, one reported case of "sweet urine disease" was effectively treated with a traditional folk remedy of alfalfa tea. The "sweet urine" was found to have both elevated glucose and manganese urinary output. The alfalfa tea was high in manganese and found to reduce the urinary glucose levels.[159] Therefore, supplementation of manganese may be efficacious as a treatment in diabetes or to regulate blood glucose levels in patients with HD where insulin-resistance and diabetes are common.[147,148] AKT signaling, insulin secretion, and blood glucose regulation are all affected by both manganese and mutant Huntingtin, but other modifiers of the AKT pathway can also be altered through this gene–environment interaction.

22.3.4.3   Protein Phosphatase 1

Protein phosphatase 1 (PP1) is an important phosphatase in many cellular signaling pathways, including the AKT pathway. In fact, PP1 has been shown to dephosphorylate AKT directly at serine 473 and p53 at ser15 (to be discussed further in Section 22.3.5).[160,161] Interestingly, both AKT and p53 phosphorylation have been shown to increase with manganese exposure; however, this would presumably increase the ability of PP1 to dephosphorylate at these sites as well.[47,154]

One of the many functions of PP1 is to regulate the function of glycogen synthase by dephosphorylation and, thereby, increase its activity. The increased glucose levels noted in HD model mice could indicate a deficiency in glycogen synthase activity. Glycogen synthase is also regulated by GSK3B, which phosphorylates and inhibits its activity. Intriguingly, GSK3B inhibitors are protective in some HD models, possibly because of deficient ability of PP1 to dephosphorylate glycogen synthase.[162] Further analysis is needed to elucidate the role of this manganese-dependent enzyme, PP1, in HD.

22.3.5   p53 Pathway

p53 is one of the most studied and interconnected cellular signaling molecules known. p53 is important for regulating nearly all cellular stress signals, including DNA damage, oxidative stress, and various other stressors, by integrating cellular stress response pathways to elicit an appropriate pro-survival or apoptotic transcriptional response.[163] p53 has been implicated as a player in HD neurodegenerative processes, and this pathway is activated under low manganese exposure conditions. Therefore, the p53 pathway may be another point of intersection between manganese homeostasis and HD pathology.

22.3.5.1   p53 and HD

Patients with HD have been shown to have brain accumulation of p53 that correlates with disease stage. Many cellular models of HD have also shown increased p53 activation by phosphorylation at serine15.[164,165] Mutant Huntingtin is thought to bind to p53 and alter its function.[166] Additionally, when HD model mice were crossed onto a p53 knockout background, nearly all of the pathogenic and behavioral phenotypes in HD were ameliorated, suggesting that p53 may play an important role in the disease-associated neurodegeneration.[164] Additionally, the mitochondrial depolarization and fragmentation phenotypes seen in HD human and mouse cell models were also shown to be diminished by using a inhibitor of p53 transcriptional activity.[164,167]

In fact, single-nucleotide polymorphism (SNP) variants of the *TP53* gene have also been found strongly to modify age at onset in HD. In fact, 12.6% of variability after taking CAG repeat length into account could be explained by p53 polymorphisms, further underscoring the importance of p53 in HD pathology.[168] Finally, human induced pluripotent stem cells from patients with HD also demonstrate increased activation of p53 as well as one of its upstream kinases, ataxia telangiectasia mutated (ATM).[169] ATM has also been implicated in abnormal activation and phosphorylation of p53.[170]

22.3.5.2 *Manganese and p53*

p53 has been shown to be a metal-binding protein that relies on a divalent metal ion, Zn^{2+}, to maintain its proper conformation.[171] The ability of Mn^{2+} to interact directly with p53 has not been explored. Cellular exposure to several other divalent metal ions (Cd^{2+}, Cu^{2+}, Zn^{2+}, Co^{2+}, Cr^{2+}) has been shown to cause the activation of p53, as monitored by phosphorylation at serine 15, but manganese had not been investigated in this context.[172–175] However, primates exposed to manganese showed increases in p53-dependent transcripts,[176] and ATM is an important kinase in the double-stranded DNA break (DSB) repair process.[177,178] In this process, breaks in genomic DNA are associated with a change in the inactive dimer form of ATM, which is converted into an active monomer that phosphorylates important downstream targets that regulate DNA repair (MRN, gamma-H2AX), cell cycle arrest (CHK2, p53), and apoptosis (p53). ATM has also been recently shown to be activated by oxidative stress through an alternative process of dimer conjugation *via* formation of a cysteine disulfide bond.[179] Manganese has been shown to be necessary for the *in vitro* enzyme activity of ATM, as demonstrated by phosphorylation of downstream targets such as p53 at ser15.[179,180] While these studies roughly describe ATM as a manganese-dependent enzyme, the exact nature of the relationship between Mn and ATM activity is poorly defined. Manganese may be necessary for proper ATM function in the context of other stressors, or the manganese itself may elicit ATM activation *via* the oxidative stress, resulting in the dimerization pathway shown by Guo *et al.* (2010).[179] This seems possible given that very high concentrations of manganese (5 mM) are used for activation of ATM and subsequent phosphorylation of p53. The nature of the activation of ATM and p53 *via* manganese exposure needs to be more clearly defined.

22.3.5.3 *p53 and HD/Manganese Phenotypes*

Owing to the broad role of p53 in regulating metabolism, DNA repair, stress response, and apoptosis *via* both transcriptional and non-transcriptional means, p53 dysregulation is also closely related to several of the HD phenotypes explored in previous sections. For example, p53 regulates the IGF-1/AKT pathway by regulating the transcription of several AKT regulatory genes (*e.g.* IGFBP3, PTEN).[181-183] AKT also regulates p53 signaling by binding and phosphorylating HDM2, a major regulatory protein of p53.[184]

There is also a role for p53 in the mitochondrial dysfunction seen in HD. The propensity for mitochondrial depolarization and fragmentation in HD has been shown in cell culture to be blocked by a p53 inhibitor (pifithrin-α) and by knockdown of p53 expression, and p53 has been found to be integral to opening of the mitochondrial membrane transition pore and the release of cytochrome c.[164,167,185,186] Interestingly, both the activity and transcription of MnSOD, which detoxifies mitochondrial ROS, are also negatively regulated by activated p53.[187,188] Therefore, increased mitochondrial susceptibility to glutamate/Ca^{2+} excitotoxic processes could be due to observed increases in p53 activity, leading to either increased propensity to open the MPTP or diminishing MnSOD expression and activity, and thereby increasing ROS production and mitochondrial dysfunction (see Sections 22.3.2.2–22.3.3.2).

Activation of p53 has also been shown to reduce glutamate uptake in astrocytes as well as increasing the level of mitochondrial glutaminase, which generates glutamate and ammonia.[189-191] Thus, activation of p53 in HD or *via* manganese exposure could lead to increased glutamate production and decreased glutamate clearance from the synapse, and increased ammonia levels, all of which have been seen in HD models (see Section 22.3.2.1).[80-100] The far-reaching effects of p53 in regulating the cellular transcriptome as well as its intimate regulation of cell signaling, mitochondrial health, and apoptosis provide a potential unifying theory for the broad effects of manganese homeostasis alterations.

22.4 Conclusions

Metal toxicity and deficiencies can have many detrimental effects on the brain, particularly the basal ganglia, which is also selectively susceptible in many neurodegenerative diseases such as HD and Parkinsonism. Iron, copper, and, more recently, manganese levels have been shown to be altered in patients and in mouse models of HD. Interestingly, manganese seems to be regionally decreased in organisms containing the mutant Huntingtin protein. The underlying manganese homeostatic dysfunction may cause reduced activity of enzymes that depend on manganese as a cofactor. Reduced manganese levels can lead to many maladaptive phenotypes found in HD models, including urea cycle dysfunction, altered synaptic glutamate regulation, decreased energy production, increased oxidative stress, and altered cell signaling that regulates energy homeostasis and cellular stress. This close relationship between manganese levels and HD indicates that deepening our understanding of the underlying defect could lead to therapeutics that, by rescuing deficient manganese levels, may ameliorate dysfunctional HD processes.

References

1. F. O. Walker, Huntington's disease, *Lancet*, 2007, **369**, 218–228.
2. C. Zuccato, M. Valenza and E. Cattaneo, Molecular mechanisms and potential therapeutic targets in Huntington's disease, *Physiol. Rev.*, 2010, **90**, 905–981.
3. S. E. Folstein, *Huntington's disease: a disorder of families*, Johns Hopkins University Press, 1989.
4. C.-A. Gutekunst, F. Norflus and S. M. Hersch, The neuropathology of Huntington's disease, *Oxford Monogr. Med. Genet.*, 2002, **45**, 251–275.
5. M. E. MacDonald, C. M. Ambrose, M. P. Duyao, R. H. Myers, C. Lin, L. Srinidhi, G. Barnes, S. A. Taylor, M. James and N. Groot, A novel gene containing a trinucleotide repeat that is expanded and unstable on Huntington's disease chromosomes, *cell*, 1993, **72**, 971–983.
6. D. C. Rubinsztein, J. Leggo, R. Coles, E. Almqvist, V. Biancalana, J.-J. Cassiman, K. Chotai, M. Connarty, D. Craufurd and A. Curtis, Phenotypic Characterization of Individuals with 30–40 CAG Repeats in the Huntington Disease (HD) Gene Reveals HD Cases with 36 Repeats and Apparently Normal Elderly Individuals with 36–39 Repeats, *Am. J. Hum. Genet.*, 1996, **59**, 16.
7. D. Langbehn, R. Brinkman, D. Falush, J. Paulsen and M. Hayden, A new model for prediction of the age of onset and penetrance for Huntington's disease based on CAG length, *Clin. Genet.*, 2004, **65**, 267–277.
8. S. E. Andrew, Y. P. Goldberg, B. Kremer, H. k. Telenius, J. Theilmann, S. Adam, E. Starr, F. Squitieri, B. Lin and M. A. Kalchman, The relationship between trinucleotide (CAG) repeat length and clinical features of Huntington's disease, *Nat. Genet.*, 1993, **4**, 398–403.
9. N. S. Wexler, Venezuelan kindreds reveal that genetic and environmental factors modulate Huntington's disease age of onset, *Proc. Natl. Acad. Sci. U. S. A.*, 2004, **101**, 3498–3503.
10. J. Gomez-Esteban, E. Lezcano, J. Zarranz, F. Velasco, I. Garamendi, T. Pérez and B. Tijero, Monozygotic twins suffering from Huntington's disease show different cognitive and behavioural symptoms, *Eur. Neurol.*, 2006, 57, 26–30.
11. N. Georgiou, J. L. Bradshaw, E. Chiu, A. Tudor, L. O'Gorman and J. G. Phillips, Differential clinical and motor control function in a pair of monozygotic twins with Huntington's disease, *Mov. Disord.*, 1999, **14**, 320–325.
12. J. H. Friedman, M. E. Trieschmann, R. H. Myers and H. H. Fernandez, Monozygotic twins discordant for Huntington disease after 7 years, *Arch. Neurol.*, 2005, **62**, 995.
13. M. Anca, E. Gazit, R. Loewenthal, O. Ostrovsky, M. Frydman and N. Giladi, Different phenotypic expression in monozygotic twins with Huntington disease, *Am. J. Med. Genet., Part A*, 2004, **124**, 89–91.

14. E. Hockly, P. M. Cordery, B. Woodman, A. Mahal, A. Van Dellen, C. Blakemore, C. M. Lewis, A. J. Hannan and G. P. Bates, Environmental enrichment slows disease progression in R6/2 Huntington's disease mice, *Ann. Neurol.*, 2002, **51**, 235–242.

15. T. L. Spires, H. E. Grote, N. K. Varshney, P. M. Cordery, A. van Dellen, C. Blakemore and A. J. Hannan, Environmental enrichment rescues protein deficits in a mouse model of Huntington's disease, indicating a possible disease mechanism, *J. Neurosci.*, 2004, **24**, 2270–2276.

16. A. van Dellen, C. Blakemore, R. Deacon, D. York and A. J. Hannan, Delaying the onset of Huntington's in mice, *Nature*, 2000, **404**, 721–722.

17. R. J. Ferrante, O. A. Andreassen, B. G. Jenkins, A. Dedeoglu, S. Kuemmerle, J. K. Kubilus, R. Kaddurah-Daouk, S. M. Hersch and M. F. Beal, Neuroprotective effects of creatine in a transgenic mouse model of Huntington's disease, *J. Neurosci.*, 2000, **20**, 4389–4397.

18. O. A. Andreassen, A. Dedeoglu, R. J. Ferrante, B. G. Jenkins, K. L. Ferrante, M. Thomas, A. Friedlich, S. E. Browne, G. Schilling and D. R. Borchelt, Creatine increases survival and delays motor symptoms in a transgenic animal model of Huntington's disease, *Neurobiol. Dis.*, 2001, **8**, 479–491.

19. W. Duan, Z. Guo, H. Jiang, M. Ware, X.-J. Li and M. P. Mattson, Dietary restriction normalizes glucose metabolism and BDNF levels, slows disease progression, and increases survival in huntingtin mutant mice, *Proc. Natl. Acad. Sci.*, 2003, **100**, 2911–2916.

20. O. A. Andreassen, R. J. Ferrante, A. Dedeoglu and M. F. Beal, Lipoic acid improves survival in transgenic mouse models of Huntington's disease, *Neuroreport*, 2001, **12**, 3371–3373.

21. T. Pang, N. Stam, J. Nithianantharajah, M. Howard and A. Hannan, Differential effects of voluntary physical exercise on behavioral and brain-derived neurotrophic factor expression deficits in Huntington's disease transgenic mice, *Neuroscience*, 2006, **141**, 569–584.

22. K. Jomova, D. Vondrakova, M. Lawson and M. Valko, Metals, oxidative stress and neurodegenerative disorders, *Mol. Cell. Biochem.*, 2010, **345**, 91–104.

23. A. B. Bowman, G. F. Kwakye, E. Herrero Hernández and M. Aschner, Role of manganese in neurodegenerative diseases, *J. Trace Elem. Med. Biol.*, 2011, **25**, 191–203.

24. R. Betarbet, T. B. Sherer, G. MacKenzie, M. Garcia-Osuna, A. V. Panov and J. T. Greenamyre, Chronic systemic pesticide exposure reproduces features of Parkinson's disease, *Nat. Neurosci.*, 2000, **3**, 1301–1306.

25. S. Zeitlin, J.-P. Liu, D. L. Chapman, V. E. Papaioannou and A. Efstratiadis, Increased apoptosis and early embryonic lethality in mice nullizygous for the Huntington's disease gene homologue, *Nat. Genet.*, 1995, **11**, 155–163.

26. J. Nasir, S. B. Floresco, J. R. O'Kusky, V. M. Diewert, J. M. Richman, J. Zeisler, A. Borowski, J. D. Marth, A. G. Phillips and M. R. Hayden,

Targeted disruption of the Huntington's disease gene results in embryonic lethality and behavioral and morphological changes in heterozygotes, *cell*, 1995, **81**, 811–823.

27. M. P. Duyao, A. B. Auerbach, A. Ryan, F. Persichetti, G. T. Barnes, S. M. McNeil, P. Ge, J.-P. Vonsattel, J. F. Gusella and A. L. Joyner, Inactivation of the mouse Huntington's disease gene homolog Hdh, *Science*, 1995, **269**, 407–410.

28. A. L. Lumsden, T. L. Henshall, S. Dayan, M. T. Lardelli and R. I. Richards, Huntingtin-deficient zebrafish exhibit defects in iron utilization and development, *Hum. Mol. Genet.*, 2007, **16**, 1905–1920.

29. J. K. White, W. Auerbach, M. P. Duyao, J.-P. Vonsattel, J. F. Gusella, A. L. Joyner and M. E. MacDonald, Huntingtin is required for neurogenesis and is not impaired by the Huntington's disease CAG expansion, *Nat. Genet.*, 1997, **17**, 404–410.

30. O. C. Stine, S.-H. Li, N. Pleasant, M. V. Wagster, J. C. Hedreen and C. A. Ross, Expression of the mutant allele of IT-15 (the HD gene) in striatum and cortex of Huntington's disease patients, *Hum. Mol. Genet.*, 1995, **4**, 15–18.

31. F. Molina-Holgado, R. C. Hider, A. Gaeta, R. Williams and P. Francis, Metals ions and neurodegeneration, *Biometals*, 2007, **20**, 639–654.

32. A. Gaeta and R. C. Hider, The crucial role of metal ions in neurodegeneration: the basis for a promising therapeutic strategy, *Br. J. Pharmacol.*, 2005, **146**, 1041–1059.

33. A. W. Dobson, K. M. Erikson and M. Aschner, Manganese neurotoxicity, *Ann. N. Y. Acad. Sci.*, 2004, **1012**, 115–128.

34. A. Ala, A. P. Walker, K. Ashkan, J. S. Dooley and M. L. Schilsky, Wilson's disease, *The Lancet*, 2007, **369**, 397–408.

35. S. Zidenberg-Cherr, C. L. Keen, B. Lonnerdal and L. S. Hurley, Superoxide dismutase activity and lipid peroxidation in the rat: developmental correlations affected by manganese deficiency, *J. Neurosci.*, 1983, **113**, 2498.

36. J. Beard, Iron deficiency alters brain development and functioning, *J. Nutr.*, 2003, **133**, 1468S–1472S.

37. T. Walter, I. De Andraca, P. Chadud and C. G. Perales, Iron deficiency anemia: adverse effects on infant psychomotor development, *Pediatrics*, 1989, **84**, 7–17.

38. L. Erway, L. S. Hurley and A. S. Fraser, Congenital ataxia and otolith defects due to manganese deficiency in mice, *J. Nutr.*, 1970, **100**, 643–654.

39. C. L. Keen, J. Y. Uriu-Hare, S. N. Hawk, M. A. Jankowski, G. P. Daston, C. L. Kwik-Uribe and R. B. Rucker, Effect of copper deficiency on prenatal development and pregnancy outcome, *Am. J. Clin. Nutr.*, 1998, **67**, 1003S–1011S.

40. R. V. Dipaolo, J. N. Kanfer and P. M. Newberne, Copper deficiency and the central nervous system: myelination in the rat: morphological and biochemical studies, *J. Neuropathol. Exp. Neurol.*, 1974, **33**, 226–236.

41. D. Dexter, A. Carayon, F. Javoy-Agid, Y. Agid, F. Wells, S. Daniel, A. Lees, P. Jenner and C. Marsden, Alterations in the levels of iron, ferritin and other trace metals in Parkinson's disease and other neurodegenerative diseases affecting the basal ganglia, *Brain*, 1991, **114**, 1953–1975.

42. H. D. Rosas, Y. Chen, G. Doros, D. H. Salat, N.-k. Chen, K. K. Kwong, A. Bush, J. Fox and S. M. Hersch, Alterations in brain transition metals in Huntington disease: an evolving and intricate story. *Arch. Neurol.*, 2012, **69**, 887–893.

43. G. Bartzokis, J. Cummings, S. Perlman, D. B. Hance and J. Mintz, Increased basal ganglia iron levels in Huntington disease, *Arch. Neurol.*, 1999, **56**, 569.

44. K. M. Erikson, T. Syversen, E. Steinnes and M. Aschner, Globus pallidus: a target brain region for divalent metal accumulation associated with dietary iron deficiency, *J. Nutr. Biochem.*, 2004, **15**, 335–341.

45. N. A. Larsen, H. Pakkenberg, E. Damsgaard and K. Heydorn, Topographical distribution of arsenic, manganese, and selenium in the normal human brain, *J. Neurol. Sci.*, 1979, **42**, 407–416.

46. J. R. Prohaska, Functions of trace elements in brain metabolism, *Physiol. Rev.*, 1987, **67**, 858–901.

47. B. B. Williams, D. Li, M. Wegrzynowicz, B. K. Vadodaria, J. G. Anderson, G. F. Kwakye, M. Aschner, K. M. Erikson and A. B. Bowman, Disease-toxicant screen reveals a neuroprotective interaction between Huntington's disease and manganese exposure, *J. Neurochem.*, 2010, **112**, 227–237.

48. F. A. Azevedo, L. R. Carvalho, L. T. Grinberg, J. M. Farfel, R. E. Ferretti, R. E. Leite, R. Lent and S. Herculano-Houzel, Equal numbers of neuronal and nonneuronal cells make the human brain an isometrically scaled-up primate brain, *J. Comp. Neurol.*, 2009, **513**, 532–541.

49. F. Wedler and R. Denman, Glutamine synthetase: the major Mn (II) enzyme in mammalian brain, *Curr. Top. Cell. Regul.*, 1984, **24**, 153–169.

50. G. Tholey, M. Ledig, P. Mandel, L. Sargentini, A. Frivold, M. Leroy, A. Grippo and F. Wedler, Concentrations of physiologically important metal ions in glial cells cultured from chick cerebral cortex, *Neurochem. Res.*, 1988, **13**, 45–50.

51. N. M. Filipov and C. A. Dodd, Role of glial cells in manganese neurotoxicity, *J. Appl. Toxicol.*, 2012, **32**, 310–317.

52. J. L. Madison, M. Wegrzynowicz, M. Aschner and A. B. Bowman, Disease-toxicant interactions in manganese exposed Huntington disease mice: early changes in striatal neuron morphology and dopamine metabolism, *PloS One*, 2012, 7, e31024.

53. C. A. Ross, E. H. Aylward, E. J. Wild, D. R. Langbehn, J. D. Long, J. H. Warner, R. I. Scahill, B. R. Leavitt, J. C. Stout and J. S. Paulsen, Huntington disease: natural history, biomarkers and prospects for therapeutics, *Nat. Rev. Neurol.*, 2014.

54. D. A. Simmons, M. Casale, B. Alcon, N. Pham, N. Narayan and G. Lynch, Ferritin accumulation in dystrophic microglia is an early event in the development of Huntington's disease, *Glia*, 2007, **55**, 1074–1084.

55. J. Chen, P. Hardy, W. Kucharczyk, M. Clauberg, J. Joshi, A. Vourlas, M. Dhar and R. Henkelman, MR of human postmortem brain tissue: correlative study between T2 and assays of iron and ferritin in Parkinson and Huntington disease, *Am. J. Neuroradiol.*, 1993, **14**, 275–281.

56. S. Levi, A. Cozzi and P. Arosio, Neuroferritinopathy: a neurodegenerative disorder associated with L-ferritin mutation, *Best Pract. Res., Clin. Haematol.*, 2005, **18**, 265–276.

57. A. R. Curtis, C. Fey, C. M. Morris, L. A. Bindoff, P. G. Ince, P. F. Chinnery, A. Coulthard, M. J. Jackson, A. P. Jackson and D. P. McHale, Mutation in the gene encoding ferritin light polypeptide causes dominant adult-onset basal ganglia disease, *Nat. Genet.*, 2001, **28**, 350–354.

58. E. Bonilla, J. Estevez, H. Suarez, L. Morales, L. C. de Bonilla, R. Villalobos and J. Davila, Serum ferritin deficiency in Huntington's disease patients, *Neurosci. Lett.*, 1991, **129**, 22–24.

59. P. F. Chinnery, D. E. Crompton, D. Birchall, M. J. Jackson, A. Coulthard, A. Lombes, N. Quinn, A. Wills, N. Fletcher and J. P. Mottershead, Clinical features and natural history of neuroferritinopathy caused by the FTL1 460InsA mutation, *Brain*, 2007, **130**, 110–119.

60. J. A. Roth, Homeostatic and toxic mechanisms regulating manganese uptake, retention, and elimination, *Biol. Res.*, 2006, **39**, 45.

61. B. B. Williams, G. F. Kwakye, M. Wegrzynowicz, D. Li, M. Aschner, K. M. Erikson and A. B. Bowman, Altered manganese homeostasis and manganese toxicity in a Huntington's disease striatal cell model are not explained by defects in the iron transport system, *Toxicol. Sci.*, 2010, **117**, 169–179.

62. P. Wardman and L. P. Candeias, Fenton chemistry: an introduction, *Radiat. Res.*, 1996, **145**, 523–531.

63. S. E. Browne, R. J. Ferrante and M. F. Beal, Oxidative stress in Huntington's disease, *Brain Pathol.*, 1999, **9**, 147–163.

64. J. H. Fox, J. A. Kama, G. Lieberman, R. Chopra, K. Dorsey, V. Chopra, I. Volitakis, R. A. Cherny, A. I. Bush and S. Hersch, Mechanisms of copper ion mediated Huntington's disease progression, *PloS One*, 2007, **2**, e334.

65. S. L. Hands, R. Mason, M. Umar Sajjad, F. Giorgini and A. Wyttenbach, Metallothioneins and copper metabolism are candidate therapeutic targets in Huntington's disease, *Biochem. Soc. Trans.*, 2010, **38**, 552–558.

66. A. Hodges, A. D. Strand, A. K. Aragaki, A. Kuhn, T. Sengstag, G. Hughes, L. A. Elliston, C. Hartog, D. R. Goldstein and D. Thu, Regional and cellular gene expression changes in human Huntington's disease brain, *Hum. Mol. Genet.*, 2006, **15**, 965–977.

67. S. W. Davies, M. Turmaine, B. A. Cozens, M. DiFiglia, A. H. Sharp, C. A. Ross, E. Scherzinger, E. E. Wanker, L. Mangiarini and G. P. Bates, Formation of neuronal intranuclear inclusions underlies the neurological dysfunction in mice transgenic for the HD mutation, *cell*, 1997, **90**, 537–548.

68. G. Bjørkøy, T. Lamark, A. Brech, H. Outzen, M. Perander, A. Øvervatn, H. Stenmark and T. Johansen, p62/SQSTM1 forms protein aggregates degraded by autophagy and has a protective effect on huntingtin-induced cell death, *J. Cell Biol.*, 2005, **171**, 603–614.

69. T. Nguyen, A. Hamby and S. M. Massa, Clioquinol down-regulates mutant huntingtin expression in vitro and mitigates pathology in a Huntington's disease mouse model, *Proc. Natl. Acad. Sci. U. S. A.*, 2005, **102**, 11840–11845.

70. G. Xiao, Q. Fan, X. Wang and B. Zhou, Huntington disease arises from a combinatory toxicity of polyglutamine and copper binding, *Proc. Natl. Acad. Sci.*, 2013, **110**, 14995–15000.

71. M.-C. Boll, M. Alcaraz-Zubeldia, S. Montes and C. Rios, Free copper, ferroxidase and SOD1 activities, lipid peroxidation and NO x content in the CSF. A different marker profile in four neurodegenerative diseases, *Neurochem. Res.*, 2008, **33**, 1717–1723.

72. V. L. Dawson, T. M. Dawson, E. D. London, D. S. Bredt and S. H. Snyder, Nitric oxide mediates glutamate neurotoxicity in primary cortical cultures, *Proc. Natl. Acad. Sci.*, 1991, **88**, 6368–6371.

73. V. Calabrese, C. Mancuso, M. Calvani, E. Rizzarelli, D. A. Butterfield and A. M. G. Stella, Nitric oxide in the central nervous system: neuroprotection versus neurotoxicity, *Nat. Rev. Neurosci.*, 2007, **8**, 766–775.

74. W. Durante, F. K. Johnson and R. A. Johnson, Arginase: a critical regulator of nitric oxide synthesis and vascular function, *Clin. Exp. Pharmacol. Physiol.*, 2007, **34**, 906–911.

75. Z. F. Kanyo, L. R. Scolnick, D. E. Ash and D. W. Christianson, Structure of a unique binuclear manganese cluster in arginase. *Nature*, 1996, **383**, 554–557.

76. A. A. Brock, S. A. Chapman, E. A. Ulman and G. Wu, Dietary manganese deficiency decreases rat hepatic arginase activity, *J. Nutr.*, 1994, **124**, 340–344.

77. J. L. Ensunsa, J. D. Symons, L. Lanoue, H. R. Schrader and C. L. Keen, Reducing arginase activity via dietary manganese deficiency enhances endothelium-dependent vasorelaxation of rat aorta, *Exp. Biol. Med.*, 2004, **229**, 1143–1153.

78. H. Yu, R. K. Iyer, R. M. Kern, W. I. Rodriguez, W. W. Grody and S. D. Cederbaum, Expression of arginase isozymes in mouse brain, *J. Neurosci. Res.*, 2001, **66**, 406–422.

79. O. Braissant, T. Gotoh, M. Loup, M. Mori and C. Bachmann, l-arginine uptake, the citrulline-NO cycle and arginase II in the rat brain: an in situ hybridization study, *Mol. Brain Res.*, 1999, **70**, 231–241.

80. M.-C. Chiang, H.-M. Chen, Y.-H. Lee, H.-H. Chang, Y.-C. Wu, B.-W. Soong, C.-M. Chen, Y.-R. Wu, C.-S. Liu and D.-M. Niu, Dysregulation of C/EBPα by mutant Huntingtin causes the urea cycle deficiency in Huntington's disease, *Hum. Mol. Genet.*, 2007, **16**, 483–498.

81. M. A. Pouladi, A. J. Morton and M. R. Hayden, Choosing an animal model for the study of Huntington's disease, *Nat. Rev. Neurosci.*, 2013, **14**, 708–721.

82. M.-C. Chiang, H.-M. Chen, H.-L. Lai, H.-W. Chen, S.-Y. Chou, C.-M. Chen, F.-J. Tsai and Y. Chern, The A2A adenosine receptor rescues the urea cycle deficiency of Huntington's disease by enhancing the activity of the ubiquitin-proteasome system, *Hum. Mol. Genet.*, 2009, **18**, 2929–2942.

83. C. Colton, Q. Xu, J. Burke, S. Bae, J. Wakefield, A. Nair, W. Strittmatter and M. Vitek, Disrupted spermine homeostasis: a novel mechanism in polyglutamine-mediated aggregation and cell death, *J. Neurosci.*, 2004, **24**, 7118–7127.

84. R. P. Shank, Ornithine as a precursor of glutamate and GABA: uptake and metabolism by neuronal and glial enriched cellular material, *J. Neurosci. Res.*, 1983, **9**, 47–57.

85. A. W. Deckel, V. Tang, D. Nuttal, K. Gary and R. Elder, Altered neuronal nitric oxide synthase expression contributes to disease progression in Huntington's disease transgenic mice, *Brain Res.*, 2002, **939**, 76–86.

86. A. W. Deckel, Nitric oxide and nitric oxide synthase in Huntington's disease, *J. Neurosci. Res.*, 2001, **64**, 99–107.

87. S. E. Browne and M. F. Beal, Oxidative damage in Huntington's disease pathogenesis, *Antioxid. Redox Signaling*, 2006, **8**, 2061–2073.

88. A. G. Estevez, M. A. Sahawneh, P. S. Lange, N. Bae, M. Egea and R. R. Ratan, Arginase 1 regulation of nitric oxide production is key to survival of trophic factor-deprived motor neurons, *J. Neurosci.*, 2006, **26**, 8512–8516.

89. K. Deng, H. He, J. Qiu, B. Lorber, J. B. Bryson and M. T. Filbin, Increased synthesis of spermidine as a result of upregulation of arginase I promotes axonal regeneration in culture and in vivo, *J. Neurosci.*, 2009, **29**, 9545–9552.

90. H. C. Ha, N. S. Sirisoma, P. Kuppusamy, J. L. Zweier, P. M. Woster and R. A. Casero, The natural polyamine spermine functions directly as a free radical scavenger, *Proc. Natl. Acad. Sci.*, 1998, **95**, 11140–11145.

91. N. d. A. Velloso, G. D. Dalmolin, G. M. Gomes, M. A. Rubin, P. M. Canas, R. A. Cunha and C. F. Mello, Spermine improves recognition memory deficit in a rodent model of Huntington's disease, *Neurobiol. Learn. Mem.*, 2009, **92**, 574–580.

92. E. J. Slow, R. K. Graham, A. P. Osmand, R. S. Devon, G. Lu, Y. Deng, J. Pearson, K. Vaid, N. Bissada and R. Wetzel, Absence of behavioral abnormalities and neurodegeneration in vivo despite widespread neuronal huntingtin inclusions, *Proc. Natl. Acad. Sci. U. S. A.*, 2005, **102**, 11402–11407.

93. M. M. Zeron, O. Hansson, N. Chen, C. L. Wellington, B. R. Leavitt, P. Brundin, M. R. Hayden and L. A. Raymond, Increased sensitivity to N-methyl-D-aspartate receptor-mediated excitotoxicity in a mouse model of Huntington's disease, *Neuron*, 2002, **33**, 849–860.

94. P. Behrens, P. Franz, B. Woodman, K. Lindenberg and G. Landwehrmeyer, Impaired glutamate transport and glutamate-glutamine cycling: downstream effects of the Huntington mutation, *Brain*, 2002, **125**, 1908–1922.

95. V. A. Fitsanakis, C. Au, K. M. Erikson and M. Aschner, The effects of manganese on glutamate, dopamine and α-aminobutyric acid regulation, *Neurochem. Int.*, 2006, **48**, 426–433.

96. L. Normandin and A. S. Hazell, Manganese neurotoxicity: an update of pathophysiologic mechanisms, *Metab. Brain Dis.*, 2002, **17**, 375–387.

97. K. M. Erikson and M. Aschner, Manganese neurotoxicity and glutamate-GABA interaction, *Neurochem. Int.*, 2003, **43**, 475–480.

98. D. R. Crooks, N. Welch and D. R. Smith, Low-level manganese exposure alters glutamate metabolism in GABAergic AF5 cells, *Neurotoxicology*, 2007, **28**, 548–554.

99. T. R. Guilarte and M.-K. Chen, Manganese inhibits NMDA receptor channel function: implications to psychiatric and cognitive effects, *Neurotoxicology*, 2007, **28**, 1147–1152.

100. J.-C. Lievens, B. Woodman, A. Mahal, O. Spasic-Boscovic, D. Samuel, L. Kerkerian-Le Goff and G. Bates, Impaired glutamate uptake in the R6 Huntington's disease transgenic mice, *Neurobiol. Dis.*, 2001, **8**, 807–821.

101. F. C. Wedler, R. B. Denman and W. G. Roby, Glutamine synthetase from ovine brain is a manganese (II) enzyme, *Biochemistry*, 1982, **21**, 6389–6396.

102. F. C. Wedler and B. W. Ley, Kinetic, ESR, and trapping evidence for in vivo binding of Mn (II) to glutamine synthetase in brain cells, *Neurochem. Res.*, 1994, **19**, 139–144.

103. J. Häeberle, B. Göerg, F. Rutsch, E. Schmidt, A. Toutain, J. Benoist, A. Gelot, A. Suc, W. Höehne and F. Schliess, Congenital glutamine deficiency with glutamine synthetase mutations, *N. Engl. J. Med.*, 2005, **353**, 1926–1933.

104. T. Eid, M. Thomas, D. Spencer, E. Runden-Pran, J. Lai, G. Malthankar, J. Kim, N. Danbolt, O. Ottersen and N. De Lanerolle, Loss of glutamine synthetase in the human epileptogenic hippocampus: possible mechanism for raised extracellular glutamate in mesial temporal lobe epilepsy, *Lancet*, 2004, **363**, 28–37.

105. T. Eid, A. Ghosh, Y. Wang, H. Beckström, H. P. Zaveri, T.-S. W. Lee, J. C. Lai, G. H. Malthankar-Phatak and N. C. de Lanerolle, Recurrent seizures and brain pathology after inhibition of glutamine synthetase in the hippocampus in rats, *Brain*, 2008, **131**, 2061–2070.

106. J. Butterworth, Changes in nine enzyme markers for neurons, glia, and endothelial cells in agonal state and Huntington's disease caudate nucleus, *J. Neurochem.*, 1986, **47**, 583–587.

107. C. Carter, Glutamine synthetase activity in Huntington's disease, *Life Sci.*, 1982, **31**, 1151–1159.
108. M. A. Nance and R. H. Myers, Juvenile onset Huntington's disease-clinical and research perspectives, *MRDD Research Reviews*, 2001, 7, 153–157.
109. A. Gambardella, M. Muglia, A. Labate, A. Magariello, A. Gabriele, R. Mazzei, D. Pirritano, F. Conforti, A. Patitucci and P. Valentino, Juvenile Huntington's disease presenting as progressive myoclonic epilepsy, *Neurology*, 2001, **57**, 708–711.
110. C. Dupont and Y. Tanaka, Blood manganese levels in children with convulsive disorder, *Biochem. Med.*, 1985, **33**, 246–255.
111. L. S. Hurley, D. E. Woolley, F. Rosenthal and P. S. Timiras, Influence of manganese on susceptibility of rats to convulsions, *Am. J. Physiol.*, 1963, **204**, 493–496.
112. P. S. Papavasiliou, H. Kutt, S. T. Miller, V. Rosal, Y. Y. Wang and R. B. Aronson, Seizure disorders and trace metals Manganese tissue levels in treated epileptics, *Neurology*, 1979, **29**, 1466.
113. D. L. Birken and W. H. Oldendorf, N-Acetyl-L-Aspartic acid: A literature review of a compound prominent in ^{1}H-NMR spectroscopic studies of brain, *Neurosci. Biobehav. Rev.*, 1989, **13**, 23–31.
114. A. J. Cooper, Role of glutamine in cerebral nitrogen metabolism and ammonia neurotoxicity, *MRDD Research Reviews*, 2001, 7, 280–286.
115. M. Tuchman, G. R. Lichtenstein, B. Rajagopal, M. T. McCann, E. E. Furth, J. Bavaria, P. B. Kaplan, J. B. Gibson and G. T. Berry, Hepatic glutamine synthetase deficiency in fatal hyperammonemia after lung transplantation, *Ann. Intern. Med.*, 1997, **127**, 446–449.
116. T. Tang, E. Slow, V. Lupu, I. G. Stavrovskaya, M. Sugimori, R. Llinás, B. S. Kristal, M. R. Hayden and I. Bezprozvanny, Disturbed Ca^{2+} signaling and apoptosis of medium spiny neurons in Huntington's disease, *Proc. Natl. Acad. Sci. U. S. A.*, 2005, **102**, 2602–2607.
117. N. Brustovetsky, T. Brustovetsky, R. Jemmerson and J. M. Dubinsky, Calcium-induced cytochrome c release from CNS mitochondria is associated with the permeability transition and rupture of the outer membrane, *J. Neurochem.*, 2002, **80**, 207–218.
118. C. Gavin, K. Gunter and T. Gunter, Manganese and calcium transport in mitochondria: implications for manganese toxicity, *Neurotoxicology*, 1998, **20**, 445–453.
119. T.-I. Peng and J. T. Greenamyre, Privileged Access to Mitochondria of Calcium Influx throughN-Methyl-d-Aspartate Receptors, *Mol. Pharmacol.*, 1998, **53**, 974–980.
120. S. Zhang, J. Fu and Z. Zhou, In vitro effect of manganese chloride exposure on reactive oxygen species generation and respiratory chain complexes activities of mitochondria isolated from rat brain, *Toxicol. In Vitro*, 2004, **18**, 71–77.
121. A. F. Schinder, E. C. Olson, N. C. Spitzer and M. Montal, Mitochondrial dysfunction is a primary event in glutamate neurotoxicity, *J. Neurosci.*, 1996, **16**, 6125–6133.

122. H. B. Fernandes, K. G. Baimbridge, J. Church, M. R. Hayden and L. A. Raymond, Mitochondrial sensitivity and altered calcium handling underlie enhanced NMDA-induced apoptosis in YAC128 model of Huntington's disease, *J. Neurosci.*, 2007, **27**, 13614–13623.

123. N. Brustovetsky, T. Brustovetsky, K. J. Purl, M. Capano, M. Crompton and J. M. Dubinsky, Increased susceptibility of striatal mitochondria to calcium-induced permeability transition, *J. Neurosci.*, 2003, **23**, 4858–4867.

124. C. D. Davis and J. Greger, Longitudinal changes of manganese-dependent superoxide dismutase and other indexes of manganese and iron status in women, *Am. J. Clin. Nutr.*, 1992, **55**, 747–752.

125. M. Aschner and J. L. Aschner, Manganese neurotoxicity: cellular effects and blood-brain barrier transport, *Neurosci. Biobehav. Rev.*, 1991, **15**, 333–340.

126. L. S. Hurley, L. L. Theriault and I. E. Dreosti, Liver mitochondria from manganese-deficient and pallid mice: function and ultrastructure, *Science*, 1970, **170**, 1316–1318.

127. O. A. Andreassen, R. J. Ferrante, A. Dedeoglu, D. W. Albers, P. Klivenyi, E. J. Carlson, C. J. Epstein and M. F. Beal, Mice with a partial deficiency of manganese superoxide dismutase show increased vulnerability to the mitochondrial toxins malonate, 3-nitropropionic acid, and MPTP, *Exp. Neurol.*, 2001, **167**, 189–195.

128. L.-s. Huang, G. Sun, D. Cobessi, A. C. Wang, J. T. Shen, E. Y. Tung, V. E. Anderson and E. A. Berry, 3-nitropropionic acid is a suicide inhibitor of mitochondrial respiration that, upon oxidation by complex II, forms a covalent adduct with a catalytic base arginine in the active site of the enzyme, *J. Biol. Chem.*, 2006, **281**, 5965–5972.

129. M. Beal, E. Brouillet, B. Jenkins, R. Ferrante, N. Kowall, J. Miller, E. Storey, R. Srivastava, B. Rosen and B. Hyman, Neurochemical and histologic characterization of striatal excitotoxic lesions produced by the mitochondrial toxin 3-nitropropionic acid, *J. Neurosci.*, 1993, **13**, 4181–4192.

130. S. Gines, I. S. Seong, E. Fossale, E. Ivanova, F. Trettel, J. F. Gusella, V. C. Wheeler, F. Persichetti and M. E. MacDonald, Specific progressive cAMP reduction implicates energy deficit in presymptomatic Huntington's disease knock-in mice, *Hum. Mol. Genet.*, 2003, **12**, 497–508.

131. P. Weydt, V. V. Pineda, A. E. Torrence, R. T. Libby, T. F. Satterfield, E. R. Lazarowski, M. L. Gilbert, G. J. Morton, T. K. Bammler and A. D. Strand, Thermoregulatory and metabolic defects in Huntington's disease transgenic mice implicate PGC-1? in Huntington's disease neurodegeneration, *Cell Metab.*, 2006, **4**, 349–362.

132. F. Mochel, B. Durant, X. Meng, J. O'Callaghan, H. Yu, E. Brouillet, V. C. Wheeler, S. Humbert, R. Schiffmann and A. Durr, Early alterations of brain cellular energy homeostasis in Huntington disease models, *J. Biol. Chem.*, 2012, **287**, 1361–1370.

133. I. S. Seong, E. Ivanova, J.-M. Lee, Y. S. Choo, E. Fossale, M. Anderson, J. F. Gusella, J. M. Laramie, R. H. Myers and M. Lesort, HD CAG repeat

implicates a dominant property of huntingtin in mitochondrial energy metabolism, *Hum. Mol. Genet.*, 2005, **14**, 2871–2880.

134. J.-M. Lee, E. V. Ivanova, I. S. Seong, T. Cashorali, I. Kohane, J. F. Gusella and M. E. MacDonald, Unbiased gene expression analysis implicates the huntingtin polyglutamine tract in extra-mitochondrial energy metabolism, *PLoS Genet.*, 2007, **3**, e135.

135. M. C. Scrutton, M. F. Utter and A. S. Mildvan, Pyruvate carboxylase VI, The presence of tightly bound manganese, *J. Biol. Chem.*, 1966, **241**, 3480–3487.

136. F. Mochel, P. DeLonlay, G. Touati, H. Brunengraber, R. P. Kinman, D. Rabier, C. R. Roe and J.-M. Saudubray, Pyruvate carboxylase deficiency: clinical and biochemical response to anaplerotic diet therapy, *Mol. Genet. Metab.*, 2005, **84**, 305–312.

137. K. Josefsen, S. Nielsen, A. Campos, T. Seifert, L. Hasholt, J. E. Nielsen, A. Nørremølle, N. H. Skotte, N. H. Secher and B. Quistorff, Reduced gluconeogenesis and lactate clearance in Huntington's disease, *Neurobiol. Dis.*, 2010, **40**, 656–662.

138. L. Harms, H. Meierkord, G. Timm, L. Pfeiffer and A. Ludolph, Decreased N-acetyl-aspartate/choline ratio and increased lactate in the frontal lobe of patients with Huntington's disease: a proton magnetic resonance spectroscopy study, *J. Neurol. Psychiatry*, 1997, **62**, 27–30.

139. B. G. Jenkins, W. J. Koroshetz, M. F. Beal and B. R. Rosen, Evidence for irnnairment of energy metabofism in vivo in Huntington's disease using localized 1H NMR spectroscopy, *Neurology*, 1993, **43**, 2689.

140. L. Chiosa, V. Niculescu, C. Bonciocat and C. Stancu, The protective action of N-acetyl and N-carbamyl derivatives of glutamic and aspartic acids against ammonia intoxication, *Biochem. Pharmacol.*, 1965, **14**, 1635–1643.

141. B. Gruber, G. Kłaczkow, M. Jaworska, J. Krzysztoń-Russjan, E. Anuszewska, D. Zielonka, A. Klimberg and J. Marcinkowski, Huntington's disease-imbalance of amino acid levels in plasma of patients and mutation carriers, *Ann. Agric. Environ. Med.*, 2013, **20**, 779.

142. M. C. Rodriguez, J. R. MacDonald, D. J. Mahoney, G. Parise, M. F. Beal and M. A. Tarnopolsky, Beneficial effects of creatine, CoQ10, and lipoic acid in mitochondrial disorders, *Muscle nerve*, 2007, **35**, 235–242.

143. D. L. Baly, C. L. Keen and L. S. Hurley, Pyruvate carboxylase and phosphoenolpyruvate carboxykinase activity in developing rats: effect of manganese deficiency, *J.Nutr.*, 1985, **115**, 872.

144. E. Colin, E. Régulier, V. r. Perrin, A. Dürr, A. Brice, P. Aebischer, N. Déglon, S. Humbert and F. Saudou, Akt is altered in an animal model of Huntington's disease and in patients, *Eur. J. Neurosci.*, 2005, **21**, 1478–1488.

145. V. Maglione, P. Marchi, A. Di Pardo, S. Lingrell, M. Horkey, E. Tidmarsh and S. Sipione, Impaired ganglioside metabolism in Huntington's disease and neuroprotective role of GM1, *J. Neurosci.*, 2010, **30**, 4072–4080.

146. S. Podolsky, N. Leopold and D. Sax, Increased frequency of diabetes mellitus in patients with Huntington's chorea, *Lancet*, 1972, **299**, 1356–1359.

147. L. A. Farrer, Diabetes mellitus in Huntington disease, *Clin. Genet.*, 1985, **27**, 62–67.

148. M. S. Hurlbert, W. Zhou, C. Wasmeier, F. Kaddis, J. Hutton and C. Freed, Mice transgenic for an expanded CAG repeat in the Huntington's disease gene develop diabetes, *Diabetes*, 1999, **48**, 649–651.

149. J. C. Mazziotta, M. E. Phelps, J. J. Pahl, S.-C. Huang, L. R. Baxter, W. H. Riege, J. M. Hoffman, D. E. Kuhl, A. B. Lanto and J. A. Wapenski, Reduced cerebral glucose metabolism in asymptomatic subjects at risk for Huntington's disease, *N. Engl. J. Med.*, 1987, **316**, 357–362.

150. S. Humbert, E. A. Bryson, F. P. Cordelières, N. C. Connors, S. R. Datta, S. Finkbeiner, M. E. Greenberg and F. d. r. Saudou, The IGF-1/Akt pathway is neuroprotective in Huntington's disease and involves Huntingtin phosphorylation by Akt, *Dev. Cell*, 2002, **2**, 831–837.

151. M. S. Clegg, S. M. Donovan, M. H. Monaco, D. L. Baly, J. L. Ensunsa and C. L. Keen, *Proceedings of the Society for Experimental Biology and Medicine*, Society for Experimental Biology and Medicine, New York, NY, 1998.

152. J. K. Hiney, V. K. Srivastava and W. Les Dees, Manganese induces IGF-1 and cyclooxygenase-2 gene expressions in the basal hypothalamus during prepubertal female development, *Toxicol. Sci.*, 2011, **121**, 389–396.

153. F. M. Cordova, A. S. Aguiar Jr, T. V. Peres, M. W. Lopes, F. M. Goncalves, A. P. Remor, S. C. Lopes, C. l. Pilati, A. S. Latini and R. D. Prediger, In vivo manganese exposure modulates erk, akt and darpp-32 in the striatum of developing rats, and impairs their motor function, *PloS One*, 2012, **7**, e33057.

154. J.-H. Bae, B.-C. Jang, S.-I. Suh, E. Ha, H. H. Baik, S.-S. Kim, M.-y. Lee and D.-H. Shin, Manganese induces inducible nitric oxide synthase (iNOS) expression via activation of both MAP kinase and PI3K/Akt pathways in BV2 microglial cells, *Neurosci. Lett.*, 2006, **398**, 151–154.

155. S. Subasinghe, A. L. Greenbaum and P. McLean, The insulin-mimetic action of Mn^{2+}: Involvement of cyclic nucleotides and insulin in the regulation of hepatic hexokinase and glucokinase, *Biochem. Med.*, 1985, **34**, 83–92.

156. N. Z. Baquer, M. Sinclair, S. Kunjara, U. C. Yadav and P. McLean, Regulation of glucose utilization and lipogenesis in adipose tissue of diabetic and fat fed animals: Effects of insulin and manganese, *J. Biosci.*, 2003, **28**, 215–221.

157. D. L. Baly, D. L. Curry, C. L. Keen and L. S. Hurley, Effect of manganese deficiency on insulin secretion and carbohydrate homeostasis in rats, *J. Nutr.*, 1984, **114**, 1438–1446.

158. M. Hassanein, H. Ghaleb, E. Haroun, M. Hegazy and M. Khayyal, Chronic manganism: preliminary observations on glucose tolerance and serum proteins, *Br. J. Ind. Med.*, 1966, **23**, 67–70.

159. C. L. Keen, J. L. Ensunsa and M. S. Clegg, Manganese metabolism in animals and humans including the toxicity of manganese, *Met. Ions Biol. Syst.*, 2000, **37**, 89–121.

160. L. Xiao, L. Gong, D. Yuan, M. Deng, X. Zeng, L. Chen, L. Zhang, Q. Yan, J. Liu and X. Hu, Protein phosphatase-1 regulates Akt1 signal transduction pathway to control gene expression, cell survival and differentiation, *Cell Death Differ.*, 2010, **17**, 1448–1462.

161. D. W. Li, J. Liu, P. Schmid, R. Schlosser, H. Feng, W. Liu, Q. Yan, L. Gong, S. Sun and M. Deng, Protein serine/threonine phosphatase-1 dephosphorylates p53 at Ser-15 and Ser-37 to modulate its transcriptional and apoptotic activities, *Oncogene*, 2006, **25**, 3006–3022.

162. J. Carmichael, K. L. Sugars, Y. P. Bao and D. C. Rubinsztein, Glycogen synthase kinase-3β inhibitors prevent cellular polyglutamine toxicity caused by the Huntington's disease mutation, *J. Biol. Chem.*, 2002, **277**, 33791–33798.

163. L. A. Carvajal and J. J. Manfredi, Another fork in the road—life or death decisions by the tumour suppressor p53, *EMBO Rep.*, 2013, **14**, 414–421.

164. B.-I. Bae, H. Xu, S. Igarashi, M. Fujimuro, N. Agrawal, Y. Taya, S. D. Hayward, T. H. Moran, C. Montell and C. A. Ross, p53 mediates cellular dysfunction and behavioral abnormalities in Huntington's disease, *Neuron*, 2005, **47**, 29–41.

165. J. L. Illuzzi, C. A. Vickers and E. B. Kmiec, Modifications of p53 and the DNA damage response in cells expressing mutant form of the protein huntingtin, *J. Mol. Neurosci.*, 2011, **45**, 256–268.

166. J. S. Steffan, A. Kazantsev, O. Spasic-Boskovic, M. Greenwald, Y.-Z. Zhu, H. Gohler, E. E. Wanker, G. P. Bates, D. E. Housman and L. M. Thompson, The Huntington's disease protein interacts with p53 and CREB-binding protein and represses transcription, *Proc. Natl. Acad. Sci.*, 2000, **97**, 6763–6768.

167. X. Guo, M.-H. Disatnik, M. Monbureau, M. Shamloo, D. Mochly-Rosen and X. Qi, Inhibition of mitochondrial fragmentation diminishes Huntington's disease – associated neurodegeneration, *J. Clin. Invest.*, 2013, **123**, 5371.

168. B. Chattopadhyay, K. Baksi, S. Mukhopadhyay and N. P. Bhattacharyya, Modulation of age at onset of Huntington disease patients by variations in TP53 and human caspase activated DNase (hCAD) genes, *Neurosci. Lett.*, 2005, **374**, 81–86.

169. C. Jung-Il, K. Dong-Wook, L. Nayeon, J. Young-Joo, J. Iksoo, K. Jihye, K. Jumi, S. Yunjo, L. Dong-Seok and S. S. Kang, Quantitative proteomic analysis of induced pluripotent stem cells derived from a human Huntington's disease patient, *Biochem. J.*, 2012, **446**, 359–371.

170. A. Grison, F. Mantovani, A. Comel, E. Agostoni, S. Gustincich, F. Persichetti and G. Del Sal, Ser46 phosphorylation and prolyl-isomerase Pin1-mediated isomerization of p53 are key events in p53-dependent apoptosis induced by mutant huntingtin, *Proc. Natl. Acad. Sci.*, 2011, **108**, 17979–17984.

171. C. Meplan, M.-J. Richard and P. Hainaut, Metalloregulation of the tumor suppressor protein p53: zinc mediates the renaturation of p53 after exposure to metal chelators in vitro and in intact cells, *Oncogene*, 2000, **19**, 5227–5236.

172. M. Matsuoka and H. Igisu, Cadmium induces phosphorylation of p53 at serine 15 in MCF-7 cells, *Biochem. Biophys. Res. Commun.*, 2001, **282**, 1120–1125.

173. E. Ostrakhovitch and M. Cherian, Differential regulation of signal transduction pathways in wild type and mutated p53 breast cancer epithelial cells by copper and zinc, *Arch. Biochem. Biophys.*, 2004, **423**, 351–361.

174. C. Stenger, T. Naves, M. Verdier and M.-H. Ratinaud, The cell death response to the ROS inducer, cobalt chloride, in neuroblastoma cell lines according to p53 status, *Int. J. Oncol.*, 2011, **39**, 601.

175. S. Wang and X. Shi, Mechanisms of Cr (VI)-induced p53 activation: the role of phosphorylation, mdm2 and ERK, *Carcinogenesis*, 2001, **22**, 757–762.

176. T. s. R. Guilarte, N. C. Burton, T. Verina, V. V. Prabhu, K. G. Becker, T. Syversen and J. S. Schneider, Increased APLP1 expression and neurodegeneration in the frontal cortex of manganese-exposed non-human primates, *J. Neurochem.*, 2008, **105**, 1948–1959.

177. C. E. Canman, D.-S. Lim, K. A. Cimprich, Y. Taya, K. Tamai, K. Sakaguchi, E. Appella, M. B. Kastan and J. D. Siliciano, Activation of the ATM kinase by ionizing radiation and phosphorylation of p53, *Science*, 1998, **281**, 1677–1679.

178. S. Banin, L. Moyal, S.-Y. Shieh, Y. Taya, C. Anderson, L. Chessa, N. Smorodinsky, C. Prives, Y. Reiss and Y. Shiloh, Enhanced phosphorylation of p53 by ATM in response to DNA damage, *Science*, 1998, **281**, 1674–1677.

179. Z. Guo, S. Kozlov, M. F. Lavin, M. D. Person and T. T. Paull, ATM activation by oxidative stress, *Science*, 2010, **330**, 517–521.

180. D. W. Chan, S.-C. Son, W. Block, R. Ye, K. K. Khanna, M. S. Wold, P. Douglas, A. A. Goodarzi, J. Pelley and Y. Taya, Purification and Characterization of ATM from Human Placenta A Manganese-Dependent, Wortmannin-Sensitive Serine/Threonine Protein Kinase, *J. Biol. Chem.*, 2000, **275**, 7803–7810.

181. A. J. Levine, Z. Feng, T. W. Mak, H. You and S. Jin, Coordination and communication between the p53 and IGF-1-AKT-TOR signal transduction pathways, *Genes Dev.*, 2006, **20**, 267–275.

182. Z. Feng and A. J. Levine, The regulation of energy metabolism and the IGF-1/mTOR pathways by the p53 protein, *Trends Cell Biol.*, 2010, **20**, 427–434.

183. Z. Feng, W. Hu, E. De Stanchina, A. K. Teresky, S. Jin, S. Lowe and A. J. Levine, The regulation of AMPK -α1, TSC2, and PTEN expression by p53: stress, cell and tissue specificity, and the role of these gene products in modulating the IGF-1-AKT-mTOR pathways, *Cancer Res.*, 2007, **67**, 3043–3053.

184. T. M. Gottlieb, J. Leal, R. Seger, Y. Taya and M. Oren, Cross-talk between Akt, p53 and Mdm2: possible implications for the regulation of apoptosis, *Oncogene*, 2002, **21**, 1299–1303.

185. M. Mihara, S. Erster, A. Zaika, O. Petrenko, T. Chittenden, P. Pancoska and U. M. Moll, p53 has a direct apoptogenic role at the mitochondria, *Mol. Cell*, 2003, **11**, 577–590.

186. A. V. Vaseva, N. D. Marchenko, K. Ji, S. E. Tsirka, S. Holzmann and U. M. Moll, p53 opens the mitochondrial permeability transition pore to trigger necrosis, *Cell*, 2012, **149**, 1536–1548.

187. G. Pani, B. Bedogni, R. Anzevino, R. Colavitti, B. Palazzotti, S. Borrello and T. Galeotti, Deregulated manganese superoxide dismutase expression and resistance to oxidative injury in p53-deficient cells, *Cancer Res.*, 2000, **60**, 4654–4660.

188. P. Drane, A. Bravard, V. Bouvard and E. May, Reciprocal downregulation of p53 and SOD2 gene expression-implication in p53 mediated apoptosis, *Oncogene*, 2001, **20**, 430–439.

189. K. Panickar, A. Jayakumar, K. Rao and M. Norenberg, Ammoniainduced activation of p53 in cultured astrocytes: role in cell swelling and glutamate uptake, *Neurochem. Int.*, 2009, **55**, 98–105.

190. W. Hu, C. Zhang, R. Wu, Y. Sun, A. Levine and Z. Feng, Glutaminase 2, a novel p53 target gene regulating energy metabolism and antioxidant function, *Proc. Natl. Acad. Sci.*, 2010, **107**, 7455–7460.

191. S. Suzuki, T. Tanaka, M. V. Poyurovsky, H. Nagano, T. Mayama, S. Ohkubo, M. Lokshin, H. Hosokawa, T. Nakayama and Y. Suzuki, Phosphate-activated glutaminase (GLS2), a p53-inducible regulator of glutamine metabolism and reactive oxygen species, *Proc. Natl. Acad. Sci.*, 2010, **107**, 7461–7466.

CHAPTER 23

Manganese and Prion Disease

HUAJUN JIN,[†] DILSHAN S. HARISCHANDRA,[†]
CHRISTOPHER CHOI, DUSTIN MARTIN,
VELLAREDDY ANANTHARAM, ARTHI KANTHASAMY AND
ANUMANTHA G. KANTHASAMY*

Parkinson's Disorder Research Laboratory, Iowa Center for Advanced
Neurotoxicology, Department of Biomedical Sciences, Iowa State
University, Ames, Iowa 50011, USA
*Email: akanthas@iastate.edu

23.1   Introduction

Prion diseases, also termed transmissible spongiform encephalopathies
(TSEs), are a group of fatal neurodegenerative disorders affecting both
humans and animals.[1] Although the prevalence of prion disease is relatively
low, with about 300 cases reported annually in the United States, the potential for the disease to be transmitted between humans and other animals
is a serious concern. The recent emergence of chronic wasting disease
(CWD), a prion disease in cervids, raises concerns over potential zoonotic
transmission of the disease. Prion diseases differ in their pathology as well
as their modes of pathogenesis.[2] Their routes of transmission are also
multiple, with the possibility of iatrogenic transmission from infected
surgical materials, human grafts, and even blood transfusions, and there
may be a hereditary component to their transmission. Typical clinical

---

[†]These authors contributed equally to this chapter.

Issues in Toxicology No. 22
Manganese in Health and Disease
Edited by Lucio G. Costa and Michael Aschner

symptoms of prion diseases include rapidly developing dementia, difficulty walking, change in gait, hallucinations, confusion, fatigue, muscle stiffness and difficulty speaking. Historically, prion diseases have been characterized neuropathologically by vacuolation of neutrophils, neuronal loss, gliosis, and by the deposition of amyloid plaques in the brains of diseased animals and humans.[3,4] The neurodegeneration in prion disease primarily occurs in the brain regions that coordinate motor functions, such as the basal ganglia, cerebral cortex, thalamus and cerebellum.[5,6] Reactive astrocytic gliosis, microglial activation and the deposition of amyloid plaques have been associated with the accumulation of the abnormal prion protein PrP^{Sc} (scrapie isoform) derived from the normal prion protein PrP^C (cellular isoform).[7] Thus far, neither effective treatment nor prevention methods have been developed for prion disease.[8,9] Most confirmed cases have been identified post mortem, so the development of antemortem tests is essential for effective detection. There are different methods of dealing with prion disease.[10,11] Given that total eradication of the diseases has not been possible, reducing their infectivity, reducing the substrate, and developing vaccinations have been explored.

Despite extensive research, the precise cause of prion disease remains unknown. However, it is generally believed that the diseased state of an organism occurs when the soluble α-helix-rich form of PrP^C has been converted to the insoluble β-sheet-rich pathogenic form of PrP^{Sc}.[12] The conversion steps required for the protein to become pathogenic remain unclear. Interestingly, different conformational isomers with identical primary sequences can display widely varying pathology, indicating that the secondary and tertiary structures of PrP^{Sc} can encode different strains of TSE.[13–15] Though the absolute pathogenesis of prion disease remains obscure, several pathogenic mechanisms have been proposed, including synaptic damage, dendrite atrophy, autophagy, microglial activation, oxidative stress, protein misfolding, endoplasmic reticulum (ER) stress and apoptosis. Further complicating the effort to uncover the etiology, a combination of multiple interlinking pathways, rather than a unifying mechanism, could contribute to the pathology of prion diseases.

Substantial evidence also indicates that the prion protein is a metalloprotein, with affinity for various cations such as copper, zinc, manganese, and nickel.[16–18] Many studies have reported significant differences in metal content in diseased humans and other animals,[19,20] suggesting the involvement of metal homeostasis in prion disease pathogenesis. The redox metals are essential trace element metals normally required for physiological processes, and are precisely regulated by the cells. Interactions of these metal ions with prion protein may have a role in protein misfolding and the neurodegenerative progression of prion diseases. Therefore, a detailed understanding of metal–prion interactions would not only expand our understanding of the pathophysiological mechanisms of prion diseases, but may also enable the development of effective treatment strategies for these debilitating diseases.

23.2 Prion Diseases

Although the biochemical process that converts PrP^C to PrP^{Sc} is not completely understood, according to a seeding-nucleation model the pre-existing or acquired PrP^{Sc} oligomers catalyze the conversion of PrP^C molecules into PrP^{Sc} fibrils. The breakage of these fibrils provides more PrP^{Sc} templates for further seeding of the conversion process.[21] This process results in increased amounts of aberrant PrP^{Sc} proteins, which are extremely resistant to proteolysis or degradation by conventional means, thereby initiating the classic prion disease state.

The earliest description of a TSE, dating back to the mid-18th century, was a disease identified as scrapie, the prototypic prion disease affecting sheep and goats.[22] In the early 20th century, a critical experiment performed by Cuille and Chelle confirmed the transmissibility of scrapie to goats, turning a new page in prion biology.[23] In 1920, the first human case of TSE, Creutzfeldt–Jakob disease (CJD), was reported,[24] and this remains the most common form of human prion disease. The number of human and animal prion diseases identified has increased steadily, and more than 15 different diseases have been described so far.[22,25] CWD in deer is considered a major form of prion disease in the US, and the disease is rapidly spreading to several states. In the following section, we will review some of the prevalent human prion diseases.

23.2.1 Creutzfeldt–Jakob Disease (CJD)

CJD was initially described as a sporadic disease (sCJD) caused by the spontaneous transformation of PrP^C into PrP^{Sc}, resulting in rapid cognitive decline, involuntary movements, blindness, weakness of the extremities, and coma. Disease onset for sCJD occurs at about 60 years old, and 90% of patients die within one year. Sporadic CJD accounts for 85% of all CJD cases, with an annual incidence rate of approximately 0.6–1.2 cases per million worldwide.[26] Epidemiologically, there are three other distinct types of CJD, in addition to sCJD: familial (fCJD), iatrogenic (iCJD), and variant (vCJD). Familial CJD develops from a heritable mutation in the *Prnp* gene, which accounts for 5–10% of all CJD cases. Several point mutations have been identified as risk factors for fCJD, including E200K and V210I.[27] A smaller number of cases (<5%) have been classified as acquired forms, accounting for both iCJD and vCJD. These acquired CJD cases are mostly transmitted iatrogenically by accidental exposure to PrP^{Sc}-infected brain or nervous system tissues during medical procedures. The first reported iCJD case occurred in 1974 following corneal transplantation from a deceased patient with undiagnosed sCJD.[28] Pituitary growth hormones and dura matter graft-transplants obtained from CJD-infected individuals account for most iCJD cases,[29] and the current worldwide total of growth hormone-associated cases of CJD is up to 226.[30] Variant CJD is a new subset of acquired CJD, with its own distinct pathological and clinical phenotypes. The first confirmed case of vCJD was reported in 1996 in the United Kingdom (UK);[31] it was suggested to be causally linked to the bovine spongiform encephalopathy (BSE) outbreak in Europe. By the end of 2013, 177 deaths in the UK (http://www.cjd. ed.ac.uk/) had been attributed to vCJD. Unlike sCJD cases, vCJD patients are relatively young (median age of 28 years) and the incubation period of the disease is relatively long. Interestingly, unlike other CJD patients, vCJD patients lack the classic CJD electroencephalogram and they develop extensive fluoride plaques in the brain.[31] Also, epidemiological and experimental data suggest that sCJD is not transmitted from person to person *via* blood transfusion, but this might not be the case for vCJD. For example, there is incidence report of one recipient developing symptoms of vCJD 6.5 years after receiving a transfusion of red blood cells donated by an individual later identified as a vCJD patient.[32] Such evidence continues to mount, as there have been four more identified cases of vCJD resulting from blood transfusions in the UK. To prevent further transmission, since April 2004 anyone who had received a blood transfusion in the UK after 1980 became ineligible to donate blood (http://www.hpa.org.uk/).

23.2.2 Kuru

Kuru is another acquired human prion disease, seen exclusively in the Fore linguistic groups and neighboring tribes in the Okapa area of the Eastern Highlands of Papua New Guinea. The disease resulted from the practice of ritualistic cannibalism among the Fore, in which relatives prepared and consumed the tissues, including the brains, of deceased family members. Kuru affected predominantly women and young children, because they were exposed to infectious brain and visceral tissues, while adult men primarily consumed muscles.[25] Although the first case was observed in about 1920, Kuru was not thoroughly documented until 1957, and since then over 2700 cases have been reported.[33] Dr Gajdusek and colleagues showed disease transmission in non-human primates by intracerebral introduction of brain homogenates from Kuru patients, proving that Kuru was caused by a transmissible agent,[34] and thus documenting for the first time the infectious nature of prion diseases in primates.

Clinically, the prodrome stage of Kuru consists of headaches and joint pain in the legs followed by three clinical stages: an ambulatory stage, a sedentary stage and a tertiary stage.[35] The symptoms seen in these clinical stages are characteristic of the term "Kuru", which means "to shiver from fever and cold" in the Fore language. Additionally, Kuru is considered largely a cerebellar syndrome with ataxia, tremors, and choreiform and athetoid movements being the prominent clinical signs of the disease;[36] dementia is a late and less prominent symptom. Interestingly, recent genome-wide studies of Kuru confirmed a strong association with a single nucleotide polymorphism (SNP) localized within codon 129 and also with two other SNPs localized within the gene's retinoic acid receptor beta (*RARB*) and in stathmin like 2 (*STMN2*; the gene encoding SCG10, a neuron-specific growth-associated protein).[37,38] Importantly, studies have shown that individuals with *PRNP* 129$^{Val/Val}$ and 129$^{Met/Met}$ genotypes are susceptible to Kuru,[39] whereas heterozygosity at codon 129 confers relative resistance to prion diseases.[40] An evolutionarily strong balancing selection for these alleles had been imposed at this locus, not only in Fore, but also in other human populations practicing cannibalism.[41]

23.2.3 Gerstmann–Sträussler–Scheinker Syndrome

Gerstmann–Sträussler–Scheinker disease (GSS) is a genetically determined adult prion disease associated with the autosomal dominant inheritance of *Prnp* mutations. GSS is also considered one of the rarest forms of prion disease, with an incidence of 2–5 per 100 million people. However, this disorder may be underdiagnosed. Many of its key clinical symptoms, such as cerebellar ataxia, dysdiadochokinesia, speech disturbance, personality changes, and dementia, are also characteristic of other neurodegenerative diseases such as Parkinson's disease (PD), amyotrophic lateral sclerosis (ALS or Lou Gherig's disease), Huntington's disease (HD), and Alzheimer's disease (AD).

Neuropathologically, GSS is characterized primarily by prominent amyloid plaques and diffuse deposits resulting from the accumulation of PrP degradation products in the cerebellum.[42] However, GSS shows great pathological heterogeneity, which often partly overlaps with AD, PD and dementia with Lewy bodies (DLB). The phenotypic differences among cases of GSS correlate to the haplotype-specific pattern. For instance, the findings on immunohistochemical analysis of the neocortex of patients with GSS are associated with the F198S,[43] Q227X,[44] and D202N haplotypes,[45] and indicate the presence of neurofibrillary tangles normally found in AD. Patients with the F198S mutation have also been found with α-synuclein immunopositive Lewy bodies typical of PD.[46] Currently, at least 16 point mutations in the *Prnp* gene have been implicated as risk factors for GSS, including P102L, P105L, A117V, Y145X, H187R, D178N, Q160X, Q217R, Y218N, Y226X, G131V, Q212P and S132I.[25,47–49] Biochemically, GSS is characterized by the presence of proteinase K (PK) resistant N- and C-terminal truncated and non-glycosylated PrP peptides ranging from ~7 to 15 kDa.[50] Interestingly, the transmissibility of GSS is widely studied, and common GSS-associated mutations (*e.g.* P102L) seem to be more efficient transmitters of the disease than less frequent GSS-associated mutations.[51]

23.2.4 Fatal Familial Insomnia

Fatal familial insomnia (FFI) is an extremely rare genetic disorder, and arguably considered the deadliest form of insomnia, which "steals one's sleep, mind and eventually life". Clinically, FFI is characterized by untreatable alterations of the sleep–wake cycle (loss of sleep spindles, slow-wave sleep, non-rapid eye movement sleep and enacted dreams),[52,53] autonomic hyperactivation, and cognitive and motor impairments such as dysarthria, myoclonus, ataxia, tremor, and pyramidal and extrapyramidal signs.[54] Thalamic atrophy is recognized as the histopathological hallmark of FFI, while pathological lesions in the neocortex and the limbic cortex are also observed. Given that the importance of the thalamus in sleep physiology has been well characterized,[55] the observed sleep disturbance is closely associated with such physiological changes. These clinical manifestations also result in dysautonomia (hyperhidrosis, hyperthermia, tachycardia, and hypertension) and endocrine disturbances (decreased adrenocorticotropic hormone secretion, increased cortisol secretion), and disturbances in growth hormone, melatonin, and prolactin secretion. FFI is the only known prion disease to exhibit these secondary health complications.[56]

Historically, FFI was first identified and characterized in 1986 by Lugaresi and colleagues, and to date at least 40 unrelated kindred cases have been identified worldwide, including families from Japan, China and Africa, as well as families with American and European ancestry.[57–59] Genetically, FFI is characterized as an autosomal dominant disease associated with a point mutation at codon 178 in the *Prnp* gene where asparagine has been substituted for aspartic acid (D178N).[60] However, the same D178N mutation is also linked to fCJD. What distinguishes the two diseases is the genotype at the polymorphic codon 129, where FFI is associated with methionine (M) (D178N-129M haplotype) and fCJD is associated with valine (V) (D178N-129V haplotype).[61] Additionally, the polymorphic codon 129 in the non-mutated allele determines the severity of the disease. In FFI, homozygous methionine at polymorphic codon 129 (Figure 23.1A) is associated with more severe insomnia and dysautonomia at onset and with thalamic damage and fewer cortical alterations, whereas heterozygosity (methionine/valine) (Figure 23.1B)

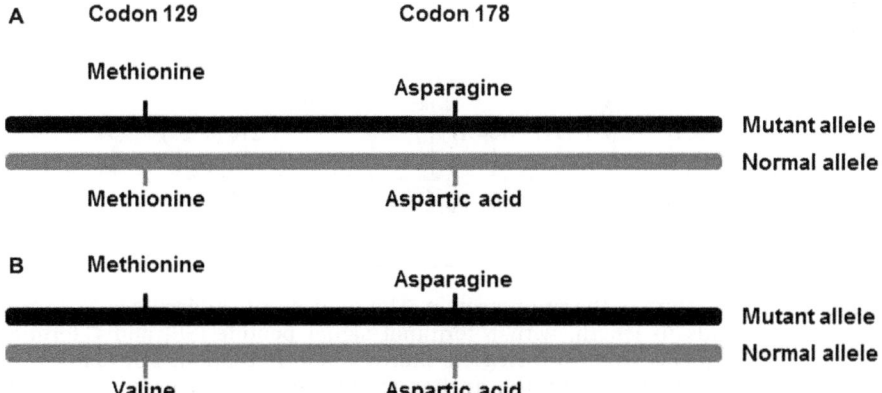

Figure 23.1   The haplotypes in fatal familial insomnia (FFI). (A) Methionine at polymorphic codon 129 is associated with thalamic damage and fewer cortical alterations, whereas (B) valine at polymorphic codon 129 is associated with prolonged disease and widespread neuropathological damage with cortical spongiosis.

is associated with ataxia and dysarthria at onset, and after prolonged disease, with widespread neuropathological damage and cortical spongiosis.[62,63]

Biochemically, FFI is quite distinct from other prion diseases: the prions involved in fCJD and sCJD give a ~ 21 kDa protein fragment after digestion with protease, while FFI produces i a shorter, ~ 19 kDa, protein fragment.[64] These results further confirm the "prion strain" hypothesis, allowing researchers to identify two different strains from their origin to their experimental transmission to laboratory animals. Further characterization of the glycosylation patterns helped to distinguish FFI from sporadic fatal insomnia (sFI), a rare sporadic form of the disease similar to FFI in its clinical symptoms. Biochemically, sFI is characterized by a predominant monoglycosylated form of the prion protein, whereas FFI is characterized by an under-representation of the unglycosylated form.[65]

23.3 Prion Protein (PrP^C)

23.3.1 Structure of PrP^C

PrP^C is an N-linked glycoprotein tethered to extracellular membranes by a glycosylphosphatidylinositol (GPI) anchor, which is ubiquitously expressed throughout the central nervous system, particularly in both neuronal and glial cells. In humans, PrP^C is encoded by the *PRNP* gene located on the short arm of chromosome 20 (20q13) as a 16 kb (kilobase) single gene copy. The human *PRNP* gene encodes a 253-residue precursor prion protein and, following translation, the first 22 N-terminal residues (signal peptide) are cleaved during transport to the cell surface[66] (Figure 23.2). The last 23 C-terminal amino acids are excised after the addition of the GPI anchor, resulting in mature PrP^C on the cell surface consisting of 208 amino acid residues.[67] Properly folded and GPI-anchored PrP^C becomes localized in detergent-resistant membranes, also known as lipid rafts.[24] This PrP^C is

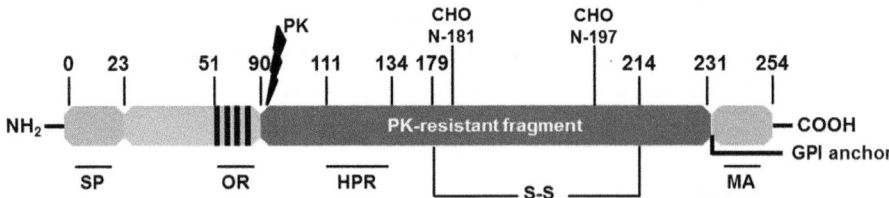

Figure 23.2   Structure of the prion protein. The mouse PrP^C molecule is 254 amino acids in length, with N-terminal signal peptide (SP) and C-terminal sequences that are cleaved shortly after translation. MA denotes the C-terminal membrane anchor region and HPR denotes the central hydrophobic domain (amino acids 111–134) of the prion protein. Toward the N-terminus is the octapeptide repeat region (OR), which is suggested to play a role in metal binding. S-S indicates the single disulfide bridge between residues 179 and 214. The approximate cutting site of PK within PrP^{Sc} is indicated by the lightening symbol, and a PK-resistant fragment is between residues 90 and 231.

rapidly internalized from the cell membrane *via* caveolae-like[68,69] or clathrin-dependent[70,71] endocytosis. Internalization is considered to be crucial for PrP^C function in regulating signal transduction pathways and neurite outgrowth.[72,73] Experimentally, this surface-bound PrP^C can be removed *in vitro* by incubating with bacterial phosphatidylinositol-specific phospholipase C (PI-PLC), which liberates PrP^C from the cell membrane by cleaving the GPI anchor.[67,74]

Structural modeling through nuclear magnetic resonance (NMR) studies on recombinant human PrP^C reveals the C-terminal globular domain to have three α helices (α$_1$, α$_2$ and α$_3$) and short anti-parallel β-sheets (β$_1$ and β$_2$).[75] The C-terminal domain of PrP^C also contains a single disulfide-bonded bridge linking the Cys residues of α$_2$ and α$_3$ at positions 179 and 214. This disulfide bridge is important for the conformational stability of PrP^C, and its removal greatly destabilizes the native PrP^C structure, which allows it to switch reversibly between the α-helical conformation and a monomeric form rich in β-sheet structure.[76] The N-terminal moiety of the prion protein is an unstructured region that characteristically interacts with a broad range of partners with contrasting capabilities, including neuroprotection and neurotoxicity. One hallmark of this region is the highly conserved tandem repeats of an eight-residue sequence (PHGGSWGQ) referred to as an octapeptide repeat domain. The number of repeats differs from species to species (see Figure 23.3 for PrP^C homology among mammalian species). For example, humans, mice, sheep, and deer each have five octapeptide repeats in the wild-type cellular protein, while bovine PrP has six repeats.[77]

23.3.2 Physiological Function of Prion Protein

Octapeptide repeat regions of prion protein have a high affinity for binding to divalent metals, with the highest affinity for copper, followed by nickel, zinc and manganese.[78] Metal binding has been suggested to play an important role in the biological function and pathogenesis of prion protein. This protein also has been speculated to act as an antioxidant. Its role in antioxidant defense was demonstrated by blocking toxic effects by treating cells with synthetic PrP^C 59–91 peptide in cells exposed to high levels of copper.[79]

Transmembrane signaling is another physiological function of prion proteins, which is regulated by different binding partners. Given that most PrP^{C}s are localized on plasma membranes, specifically on cholesterol- and glycosphingolipid-rich lipid raft domains serving as scaffolds for signal transduction,[80] numerous studies have identified binding ligands of prion proteins. Indeed, PrP^C interacts with various macromolecules at the cell membrane to activate transmembrane signaling pathways involved in several different phenomena, including neuronal survival, neurite outgrowth and neurotoxicity. It has been demonstrated that both the laminin precursor protein (LRP), *via* the yeast two hybrid system,[81] and the neuronal cell adhesion molecule (N-CAM), *via* formaldehyde cross-linking studies,[82]

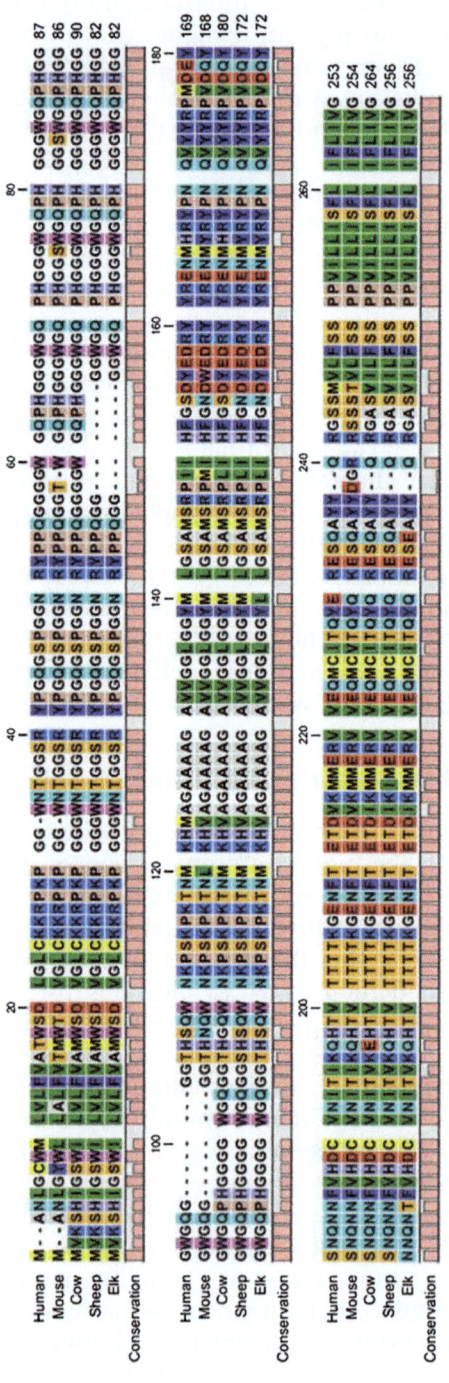

Figure 23.3 Prion protein (PrP^C) sequence homology among various mammalian species. Peptide sequences were obtained from NCBI Entrez protein database and aligned using CLC DNA workbench 6 software.

interact with PrP^C on cell surfaces, promoting diverse transduction pathways involved in differentiation and neurite outgrowth.[83,84] Moreover, Santuccione and colleagues have shown that heterophilic *cis* and *trans* interactions between N-CAM and PrP at the neuronal surface promote N-CAM recruitment to lipid rafts for activation of $p59^{Fyn}$, a member of the Src family of non-receptor tyrosine protein kinases.[85] Furthermore, studies carried out with the 1C11 neuronal differentiation model with antibody-mediated cross-linking have shown $p59^{Fyn}$ activation in a caveolin-1-dependent manner.[86] In subsequent studies, as a downstream event, they also report nicotinamide adenine dinucleotide phosphate (NADPH) oxidase-dependent reactive oxygen species (ROS) production and extracellularly regulated kinase 1/2 (ERK1/2) phosphorylation in fully differentiated progenies, identifying NADPH oxidase and ERK1/2 as targets of PrP^C-mediated signaling in neuronal and non-neuronal cells.[87]

23.4 Metals and Prion Diseases

Our understanding of the role of metals in key neurobiological processes as well as in the pathogenesis of various neurodegenerative diseases has continued to expand over the last two decades. Although it is well known that elemental metals are required for cells to function normally, the degree to which the central nervous system (CNS) uses metals in synaptic signaling, and the loss of metal homeostasis during neurodegenerative diseases, was, until recently, unknown. Dyshomeostasis of transition metals such as manganese, iron, copper, and zinc has been implicated in major neurodegenerative conditions such as PD, AD, ALS, HD, and prion disease.[88–91] Another common pathological feature of neurodegenerative diseases is the aggregation of proteins rich in β-sheets associated with each specific disease.[92] Alarmingly, exposing amyloidogenic proteins or their cleavage products to metals impacts the protein misfolding and progression of neurodegenerative processes.[88] Specifically, manganese has been shown to bind to the cellular form of prion protein, and this interaction has been implicated in the aggregation and misfolding of PrP^C. In this section, we will summarize the current evidence on manganese and prion interaction and its functional consequences in prion disease.

23.4.1 Manganese

Manganese (Mn) is one of the most abundant transitional metals on earth, comprising approximately 0.1% of the earth's crust. It may exist in both inorganic and organic forms, and has been heavily used in welding, mining, the manufacturing of batteries, glass, fireworks, chemicals, pesticides and fertilizers and in other industrial settings.[93] In nature, Mn exhibits 11 oxidation states ranging from -3 to $+7$, while in biological systems[94] it occurs primarily as Mn^{2+}, Mn^{3+}, and Mn^{4+}. In the human body, Mn is an essential trace element that functions as a key cofactor for numerous metalloenzymes, such as manganese superoxide dismutase, pyruvate carboxylase, arginase, phosphoenolpyruvate decarboxylase, and glutamine synthetase.[95] As such, Mn is involved in various biochemical and cellular functions, including blood clotting, adenosine triphosphate (ATP) production, immune responsiveness, digestion, and reproduction.[96] It also plays a key role in the development and normal functioning of the brain. Transport of Mn ions into the brain can be mediated through both the blood–brain and the blood–cerebrospinal fluid barriers,[97,98] and studies have documented various transporting mechanisms in this process.[99–101] Recommended intake levels of Mn for men and women have been established at 2.3 mg per day and 1.8 mg per day, respectively.[102] Although dietary Mn deficiencies are exceedingly rare in humans, diets low in Mn may cause various developmental defects. On the other hand, excess exposure to Mn results in a severe and degenerative neurological condition, known as manganism or Mn-induced Parkinsonism. Early signs of manganism include a variety of psychiatric disturbances, such as emotional instability, mania, and hallucinations, while motor symptoms including bradykinesia, rigidity, and dystonia are late manifestations of this disorder.[103] Unlike PD, manganism is pathologically characterized by the loss of neurons in the globus pallidus, cortex and hypothalamus, without the formation of Lewy bodies.[104,105] Although the pathogenic mechanisms underlying manganism are poorly understood, several lines of evidence suggest that Mn-induced neurotoxicity is associated with increased oxidative stress, impairment of energy metabolism and antioxidant systems, attenuation of astrocytic glutamate uptake, upregulation of binding sites for peripheral benzodiazepine receptor ligands, and alterations in various cell signaling pathways.[106–108] We have previously reported that the proapoptotic kinase PKCδ plays a crucial role in mediating Mn-induced dopaminergic neurodegeneration.[109–112] Interestingly, a role for Mn in the pathogenesis of prion disease has been emerging in recent years. Particularly, evidence has indicated that Mn-bound PrP^{Sc} can be isolated from both human and animal prion diseases.[113]

23.4.2 Manganese Binding to PrP^{C}

As mentioned previously, PrP^{C} is a putative metalloprotein because the octapeptide repeat sequences at the N-terminus of the protein have a high affinity for divalent cations including copper (Cu), Mn, and zinc (Zn). The possible complexation of Mn by PrP^{C} first came to light in 2000 when total reflection X-ray fluorescence spectrometry (TXRF) was applied.[114] In this study, recombinant full-length PrP^{C} was discovered to bind Mn *in vitro* followed by the refolding of this protein in the presence of high concentrations of Mn. However, deletion of the octapeptide repeat sequences completely abolished this Mn binding. The authors also documented that Mn can equivalently substitute for Cu in the octameric repeat region. These facts highlight the importance of the octameric repeat domain in mediating Mn and prion interactions. Later on, a few other studies on the affinity of Mn for prion were carried out using different amino acid sequences of recombinant prion protein; however, the results varied. Although evidence has indicated that Mn does have a binding affinity to the octameric repeat region, a spectroscopic study by Garnett and Viles showed that Mn does not bind to the PrP octameric repeat region.[115] Similarly, Treiber *et al.*, using the surface plasmon resonance (SPR) technique, found that peptides covering the octameric repeat sequences of prion were not able to bind Mn and, on the contrary, full-length PrP and the mutant PrP lacking the octameric repeat region bound to Mn with a nanomolar dissociation constant.[116] They concluded that the octameric repeat region is not involved in Mn binding, and proposed a conformational binding site for Mn involving the PrP residues 91–230. In contrast, analysis involving nuclear magnetic resonance (NMR) and circular dichroism (CD) spectroscopy in the presence of glycine has confirmed the binding of Mn to the octameric repeat region, despite an affinity at least three orders of magnitude less than that of Cu.[78] These divergent results might be due to variations in detection techniques and the use of different peptide sequences of PrP. Additional Mn-binding sites were identified at His 95 and 110 in the so-called "fifth site".[78,117] However, another report revealed that His 95 is the preferential binding site for Mn in this region, and His 110 plays no role in Mn binding.[88] Interestingly, further work by the same group suggests that the higher-affinity Mn binding site is not the octapeptide repeat motif, but the His 95 in the fifth site.[118] Using isothermal titration calorimetry, they identified two Mn binding sites for prion protein. The principle one is located at histidine residue 95 and the second, low-affinity, site is associated with the octameric repeat region, with dissociation constants of 63 µM and 200 µM, respectively.[118] This study also revealed an optimum pH of 5.5 for Mn binding at both sites. Additionally, the authors determined that PrP^C binds two molecules of Mn at these sites, while it was originally thought to bind up to four molecules of Mn at the octapeptide repeat region.[114] This study went on to show that Mn is able to replace Cu in Cu-saturated prion, even though PrP has a higher affinity for Cu at both binding sites. It should be noted that the micromolar range of affinity values for Mn binding to prion is in the range of other known Mn-binding proteins.[119] Although the research relating to PrP and Mn interaction has mostly been conducted *in vitro*, prion clearly is a Mn-binding protein. Given the important roles Mn and prion protein play in normal and disease states, it is logical to assume that the binding of Mn to prion protein has significant structural and functional consequences.

23.4.3   Role of Manganese in the Pathogenesis of Prion Disease

Over the past decade, evidence has been emerging on a possible role for Mn binding to PrP in the pathogenesis of prion diseases. Limited proteolytic digestion experiments using proteinase K and recombinant prion protein have revealed that Mn-loaded PrP gains partial protease resistance *in vitro*.[114] Further cell-free studies involving the protein misfolding cyclic amplification (PMCA) technique supported the idea that Mn acts as a cofactor in the conversion of PrP^C into the protease-resistant PrP^{Sc}-like form PrP^{res}, and determined that this conversion ability was similar to, but less profound than, the diagnostic proteolytic resistance characteristic of PrP^{Sc}.[120] Using a PrP-expressing yeast cell system, PrP^{res} formation was induced *in vivo* after the supplementation of Mn-containing media, suggesting that environmental Mn could be a risk factor for prion disease.[121] Similarly, *in vivo* PrP^{res} formation was detected in rat astrocytes when incubated with Mn for a prolonged period of time.[114] These changes in protease resistance of PrP^C have been suggested to be related to an altered conformation of PrP when Mn is bound. Indeed, analysis of full-length recombinant PrP using CD indicated that Mn-bound PrP has increased β-sheet content.[114,122] A near-infrared spectroscopy (NIRS) study on metal binding of prion protein in aqueous solutions documented that Mn-bound PrP undergoes highly different structural changes leading to fibril formation.[123] Interestingly, experiments using a method combining Raman optical activity (ROA) and ultraviolet circular dichroism (UV CD) demonstrated a very different impact of Cu and Mn on prion protein structure. Cu binding to prion protein destroyed its folded α-helical structure in the N-terminus; however, upon binding to Mn, the secondary structure became more organized, gaining more α-helices.[124] Another study with recombinant PrP further showed that Mn can replace Cu bound to PrP, resulting in an altered protein conformation with fewer helices.[118] In this study, cyclic voltammetric measurements indicated that the oxidation of Mn bound to PrP^C rendered the PrP^C binding irreversible, which is not seen with Cu-bound PrP^C. In addition to the resulting conformational changes in prion structure, Mn-bound PrP^C has been shown to initiate PrP aggregation and seed polymerization of soluble PrP^C. The protease-resistant PrP^{res} with Mn was shown to propagate and form more PrP^{res} in the presence of normal hamster brain homogenate by a standard PMCA technique, whereas treatment with the Mn chelator ethylenediaminetetraacetic acid (EDTA) inhibited this process, indicating reversible intermolecular Mn binding with PrP.[120] Other studies confirmed that, once bound to Mn, the β-sheet-rich PrP was able to seed polymerization of soluble metal-free PrP.[118,125] Additionally, Hesketh *et al.* characterized the Mn-bound PrP seed in a non-denature polymerization assay.[126] In this assay, a ~200 kDa oligomeric form of PrP seed capable of catalyzing PrP aggregation was generated by exposing recombinant prion protein to Mn. Using mutant recombinant PrP molecules, the authors further showed that Mn binding to PrP is essential for seed formation but not for polymerization. Another interesting finding from this report is that prion protein from chickens, in which no known prion disease has been found, was able to generate PrP seed after treatment with Mn.

Despite extensive *in vitro* studies on prion–Mn interaction and its subsequent effects on prion aggregation, the *in vivo* consequences of Mn binding to prion in terms of neuronal loss have not yet been well studied. An earlier report showed that recombinant PrP refolding in the presence of Mn was toxic to PrP-expressing cell lines and primary neuronal cultures.[127] More recently, treatment with an effective and relatively selective chelator for Mn, cyclohexanediaminetetraacetic acid (Na$_2$CaCDTA), significantly extended survival time in an animal model of prion disease in which mice were infected with a low dose of prion protein.[128] In agreement with these findings, Hortells *et al.* demonstrated that a Mn-rich diet given to scrapie prion-inoculated mice increased neuronal loss and the levels of PrP-containing plaques,[129] although this has not been replicated by other workers.[130] Interestingly, elevated Mn levels have been observed in the brains and blood of humans and other animals afflicted with TSE. In particular, altered Mn content has been observed in the blood and brains of humans infected with CJD,[113,131] mice infected with scrapie,[120,132] cattle infected with BSE,[133] and elk infected with CWD.[134] Of even greater interest is that the elevated blood Mn levels in scrapie and BSE were detected prior to the onset of disease symptoms, suggesting that altered metal levels might be a biomarker for diagnosis in the early stages of prion disease. Thus far, whether alteration in Mn levels in prion disease is a primary cause leading to infection or a secondary effect due to the infection itself remains unclear. Indeed, we lack clear evidence as to why Mn levels were elevated in these prion disease cases. At the cellular level, these increases may be linked to altered Mn homeostasis and signaling. Several cell culture studies, including ours, have shown that prion protein expression and infection can modulate the expression of Mn-transporting proteins and cause cellular retention of Mn.[135,136] On the other hand, increased Mn may in turn result in more PrP, owing to its inhibitory effects on proteasome degradation.[137,138] Therefore, a feed-forward mechanism may be involved in this process. However, the consequences of increased Mn in prion disease are likely to be abnormalities in iron metabolism because both Mn and iron ions are generally complexed and transported to the brain through a transferrin–transferrin receptor pathway.[139] In fact, iron abnormalities have been found in the brains of humans and other animals with prion diseases.[140,141] Additionally, PrP^{Sc}-infected cells are more susceptible to oxidative stress,[142,143] suggesting that Mn toxicity might be responsible for TSE-related neuronal loss.

To date, the vast majority of work done to elucidate the mechanisms of TSE pathogenesis has focused on the genetic determinants and biophysical kinetics of protein aggregation. Despite the fact that essential trace minerals are not manufactured by the body and that foreign (ingested) PrP^{Sc} can propagate, their environmental contribution to TSE etiology has not received the same kind of attention. However, the evidence continues to build for an environmental role in the initiation or development of the disease. In particular, the environmental distribution of metals correlates with the incidence of TSE.[144,145] Likewise, a correlation between soil clay content and the incidence of CWD in elk further indicates that soil constituents may affect the persistence of PrP^{Sc} in the environment.[146] A pair of monozygotic human twins with a pathogenic hereditary mutation to PrP developed TSE pathology seven years apart from each other,[147] providing more evidence for an environmental trigger. The recurrence of TSEs in livestock, despite multiple governmental programs designed to eradicate the diseases, argues for the persistence of a pathogen or trigger in the environment. Interestingly, a recent study strongly argues in favor of the notion that environmental Mn levels could be relevant to prion disease transmission. In this study, it was found that infectious PrP^{Sc} can persist in soil for at least two years.[148] Additionally, the presence of high levels of Mn in soil not only protects the protein from degradation, but may actually increase infectivity by <100-fold. These findings provide a route whereby PrP^{Sc} derived from carcasses or farm runoff can enter and persist in the environment. Thus, oral inoculation can occur in ruminants ingesting soil microparticles while grazing. The retention of infectivity in the environment seems to depend greatly on the presence of Mn in the soil. However, it should be noted that soil is a very complex matrix comprising many other components that could influence the incidence of prion disease. A better understanding of environmental determinants would greatly help in assessing the risk factors for TSEs.

23.4.4   Role of Manganese in the Physiological Function and Expression of PrP^{C}

Current evidence suggests a role for prion protein in modulating metal homeostasis as well as antioxidant levels.[149,150] Previously, we examined the role of PrP^{C} in regulating Mn-induced neurotoxicity.[151] Using mouse neuronal cell lines, we demonstrated that cells expressing prion protein (PrP^{C}) were more protected against Mn-induced neurotoxicity than were prion-knockout cells (PrP^{KO}). Inductively coupled plasma–mass spectrometry (ICP-MS) revealed that a lack of prion protein expression caused significantly lower basal Mn levels in PrP^{KO} cells, and upon Mn treatment PrP^{C} cells internalized significantly less Mn than did PrP^{KO} cells. Examination of ROS formation, caspase activation, and cell death all revealed that the mouse neuronal cell line lacking prion protein expression was more susceptible to Mn neurotoxicity. This increased susceptibility likely resulted from the homeostatic imbalance of Mn, as evidenced by the significantly higher amount of cellular Mn in these cells following treatment. Similar findings were achieved with hydrogen peroxide, which also increased the susceptibility of PrP^{KO} cells (unpublished data). Our findings suggest that prion might act as a metal sink, thereby preventing Mn from entering the cells and exerting its neurotoxic effect. In another recent study,[152] we further examined the fate of cellular prion protein in a Mn-treated PrP^{C}-expressing mouse neuronal cell line. Interestingly, our results indicated that with, Mn treatment, prion protein levels significantly increased over time. A previous study showed increased prion protein expression in a particular cell line treated with Cu, which elevated the activity of the reporter vector with Prnp.[153] Surprisingly, this was not the case for the Mn-mediated upregulation of cellular prion protein. We eliminated the possibility that Mn increased the transcription or inhibition of the ubiquitin proteasomal system (UPS), and further verified that, in our system, Mn treatment significantly altered the turnover of PrP^C. Pulse chase experiments confirmed that the half-life of PrP^C in Mn-treated cells was significantly increased. In contrast, we showed that the cation cadmium can significantly impair proteasomal activity, leading to increased oligomer formation and ubiquitinated prion proteins in neural cells, suggesting a metal-specific mechanism.[91] Together, these findings suggest that prion-mediated alterations in cellular Mn uptake and Mn-induced upregulation of PrP^C levels through increasing global protein stability may contribute to neurodegeneration in prion and other neurodegenerative diseases.

23.4.5 Role of other Metals in Prion Disease

Additional metals such as Cu, Zn, and nickel (Ni) may have the potential to bind to prion protein. A reduced Zn content in brains with prion disease has been revealed,[20,113] and a few other studies suggest that Zn binds to the octapeptide repeats in PrP^C, albeit at an apparently lower affinity.[154–156] Evidence also exists that the interaction promotes the endocytosis of PrP^c.[157,158] Although Spevacek *et al.* showed that Zn binding changes the structure of murine PrP^c,[159] the structural and functional consequences of Zn–prion interaction are still largely unknown. Interestingly, a recent study pointed out a role for cellular prion protein in facilitating the uptake of Zn into neurons, which was not observed in prion disease.[160,161] The mechanism of prion-mediated metal uptake seems to be metal-specific, because our study showed that prion protein reduced Mn uptake in neurons.[151] Ni has been used to isolate PrP in affinity columns,[162] but its potential interaction with PrP has been neglected because Ni binds to PrP^c with quite low affinity.[114] In contrast, it is widely accepted that PrP binds Cu. Many lines of *in vitro* evidence indicate that the interaction between Cu and PrP^C primarily occurs within the octapeptide repeat region or at a second site at the His 95 and 110 residues in the fifth site with a dissociation constant in the nanomolar range.[78,117,163] Further studies on Cu coordination with the octapeptide repeat domain have shown that there are three distinct modes at physiological pH.[164] Similar to Zn, Cu binding governs PrP^c endocytosis, and this requires the octapeptide repeat region.[157,158,165] The binding of Cu to PrP^c appears to be crucial to the normal function of the protein. For instance, the antioxidant activity of PrP^C requires Cu bound within the octapeptide repeat domain.[166–169] Cu bound by the octapeptide repeat domain of PrP^C undergoes full and reversible redox chemistry and can detoxify superoxide and reduce hydroxyl radicals.[170] However, the concept that the Cu–PrP complex acts as an antioxidant is controversial, based on conflicting results of several studies.[171,172] Loss of Cu accompanied by increased Mn levels has been described in prion-infected brains,[20,113,132] suggesting that a Cu deficiency or Cu displacement from the PrP^C might contribute to the pathology of prion disease. Unfortunately, experiments on the role of Cu in the pathology of prion disease have generated contradictory viewpoints. Some studies reported that Cu-bound PrP^C undergoes conformational changes and increases protease resistance,[173–175] while others showed that, when refolded in the presence of Cu, PrP^C decreases its protease resistance and its level of aggregation.[176,177] In addition, reducing Cu levels in the brain and blood using a Cu chelator extended the incubation period of the diseases,[178] whereas another report indicated that increased Cu in the diet delayed the onset of prion disease.[179] Overall, these findings and those of others suggest that the contributions of Cu to the pathogenesis of prion disease tend to be far more complex than originally expected.

23.5   Conclusions

Prion protein readily binds Mn despite an apparent low affinity. Once bound to Mn, it adopts a conformational change and converts into the protease-resistant PrP^{Sc}-like form that is essential for seed formation. Chelation of Mn in a prion animal model extended the incubation period of the disease, indicating that Mn could be a significant risk factor for prion disease. The discovery that Mn levels increased during the course of TSE in both humans and other animals, and that it could stabilize prions in the soil, thereby increasing PrP^{Sc} availability, further implicated Mn in the pathogenesis of the disease. However, many questions remain unanswered; for example, What is the impact of Mn binding on the normal functioning of PrP^C? What is the molecular basis behind the Mn-induced conformational conversion of prion protein? Furthermore, why are Mn levels altered in prion diseases? Additionally, is there any role for divalent metals in transmitting protein aggregates between cells?

Although Mn could be an important component in the pathogenesis of prion disease, by no means is it the only causative factor in disease development and progression. Most likely, development of a prion disease results from the culmination of various environmental, genetic, and even sporadic conditions. Therefore, developing cures for prion disease may require multifaceted strategies to combat its progression and even its development effectively.

Acknowledgements

This work was supported by National Institutes of Health Grants ES19276 and ES10586. The W. Eugene and Linda Lloyd Endowed Chair for AGK is also acknowledged. We thank Gary Zenitsky for assistance in preparing this chapter.

References

1. S. B. Prusiner, Novel properties and biology of scrapie prions, *Curr. Top. Microbiol. Immunol.*, 1991, **172**, 233–257.

2. M. W. Head, Human prion diseases: molecular, cellular and population biology, *Neuropathology*, 2013, **33**, 221–236.
3. E. D. Belay and L. B. Schonberger, The public health impact of prion diseases, *Annu. Rev. Public Health*, 2005, **26**, 191–212.
4. J. D. Wadsworth and J. Collinge, Molecular pathology of human prion disease, *Acta Neuropathol.*, 2011, **121**, 69–77.
5. B. Jang, A. Ishigami, N. Maruyama, R. I. Carp, Y. S. Kim and E. K. Choi, Peptidylarginine deiminase and protein citrullination in prion diseases: strong evidence of neurodegeneration, *Prion*, 2013, 7, 42–46.
6. G. R. Mallucci, Prion neurodegeneration: starts and stops at the synapse, *Prion*, 2009, **3**, 195–201.
7. S. J. DeArmond, W. C. Mobley, D. L. DeMott, R. A. Barry, J. H. Beckstead and S. B. Prusiner, Changes in the localization of brain prion proteins during scrapie infection, *Neurology*, 1987, **37**, 1271–1280.
8. T. Wisniewski and F. Goni, Could immunomodulation be used to prevent prion diseases?, *Expert Rev. Anti-Infect. Ther.*, 2012, **10**, 307–317.
9. Y. Roettger, Y. Du, M. Bacher, I. Zerr, R. Dodel and J. P. Bach, Immunotherapy in prion disease, *Nat. Rev. Neurol.*, 2013, **9**, 98–105.
10. V. L. Sim and B. Caughey, Recent advances in prion chemotherapeutics, *Infect. Disord.: Drug Targets*, 2009, **9**, 81–91.
11. G. Forloni, V. Artuso, I. Roiter, M. Morbin and F. Tagliavini, Therapy in prion diseases, *Curr. Top. Med. Chem.*, 2013, **13**, 2465–2476.
12. J. Collinge, Molecular neurology of prion disease, *J. Neurol. Neurosurg. Psychiatry*, 2005, **76**, 906–919.
13. P. Gambetti, I. Cali, S. Notari, Q. Kong, W. Q. Zou and W. K. Surewicz, Molecular biology and pathology of prion strains in sporadic human prion diseases, *Acta Neuropathol.*, 2011, **121**, 79–90.
14. P. Parchi, M. Cescatti, S. Notari, W. J. Schulz-Schaeffer, S. Capellari, A. Giese, W. Q. Zou, H. Kretzschmar, B. Ghetti and P. Brown, Agent strain variation in human prion disease: insights from a molecular and pathological review of the National Institutes of Health series of experimentally transmitted disease, *Brain*, 2010, **133**, 3030–3042.
15. C. Weissmann, J. Li, S. P. Mahal and S. Browning, Prions on the move, *EMBO Rep.*, 2011, **12**, 1109–1117.
16. G. Di Natale, G. Grasso, G. Impellizzeri, D. La Mendola, G. Micera, N. Mihala, Z. Nagy, K. Osz, G. Pappalardo, V. Rigo, E. Rizzarelli, D. Sanna and I. Sovago, Copper(II) interaction with unstructured prion domain outside the octarepeat region: speciation, stability, and binding details of copper(II) complexes with PrP106-126 peptides, *Inorg. Chem.*, 2005, **44**, 7214–7225.
17. D. R. Brown, Metallic prions, *Biochem. Soc. Symp.*, 2004, 193–202.
18. M. P. Hornshaw, J. R. McDermott and J. M. Candy, Copper binding to the N-terminal tandem repeat regions of mammalian and avian prion protein, *Biochem. Biophys. Res. Commun.*, 1995, **207**, 621–629.

19. S. Hesketh, J. Sassoon, R. Knight, J. Hopkins and D. R. Brown, Elevated Manganese Levels in Blood and CNS Occur Prior to Onset of Clinical Signs in Scrapie and BSE, *J. Anim. Sci.*, 2007, **85**, 1596–1609.

20. B. S. Wong, D. R. Brown, T. Pan, M. Whiteman, T. Liu, X. Bu, R. Li, P. Gambetti, J. Olesik, R. Rubenstein and M. S. Sy, Oxidative impairment in scrapie-infected mice is associated with brain metals perturbations and altered antioxidant activities, *J. Neurochem.*, 2001, **79**, 689–698.

21. S. B. Prusiner, Novel proteinaceous infectious particles cause scrapie, *Science*, 1982, **216**, 136–144.

22. A. Aguzzi, Prion diseases of humans and farm animals: epidemiology, genetics, and pathogenesis, *J. Neurochem.*, 2006, **97**, 1726–1739.

23. P. L. Chelle and J. Cuille, Experimental transmission of trembling to the goat, *C.R. Seances Acad. Sci.*, 1939, 1058–1160.

24. A. Aguzzi, F. Baumann and J. Bremer, The prion's elusive reason for being, *Annu. Rev. Neurosci.*, 2008, **31**, 439–477.

25. M. Imran and S. Mahmood, An overview of human prion diseases, *Virol. J.*, 2011, **8**, 559.

26. A. Ladogana, M. Puopolo, E. A. Croes, H. Budka, C. Jarius, S. Collins, G. M. Klug, T. Sutcliffe, A. Giulivi, A. Alperovitch, N. Delasnerie-Laupretre, J. P. Brandel, S. Poser, H. Kretzschmar, I. Rietveld, E. Mitrova, P. Cuesta Jde, P. Martinez-Martin, M. Glatzel, A. Aguzzi, R. Knight, H. Ward, M. Pocchiari, C. M. van Duijn, R. G. Will and I. Zerr, Mortality from Creutzfeldt-Jakob disease and related disorders in Europe, Australia, and Canada, *Neurology*, 2005, **64**, 1586–1591.

27. S. B. Prusiner and S. J. DeArmond, Prion diseases and neurodegeneration, *Annu. Rev. Neurosci.*, 1994, **17**, 311–339.

28. P. Duffy, J. Wolf, G. Collins, A. G. DeVoe, B. Streeten and D. Cowen, Letter: Possible person-to-person transmission of Creutzfeldt-Jakob disease, *N. Engl. J. Med.*, 1974, **290**, 692–693.

29. C. T. Gay, W. A. Marks, H. D. Riley, Jr., J. B. Bodensteiner, M. Hamza, P. A. Noorani and G. B. Bobele, Infantile botulism, *South. Med. J.*, 1988, **81**, 457–460.

30. C. J. Gibbs, Jr., A. Joy, R. Heffner, M. Franko, M. Miyazaki, D. M. Asher, J. E. Parisi, P. W. Brown and D. C. Gajdusek, Clinical and pathological features and laboratory confirmation of Creutzfeldt-Jakob disease in a recipient of pituitary-derived human growth hormone, *N. Engl. J. Med.*, 1985, **313**, 734–738.

31. R. G. Will, J. W. Ironside, M. Zeidler, S. N. Cousens, K. Estibeiro, A. Alperovitch, S. Poser, M. Pocchiari, A. Hofman and P. G. Smith, A new variant of Creutzfeldt-Jakob disease in the UK, *Lancet*, 1996, **347**, 921–925.

32. C. A. Llewelyn, P. E. Hewitt, R. S. Knight, K. Amar, S. Cousens, J. Mackenzie and R. G. Will, Possible transmission of variant Creutzfeldt-Jakob disease by blood transfusion, *Lancet*, 2004, **363**, 417–421.

33. R. G. Will, Acquired prion disease: iatrogenic CJD, variant CJD, kuru, *Br. Med. Bull.*, 2003, **66**, 255–265.

34. D. C. Gajdusek, C. J. Gibbs and M. Alpers, Experimental transmission of a Kuru-like syndrome to chimpanzees, *Nature*, 1966, **209**, 794–796.

35. M. P. Kaufman, G. P. Kozlowski and K. J. Rybicki, Attenuation of the reflex pressor response to muscular contraction by a substance P antagonist, *Brain Res.*, 1985, **333**, 182–184.

36. M Alpers, Epidemiology and clinical aspects of kuru, in Prions, ed. M. McKinley and S. B Prusiner, Academic Press, New York, 1987, vol. 1987.

37. A. H. Lockwood, Medical problems of musicians, *N. Engl. J. Med.*, 1989, **320**, 221–227.

38. J. M. Thomson, V. Bowles, J. W. Choi, U. Basu, Y. Meng, P. Stothard and S. Moore, The identification of candidate genes and SNP markers for classical bovine spongiform encephalopathy susceptibility, *Prion*, 2012, **6**, 461–469.

39. L. Cervenakova, L. G. Goldfarb, R. Garruto, H. S. Lee, D. C. Gajdusek and P. Brown, Phenotype-genotype studies in kuru: implications for new variant Creutzfeldt-Jakob disease, *Proc. Natl. Acad. Sci. U. S. A.*, 1998, **95**, 13239–13241.

40. S. Mead, M. P. Stumpf, J. Whitfield, J. A. Beck, M. Poulter, T. Campbell, J. B. Uphill, D. Goldstein, M. Alpers, E. M. Fisher and J. Collinge, Balancing selection at the prion protein gene consistent with prehistoric kurulike epidemics, *Science*, 2003, **300**, 640–643.

41. P. P. Liberski, Kuru: A Journey Back in Time from Papua New Guinea to the Neanderthals' Extinction, *Pathogens*, 2013, 472–505.

42. C. L. Masters, D. C. Gajdusek and C. J. Gibbs, Jr., Creutzfeldt-Jakob disease virus isolations from the Gerstmann-Straussler syndrome with an analysis of the various forms of amyloid plaque deposition in the virus-induced spongiform encephalopathies, *Brain: J. Neurol.*, 1981, **104**, 559–588.

43. B. Ghetti, F. Tagliavini, C. L. Masters, K. Beyreuther, G. Giaccone, L. Verga, M. R. Farlow, P. M. Conneally, S. R. Dlouhy, B. Azzarelli, *et al.*, Gerstmann-Straussler-Scheinker disease. II. Neurofibrillary tangles and plaques with PrP-amyloid coexist in an affected family, *Neurology*, 1989, **39**, 1453–1461.

44. P. P. Liberski, M. Barcikowska, L. Cervenakova, J. Bratosiewicz, M. Marczewska, P. Brown and D. C. Gajdusek, A case of sporadic Creutzfeldt-Jakob disease with a Gerstmann-Straussler-Scheinker phenotype but no alterations in the PRNP gene, *Acta Neuropathol.*, 1998, **96**, 425–430.

45. P. Piccardo, S. R. Dlouhy, P. M. Lievens, K. Young, T. D. Bird, D. Nochlin, D. W. Dickson, H. V. Vinters, T. R. Zimmerman, I. R. Mackenzie, S. J. Kish, L. C. Ang, C. De Carli, M. Pocchiari, P. Brown, C. J. Gibbs, Jr., D. C. Gajdusek, O. Bugiani, J. Ironside, F. Tagliavini and B. Ghetti, Phenotypic variability of Gerstmann-Straussler-Scheinker disease is associated with prion protein heterogeneity, *J. Neuropathol. Exp. Neurol.*, 1998, **57**, 979–988.

46. S. Mirra, P. Piccardo, K. Young, M. Gearing, S. R. Dlouhy and B. Ghetti, α-Synuclein accumulation in Gerstmann-Straussler-Schenker disease (GSS) with prion protein gene (PRNP) F198S, *Neurobiol. Aging*, 1998, S724.

47. K. Doh-ura, J. Tateishi, H. Sasaki, T. Kitamoto and Y. Sakaki, Pro–leu change at position 102 of prion protein is the most common but not the sole mutation related to Gerstmann-Straussler syndrome, *Biochem. Biophys. Res. Commun.*, 1989, **163**, 974–979.

48. K. Hsiao, S. R. Dlouhy, M. R. Farlow, C. Cass, M. Da Costa, P. M. Conneally, M. E. Hodes, B. Ghetti and S. B. Prusiner, Mutant prion proteins in Gerstmann-Straussler-Scheinker disease with neurofibrillary tangles, *Nat. Genet.*, 1992, **1**, 68–71.

49. C. Jansen, P. Parchi, S. Capellari, A. J. Vermeij, P. Corrado, F. Baas, R. Strammiello, W. A. van Gool, J. C. van Swieten and A. J. Rozemuller, Prion protein amyloidosis with divergent phenotype associated with two novel nonsense mutations in PRNP, *Acta Neuropathol.*, 2010, **119**, 189–197.

50. F. Tagliavini, F. Prelli, M. Porro, G. Rossi, G. Giaccone, M. R. Farlow, S. R. Dlouhy, B. Ghetti, O. Bugiani and B. Frangione, Amyloid fibrils in Gerstmann-Straussler-Scheinker disease (Indiana and Swedish kindreds) express only PrP peptides encoded by the mutant allele, *Cell*, 1994, **79**, 695–703.

51. J. Tateishi, T. Kitamoto, H. Hashiguchi and H. Shii, Gerstmann-Straussler-Scheinker disease: immunohistological and experimental studies, *Ann. Neurol.*, 1988, **24**, 35–40.

52. D. R. Brown, Don't lose sleep over prions: role of prion protein in sleep regulation, *NeuroReport*, 2002, **13**, A1.

53. R. Huber, T. Deboer and I. Tobler, Prion protein: a role in sleep regulation?, *J. sleep Res.*, 1999, **8**(Suppl 1), 30–36.

54. P. Montagna, P. Gambetti, P. Cortelli and E. Lugaresi, Familial and sporadic fatal insomnia, *Lancet Neurol.*, 2003, **2**, 167–176.

55. R. E. Brown, R. Basheer, J. T. McKenna, R. E. Strecker and R. W. McCarley, Control of sleep and wakefulness, *Physiol. Rev.*, 2012, **92**, 1087–1187.

56. C. D. Engleberg, V. J. Dermody, S. Terence, Schaechter's Mechanisms of Microbial Disease, *Lippincott Williams & Wilkins*, 5th edn, 2012.

57. A. Harder, K. Jendroska, F. Kreuz, T. Wirth, C. Schafranka, N. Karnatz, A. Theallier-Janko, J. Dreier, K. Lohan, D. Emmerich, J. Cervos-Navarro, O. Windl, H. A. Kretzschmar, P. Nurnberg and R. Witkowski, Novel twelve-generation kindred of fatal familial insomnia from germany representing the entire spectrum of disease expression, *Am. J. Med. Genet.*, 1999, **87**, 311–316.

58. A. Padovani, M. D'Alessandro, P. Parchi, P. Cortelli, G. P. Anzola, P. Montagna, L. A. Vignolo, R. Petraroli, M. Pocchiari, E. Lugaresi and P. Gambetti, Fatal familial insomnia in a new Italian kindred, *Neurology*, 1998, **51**, 1491–1494.

59. E. Lugaresi, R. Medori, P. Montagna, A. Baruzzi, P. Cortelli, A. Lugaresi, P. Tinuper, M. Zucconi and P. Gambetti, Fatal familial insomnia and dysautonomia with selective degeneration of thalamic nuclei, *New Engl. J. Med.*, 1986, **315**, 997–1003.

60. R. Medori, H. J. Tritschler, A. LeBlanc, F. Villare, V. Manetto, H. Y. Chen, R. Xue, S. Leal, P. Montagna, P. Cortelli, *et al.*, Fatal familial insomnia, a prion disease with a mutation at codon 178 of the prion protein gene, *N. Engl. J. Med.*, 1992, **326**, 444–449.

61. Y. Taniwaki, H. Hara, K. Doh-Ura, I. Murakami, H. Tashiro, T. Yamasaki, H. Shigeto, K. Arakawa, E. Araki, T. Yamada, T. Iwaki and J. Kira, Familial Creutzfeldt-Jakob disease with D178N-129M mutation of PRNP presenting as cerebellar ataxia without insomnia, *J. Neurol. Neurosurg. Psychiatry*, 2000, **68**, 388.

62. R. Medori and H. J. Tritschler, Prion protein gene analysis in three kindreds with fatal familial insomnia (FFI): codon 178 mutation and codon 129 polymorphism, *Am. J. Hum. Genet.*, 1993, **53**, 822–827.

63. C. Tabernero, J. M. Polo, M. D. Sevillano, R. Munoz, J. Berciano, A. Cabello, B. Baez, J. R. Ricoy, R. Carpizo, J. Figols, N. Cuadrado and L. E. Claveria, Fatal familial insomnia: clinical, neuropathological, and genetic description of a Spanish family, *J. Neurol. Neurosurg. Psychiatry*, 2000, **68**, 774–777.

64. S. B. Prusiner, Prions and Brain Diseases in Animals and Humans, *Springer*, 1998.

65. G. A. Broderick, Effect of processing on protein utilization by ruminants, *Adv. Exp. Med. Biol.*, 1977, **86B**, 531–544.

66. H. A. Kretzschmar, L. E. Stowring, D. Westaway, W. H. Stubblebine, S. B. Prusiner and S. J. Dearmond, Molecular cloning of a human prion protein cDNA, *DNA*, 1986, **5**, 315–324.

67. N. Stahl, D. R. Borchelt, K. Hsiao and S. B. Prusiner, Scrapie prion protein contains a phosphatidylinositol glycolipid, *Cell*, 1987, **51**, 229–240.

68. M. Marella, S. Lehmann, J. Grassi and J. Chabry, Filipin prevents pathological prion protein accumulation by reducing endocytosis and inducing cellular PrP release, *J. Biol. Chem.*, 2002, **277**, 25457–25464.

69. P. J. Peters, A. Mironov, Jr., D. Peretz, E. van Donselaar, E. Leclerc, S. Erpel, S. J. DeArmond, D. R. Burton, R. A. Williamson, M. Vey and S. B. Prusiner, Trafficking of prion proteins through a caveolae-mediated endosomal pathway, *J. Cell Biol.*, 2003, **162**, 703–717.

70. M. A. Prado, J. Alves-Silva, A. C. Magalhaes, V. F. Prado, R. Linden, V. R. Martins and R. R. Brentani, PrPc on the road: trafficking of the cellular prion protein, *J. Neurochem.*, 2004, **88**, 769–781.

71. D. R. Taylor, N. T. Watt, W. S. Perera and N. M. Hooper, Assigning functions to distinct regions of the N-terminus of the prion protein that are involved in its copper-stimulated, clathrin-dependent endocytosis, *J. Cell Sci.*, 2005, **118**, 5141–5153.

72. G. M. Di Guglielmo, C. Le Roy, A. F. Goodfellow and J. L. Wrana, Distinct endocytic pathways regulate TGF-beta receptor signalling and turnover, *Nat. Cell Biol.*, 2003, **5**, 410–421.

73. R. D. York, D. C. Molliver, S. S. Grewal, P. E. Stenberg, E. W. McCleskey and P. J. Stork, Role of phosphoinositide 3-kinase and endocytosis in nerve growth factor-induced extracellular signal-regulated kinase activation via Ras and Rap1, *Mol. Cell. Biol.*, 2000, **20**, 8069–8083.

74. C. Weissmann, The state of the prion, *Nat. Rev. Microbiol.*, 2004, **2**, 861–871.

75. R. Zahn, A. Liu, T. Luhrs, R. Riek, C. von Schroetter, F. Lopez Garcia, M. Billeter, L. Calzolai, G. Wider and K. Wuthrich, NMR solution structure of the human prion protein, *Proc. Natl. Acad. Sci. U. S. A.*, 2000, **97**, 145–150.

76. N. R. Maiti and W. K. Surewicz, The role of disulfide bridge in the folding and stability of the recombinant human prion protein, *J. Biol. Chem.*, 2001, **276**, 2427–2431.

77. P. Mastrangelo and D. Westaway, Biology of the prion gene complex, Biochemistry and cell biology, *Biochem. Cell Biol.*, 2001, **79**, 613–628.

78. G. S. Jackson, I. Murray, L. L. Hosszu, N. Gibbs, J. P. Waltho, A. R. Clarke and J. Collinge, Location and properties of metal-binding sites on the human prion protein, *Proc. Natl. Acad. Sci. U. S. A.*, 2001, **98**, 8531–8535.

79. D. R. Brown, B. Schmidt and H. A. Kretzschmar, Effects of copper on survival of prion protein knockout neurons and glia, *J. Neurochem.*, 1998, **70**, 1686–1693.

80. D. R. Taylor and N. M. Hooper, The prion protein and lipid rafts, *Mol. Membr. Biol.*, 2006, **23**, 89–99.

81. R. Rieger, F. Edenhofer, C. I. Lasmezas and S. Weiss, The human 37-kDa laminin receptor precursor interacts with the prion protein in eukaryotic cells, *Nat. Med.*, 1997, **3**, 1383–1388.

82. G. Schmitt-Ulms, G. Legname, M. A. Baldwin, H. L. Ball, N. Bradon, P. J. Bosque, K. L. Crossin, G. M. Edelman, S. J. DeArmond, F. E. Cohen and S. B. Prusiner, Binding of neural cell adhesion molecules (N-CAMs) to the cellular prion protein, *J. Mol. Biol.*, 2001, **314**, 1209–1225.

83. H. Colognato and P. D. Yurchenco, Form and function: the laminin family of heterotrimers, *Dev. Dyn.*, 2000, **218**, 213–234.

84. P. F. Maness and M. Schachner, Neural recognition molecules of the immunoglobulin superfamily: signaling transducers of axon guidance and neuronal migration, *Nat. Neurosci.*, 2007, **10**, 19–26.

85. A. Santuccione, V. Sytnyk, I. Leshchyns'ka and M. Schachner, Prion protein recruits its neuronal receptor NCAM to lipid rafts to activate p59fyn and to enhance neurite outgrowth, *J. Cell Biol.*, 2005, **169**, 341–354.

86. S. Mouillet-Richard, M. Ermonval, C. Chebassier, J. L. Laplanche, S. Lehmann, J. M. Launay and O. Kellermann, Signal transduction through prion protein, *Science*, 2000, **289**, 1925–1928.

87. B. Schneider, V. Mutel, M. Pietri, M. Ermonval, S. Mouillet-Richard and O. Kellermann, NADPH oxidase and extracellular regulated kinases 1/2 are targets of prion protein signaling in neuronal and nonneuronal cells, *Proc. Natl. Acad. Sci. U. S. A.*, 2003, **100**, 13326–13331.

88. D. R. Brown, Brain proteins that mind metals: a neurodegenerative perspective, *Dalton Trans.*, 2009, 4069–4076.

89. A. I. Bush and C. C. Curtain, Twenty years of metallo-neurobiology: where to now?, *Eur. Biophys. J.*, 2008, **37**, 241–245.

90. F. Molina-Holgado, R. C. Hider, A. Gaeta, R. Williams and P. Francis, Metals ions and neurodegeneration, *BioMetals*, 2007, **20**, 639–654.

91. A. G. Kanthasamy, C. Choi, H. Jin, D. S. Harischandra, V. Anantharam and A. Kanthasamy, Effect of divalent metals on the neuronal proteasomal system, prion protein ubiquitination and aggregation, *Toxicol. Lett.*, 2012, **214**, 288–295.

92. J. Tyedmers, A. Mogk and B. Bukau, Cellular strategies for controlling protein aggregation, *Nat. Rev. Mol. Cell Biol.*, 2010, **11**, 777–788.

93. J. D. Meeker, P. Susi and M. R. Flynn, Manganese and welding fume exposure and control in construction, *J. Occup. Environ. Hyg.*, 2007, **4**, 943–951.

94. J. Su, P. Bao, T. Bai, L. Deng, H. Wu, F. Liu and J. He, a multicopper oxidase from Bacillus pumilus WH4, exhibits manganese-oxidase activity, *PLoS One*, 2013, **8**, e60573.

95. J. L. Aschner and M. Aschner, Nutritional aspects of manganese homeostasis, *Mol. Aspects Med.*, 2005, **26**, 353–362.

96. K. M. Erikson, K. Thompson, J. Aschner and M. Aschner, Manganese neurotoxicity: a focus on the neonate, *Pharmacol. Ther.*, 2007, **113**, 369–377.

97. J. S. Crossgrove and R. A. Yokel, Manganese distribution across the blood-brain barrier III. The divalent metal transporter-1 is not the major mechanism mediating brain manganese uptake, *Neurotoxicology*, 2004, **25**, 451–460.

98. R. A. Yokel, Brain uptake, retention, and efflux of aluminum and manganese, *Environ. Health Perspect.*, 2002, **110**(Suppl 5), 699–704.

99. M. Aschner and M. Gannon, Manganese (Mn) transport across the rat blood-brain barrier: saturable and transferrin-dependent transport mechanisms, *Brain Res. Bull.*, 1994, **33**, 345–349.

100. M. Aschner, K. E. Vrana and W. Zheng, Manganese uptake and distribution in the central nervous system (CNS), *Neurotoxicology*, 1999, **20**, 173–180.

101. E. J. Martinez-Finley, C. E. Gavin, M. Aschner and T. E. Gunter, Manganese neurotoxicity and the role of reactive oxygen species, *Free Radical Biol. Med.*, 2013, **62**, 65–75.

102. P. Trumbo, A. A. Yates, S. Schlicker and M. Poos, Dietary reference intakes: vitamin A, vitamin K, arsenic, boron, chromium, copper, iodine, iron, manganese, molybdenum, nickel, silicon, vanadium, and zinc, *J. Am. Diet. Assoc.*, 2001, **101**, 294–301.

103. A. W. Dobson, K. M. Erikson and M. Aschner, Manganese neurotoxicity, *Ann. N. Y. Acad. Sci.*, 2004, **1012**, 115–128.

104. T. Verina, J. S. Schneider and T. R. Guilarte, Manganese exposure induces alpha-synuclein aggregation in the frontal cortex of non-human primates, *Toxicol. Lett.*, 2013, **217**, 177–183.

105. M. Aschner, K. M. Erikson, E. Herrero Hernandez and R. Tjalkens, Manganese and its role in Parkinson's disease: from transport to neuropathology, *NeuroMol. Med.*, 2009, **11**, 252–266.

106. J. A. Roth and M. D. Garrick, Iron interactions and other biological reactions mediating the physiological and toxic actions of manganese, *Biochem. Pharmacol.*, 2003, **66**, 1–13.

107. K. M. Erikson and M. Aschner, Manganese neurotoxicity and glutamate-GABA interaction, *Neurochem. Int.*, 2003, **43**, 475–480.

108. M. Kitazawa, J. R. Wagner, M. L. Kirby, V. Anantharam and A. G. Kanthasamy, Oxidative stress and mitochondrial-mediated apoptosis in dopaminergic cells exposed to methylcyclopentadienyl manganese tricarbonyl, *J. Pharmacol. Exp. Ther.*, 2002, **302**, 26–35.

109. M. Kitazawa, V. Anantharam, Y. Yang, Y. Hirata, A. Kanthasamy and A. G. Kanthasamy, Activation of protein kinase C delta by proteolytic cleavage contributes to manganese-induced apoptosis in dopaminergic cells: protective role of Bcl-2, *Biochem. Pharmacol.*, 2005, **69**, 133–146.

110. V. Anantharam, M. Kitazawa, J. Wagner, S. Kaul and A. G. Kanthasamy, Caspase-3-dependent proteolytic cleavage of protein kinase Cdelta is essential for oxidative stress-mediated dopaminergic cell death after exposure to methylcyclopentadienyl manganese tricarbonyl, *J. Neurosci.*, 2002, **22**, 1738–1751.

111. C. Latchoumycandane, V. Anantharam, M. Kitazawa, Y. Yang, A. Kanthasamy and A. G. Kanthasamy, Protein kinase Cdelta is a key downstream mediator of manganese-induced apoptosis in dopaminergic neuronal cells, *J. Pharmacol. Exp. Ther.*, 2005, **313**, 46–55.

112. A. Kanthasamy, H. Jin, S. Mehrotra, R. Mishra, A. Kanthasamy and A. Rana, Novel cell death signaling pathways in neurotoxicity models of dopaminergic degeneration: relevance to oxidative stress and neuroinflammation in Parkinson's disease, *Neurotoxicology*, 2010, **31**, 555–561.

113. B. S. Wong, S. G. Chen, M. Colucci, Z. Xie, T. Pan, T. Liu, R. Li, P. Gambetti, M. S. Sy and D. R. Brown, Aberrant metal binding by prion protein in human prion disease, *J. Neurochem.*, 2001, **78**, 1400–1408.

114. D. R. Brown, F. Hafiz, L. L. Glasssmith, B. S. Wong, I. M. Jones, C. Clive and S. J. Haswell, Consequences of manganese replacement of copper for prion protein function and proteinase resistance, *EMBO J.*, 2000, **19**, 1180–1186.

115. A. P. Garnett and J. H. Viles, Copper binding to the octarepeats of the prion protein. Affinity, specificity, folding, and cooperativity: insights from circular dichroism, *J. Biol. Chem.*, 2003, **278**, 6795–6802.

116. C. Treiber, A. R. Thompsett, R. Pipkorn, D. R. Brown and G. Multhaup, Real-time kinetics of discontinuous and highly conformational

metal-ion binding sites of prion protein, *JBIC, J. Biol. Inorg. Chem.*, 2007, **12**, 711–720.

117. C. E. Jones, S. R. Abdelraheim, D. R. Brown and J. H. Viles, Preferential Cu^{2+} coordination by His96 and His111 induces beta-sheet formation in the unstructured amyloidogenic region of the prion protein, *J. Biol. Chem.*, 2004, **279**, 32018–32027.

118. M. W. Brazier, P. Davies, E. Player, F. Marken, J. H. Viles and D. R. Brown, Manganese binding to the prion protein, *J. Biol. Chem.*, 2008, **283**, 12831–12839.

119. D. R. Brown, Prions and manganese: A maddening beast, *Metallomics*, 2011, **3**, 229–238.

120. N. H. Kim, J. K. Choi, B. H. Jeong, J. I. Kim, M. S. Kwon, R. I. Carp and Y. S. Kim, Effect of transition metals (Mn, Cu, Fe) and deoxycholic acid (DA) on the conversion of PrPC to PrPres, *FASEB J.*, 2005, **19**, 783–785.

121. C. Treiber, A. Simons and G. Multhaup, Effect of copper and manganese on the de novo generation of protease-resistant prion protein in yeast cells, *Biochemistry*, 2006, **45**, 6674–6680.

122. A. Giese, J. Levin, U. Bertsch and H. Kretzschmar, Effect of metal ions on de novo aggregation of full-length prion protein, *Biochem. Biophys. Res. Commun.*, 2004, **320**, 1240–1246.

123. R. N. Tsenkova, I. K. Iordanova, K. Toyoda and D. R. Brown, Prion protein fate governed by metal binding, *Biochem. Biophys. Res. Commun.*, 2004, **325**, 1005–1012.

124. F. Zhu, P. Davies, A. R. Thompsett, S. M. Kelly, G. E. Tranter, L. Hecht, N. W. Isaacs, D. R. Brown and L. D. Barron, Raman optical activity and circular dichroism reveal dramatic differences in the influence of divalent copper and manganese ions on prion protein folding, *Biochemistry*, 2008, **47**, 2510–2517.

125. T. Lekishvili, J. Sassoon, A. R. Thompsett, A. Green, J. W. Ironside and D. R. Brown, BSE and vCJD cause disturbance to uric acid levels, *Exp. Neurol.*, 2004, **190**, 233–244.

126. S. Hesketh, A. R. Thompsett and D. R. Brown, Prion protein polymerisation triggered by manganese-generated prion protein seeds, *J. Neurochem.*, 2012, **120**, 177–189.

127. K. M. Uppington and D. R. Brown, Resistance of cell lines to prion toxicity aided by phospho-ERK expression, *J. Neurochem.*, 2008, **105**, 842–852.

128. M. W. Brazier, I. Volitakis, M. Kvasnicka, A. R. White, J. R. Underwood, J. E. Green, S. Han, A. F. Hill, C. L. Masters and S. J. Collins, Manganese chelation therapy extends survival in a mouse model of M1000 prion disease, *J. Neurochem.*, 2010, **114**, 440–451.

129. P. Hortells, E. Monleon, C. Acin, A. Vargas, V. Vasseur, A. Salomon, B. Ryffel, J. Y. Cesbron, J. J. Badiola and M. Monzon, The effect of metal imbalances on scrapie neurodegeneration, *Zoonoses Public Health*, 2010, **57**, 358–366.

130. L. R. Legleiter, H. C. Liu, K. E. Lloyd, S. L. Hansen, R. S. Fry and J. W. Spears, Exposure to low dietary copper or low copper coupled with high dietary manganese for one year does not alter brain prion protein characteristics in the mature cow, *J. Anim. Sci.*, 2007, **85**, 2895–2903.

131. S. Hesketh, J. Sassoon, R. Knight and D. R. Brown, Elevated manganese levels in blood and CNS in human prion disease, *Mol. Cell. Neurosci.*, 2008, **37**, 590–598.

132. A. M. Thackray, R. Knight, S. J. Haswell, R. Bujdoso and D. R. Brown, Metal imbalance and compromised antioxidant function are early changes in prion disease, *Biochem. J.*, 2002, **362**, 253–258.

133. S. Hesketh, J. Sassoon, R. Knight, J. Hopkins and D. R. Brown, Elevated manganese levels in blood and central nervous system occur before onset of clinical signs in scrapie and bovine spongiform encephalopathy, *J. Anim. Sci.*, 2007, **85**, 1596–1609.

134. S. N. White, K. I. O'Rourke, T. Gidlewski, K. C. VerCauteren, M. R. Mousel, G. E. Phillips and T. R. Spraker, Increased risk of chronic wasting disease in Rocky Mountain elk associated with decreased magnesium and increased manganese in brain tissue, *Can. J. Vet. Res.*, 2010, **74**, 50–53.

135. S. Kralovicova, S. N. Fontaine, A. Alderton, J. Alderman, K. V. Ragnarsdottir, S. J. Collins and D. R. Brown, The effects of prion protein expression on metal metabolism, *Mol. Cell. Neurosci.*, 2009, **41**, 135–147.

136. D. P. Martin, V. Anantharam, H. Jin, T. Witte, R. Houk, A. Kanthasamy and A. G. Kanthasamy, Infectious prion protein alters manganese transport and neurotoxicity in a cell culture model of prion disease, *Neurotoxicology*, 2011, **32**, 554–562.

137. Y. Zhou, F. S. Shie, P. Piccardo, T. J. Montine and J. Zhang, Proteasomal inhibition induced by manganese ethylene-bis-dithiocarbamate: relevance to Parkinson's disease, *Neuroscience*, 2004, **128**, 281–291.

138. T. Cai, T. Yao, Y. Li, Y. Chen, K. Du, J. Chen and W. Luo, Proteasome inhibition is associated with manganese-induced oxidative injury in PC12 cells, *Brain Res.*, 2007, **1185**, 359–365.

139. E. A. Heilig, K. J. Thompson, R. M. Molina, A. R. Ivanov, J. D. Brain and M. Wessling-Resnick, Manganese and iron transport across pulmonary epithelium, *Am. J. Physiol.: Lung Cell. Mol. Physiol.*, 2006, **290**, L1247–1259.

140. A. Singh, A. O. Isaac, X. Luo, M. L. Mohan, M. L. Cohen, F. Chen, Q. Kong, J. Bartz and N. Singh, Abnormal brain iron homeostasis in human and animal prion disorders, *PLoS Pathog.*, 2009, **5**, e1000336.

141. A. Singh, S. Haldar, K. Horback, C. Tom, L. Zhou, H. Meyerson and N. Singh, Prion protein regulates iron transport by functioning as a ferrireductase, *J. Alzheimer's Dis.*, 2013, **35**, 541–552.

142. O. Milhavet, H. E. McMahon, W. Rachidi, N. Nishida, S. Katamine, A. Mange, M. Arlotto, D. Casanova, J. Riondel, A. Favier and

S. Lehmann, Prion infection impairs the cellular response to oxidative stress, *Proc. Natl. Acad. Sci. U. S. A.*, 2000, **97**, 13937–13942.

143. S. Fernaeus, K. Reis, K. Bedecs and T. Land, Increased susceptibility to oxidative stress in scrapie-infected neuroblastoma cells is associated with intracellular iron status, *Neurosci. Lett.*, 2005, **389**, 133–136.

144. M. Purdey, Ecosystems supporting clusters of sporadic TSEs demonstrate excesses of the radical-generating divalent cation manganese and deficiencies of antioxidant co factors Cu, Se, Fe, Zn. Does a foreign cation substitution at prion protein's Cu domain initiate TSE?, *Med. Hypotheses*, 2000, **54**, 278–306.

145. M. Polano, C. Anselmi, L. Leita, A. Negro and M. De Nobili, Organic polyanions act as complexants of prion protein in soil, *Biochem. Biophys. Res. Commun.*, 2008, **367**, 323–329.

146. C. J. Johnson, J. A. Pedersen, R. J. Chappell, D. McKenzie and J. M. Aiken, Oral transmissibility of prion disease is enhanced by binding to soil particles, *PLoS Pathog.*, 2007, **3**, e93.

147. A. B. Bowman, G. F. Kwakye, E. H. Hernandez and M. Aschner, Role of manganese in neurodegenerative diseases, *J. Trace Elem. Med. Biol.*, 2011, **25**, 191–203.

148. P. Davies and D. R. Brown, Manganese enhances prion protein survival in model soils and increases prion infectivity to cells, *PLoS One*, 2009, **4**, e7518.

149. D. R. Brown, R. S. Nicholas and L. Canevari, Lack of prion protein expression results in a neuronal phenotype sensitive to stress, *J. Neurosci. Res.*, 2002, **67**, 211–224.

150. B. S. Wong, T. Liu, R. Li, T. Pan, R. B. Petersen, M. A. Smith, P. Gambetti, G. Perry, J. C. Manson, D. R. Brown and M. S. Sy, Increased levels of oxidative stress markers detected in the brains of mice devoid of prion protein, *J. Neurochem.*, 2001, **76**, 565–572.

151. C. J. Choi, V. Anantharam, N. J. Saetveit, R. S. Houk, A. Kanthasamy and A. G. Kanthasamy, Normal cellular prion protein protects against manganese-induced oxidative stress and apoptotic cell death, *Toxicol. Sci.*, 2007, **98**, 495–509.

152. C. J. Choi, V. Anantharam, D. P. Martin, E. M. Nicholson, J. A. Richt, A. Kanthasamy and A. G. Kanthasamy, Manganese upregulates cellular prion protein and contributes to altered stabilization and proteolysis: relevance to role of metals in pathogenesis of prion disease, *Toxicol. Sci.*, 2010, **115**, 535–546.

153. L. Varela-Nallar, E. M. Toledo, L. F. Larrondo, A. L. Cabral, V. R. Martins and N. C. Inestrosa, Induction of cellular prion protein (PrPC) gene expression by copper in neurons, *Am. J. Physiol.: Cell Physiol.*, 2006, **290**, C271–C281.

154. E. D. Walter, D. J. Stevens, M. P. Visconte and G. L. Millhauser, The prion protein is a combined zinc and copper binding protein: Zn^{2+}

alters the distribution of Cu^{2+} coordination modes, *J. Am. Chem. Soc.*, 2007, **129**, 15440–15441.

155. F. Stellato, A. Spevacek, O. Proux, V. Minicozzi, G. Millhauser and S. Morante, Zinc modulates copper coordination mode in prion protein octa-repeat subdomains, *Eur. Biophys. J.*, 2011, **40**, 1259–1270.

156. A. G. Kenward, L. J. Bartolotti and C. S. Burns, Copper and zinc promote interactions between membrane-anchored peptides of the metal binding domain of the prion protein, *Biochemistry*, 2007, **46**, 4261–4271.

157. W. S. Perera and N. M. Hooper, Ablation of the metal ion-induced endocytosis of the prion protein by disease-associated mutation of the octarepeat region, *Curr. Biol.*, 2001, **11**, 519–523.

158. P. C. Pauly and D. A. Harris, Copper stimulates endocytosis of the prion protein, *J. Biol. Chem.*, 1998, **273**, 33107–33110.

159. A. R. Spevacek, E. G. Evans, J. L. Miller, H. C. Meyer, J. G. Pelton and G. L. Millhauser, Zinc drives a tertiary fold in the prion protein with familial disease mutation sites at the interface, *Structure*, 2013, **21**, 236–246.

160. N. T. Watt, D. R. Taylor, T. L. Kerrigan, H. H. Griffiths, J. V. Rushworth, I. J. Whitehouse and N. M. Hooper, Prion protein facilitates uptake of zinc into neuronal cells, *Nat. Commun.*, 2012, **3**, 1134.

161. N. T. Watt, H. H. Griffiths and N. M. Hooper, Neuronal zinc regulation and the prion protein, *Prion*, 2013, **7**, 203–208.

162. C. J. Choi, A. Kanthasamy, V. Anantharam and A. G. Kanthasamy, Interaction of metals with prion protein: possible role of divalent cations in the pathogenesis of prion diseases, *Neurotoxicology*, 2006, **27**, 777–787.

163. E. Aronoff-Spencer, C. S. Burns, N. I. Avdievich, G. J. Gerfen, J. Peisach, W. E. Antholine, H. L. Ball, F. E. Cohen, S. B. Prusiner and G. L. Millhauser, Identification of the Cu2+ binding sites in the N-terminal domain of the prion protein by EPR and CD spectroscopy, *Biochemistry*, 2000, **39**, 13760–13771.

164. M. Chattopadhyay, E. D. Walter, D. J. Newell, P. J. Jackson, E. Aronoff-Spencer, J. Peisach, G. J. Gerfen, B. Bennett, W. E. Antholine and G. L. Millhauser, The octarepeat domain of the prion protein binds Cu(II) with three distinct coordination modes at pH 7.4, *J. Am. Chem. Soc.*, 2005, **127**, 12647–12656.

165. C. L. Haigh, K. Edwards and D. R. Brown, Copper binding is the governing determinant of prion protein turnover, *Mol. Cell. Neurosci.*, 2005, **30**, 186–196.

166. D. R. Brown, B. S. Wong, F. Hafiz, C. Clive, S. J. Haswell and I. M. Jones, Normal prion protein has an activity like that of superoxide dismutase, *Biochem. J.*, 1999, **344**(Pt 1), 1–5.

167. D. R. Brown, C. Clive and S. J. Haswell, Antioxidant activity related to copper binding of native prion protein, *J. Neurochem.*, 2001, **76**, 69–76.

168. C. Treiber, R. Pipkorn, C. Weise, G. Holland and G. Multhaup, Copper is required for prion protein-associated superoxide dismutase-I activity in Pichia pastoris, *FEBS J.*, 2007, **274**, 1304–1311.

169. E. Gaggelli, E. Jankowska, H. Kozlowski, A. Marcinkowska, C. Migliorini, P. Stanczak, D. Valensin and G. Valensin, Structural characterization of the intra- and inter-repeat copper binding modes within the N-terminal region of "prion related protein" (PrP-rel-2) of zebrafish, *J. Phys. Chem. B*, 2008, **112**, 15140–15150.

170. R. C. Nadal, S. R. Abdelraheim, M. W. Brazier, S. E. Rigby, D. R. Brown and J. H. Viles, Prion protein does not redox-silence Cu2+, but is a sacrificial quencher of hydroxyl radicals, *Free Radic. Biol. Med.*, 2007, **42**, 79–89.

171. G. Hutter, F. L. Heppner and A. Aguzzi, No superoxide dismutase activity of cellular prion protein in vivo, *Biol. Chem.*, 2003, **384**, 1279–1285.

172. S. Jones, M. Batchelor, D. Bhelt, A. R. Clarke, J. Collinge and G. S. Jackson, Recombinant prion protein does not possess SOD-1 activity, *Biochem. J.*, 2005, **392**, 309–312.

173. K. Qin, D. S. Yang, Y. Yang, M. A. Chishti, L. J. Meng, H. A. Kretzschmar, C. M. Yip, P. E. Fraser and D. Westaway, Copper(II)-induced conformational changes and protease resistance in recombinant and cellular PrP. Effect of protein age and deamidation, *J. Biol. Chem.*, 2000, **275**, 19121–19131.

174. E. Quaglio, R. Chiesa and D. A. Harris, Copper converts the cellular prion protein into a protease-resistant species that is distinct from the scrapie isoform, *J. Biol. Chem.*, 2001, **276**, 11432–11438.

175. J. Stockel, J. Safar, A. C. Wallace, F. E. Cohen and S. B. Prusiner, Prion protein selectively binds copper(II) ions, *Biochemistry*, 1998, **37**, 7185–7193.

176. O. V. Bocharova, L. Breydo, V. V. Salnikov and I. V. Baskakov, Copper(II) inhibits in vitro conversion of prion protein into amyloid fibrils, *Biochemistry*, 2005, **44**, 6776–6787.

177. B. S. Wong, C. Venien-Bryan, R. A. Williamson, D. R. Burton, P. Gambetti, M. S. Sy, D. R. Brown and I. M. Jones, Copper refolding of prion protein, *Biochem. Biophys. Res. Commun.*, 2000, **276**, 1217–1224.

178. E. M. Sigurdsson, D. R. Brown, M. A. Alim, H. Scholtzova, R. Carp, H. C. Meeker, F. Prelli, B. Frangione and T. Wisniewski, Copper chelation delays the onset of prion disease, *J. Biol. Chem.*, 2003, **278**, 46199–46202.

179. N. Hijazi, Y. Shaked, H. Rosenmann, T. Ben-Hur and R. Gabizon, Copper binding to PrPC may inhibit prion disease propagation, *Brain Res.*, 2003, **993**, 192–200.

DNA Damage Induced by Manganese

JULIA BORNHORST* AND TANJA SCHWERDTLE*

Institute of Nutritional Sciences, University of Potsdam,
Arthur-Scheunert-Allee 114-116, 14558, Nuthetal, Germany
*Email: julia.bornhorst@uni-potsdam.de; tanja.schwerdtle@uni-potsdam.de

24.1 Genotoxic Lesions Induced by Manganese

DNA damage at the molecular and chromosomal levels is an essential part of genetic toxicology. Thus, this chapter deals with genotoxic lesions induced by various manganese (Mn) species at both the chromosomal and the DNA level. Given that genotoxic events can result in mutations, studies investigating the mutagenic potential of Mn species are additionally summarized in this chapter.

24.1.1 Damage Induced by Manganese at the Chromosomal Level

Genetic instability at the chromosomal level can be analyzed using cytogenetic parameters such as chromosomal aberrations (CAs), sister chromatid exchanges (SCEs), and micronuclei (MN). CAs are abnormalities resulting either from a variation in the chromosome number or from structural changes. SCE represent symmetrical exchanges between newly replicated chromatids and their sisters, which can be visualized by labelling the DNA of one chromatid with 5-bromodeoxyuridine (BrdU) during

Issues in Toxicology No. 22
Manganese in Health and Disease
Edited by Lucio G. Costa and Michael Aschner
© The Royal Society of Chemistry 2015
Published by the Royal Society of Chemistry, www.rsc.org

synthesis.[1] MN are derived from acentric chromosome fragments, acentric chromatid fragments or whole chromosomes that fail to be included in the daughter nuclei during mitosis. The cytokinesis-block micronucleus assay (CBMN) has emerged as one of the preferred methods for measuring MN. Here, scoring is specifically restricted to once-divided binucleate cells, which are the cells that can express MN.[2,3]

There are a few *in vitro* and *in vivo* studies in the literature dealing with the genotoxic action of various Mn species at the chromosomal level. The most important studies, ordered by the respective Mn species, are described below and are additionally summarized in Table 24.1.

The genotoxic potential of Mn-containing dithiocarbamate fungicides, which are used in agriculture to control many diseases of major food crops, has been demonstrated by an increased number of CAs, SCEs and MN. In cultured human lymphocytes, isolated from the peripheral blood of healthy non-smoking donors, exposure to Mancozeb ([1,2-ethanediylbis[carbamodithioato]](2-)]manganese mixture with [[1,2-ethanediylbis[carbamodithioato]](2-)]zinc) resulted in a significant and dose-dependent induction of CAs and MN.[4] In chick embryos, Maneb incubation significantly increased SCE values at 13.5 g l^{-1} and at 27 g l^{-1} when compared with untreated controls. The concentration of 27 g l^{-1}, which corresponds to 10.8 times the recommended maximum application level for use in the field, suggests a low genotoxic potential in this study.[5] Besides these *in vitro* and *in vivo* data there is also limited epidemiological evidence for the genotoxic potential of Mn-containing pesticides. CAs and SCEs were analyzed in short-term cultures of peripheral lymphocytes from 44 workers occupationally exposed to Mancozeb during the production of the pesticide Novozir Mn80 (80–85% Mancozeb) and control persons. Increased frequencies of cells with structural CAs and an increased number of SCEs per cell suggested that the exposed workers may be considered as a slight risk group on the basis of CAs.[6] Farm workers exposed to pesticides containing Mn showed significantly higher CAs, even at a low level of exposure.[7]

Manganese dioxide has been reported to cause CAs and MN in cultured human peripheral lymphocytes.[8] However, no CAs could be observed in the nasal mucosa of mammals exposed to Mn dioxide aerosol.[9]

Given that manganese oxide nanoparticles are promising materials for use as contrast agents for magnetic resonance imaging (MRI), and in drug delivery, waste water treatment and batteries, concerns are rising about intentional but also accidental human and environmental exposure, which may lead to significant genotoxic effects.[10,11] A study in Wistar rats demonstrated a significant increase in MN and CAs in bone marrow cells following 28 days of repeated oral exposure to MnO_2 nanoparticles or MnO_2 microparticles.[10]

In vitro and *in vivo* assays in mammalian cells have given conflicting results concerning the genotoxic effects of manganese salts. Manganese sulfate ($MnSO_4$) induced SCEs in Chinese hamster ovary (CHO) cells in both the presence and the absence of S9 from Aroclor 1254-induced rat liver.[12]

Table 24.1 Summary of important studies showing genotoxic lesions induced by manganese species. bw, body weight; CA, chromosomal aberrations; MN, micronuclei; SCE, sister chromatid exchange; 8-oxoG: 7,8-dihydro-8-oxoguanine.

Genotoxic endpoint	Mn species	Model	Exposure range	Effective doses	Exposure conditions	Reference
CA, MN	Mancozeb	Human lymphocytes	0, 0.5, 2, 5 µg ml^{-1}	0.5, 2, 5 µg ml^{-1}	24 h	4
SCE	Maneb	Chick embryos	0, 0.5, 1.5, 4.5, 13.5, 27 g l^{-1}	13.5, 27 g l^{-1}	Eggs were dipped in Maneb aqueous solutions (30 s), then incubated for 4 days	5
CA, MN	MnO$_2$	Human lymphocytes	0, 1.2 mmol l^{-1}	1.2 mmol l^{-1}	3 h	8
CA, MN	MnO$_2$–nanoparticles	Wistar rats	0, 30, 300, 1000 mg kg^{-1} bw per day	300, 1000 mg kg^{-1} bw per day	28 days	11
SCE	MnSO$_4$	Chinese hamster ovary cells	0, 5–50 µg ml^{-1} (−S9); 0, 200–350 µg ml^{-1} (+S9)	<5 µg ml^{-1} (−S9); 298/300 µg ml^{-1} (+S9)	24 h	12
CA	MnSO$_4$	Chinese hamster ovary cells	0, 141–300 µg ml^{-1} (−S9); 0, 400–500 µg ml^{-1} (+S9)	200/180 µg ml^{-1} (−S9); –	24 h	12
CA, MN	MnSO$_4$	Swiss albino mice	0, 10.25, 20.5, 61 mg 100 g^{-1} bw per day	10.25, 20.5, 61 mg 100 g^{-1} bw per day	7, 14 or 21 days	13
CA, MN	KMnO$_4$	Swiss albino mice	0, 6.5, 13, 38 mg 100 g^{-1} bw per day	6.5, 13, 38 mg 100 g^{-1} bw per day	7, 14 or 21 days	13
CA	MnCl$_2$	Male albino rats	0, 50 µg kg^{-1} bw per day	—	180 days	14
CA	MnCl$_2$	Human lymphocytes	0, 15, 20, 25 µM	25 µM (G2 phase)	3 h	15
CA	KMnO$_4$	FM3A cells	0, 6.4×10^{-4}, 1.0×10^{-3}, 2.0×10^{-3}, 3.2×10^{-3} M	>10^{-3} M	24, 48 h	16

DNA damage	MnCl$_2$	Human lymphocytes	0, 1.5, 3, 4.5 mmol L^{-1}	1.5, 3, 4.5 mmol L^{-1}	60 min	19
DNA strand breaks	MnCl$_2$	Human lymphocytes	0, 15, 20, 25 μM	25 μM	3 h	15
Single strand breaks	MnCl$_2$	Neuroblastoma SH-SY5Y cells	0, 2, 62 μM	2, 62 μM	24 h	20
DNA damage	KMnO$_4$	Human lymphocytes	0, 10, 20, and 30 μl of KMnO$_4$ + 6NH$_2$SO$_4$	10, 20, and 30 μl of KMnO$_4$ + 6NH$_2$SO$_4$	60 min	19
Single strand breaks	MnCl$_2$	Rat hepatic mitochondria	0, 0.1, 0.2, 0.5, 1 mmol l^{-1}	0.5, 1 mmol l^{-1}	1 h	23
Single strand breaks	MnCl$_2$	Sprague–Dawley rats	0, 5, 10, 20 mg kg^{-1} bw	5, 10, 20 mg kg^{-1} bw	Daily injections over 3 months	23
Single strand breaks	MnCl$_2$	Human astrocytes	0, 10, 50, 100, 250 μM	—	2, 24, 48 h	22
8-oxoG	MnCl$_2$	PC12-derived neurons	0, 100 ng ml^{-1}	—	72 h	27
8-oxoG	MnCl$_2$	PC12-derived neurons	0, 200 μM	—	12 h	28
8-oxoG	MnCl$_2$	PC12-derived neurons	0, 200 μM Mn + 0, 20, 50 μM dopamine	200 μM Mn + 20, 50 μM dopamine	12 h	28
Thymine-derived lesions	MnCl$_2$	Neuroblastoma SH-SY5Y cells	0, 2, 62, 125 μM	62, 125 μM	24 h	20

In a separate assay, MnSO$_4$ likewise induced CAs in CHO cells in the absence of S9 but not in its presence.[12] *In vivo* assays in mice showed that oral application of manganese sulfate, as well as potassium permanganate, caused MN and CAs in bone marrow.[13]

In contrast, oral uptake of MnCl$_2$ produced no significant chromosomal damage in the bone marrow or spermatogonia of rats.[14] In cultured human lymphocytes, CAs were generated after MnCl$_2$ incubation exclusively in the G2 phase but not in other phases of the cell cycle.[15]

Potassium permanganate has been reported to cause CAs in FM3A cells.[16] Epidemiological evidence from a study with occupationally exposed welders may suggest genotoxic effects following inhalative exposure to manganese. Within this study the incidence of CAs was established in three groups of welders with occupational exposure (10–24 years) to metals including manganese, nickel and chromium. A significant increase in CAs was found in the group of welders who used a semi-automatic metal active gas welding process with cored wire containing nickel for welding mild steel. The authors could not attribute the results to a single metal owing to the mixed exposure.[17]

24.1.2 Damage Induced by Manganese at the DNA Level

At the molecular level, the findings regarding the genotoxic potential of manganese are controversial. Data indicate that incubation of various cell cultures with Mn species results in an increased number of DNA strand breaks and oxidatively modified bases. Most of these studies applied MnCl$_2$ and used the comet assay, which has become one of the standard methods for assessing DNA strand breaks in eukaryotic cells. The comet assay is formally named single gel electrophoresis and is also frequently used in modified form to quantify oxidatively modified DNA bases.[18]

MnCl$_2$ induced DNA damage in human lymphocytes in a dose-dependent manner in the absence of metabolic activation, but caused no DNA damage when S9 was present.[19] In a further study, with cultured human lymphocytes, Mn incubation only generated DNA strand breaks at the highest applied dose of 25 μM MnCl$_2$, which was shown to be cytotoxic to the lymphocytes.[15] Exposure of neuroblastoma SH-SY5Y cells to low, subcytotoxic, MnCl$_2$ concentrations (2 μM) resulted in an eight-fold increase in DNA damage.[20] In human leukocytes, Mn exposure did not increase the amount of strand breaks or cross-links.[21] Potassium permanganate has been shown to cause DNA strand breaks in human peripheral blood lymphocytes.[19] However, in cultured human astrocytes, no induction of single strand breaks (SSB) could be observed following MnCl$_2$ incubation. Even though in this study no direct genotoxic effect could be indicated, the suggestion of a possible indirect mechanism has been pointed out[22] by examining the effect of MnCl$_2$ on the induction of DNA strand breaks by H$_2$O$_2$.

Mn-induced SSBs have been analyzed *in vitro* and *in vivo* in hepatic and brain mitochondrial DNA (mtDNA). During *in vitro* experiments on rat hepatic mitochondria, the increased formation of SSBs was evaluated in a time-dependent manner. In the *in vivo* study, $MnCl_2$ exposure (intraperitoneal) for three months increased SSB formation in rat hepatic and brain mtDNA in a dose-dependent manner.[23]

7,8-Dihydro-8-oxoguanine (8-oxoG) is a prominent indicator of oxidative damage and is associated with neurodegeneration.[24,25] Several publications have reported that patients with Parkinson's disease show a significant increase in 8-oxoG in mitochondrial DNA or cytoplasmic RNA in their dopaminergic neurons.[26] While incubation with $MnCl_2$ alone did not increase the 8-oxoG content in PC12-derived neurons,[27,28] $MnCl_2$ increased the formation of 8-oxoG in dopamine-treated PC12 cells.[28] In human neuroblastoma SH-SY5Y cells, $MnCl_2$ promoted both the formation and the accumulation of thymine-derived oxidative DNA damage. Thus, Mn exposure resulted in a significant accumulation of thymine-derived (5-OH-5-MetHyd) lesions, whereas no statistically significant accumulation of guanine-derived (8-OH-Gua, FapyGua) lesions could be observed. Moreover, the accumulated levels of DNA damage could be abrogated by the addition of exogenous chemical antioxidants including *N*-acetylcysteine and glutathione.[28] In calf thymus DNA, the site specificity of DNA damage by dopamine and Mn(II) in the presence of Cu(II) has been analyzed. Mn(II)-treated dopamine formed piperidine-labile and Fpg-sensitive lesions at T and G of the 5′-TG-3′ sequence, indicating that Mn(II)-treated dopamine can cause double base lesions in the presence of Cu(II).[28]

24.1.3 Mutations Induced by Manganese

The mutagenic potential of manganese is equivocal and has been studied by *in vitro* tests in bacteria as well as by *in vivo/in vitro* tests in *Drosophila melanogaster* and mammalian cells.[29] Most mutagenicity studies have been carried out by applying $MnCl_2$ and $MnSO_4$.

To assess mutations in bacteria three principal assays are frequently used: the Rec-assay in *Bacillus subtilis*, the formation of mutations in the lac I gene of *Escherichia coli* (*E. coli*) and the reversion assay in *Salmonella typhimurium* (*S. typhimurium*) (Ames test). Mn(II) gave positive results in the Rec-assay[30] and showed mutations in the lac I gene of *E. coli*,[31] albeit not scoring as a direct mutagen in the Ames test owing to equivocal results and the assumption that the conventional Ames test does not represent a reliable assay when applying metal salts.[32,33] Manganese sulfate did not show positive results in *S. typhimurium* strains TA97, TA98, TA100, TA1535 and TA1537 with and without exogenous metabolic activation (S9),[34] although it was reported to be mutagenic in strain TA97 elsewhere.[35] Although $MnCl_2$ failed to induce mutagenicity in *S. typhimurium* strains TA98, TA100, and TA1535, it was mutagenic in strain TA1537, and contradictory results were achieved for TA102.[19,36]

In yeast (*Saccharomyces cerevisiae* strain D7), manganese sulfate has been reported to induce gene conversion/reverse mutation.[37]

In *Drosophila melanogaster*, MnCl$_2$ ($\geq$20 mM) was clearly effective in inducing spots with one or two mutant hairs (small spots), as measured by the wing spot test.[38] Moreover, MnCl$_2$ has been reported to induce forward mutations at the thymidine kinase locus in L5178Y mouse lymphoma cells (mutants resistant to 6-thioguanine)[39] and in Chinese hamster V79 mammalian cells (mutants resistant to 8-azaguanine).[40] MnDPDP [manganese(II) *N,N'*-dipyridoxylethylenediamine-*N,N'*-diacetate-5,5'-bis(phosphate)], which is used as a contrast medium for MRI, can cause gene mutations in *E. coli* and *S. typhimurium* strains TA100, TA1535, TA98 and TA1537. Additionally, MnDPDP results in forward mutation in CHO cells.[41]

24.2   Sources of the Genotoxic Potential of Manganese

This section deals with possible underlying mechanisms of the genotoxic potential of manganese. From a chemical point of view Mn is unlikely to bind covalently to DNA, but different Mn species are reported to cause oxidative stress, which subsequently might cause DNA strand breaks and oxidative base modifications (see Section 24.1.2). Given that Mn-induced oxidative stress is extensively summarized in Chapter 8, this will not be further discussed here. Other probable mechanisms for the observed increase in DNA damage in the presence of high Mn concentrations include the interaction of Mn with DNA replication/DNA polymerases as well as with DNA damage response pathways.

The DNA polymerases are classified into six main families on the basis of phylogenetic relationships with *E. coli* Pol I (class A), *E. coli* Pol II (class B), *E. coli* Pol III (class C), *Euryarchaeotic* Pol II (class D), human Pol ß (class X), and *E. coli* UmuC/DinB and eukaryotic RAD30/xeroderma pigmentosum variant (class Y).[42] It is generally assumed that Mg(II) is the physiological cofactor for replicative DNA polymerases *in vivo*. However, the effect of replacing Mg with Mn on the activity of A and B family polymerases has been intensively studied.[43–46] Exposing *E. coli* DNA polymerase I to both poly[d(A–T)] and natural DNA templates for Mn(II) showed a substantially Mn-induced decrease of the fidelity of DNA replication resulting from alteration of the activity of the DNA polymerase.[45] Moreover, this effect has been confirmed by a study reporting that Mn(II) promotes the B–Z transition of poly[d(G-m5C)] at substoichiometric concentrations with respect to DNA nucleotides.[47] Other studies have revealed that, within a number of DNA polymerases, the replacement of Mg with Mn influences both the fidelity[48] and the lesion bypass capacity.[49] Mn has also been reported to promote translesion synthesis by certain replicative polymerases, including herpes simplex virus-1 UL30 protein, but not by others such as T4 DNA polymerase or σ-polymerases.[50] In an early study, the authors demonstrated that MnCl$_2$ inhibited to a greater extent the activity of the proofreading exonuclease associated with the HSV-1 DNA polymerase than that allied to T4 DNA polymerase. Based on the results with the T4 DNA polymerase, they suggested that the mutagenic effects of Mn(ıı) are caused by increasing the binding of mispaired nucleotides.[50] The DNA polymerase iota (Pol iota), which has some peculiar features and is characterized by extremely error-prone DNA synthesis, was shown to utilize Mn(ıı) in DNA synthesis. Interestingly, Pol iota is inducible by Mn(ıı) and may in part contribute to the genotoxic potential of Mn.[51] Gene expression profiling of human primary astrocytes exposed to Mn indicated 15 genes, encoding functions involved in DNA replication and repair as well as cell cycle checkpoint control, which were downregulated. Among these, ATM kinase (ataxia telangiectasia mutated), the checkpoint kinase BUB1, the Fanconi anemia FANCF protein, two subunits of DNA primase PRIM1 and PRIM2A, and the replication factor RFC5 are of particular interest.[52]

Damage to genomic DNA triggers an immediate set of signaling events known as the DNA damage response (DDR). The cellular DDR coordinates DNA repair, cell cycle arrest and ultimately cell death pathways.

DNA repair mechanisms belong to an elaborate genomic maintenance apparatus, which minimizes the consequences of DNA damage. DNA repair genes can be sub-grouped into genes associated with signaling and regulation of DNA repair on the one hand and genes associated with distinct repair mechanisms such as base excision repair (BER), SSB repair, nucleotide excision repair (NER), and DNA double-strand break (DSB) repair on the other.[53] BER removes subtle modifications of DNA, including oxidative lesions, small alkylation products and different kinds of single-strand break, while NER eliminates helix-distorting DNA damage, a broad category of damage that affects one of the two DNA strands.[54] Deficiency in repair of nuclear and mitochondrial DNA damage has been linked to several neurodegenerative disorders including Huntington's, Alzheimer's and Parkinson's diseases, underscoring the critical importance of DNA repair for neural homeostasis.[55-58]

An indication that DNA repair might play a major role in the genotoxic potential of manganese has been provided by a study in which genotoxicity, observed at G2 phase (CAs) and in the comet assay, may be related to a lack of time for the cellular repair system to act.[15] Among theoretical studies the interaction of Mn with the human DNA polymerase lambda (Polλ) has been discussed, indicating the interaction of Mn with DNA repair pathways. Polλ is a DNA repair polymerase that fills short gaps during DNA repair. It plays a role in base excision repair as well as DNA double-strand break repair by non-homologous end joining.[59,60] Polλ has been suggested to have a slight preference for Mn(ıı) over Mg(ıı) as the activating metal, based on lower activation energy for the former metal.[61,62]

That base excision repair is affected by Mn has further been indicated by the effect of Mn on the enzymatic activity of 8oxo-G DNA glycosylase 1 (OGG1). In the BER pathway, OGG1 is responsible for the excision of 8-oxoguanine. After recognizing and removing the damaged base, OGG1 remains tightly bound to the apurinic/apyrimidinic (AP) site until displaced by AP endonuclease 1 (APE1).[63] Even though the bifunctional glycosylase OGG1 contains intrinsic AP lyase activity, under physiological conditions OGG1 might act as monofunctional glycosylase, whereupon the AP site is cleaved by APE1. A study conducted in PC12-derived neuronal cells examined diminished activity of OGG1 caused by Mn after three days of exposure; in contrast, following one day of exposure no reduction could be observed.[27] In OGG1 knockout mice, the lack of OGG1, was sensitizing dopaminergic neurons towards manganese induced neurotoxicity during developmental stages.[64] Moreover, the contribution of various DNA repair pathways to the survival of yeast cells (*Saccharomyces cerevisiae*) exposed to Mn toxicity has been studied recently. The strains most sensitive to Mn toxicity were those defective in base excision repair; for example, mutants *apn1*, *rad27*, and *ntg1* were more than four-fold more sensitive to Mn(ɪɪ) than wild type, and *ntg1* was the most sensitive (7.5-fold). *Ntg1* is a DNA-glycosylase which removes an oxidized damaged base. The high sensitivity of the strains *ubc13*, containing a DNA-damage-inducible gene, a member of the errorfree post-replication repair pathway, and *rad30* mutants, which are defective in translesion synthesis DNA polymerase eta, indicated the interaction of Mn with DNA replication. However, NER and double strand break repair seem not to play a major role in the repair of Mn-induced DNA damage. Use of two distinct mutator assays, *CAN1* and *lys2-10A*, demonstrated a significant dose-dependent Mn-induced increase in the accumulation of mutations. Data from yeast cells observed in this study suggested that Mn causes oxidative DNA damage that requires base excision repair for processing, and that Mn interferes with polymerase fidelity. The authors proposed further that the status of BER might provide a biomarker for the Mn sensitivity of individuals.[65] Overexpression of RAD23, a well-conserved protein involved in DNA repair (NER) and proteosomal degradation, in *S. cerevisiae* resulted in a reduction of Mn toxicity by lowering Mn levels. However, data suggest that the RAD23-induced reversal of manganese toxicity reflects its role in protein quality control, not DNA repair.[66]

The genomic instability caused by Mn might also be due to an inhibitory effect on an essential signaling reaction related to the response to DNA damage, as indicated *in vitro* by strongly inhibited DNA damage-stimulated poly(ADP-ribosyl)ation following Mn exposure.[22,67] Poly(ADP-ribosyl)ation is a post-translational modification which is catalyzed by the poly(ADP-ribose) polymerase (PARP) family of proteins. In response to DNA strand breaks, two members of the PARP superfamily, PARP-1 and PARP-2, are rapidly activated and transfer ADP-ribosyl units from NAD^+ onto themselves and other target proteins, thus producing protein-coupled ADP-ribose polymers of up to 200 units. PARP-1 is responsible for about 90% of cellular poly(ADP-ribose) (PAR) formation and is involved in several biological pathways. Thus poly(ADP-ribosyl)ation affects proteins involved in transcription, replication, telomere maintenance, genomic stability, chromatin organization and DNA repair.[68,69] In cultured human astrocytes $MnCl_2$ efficiently (≥ 1 µM) disturbed DNA damage-induced poly(ADP-ribosyl)ation. This sensitive effect resulted neither from a delay in the signaling reaction nor from diminished formation of DNA strand breaks after co-incubation with the damage-inducing H_2O_2 and $MnCl_2$. Moreover, Mn did not decrease *PARP-1* gene expression or PARP-1 protein level, or affect the activity of isolated PARP-1. Although the underlying mechanism remains unclear, concerns have been raised about a direct interaction of Mn with the zinc finger structures of PARP-1, especially because evidence suggests a replacement of Mn(II) by Zn(II) in a Mn(II)-dependent lactonase ULaG.[70] Moreover, it cannot be excluded that the observed inhibition of H_2O_2-induced poly(ADP-ribosyl)ation is partly due to an increase of poly(ADP-ribose) glycohydrolase (PARG) expression. PARG contributes to poly(ADP-ribose) degradation.

24.3 Consequences of DNA Damage Induced by Manganese

In summary, Mn species have been shown to cause DNA damage, to interfere with the fidelity of DNA replication and to affect the cellular DNA damage response. Within this section, possible consequences of the Mn-induced effect on the previously mentioned pathways are discussed.

Consequences of DNA damage are diverse and generally adverse. Evidence suggests that an acute consequence is triggering cell-cycle arrest or cell death. Cell cycle arrest at specific checkpoints in G1, S, G2 and M phases averts genome injury by allowing DNA repair and preventing conversion into permanent mutations.[71] $MnCl_2$ has been reported to cause G(0)/G(1) phase cell cycle arrest and S phase arrest in cultured rat astrocytes and A549 cells.[72,73] Moreover, in PC12-derived neurons, $MnCl_2$ increased the population of cells in the G2/M phase, but reduced the population of cells in the S phase, leading to a significant increase in the G2/S ratio.[74] The consequence of these observed arrests might be persistence of DNA damage. In the case of significant damage, a cell may opt for the ultimate mode of rescue by initiating apoptosis. In particular, neuronal apoptosis is of great concern because neuronal cell death is a hallmark of many neurodegenerative disorders. Mn has been shown to induce apoptosis in several studies. Long-term consequences of damaged DNA may result in permanent changes in the DNA sequence (mutations) as well as damage at the chromosomal level, which have both been observed following Mn exposure (summarized in Section 24.1). Moreover, concern has been raised about possible consequences of the inhibitory effect of Mn on DNA repair pathways and DNA damage response-related signaling reactions, especially DNA damage-induced poly(ADP-ribosyl)ation. Deficiency in repair of nuclear and mitochondrial DNA damage has been linked to several neurodegenerative disorders, including Huntington's, Alzheimer's and Parkinson's diseases.[55,56] PARP-1 contributes to several DNA repair pathways, including SSB repair, BER and DSB repair, and inhibition is well known to result in diminished DNA repair, thereby causing genomic instability at the DNA and chromosomal level.[75,76] Thus, the Mn-induced inhibition of DNA damage-induced poly(ADP-ribosyl)ation indicates sensitization of cells to genotoxic treatment, and inhibition may be an underlying mechanism for the genotoxic potential of Mn. Additionally, PARP-1 activation is associated with neurite outgrowth and long-term memory. Therefore, chronic PARP-1 inhibition may compromise neurogenesis and learning ability.[77,78] Whether there is a correlation with the recent epidemiological studies showing associations between elevated dietary Mn exposure and neurobehavioral and neurocognitive deficits in children has to be elucidated in further studies.[79–81]

24.4   Conclusions

Overall, given that the results of *in vitro* and *in vivo* studies regarding the genotoxic potential of manganese are inconsistent, further studies are needed to clarify whether Mn species are genotoxic. Furthermore, it seems probable that the potential for DNA damage observed in some studies is not due to an intrinsic, directly genotoxic mode of action, but likely results from indirect genotoxic mechanisms. Thus, recent evidence from *in vitro* studies suggests that Mn can disturb cellular DNA damage response pathways under conditions of either overload due to high exposure or disturbed homeostasis, and thereby causes genomic instability.

References

1. F. Palitti, C. Tanzarella, R. Cozzi, R. Ricordy, E. Vitagliano and M. Fiore, Comparison of the Frequencies of Sces Induced by Chemical Mutagens in Bone-Marrow, Spleen and Spermatogonial Cells of Mice, *Mutat. Res.*, 1982, **103**, 191–195.
2. M. Fenech, Micronuclei and their association with sperm abnormalities, infertility, pregnancy loss, pre-eclampsia and intra-uterine growth restriction in humans, *Mutagenesis*, 2011, **26**, 63–67.
3. M. Fenech, Cytokinesis-block micronucleus cytome assays, *Nat. Protoc.*, 2007, **2**, 1084–1104.
4. A. K. Srivastava, W. Ali, R. Singh, K. Bhui, S. Tyagi, A. A. Al-Khedhairy, P. K. Srivastava, J. Musarrat and Y. Shukla, Mancozeb-induced genotoxicity and apoptosis in cultured human lymphocytes, *Life Sci.*, 2012, **90**, 815–824.
5. E. Arias, Sister-chromatid exchanges and chromosomal aberrations in chick embryos after treatment with the fungicide maneb, *Mutat. Res.*, 1988, **206**, 271–273.
6. A. Jablonicka, H. Polakova, J. Karelova and M. Vargova, Analysis of chromosome aberrations and sister-chromatid exchanges in peripheral blood lymphocytes of workers with occupational exposure to the mancozeb-containing fungicide Novozir Mn80, *Mutat. Res.*, 1989, **224**, 143–146.

7. S. M. Brega, I. Vassilieff, A. Almeida, A. Mercadante, D. Bissacot, P. R. Cury and D. V. Freire-Maia, Clinical, cytogenetic and toxicological studies in rural workers exposed to pesticides in Botucatu, Sao Paulo, Brazil, *Cad. Saude Publica*, 1998, **14**(Suppl 3), 109–115.

8. D. Dutta, S. S. Devi, K. Krishnamurthi and T. Chakrabarti, Anti-clastogenic effect of redistilled cow's urine distillate in human peripheral lymphocytes challenged with manganese dioxide and hexavalent chromium, *Biomed. Environ. Sci.*, 2006, **19**, 487–494.

9. O. I. Timchenko, N. M. Paran'ko, E. E. Shantyr and S. D. Kuz'menko, The cytogenetic effects of separate and combined exposures to a manganese dioxide aerosol and wide-band noise, *Gig. Sanit.*, 1991, 70–72.

10. S. P. Singh, M. Kumari, S. I. Kumari, M. F. Rahman, M. Mahboob and P. Grover, Toxicity assessment of manganese oxide micro and nano-particles in Wistar rats after 28 days of repeated oral exposure, *J. Appl. Toxicol.*, 2013, **33**, 1165–1179.

11. S. P. Singh, M. Kumari, S. I. Kumari, M. F. Rahman, S. S. Kamal, M. Mahboob and P. Grover, Genotoxicity of nano- and micron-sized manganese oxide in rats after acute oral treatment, *Mutat. Res.*, 2013, **754**, 39–50.

12. S. M. Galloway, M. J. Armstrong, C. Reuben, S. Colman, B. Brown, C. Cannon, A. D. Bloom, F. Nakamura, M. Ahmed, S. Duk, *et al.*, Chromosome aberrations and sister chromatid exchanges in Chinese hamster ovary cells: evaluations of 108 chemicals, *Environ. Mol. Mutagen.*, 1987, **10**, 1–175.

13. M. Joardar and A. Sharma, Comparison of clastogenicity of inorganic Mn administered in cationic and anionic forms in vivo, *Mutat. Res.*, 1990, **240**, 159–163.

14. T. S. Dikshith and S. V. Chandra, Cytological studies in albino rats after oral administration of manganese chloride, *Bull. Environ. Contam. Toxicol.*, 1978, **19**, 741–746.

15. P. D. Lima, M. C. Vasconcellos, M. O. Bahia, R. C. Montenegro, C. O. Pessoa, L. V. Costa-Lotufo, M. O. Moraes and R. R. Burbano, Genotoxic and cytotoxic effects of manganese chloride in cultured human lymphocytes treated in different phases of cell cycle, *Toxicol. In Vitro*, 2008, **22**, 1032–1037.

16. M. Umeda and M. Nishimura, Inducibility of chromosomal aberrations by metal compounds in cultured mammalian cells, *Mutat. Res.*, 1979, **67**, 221–229.

17. Z. Elias, J. M. Mur, F. Pierre, S. Gilgenkrantz, O. Schneider, F. Baruthio, M. C. Daniere and J. M. Fontana, Chromosome aberrations in peripheral blood lymphocytes of welders and characterization of their exposure by biological samples analysis, *J. Occup. Med.*, 1989, **31**, 477–483.

18. A. Azqueta and A. R. Collins, The essential comet assay: a comprehensive guide to measuring DNA damage and repair, *Arch. Toxicol.*, 2013, **87**, 949–968.

19. M. De Meo, M. Laget, M. Castegnaro and G. Dumenil, Genotoxic activity of potassium permanganate in acidic solutions, *Mutat. Res.*, 1991, **260**, 295–306.

20. A. P. Stephenson, J. A. Schneider, B. C. Nelson, D. H. Atha, A. Jain, K. F. Soliman, M. Aschner, E. Mazzio and R. R. Reams, Manganese-induced oxidative DNA damage in neuronal SH-SY5Y cells: attenuation of thymine base lesions by glutathione and N-acetylcysteine, *Toxicol. Lett.*, 2013, **218**, 299–307.

21. J. R. McLean, R. S. McWilliams and J. G. Kaplan, Rapid detection of DNA strand breaks in human peripheral blood cells and animal organs following treatment with physical and chemical agents, *Prog. Mutat. Res.*, 1982, **3**, 137–141.

22. J. Bornhorst, S. Meyer, T. Weber, C. Boker, T. Marschall, A. Mangerich, S. Beneke, A. Burkle and T. Schwerdtle, Molecular mechanisms of Mn induced neurotoxicity: RONS generation, genotoxicity, and DNA-damage response, *Mol. Nutr. Food Res.*, 2013, **57**, 1255–1269.

23. J. Jiao, Y. Qi, J. Fu and Z. Zhou, Manganese-induced single strand breaks of mitochondrial DNA in vitro and in vivo, *Environ. Toxicol. Pharmacol.*, 2008, **26**, 123–127.

24. Z. Sheng, S. Oka, D. Tsuchimoto, N. Abolhassani, H. Nomaru, K. Sakumi, H. Yamada and Y. Nakabeppu, 8-Oxoguanine causes neurodegeneration during MUTYH-mediated DNA base excision repair, *J. Clin. Invest.*, 2012, **122**, 4344–4361.

25. M. Sekiguchi and T. Tsuzuki, Oxidative nucleotide damage: consequences and prevention, *Oncogene*, 2002, **21**, 8895–8904.

26. Y. Nakabeppu, D. Tsuchimoto, H. Yamaguchi and K. Sakumi, Oxidative damage in nucleic acids and Parkinson's disease, *J. Neurosci. Res.*, 2007, **85**, 919–934.

27. V. Sava, D. Mosquera, S. Song, F. Cardozo-Pelaez and J. R. Sanchez-Ramos, Effects of melanin and manganese on DNA damage and repair in PC12-derived neurons, *Free Radical Biol. Med.*, 2004, **36**, 1144–1154.

28. S. Oikawa, I. Hirosawa, S. Tada-Oikawa, A. Furukawa, K. Nishiura and S. Kawanishi, Mechanism for manganese enhancement of dopamine-induced oxidative DNA damage and neuronal cell death, *Free Radical Biol. Med.*, 2006, **41**, 748–756.

29. G. B. Gerber, A. Leonard and P. Hantson, Carcinogenicity, mutagenicity and teratogenicity of manganese compounds, *Crit. Rev. Oncol. Hematol.*, 2002, **42**, 25–34.

30. H. Nishioka, Mutagenic activities of metal compounds in bacteria, *Mutat. Res.*, 1975, **31**, 185–189.

31. R. A. Zakour and B. W. Glickman, Metal-induced mutagenesis in the lacI gene of Escherichia coli, *Mutat. Res.*, 1984, **126**, 9–18.

32. A. Leonard, Mechanisms in metal genotoxicity: the significance of in vitro approaches, *Mutat. Res.*, 1988, **198**, 321–326.

33. D. R. Marzin and H. V. Phi, Study of the mutagenicity of metal derivatives with Salmonella typhimurium TA102, *Mutat. Res.*, 1985, **155**, 49–51.

34. K. Mortelmans, S. Haworth, T. Lawlor, W. Speck, B. Tainer and E. Zeiger, Salmonella mutagenicity tests: II. Results from the testing of 270 chemicals, *Environ. Mutagen.*, 1986, **8**(Suppl 7), 1–119.

35. D. A. Pagano and E. Zeiger, Conditions for detecting the mutagenicity of divalent metals in Salmonella typhimurium, *Environ. Mol. Mutagen.*, 1992, **19**, 139–146.

36. P. K. Wong, Mutagenicity of heavy metals, *Bull. Environ. Contam. Toxicol.*, 1988, **40**, 597–603.

37. I. Singh, Induction of gene conversion and reverse mutation by manganese sulphate and nickel sulphate in Saccharomyces cerevisiae, *Mutat. Res.*, 1984, **137**, 47–49.

38. H. I. Ogawa, T. Shibahara, H. Iwata, T. Okada, S. Tsuruta, K. Kakimoto, K. Sakata, Y. Kato, H. Ryo, T. Itoh, *et al.*, Genotoxic activities in vivo of cobaltous chloride and other metal chlorides as assayed in the Drosophila wing spot test, *Mutat. Res.*, 1994, **320**, 133–140.

39. T. J. Oberly, C. E. Piper and D. S. McDonald, Mutagenicity of metal salts in the L5178Y mouse lymphoma assay, *J. Toxicol. Environ. Health*, 1982, **9**, 367–376.

40. M. Miyaki, N. Akamatsu, T. Ono and H. Koyama, Mutagenicity of Metal-Cations in Cultured-Cells from Chinese-Hamster, *Mutat. Res.*, 1979, **68**, 259–263.

41. L. E. Larsen and D. Grant, General toxicology of MnDPDP, *Acta Radiol.*, 1997, **38**, 770–779.

42. P. M. Burgers, E. V. Koonin, E. Bruford, L. Blanco, K. C. Burtis, M. F. Christman, W. C. Copeland, E. C. Friedberg, F. Hanaoka, D. C. Hinkle, C. W. Lawrence, M. Nakanishi, H. Ohmori, L. Prakash, S. Prakash, C. A. Reynaud, A. Sugino, T. Todo, Z. Wang, J. C. Weill and R. Woodgate, Eukaryotic DNA polymerases: proposal for a revised nomenclature, *J. Biol. Chem.*, 2001, **276**, 43487–43490.

43. M. F. Goodman, S. Keener, S. Guidotti and E. W. Branscomb, On the Enzymatic Basis for Mutagenesis by Manganese, *J. Biol. Chem.*, 1983, **258**, 3469–3475.

44. W. S. El-Deiry, K. M. Downey and A. G. So, Molecular mechanisms of manganese mutagenesis, *Proc. Natl. Acad. Sci. U. S. A.*, 1984, **81**, 7378–7382.

45. R. A. Beckman, A. S. Mildvan and L. A. Loeb, On the fidelity of DNA replication: manganese mutagenesis in vitro, *Biochemistry*, 1985, **24**, 5810–5817.

46. E. Zakharova, J. Wang and W. Konigsberg, The activity of selected RB69 DNA polymerase mutants can be restored by manganese ions: the existence of alternative metal ion ligands used during the polymerization cycle, *Biochemistry*, 2004, **43**, 6587–6595.

47. J. H. Vandesande, L. P. Mcintosh and T. M. Jovin, Mn-2+ and Other Transition-Metals at Low Concentration Induce the Right-to-Left Helical Transformation of Poly[D(G-C)], *EMBO J.*, 1982, **1**, 777–782.

48. M. A. Sirover and L. A. Loeb, Infidelity of DNA-Synthesis Invitro - Screening for Potential Metal Mutagens or Carcinogens, *Science*, 1976, **194**, 1434–1436.

49. P. D. Moore, K. K. Bose, S. D. Rabkin and B. S. Strauss, Sites of termination of in vitro DNA synthesis on ultraviolet- and N-acetylamino-fluorene-treated phi X174 templates by prokaryotic and eukaryotic DNA polymerases, *Proc. Natl. Acad. Sci. U. S. A.*, 1981, **78**, 110–114.

50. G. Villani, N. Tanguy Le Gac, L. Wasungu, D. Burnouf, R. P. Fuchs and P. E. Boehmer, Effect of manganese on in vitro replication of damaged DNA catalyzed by the herpes simplex virus type-1 DNA polymerase, *Nucleic Acids Res.*, 2002, **30**, 3323–3332.

51. A. V. Lakhin, A. S. Efremova, I. V. Makarova, E. E. Grishina, S. I. Shram, V. Z. Tarantul and L. V. Gening, Effect of Mn(II) on the error-prone DNA polymerase iota activity in extracts from human normal and tumor cells, *Mol. Genet., Mikrobiol. Virusol.*, 2013, 14–20.

52. A. Sengupta, S. M. Mense, C. Lan, M. Zhou, R. E. Mauro, L. Kellerman, G. Bentsman, D. J. Volsky, E. D. Louis, J. H. Graziano and L. Zhang, Gene expression profiling of human primary astrocytes exposed to manganese chloride indicates selective effects on several functions of the cells, *Neurotoxicology*, 2007, **28**, 478–489.

53. M. Christmann, M. T. Tomicic, W. P. Roos and B. Kaina, Mechanisms of human DNA repair: an update, *Toxicology*, 2003, **193**, 3–34.

54. J. H. Hoeijmakers, DNA damage, aging, and cancer, *N. Engl. J. Med.*, 2009, **361**, 1475–1485.

55. D. K. Jeppesen, V. A. Bohr and T. Stevnsner, DNA repair deficiency in neurodegeneration, *Prog. Neurobiol.*, 2011, **94**, 166–200.

56. S. Katyal and P. J. McKinnon, DNA strand breaks, neurodegeneration and aging in the brain, *Mech. Ageing Dev.*, 2008, **129**, 483–491.

57. A. Bender, K. J. Krishnan, C. M. Morris, G. A. Taylor, A. K. Reeve, R. H. Perry, E. Jaros, J. S. Hersheson, J. Betts, T. Klopstock, R. W. Taylor and D. M. Turnbull, High levels of mitochondrial DNA deletions in substantia nigra neurons in aging and Parkinson disease, *Nat. Genet.*, 2006, **38**, 515–517.

58. L. Weissman, D. G. Jo, M. M. Sorensen, N. C. de Souza-Pinto, W. R. Markesbery, M. P. Mattson and V. A. Bohr, Defective DNA base excision repair in brain from individuals with Alzheimer's disease and amnestic mild cognitive impairment, *Nucleic Acids Res.*, 2007, **35**, 5545–5555.

59. M. Garcia-Diaz, K. Bebenek, G. Gao, L. C. Pedersen, R. E. London and T. A. Kunkel, Structure-function studies of DNA polymerase lambda, *DNA Repair*, 2005, **4**, 1358–1367.

60. M. Garcia-Diaz, K. Bebenek, J. M. Krahn, L. C. Pedersen and T. A. Kunkel, Structural analysis of strand misalignment during DNA synthesis by a human DNA polymerase, *Cell*, 2006, **124**, 331–342.

61. G. Blanca, I. Shevelev, K. Ramadan, G. Villani, S. Spadari, U. Hubscher and G. Maga, Human DNA polymerase lambda diverged in evolution

from DNA polymerase beta toward specific Mn(++) dependence: a kinetic and thermodynamic study, *Biochemistry*, 2003, **42**, 7467–7476.

62. G. A. Cisneros, L. Perera, M. Garcia-Diaz, K. Bebenek, T. A. Kunkel and L. G. Pedersen, Catalytic mechanism of human DNA polymerase lambda with Mg2+ and Mn2+ from ab initio quantum mechanical/molecular mechanical studies, *DNA Repair*, 2008, 7, 1824–1834.

63. P. Fortini and E. Dogliotti, Base damage and single-strand break repair: mechanisms and functional significance of short- and long-patch repair subpathways, *DNA Repair*, 2007, **6**, 398–409.

64. F. Cardozo-Pelaez, D. P. Cox and C. Bolin, Lack of the DNA repair enzyme OGG1 sensitizes dopamine neurons to manganese toxicity during development, *Gene Expression*, 2005, **12**, 315–323.

65. A. P. Stephenson, T. K. Mazu, J. S. Miles, M. D. Freeman, R. R. Reams and H. Flores-Rozas, Defects in base excision repair sensitize cells to manganese in S. cerevisiae, *Biomed. Res. Int.*, 2013, **2013**, 295635.

66. L. Rosenfeld and V. C. Culotta, Phosphate disruption and metal toxicity in Saccharomyces cerevisiae: effects of RAD23 and the histone chaperone HPC2, *Biochem. Biophys. Res. Commun.*, 2012, **418**, 414–419.

67. J. Bornhorst, F. Ebert, A. Hartwig, B. Michalke and T. Schwerdtle, Manganese inhibits poly(ADP-ribosyl)ation in human cells: a possible mechanism behind manganese-induced toxicity?, *J. Environ. Monit.*, 2010, **12**, 2062–2069.

68. S. Beneke and A. Buerkle, Poly(ADP-ribosyl)ation in mammalian ageing, *Nucleic Acids Res.*, 2007, **35**, 7456–7465.

69. J. Yelamos, J. Farres, L. Llacuna, C. Ampurdanes and J. Martin-Caballero, PARP-1 and PARP-2: New players in tumour development, *Am. J. Cancer Res.*, 2011, **1**, 328–346.

70. F. Garces, F. J. Fernandez, C. Montella, E. Penya-Soler, R. Prohens, J. Aguilar, L. Baldoma, M. Coll, J. Badia and M. C. Vega, Molecular architecture of the Mn2+-dependent lactonase UlaG reveals an RNase-like metallo-beta-lactamase fold and a novel quaternary structure, *J. Mol. Biol.*, 2010, **398**, 715–729.

71. B. B. Zhou and S. J. Elledge, The DNA damage response: putting checkpoints in perspective, *Nature*, 2000, **408**, 433–439.

72. Y. Deng, D. Xu, B. Xu, Z. Xu, Y. Tian, W. Feng, W. Liu and H. Yang, G0/G1 phase arrest and apoptosis induced by manganese chloride on cultured rat astrocytes and protective effects of riluzole, *Biol. Trace Elem. Res.*, 2011, **144**, 832–842.

73. P. Zhao, W. Zhong, X. Ying, Z. Yuan, J. Fu and Z. Zhou, Manganese chloride-induced G0/G1 and S phase arrest in A549 cells, *Toxicology*, 2008, **250**, 39–46.

74. F. Zhao, J. B. Zhang, T. J. Cai, X. Q. Liu, M. C. Liu, T. Ke, J. Y. Chen and W. J. Luo, Manganese induces p21 expression in PC12 cells at the transcriptional level, *Neuroscience*, 2012, **215**, 184–195.

75. T. M. Kauppinen, Multiple roles for poly(ADP-ribose)polymerase-1 in neurological disease, *Neurochem. Int.*, 2007, **50**, 954–958.

76. S. W. Yu, H. Wang, T. M. Dawson and V. L. Dawson, Poly(ADP-ribose) polymerase-1 and apoptosis inducing factor in neurotoxicity, *Neurobiol. Dis.*, 2003, **14**, 303–317.

77. L. Visochek, R. A. Steingart, I. Vulih-Shultzman, R. Klein, E. Priel, I. Gozes and M. Cohen-Armon, PolyADP-ribosylation is involved in neurotrophic activity, *J. Neurosci.*, 2005, **25**, 7420–7428.

78. M. Cohen-Armon, L. Visochek, A. Katzoff, D. Levitan, A. J. Susswein, R. Klein, M. Valbrun and J. H. Schwartz, Long-term memory requires polyADP-ribosylation, *Science*, 2004, **304**, 1820–1822.

79. D. Hernandez-Bonilla, A. Schilmann, S. Montes, Y. Rodriguez-Agudelo, S. Rodriguez-Dozal, R. Solis-Vivanco, C. Rios and H. Riojas-Rodriguez, Environmental exposure to manganese and motor function of children in Mexico, *Neurotoxicology*, 2011, **32**, 615–621.

80. M. F. Bouchard, S. Sauve, B. Barbeau, M. Legrand, M. E. Brodeur, T. Bouffard, E. Limoges, D. C. Bellinger and D. Mergler, Intellectual impairment in school-age children exposed to manganese from drinking water, *Environ. Health Perspect.*, 2011, **119**, 138–143.

81. J. A. Menezes-Filho, O. Novaes Cde, J. C. Moreira, P. N. Sarcinelli and D. Mergler, Elevated manganese and cognitive performance in school-aged children and their mothers, *Environ. Res.*, 2011, **111**, 156–163.

Post-face

Enormous strides have been made in the understanding of the biology of manganese (Mn) since 1774 when Johan Gottlieb Gahn first isolated an impure sample of Mn metal by reducing the dioxide with carbon. As discussed in the preceding chapters, adequate Mn intake ensures optimal functioning of multiple biological processes; yet, excessive exposure to this metal may contribute to infertility, bone malformations, and predominantly neurological effects. The most significant Mn exposure for the general population is from food, largely from plant-based diets, which tend to be rich sources of Mn. Even though gastrointestinal absorption is tightly controlled (3–5%), oral absorption is the primary source for Mn. People ingesting food high in Mn content have potential for higher-than-usual exposures. This group would include vegetarians who ingest a larger proportion of grains, legumes and nuts in their diets than the average population, as well as heavy tea drinkers. In addition, iron deficiency is associated with increased rates of Mn absorption. Workplace exposure to Mn is most likely to occur by inhalation and represents a major concern for the ferromanganese, iron and steel, dry-cell battery, welding, mining, and ore processes industries. In addition, individuals living in the vicinity of ferromanganese or iron and steel manufacturing facilities or hazardous waste sites may be exposed to elevated Mn particulate matter in air or water, although this exposure is likely to be much lower than in the workplace. Mn derived from the combustion of methylcyclopentadienyl manganese tricarbonyl (MMT) must also be considered as a risk in those countries where it is used as an anti-knock agent in gasoline.

Several sensitive populations must also be considered. Mn exposure of children is of particular concern. Similarly, the elderly are at greater risk because Mn exposure may compound normal aging processes or disease states (*e.g.* preclinical Parkinsonism), with attenuated ability to compensate

Issues in Toxicology No. 22
Manganese in Health and Disease
Edited by Lucio G. Costa and Michael Aschner
© The Royal Society of Chemistry 2015
Published by the Royal Society of Chemistry, www.rsc.org

for declines in function. Sub-clinical impairment of liver function should also be considered because the majority of Mn is excreted via the biliary system.

While attempts have been made to develop reliable exposure biomarkers or specific early biomarkers of effects, at present no such biomarkers exist. Plasma and serum analysis may not reflect exposures, and urine is deemed an inadequate assessment medium, given the low rate of Mn excretion by this route.

While decades of research have significantly improved our appreciation for the role of Mn in health and disease, further research is clearly warranted. Attempts have been made throughout the book to highlight those research needs. A select number of these needs are identified below:

- Dietary Mn is tightly regulated by changes in absorption and excretion. Comparative data for inhaled Mn need to be established to better understand the health risk associated with exposure to Mn via this route.
- Prospective imaging studies in occupationally exposed Mn cohorts will advance the understanding on the pathophysiology associated with excessive exposure to this metal.
- The route-dependent form in which Mn is transported into the brain, and the mechanisms that govern its accumulation in basal ganglia have yet to be fully understood. The potential for human Mn accumulation in basal ganglia via olfactory transport as a risk factor merits further scrutiny.
- As of yet, specific biomarkers of Mn exposure have not been reliably identified.
- The long-term effects of low-level Mn exposure, specifically increased risk for neurological disease, have yet to be identified.
- Better understanding of the molecular and cellular mechanisms of Mn-induced toxicity is warranted.
- Studies are necessary to determine differences in Mn metabolism and toxicity between subpopulations and susceptible groups, focusing on genetic susceptibility.
- The interaction between dietary and inhaled Mn remains unknown.
- Additional studies are necessary to set safe limits for Mn exposure, both from the diet and via inhalation.

We thank the publisher and the international experts who contributed to the book. The goal of our efforts was to produce a comprehensive, yet easy-to-follow authoritative book on current knowledge in this ever-growing field on Mn. No book to date has been wholly dedicated to the biology of Mn, thus we felt a book of this nature would be greatly desired and appreciated. We attempted to bridge basic research with clinical, epidemiological, regulatory, and translational research, conveying both an introductory understanding and the latest development in the field. Utmost, it was designed to serve both the novice student and the expert scientist and physician. With a comprehensive and integrated view on the biology of Mn, we hope that we are successful in stimulating a deeper understanding and promoting renewed collaborations on this multifaceted topic, leading to improved understanding of the role of this unique metal, Mn, both in health and disease.

Michael Aschner
Lucio G. Costa

Subject Index